Mathematical Analysis

THE WILEY BICENTENNIAL–KNOWLEDGE FOR GENERATIONS

Each generation has its unique needs and aspirations. When Charles Wiley first opened his small printing shop in lower Manhattan in 1807, it was a generation of boundless potential searching for an identity. And we were there, helping to define a new American literary tradition. Over half a century later, in the midst of the Second Industrial Revolution, it was a generation focused on building the future. Once again, we were there, supplying the critical scientific, technical, and engineering knowledge that helped frame the world. Throughout the 20th Century, and into the new millennium, nations began to reach out beyond their own borders and a new international community was born. Wiley was there, expanding its operations around the world to enable a global exchange of ideas, opinions, and know-how.

For 200 years, Wiley has been an integral part of each generation's journey, enabling the flow of information and understanding necessary to meet their needs and fulfill their aspirations. Today, bold new technologies are changing the way we live and learn. Wiley will be there, providing you the must-have knowledge you need to imagine new worlds, new possibilities, and new opportunities.

Generations come and go, but you can always count on Wiley to provide you the knowledge you need, when and where you need it!

WILLIAM J. PESCE
PRESIDENT AND CHIEF EXECUTIVE OFFICER

PETER BOOTH WILEY
CHAIRMAN OF THE BOARD

Mathematical Analysis

A Concise Introduction

Bernd S. W. Schröder
Louisiana Tech University
Program of Mathematics and Statistics
Ruston, LA

WILEY-INTERSCIENCE
A John Wiley & Sons, Inc., Publication

Published by John Wiley & Sons, Inc , Hoboken, New Jersey
Published simultaneously in Canada.

For general information on our other products and services or for technical support, please contact our Customer Care Department within the United States at (800) 762-2974, outside the United States at (317) 572-3993 or fax (317) 572-4002

Wiley also publishes its books in a variety of electronic formats Some content that appears in print may not be available in electronic format. For information about Wiley products, visit our web site at www.wiley.com.

Wiley Bicentennial Logo: Richard J Pacifico

Library of Congress Cataloging-in-Publication Data:

Schroder, Bernd S. W (Bernd Siegfried Walter), 1966–
Mathematical analysis : a concise introduction / Bernd S.W Schröder
p cm.
ISBN 978-0-470-10796-6 (cloth)
1 Mathematical analysis. I Title
QA300.S376 2007
515—dc22 2007024690

10 9 8 7 6 5 4

Contents

Part II: Analysis in Abstract Spaces

Part III: Applied Analysis

Appendices

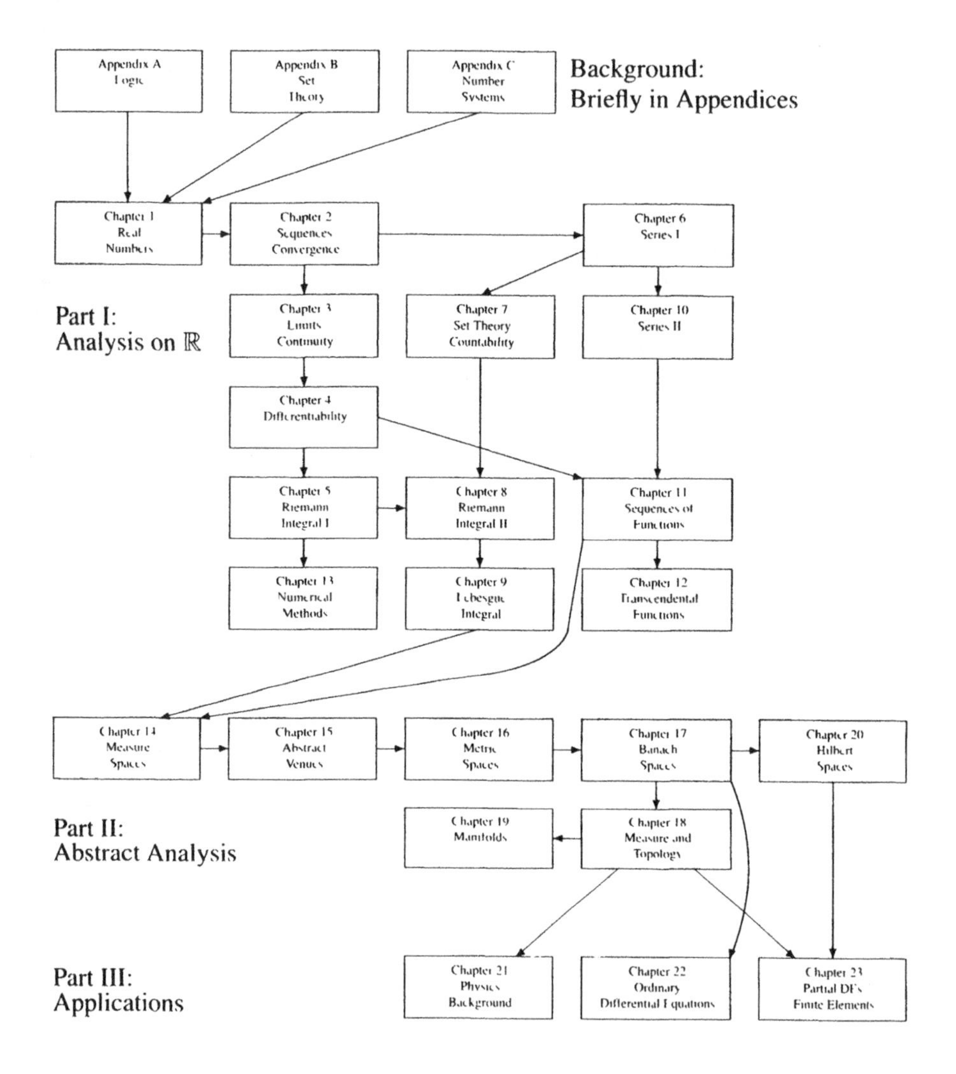

Figure I: Content dependency chart with minimum prerequisites indicated by arrows. Some remarks, examples, and exercises in the later chapter might still depend on other earlier chapters, but this problem typically can be resolved by quoting a single result. Details about where and how the reader can "branch out" are given in **boxes** in the text.

Preface

This text is a self-contained introduction to the fundamentals of analysis. The only prerequisite is some experience with mathematical language and proofs. That is, it helps to be familiar with the structure of mathematical statements and with proof methods, such as direct proofs, proofs by contradiction, or induction. With some support in the right places, mostly in the early chapters, this text can also be used without prerequisites in a first proof class.

Mastering proofs in analysis is one of the key steps toward becoming a mathematician. To develop sound proof writing techniques, standard proof techniques are discussed early in the text and for a while they are pointed out explicitly. Throughout, proofs are presented with as much detail and as little hand waving as possible. This makes some proofs (for example, the density of $C[a, b]$ in $L^p[a, b]$ in Part II) notationally a bit complicated. With computers now being a regular tool in mathematics, the author considers this appropriate. When code is written for a problem, all details must be implemented, even those that are omitted in proofs. Seeing a few highly detailed proofs is reasonable preparation for such tasks. Moreover, to facilitate the transition to more abstract settings, such as measure, inner product, normed, and metric spaces, the results for single variable functions are proved using methods that translate to these abstract settings. For example, early proofs rely extensively on sequences and we also use the completeness of the real numbers rather than their order properties.

Analysis is important for applications, because it provides the abstract background that allows us to apply the full power of mathematics to scientific problems. This text shows that all abstractions are well motivated by the desire to build a strong theory that connects to specific applications. Readers who complete this text will be ready for *all* analysis-based and analysis-related subjects in mathematics, including complex analysis, differential equations, differential geometry, functional analysis, harmonic analysis, mathematical physics, measure theory, numerical analysis, partial differential equations, probability theory, and topology. Readers interested in motivation from physics are advised to browse Chapter 21, even if they have not read any of the earlier chapters.

Aside from the topics covered, readers interested in applications should note that the axiomatic approach of mathematics is similar to problem solving in other fields. In mathematics, theories are built on axioms. Similarly, in applications, models are subject to constraints. Neither the axioms, nor the constraints can be violated by the theory or model. Building a theory based on axioms fosters the reader's discipline to not make unwarranted assumptions.

Organization of the content. The text consists of three large parts. Part I, comprised of Chapters 1–13, presents the analysis of functions of one real variable, including a motivated introduction to the Lebesgue integral. Chapters 1–6 and 10–13 could be called "single variable calculus with proofs." For a smooth transition from calculus and a gradual increase in abstraction, Chapters 1–6 require very little set theory. Chapter 1 presents the properties of the real line and limits of sequences are introduced in Chapter 2. Chapters 3–5 present the fundamentals on continuity, differentiation, and (Riemann) integration in this order and Chapter 6 gives a first introduction to series.

Chapters 6–8 are motivated by the desire to further explore the Riemann integral while avoiding the excessive use of Riemann sums. This exploration is done with the Lebesgue criterion for Riemann integrability. Although this criterion requires the Lebesgue measure, the payoff is that many proofs become simpler. To quickly reach this criterion, the first presentation of series in Chapter 6 is deliberately kept short. It presents enough about series to allow the definition of Lebesgue measure. Chapter 7 presents fundamental notions of set theory. Most of these ideas are needed for Lebesgue measure, but, overall, Chapter 7 contains all the set theory needed in the remainder of the text. Chapter 8 finishes the presentation of the Riemann integral. With Lebesgue measure available, it is natural to investigate the Lebesgue integral in Chapter 9. This chapter could also be delayed to the end of Part I, but the author believes that early exposure to the crucial ideas will ease the later transition to measure spaces.

The analysis of single variable functions is finished with the rigorous introduction of the transcendental functions. The necessary background on power series is explored in Chapter 10. Chapter 11 presents some fundamentals on the convergence of sequences of functions and Chapter 12 is devoted to the transcendental functions themselves. Chapter 13 discusses general numerical methods, but transcendental functions provide a rich test bed for the methods presented.

Part I of the text can be read or presented in many orders. Figure 1 shows the prerequisite structure of the text. Prerequisites for each chapter have deliberately been kept minimal. In this fashion, the order of topics in the reader's first contact with proofs in analysis can be adapted to many readers' preferences. Most notably, the intentionally early presentation of Lebesgue integration can be postponed to the end of Part I if so desired. Throughout, the author intends to keep the reader engaged by providing motivation for all abstractions. Consequently, as Figure 1 and the table of contents indicate, some concepts and results are presented in a "just-in-time" fashion rather than in what may be considered their traditional place. If a concept is needed in an exercise before the concept is "officially" defined in the text, the concept will be defined in the exercise and in the text.

Part II, comprised of Chapters 14–20, explores how the appropriate abstractions lead to a powerful and widely applicable theoretical foundation for all branches of applied mathematics. The desire to define an integral in d-dimensional space provides a natural motivation to introduce measure spaces in Chapter 14. This chapter facilitates the transition to more abstract mathematics by frequently referring back to corresponding results for the one dimensional Lebesgue integral. The proofs of these results usually are verbatim the same as in the one-dimensional setting. Moreover, this early introduction makes L^p spaces available as examples for the rest of the text. The abstract venues of analysis are then presented in Chapter 15, which provides all examples for the rest of Part II.

The fundamentals on metric spaces and continuity are presented in Chapter 16. As with measure spaces, for several results on metric spaces the reader is referred back to the corresponding proof for single variable functions. Proofs are no longer verbatim the same and abstraction is facilitated by translating proofs from a familiar setting to the new setting while analyzing similarities and differences. In a class, the author suggests that the teacher fill in some of these proofs to demonstrate the process.

Chapter 17 presents the fundamentals on normed spaces and differentiation. Again, ideas are similar to those for functions of a single variable, but this time the abstraction goes beyond translation. With all three fundamental concepts (integration, continuity, and differentiation) available in the abstract setting, Chapter 18 shows the interrelationship between concepts presented separately before, culminating in the Multivariable Substitution Formula.

The second part is completed by a presentation of the fundamentals of analysis on manifolds, together with a physical interpretation of key concepts in Chapter 19 and by an introduction to Hilbert spaces in Chapter 20.

The remaining chapters give a brief outlook to applied subjects in which analysis is used, specifically, physics in Chapter 21, ordinary differential equations in Chapter 22, and partial differential equations and the finite element method in Chapter 23. Each of these chapters can only give a taste of its subject and I encourage the reader to go deeper into the utterly fascinating applications that lie behind part III. The mathematical preparation through this text should facilitate the transition.

It should be possible to cover the bulk of the text in a two course sequence. Although Chapters 14-16 should be read in order, depending on the available time, the pace and the choice of topics, any of Chapters 17-23 can serve as a capstone experience.

How to read this text. Mathematics in general, and analysis in particular, is not a spectator sport. It is learned by doing. To allow the reader to "do" mathematics, each section has exercises of varying degrees of difficulty. Some exercises require the adaptation of an argument in the text. These exercises are also intended to make the reader critically analyze the argument before adapting it. This is the first step toward being able to write proofs. Of course the need for very critical (and slow) reading of mathematics is nicely summed up in the old quote that "To read without a pencil is daydreaming." The reader should ask him/herself after every sentence "What does this mean? Why is this justified?" Making notes in the margin to explain the harder steps will allow the reader to answer these questions more easily in the second and third readings of a proof. So it is important to read thoroughly and slowly, to make notes and to reread as often as needed. The extensive index should help with unknown or forgotten terminology as necessary. Other exercises have hints on how to create a proof that the reader has not seen before. These exercises require the use of proof techniques in a new setting. Finally, there are also exercises without hints. Being able to create the proof with nothing but the result given is the deepest task in a mathematics course. This is not to say that exercises without hints are always the hardest and adaptations are always the easiest, but in many cases this is true. Finally, some exercises give a sequence of hints and intermediate results leading up to a famous theorem or a specific example. These exercises could also be used as mini-projects. In a class, some of them could be the basis for separate lectures that spotlight a particular theorem or example.

To get the most out of this text, the reader is encouraged to *not* look for hints and solutions in other background materials. In fact, even for proofs that are adaptations of proofs in this text, it is advantageous to try to create the proof *without* looking up the proof that is to be adapted. There is evidence that the struggle to solve a problem, which can take days for a single proof, is exactly what ultimately contributes to the development of strong skills. "Shortcuts," while pleasant, can actually diminish this development. Readers interested in quantitative evidence that shows how the struggle to acquire a skill actually can lead to deeper learning may find the article [4] quite enlightening. A better survival mechanism than shortcuts is the development of connections between newly learned content and existing knowledge. The reader will need to find these connections to his/her existing knowledge, but the structure of the text is intended to help by motivating all abstractions. Readers interested in how knowledge is activated more easily when it was learned in a known context may be interested in the article [5].

Acknowledgments. Strange as it may sound, I started writing this text in the spring of 1987, as I prepared for my oral final examination in the traditional Analysis I–III sequence in Germany. Basically, I took all topics in the sequence and arranged them in what was the most logical fashion to me at the time. Of course, these notes are, in retrospect, immature. But they did a lot to shape my abilities and they were a good source of ideas and exercises. In this respect, I am indebted to my teachers for this sequence: Professor Wegener and teaching assistant Ms. Lange for Analysis I, Professor Kutzler and teaching assistant *Herr* Böttger for Analysis II-III as well as Professor Herz in whose Differential Equations class I first saw analysis "at work." With all due respect to the other individuals, to me and many of my fellow students, the force that drove us in analysis (and beyond) was Herr Böttger. This gentleman was uncompromising in his pursuit of mathematical excellence and we feared as well as looked forward to his demanding exercise sets. He was highly respected because he was ready to spend hours with anyone who wanted to talk mathematics. Those who kept up with him were extremely well prepared for their mathematical careers. Incidentally, Dr. Ansgar Jüngel, whose notes I used for the chapter on the finite element method, took the above mentioned classes with me. The thorough preparation through these classes is the main reason why most of this text was comparatively easy to write. If this text does half as good a job as Herr Böttger did with us, it has more than achieved its purpose.

It was thrilling to test my limitations, it was humbling to find them and ultimately I was left awed once more by the beauty of mathematics. When my abilities were insufficient to proceed, I used the texts listed in the bibliography for proofs, hints or to structure the presentation. To make the reader fully concentrate on matters at hand, and to force myself to make the exposition self-contained, outside references are limited to places where results were beyond the scope of this exposition. A solid foundation will allow readers to judiciously pick their own resources for further study. Nonetheless, it is appropriate to recognize the influence of the works of a number of outstanding individuals. I used Adams [2], Renardy and Rogers [23], Yosida [33] and Zeidler [34] for Sobolev spaces, Aris [3], Cramer's `http://www.navier-stokes.net/`, and

Welty, Wicks and Wilson [31] for fluid dynamics, Chapman [6] for heat transfer, Cohn [7] for measure theory, Dieudonné [8] for differentiation in Banach spaces, Dodge [9] and Halmos [13] for set theory, Ferguson [10], Sandefur [24] and Stoer and Bulirsch [28] for numerical analysis, Halliday, Resnick and Walker [12] for elementary physics, Hewitt and Stromberg [14], Heuser [15], [16], Johnsonbaugh and Pfaffenberger [20], Lehn [22] and Stromberg [29] for general background on analysis, Heuser [17] for functional analysis, Hurd and Loeb [18] for the use of quantifiers in logic, Jüngel [21] and Šolín [25] for the finite element method, Spivak [26], [27] for manifolds, Torchinsky [30] for Fourier series, Willard [32] for topology, and the Online Encyclopaedia of Mathematics `http://eom.springer.de/` for quick checks of notation and definitions. Readers interested in further study of these subjects may wish to start with the above references.

The first draft of the manuscript was used in my analysis classes in the Winter and Spring quarters of 2007. The first class covered Chapters 1–9, the second covered Chapters 11 and 14–18 (with some strategic "fast forwards"). This setup assured that graduating students would have full exposure to the essentials of analysis on the real line and to as much abstract analysis as possible without "handwaving arguments." I am grateful to the students in these classes for keeping up with the pace, solving large numbers of homework problems, being patient with the typos we found and also for suggesting at least one order in which to present the material that I had not considered. The students' evaluations (my best ever) also reaffirmed for me that people will enjoy, or at least accept and honor, a challenge, and that an ambitious, motivated course should be the way to go. Devery Rowland once more did an excellent job printing drafts of the text for the classes.

Aside from the referees, several colleagues also commented on this text and I owe them my thanks for making it a better product. In particular, I would like to thank Natalia Zotov for some comments on an early version that significantly improved the presentation, and Ansgar Jüngel for pointing out some key references on Sobolev spaces. Although I hope that we have found all remaining errors and typos, any that remain are my responsibility and mine alone. I request readers to report errors and typos to me so I can post an errata. My contacts at Wiley, Susanne Steitz, Jacqueline Palmieri, and Melissa Yanuzzi bore with me when the stress level rose and their patience made the publishing process very smooth.

As always, this work would not have been possible without the love of my family. It is truly wonderful to be supported by individuals who accept your decision to spend large amounts of time reliving your formative years.

Finally, I was sad to learn that Herr Böttger died unexpectedly a few years after I had my last class with him. Sir, this one's for you.

Ruston, LA, August 30, 2007

Bernd Schröder

Part I

Analysis of Functions of a Single Real Variable

Chapter 1

The Real Numbers

This investigation of analysis starts with minimal prerequisites. Regarding set theory, the terms "set" and "element" will remain undefined, as is customary in mathematics to avoid paradoxes. The **empty set** $\emptyset$ is the set that has no elements. The statement "$e \in S$" says that e is an element of the set S. The statement "$A \subseteq B$" says that every element of A is an element of B. Sets A and B are equal if and only if $A \subseteq B$ and $B \subseteq A$. The statement "$A \subset B$" says that $A \subseteq B$ and $A \neq B$. Subsets will be defined as "$A = \{x \in S : \langle\text{property}\rangle\}$," that is, with a statement from which set S the elements of A are taken and a property describing them. The **union** of two sets A and B is $A \cup B = \{x : x \in A \text{ or } x \in B\}$, the **intersection** is $A \cap B = \{x : x \in A \text{ and } x \in B\}$. Union and intersection of finitely many sets are denoted $\bigcup_{j=1}^{n} A_j$ and $\bigcap_{j=1}^{n} A_j$, respectively, and the **relative complement** of B in A is $A \setminus B = \{x \in A : x \notin B\}$. Further details on set theory are purposely delayed until Section 7.1. Until then, we focus on analytical techniques. Any required notions of set theory will be clarified on the spot.

To define properties, sometimes the **universal quantifier** "$\forall$" (read "for all") or the **existential quantifier** "$\exists$" (read "there exists") are used. Formal logic is described in more detail in Appendix A. Finally, the reader needs an intuitive idea what a function, a relation and a binary operation are. Details are relegated to Appendices B.2 and C.2.

The real numbers $\mathbb{R}$ are the "staging ground" for analysis. They can be characterized as the unique (up to isomorphism) mathematical entity that satisfies Axioms 1.1, 1.6, and 1.19. That is, they are the unique linearly ordered, complete field (see Exercise 1-30). In this chapter, we introduce the axioms for the real numbers and some fundamental consequences. These results assure that the real numbers indeed have the properties that we are familiar with from algebra and calculus.

1.1 Field Axioms

The description of the real numbers starts with their algebraic properties.

Axiom 1.1 *The* **real numbers** $\mathbb{R}$ *are a* **field**. *That is,* $\mathbb{R}$ *has at least two elements and there are two binary operations,* **addition** $+ : \mathbb{R} \times \mathbb{R} \to \mathbb{R}$ *and* **multiplication** $\cdot : \mathbb{R} \times \mathbb{R} \to \mathbb{R}$, *so that*

1. *Addition is* **associative**, *that is, for all* $x, y, z \in \mathbb{R}$ *we have*
   $(x + y) + z = x + (y + z)$.
2. *Addition is* **commutative**, *that is, for all* $x, y \in \mathbb{R}$ *we have*
   $x + y = y + x$.
3. *There is a* **neutral element** 0 *for addition, that is, there is an element* $0 \in \mathbb{R}$ *so that for all* $x \in \mathbb{R}$ *we have* $x + 0 = x$.
4. *For every element* $x \in \mathbb{R}$ *there is an* **additive inverse** *element* $(-x)$ *so that*
   $x + (-x) = 0$.
5. *Multiplication is* **associative**, *that is, for all* $x, y, z \in \mathbb{R}$ *we have*
   $(x \cdot y) \cdot z = x \cdot (y \cdot z)$.
6. *Multiplication is* **commutative**, *that is, for all* $x, y \in \mathbb{R}$ *we have*
   $x \cdot y = y \cdot x$.
7. *There is a* **neutral element** 1 *for multiplication, that is, there is an element* $1 \in \mathbb{R}$ *so that for all* $x \in \mathbb{R}$ *we have* $1 \cdot x = x$.
8. *For every element* $x \in \mathbb{R} \setminus \{0\}$ *there is a* **multiplicative inverse** *element* x^{-1} *so that* $x \cdot x^{-1} = 1$.
9. *Multiplication is* **(left) distributive** *over addition, that is, for all* $\alpha, x, y \in \mathbb{R}$ *we have* $\alpha \cdot (x + y) = \alpha \cdot x + \alpha \cdot y$.

As is customary for multiplication, the dot between factors is usually omitted.

Fields are investigated in detail in abstract algebra. For analysis, it is most effective to remember that the field axioms guarantee the properties needed so that we can perform algebra and arithmetic "as usual." Some of these properties are exhibited in this section and in the exercises. The exercises also include examples that show that not every field needs to be infinite (see Exercises 1-7–1-9).

Theorem 1.2 *The following are true in* $\mathbb{R}$:

1. *For all* $x \in \mathbb{R}$, *we have* $0x = 0$.
2. $0 \neq 1$.
3. *Additive inverses are unique. That is, if* $x \in \mathbb{R}$ *and* x' *and* $\overline{x}$ *both have the property in part 4 of Axiom 1.1, then* $x' = \overline{x}$.
4. *For all* $x \in \mathbb{R}$, *we have* $(-1)x = -x$.

Proof. *Early in the text, proofs will sometimes be interrupted by comments in italics to point out standard formulations and proof techniques.*

To prove part 1, let $x \in \mathbb{R}$. Then the axioms allow us to obtain the following equation. $0x \overset{\text{Ax.3}}{=} (0+0)x \overset{\text{Ax.6}}{=} x(0+0) \overset{\text{Ax 9}}{=} x0 + x0 \overset{\text{Ax.6}}{=} 0x + 0x$. This implies

$$0 \overset{\text{Ax.4}}{=} 0x + (-0x) \overset{\text{above}}{=} (0x + 0x) + (-0x) \overset{\text{Ax 1}}{=} 0x + \big(0x + (-0x)\big) \overset{\text{Ax.4}}{=} 0x + 0 \overset{\text{Ax 3}}{=} 0x$$

as was claimed. *The proof of part 1 shows how every step in a proof needs to be justified. Usually we will not explicitly justify each step in a computation with an axiom or a previous result. However, the reader should always mentally fill in the justification. The practice of filling in these justifications should be started in the computations in the remainder of this proof.*

To prove part 2, first note that, because $\mathbb{R}$ has at least two elements, there is an $x \in \mathbb{R} \setminus \{0\}$. Now suppose for a contradiction *(see Standard Proof Technique 1.4 below)* that $0 = 1$. Then $x = 1 \cdot x = 0 \cdot x = 0$ is a contradiction to $x \in \mathbb{R} \setminus \{0\}$.

For part 3, note that if x' and $\overline{x}$ both have the property in part 4 of Axiom 1.1, then $x' = x' + 0 = x' + (x + \overline{x}) = \big(x' + x\big) + \overline{x} = \big(x + x'\big) + \overline{x} = 0 + \overline{x} = \overline{x} + 0 = \overline{x}$. *Note that the statement of part 3 already encodes the typical approach to a uniqueness proof (see Standard Proof Technique 1.5 below).*

Finally, for part 4 note that $x + (-1)x = 1x + (-1)x = \big(1 + (-1)\big)x = 0x = 0$. Because by part 3 additive inverses are unique, $(-1)x$ must be the additive inverse $-x$ of x. *The last step is a typical application of modus ponens, see Standard Proof Technique 1.3 below.* ■

To familiarize the reader with standard proof techniques, these techniques will be pointed out explicitly in the early part of the text. The techniques presented in Chapter 1 are general proof techniques applicable throughout mathematics. Techniques presented in later chapters are mostly specific to analysis.

Standard Proof Technique 1.3 The simplest mathematical proof technique is a **direct proof** in which a result that says "A implies B" is applied after we have proved that A is true. Truth of A and of "A implies B" guarantees truth of B. This technique is also called **modus ponens**. An example is in the proof of part 4 of Theorem 1.2. □

Standard Proof Technique 1.4 In a proof by **contradiction**, we suppose the contrary (the negation, also see Appendix A.2) of what is claimed is true and then we derive a contradiction. Typically, we derive a statement and its negation, which is a contradiction, because they cannot both be true. For an example, see the proof of part 2 of Theorem 1.2 above. Given that the reasoning that led to the contradiction is correct, the contradiction must be caused by the assumption that the contrary of the claim is true. Hence, the contrary of the claim must be false, because true statements cannot imply false statements like contradictions (see part 3 of Definition A.2 in Appendix A). But this means the claim must be true.

We will usually indicate proofs by contradiction with a starting statement like "suppose for a contradiction." □

Standard Proof Technique 1.5 For many mathematical objects it is important to assure that they are the *only* object that has certain properties. That is, we want to assure that the object is unique. In a typical **uniqueness proof**, we assume that there is more than one object with the properties under investigation and we prove that any two of these objects must be equal. Part 3 of Theorem 1.2 shows this approach. □

Exercises

1-1. Prove that $(-1) \cdot (-1) = 1$.

1-2. $\cdot$ **is right distributive over** $+$. Prove that for all $x, y, z \in \mathbb{R}$ we have $(x + y)z = xz + yz$.

1-3. **Multiplicative inverses are unique** Prove that if $x \in \mathbb{R}$ and x' and $\overline{x}$ both have the property in part 8 of Axiom 1.1 then $x' = \overline{x}$.

1-4. Prove that 0 does not have a multiplicative inverse

1-5. Prove that if $x, y \neq 0$, then $(xy)^{-1} = y^{-1}x^{-1}$. Conclude in particular that $xy \neq 0$.

1-6. Prove each of the **binomial formulas** below. Justify each step with the appropriate axiom.

(a) $(a + b)^2 = a^2 + 2ab + b^2$ (b) $(a - b)^2 = a^2 - 2ab + b^2$

(c) $(a + b)(a - b) = a^2 - b^2$

1-7. Prove that the set $\{0, 1\}$ with the usual multiplication and the usual addition, except that $1 + 1 := 0$, is a field. That is, prove that the set and addition and multiplication as stated have the properties listed in Axiom 1.1.

1-8. Prove that the set $\{0, 1, 2\}$ with the sum and product of two elements being the remainder obtained when dividing the regular sum and product by 3 is a field

1-9. A property and some finite fields

(a) Let F be a field and let $x, y \in F$ Prove that $x \cdot y = 0$ if and only if $x = 0$ or $y = 0$.

(b) Prove that the set $\{0, 1, 2, 3\}$ with the sum and product of two elements being the remainder obtained when dividing the regular sum and product by 4 is *not* a field.

(c) Prove that the set $\{0, 1, \ldots, p - 1\}$ with the sum and product of two elements being the remainder obtained when dividing the regular sum and product by p is a field if and only if p is a prime number

1.2 Order Axioms

Exercises 1-7–1-9c show that the field axioms alone are not enough to describe the real numbers. In fact, fields need not even be infinite. However, aside from executing the familiar algebraic operations, we can also compare real numbers. This section presents the order relation on the real numbers and its properties.

Axiom 1.6 *The real numbers* $\mathbb{R}$ *contain a subset* $\mathbb{R}^+$, *called the* **positive real numbers** *such that*

1. *For all* $x, y \in \mathbb{R}^+$, *we have* $x + y \in \mathbb{R}^+$ *and* $xy \in \mathbb{R}^+$,
2. *For all* $x \in \mathbb{R}$, *exactly one of the following three properties holds.*

   Either $x \in \mathbb{R}^+$ *or* $-x \in \mathbb{R}^+$ *or* $x = 0$.

A real number x is called **negative** if and only if $-x \in \mathbb{R}^+$.

Once positive numbers are defined, we can define an order relation. As usual, instead of writing $y + (-x)$ we write $y - x$ and call it the **difference** of x and y. The binary operation "$-$" is called **subtraction**.

The phrase "if and only if," which is used in definitions and biconditionals, is normally abbreviated with the artificial word "**iff**."

Definition 1.7 *For $x, y \in \mathbb{R}$, we say x is* **less than** *y, in symbols $x < y$, iff $y - x \in \mathbb{R}^+$. We say x is* **less than or equal to** *y, denoted $x \leq y$, iff $x < y$ or $x = y$. Finally, we say x is* **greater than** *y, denoted $x > y$, iff $y < x$, and we say x is* **greater than or equal to** *y, denoted $x \geq y$, iff $y \leq x$.*

The relation $\leq$ satisfies the properties that define an order relation.

Proposition 1.8 *The relation $\leq$ is an* **order** *relation on $\mathbb{R}$. That is,*

1. *$\leq$ is* **reflexive**. *For all $x \in \mathbb{R}$ we have $x \leq x$,*
2. *$\leq$ is* **antisymmetric**. *For all $x, y \in \mathbb{R}$ we have that $x \leq y$ and $y \leq x$ implies $x = y$,*
3. *$\leq$ is* **transitive**. *For all $x, y, z \in \mathbb{R}$, we have that $x \leq y$ and $y \leq z$ implies $x \leq z$.*

Moreover, the relation $\leq$ is a **total order** *relation, that is, for any two $x, y \in \mathbb{R}$ we have that $x \leq y$ or $y \leq x$.*

Proof. The relation $\leq$ is reflexive, because it includes equality.

For antisymmetry, let $x \leq y$ and $y \leq x$ and suppose for a contradiction that $x \neq y$. Then $x - y \in \mathbb{R}^+$ and $-(x - y) = y - x \in \mathbb{R}^+$, which cannot be by Axiom 1.6. Thus $\leq$ must be antisymmetric.

For transitivity, let $x \leq y$ and $y \leq z$. There is nothing to prove if one of the inequalities is an equality. Thus we can assume that $x < y$ and $y < z$, which means $y - x \in \mathbb{R}^+$ and $z - y \in \mathbb{R}^+$. But then $\mathbb{R}^+$ contains $(z - y) + (y - x) = z - x$, and hence $x < z$. We have shown that for all $x, y, z \in \mathbb{R}$ the inequalities $x \leq y$ and $y \leq z$ imply $x \leq z$, which means that $\leq$ is transitive.

For the "moreover" part note that if $x, y \in \mathbb{R}$, then $y - x \in \mathbb{R}$ and we have either $y - x \in \mathbb{R}^+$, which means $x < y$, or $y - x = 0$, which means $y = x$, or $x - y = -(y - x) \in \mathbb{R}^+$, which means $y < x$. Therefore for all $x, y \in \mathbb{R}$ one of $x \leq y$ or $y \leq x$ holds, and hence $\leq$ is a total order. ■

Once an order relation is established, we can define intervals.

Definition 1.9 *An* **interval** *is a set $I \subseteq \mathbb{R}$ so that for all $c, d \in I$ and $x \in \mathbb{R}$ the inequalities $c < x < d$ imply $x \in I$. In particular for $a, b \in \mathbb{R}$ with $a < b$ we define*

1. $[a, b] := \{x \in \mathbb{R} : a \leq x \leq b\}$,
2. $(a, b) := \{x \in \mathbb{R} : a < x < b\}$, $(a, \infty) := \{x \in \mathbb{R} : a < x\}$, $(-\infty, b) := \{x \in \mathbb{R} : x < b\}$, $(-\infty, \infty) := \mathbb{R}$,

3. $[a, b) := \{x \in \mathbb{R} : a \leq x < b\}$, $[a, \infty) := \{x \in \mathbb{R} : a \leq x\}$,

4. $(a, b] := \{x \in \mathbb{R} : a < x \leq b\}$, $(-\infty, b] := \{x \in \mathbb{R} : x \leq b\}$.

The points a and b are also called the **endpoints** *of the interval. An interval that does not contain either of its endpoints (where $\pm\infty$ are also considered to be "endpoints") is called* **open**. *An interval that contains exactly one of its endpoints is called* **half-open** *and an interval that contains both its endpoints is called* **closed**.

For the first part of this text, the domains of functions will almost exclusively be intervals. Because analysis requires extensive work with inequalities, we need to investigate how the order relation relates to the algebraic operations.

Theorem 1.10 *Properties of the order relation. Let $x, y, z \in \mathbb{R}$.*

1. *The number x is positive iff $x > 0$ and x is negative iff $x < 0$.*

2. *If $x \leq y$, then $x + z \leq y + z$.*

3. *If $x \leq y$ and $z > 0$, then $xz \leq yz$.*

4. *If $x \leq y$ and $z < 0$, then $xz \geq yz$.*

5. *If $0 < x \leq y$, then $y^{-1} \leq x^{-1}$.*

Similar results can be proved for other combinations of strict and nonstrict inequalities. We will not state these here, but instead trust that the reader can make the requisite translation from the statements in this theorem.

Proof. Parts 1 and 2 are left to the reader as Exercises 1-10a and 1-10b. *Throughout this text, parts of proofs will be delegated to the reader to facilitate a better connection to the material presented.*

For part 3, let $x \leq y$ and let $z > 0$. Then, $y - x \in \mathbb{R}^+$ or $y = x$. In case $y = x$, we obtain $yz = xz$ and thus, in particular, $xz \leq yz$. In case $y - x \in \mathbb{R}^+$, note that $z > 0$ means $z \in \mathbb{R}^+$, and hence $yz - xz = (y - x)z \in \mathbb{R}^+$. By definition, this implies $xz < yz$, and in particular $xz \leq yz$. Because we have shown $xz \leq yz$ in each case, the result is established. *All proofs in this section are done with the above kind of case distinction (see Standard Proof Technique 1.11).*

For part 4, let $x \leq y$ and let $z < 0$. Then, $y - x \in \mathbb{R}^+$ or $y = x$. In case $y = x$, we obtain $yz = xz$, and hence $xz \geq yz$. In case $y - x \in \mathbb{R}^+$, note that $z < 0$ means $-z \in \mathbb{R}^+$, and hence $xz - yz = (x - y)z = (y - x)(-z) \in \mathbb{R}^+$. By definition, this implies $yz < xz$, and hence $yz \leq xz$, which establishes the result.

For part 5, first note that there is nothing to prove if $x = y$. Hence, we can assume that $x < y$. Suppose for a contradiction that $x^{-1} < y^{-1}$. Then by part 3 we have that $1 = x^{-1}x < y^{-1}x$, and hence $x < y \cdot 1 < yy^{-1}x = x$, contradiction. ■

Standard Proof Technique 1.11 When several possibilities must be considered in a proof, the proof usually continues with separate arguments for each possibility. The proof is complete when each separate argument has led to the desired conclusion. This type of proof is also called a proof by **case distinction**. □

We conclude this section by introducing the absolute value function and some of its properties.

Definition 1.12 *For $x \in \mathbb{R}$, we set* $|x| = \begin{cases} x; & \text{if } x \geq 0, \\ -x; & \text{if } x < 0, \end{cases}$ *and we call it the* **absolute value** *of x.*

Theorem 1.13 summarizes the properties of the absolute value. The numbering is adjusted so that properties 1, 2, and 3 correspond to the analogous properties for norms (see Definition 15.38). *We will formulate many results in the first part of the text to be analogous or easily generalizable to more abstract settings, but we will usually do so without explicit forward references. In this fashion many abstract situations will be more familiar because of similarities to situations investigated in the first part.*

Theorem 1.13 *Properties of the absolute value.*

0. *For all $x \in \mathbb{R}$, we have $|x| \geq 0$,*
1. *For all $x \in \mathbb{R}$, we have $|x| = 0$ iff $x = 0$,*
2. *For all $x, y \in \mathbb{R}$, we have $|xy| = |x||y|$,*
3. **Triangular inequality**. *For all $x, y \in \mathbb{R}$, we have $|x + y| \leq |x| + |y|$.*
4. **Reverse triangular inequality**. *For all $x, y \in \mathbb{R}$, we have $\big||x| - |y|\big| \leq |x - y|$.*

Proof. For part 0, let $x \in \mathbb{R}$. In case $x \geq 0$, by Definition 1.12 we have $|x| = x \geq 0$. In case $x < 0$, we have $x \notin \mathbb{R}^+$ and by part 2 of Axiom 1.6 we conclude $-x > 0$. Because in this case $|x| = -x > 0$, part 0 follows.

Throughout the text, the two implications of a biconditional "A iff B" will be referred to as "$\Rightarrow$," denoting "if A, then B" and "$\Leftarrow$," denoting "if B, then A."

For part 1, note that the direction "$\Leftarrow$" is trivial, because $|0| = 0$. For the direction "$\Rightarrow$," let $x \in \mathbb{R}$ be so that $|x| = 0$ and suppose for a contradiction that $x \neq 0$. If $x > 0$, then $0 < x = |x| = 0$, a contradiction. *(Note that the previous sentence is a short proof by contradiction that is part of a longer proof by contradiction.)* Therefore $x < 0$. But then $0 < -x = |x| = 0$, a contradiction. Hence, x must be equal to 0.

For part 2, let $x, y \in \mathbb{R}$. If $x \geq 0$ and $y \geq 0$, then by part 3 of Theorem 1.10 $xy \geq 0$, and hence $|xy| = xy = |x||y|$. If $x \geq 0$ and $y < 0$, then by part 4 of Theorem 1.10 we infer $xy \leq 0$. Hence, $|xy| = -xy = x(-y) = |x||y|$. The case $x < 0$ and $y \geq 0$ is similar and the reader will produce it in Exercise 1-11a. Finally, if $x < 0$ and $y < 0$, then by part 4 of Theorem 1.10 we obtain $xy > 0$. Hence, $|xy| = xy = (-1)(-1)xy = (-x)(-y) = |x||y|$.

To prove the triangular inequality, first note that for all $x \in \mathbb{R}$ we have that $x \leq |x|$. This is clear for $x \geq 0$ and for $x < 0$ we simply note $x < 0 < -x = |x|$. Moreover, (see Exercise 1-11b) for all $x \in \mathbb{R}$ we have $-x \leq |x|$. Now let $x, y \in \mathbb{R}$. If the inequality $x + y \geq 0$ holds, then by part 2 of Theorem 1.10 at least one of x, y is greater than or equal to 0. (Otherwise $x < 0$ and $y < 0$ would imply $x + y < 0$.) Hence, by part 2 of Theorem 1.10 $|x+y| = x+y \leq |x|+y \leq |x|+|y|$. If $x+y < 0$, then at least one of x and y is less than 0. Hence, by part 2 of Theorem 1.10 we obtain $|x+y| = -(x+y) = -x+(-y) \leq |-x|+(-y) \leq |-x|+|-y| = |x|+|y|$.

Finally, for the reverse triangular inequality, let $x, y \in \mathbb{R}$. Without loss of generality *(see Standard Proof Technique 1.14)* assume that $|x| \geq |y|$. (The proof for the case $|x| < |y|$ is left as Exercise 1-11c.) Then $|x| = |x-y+y| \leq |x-y|+|y|$, which implies $\big||x|-|y|\big| = |x|-|y| \leq |x-y|$. ■

Standard Proof Technique 1.14 If the proofs for the cases in a case distinction are very similar, it is customary to assume **without loss of generality** that one of these similar cases is true. This is not a loss of generality, because it is assumed that what is presented enables the reader to fill in the proof(s) for the other case(s). In this text, the omitted part is sometimes included as an explicit exercise for the reader. □

Exercises

1-10 Finishing the proof of Theorem 1.10

(a) Prove part 1 of Theorem 1.10.

(b) Prove part 2 of Theorem 1.10.

1-11 Finishing the proof of Theorem 1.13.

(a) Let $x, y \in \mathbb{R}$. Prove that if $x \geq 0$ and $y < 0$, then $|xy| = |x||y|$

(b) Prove that for all $x \in \mathbb{R}$ we have $-x \leq |x|$

(c) Prove that if $|x| < |y|$, then $\big||x|-|y|\big| \leq |x-y|$.

1-12 Let $I, J \subseteq \mathbb{R}$ be intervals Prove that $I \cap J = \{x \in \mathbb{R} : x \in I \text{ and } x \in J\}$ is again an interval.

1-13 Let $a < b$ and let $x, y \in [a, b]$ Prove that $|x-y| \leq b-a$.

1-14 Prove that none of the fields from Exercise 1-9c can satisfy Axiom 1.6 by showing that for these fields part 2 of Axiom 1.6 fails for $x = 1$.

Note. This result shows that Axiom 1.6 distinguishes $\mathbb{R}$ from the finite fields of Exercise 1-9c.

1.3 Lowest Upper and Greatest Lower Bounds

A structure that has the properties outlined in Axioms 1.1 and 1.6 is also called a **linearly ordered field**. The rational numbers satisfy these properties just as well as the real numbers. Thus we are not done with our characterization of $\mathbb{R}$. The final axiom for the real numbers addresses upper and lower bounds of sets.

Definition 1.15 *Let A be a subset of $\mathbb{R}$.*

1. *The number $u \in \mathbb{R}$ is called an* **upper bound** *of A iff $u \geq a$ for all $a \in A$. If A has an upper bound, it is also called* **bounded above**.

2. *The number $l \in \mathbb{R}$ is called a* **lower bound** *of A iff $l \leq a$ for all $a \in A$. If A has a lower bound, it is also called* **bounded below**.

A subset $A \subseteq \mathbb{R}$ that is bounded above and bounded below is also called **bounded**.

Among all upper bounds of a set, the smallest one (if it exists) plays a special role. Similarly, the greatest lower bound plays a special role if it exists.

Definition 1.16 *Let $A \subseteq \mathbb{R}$.*

1. *The number $s \in \mathbb{R}$ is called* **lowest upper bound** *of A or* **supremum** *of A, denoted* $\sup(A)$, *iff s is an upper bound of A and for all upper bounds u of A we have that $s \leq u$.*

2. *The number $i \in \mathbb{R}$ is called* **greatest lower bound** *of A or* **infimum** *of A, denoted* $\inf(A)$, *iff i is a lower bound of A and for all lower bounds l of A we have that $l \leq i$.*

Formally, it is not guaranteed that suprema and infima are unique, but the next result shows that this is indeed the case. Note that the statement of Proposition 1.17 follows the standard pattern for a uniqueness statement.

Proposition 1.17 Suprema are unique. *That is, if the set $A \subseteq \mathbb{R}$ is bounded above and $s, t \in \mathbb{R}$ both are suprema of A, then $s = t$.*

Proof. Let $A \subseteq \mathbb{R}$ and $s, t \in \mathbb{R}$ be as indicated. Then s is an upper bound of A and, because t is a supremum of A, we infer $s \geq t$. Similarly, t is an upper bound of A and, because s is a supremum of A, we infer $t \geq s$. This implies $s = t$. ■

Standard Proof Technique 1.18 (Also compare with Standard Proof Technique 1.14.) When, as in the proof of Proposition 1.17, two parts of a proof are very similar, it is common to only prove one part and state that the other part is similar. Throughout the text, the reader will become familiar with this idea through exercises that require the construction of proofs that are similar to proofs given in the narrative. □

The proof that infima are unique is similar (see Exercise 1-15). Because suprema and infima are unique if they exist, we speak of *the* supremum and *the* infimum.

The final axiom for the real numbers now states that suprema and infima exist under mild hypotheses.

Axiom 1.19 Completeness Axiom. *Every nonempty subset S of $\mathbb{R}$ that has an upper bound has a lowest upper bound.*

Although the Completeness Axiom formally only guarantees that nonempty subsets of $\mathbb{R}$ that are bounded above have suprema, existence of infima is a consequence.

Proposition 1.20 *Let $S \subseteq \mathbb{R}$ be nonempty and bounded below. Then S has a greatest lower bound.*

Proof. Let $L := \{x \in \mathbb{R} : x \text{ is a lower bound of } S\}$. Then $L \neq \emptyset$. Let $s \in S$. Then for all $l \in L$ we have that $l \leq s$. Because $S \neq \emptyset$ this means that L is bounded above. Because $L \neq \emptyset$, by the Completeness Axiom, L has a supremum $\sup(L)$. Every $s \in S$ is an upper bound of L, which means that $s \geq \sup(L)$ and so $\sup(L)$ is a lower bound of S. By definition of suprema, $\sup(L)$ is greater than or equal to all elements of L, that is, it is greater than or equal to all lower bounds of S. By definition of infima, this means that $\sup(L) = \inf(S)$. ■

We will see that suprema and infima are valuable tools in analysis on the real line. The next result shows that in any set with a supremum we can find numbers that are arbitrarily close to the supremum. This fact is important, because analysis ultimately is about objects "getting close to each other."

Proposition 1.21 *Let $S \subset \mathbb{R}$ be a nonempty subset of $\mathbb{R}$ that is bounded above and let $s := \sup(S)$. Then for every $\varepsilon > 0$ there is an element $x \in S$ so that $s - x < \varepsilon$.*

Proof. Suppose for a contradiction that there is an $\varepsilon > 0$ so that for all $x \in S$ we have that $s - x \geq \varepsilon$. Then for all $x \in S$ we would obtain $s - \varepsilon \geq x$, that is, $s - \varepsilon$ would be an upper bound of S. But $s - \varepsilon < s$ contradicts the fact that s is the lowest upper bound of S. ■

Although the supremum and infimum of a set need not be elements of the set, we have different names for them in case they are in the set.

Definition 1.22 *Let A be a subset of $\mathbb{R}$.*

1. *If A is bounded above and $\sup(A) \in A$, then the supremum of A is also called the* **maximum** *of A, denoted $\max(A)$.*

2. *If A is bounded below and $\inf(A) \in A$, then the infimum of A is also called the* **minimum** *of A, denoted $\min(A)$.*

Although the distinctions between suprema and maxima and between infima and minima are small, the notions are distinct. For example, the open interval $(0, 1)$ has a supremum (1) and an infimum (0), but it has neither a maximum, nor a minimum.

Exercises

1-15. Let $A \subseteq \mathbb{R}$ be nonempty and bounded below and let $s, t \in \mathbb{R}$ both be infima of A. Prove that $s = t$.

1-16. Approaching infima. State and prove a version of Proposition 1.21 that applies to infima. Is the proof significantly different from that of Proposition 1.21?

1-17. Let $S \subseteq \mathbb{R}$ be nonempty and bounded above. Prove that $s \in \mathbb{R}$ is the supremum of S iff s is an upper bound of S and for all $\varepsilon > 0$ there is an $x \in S$ so that $|s - x| < \varepsilon$.

1-18. Suprema and infima vs. containment of sets.

(a) Let $A, B \subseteq \mathbb{R}$ be nonempty and bounded above. Prove that $A \subseteq B$ implies $\sup(A) \leq \sup(B)$.

(b) Let $A, B \subseteq \mathbb{R}$ be nonempty and bounded below. Prove that $A \subseteq B$ implies $\inf(A) \geq \inf(B)$.

1-19. Let $A \subseteq \mathbb{R}$ be nonempty and bounded above. Prove that $\inf\{x \in \mathbb{R} : -x \in A\} = -\sup(A)$.

This section concludes the introduction of the axioms for the real numbers. Exercise 1-30 after the next section shows that the axioms *uniquely* determine the real numbers. We will not explicitly construct a mathematical entity that satisfies these axioms. Readers interested in the construction of $\mathbb{R}$ from $\mathbb{Q}$ can revisit this idea after Theorem 16.89 (see Exercise 16-93). The construction of the rational numbers from the axioms of set theory is sketched in Appendix C.

1.4 Natural Numbers, Integers, and Rational Numbers

Although Axioms 1.1, 1.6 and 1.19 uniquely describe the real numbers, they do not mention familiar subsets, such as natural numbers, integers, and rational numbers. This is because these sets can be constructed from the axioms as subsets of the real numbers. We start with the natural numbers, which are the unique subset with properties as stated in Theorem 1.23. While their existence is easy to establish, the uniqueness of the natural numbers can only be proved in Theorem 1.28 after some more machinery has been developed.

Theorem 1.23 *There is a subset* $\mathbb{N} \subseteq \mathbb{R}$*, called the* **natural numbers**, *so that*

1. $1 \in \mathbb{N}$.

2. *For each* $n \in \mathbb{N}$ *the number* $n + 1$ *is also in* $\mathbb{N}$.

3. **Principle of Induction**. *If* $S \subseteq \mathbb{N}$ *is such that* $1 \in S$ *and for each* $n \in S$ *we also have* $n + 1 \in S$, *then* $S = \mathbb{N}$.

Proof. Call a subset $A \subseteq \mathbb{R}$ a **successor set** iff $1 \in A$ and for all $a \in A$ we also have $a + 1 \in A$. Successor sets exist, because, for example, $\mathbb{R}$ itself is a successor set. Let $\mathbb{N}$ be the set of all elements of $\mathbb{R}$ that are in all successor sets. Because 1 is an element of every successor set, we infer $1 \in \mathbb{N}$. Moreover, if $n \in \mathbb{N}$, then n is in every successor set, which means $n + 1$ is in every successor set, and hence $n + 1 \in \mathbb{N}$. Finally, any subset $S \subseteq \mathbb{N}$ as given in the Principle of Induction is a successor set. Because the elements of $\mathbb{N}$ are contained in all successor sets, we conclude that $\mathbb{N} \subseteq S$, and hence $\mathbb{N} = S$. ■

Of course, we will denote the natural numbers by their usual names $1, 2, 3, \ldots$ As algebraic objects, natural numbers are suited for addition and multiplication (see Proposition 1.24), but they are not so well suited for subtraction (see Proposition 1.25). Although all results until Theorem 1.28 are stated for $\mathbb{N}$, they hold "for every subset of $\mathbb{R}$ that satisfies the properties in Theorem 1.23." The reader should keep this in mind and double check, because we will need it in the proof of Theorem 1.28. To avoid awkward formulations, the results up to Theorem 1.28 are formulated for $\mathbb{N}$, however.

Proposition 1.24 *The natural numbers are closed under addition and multiplication. That is, if* $m, n \in \mathbb{N}$, *then* $m + n$ *and* mn *are in* $\mathbb{N}$ *also.*

Proof. The key to this result is the Principle of Induction. Let $m \in \mathbb{N}$ be arbitrary and let $S_m := \{n \in \mathbb{N} : m+n \in \mathbb{N}\}$. Then $m \in \mathbb{N}$ implies $m+1 \in \mathbb{N}$, and hence $1 \in S_m$. Moreover, if $n \in S_m$, then $m + n \in \mathbb{N}$, and hence $m + (n + 1) = (m + n) + 1 \in \mathbb{N}$, which means that $n + 1 \in S_m$. By the Principle of Induction we conclude that $S_m = \mathbb{N}$. Because $m \in \mathbb{N}$ was arbitrary, this means that for any $m, n \in \mathbb{N}$ we have $m + n \in \mathbb{N}$.

The proof for products is similar and left to the reader as Exercise 1-20. ■

Readers familiar with induction recognize the part "$1 \in S_m$" of the preceding proof as the **base step** of an induction and the part "$n \in S_m \Rightarrow n + 1 \in S_m$" as the **induction step**. In this section, we use the "induction on sets" as done in the preceding proof. The more commonly known Principle of Induction is introduced in Theorem 1.39.

Proposition 1.25 *Let $m, n \in \mathbb{N}$ be such that $m > n$. Then $m - n \in \mathbb{N}$.*

Proof. We first show that if $m \in \mathbb{N}$, then $m - 1 \in \mathbb{N}$ or $m - 1 = 0$. To do this, let $A := \{m \in \mathbb{N} : m - 1 \in \mathbb{N} \text{ or } m - 1 = 0\}$. Then $1 \in A$ and if $m \in A$, then $(m + 1) - 1 = m \in A \subseteq \mathbb{N}$, which means $m + 1 \in A$. Hence, $A = \mathbb{N}$ by the Principle of Induction.

Now let $S := \{n \in \mathbb{N} : (\forall m \in \mathbb{N} : m > n \text{ implies } m - n \in \mathbb{N})\}$. If $n = 1$ and $m \in \mathbb{N}$ satisfies $m > 1$, then $m - 1 > 0$ and so by the above $m - 1 \in \mathbb{N}$, which means $1 \in S$. Let $n \in S$. If $m > n + 1$, then $m - 1 > n$, and hence $m - (n + 1) = (m - 1) - n \in \mathbb{N}$, which means $n + 1 \in S$. By the Principle of Induction we conclude that $S = \mathbb{N}$, and hence for all $m, n \in \mathbb{N}$ we have proved that $m > n$ implies $m - n \in \mathbb{N}$. ■

Proposition 1.26 shows that the natural numbers are positive and the smallest difference between any two of them is 1.

Proposition 1.26 *For all $n \in \mathbb{N}$, the inequality $n \geq 1$ holds and there is no $m \in \mathbb{N}$ so that the inequalities $n < m < n + 1$ hold.*

Proof. The proof that all natural numbers are greater than or equal to 1 is left to Exercise 1-21.

Now suppose for a contradiction that there is an $n \in \mathbb{N}$ and an $m \in \mathbb{N}$ so that $n < m < n + 1$. Then $m - n \in \mathbb{N}$ and $m - n < 1$, a contradiction. ■

The Well-ordering Theorem turns out to be equivalent to the Principle of Induction (see Exercise 1-22).

Theorem 1.27 Well-ordering Theorem. *Every nonempty subset of $\mathbb{N}$ has a smallest element.*

Proof. Suppose for a contradiction that $B \subseteq \mathbb{N}$ is not empty and does not have a smallest element. Let $S := \{n \in \mathbb{N} : (\forall m \in \mathbb{N} : m \leq n \text{ implies } m \notin B)\}$. By Proposition 1.26, 1 is less than or equal to all elements of $\mathbb{N}$, so $1 \notin B$, and hence $1 \in S$. Now let $n \in S$. Then all $m \in \mathbb{N}$ with $m \leq n$ are not in B. But then $n + 1 \in B$ would by Proposition 1.26 imply that $n + 1$ is the smallest element of B. Hence, $n + 1 \notin B$ and we conclude $n + 1 \in S$. By the Principle of Induction, $S = \mathbb{N}$ and consequently $B = \emptyset$, a contradiction. ■

Now we are finally ready to show that the natural numbers are unique.

Theorem 1.28 *The natural numbers $\mathbb{N}$ are the* unique *subset of $\mathbb{R}$ that satisfies the properties in Theorem 1.23.*

Proof. Examination of the proofs of all results since Theorem 1.23 reveals that any set $S \subseteq \mathbb{R}$ that satisfies the properties in Theorem 1.23 must also have the properties given in these results.

It may feel tedious to go back and verify the above statement. However, mathematical presentations more often than not will ask a reader to use a modification of a known proof to prove a result (also see Standard Proof Technique 1.14). When this occurs, the reader is expected to verify that the result(s) can indeed be proved with similar methods as were used for earlier results.

Now suppose for a contradiction that there is a set $S \neq \mathbb{N}$ with properties as in Theorem 1.23. Then S is a successor set, so $\mathbb{N} \subseteq S$. Let $B := S \setminus \mathbb{N} = \{s \in S : s \notin \mathbb{N}\}$. Then $B \neq \emptyset$, and hence by the Well-ordering Theorem, which is valid for S, B has a smallest element b. Because $1 \in \mathbb{N}$ we infer $b > 1$, and hence by Proposition 1.25, which is valid for S, we have $b - 1 \in S$. But then $b - 1 \notin \mathbb{N}$, because this would imply $b = (b-1) + 1 \in \mathbb{N}$. Hence, $b - 1 \in B$, which is a contradiction to the fact that b is the smallest element of B. ■

Once we have constructed the natural numbers, the next number system to consider are the integers.

Definition 1.29 *The set* $\mathbb{Z} := \{m \in \mathbb{R} : m \in \mathbb{N} \text{ or } m = 0 \text{ or } -m \in \mathbb{N}\}$ *is called the set of* **integers**.

We leave several proofs of natural properties of the integers to the reader.

Proposition 1.30 *The integers are closed under addition, subtraction and multiplication. Moreover, for any two integers k, l with $k > l$ we have that $k - l \geq 1$, every nonempty set $A \subseteq \mathbb{Z}$ that is bounded below has a minimum, and every nonempty set $A \subseteq \mathbb{Z}$ that is bounded above has a maximum.*

Proof. To prove that $\mathbb{Z}$ is closed under addition, let $m, n \in \mathbb{Z}$. In case both are natural numbers or in case one of them is zero, there is nothing to prove. Moreover, in case $-m, -n \in \mathbb{N}$ we have $m + n = -\big((-m) + (-n)\big)$, which is in $\mathbb{Z}$, because $(-m) + (-n) \in \mathbb{N}$. Now consider the case $m \in \mathbb{N}$ and $-n \in \mathbb{N}$. If $m = -n$, we obtain $m + n = 0 \in \mathbb{Z}$. If $m > -n$, then by Proposition 1.25 we conclude that $m + n = m - (-n) \in \mathbb{N} \subseteq \mathbb{Z}$. Finally, if $m < -n$ again by Proposition 1.25 we conclude that $-(m + n) = (-n) - m \in \mathbb{N}$, which means by definition of $\mathbb{Z}$ that $m + n \in \mathbb{Z}$. The case $-m \in \mathbb{N}$ and $n \in \mathbb{N}$ is treated similarly (see Exercise 1-23a).

Closedness under subtraction and multiplication as well as the claim about differences are left to Exercises 1-23b–1-23d.

Now let $A \subseteq \mathbb{Z}$ be nonempty and bounded below. Then, because $A \subseteq \mathbb{R}$, it has an infimum a. By the version of Proposition 1.21 for infima, there is an integer $m \in A$ with $m - a < 1$. Because the absolute value of the difference between any two distinct integers is at least 1, m is the only integer in $[a, a+1)$. Hence, m is below all elements of A that are not in $[a, a+1)$. Because m is the only element of A in $[a, a+1)$, m must be the minimum of A.

The proof of the corresponding result for nonempty subsets $A \subseteq \mathbb{Z}$ that are bounded above is left to Exercise 1-23e. ■

A key property of the natural numbers is that any real number is exceeded by a natural number. To prove this, we need the usual fractions, which are easily introduced.

Definition 1.31 *For all $a \in \mathbb{R} \setminus \{0\}$ we set $\frac{1}{a} := a^{-1}$ and call it the* **reciprocal** *of a. For $b \in \mathbb{R}$ and $a \in \mathbb{R} \setminus \{0\}$ we set $\frac{b}{a} := b \cdot \frac{1}{a} = ba^{-1}$ and call it a* **fraction**.

Because $\frac{1}{2}+\frac{1}{2}=2^{-1}+2^{-1}=(1+1)\cdot 2^{-1}=2\cdot 2^{-1}=1$ we can now prove the following.

Theorem 1.32 *For every $x \in \mathbb{R}$, there is an $n \in \mathbb{N}$ so that $n \geq x$.*

Proof. For a contradiction, suppose that x is such that for all $n \in \mathbb{N}$ we have that $n < x$. Then $B := \{y \in \mathbb{R} : (\forall n \in \mathbb{N} : n < y)\}$ is not empty. Moreover, B is bounded below by all $n \in \mathbb{N}$. By the Completeness Axiom, B has an infimum, call it b. Then $b - \frac{1}{2} \notin B$, which means there is an $n \in \mathbb{N}$ with $n \geq b - \frac{1}{2}$. But then $n+1 \geq b + \frac{1}{2}$ is a lower bound of B, a contradiction to $b = \inf(B)$. ■

Because $\mathbb{N} \subseteq \mathbb{Z}$ and because subsets of $\mathbb{Z}$ that are bounded below have a minimum, we infer that for every real number x there is a unique smallest integer that is greater than or equal to x. Similarly there is a unique largest integer that is less than or equal to x. These numbers are useful when we need integers instead of real numbers, so we define the following.

Definition 1.33 *For every $x \in \mathbb{R}$, let $\lceil x \rceil$ be the smallest integer greater than or equal to x. Moreover, let $\lfloor x \rfloor$ be the largest integer less than or equal to x. As functions from $\mathbb{R}$ to $\mathbb{Z}$, $\lceil \cdot \rceil$ is called the* **ceiling function** *and $\lfloor \cdot \rfloor$ is called the* **floor function**.

The last subset of $\mathbb{R}$ that we introduce is the set of rational numbers. Rational numbers are naturally defined as fractions.

Definition 1.34 *The set $\mathbb{Q} := \left\{\frac{n}{d} : n \in \mathbb{Z}, d \in \mathbb{N}\right\}$ is called the set of* **rational numbers**. *The set $\mathbb{R} \setminus \mathbb{Q} := \{x \in \mathbb{R} : x \notin \mathbb{Q}\}$ is called the set of* **irrational numbers**.

Proposition 1.35 *The rational numbers are closed under addition, subtraction and multiplication. Moreover, if $q, r \in \mathbb{Q}$ and $r \neq 0$, then $\frac{q}{r} \in \mathbb{Q}$.*

Proof. Let $m, n \in \mathbb{Z}$, let $c, d \in \mathbb{N}$ and consider the rational numbers $\frac{m}{c}$ and $\frac{n}{d}$. Then $\mathbb{Q}$ is closed under addition because

$$\begin{aligned} \frac{m}{c}+\frac{n}{d} &= mc^{-1}+nd^{-1} = mdd^{-1}c^{-1}+ncc^{-1}d^{-1} \\ &= (md+nc)c^{-1}d^{-1} = \frac{md+nc}{cd}. \end{aligned}$$

For multiplication, note that $\frac{m}{c}\frac{n}{d} = mc^{-1}nd^{-1} = mnc^{-1}d^{-1} = \frac{mn}{cd}$. The remainder is left to Exercise 1-24. ■

Rational numbers can be found between any two real numbers and Exercise 1-45 will establish a similar result for irrational numbers.

Theorem 1.36 *Let $a, b \in \mathbb{R}$ with $a < b$. Then there is a rational number $q \in \mathbb{Q}$ such that $a < q < b$.*

Proof. By Theorem 1.32, there is an $n \in \mathbb{N}$ so that $0 < \frac{1}{b-a} < n$. By part 5 of Theorem 1.10, we obtain $\frac{1}{n} < b - a$. Now let $u := \min\left\{m \in \mathbb{Z} : \frac{m}{n} \geq b\right\}$ and similarly let $l := \max\left\{m \in \mathbb{Z} : \frac{m}{n} \leq a\right\}$. Then $\frac{u}{n} - \frac{l}{n} \geq b - a > \frac{1}{n}$, which means $\frac{l+1}{n} < \frac{u}{n}$. Hence, by definition of l and u we infer $a < \frac{l+1}{n} < b$. ■

We conclude with a simple looking result that is actually at the heart of a standard proof technique (see Standard Proof Technique 2.7). Exercise 1-25 extends Theorem 1.37 to inequalities.

Theorem 1.37 *Let $x \in \mathbb{R}$. If $x \geq 0$ and for all $\varepsilon > 0$ we have $x \leq \varepsilon$, then $x = 0$.*

Proof. Let x be as indicated and suppose for a contradiction that $x > 0$. Then $\varepsilon := \frac{x}{2}$ is positive and $x \leq \varepsilon = \frac{x}{2}$ implies $1 \leq \frac{1}{2}$, a contradiction. ■

Exercises

1-20 Prove that if $m, n \in \mathbb{N}$, then $mn \in \mathbb{N}$.

Hint. Same idea as the first part of the proof of Proposition 1.24 with sets $S_m := \{n \in \mathbb{N} : mn \in \mathbb{N}\}$.

1-21 Prove that if $n \in \mathbb{N}$, then $n \geq 1$.

Hint. Use $S := \{n \in \mathbb{N} : n \geq 1\}$.

1-22. Use the Well-ordering Theorem to prove the Principle of Induction.

1-23 Finish the proof of Proposition 1.30 by proving the following.

(a) Finish the proof that $\mathbb{Z}$ is closed under addition. That is, prove that if $-m \in \mathbb{N}$ and $n \in \mathbb{N}$, then $m + n \in \mathbb{Z}$.

(b) Prove that $\mathbb{Z}$ is closed under subtraction. That is, prove that $m - n \in \mathbb{Z}$ for all $m, n \in \mathbb{Z}$.

(c) Prove that $\mathbb{Z}$ is closed under multiplication. That is, prove that $mn \in \mathbb{Z}$ for all $m, n \in \mathbb{Z}$.

(d) Prove that for any two integers m, n with $m > n$ we have $m - n \geq 1$.

Hint. Find a contradiction to Proposition 1.26.

(e) Prove that every nonempty set $A \subseteq \mathbb{Z}$ that is bounded above has a maximum.

1-24. Finish the proof of Proposition 1.35. That is,

(a) Prove that $\mathbb{Q}$ is closed under subtraction.

(b) Prove that if $q, r \in \mathbb{Q}$ and $r \neq 0$, then $\frac{q}{r} \in \mathbb{Q}$.

Hint. First show that for $n \in \mathbb{Z} \setminus \{0\}$ and $d \in \mathbb{N}$ we have that $\left(\frac{n}{d}\right)^{-1} = \frac{d}{n}$.

1-25. Prove that if $a, b \in \mathbb{R}$ are such that for all $\varepsilon > 0$ we have $a \leq b + \varepsilon$, then $a \leq b$.

1-26. Prove that for every real number x there is an integer n so that $n \leq x$.

1-27. Prove that for any real numbers $x, \varepsilon > 0$ there is an $n \in \mathbb{N}$ so that $\frac{x}{n} < \varepsilon$.

Hint. Theorem 1.32.

1-28. Prove that $\frac{1}{3} + \frac{1}{3} + \frac{1}{3} = 1$.

1-29. A rational number r is called a **dyadic rational number** iff there are $p \in \mathbb{Z}$ and $n \in \mathbb{N}$ so that $r = \frac{p}{2^n}$. Dyadic rational numbers are useful in analysis because they can provide a sequence of "grids" such that each new grid contains the old one (see part 1-29a below), the whole set is the union of the "grids" (see part 1-29b) and between any two real numbers there is a dyadic rational number (see part 1-29c).

Let D be the set of dyadic rational numbers and for each $n \in \mathbb{N}$ let $D_n := \left\{ \frac{p}{2^n} : p \in \mathbb{Z} \right\}$.

(a) Prove that for all $n \in \mathbb{N}$ we have $D_n \subset D_{n+1}$.

(b) Let $\bigcup_{n=1}^{\infty} D_n := \{ x \in \mathbb{R} : (\exists n \in \mathbb{N} : x \in D_n) \}$ and prove that $D = \bigcup_{n=1}^{\infty} D_n$.

(c) Prove that for any $x, y \in \mathbb{R}$ with $x < y$ there is a dyadic rational number d so that $x < d < y$.

1-30 In this exercise, we will prove that the **real numbers are the (up to isomorphism) unique linearly ordered complete field**. That is, we will prove that every mathematical object that satisfies Axioms 1.1, 1.6, and 1.19 is in a certain sense (defined below) "the same as $\mathbb{R}$."

First notice that, similar to the proof of Theorem 1.28, *all* results proved so far hold for *any* object that satisfies Axioms 1.1, 1.6, and 1.19 (because the results are derived from these axioms). That is, every set $\tilde{\mathbb{R}}$ that satisfies Axioms 1.1, 1.6, and 1.19 contains subsets $\tilde{\mathbb{N}}$, $\tilde{\mathbb{Z}}$, and $\tilde{\mathbb{Q}}$ that have the properties that we have proved up to now for the natural numbers, the integers and the rational numbers.

(a) Prove that for all $x \in \mathbb{R}$ we have that $x = \sup\{r \in \mathbb{Q} : r \leq x\}$.

(b) Now let $\tilde{\mathbb{R}}$ be a set that satisfies Axioms 1.1, 1.6, and 1.19 and let $\tilde{\mathbb{N}}$, $\tilde{\mathbb{Z}}$, and $\tilde{\mathbb{Q}}$ be subsets of $\tilde{\mathbb{R}}$ that have the properties that we have proved up to now for the natural numbers, the integers and the rational numbers, including Exercise 1-30a.

i. Define a function $f : \mathbb{Q} \to \tilde{\mathbb{R}}$ as follows. For $n \in \mathbb{N}$, let $f(1) := \tilde{1}$ and once $f(n)$ is defined let $f(n+1) := f(n)\tilde{+}\tilde{1}$. Also let $f(-n) = \tilde{-}\tilde{n}$. For $n \in \mathbb{Z}$ and $d \in \mathbb{N}$ let $f\left(\frac{n}{d}\right) := \frac{\tilde{n}}{\tilde{d}}$. Prove that for all $x \in \mathbb{Q}$ the above definition is not self-contradictory by proving that it assigns exactly one value to each $x \in \mathbb{Q}$. Then prove that $f(x) \in \tilde{\mathbb{Q}}$ for each $x \in \mathbb{Q}$ and that f preserves the order, that is, if $x < z$, then $f(x) < f(z)$.

ii. For $x \in \mathbb{R}$ let $f(x) = \sup \{ f(r) : r \in \mathbb{Q} \text{ and } r \leq x \}$. Prove that for all $x \in \mathbb{R}$ the above definition is not self-contradictory by proving it assigns exactly one value to each $x \in \mathbb{R}$.
(Formally this says that f is **well-defined**.)

iii. Prove that the above function does not map any two points to the same image by proving that for all $x, y \in \mathbb{R}$ the inequality $x \neq y$ implies that $f(x) \neq f(y)$.
(Formally, this says that the function f is **one-to-one** or **injective**.)

iv. Prove that the above function "reaches" every element of $\tilde{\mathbb{R}}$ by proving that for all $\tilde{x} \in \tilde{\mathbb{R}}$ there is an $x \in \mathbb{R}$ so that $f(x) = \tilde{x}$.
(Formally, this says that the function f is **onto** or **surjective**.)

v. Prove that the above function is consistent with the algebraic operations by proving that for all $x, y \in \mathbb{R}$ we have that $f(x+y) = f(x)\tilde{+}f(y)$ and $f(x \cdot y) = f(x)\tilde{\cdot}f(y)$.
(Formally, this says that f is a **field isomorphism**.)

vi. Prove that the above function is consistent with the order relation by proving that for all $x, y \in \mathbb{R}$ we have that $x \leq y$ implies that $f(x)\tilde{\leq}f(y)$.
(Formally, this says that f is an **order isomorphism**.)

The above steps show that the points and operations in $\mathbb{R}$ and in $\tilde{\mathbb{R}}$ can be identified with each other in such a way that it does not matter if we are working in $\mathbb{R}$ or in $\tilde{\mathbb{R}}$. Thus for all intents and purposes, $\mathbb{R}$ and $\tilde{\mathbb{R}}$ are "the same." This is the essence of saying that the real numbers are up to isomorphism the unique linearly ordered, complete field.

1.5 Recursion, Induction, Summations, and Products

A recursive definition defines an entity X_n that depends on a natural number n first for $n = 1$ and then it defines X_{n+1} in terms of X_n. By the Principle of Induction the set $S = \{n \in \mathbb{N} : X_n \text{ is defined}\}$ is equal to $\mathbb{N}$, which means that a recursive definition defines the entity X_n for all natural numbers n. In this fashion, the sum of finitely many numbers can be defined.

Definition 1.38 *For each $j \in \mathbb{N}$ let $a_j \in \mathbb{R}$. Define the* **sum** $\sum_{j=1}^{1} a_j := a_1$ *and for $n \in \mathbb{N}$ define the* **sum** $\sum_{j=1}^{n+1} a_j := a_{n+1} + \sum_{j=1}^{n} a_j$. *For $m \in \mathbb{N} \cup \{0\}$, set* $\sum_{j=1}^{-m} a_j := 0$. *The parameter j is also called the* **summation index**.

In particular, note that a sum whose index starts at 1 and ends at a number smaller than 1 is always zero. It is also called an **empty sum**. Summations that start at numbers other than 1 are defined similarly (Exercise 1-31). By their nature, recursive definitions are closely linked to induction. Unlike what is stated in Theorem 1.23, induction normally is used to prove statements about natural numbers. This is possible, because a proof that a statement is true for all natural numbers is the same as a proof that a certain set is equal to $\mathbb{N}$.

Theorem 1.39 Principle of Induction. *Let $P(n)$ be a statement about the natural number n. If $P(1)$ is true and if for all $n \in \mathbb{N}$ truth of $P(n)$ implies truth of $P(n+1)$, then $P(n)$ holds for all natural numbers.*

Proof. Let P be as indicated and consider the set $S := \{n \in \mathbb{N} : P(n) \text{ is true}\}$. Then $1 \in S$. For every $n \in S$ the statement $P(n)$ is true, hence $P(n+1)$ is true, which means $n + 1 \in S$. By Theorem 1.23 we conclude $S = \mathbb{N}$ and thus $P(n)$ is true for all $n \in \mathbb{N}$. ■

Standard Proof Technique 1.40 In the form of Theorem 1.39, **induction** is a standard proof technique. It involves a two-step process. In the first step, called the **base step**, $P(1)$ is proved. Then, in the **induction step**, $P(n)$ is used to prove $P(n+1)$. In this context, $P(n)$ is also called the **induction hypothesis**. All proofs in this section rely on induction. Moreover, Exercise 1-32 exhibits another way to carry out an induction (sometimes called strong induction). □

Example 1.41 *For all $n \in \mathbb{N}$, the* **summation formula** $\sum_{j=1}^{n} j = \frac{1}{2}n(n+1)$ *holds.*

Proof. The statement is $P(n) = \left[\sum_{j=1}^{n} j = \frac{1}{2}n(n+1)\right]$.

Base step. We prove $P(n)$ for $n = 1$. $\sum_{j=1}^{1} j = 1 = \frac{1}{2}1(1+1)$, so $P(1)$ holds.

Induction step. Under the induction hypothesis $\sum_{j=1}^{n} j = \frac{1}{2}n(n+1)$ we must prove $\sum_{j=1}^{n+1} j = \frac{1}{2}(n+1)\big((n+1)+1\big)$. A standard step in induction for recursively defined quantities is to split off the last term. This is done in the first step here.

$$\begin{aligned}\sum_{j=1}^{n+1} j &= (n+1)+\sum_{j=1}^{n} j\\ &= (n+1)+\frac{1}{2}n(n+1) = \frac{1}{2}2(n+1)+\frac{1}{2}n(n+1)\\ &= \frac{1}{2}(n+2)(n+1).\end{aligned}$$

Now we can apply the induction hypothesis to the sum and the rest is algebra.

Further examples of similar inductions can be found in Exercise 1-33. □

Similar to sums we can define products. Although products occur less frequently than sums, they are useful to define powers.

Definition 1.42 *For each $j \in \mathbb{N}$, let $a_j \in \mathbb{R}$. Define the* **product** $\prod_{j=1}^{1} a_j := a_1$ *and for all $n \in \mathbb{N}$ define the* **product** $\prod_{j=1}^{n+1} a_j := a_{n+1} \cdot \prod_{j=1}^{n} a_j$. *For all $m \in \mathbb{N} \cup \{0\}$, set $\prod_{j=1}^{-m} a_j := 1$. The parameter j is also called the* **product index**.

Products that start at numbers other than 1 are defined similarly (Exercise 1-31). Products that end at an index that is smaller than the starting index are set to 1 and are also called **empty products**.

Definition 1.43 *For all $a \in \mathbb{R}$, and all $n \in \mathbb{N}\cup\{0\}$, we define the nth* **power** $a^n := \prod_{j=1}^{n} a$.

Aside from integer powers of numbers, we want to work with rational powers. To define rational powers, we need n^{th} roots of nonnegative real numbers. To formally prove their existence, we need the Binomial Theorem. As a start we need binomial coefficients and one of their key properties.

Definition 1.44 *For all $n \in \mathbb{N} \cup \{0\}$, we define $n! := \prod_{j=1}^{n} j$ and call it the* **factorial** *of n. For all $n, k \in \mathbb{N} \cup \{0\}$ with $k \leq n$, we define the* **binomial coefficient** *as $\binom{n}{k} := \frac{n!}{k!(n-k)!}$.*

Theorem 1.45 *The equation* $\binom{n}{k-1}+\binom{n}{k}=\binom{n+1}{k}$ *holds for all* $n, k \in \mathbb{N}$ *with* $k \leq n$.

Proof. This result can be proved by direct computation.

$$\begin{aligned}
\binom{n}{k-1}+\binom{n}{k} &= \frac{n!}{(k-1)!\big(n-(k-1)\big)!}+\frac{n!}{k!(n-k)!} \\
&= \frac{n!k}{k!(n+1-k)!}+\frac{n!(n+1-k)}{k!(n+1-k)!}=\frac{n!(k+n+1-k)}{k!(n+1-k)!} \\
&= \frac{(n+1)!}{k!(n+1-k)!}.
\end{aligned}$$

■

Now we are ready to prove the Binomial Theorem.

Theorem 1.46 *The* **Binomial Theorem**. *For all real numbers* $a, b \in \mathbb{R}$, *and all* $n \in \mathbb{N}$, *we have* $(a+b)^n=\sum_{k=0}^{n}\binom{n}{k}a^k b^{n-k}$.

Proof. Throughout the proof we will freely use the properties of sums proved in Exercise 1-34. The proof is by induction on n, with $P(n)$ being the statement about $(a+b)^n$.

Base step. For $n=1$, note that

$$(a+b)^1=a+b=\binom{1}{0}a^0b^{1-0}+\binom{1}{1}a^1b^{1-1}=\sum_{k=0}^{1}\binom{1}{k}a^kb^{1-k},$$

which proves the base step.

Induction step. Assuming that the result holds for n, we must prove it for $n+1$. First note that it follows easily from Definition 1.43 that for all $x \in \mathbb{R}$ and all $m \in \mathbb{N}$ we have $x \cdot x^m = x^{m+1}$.

$(a+b)^{n+1}$

The first step is to split off the last term of the power.

$= (a+b)(a+b)^n$

Now we can apply the induction hypothesis to $(a+b)^n$.

$$= (a+b)\sum_{k=0}^{n}\binom{n}{k}a^kb^{n-k}$$

$$= \sum_{k=0}^{n}\binom{n}{k}a^{k+1}b^{n-k}+\sum_{k=0}^{n}\binom{n}{k}a^kb^{n+1-k}$$

After multiplying out parentheses, we want to combine the sums. In order to do this we shift the indices to obtain similar terms in both sums. In the first sum we set $j := k + 1$ and in the second sum we set $j := k$.

$$= \sum_{j=1}^{n+1} \binom{n}{j-1} a^j b^{n+1-j} + \sum_{j=0}^{n} \binom{n}{j} a^j b^{n+1-j}$$

To combine the sums, the indices must start and end at the same numbers. Thus we split off the last term of the first sum and the first term of the second sum. Then we can combine the sums.

$$= \binom{n}{n} a^{n+1} b^{n+1-(n+1)} + \binom{n}{0} a^0 b^{n+1-0} + \sum_{j=1}^{n} \left[\binom{n}{j-1} + \binom{n}{j} \right] a^j b^{n+1-j}$$

Now we can apply Theorem 1.45. Moreover, by rewriting the terms outside the sum, we see that they fit the requisite pattern and can be absorbed into the sum.

$$= \binom{n+1}{n+1} a^{n+1} b^{n+1-(n+1)} + \binom{n+1}{0} a^0 b^{n+1-0} + \sum_{j=1}^{n} \binom{n+1}{j} a^j b^{n+1-j}$$

$$= \sum_{j=0}^{n+1} \binom{n+1}{j} a^j b^{n+1-j}.$$

■

With the Binomial Theorem, we can prove that n^{th} roots exist. The proof of Theorem 1.47 is the first proof in this text in which we have to choose a number to make another number smaller than a given bound. That is, this is our first proof with a distinct analytical flavor.

Theorem 1.47 *Let $n \in \mathbb{N}$. For every nonnegative real number a, there exists a unique nonnegative real number r such that $r^n = a$.*

Proof. We first prove the existence of r. Let $R := \{x \in \mathbb{R} : x \geq 0 \text{ and } x^n \leq a\}$. Then $0 \in R$ and R is bounded above by $\max\{1, a\}$. Let $r := \sup(R)$. To show that $r^n = a$, we will show that $r^n \not< a$ and $r^n \not> a$. First, suppose for a contradiction that $r^n < a$. Then there is a $\delta > 0$ so that $r^n + \delta < a$. By Theorem 1.32 (or Exercise 1-27), for each $k \in \{1, \ldots, n\}$ we can find an $m_k \in \mathbb{N}$ so that $\binom{n}{k} r^{n-k} \frac{1}{m_k^k} < \frac{\delta}{n}$. Let $m := \max\{m_1, \ldots, m_n\}$. Then by the Binomial Theorem we conclude

$$\left(r + \frac{1}{m}\right)^n = \sum_{k=0}^{n} \binom{n}{k} r^{n-k} \left(\frac{1}{m}\right)^k$$

Split off the zeroth term.

$$= r^n + \sum_{k=1}^{n} \binom{n}{k} r^{n-k} \frac{1}{m^k}$$

Now use the definition of m.

$$< \quad r^n + \sum_{k=1}^{n} \frac{\delta}{n} = r^n + \delta < a.$$

The above shows that $r + \frac{1}{m} \in R$, contradicting the fact that $r = \sup(R)$. Hence, $r^n \not< a$. The proof that $r^n \not> a$ is similar and left to the reader as Exercise 1-36.

For uniqueness, suppose for a contradiction that there is another $b \geq 0$ with $b^n = a$. Then $b < r$ or $b > r$. But if $b > r$, then with $\delta := b - r$ we obtain $a = b^n = (r+\delta)^n = r^n + \sum_{k=1}^{n} \binom{n}{k} r^{n-k}\delta^k > a$, a contradiction. Hence, $b < r$. But then with $\delta := r - b$ we have $a = r^n = (b+\delta)^n = b^n + \sum_{k=1}^{n} \binom{n}{k} b^{n-k}\delta^k > a$, a contradiction. Therefore r is unique. ■

We conclude by defining rational powers of nonnegative numbers and by proving some of their properties.

Definition 1.48 *Let $n \in \mathbb{N}$ and let $a \in \mathbb{R}$ be nonnegative. The unique nonnegative real number r such that $r^n = a$ is called the nth* **root of** *a, denoted $\sqrt[n]{a}$. For $n = 2$ the root is called the* **square root**, *denoted $\sqrt{a}$.*

Existence of nth roots is another property that distinguishes $\mathbb{R}$ from $\mathbb{Q}$. Although Theorem 1.36 indicates that there are "many" rational numbers, the rational number system has some shortcomings when it comes to powers.

Proposition 1.49 *There is no rational number r such that $r^2 = 2$.*

Proof. We first prove by induction as stated in Exercise 1-32 (strong induction) that if $n^2 = 2z$ for some $z \in \mathbb{N}$, then $n = 2z'$ for some $z' \in \mathbb{N}$. The base step for $n = 1$ is **vacuously true**. That is, because the hypothesis $1^2 = 2z$ leads to the contradiction $1 = 1^2 = 2z = z + z > 1$, the hypothesis is never true, which means that the implication is automatically true (see Definition A.2 in Appendix A).

For the induction step, first note that the result is trivial for $n = 2$, because $2 = 2 \cdot 1$. Now assume that $n > 2$ and the statement has been proved for all natural numbers less than n. Then $2z = n^2 = (n - 2 + 2)^2 = (n-2)^2 + 4(n-2) + 4$ implies that $(n-2)^2 = 2\bar{z}$ for some $\bar{z} \in \mathbb{N}$. By induction hypothesis, we conclude that $n - 2 = 2\tilde{z}$ for some $\tilde{z} \in \mathbb{N}$, and hence $n = 2\tilde{z} + 2 = 2z'$ for some $z' \in \mathbb{N}$. This proves that if $n^2 = 2z$ for some $z \in \mathbb{N}$, then $n = 2z'$ for $z' = \tilde{z} + 1 \in \mathbb{N}$.

Now suppose for a contradiction that there are $n \in \mathbb{Z}$ and $d \in \mathbb{N}$ so that $\left(\frac{n}{d}\right)^2 = 2$ and such that there is no $k \in \mathbb{N} \setminus \{1\}$ such that $n = n_k \cdot k$ and $d = d_k \cdot k$. But by the above $n^2 = 2d^2$ implies $n = n_2 \cdot 2$. Consequently, $2d^2 = (n_2 \cdot 2)^2$, that is, $d^2 = n_2^2 \cdot 2$, which implies $d = d_2 \cdot 2$, a contradiction. ■

We conclude from Theorem 1.47 and Proposition 1.49 that $\sqrt{2}$ is irrational.

For odd natural numbers (that is, natural numbers of the form $n = 2k + 1$), it is possible to define the nth root of a negative number $a < 0$ as $\sqrt[n]{a} := -\sqrt[n]{|a|}$. For the most part, powers are considered for nonnegative numbers, though.

Definition 1.50 *For all real numbers $a \geq 0$, all $m \in \mathbb{N} \cup \{0\}$, $n \in \mathbb{N}$ and all $q \in \mathbb{Q}$ with $q > 0$ we define*

1. $a^{\frac{1}{n}} := \sqrt[n]{a}$. *That is, the* $\left(\frac{1}{n}\right)^{\text{th}}$ **power** *of a is the nth root of a.*

2. $a^{\frac{m}{n}} := \left(a^m\right)^{\frac{1}{n}}$.

3. $a^{-q} := \left(a^q\right)^{-1} = \frac{1}{a^q}$ *for $a \neq 0$.*

Theorem 1.51 *For all positive numbers a and b and all rational numbers x and y, the following* **power laws** *hold:*

$$a^x a^y = a^{x+y} \qquad (ab)^x = a^x b^x$$

$$\frac{a^x}{a^y} = a^{x-y} = \frac{1}{a^{y-x}} \qquad \left(\frac{a}{b}\right)^x = \frac{a^x}{b^x}$$

$$\left(a^x\right)^y = a^{xy}$$

Proof. We first prove $(ab)^x = a^x b^x$. For exponents $n \in \mathbb{N}$, this is an easy induction. The base step $n = 1$ is trivial and the induction step from n to $n+1$ is $(ab)^{n+1} = ab(ab)^n = aba^n b^n = aa^n bb^n = a^{n+1}b^{n+1}$.

For rational exponents $\frac{n}{d}$ with $n, d \in \mathbb{N}$, we have that $\left((ab)^{\frac{n}{d}}\right)^d = (ab)^n$ and $\left(a^{\frac{n}{d}} b^{\frac{n}{d}}\right)^d = \left(a^{\frac{n}{d}}\right)^d \left(b^{\frac{n}{d}}\right)^d = a^n b^n = (ab)^n$. *Note that in both equalities we used the definition of fractional powers,* not *the power law that we are currently proving.* Because all numbers involved are positive and d^{th} roots are unique, we conclude that $(ab)^{\frac{n}{d}} = a^{\frac{n}{d}} b^{\frac{n}{d}}$.

For $x = 0$, the equality $(ab)^x = a^x b^x$ is trivial. Finally, for all positive $x \in \mathbb{Q}$ we note $(ab)^{-x} a^x b^x = (ab)^{-x}(ab)^x = 1$. Therefore $(ab)^{-x}$ is the multiplicative inverse of $a^x b^x$, that is, $(ab)^{-x} = a^{-x} b^{-x}$. Thus $(ab)^x = a^x b^x$ for all $a, b > 0$ and all $x \in \mathbb{Q}$.

To prove that $a^{x+y} = a^x a^y$ we proceed similarly. For exponents $m, n \in \mathbb{N}$, the proof for arbitrary m is an induction on n. The base step $a^m a^1 = a^m a = a^{m+1}$ follows straight from the definition of powers with natural exponents. For the induction step from n to $n+1$, note that $a^m a^{n+1} = a^m a^n a = a^{m+n} a = a^{m+(n+1)}$, which proves the result for exponents $m, n \in \mathbb{N}$.

For positive rational exponents x and y, note that there are $m, n, d \in \mathbb{N}$ so that $x = \frac{m}{d}$ and $y = \frac{n}{d}$. Then, using the equality we already proved, we obtain

$$a^x a^y = a^{\frac{m}{d}} a^{\frac{n}{d}} = \left(a^m\right)^{\frac{1}{d}} \left(a^n\right)^{\frac{1}{d}} = \left(a^m a^n\right)^{\frac{1}{d}} = \left(a^{m+n}\right)^{\frac{1}{d}} = a^{\frac{m+n}{d}} = a^{x+y}.$$

The equality is trivial if one of x and y is zero.

In case both exponents are negative, note that for all positive $x, y \in \mathbb{Q}$ we have $a^{x+y} a^{-x} a^{-y} = a^x a^y a^{-x} a^{-y} = 1$, which means $a^{-x} a^{-y} = a^{-x+(-y)}$ as was to be proved. This leaves the case in which one exponent is positive and the other is negative. Let $x, y \in \mathbb{Q}$ be positive and consider a^{x-y}. If the inequality $|x| > |y|$ holds we have $a^{x-y}a^y a^{-x} = a^x a^{-x} = 1$, which means that $a^{x-y} = a^x a^{-y}$. If $|x| < |y|$ we have $a^{y-x}a^x a^{-y} = a^y a^{-y} = 1$, which means that $a^{y-x} = a^y a^{-x}$. If $|x| = |y|$ the claim is trivial. Thus $a^{x+y} = a^x a^y$ for all $a > 0$ and all $x, y \in \mathbb{Q}$.

We leave the remaining three equalities as Exercise 1-37. ■

Power laws for $a \leq 0$ and $b \leq 0$ (as applicable) can be proved similarly. To conclude, note that the results presented in this chapter guarantee that the real numbers have the properties we expect them to have. We will therefore use the usual notation (fractions, etc.) and laws of algebra throughout this text without further qualms about the need to justify that we are indeed allowed to do so.

Exercises

1-31. Let $k, m \in \mathbb{Z}$ and for each $j \in \mathbb{Z}$ let $a_j \in \mathbb{R}$. Define the sum $\sum_{j=k}^{m} a_j$ and the product $\prod_{j=k}^{m} a_j$.

1-32. Let $P(n)$ be a statement about the natural number n. Prove that if $P(1)$ is true and if for all $n \in \mathbb{N}\setminus\{1\}$ truth of $P(1), \ldots, P(n-1)$ implies truth of $P(n)$, then $P(n)$ holds for all natural numbers. This type of induction is sometimes called **strong induction**.

Hint. Consider $S := \left\{n \in \mathbb{N} : \left(\forall k < n : P(k) \text{ holds }\right)\right\}$.

1-33. Prove each of the following by induction.

(a) $\sum_{j=1}^{n} j^2 = \frac{n}{6}(n+1)(2n+1)$

(b) $\sum_{j=1}^{n} j^3 = \frac{1}{4}n^2(n+1)^2$

(c) $\sum_{j=1}^{n} j^4 = \frac{1}{30}n(n+1)(2n+1)\left(3n^2+3n-1\right)$

(d) **Bernoulli's inequality** Prove that for all real numbers $x > -1$, $x \neq 0$ and $n \geq 2$ we have that $(1+x)^n > 1 + nx$.

1-34. **Properties of sums and products**. Let $c \in \mathbb{R}$ and for all $j \in \mathbb{N}$ let a_j and b_j be real numbers.

(a) Prove that for all $n \in \mathbb{N}$ we have $\sum_{j=1}^{n}(a_j + b_j) = \sum_{j=1}^{n} a_j + \sum_{j=1}^{n} b_j$.

(b) Prove that for all $n \in \mathbb{N}$ we have $\sum_{j=1}^{n}(ca_j) = c\sum_{j=1}^{n} a_j$.

(c) Prove that for all $n \in \mathbb{N}$ we have $\sum_{j=1}^{n} 1 = n$.

(d) Prove that for all $n \in \mathbb{N}$ we have $\prod_{j=1}^{n}(a_j \cdot b_j) = \left(\prod_{j=1}^{n} a_j\right)\left(\prod_{j=1}^{n} b_j\right)$.

1-35. **Reindexing sums**. Let $s \in \mathbb{Z}$, $n \in \mathbb{N}$ and for $j \in \mathbb{Z}$ let $a_j \in \mathbb{R}$. Prove that $\sum_{j=s}^{s+n} a_j = \sum_{k=1}^{n+1} a_{k+s-1}$

1-36 Finish the proof of Theorem 1.47 by showing that $r^n \not> a$
Hint. Suppose $r^n > a$ and prove that then for some $\varepsilon > 0$ and *all* $\delta \in (0, \varepsilon)$ we have $r^n - \delta \notin R$.

1-37. Finish the proof of Theorem 1.51 That is, let a and b be positive real numbers, let $x, y \in \mathbb{Q}$ and prove each of the following.

(a) $\frac{a^x}{a^y} = a^{x-y} = \frac{1}{a^{y-x}}$ (b) $\left(\frac{a}{b}\right)^x = \frac{a^x}{b^x}$ (c) $\left(a^x\right)^y = a^{xy}$

1-38. Let $0 \leq a < b$ and let $q > 0$ be rational Prove that $a^q < b^q$

1-39. Let $a, x \in (0, \infty)$ and let x be a rational number.

(a) Prove that if $a > 1$ and $x > 1$, then $a^x > a$
Hint. Let $p, q \in \mathbb{N}$ be so that $x = \frac{p}{q}$ and compare a^p and a^q.

(b) Prove that if $a < 1$ and $x < 1$, then $a^x > a$

(c) Prove that if $a > 1$ and $x < 1$, then $a^x < a$.

(d) Prove that if $a < 1$ and $x > 1$, then $a^x < a$.

1-40. Let $n \in \mathbb{N}$. Prove that $\binom{n}{0} = 1$ and that $\binom{n}{n} = 1$.

1-41 Prove that there is no rational number r such that $r^2 = 3$.

1-42. Prove that for any n real numbers $x_1, \ldots, x_n$ the inequality $\left|\sum_{i=1}^{n} x_i\right| \leq \sum_{i=1}^{n} |x_i|$ holds

1-43 Prove that for all $a, b \geq 0$ the inequality $\sqrt{ab} \leq \frac{a+b}{2}$ holds

1-44 (a) Prove that for all $a, b \in \mathbb{R}$ we have $\sqrt{a^2 + b^2} \leq |a| + |b|$.

(b) Prove that for any $a_1, \ldots, a_n \in \mathbb{R}$ we have $\sqrt{\sum_{j=1}^{n} a_j^2} \leq \sum_{j=1}^{n} |a_j|$.

1-45. Let $a, b \in \mathbb{R}$ with $a < b$. Prove that there is an irrational number $x \in \mathbb{R} \setminus \mathbb{Q}$ such that $a < x < b$
Hint Use that $\sqrt{2}$ is irrational and Exercise 1-27 and mimic the proof of Theorem 1.36

Chapter 2

Sequences of Real Numbers

Convergence is the fundamental concept of analysis. It explores what happens when two quantities get close to each other, or when a quantity grows beyond all bounds. This chapter exhibits these ideas for sequences, with special emphasis on standard proof techniques.

2.1 Limits

We start by defining sequences.

Definition 2.1 *A* **sequence** *of real numbers is a function f from the natural numbers to the real numbers. To emphasize their discrete nature, we denote sequences as $\{a_n\}_{n=1}^{\infty}$ with the understanding that $a_n = f(n)$ for all $n \in \mathbb{N}$.*

Similar to sums and products, a sequence can actually start at any integer k (Exercise 2-1). The limit of a sequence should be the place where the sequence "stabilizes" for large n. Definition 2.2 encodes this property by demanding that for every given tolerance ε, there is a threshold N so that once the running index n has gone past the threshold N, the sequence can only deviate from the limit by less than the tolerance ε.

Definition 2.2 *Let $\{a_n\}_{n=1}^{\infty}$ be a sequence of real numbers. Then $L \in \mathbb{R}$ is called* **limit** *of $\{a_n\}_{n=1}^{\infty}$ iff for all $\varepsilon > 0$ there is an $N \in \mathbb{N}$ so that for all $n \geq N$ we have that $|a_n - L| < \varepsilon$ (see Figure 2). A sequence that has a limit will be called* **convergent**, *a sequence that does not have a limit will be called* **divergent**.

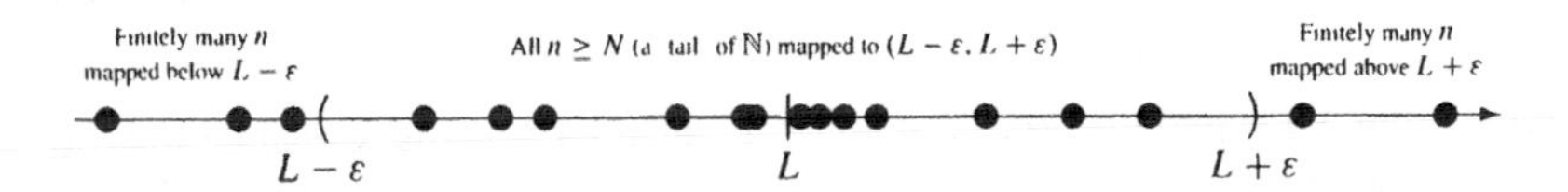

Figure 2: Visualization of convergence to L. For every $\varepsilon > 0$ a "tail" of the sequence is in $(L - \varepsilon, L + \varepsilon)$.

Remark 2.3 It can be helpful to restate the definition using **quantifiers** (see Definition A.3 in Appendix A).

$$L \in \mathbb{R} \text{ is a limit of } \{a_n\}_{n=1}^{\infty} \text{ iff } \forall \varepsilon > 0 : \exists N \in \mathbb{N} : \forall n \geq N : |a_n - L| < \varepsilon.$$

Formal statements with quantifiers enforce the rule that a variable must be defined or quantified before it can be used. We will usually enforce the same rule in natural language. Although the prose becomes a bit rigid this way, in nested quantifications like the definition of limits it is clearer to say "...for all x we have $P(x)$ ..." than to say "... $P(x)$ holds for all x ..." □

Note that we did not speak of *the* limit of a sequence in Definition 2.2. This is because we have not proved yet that every convergent sequence has only one limit. The next theorem shows that this indeed the case.

Proposition 2.4 *Limits of sequences of real numbers are unique. That is, if* $\{a_n\}_{n=1}^{\infty}$ *is a sequence of real numbers and both* L *and* M *are limits of* $\{a_n\}_{n=1}^{\infty}$, *then* $L = M$.

Proof. Let $\{a_n\}_{n=1}^{\infty}$ be a sequence and let L and M be limits of $\{a_n\}_{n=1}^{\infty}$. We need to prove that $L = M$. By Theorem 1.37 we can do so by showing that for all $\varepsilon > 0$ the inequality $|L - M| < \varepsilon$ holds.

Let $\varepsilon > 0$ be arbitrary but fixed. Then there is an $N_1 \in \mathbb{N}$ such that for all $n \geq N_1$ we have that $|a_n - L| < \frac{\varepsilon}{2}$. There also is an $N_2 \in \mathbb{N}$ such that for all $n \geq N_2$ we have $|a_n - M| < \frac{\varepsilon}{2}$. Let $N := \max\{N_1, N_2\}$. Because $N \geq N_1$, for all $n \geq N$ we have $|a_n - L| < \frac{\varepsilon}{2}$. Because $N \geq N_2$, for all $n \geq N$ we have $|a_n - M| < \frac{\varepsilon}{2}$. Then by adding and subtracting a_N and applying the triangular inequality we obtain (with $n = N$)

$$|L - M| = |L - a_N + a_N - M| \leq |L - a_N| + |a_N - M| < \frac{\varepsilon}{2} + \frac{\varepsilon}{2} = \varepsilon.$$

Because for arbitrary $\varepsilon > 0$ we have the inequality $|L - M| < \varepsilon$, by Theorem 1.37 we conclude that $|L - M| = 0$, and hence $L = M$. ■

We will ultimately read and produce proofs that are much more complicated than the proof of Proposition 2.4. Therefore it is only appropriate to analyze how such proofs can be conceived. The standard proof techniques discussed later in this section reveal that certain details are indeed standard techniques which simply need to be internalized and used at the right time. Other than that, the novice usually is impressed by the sometimes "strange" choices for ε. The reason why $\frac{\varepsilon}{2}$ is chosen in the proof of Proposition 2.4 is that the proof is actually created backwards. Consider the following.

To show that $L = M$ we first note that because the a_n are eventually close to L and close to M, we can put an a_n with a sufficiently large index between L and M. After applying the triangular inequality

$$|L - M| = |L - a_n + a_n - M| \leq |L - a_n| + |a_n - M|,$$

the resulting differences should be small. By Theorem 1.37, if for all $\varepsilon > 0$ we can make the difference $|L - M|$ less than or equal to ε then $|L - M| = 0$, that is, $L = M$. So we want to make the sum of the differences $|L - a_n|$ and $|a_n - M|$ smaller than ε. It is most natural to make each of the two terms smaller than $\frac{\varepsilon}{2}$ to obtain

$$|L - M| = |L - a_N + a_N - M| \leq |L - a_N| + |a_N - M| < \frac{\varepsilon}{2} + \frac{\varepsilon}{2} = \varepsilon.$$

This argument provides the final few lines of the proof. Note that up to here we have not chosen any N_1, N_2, or N. However, now that we have the "meat" of the argument, it is easy to create the "header."

To make $|a_n - L|$ and $|a_n - M|$ smaller than $\frac{\varepsilon}{2}$, we use that by the definition of convergence there are N_1 and N_2 as mentioned in the proof and choose N to be their maximum so that both required inequalities hold for indices beyond N. Note that even though the "header" is the first thing we encounter, it is often the last thing that materializes as a proof is created. So, to set up an analysis proof it is standard practice to start by working with inequalities. Once the inequalities work, we create a "header" with the appropriate choices for ε, n, and so on.

Standard proof techniques in analysis. Certain steps occur so frequently in analysis proofs that they should become second nature. In this fashion, communication becomes more effective because memory is less strained to recall details of proofs. This is a cognitive technique commonly known as "chunking" of data. By internalizing certain standard "chunks," larger amounts of data can be recalled, because we only need to recall which chunks are involved rather than all details. Unlike the standard proof techniques listed so far, from here on most standard proof techniques will be specific to analysis. The standard techniques used in the proof of Proposition 2.4 are listed below.

Standard Proof Technique 2.5 It is common practice to rearrange terms and to **add and subtract the same term** to obtain more manageable expressions. When working with absolute values, this is often done in conjunction with the **triangular inequality**. In Proposition 2.4 this is the step

$$|L - M| = |L - a_N + a_N - M| \leq |L - a_N| + |a_N - M|.$$

Such a step is usually abbreviated as $|L - M| \leq |L - a_N| + |a_N - M|$. In other computations, we will see that it can also be useful to multiply and divide by the same nonzero term. □

Standard Proof Technique 2.6 If finitely many numbers $N_1, \ldots, N_k \in \mathbb{N}$ are such that for all $n \geq N_i$ a certain inequality holds, we can choose $N := \max\{N_1, \ldots, N_k\}$. Then for all $n \geq N$ *all* these inequalities hold. We usually claim directly that such an N exists, skipping the intermediate N_i. □

Standard Proof Technique 2.7 To prove that two quantities are equal, we can prove that for any $\varepsilon > 0$ the absolute value of the difference is less than ε. This is usually done without explicit reference to Theorem 1.37. To prove an inequality $a \leq b$ we often prove that for all $\varepsilon > 0$ the inequality $a \leq b + \varepsilon$ holds (see Exercise 1-25). □

Standard Proof Technique 2.8 In many analysis proofs we prove results about a universally quantified variable, often denoted ε. To prove such results, we pick one such ε that is "arbitrary, but fixed," throughout the proof. It must be fixed throughout the proof so we can uniquely define quantities that depend on it, and it must be arbitrary so that we really prove something about all variables in the scope of the universal quantification. Once the result is proved, we can conclude that "Because ε was arbitrary we have proved ... for all such ε." This final statement reiterates that, even though we made specific choices for the ε in the proof, we can indeed make these choices for all ε, which proves the universally quantified statement.

Because this approach is so common, the bracketing statements about the variable ε are usually left out or abbreviated. □

Standard Proof Technique 2.9 Finally, statements like "We need to prove $L = M$," that are put at the start to remind the reader what we will prove are often left out. Similarly, statements put at the end to reiterate what we have proved are often left out, too. □

To phase in these techniques, we will first carry them out explicitly and give a reference to the above list. Then we will omit the explicit step, but still refer to the appropriate entry in the above list. Eventually a proof for something like Proposition 2.4 will condense to the following.

"Expert Proof" of Proposition 2.4. Let $\varepsilon > 0$. Then there is an $N \in \mathbb{N}$ such that for all $n \geq N$ the inequalities $|a_n - L| < \frac{\varepsilon}{2}$ and $|a_n - M| < \frac{\varepsilon}{2}$ hold. Therefore

$$|L - M| \leq |L - a_N| + |a_N - M| < \frac{\varepsilon}{2} + \frac{\varepsilon}{2} = \varepsilon,$$

which implies $L = M$. ■

The "expert proof" contains all relevant information needed to communicate the idea. It does not list the details of standard techniques, which are assumed to be known. In this fashion the reader may have to fill in more "steps between the lines," but as long as these steps can be considered manageable, communication becomes more effective.

Definition 2.10 *Because the limit is unique, we speak of* the *limit of a sequence. The notation* $\lim_{n\to\infty} a_n = L$ *will indicate that the limit of* $\{a_n\}_{n=1}^{\infty}$ *exists and is equal to the number* L.

The next result shows that the first finitely many values of a sequence do not affect its convergence behavior.

Theorem 2.11 *Let* $\{a_n\}_{n=1}^{\infty}$ *and* $\{b_n\}_{n=1}^{\infty}$ *be sequences of real numbers and let* $K \in \mathbb{N}$ *be so that for all* $n \geq K$ *we have* $a_n = b_n$. *Then* $\{a_n\}_{n=1}^{\infty}$ *converges iff* $\{b_n\}_{n=1}^{\infty}$ *converges and in this case the equality* $\lim_{n\to\infty} a_n = \lim_{n\to\infty} b_n$ *holds.*

Proof. Left to Exercise 2-2. ■

We conclude this section with an explicit proof that a limit is a certain number. This exercise is good for learning how to choose more exotic values for N. Remember that while the proof is presented in linear fashion, it is actually created by working with the inequalities that we see at the end. In fact, N was found by solving the inequality $\frac{17}{4n+10} < \varepsilon$, which arises naturally from the computation at the end.

Example 2.12 *Prove that* $\lim_{n\to\infty} \frac{3n-1}{2n+5} = \frac{3}{2}$.

Let $\varepsilon > 0$ be arbitrary but fixed, and let $N \in \mathbb{N}$ be such that $N > \frac{\frac{17}{\varepsilon}-10}{4}$. Then for all $n \geq N$ the following holds.

$$\begin{aligned} \left|\frac{3n-1}{2n+5} - \frac{3}{2}\right| &= \left|\frac{6n-2}{4n+10} - \frac{6n+15}{4n+10}\right| = \left|\frac{-17}{4n+10}\right| = \frac{17}{4n+10} \\ &< \frac{17}{4\frac{\frac{17}{\varepsilon}-10}{4}+10} = \frac{17}{\frac{17}{\varepsilon}-10+10} = \frac{17}{\frac{17}{\varepsilon}} = \varepsilon. \end{aligned}$$

Because $\varepsilon > 0$ was arbitrary, this proves that $\lim_{n\to\infty} \frac{3n-1}{2n+5} = \frac{3}{2}$. □

Exercises

2-1 Let $k \in \mathbb{Z}$ and for each $n \in \mathbb{Z}$ with $n \geq k$ let $a_n \in \mathbb{R}$. Define the sequence $\{a_n\}_{n=k}^{\infty}$ and define what it means for $L \in \mathbb{R}$ to be its limit.

2-2 Prove Theorem 2.11

2-3 Write out the argument that produced the choice for N in Example 2.12.

2-4 Let $\{a_n\}_{n=1}^{\infty}$ and $\{b_n\}_{n=1}^{\infty}$ be sequences such that for all $n \in \mathbb{N}$ we have $|a_n - b_n| < \frac{1}{n}$ Prove that if $\{a_n\}_{n=1}^{\infty}$ converges then so does $\{b_n\}_{n=1}^{\infty}$ and $\lim_{n\to\infty} a_n = \lim_{n\to\infty} b_n$.

2-5. Let $\{a_n\}_{n=1}^{\infty}$ be a sequence. Prove that $\lim_{n\to\infty} a_n = L$ iff $\lim_{n\to\infty} |a_n - L| = 0$.

2-6 Prove each of the following.

(a) $\lim_{n\to\infty} \frac{4n-9}{5n+1} = \frac{4}{5}$

(b) $\lim_{n\to\infty} \frac{6n^2+1}{n^2-5} = 6$

(c) $\lim_{n\to\infty} \frac{2n+4}{5n^2-11} = 0$

(d) $\lim_{n\to\infty} \sqrt{n+1} - \sqrt{n} = 0$

2-7. The usefulness of quantifiers

(a) State the negation of the statement "L is the limit of the sequence $\{a_n\}_{n=1}^{\infty}$" by finding the negation of Definition 2.2.

(b) State the negation of the statement "L is the limit of the sequence $\{a_n\}_{n=1}^{\infty}$" by finding the negation of the quantifier version of Definition 2.2 in Remark 2.3

Hint. Appendix A.2.

2.2 Limit Laws

Example 2.12 shows how to prove that a limit is a certain number. However, to write such a proof we must know the limit in advance. It would be nice to have a computational tool that provides the limit. It would be even nicer if the computation somehow could justify (prove) that the result of the computation really is the limit. This can be done with the limit laws discussed in this section. Because we are investigating the theory, it will be important to understand how the limit laws come into being. Applying the limit laws is fairly simple (see Exercise 2-8).

To use the limit laws, we must know the limits of certain standard sequences.

Theorem 2.13 $\lim_{n\to\infty} \frac{1}{n} = 0.$

Proof. Exercise 2-9. ■

The limit laws state that limits can be moved into the familiar algebraic operations. The proof consists of several similar arguments. The argument for sums is spelled out in great detail. The argument for differences is almost verbatim the same argument as for sums and it is left to the reader in Exercise 2-10. The argument for products is proved closer to the fashion of the "expert proof" on page 28, but with remarks interspersed. For the argument for quotients, we only present the final inequalities, leaving the remaining parts to the reader in Exercise 2-11.

Theorem 2.14 Limit laws *for sequences. Let $\{a_n\}_{n=1}^{\infty}$ and $\{b_n\}_{n=1}^{\infty}$ be two convergent sequences. Then the following hold.*

1. *The sum $\{a_n + b_n\}_{n=1}^{\infty}$ converges and $\lim_{n\to\infty} a_n + b_n = \lim_{n\to\infty} a_n + \lim_{n\to\infty} b_n$.*
2. *The difference $\{a_n - b_n\}_{n=1}^{\infty}$ converges and $\lim_{n\to\infty} a_n - b_n = \lim_{n\to\infty} a_n - \lim_{n\to\infty} b_n$.*
3. *The product $\{a_n b_n\}_{n=1}^{\infty}$ converges and $\lim_{n\to\infty} a_n \cdot b_n = \lim_{n\to\infty} a_n \cdot \lim_{n\to\infty} b_n$.*
4. *If all $b_n \neq 0$ and $\lim_{n\to\infty} b_n \neq 0$, then the quotient $\left\{\frac{a_n}{b_n}\right\}_{n=1}^{\infty}$ converges and $\lim_{n\to\infty} \frac{a_n}{b_n} = \frac{\lim_{n\to\infty} a_n}{\lim_{n\to\infty} b_n}$.*

Proof. Throughout this proof let $L := \lim_{n\to\infty} a_n$ and let $M := \lim_{n\to\infty} b_n$.

To show for part 1 that the sum $\{a_n+b_n\}_{n=1}^{\infty}$ converges to the limit $L+M$, we must show that for all $\varepsilon > 0$ there is an $N \in \mathbb{N}$ so that for all $n \geq N$ we have the inequality $\big|(a_n + b_n) - (L + M)\big| < \varepsilon$.

Let $\varepsilon > 0$ be arbitrary but fixed (see Standard Proof Technique 2.8). Because $\lim_{n\to\infty} a_n = L$ there is an $N_L \in \mathbb{N}$ so that for all $n \geq N_L$ we have $|a_n - L| < \frac{\varepsilon}{2}$. Similarly, because $\lim_{n\to\infty} b_n = M$ there is an $N_M \in \mathbb{N}$ so that for all $n \geq N_M$ we have $|b_n - M| < \frac{\varepsilon}{2}$. Let $N := \max\{N_L, N_M\}$. Then (compare with Standard Proof Technique 2.6) for all $n \geq N$ the inequalities $|a_n - L| < \frac{\varepsilon}{2}$ and $|b_n - M| < \frac{\varepsilon}{2}$ hold. By rearranging the terms and applying the triangular inequality (compare with Standard Proof Technique 2.5) we obtain the following for all $n \geq N$.

$$\begin{aligned} \left|(a_n + b_n) - (L + M)\right| &= |a_n - L + b_n - M| \leq |a_n - L| + |b_n - M| \\ &< \frac{\varepsilon}{2} + \frac{\varepsilon}{2} = \varepsilon. \end{aligned}$$

We have proved that for arbitrary $\varepsilon > 0$ we can find an $N \in \mathbb{N}$ so that for all $n \geq N$ we have that $\left|(a_n + b_n) - (L + M)\right| < \varepsilon$ (see Standard Proof Technique 2.9). Therefore by the definition of the limit, $\lim_{n\to\infty} a_n + b_n = L + M$.

Part 2 is Exercise 2-10.

To show how an abbreviated proof still contains the standard statement of the definition of the limit, the key parts of the definition are **boxed** *in the proof of part 3. We also intersperse comments in italics on how certain choices arise.*

For part 3, let $\boxed{\varepsilon > 0}$.

The choices for ε look a bit strange. They are made so that the inequalities later on work out and they can be motivated by reading the final estimates. Recall the discussion after the proof of Proposition 2.4. The first typed part of this proof were the final inequalities.

Because $\lim_{n\to\infty} a_n = L$ there is an $N_L \in \mathbb{N}$ so that for all $n \geq N_L$ we have that $|a_n - L| < \dfrac{\varepsilon}{2(|M| + 1)}$. *(We cannot divide by $|M|$ alone, because it could be zero.)* Similarly, because $\lim_{n\to\infty} b_n = M$ there is an $N_M \in \mathbb{N}$ so that for all $n \geq N_M$ we have $|b_n - M| < \dfrac{\varepsilon}{2(|L| + 1)}$. Finally, because $\lim_{n\to\infty} a_n = L$ there is a $K_1 \in \mathbb{N}$ so that for all $n \geq K_1$ we have $|a_n - L| < 1$. Consequently, for all $n \geq K_1$ we obtain by the reverse triangular inequality $|a_n| - |L| \leq \big||a_n| - |L|\big| \leq |a_n - L| < 1$ and thus $|a_n| < |L| + 1$. Let $\boxed{N := \max\{N_L, N_M, K_1\}}$. Then (compare with Standard Proof Technique 2.6) $\boxed{\text{for all } n \geq N}$ the inequalities $|a_n - L| < \dfrac{\varepsilon}{2(|M| + 1)}$, $|b_n - M| < \dfrac{\varepsilon}{2(|L| + 1)}$ and $|a_n| < |L| + 1$ hold.

To "connect" the $a_n b_n$ and the LM, we add and subtract the term $a_n M$. This allows us to apply the triangular inequality (compare with Standard Proof Technique 2.5) to split the original difference $|a_n b_n - LM|$ into two summands. Each of these summands contains one factor that can be made small.

For all $n \geq N$, we obtain the following.

$$\begin{aligned} \boxed{|a_n b_n - LM|} &\leq |a_n b_n - a_n M| + |a_n M - LM| \\ &= |a_n||b_n - M| + |a_n - L||M| \end{aligned}$$

The inequalities for $|a_n - L|$ and $|b_n - M|$ will only help if the terms $|a_n - L|$ and $|b_n - M|$ are not multiplied by quantities that depend on n. To remove the dependency on n represented by the term $|a_n|$, we replace it with $|L| + 1$ before proceeding.

$$\leq (|L| + 1)|b_n - M| + |a_n - L||M|$$

$$< \left(|L|+1\right)\frac{\varepsilon}{2(|L|+1)} + \frac{\varepsilon}{2(|M|+1)}|M|$$
$$\boxed{<\varepsilon},$$

which proves part 3 (see Standard Proof Technique 2.9).

For part 4, the final estimates, which are the hard part to figure out, are

$$\begin{aligned}
\left|\frac{a_n}{b_n} - \frac{L}{M}\right| &= \left|\frac{a_nM - Lb_n}{b_nM}\right| \leq \frac{|a_nM - LM| + |LM - Lb_n|}{|b_n||M|} \\
&= \frac{|a_n - L||M| + |L||M - b_n|}{|b_n||M|} \\
&\leq \frac{|a_n - L||M| + |L||M - b_n|}{\left|\frac{M}{2}\right||M|} \\
&= |a_n - L|\frac{2}{|M|} + \frac{2|L|}{M^2}|M - b_n| \\
&< \frac{\varepsilon|M|}{4}\frac{2}{|M|} + \frac{2|L|}{M^2}\frac{\varepsilon M^2}{2(2|L|+1)} < \varepsilon.
\end{aligned}$$

To train the reader in generating the appropriate "header," the complete proof of part 4 (including final estimates) is left to Exercise 2-11. ■

Standard Proof Technique 2.15 In an analysis proof, it can be necessary to divide by a nonnegative quantity $|a|$. Usually the quotient will be multiplied by that same quantity later in an estimate and the goal is to cancel it (consider the proofs of parts 3 and 4 of Theorem 2.14). However, we cannot divide by zero. To avoid any undue distractions here, when defining certain quotients, we will usually add 1 to nonnegative quantities in denominators. In this fashion, the fact that $\frac{|a|}{|a|+1} < 1$ still allows us to "cancel" the term in an estimate. This is more effective than to separately consider the case that a quantity is equal to zero. □

The limit laws allow us to efficiently establish convergence and compute the limits for many complicated looking sequences. For instance, finding the limit of the sequence in Example 2.12 reduces to the following.

Example 2.16 *Find the limit of* $\left\{\frac{3n-1}{2n+5}\right\}_{n=1}^{\infty}$.

Using Theorem 2.13 and the limit laws we obtain the following:

$$\lim_{n\to\infty}\frac{3n-1}{2n+5} = \lim_{n\to\infty}\frac{3n-1}{2n+5}\frac{\frac{1}{n}}{\frac{1}{n}} = \lim_{n\to\infty}\frac{3-\frac{1}{n}}{2+\frac{5}{n}} = \frac{3-\lim_{n\to\infty}\frac{1}{n}}{2+\lim_{n\to\infty}\frac{5}{n}} = \frac{3}{2}.$$

Note that by the limit laws, the computation also *proves* that the limit is $\frac{3}{2}$. That is, we will not need to perform the tedious argument given in Example 2.12. □

For limits of powers, we will consider integer powers here, leaving rational powers to Theorem 3.42 and real powers to Theorem 12.11.

Theorem 2.17 *Let $\{a_n\}_{n=1}^{\infty}$ be a convergent sequence. Then for all $k \in \mathbb{N}$ the sequence $\left\{a_n^k\right\}_{n=1}^{\infty}$ converges and $\lim_{n\to\infty} a_n^k = \left(\lim_{n\to\infty} a_n\right)^k$. If none of the terms are zero and the limit is not equal to zero, the result holds for all $k \in \mathbb{Z}$.*

Proof. For $k \in \mathbb{N}$ the proof is an induction on k.

The base step for $k = 1$ is trivial.

For the induction step $k \to k+1$, assume that for a given $k \geq 1$ the sequence $\left\{a_n^k\right\}_{n=1}^{\infty}$ converges and that $\lim_{n\to\infty} a_n^k = \left(\lim_{n\to\infty} a_n\right)^k$. We must prove that $\left\{a_n^{k+1}\right\}_{n=1}^{\infty}$ converges with $\lim_{n\to\infty} a_n^{k+1} = \left(\lim_{n\to\infty} a_n\right)^{k+1}$. By part 3 of Theorem 2.14, we can argue as follows.

$$\begin{aligned}
\left(\lim_{n\to\infty} a_n\right)^{k+1} &= \lim_{n\to\infty} a_n \cdot \left(\lim_{n\to\infty} a_n\right)^k = \lim_{n\to\infty} a_n \cdot \lim_{n\to\infty} a_n^k \\
&= \lim_{n\to\infty} a_n \cdot a_n^k = \lim_{n\to\infty} a_n^{k+1},
\end{aligned}$$

where each expression in the chain of equations is guaranteed to exist because the expression preceding it exists. This completes the induction step and the proof of the result for $k \in \mathbb{N}$.

Now for $k \in \mathbb{Z}$, let all terms and the limit be nonzero. For $k > 0$, we just proved the result and for $k = 0$ the result is trivial. For $k < 0$, note that

$$\left(\lim_{n\to\infty} a_n\right)^k = \frac{1}{(\lim_{n\to\infty} a_n)^{-k}} = \frac{1}{\lim_{n\to\infty} a_n^{-k}} = \lim_{n\to\infty} \frac{1}{a_n^{-k}} = \lim_{n\to\infty} a_n^k,$$

where again each expression in the chain of equations is guaranteed to exist because the expression preceding it exists. ■

Standard Proof Technique 2.18 Many statements about limits implicitly assert two things. First, that the limit exists, and second, what the limit is. This means a proof will need to verify both these claims. To simplify the language and to shorten proofs, it is customary to dispense with an explicit existence proof for the limit. Instead, the computation of the limit can be used to establish the existence of the limit as in the proof of Theorem 2.17. If we start with an existing quantity and finish with the quantity in question, then the existence claim in known limit laws establishes the existence of all intermediate quantities from left to right. *Exercise 2-23 shows that the order of progression is important. The limit of a sum, a difference, a product, or a quotient can exist even when the sequences of the individual terms diverge.* □

From here on, we will interpret some equations for which only one side is guaranteed to exist as implicitly asserting that the other side exists.

Standard Proof Technique 2.19 An induction as in the first part of the proof of Theorem 2.17, which takes a result for an operation on two objects and turns it into a result for an operation with $k \in \mathbb{N}$ objects, is often omitted by saying **"a simple induction argument shows ..."** The proofs will all be similar. The base step is usually simple. For the induction step, split off the last element, apply the induction hypothesis to the first k elements, then apply the original result and thus conclude the proof. □

Although it would feel natural to prove that Theorem 2.17 holds for real exponents, real exponents must be postponed until we define real exponentiation in Definition 12.8. Our main focus is on the abstract underpinnings and fundamentals of analysis, not on certain specifics. Hence, we postpone the introduction of transcendental functions until we have all the machinery to introduce them very efficiently. Their introduction will also make us more fully appreciate the power of the abstract tools we build. We could at least prove Theorem 2.17 for rational exponents, but at this stage the proof would be cumbersome. We will obtain the result easily in Theorem 3.42, once we have developed some more tools. We will use this approach throughout. *If a result can be obtained easily later, then (unless an important technique needs to be introduced) we will postpone the result rather than set up an unnecessarily complicated argument.*

We conclude this section by investigating the relationship between limits and inequalities.

Theorem 2.20 *Let $\{a_n\}_{n=1}^{\infty}$ and $\{b_n\}_{n=1}^{\infty}$ be two convergent sequences of real numbers so that for all $n \in \mathbb{N}$ the inequality $a_n \leq b_n$ holds. Then $\lim_{n\to\infty} a_n \leq \lim_{n\to\infty} b_n$.*

Proof. Let $L := \lim_{n\to\infty} a_n$ and let $M := \lim_{n\to\infty} b_n$. We will prove (see Standard Proof Technique 2.7) that for every $\varepsilon > 0$ the inequality $L - M < \varepsilon$ holds, which proves that $L - M \leq 0$, that is, $L \leq M$.

Let $\varepsilon > 0$. Then (see Standard Proof Technique 2.6) there is an $N \in \mathbb{N}$ so that for all $n \geq N$ we have $|a_n - L| < \frac{\varepsilon}{2}$ and $|b_n - M| < \frac{\varepsilon}{2}$. But, using Standard Proof Technique 2.5, this implies

$$\begin{aligned} L - M &= L - a_n + b_n - M + a_n - b_n \leq |L - a_n| + |b_n - M| + a_n - b_n \\ &< \frac{\varepsilon}{2} + \frac{\varepsilon}{2} + 0 = \varepsilon. \end{aligned}$$

We have proved that for all $\varepsilon > 0$ we have $L - M < \varepsilon$, which by Exercise 1-25 proves that $\lim_{n\to\infty} a_n = L \leq M = \lim_{n\to\infty} b_n$. ■

The next theorem is a refinement of the fact that limits preserve existing inequalities. If a sequence $\{b_n\}_{n=1}^{\infty}$ is "trapped" between two convergent sequences with the same limit, then $\{b_n\}_{n=1}^{\infty}$ must converge to that limit also. The novelty is that we need not assume that $\{b_n\}_{n=1}^{\infty}$ converges.

Theorem 2.21 *The* **Squeeze Theorem** *for sequences. Let $\{a_n\}_{n=1}^{\infty}$, $\{b_n\}_{n=1}^{\infty}$, $\{c_n\}_{n=1}^{\infty}$ be sequences of real numbers so that for all $n \in \mathbb{N}$ the inequalities $a_n \leq b_n \leq c_n$ hold and so that $\lim_{n\to\infty} a_n = \lim_{n\to\infty} c_n$. Then $\{b_n\}_{n=1}^{\infty}$ converges and the limits are equal, that is, $\lim_{n\to\infty} b_n = \lim_{n\to\infty} a_n = \lim_{n\to\infty} c_n$.*

Proof. Let $L := \lim_{n\to\infty} a_n = \lim_{n\to\infty} c_n$. We need to prove that for all $\varepsilon > 0$ there is an $N \in \mathbb{N}$ so that for all $n \geq N$ the inequality $|b_n - L| < \varepsilon$ holds.

Let $\varepsilon > 0$. Because $\{a_n\}_{n=1}^{\infty}$ and $\{c_n\}_{n=1}^{\infty}$ both converge to L, there are $N_a \in \mathbb{N}$ and $N_c \in \mathbb{N}$ so that for all $n \geq N_a$ we have $|a_n - L| < \varepsilon$ and so that for all $n \geq N_c$ we have $|c_n - L| < \varepsilon$. Let $N := \max\{N_a, N_c\}$. Then for all $n \geq N$ we obtain $L - \varepsilon < a_n$ and $c_n < L + \varepsilon$. Therefore, for all $n \geq N$ we conclude $L - \varepsilon < a_n \leq b_n \leq c_n < L + \varepsilon$, which means that $|b_n - L| < \varepsilon$.

We have proved that for all $\varepsilon > 0$ there is an $N \in \mathbb{N}$ so that for all $n \geq N$ we have the inequality $|b_n - L| < \varepsilon$. This means that $\{b_n\}_{n=1}^{\infty}$ converges to L. ■

Standard Proof Technique 2.22 To prove that the limit of a sequence $\{b_n\}_{n=1}^{\infty}$ is zero, it is common to prove that $|b_n|$ is bounded by some sequence, such as a multiple of $\frac{1}{n}$, that goes to zero itself. The Squeeze Theorem then guarantees that the limit of $\{b_n\}_{n=1}^{\infty}$ also is zero.

In particular (compare with Exercise 2-5), we often prove that $\lim_{n\to\infty} a_n = L$ by proving that $|a_n - L|$ is bounded by a sequence that goes to zero. □

Exercises

2-8. Use the limit laws to compute the limit of the sequence.

(a) $\left\{\frac{n+3}{4n-7}\right\}_{n=1}^{\infty}$ (b) $\left\{\frac{3n+2}{8n^2-1}\right\}_{n=1}^{\infty}$ (c) $\left\{\sqrt{n}-\sqrt{n+4}\right\}_{n=1}^{\infty}$

2-9. Prove Theorem 2.13.

Hint Exercise 1-27

2-10. Prove part 2 of Theorem 2.14.

2-11. Prove part 4 of Theorem 2 14 Explain all choices for N and for upper or lower bounds.

Hint. There are three inequalities that must hold for all $n \geq N$.

2-12. Prove that if $\{a_n\}_{n=1}^{\infty}$ converges, then $\lim_{n\to\infty} |a_n| = \left|\lim_{n\to\infty} a_n\right|$.

2-13. Prove that if $a_n \geq 0$ for all $n \in \mathbb{N}$ and $\{a_n\}_{n=1}^{\infty}$ converges, then $\lim_{n\to\infty} \sqrt{a_n} = \sqrt{\lim_{n\to\infty} a_n}$.

2-14. Let $q > 0$. Prove that $\lim_{n\to\infty} \frac{q^n}{n!} = 0$.

2-15. Conjecture the value of $\lim_{n\to\infty} \frac{n!}{n^n}$ and then prove your conjecture

2-16. Let $\{a_n\}_{n=1}^{\infty}$ be a sequence of real numbers and let $\{p_n\}_{n=1}^{\infty}$ be a sequence of positive real numbers so that $\lim_{n\to\infty} \frac{1}{\sum_{k=1}^{n} p_k} = 0$.

(a) Prove that if $\{a_n\}_{n=1}^{\infty}$ converges to $a \in \mathbb{R}$, then $\lim_{n\to\infty} \frac{\sum_{k=1}^{n} p_k a_k}{\sum_{k=1}^{n} p_k} = a$

(b) Give an example to show that the convergence of $\left\{\frac{\sum_{k=1}^{n} p_k a_k}{\sum_{k=1}^{n} p_k}\right\}_{n=1}^{\infty}$ need not imply the convergence of $\{a_n\}_{n=1}^{\infty}$.

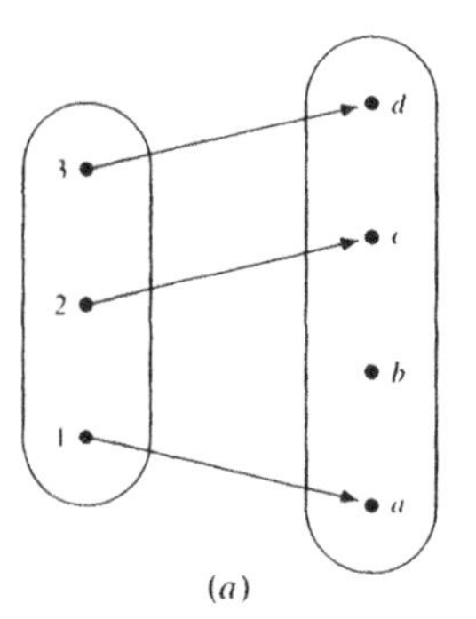

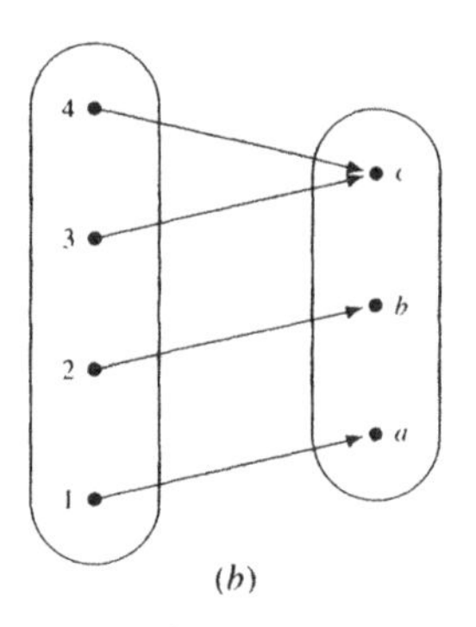

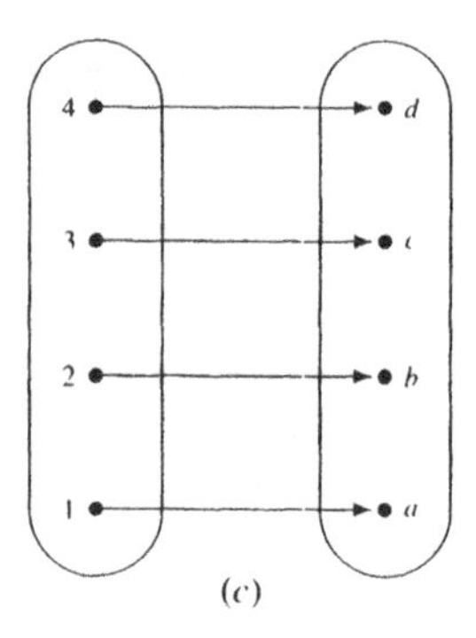

Figure 3: An injective function maps all elements of the domain to distinct images, but some elements of the range may not have a preimage (a). For a surjective function, every element of the range has a preimage, but some elements of the domain may be mapped to the same image (b). A bijective function maps all elements of the domain to distinct images and each element of the range has a preimage (c). This is why the existence of a bijection between two sets indicates that the two sets are "of the same size" (see Definitions 2.25 and 7.11).

2.3 Cauchy Sequences

A sequence so that for all $\varepsilon > 0$ all elements with a sufficiently large index are within ε of each other should converge. Indeed, this condition guarantees that the elements with large indices cluster ever more tightly. However, the number system may have a hole just where these elements cluster. For example, the sequence 1.4, 1.41, 1.414, 1.4142, ... of successively better decimal approximations of $\sqrt{2}$ does not converge in $\mathbb{Q}$ because the value that the sequence approaches is not in $\mathbb{Q}$ (see Exercise 2-17).

This section shows that this problem does not arise in the real numbers. Sequences for which elements with large indices cluster ever more tightly play an important role in analysis. They are called Cauchy sequences.

Definition 2.23 *Let* $\{a_n\}_{n=1}^{\infty}$ *be a sequence of real numbers. Then* $\{a_n\}_{n=1}^{\infty}$ *is called a* **Cauchy sequence** *iff for all* $\varepsilon > 0$ *there is an* $N \in \mathbb{N}$ *so that for all* $m, n \geq N$ *we have that* $|a_n - a_m| < \varepsilon$.

In the real numbers, convergence and being a Cauchy sequence are equivalent. Before we can prove this result, we need to define finite and infinite sets.

Definition 2.24 *Let* A, B *be sets and let* $f : A \to B$ *be a function. Then* f *is called* **injective** *or* **one-to-one** *iff for all* $x, y \in A$ *the inequality* $x \neq y$ *implies* $f(x) \neq f(y)$. f *is called* **surjective** *or* **onto** *iff for all* $b \in B$ *there is an* $a \in A$ *with* $f(a) = b$. *Finally,* f *is called* **bijective** *iff* f *is both injective and surjective.*

Figure 3 gives a visualization of injective, surjective, and bijective functions and some properties of injective and surjective functions are investigated in Exercises 2-18 and 2-19. Once we have bijective functions, we can define finite sets as "sets of size n," where $n \in \mathbb{N} \cup \{0\}$.

Definition 2.25 *A set F is called* **finite** *iff F is empty or there is an $n \in \mathbb{N}$ and a bijective function $f : \{1, \ldots, n\} \to F$. Sets that are not finite are called* **infinite**. *For finite sets $F \neq \emptyset$ we set $|F| := n$ with n as above and we set $|\emptyset| := 0$. For infinite sets I we set $|I| := \infty$.*

Lemma 2.26 *Let F be a finite set and let I be an infinite set. Then $I \setminus F$ is infinite.*

Proof. In case F is empty, there is nothing to prove. In case F is not empty let $n \in \mathbb{N}$ be so that there is a bijective function $f : \{1, \ldots, n\} \to F$. Suppose for a contradiction that $I \setminus F$ is finite. Then there are a natural number $m \in \mathbb{N}$ and a bijective function $g : \{1, \ldots, m\} \to I \setminus F$. Define the function $h : \{1, \ldots, m + n\} \to I$ by $h(j) := \begin{cases} f(j); & \text{if } j \leq n, \\ g(j-n); & \text{if } j > n. \end{cases}$ Then it is easy to show that h is bijective (Exercise 2-20). But this means that I is finite, a contradiction. ■

Theorem 2.27 *A sequence $\{a_n\}_{n=1}^{\infty}$ of real numbers converges iff it is a Cauchy sequence.*

Proof. For the direction "$\Rightarrow$," let $L := \lim_{n\to\infty} a_n$. We need to prove that for all $\varepsilon > 0$ there is an $N \in \mathbb{N}$ so that for all $m, n \geq N$ we have $|a_n - a_m| < \varepsilon$.

Let $\varepsilon > 0$. Then there is an $N \in \mathbb{N}$ so that for all $n \geq N$ we have $|a_n - L| < \frac{\varepsilon}{2}$. Therefore for all $m, n \geq N$ we obtain

$$|a_m - a_n| = |a_m - L + L - a_n| \leq |a_m - L| + |L - a_n| < \frac{\varepsilon}{2} + \frac{\varepsilon}{2} = \varepsilon.$$

We have proved that for all $\varepsilon > 0$ there is an $N \in \mathbb{N}$ so that for all $m, n \geq N$ we have $|a_n - a_m| < \varepsilon$. Hence, $\{a_n\}_{n=1}^{\infty}$ is a Cauchy sequence.

Once more we have used Standard Proof Technique 2.5. It is so common, and we have used it often enough, that we will no longer explicitly refer back to it.

"$\Leftarrow$:" Let $\{a_n\}_{n=1}^{\infty}$ be a Cauchy sequence. We need to prove that there is an $L \in \mathbb{R}$ so that for all $\varepsilon > 0$ there is an $N \in \mathbb{N}$ so that for all $n \geq N$ we have $|a_n - L| < \varepsilon$.

We first need to find a suitable number L. Because $\{a_n\}_{n=1}^{\infty}$ is a Cauchy sequence, for $\varepsilon := 1$ there is an $N \in \mathbb{N}$ such that for all $m, n \geq N$ we have $|a_n - a_m| < 1$. In particular, for all $n \geq N$ we obtain $|a_n - a_N| < 1$, and hence $a_n < a_N + 1$. Therefore the set $\{n \in \mathbb{N} : a_n \leq a_N + 1\}$ is infinite and thus $\big\{x \in \mathbb{R} : \{n \in \mathbb{N} : a_n \leq x\} \text{ is infinite}\big\} \neq \emptyset$. For all $n \geq N$ we also have that $a_n > a_N - 1$, so that for all $x \leq a_N - 1$ the set $\{n \in \mathbb{N} : a_n \leq x\}$ is finite. Therefore $\big\{x \in \mathbb{R} : \{n \in \mathbb{N} : a_n \leq x\} \text{ is infinite}\big\}$ is bounded below. This means $L := \inf\big\{x \in \mathbb{R} : \{n \in \mathbb{N} : a_n \leq x\} \text{ is infinite}\big\}$ exists by Proposition 1.20. The idea how to obtain L is visualized in Figure 4.

To prove that L is the limit of $\{a_n\}_{n=1}^{\infty}$ we need to prove that for all $\varepsilon > 0$ there is an $N \in \mathbb{N}$ so that for all $n \geq N$ we have $|a_n - L| < \varepsilon$.

Let $\varepsilon > 0$. By definition of L, the set $H_- := \left\{n \in \mathbb{N} : a_n \leq L - \frac{\varepsilon}{2}\right\}$ is finite and the set $H_+ := \left\{n \in \mathbb{N} : a_n \leq L + \frac{\varepsilon}{2}\right\}$ is infinite. Therefore, by Lemma 2.26 the relative complement $H_+ \setminus H_- = \left\{n \in \mathbb{N} : a_n \in \left(L - \frac{\varepsilon}{2}, L + \frac{\varepsilon}{2}\right]\right\}$ is infinite. Because

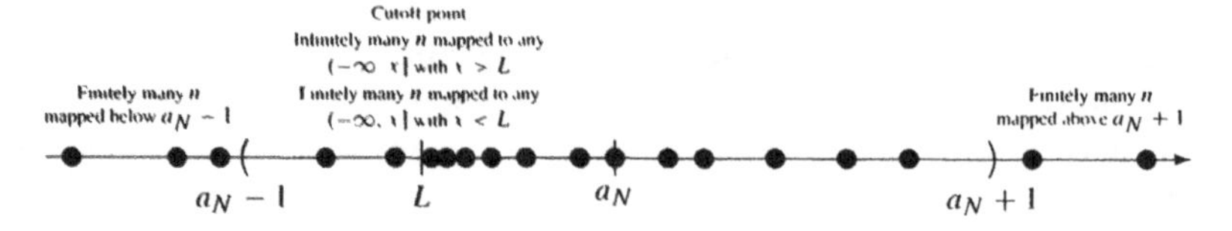

Figure 4: Visualization of the construction of L in the proof of Theorem 2.27.

$\{a_n\}_{n=1}^{\infty}$ is a Cauchy sequence, there is an $N \in \mathbb{N}$ such that for all $m, n \geq N$ we have $|a_m - a_n| < \frac{\varepsilon}{2}$. Moreover, because $\left\{n \in \mathbb{N} : a_n \in \left(L - \frac{\varepsilon}{2}, L + \frac{\varepsilon}{2}\right]\right\}$ is infinite, there is a $k \geq N$ so that $|a_k - L| \leq \frac{\varepsilon}{2}$. Therefore for all $n \geq N$ we obtain

$$|a_n - L| \quad = \quad |a_n - a_k + a_k - L| \leq |a_n - a_k| + |a_k - L| < \frac{\varepsilon}{2} + \frac{\varepsilon}{2} = \varepsilon.$$

Because ε was arbitrary we have proved that for all $\varepsilon > 0$ there is an $N \in \mathbb{N}$ so that for all $n \geq N$ we have $|a_n - L| < \varepsilon$. Hence, $\lim_{n\to\infty} a_n = L$. ■

Standard Proof Technique 2.28 Application of the **Completeness Axiom** to get the infimum or supremum of the "right" set is a standard technique on the real line. We will also see this approach in the proofs of Theorems 2.37, 2.41, 3.34, and 8.4. In abstract spaces, this technique is not available and we usually substitute compactness (see Definition 16.57), a property which, for closed and bounded intervals on the real line, is a consequence of the Completeness Axiom (see Theorems 2.41 and 8.4). □

Convergence of Cauchy sequences is a fundamental analytical property called "completeness," which is introduced in Section 16.2. Another way of formulating Theorem 2.27 is to say that the real numbers are complete. Although Axiom 1.19 is already called the Completeness Axiom of the real numbers, this terminology makes sense, because Exercise 2-25 shows that Theorem 2.27 and Axiom 1.19 are equivalent. This means that either one of them could rightly be called the Completeness Axiom.

There are many other equivalent formulations of the Completeness Axiom. We will encounter some more of them in Theorems 2.37, 2.41, and 8.4 as well as in Exercise 2-50c. Whenever we encounter one of these formulations, there will be an exercise similar to Exercise 2-25 to show that the new result is equivalent to one of the equivalent formulations of the Completeness Axiom that we already know.

Aside from the fundamental importance of completeness, Theorem 2.27 has immediate value for showing if a sequence is divergent. By Definition 2.2 a sequence is divergent iff (using negations as stated in Appendix A.2) for every real number L there is an $\varepsilon > 0$ such that for all $N \in \mathbb{N}$ there is an $n \geq N$ so that $|a_n - L| \geq \varepsilon$. This is a four times nested quantification that would require us to show for every real number that it is not the limit of the sequence. Theorem 2.27 reduces proofs of divergence to showing the sequence is not a Cauchy sequence.

Example 2.29 *The sequence* $\left\{(-1)^n\right\}_{n=1}^{\infty}$ *diverges.*

To prove that the sequence diverges, we need to prove that it is not a Cauchy sequence. This means (see Appendix A.2) we must find an $\varepsilon > 0$ so that for all $N \in \mathbb{N}$ there are $m, n \geq N$ so that $|a_n - a_m| > \varepsilon$.

But with $\varepsilon := 1$ for every $N \in \mathbb{N}$ we have that $\left|(-1)^N - (-1)^{N+1}\right| = 2 > 1$. Hence, $\left\{(-1)^n\right\}_{n=1}^{\infty}$ is not a Cauchy sequence and therefore it diverges. □

Standard Proof Technique 2.30 In Example 2.29, we had to negate the statement "for all $\varepsilon > 0$ there is an $N \in \mathbb{N}$ such that for all $m, n \geq N$ we have $|a_n - a_m| < \varepsilon$." When negating such a complicated statement, it is helpful to write the statement in quantifiers and then negate it. In this fashion, the definition of a Cauchy sequence is

$$\forall \varepsilon > 0 : \exists N \in \mathbb{N} : \forall m, n \geq N : |a_n - a_m| < \varepsilon,$$

and the negation becomes (see Appendix A.2)

$$\exists \varepsilon > 0 : \forall N \in \mathbb{N} : \exists m, n \geq N : |a_n - a_m| \geq \varepsilon,$$

which is what was needed in Example 2.29. The schematic way in which quantified statements can be negated is very helpful, especially the first few times one negates a complicated statement. However, the quantifiers must not become a crutch. It is advisable to first try the negation verbally and then double check with quantifiers. □

Standard Proof Technique 2.31 To prove that a sequence of real numbers converges, we often simply prove that it is a **Cauchy sequence**. To prove that a sequence of real numbers diverges, we prove that it is not a Cauchy sequence. □

Exercises

2-17 Prove that $\left\{\dfrac{\lfloor 10^n\sqrt{2}\rfloor}{10^n}\right\}_{n=1}^{\infty}$ is a Cauchy sequence of rational numbers whose limit is not a rational number.

2-18. Let A, B be sets and let $f : A \to B$ be a function. Prove that f is injective iff for all $b_1, b_2 \in B$ we have that $f(b_1) = f(b_2)$ implies $b_1 = b_2$

2-19. **Compositions** of injective and surjective functions. Let A, B, C be sets and let $f : B \to C$ and $g : A \to B$ be functions. The **composition** of f and g is defined by $f \circ g(a) := f\left(g(a)\right)$ for all $a \in A$

(a) Prove that if f and g are injective, then so is $f \circ g$

(b) Prove that if f and g are surjective, then so is $f \circ g$.

2-20 Prove that the function h in the proof of Lemma 2 26 is bijective.

2-21. State the definition of a convergent sequence using quantifiers.

2-22 For each of the following sequences, prove that it converges or prove that it diverges.

(a) $\left\{\dfrac{(-1)^n}{n}\right\}_{n=1}^{\infty}$ (b) $\{n\}_{n=1}^{\infty}$ (c) $\left\{\sqrt{n}\right\}_{n=1}^{\infty}$ (d) $\left\{\dfrac{2n+1}{3n-2}\right\}_{n=1}^{\infty}$

2-23 Existence of the limit on the left side of a limit law as in Theorem 2.14 does not imply the existence of the limits on the right side.

(a) Use $\{a_n\}_{n=1}^{\infty} = \left\{(-1)^n\right\}_{n=1}^{\infty}$ and $\{b_n\}_{n=1}^{\infty} := \left\{(-1)^{n+1}\right\}_{n=1}^{\infty}$ to show that $\lim_{n\to\infty} a_n + b_n$ can exist without either sequence being convergent.

(b) Show that $\lim_{n\to\infty} a_n - b_n$ can exist without either sequence being convergent.

(c) Show that $\lim_{n\to\infty} a_n \cdot b_n$ can exist without either sequence being convergent.

(d) Show that $\lim_{n\to\infty} \frac{a_n}{b_n}$ can exist without either sequence being convergent.

2-24. Can a Cauchy sequence have two limits? Explain your answer.

2-25 Use the fact that Cauchy sequences converge in the real numbers and the axioms for $\mathbb{R}$ except for Axiom 1.19 to prove Axiom 1.19.

Hint. Let $S \subseteq \mathbb{R}$ be bounded above and not empty. Construct a Cauchy sequence $\{a_n\}_{n=1}^{\infty}$ so that for all $x \in S$ there is an $N \in \mathbb{N}$ so that for all $n \geq N$ the inequality $a_n \geq x$ holds.

2.4 Bounded Sequences

If the elements of a sequence cannot become arbitrarily large, we speak of a bounded sequence. Unlike being a Cauchy sequence, boundedness is not equivalent to convergence, but it still has some important consequences.

Definition 2.32 *A sequence $\{a_n\}_{n=1}^{\infty}$ is called* **bounded above** *iff there is a number $A \in \mathbb{R}$ such that for all $n \in \mathbb{N}$ the inequality $a_n \leq A$ holds. In this case, A is also called an* **upper bound** *of the sequence. A sequence $\{a_n\}_{n=1}^{\infty}$ is called* **bounded below** *iff there is a $B \in \mathbb{R}$ such that for all $n \in \mathbb{N}$ the inequality $a_n \geq B$ holds. In this case, B is also called a* **lower bound** *of the sequence. Finally, a sequence is called* **bounded** *iff it is bounded above and bounded below and we call it* **unbounded** *if not.*

Example 2.33 The sequence $\left\{\frac{1}{n}\right\}_{n=1}^{\infty}$ is bounded, while the sequence $\{n\}_{n=1}^{\infty}$ is not bounded. □

Proposition 2.34 *Any convergent sequence of real numbers is bounded.*

Proof. Let $\{a_n\}_{n=1}^{\infty}$ be a convergent sequence and let $L := \lim_{n\to\infty} a_n$. We need to prove that there is a number $M \in \mathbb{R}$ so that for all $n \in \mathbb{N}$ the inequality $|a_n| \leq M$ holds.

Let $\varepsilon > 0$. Then there is an $N \in \mathbb{N}$ so that for all $n \geq N$ we have $|a_n - L| < \varepsilon$. Let $M := \max\left\{|L| + \varepsilon, |a_1|, \ldots, |a_{N-1}|\right\}$. Then for all $n < N$ we trivially have $|a_n| \leq M$ and for $n \geq N$ we obtain $|a_n| \leq |a_n - L| + |L| < \varepsilon + |L| \leq M$.

We have proved that $\{a_n\}_{n=1}^{\infty}$ is bounded below by $-M$ and above by M. ■

In general, the converse of Proposition 2.34 is not true as the next example shows.

Example 2.35 The sequence $\left\{(-1)^n\right\}_{n=1}^{\infty}$ is bounded, but it does not converge (see Example 2.29). □

For monotone sequences, however, boundedness does imply convergence.

Definition 2.36 *Let $\{a_n\}_{n=1}^{\infty}$ be a sequence. Then $\{a_n\}_{n=1}^{\infty}$ is called* **nondecreasing** *iff for all $n \in \mathbb{N}$ we have $a_n \leq a_{n+1}$. It is called* **nonincreasing** *iff for all $n \in \mathbb{N}$ we have $a_n \geq a_{n+1}$. If $\{a_n\}_{n=1}^{\infty}$ is either nonincreasing or nondecreasing, then it is called* **monotone**. *Moreover, $\{a_n\}_{n=1}^{\infty}$ is called* **(strictly) increasing** *iff for all $n \in \mathbb{N}$ we have $a_n < a_{n+1}$ and it is called* **(strictly) decreasing** *iff for all $n \in \mathbb{N}$ we have $a_n > a_{n+1}$.*

The sequence $\{n\}_{n=1}^{\infty}$ shows that nondecreasing sequences can grow beyond all bounds. But if this is not the case, a monotone sequence converges. The key to the proof is Standard Proof Technique 2.28.

Theorem 2.37 Monotone Sequence Theorem. *If $\{a_n\}_{n=1}^{\infty}$ is bounded and monotone, then $\{a_n\}_{n=1}^{\infty}$ converges. More precisely*

1. *If $\{a_n\}_{n=1}^{\infty}$ is bounded above and nondecreasing, then it converges.*
2. *If $\{a_n\}_{n=1}^{\infty}$ is bounded below and nonincreasing, then it converges.*

Proof. We only prove part 1. The proof of part 2 is similar (Exercise 2-30).

To prove part 1, let $\{a_n\}_{n=1}^{\infty}$ be bounded above and nondecreasing. Then the set $\{a_n : n \in \mathbb{N}\}$ is bounded above, and hence by Axiom 1.19 it has a supremum L. To prove that L is the limit of the sequence, we must prove that for every $\varepsilon > 0$ there is an $N \in \mathbb{N}$ so that for all $n \geq N$ the inequality $|a_n - L| < \varepsilon$ holds.

Let $\varepsilon > 0$. By Proposition 1.21 there is an $N \in \mathbb{N}$ with $a_N > L - \varepsilon$. But then for all $n \geq N$ we have $L - \varepsilon < a_N \leq a_n \leq L$, and hence $|a_N - L| < \varepsilon$.

We have proved that for every $\varepsilon > 0$ there is an $N \in \mathbb{N}$ so that for all $n \geq N$ we have $|a_n - L| < \varepsilon$. Hence, $\lim_{n\to\infty} a_n = L$. ∎

Although boundedness does not imply convergence, it forces the sequence to "cluster" in some places. To make this idea more precise, we need the notion of a subsequence.

Definition 2.38 *Let A, B, C be sets and let $f : B \to C$ and $g : A \to B$ be functions. The* **composition** *of f and g is defined by $f \circ g(a) := f\big(g(a)\big)$ for all $a \in A$.*

Definition 2.39 *Let $\{a_n\}_{n=1}^{\infty}$ be a sequence of real numbers and let $\{n_k\}_{k=1}^{\infty}$ be a strictly increasing sequence of natural numbers. Then $\left\{a_{n_k}\right\}_{k=1}^{\infty}$ is called a* **subsequence** *of $\{a_n\}_{n=1}^{\infty}$. Formally, a subsequence is the composition of the function that maps k to n_k and the function that maps n to a_n.*

Convergence is what happens when the indices get large. To obtain a notion that is useful to analyze convergence behavior, in the definition of a subsequence we had to specifically demand that the n_k are *strictly* increasing. If we had allowed $n_k = n_{k+1}$, then by choosing $\{n_k\}_{k=1}^{\infty}$ to be a constant sequence, any sequence would have infinitely many convergent "subsequences." This would be counterintuitive, because a sequence such as $\left\{\frac{1}{n}\right\}_{n=1}^{\infty}$ would have a "subsequence" that converges to 1 even though the sequence itself converges to 0.

With the definition as it is, subsequences behave sensibly when the sequence converges.

Proposition 2.40 *Let $\{a_n\}_{n=1}^{\infty}$ be a convergent sequence of real numbers with limit L. Then every subsequence $\left\{a_{n_k}\right\}_{k=1}^{\infty}$ also converges to L.*

Proof. Let $\{a_n\}_{n=1}^{\infty}$ be a convergent sequence of real numbers with limit L and let $\left\{a_{n_k}\right\}_{k=1}^{\infty}$ be a subsequence. We must prove that for all $\varepsilon > 0$ there is a $K \in \mathbb{N}$ so that for all $k \geq K$ we have $\left|a_{n_k} - L\right| < \varepsilon$.

Let $\varepsilon > 0$. Because $n_k < n_{k+1}$ for all $k \in \mathbb{N}$, an easy induction shows that $n_k \geq k$ for all $k \in \mathbb{N}$ (Exercise 2-31). Because $\{a_n\}_{n=1}^{\infty}$ converges, there is an $N \in \mathbb{N}$ so that for all $n \geq N$ we have $|a_n - L| < \varepsilon$. Therefore for all $k \geq N$ we obtain $n_k \geq k \geq N$, and hence $\left|a_{n_k} - L\right| < \varepsilon$.

We have proved that for all $\varepsilon > 0$ there is a $K \in \mathbb{N}$ so that for all $k \geq K$ we have $\left|a_{n_k} - L\right| < \varepsilon$. Hence, $\left\{a_{n_k}\right\}_{k=1}^{\infty}$ converges to L, too. ■

The precise statement of the idea that a bounded sequence of real numbers must "cluster" somewhere is that a bounded sequence of real numbers must have a convergent subsequence. This is an important property of the real numbers, which is ultimately encoded in the notion of compactness (see Section 16.5). The proof of the Bolzano-Weierstrass Theorem utilizes Standard Proof Technique 2.28.

Theorem 2.41 Bolzano-Weierstrass Theorem. *Any bounded sequence $\{a_n\}_{n=1}^{\infty}$ of real numbers has a convergent subsequence.*

Proof. Let $\{a_n\}_{n=1}^{\infty}$ be a bounded sequence and let $b > 0$ be such that for all $n \in \mathbb{N}$ we have $|a_n| < b$. Then b is a real number so that $\{n \in \mathbb{N} : a_n \leq b\} = \mathbb{N}$ and $\{n \in \mathbb{N} : a_n \leq -b\} = \emptyset$. Therefore b is contained in the set of real numbers $\left\{x \in \mathbb{R} : \{n \in \mathbb{N} : a_n \leq x\} \text{ is infinite}\right\}$. Moreover, this set is bounded below by $-b$. Hence, the infimum $L := \inf\left\{x \in \mathbb{R} : \{n \in \mathbb{N} : a_n \leq x\} \text{ is infinite}\right\}$ exists by Proposition 1.20.

We will prove that L is the limit of a subsequence of $\{a_n\}_{n=1}^{\infty}$. To do this, we will employ the Standard Proof Technique 2.22 and find a subsequence $\left\{a_{n_k}\right\}_{k=1}^{\infty}$ so that $\left|a_{n_k} - L\right|$ is bounded by a sequence that goes to zero. For each $k \in \mathbb{N}$ the set $H_k^+ := \left\{n \in \mathbb{N} : a_n \leq L + \frac{1}{k}\right\}$ is infinite and the set $H_k^- := \left\{n \in \mathbb{N} : a_n \leq L - \frac{1}{k}\right\}$ is finite. Therefore the set $T_k := \left\{n \in \mathbb{N} : a_n \in \left(L - \frac{1}{k}, L + \frac{1}{k}\right]\right\} = H_k^+ \setminus H_k^-$ is infinite for each $k \in \mathbb{N}$ (see Lemma 2.26). Construct $\{n_k\}_{k=1}^{\infty}$ inductively as follows. Because T_1 is infinite, it is not empty and we let $n_1 \in T_1$. Once n_k was chosen, let n_{k+1} be any natural number in T_{k+1} that is greater than n_k. Such a natural number exists, because T_{k+1} is infinite. Then $\left\{a_{n_k}\right\}_{k=1}^{\infty}$ is a subsequence of $\{a_n\}_{n=1}^{\infty}$ and for all $k \in \mathbb{N}$ we have $\left|a_{n_k} - L\right| \leq \frac{1}{k}$, because $n_k \in T_k$. By the Squeeze Theorem this means $\lim_{k\to\infty} \left|a_{n_k} - L\right| = 0$, and hence $\lim_{k\to\infty} a_{n_k} = L$. ■

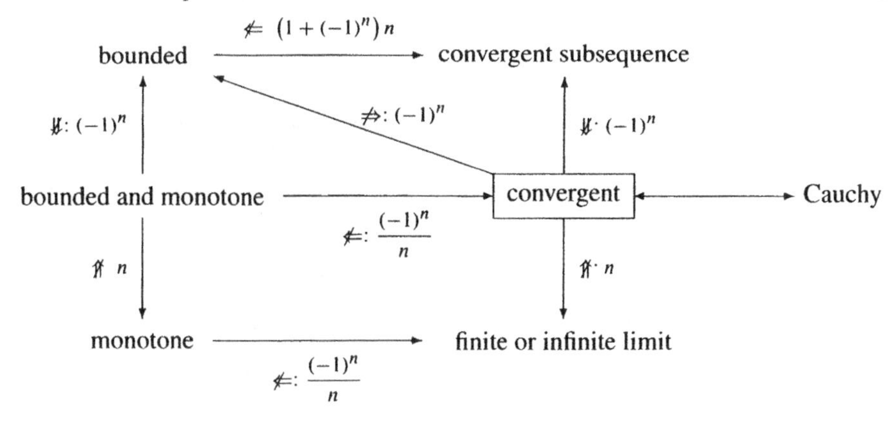

Figure 5: Implications between the various notions related to convergence that are introduced in this chapter. Implications are indicated with arrows and the examples near the arrows indicate that the opposite implication does not hold.

The Bolzano-Weierstrass Theorem is a useful tool in single variable analysis. We will see examples of its use in the proofs of Theorem 3.44 and Lemma 5.19.

For many sets of properties, it is instructive to explore which property implies which other properties and which properties are equivalent. Figure 5 summarizes the properties introduced in this chapter (including those from the next section) and how they are related to each other.

Exercises

2-26 For each of the sequences below determine if it is bounded, and then prove your claim

(a) $\left\{\frac{1}{n}\right\}_{n=1}^{\infty}$ (b) $\{n\}_{n=1}^{\infty}$ (c) $\left\{\frac{2n+1}{3n-5}\right\}_{n=1}^{\infty}$ (d) $\left\{\frac{n^2-1}{n+12}\right\}_{n=1}^{\infty}$

2-27. For the given sequence $\{a_n\}_{n=1}^{\infty}$, find an expression for the terms of the subsequence $\{a_{n_k}\}_{k=1}^{\infty}$ where n_k is as indicated.

(a) $\left\{\frac{n+4}{n^2+4}\right\}_{n=1}^{\infty}$, $n_k = 4k$ (b) $\{(-1)^n\}_{n=1}^{\infty}$, $n_k = 2k$

(c) $\{(1+(-1)^n)n\}_{n=1}^{\infty}$, $n_k = 2k$ and $n_m = 2m+1$

2-28. Use Proposition 2.40 to prove that each of the sequences below diverges

(a) $\{(-1)^n\}_{n=1}^{\infty}$ (b) $\{(1+(-1)^n)n\}_{n=1}^{\infty}$

2-29. Explain why we could have chosen $\varepsilon = 1$ in the proof of Proposition 2.34. Then explain why we cannot choose $\varepsilon = 1$ in a general convergence proof.

2-30. Prove part 2 of the Monotone Sequence Theorem.

2-31 Perform the induction mentioned in the proof of Proposition 2.40 First state exactly what it is that you prove, then execute the proof.

2-32. Sketch a visualization for the construction of L in the proof of the Bolzano-Weierstrass Theorem that is similar to Figure 4.

2-33. Let $x \in \mathbb{R}$ and let $\{x_n\}_{n=1}^{\infty}$ be a sequence of real numbers that does not converge to x. Prove that there is an $\varepsilon > 0$ and a subsequence $\{x_{n_k}\}_{k=1}^{\infty}$ so that for all $k \in \mathbb{N}$ we have $|x_{n_k} - x| \geq \varepsilon$.

2-34. Let $\{x_n\}_{n=1}^{\infty}$ be a sequence of real numbers that has no convergent subsequence. Prove that for each $x \in \mathbb{R}$ there is an $\varepsilon_x > 0$ so that $\{n \in \mathbb{N} : |x_n - x| < \varepsilon_x\}$ is finite.

2-35 Let $L \in \mathbb{R}$ and let $\{x_n\}_{n=1}^{\infty}$ be a sequence of real numbers such that every subsequence has a subsequence that converges to L Prove that $\{x_n\}_{n=1}^{\infty}$ converges.

2-36 A well-known convergent sequence

(a) Let $a, b \in \mathbb{R}$ and $n \in \mathbb{N}$. Prove that $b^{n+1} - a^{n+1} = (b-a)\sum_{j=0}^{n} a^j b^{n-j}$.

(b) Let $a, b \in \mathbb{R}$ with $0 \leq a < b$ and let $n \in \mathbb{N}$ Prove that $\dfrac{b^{n+1} - a^{n+1}}{b-a} < (n+1)b^n$.

(c) Prove that $\left\{\left(1 + \frac{1}{n}\right)^n\right\}_{n=1}^{\infty}$ is increasing and that it converges

Hint. To prove it is increasing, bring a^{n+1} to the right in part 2-36b and use $a = 1 + \dfrac{1}{n+1}$ and $b = 1 + \dfrac{1}{n}$. To prove the sequence is bounded above, use $a = 1$ and $b = 1 + \dfrac{1}{2n}$.

2-37. Use the Monotone Sequence Theorem and the axioms for $\mathbb{R}$ except Axiom 1.19 to prove Axiom 1.19.

2-38. Use the Bolzano-Weierstrass Theorem and the axioms for $\mathbb{R}$ except Axiom 1 19 to prove Axiom 1 19.

2-39. Use the Bolzano-Weierstrass Theorem and the axioms for $\mathbb{R}$ except for Axiom 1 19 to prove directly (that is, without using Theorem 2 27 or its proof) that every Cauchy sequence of real numbers converges.

2-40. Use the Bolzano-Weierstrass Theorem and the axioms for $\mathbb{R}$ except for Axiom 1.19 to prove the Monotone Sequence Theorem

2-41 Let $\{a_n\}_{n=1}^{\infty}$ and $\{b_n\}_{n=1}^{\infty}$ be bounded sequences.

(a) Prove that $\{a_n + b_n\}_{n=1}^{\infty}$ is bounded.

Hint. Unlike for convergence, the bound need not be one number M An upper bound of the form $M_a + M_b$ would work just fine

(b) Prove that $\{a_n b_n\}_{n=1}^{\infty}$ is bounded

(c) Prove that if there is a $\delta > 0$ so that $b_n > \delta$ for all $n \in \mathbb{N}$ then $\left\{\dfrac{a_n}{b_n}\right\}_{n=1}^{\infty}$ is bounded

2-42. Let $x \in [0, 1]$. Prove that the sequence $\{a_n\}_{n=0}^{\infty}$ defined recursively by $a_0 := 0$ and the recurrence relation $a_{n+1} := a_n + \frac{1}{2}\left(x - a_n^2\right)$ converges to $\sqrt{x}$.

Hint Prove by induction that the sequence is bounded above by $\sqrt{x}$ Then prove that it is nondecreasing.

2.5 Infinite Limits

Although unbounded sequences diverge, they can display some types of regular behavior, which are explored in this section.

Definition 2.42 *Let $\{a_n\}_{n=1}^{\infty}$ be a sequence of real numbers. Then we say that the limit of $\{a_n\}_{n=1}^{\infty}$ is* **infinity** *iff for every $M \in \mathbb{R}$ there is an $N \in \mathbb{N}$ so that for all $n \geq N$ the inequality $a_n \geq M$ holds. In this case, we write $\lim_{n\to\infty} a_n = \infty$. A limit of* **negative infinity** *is defined similarly (Exercise 2-43) and denoted $\lim_{n\to\infty} a_n = -\infty$.*

Intuitively an infinite limit should mean that eventually the sequence gets close to infinity. Similar to Definition 2.2, this idea is encoded in Definition 2.42 by saying that for any given bound M, there is a threshold N so that once the running index n goes past the threshold N, the sequence will not drop below the bound M any more. It is also said that a sequence with $\lim_{n\to\infty} a_n = \infty$ **grows beyond all bounds**.

Example 2.43 *Let $x > 1$. Then $\lim_{n\to\infty} x^n = \infty$.*

Clearly, for all $n \in \mathbb{N}$ the inequality $x^{n+1} > x^n$ holds. Suppose for a contradiction that $\{x^n\}_{n=1}^{\infty}$ does not go to infinity. Then (refer to Standard Proof Technique 2.30 as necessary) there is a $B > 0$ so that for all $N \in \mathbb{N}$ there is an $m > N$ with $x^m \leq B$. Let $n \in \mathbb{N}$. By the above, there is an $m \geq n$ so that $x^n \leq x^m \leq B$. Hence, the sequence $\{x^n\}_{n=1}^{\infty}$ is bounded above. Let $M := \sup\{x^n : n \in \mathbb{N}\}$. Now by Proposition 1.21 because $\frac{M}{x} < M$, there is an $n \in \mathbb{N}$ so that $x^n > \frac{M}{x}$. But then $x^{n+1} = x \cdot x^n > x\frac{M}{x} = M$, a contradiction. Thus $\lim_{n\to\infty} x^n = \infty$. ■

With some exceptions, discussed at the end of this section, the limit laws for infinite limits are similar to those for convergent sequences.

Theorem 2.44 **Limit laws** *involving ∞. Let $\{a_n\}_{n=1}^{\infty}$ be such that $\lim_{n\to\infty} a_n = \infty$.*

1. *If $\{b_n\}_{n=1}^{\infty}$ is a bounded sequence, then $\lim_{n\to\infty} a_n + b_n = \infty$ and if all a_n are nonzero, then $\lim_{n\to\infty} \frac{b_n}{a_n} = 0$.*
2. *If $\lim_{n\to\infty} b_n = \infty$, then $\lim_{n\to\infty} a_n + b_n = \infty$ and $\lim_{n\to\infty} a_n b_n = \infty$.*
3. *If $c > 0$ is a real number, then $\lim_{n\to\infty} ca_n = \infty$.*
4. *If $c < 0$ is a real number, then $\lim_{n\to\infty} ca_n = -\infty$.*

Proof. To prove part 1 let $\lim_{n\to\infty} a_n = \infty$ and let $\{b_n\}_{n=1}^{\infty}$ be bounded. Let $B \in \mathbb{R}$ be such that for all $n \in \mathbb{N}$ the inequality $|b_n| < B$ holds.

First consider the sum. We need to prove that for all $M \in \mathbb{R}$ there is an $N \in \mathbb{N}$ so that for all $n \geq N$ we have $a_n + b_n \geq M$.

Let $M \in \mathbb{R}$. There is an $N \in \mathbb{N}$ so that for all $n \geq N$ we have $a_n \geq M + B$. But then for all $n \geq N$ we obtain $a_n + b_n \geq M + B - B = M$, and hence $\lim_{n\to\infty} a_n + b_n = \infty$.

Now consider the quotient. We need to prove that for all $\varepsilon > 0$ there is an $N \in \mathbb{N}$ so that for all $n \geq N$ the inequality $\left|\frac{b_n}{a_n}\right| < \varepsilon$ holds.

Let $\varepsilon > 0$. By Theorem 1.32 there is an $M \in \mathbb{N} \subseteq \mathbb{R}$ so that $\frac{|B|}{\varepsilon} < M$, which means $\frac{|B|}{M} < \varepsilon$. Now there is an $N \in \mathbb{N}$ so that for all $n \geq N$ we have $a_n \geq M$. Thus for all $n \geq N$ we obtain $\left|\frac{b_n}{a_n}\right| \leq \frac{|B|}{M} < \varepsilon$, and hence $\lim_{n\to\infty} \frac{b_n}{a_n} = 0$.

To prove part 2 let $\lim_{n\to\infty} a_n = \lim_{n\to\infty} b_n = \infty$.

For the sum, we need to prove that for all $M \in \mathbb{R}$ there is an $N \in \mathbb{N}$ so that for all $n \geq N$ the inequality $a_n + b_n \geq M$ holds.

Let $M \in \mathbb{R}$ and note that there is an $N \in \mathbb{N}$ so that for all $n \geq N$ we have the inequalities $a_n \geq \frac{M}{2}$ and $b_n \geq \frac{M}{2}$ (see Standard Proof Technique 2.6). But then for all $n \geq N$ we have $a_n + b_n \geq \frac{M}{2} + \frac{M}{2} = M$. Hence, $\lim_{n\to\infty} a_n + b_n = \infty$.

For the product, we need to prove that for all $M \in \mathbb{R}$ there is an $N \in \mathbb{N}$ so that for all $n \geq N$ the inequality $a_n b_n \geq M$ holds.

Let $M \in \mathbb{R}$ and note that there is an $N \in \mathbb{N}$ so that for all $n \geq N$ we have the inequalities $a_n \geq \sqrt{|M|}$ and $b_n \geq \sqrt{|M|}$. But then for all $n \geq N$ the product exceeds M because $a_n b_n \geq \sqrt{|M|}\sqrt{|M|} = |M| \geq M$. Hence, $\lim_{n\to\infty} a_n b_n = \infty$.

To prove part 3, let $\lim_{n\to\infty} a_n = \infty$ and let $c > 0$. We need to prove that for all $M \in \mathbb{R}$ there is an $N \in \mathbb{N}$ so that for all $n \geq N$ the inequality $ca_n \geq M$ holds.

Let $M \in \mathbb{R}$. There is an $N \in \mathbb{N}$ so that for all $n \geq N$ we have $a_n \geq \frac{M}{c}$. But then for all $n \geq N$ we obtain $ca_n \geq c\frac{M}{c} = M$. Hence, $\lim_{n\to\infty} ca_n = \infty$.

Finally, to prove part 4 let $\lim_{n\to\infty} a_n = \infty$ and let $c < 0$. We need to prove that for all $M \in \mathbb{R}$ there is an $N \in \mathbb{N}$ so that for all $n \geq N$ the inequality $ca_n \leq M$ holds.

Let $M \in \mathbb{R}$. There is an $N \in \mathbb{N}$ so that for all $n \geq N$ we have $a_n \geq -\frac{M}{|c|}$. But then for all $n \geq N$ we obtain $ca_n \leq c\left(-\frac{M}{|c|}\right) = M$. Hence, $\lim_{n\to\infty} ca_n = -\infty$. ■

Infinite limits can also help indirectly to establish the existence of finite limits.

Example 2.45 *Let $x > 0$. Then $\lim_{n\to\infty} x^{\frac{1}{n}} = 1$.*

The result is trivial for $x = 1$. We first consider $x > 1$. Suppose for a contradiction that $\lim_{n\to\infty} x^{\frac{1}{n}} \neq 1$. For every $n \in \mathbb{N}$ we have $1 < x^{\frac{1}{n+1}} = \left(x^{\frac{1}{n}}\right)^{\frac{n}{n+1}} < x^{\frac{1}{n}}$. Thus if $\left\{x^{\frac{1}{n}}\right\}_{n=1}^{\infty}$ does not converge to 1, then (refer to Standard Proof Technique 2.30 as necessary) because $x^{\frac{1}{n}} > 1$ for all $n \in \mathbb{N}$, there is an $\varepsilon > 0$ so that for every $N \in \mathbb{N}$ there is an $n \geq N$ with $x^{\frac{1}{n}} > 1 + \varepsilon$. Because $\left\{x^{\frac{1}{n}}\right\}_{n=1}^{\infty}$ is decreasing this would mean that $x^{\frac{1}{n}} > 1 + \varepsilon$ for all $n \in \mathbb{N}$. But then for all $n \in \mathbb{N}$ we would have $(1+\varepsilon)^n < \left(x^{\frac{1}{n}}\right)^n = x$, contradicting the fact that $\lim_{n\to\infty} (1+\varepsilon)^n = \infty$ (see Example 2.43). Thus for all $x > 1$ we have $\lim_{n\to\infty} x^{\frac{1}{n}} = 1$.

The proof for $0 < x < 1$ is deferred to Exercise 2-46. □

Just as for infinite limits, there are limit laws for limits that equal negative infinity.

Theorem 2.46 Limit laws *involving* $-\infty$. *Let* $\{a_n\}_{n=1}^{\infty}$ *be such that* $\lim_{n\to\infty} a_n = -\infty$.

1. *If* $\{b_n\}_{n=1}^{\infty}$ *is a bounded sequence, then* $\lim_{n\to\infty} a_n + b_n = -\infty$ *and if all* a_n *are nonzero, then* $\lim_{n\to\infty} \frac{b_n}{a_n} = 0$.
2. *If* $\lim_{n\to\infty} b_n = -\infty$, *then* $\lim_{n\to\infty} a_n + b_n = -\infty$ *and* $\lim_{n\to\infty} a_n b_n = \infty$.
3. *If* $\lim_{n\to\infty} b_n = \infty$, *then* $\lim_{n\to\infty} a_n b_n = -\infty$.
4. *If* $c > 0$ *is a real number, then* $\lim_{n\to\infty} ca_n = -\infty$.
5. *If* $c < 0$ *is a real number, then* $\lim_{n\to\infty} ca_n = \infty$.

Proof. The proof of Theorem 2.46 is similar to that of Theorem 2.44. It is thus left to the reader as Exercise 2-44. ■

The addition of two sequences such that one has limit ∞ and the other has limit $-\infty$ is absent from Theorems 2.44 and 2.46. This is because by Exercise 2-48b the sum need not converge. Exercise 2-48c shows that even if there is a limit, it is not the same number in all cases. The situation is similar for the product of a sequence with infinite limit and a sequence with limit zero (see Exercises 2-48d and 2-48e), as well as for the quotient of two sequences with infinite limits (see Exercises 2-48f and 2-48g). These types of limits are called **indeterminate forms** and they are discussed in more detail in Section 12.3.

Exercises

2-43 State the definition of a sequence whose limit is negative infinity.

2-44. Prove Theorem 2.46. That is, prove each of the following

(a) If $\lim_{n\to\infty} a_n = -\infty$ and $\{b_n\}_{n=1}^{\infty}$ is a bounded sequence, then $\lim_{n\to\infty} a_n + b_n = -\infty$ and if all a_n are nonzero, then $\lim_{n\to\infty} \frac{b_n}{a_n} = 0$

(b) If $\lim_{n\to\infty} a_n = \lim_{n\to\infty} b_n = -\infty$, then $\lim_{n\to\infty} a_n + b_n = -\infty$ and $\lim_{n\to\infty} a_n b_n = \infty$

(c) If $\lim_{n\to\infty} a_n = -\infty$ and $\lim_{n\to\infty} b_n = \infty$, then $\lim_{n\to\infty} a_n b_n = -\infty$.

(d) If $c > 0$ is a real number and $\lim_{n\to\infty} a_n = -\infty$, then $\lim_{n\to\infty} ca_n = -\infty$.

(e) If $c < 0$ is a real number and $\lim_{n\to\infty} a_n = -\infty$, then $\lim_{n\to\infty} ca_n = \infty$

2-45 Let $0 < x < 1$.

(a) Prove that $\lim_{n\to\infty} x^n = 0$ by using the limit laws.

(b) Prove that $\lim_{n\to\infty} x^n = 0$ by mimicking the proof in Example 2 43

2-46. Let $0 < x < 1$. Prove that $\lim_{n\to\infty} x^{\frac{1}{n}} = 1$.

2-47. Prove that $\left\{ \sqrt[n]{n} \right\}_{n=1}^{\infty}$ is decreasing and that the limit is 1.

Hint. Use Exercise 2-36c to show that $\left(1+\frac{1}{n}\right)^n \le n$ for all large enough n and derive the inequality $(n+1)^{\frac{1}{n+1}} < n^{\frac{1}{n}}$ from this. For the limit L, use the result from Example 2.45 to show that $L = L^2$.

2-48. A first encounter with indeterminate forms.

(a) For $\{a_n\}_{n=1}^{\infty} := \{n\}_{n=1}^{\infty}$ and $\{b_n\}_{n=1}^{\infty} := \left\{-n^2\right\}_{n=1}^{\infty}$, prove that $\lim_{n\to\infty} a_n = \infty$, $\lim_{n\to\infty} b_n = -\infty$ and that the sequence $\{a_n + b_n\}_{n=1}^{\infty}$ diverges.

(b) For $\{a_n\}_{n=1}^{\infty} := \{n\}_{n=1}^{\infty}$ and $\{b_n\}_{n=1}^{\infty} := \{-n\}_{n=1}^{\infty}$, prove that $\lim_{n\to\infty} a_n = \infty$, $\lim_{n\to\infty} b_n = -\infty$ and that the sequence $\{a_n + b_n\}_{n=1}^{\infty}$ converges.

(c) Let $c \in \mathbb{R}$. Find sequences $\{a_n\}_{n=1}^{\infty}$ and $\{b_n\}_{n=1}^{\infty}$ so that $\lim_{n\to\infty} a_n = \infty$, $\lim_{n\to\infty} b_n = -\infty$ and $\lim_{n\to\infty} a_n + b_n = c$.

(d) Find sequences $\{a_n\}_{n=1}^{\infty}$ and $\{b_n\}_{n=1}^{\infty}$ so that $\lim_{n\to\infty} a_n = \infty$, $\lim_{n\to\infty} b_n = 0$ and $\{a_n b_n\}_{n=1}^{\infty}$ diverges.

(e) Let $c \in \mathbb{R}$. Find sequences $\{a_n\}_{n=1}^{\infty}$ and $\{b_n\}_{n=1}^{\infty}$ so that $\lim_{n\to\infty} a_n = \infty$, $\lim_{n\to\infty} b_n = 0$ and $\lim_{n\to\infty} a_n b_n = c$.

(f) Find sequences $\{a_n\}_{n=1}^{\infty}$ and $\{b_n\}_{n=1}^{\infty}$ so that $\lim_{n\to\infty} a_n = \lim_{n\to\infty} b_n = \infty$ and $\left\{\frac{a_n}{b_n}\right\}_{n=1}^{\infty}$ diverges.

(g) Let $c \in \mathbb{R}$. Find sequences $\{a_n\}_{n=1}^{\infty}$ and $\{b_n\}_{n=1}^{\infty}$ so that $\left|\lim_{n\to\infty} a_n\right| = \left|\lim_{n\to\infty} b_n\right| = \infty$ and $\lim_{n\to\infty} \frac{a_n}{b_n} = c$.

2-49. Prove that if $\{a_n\}_{n=1}^{\infty}$ and $\{b_n\}_{n=1}^{\infty}$ are sequences so that $\lim_{n\to\infty} a_n = \infty$ and there are an $\varepsilon > 0$ and an $N \in \mathbb{N}$ so that $b_n \ge \varepsilon$ for all $n \ge N$, then $\lim_{n\to\infty} a_n b_n = \infty$.

2-50. A characterization of divergent sequences.

(a) Let $\{a_n\}_{n=1}^{\infty}$ be an unbounded sequence. Prove that there is a subsequence $\left\{a_{n_k}\right\}_{k=1}^{\infty}$ so that $\lim_{k\to\infty} a_{n_k} = \infty$ or $\lim_{k\to\infty} a_{n_k} = -\infty$.

(b) Let $\{a_n\}_{n=1}^{\infty}$ be a bounded divergent sequence. Prove that there are two convergent subsequences $\left\{a_{l_m}\right\}_{m=1}^{\infty}$ and $\left\{a_{n_k}\right\}_{k=1}^{\infty}$ such that $\lim_{m\to\infty} a_{l_m}$ and $\lim_{k\to\infty} a_{n_k}$ exist, but are not equal.

(c) Let $\{a_n\}_{n=1}^{\infty}$ be a sequence. Prove that $\{a_n\}_{n=1}^{\infty}$ diverges if and only if there is a subsequence $\left\{a_{n_k}\right\}_{k=1}^{\infty}$ such that $\lim_{k\to\infty} a_{n_k} = \infty$ or $\lim_{k\to\infty} a_{n_k} = -\infty$, *or* there are two subsequences $\left\{a_{l_m}\right\}_{m=1}^{\infty}$ and $\left\{a_{n_k}\right\}_{k=1}^{\infty}$ such that $\lim_{m\to\infty} a_{l_m}$ and $\lim_{k\to\infty} a_{n_k}$ exist, but are not equal.

(d) Use the characterization of divergent sequences in part 2-50c and the axioms for $\mathbb{R}$ except for Axiom 1.19 to prove the Bolzano-Weierstrass Theorem.

2-51. Prove **Cauchy's Limit Theorem**. That is, let $\{b_n\}_{n=1}^{\infty}$ be a strictly increasing sequence of positive numbers that goes to infinity and let $\{a_n\}_{n=1}^{\infty}$ be a sequence. Prove that if the sequence $\left\{\frac{a_n - a_{n-1}}{b_n - b_{n-1}}\right\}_{n=2}^{\infty}$ converges to c, then $\lim_{n\to\infty} \frac{a_n}{b_n} = c$.

Hint. Exercise 2-16 with $p_n := b_n - b_{n-1}$ and another appropriate sequence.

Readers interested in series of numbers could read Chapters 6 and 10 before continuing.

Chapter 3

Continuous Functions

Functions are the central objects of analysis. This chapter defines limits and continuity for functions of a real variable and it presents some consequences of continuity. To avoid problems with complicated domains (see Exercise 3-32 and the end of Section 16.3.1 for some details), in this chapter functions are usually considered on intervals or on intervals from which at most finitely many points were removed. These domains are sufficient to build the traditional calculus of functions of one variable. More complicated domains are handled in metric spaces in the second part of the text.

3.1 Limits of Functions

The limit of a function at a point x is supposed to express what happens *near* x, but not necessarily *at* x. This is similar to the running index n of a sequence never actually *becoming* ∞. While ∞ is not in the domain of a sequence, a real number x can be in the domain of a function. Hence, we must explicitly remove x from consideration. In this section, functions are defined on an open interval from which a point x has been removed. In this fashion, we assure that each function is defined "close to the left of x" *and* "close to the right of x," which is what we need to investigate for (two-sided) limits. Because convergence of sequences is already defined, we can use sequences to define convergence of functions.

Definition 3.1 Sequence formulation of the limit of a function. *Let $I \subseteq \mathbb{R}$ be an open interval and let $x \in I$. The number $L \in \mathbb{R}$ is called the* **limit** *of the function $f : I \setminus \{x\} \to \mathbb{R}$ at x iff for all sequences $\{z_n\}_{n=1}^{\infty}$ with $z_n \in I \setminus \{x\}$ for all $n \in \mathbb{N}$ and $\lim_{n\to\infty} z_n = x$ we have $\lim_{n\to\infty} f(z_n) = L$. In this case, we denote $\lim_{z\to x} f(z) := L$ and we also say that f* **converges (to L) at** *x.*

Similar to Theorem 2.11, the limit of a function at x is only affected by the values of the function near x.

Theorem 3.2 *Let $I \subseteq \mathbb{R}$ be an open interval, $x \in I$, and let $f, g : I \setminus \{x\} \to \mathbb{R}$ be functions. If there is a number $\delta > 0$ so that f and g are equal on the subset $\{z \in I \setminus \{x\} : |z - x| < \delta\}$ of $I \setminus \{x\}$, then f converges at x iff g converges at x and in this case the equality $\lim_{z \to x} f(z) = \lim_{z \to x} g(z)$ holds.*

Proof. Exercise 3-2. ■

By Theorem 3.2, Definition 3.1 also defines the limit at x for any function that is defined on a set D that contains a set $I \setminus \{x\}$, where I is an open interval that contains x. Formally, we define the following.

Definition 3.3 *Let D, R, S be sets with $R \subseteq D$ and let $f : D \to S$ be a function. The* **restriction** *of the function f to R, denoted $f|_R$, is defined by $f|_R(x) := f(x)$ for all $x \in R$.*

Definition 3.4 *Let $f : D \to \mathbb{R}$ be a function and let $x \in \mathbb{R}$ be so that there is an open interval I with $x \in I$ and $I \setminus \{x\} \subseteq D$. We define the* **limit** *of f at x as the limit of the restriction $f|_{I \setminus \{x\}}$ at x and denote it $\lim_{z \to x} f(z) := \lim_{z \to x} f|_{I \setminus \{x\}}(z)$.*

All the following results on functions $f : I \setminus \{x\} \to \mathbb{R}$ also apply to functions with larger domains. Strictly speaking we would need to apply all definitions and results to the restriction of the function to a set $I \setminus \{x\}$, where I is an open interval and $x \in I$. We will usually avoid this simple formality. Ultimately, Definition 16.33 will encompass this situation as well as some situations in which, for single variable functions, we use one-sided limits (see Definition 3.15 below).

Example 3.5

1. *For all $x \in \mathbb{R}$, we have $\lim_{z \to x} z = x$.*

2. *For all $x \in \mathbb{R}$, we have $\lim_{z \to x} |z| = |x|$.*

3. *The function $f(x) := \begin{cases} 1; & \text{for } x > 0, \\ 0; & \text{for } x = 0, \\ -1; & \text{for } x < 0, \end{cases}$ does not have a limit at $x = 0$.*

Part 1 is trivial and part 2 follows from Exercise 2-12. To see that the function in part 3 does not have a limit at $x = 0$, it suffices to produce two sequences $\{y_n\}_{n=1}^{\infty}$ and $\{z_n\}_{n=1}^{\infty}$ that converge to zero so that $y_n \neq 0$ and $z_n \neq 0$ for all $n \in \mathbb{N}$ and $\{f(y_n)\}_{n=1}^{\infty}$ and $\{f(z_n)\}_{n=1}^{\infty}$ converge and $\lim_{n \to \infty} f(y_n) \neq \lim_{n \to \infty} f(z_n)$. With $y_n := \frac{1}{n}$ and $z_n := -\frac{1}{n}$ for all $n \in \mathbb{N}$ we obtain $\lim_{n \to \infty} f\left(\frac{1}{n}\right) = 1 \neq -1 = \lim_{n \to \infty} f\left(-\frac{1}{n}\right)$, which completes the argument. □

Standard Proof Technique 3.6 If a sequence that converges to a number x from the right is needed, $\left\{x + \frac{1}{n}\right\}_{n=1}^{\infty}$ is usually a good choice. For a sequence that converges to x from the left, $\left\{x - \frac{1}{n}\right\}_{n=1}^{\infty}$ is usually a good choice. This idea is extended in Standard Proof Technique 3.8. □

There are at least two ways to define the limit of a function. We have already seen the formulation with sequences and we give the formulation with ε and δ below. The two formulations are equivalent, and hence either one could serve as the definition. With both formulations available, we can choose which one to use. Depending on the situation, one formulation may be preferable over the other to produce a simpler statement or proof. The proof of Theorem 3.19 is a good example of how each formulation is better suited for certain settings than the other.

Theorem 3.7 **ε-δ formulation of the limit of a function.** *Let $I \subseteq \mathbb{R}$ be an open interval and let $x \in I$. Then $L \in \mathbb{R}$ is the* **limit** *of the function $f : I \setminus \{x\} \to \mathbb{R}$ at x iff for all $\varepsilon > 0$ there is a $\delta > 0$ such that for all $z \in I \setminus \{x\}$ with $|z - x| < \delta$ we have that $\left|f(z) - L\right| < \varepsilon$.*

Proof. For "$\Rightarrow$," let $\lim_{z \to x} f(z) = L$. Suppose, for a contradiction, the statement on the right is false. Then there is an $\varepsilon > 0$ so that for each $\delta > 0$ there is a $z \in I \setminus \{x\}$ with $|z - x| < \delta$ and $\left|f(z) - L\right| \geq \varepsilon$ (if necessary, use Standard Proof Technique 2.30 for the negation). Then for $\delta := \frac{1}{n}$ there is a $z_n \in I \setminus \{x\}$ with $|z_n - x| < \frac{1}{n}$ and $\left|f(z_n) - L\right| \geq \varepsilon$. But then $\lim_{n \to \infty} z_n = x$, while $\lim_{n \to \infty} f(z_n)$ either does not exist, or if it exists, then $\lim_{n \to \infty} f(z_n) \neq L$ (see Exercise 2-5). Either way we have arrived at a contradiction.

For "$\Leftarrow$," let $f : I \setminus \{x\} \to \mathbb{R}$ be such that for all $\varepsilon > 0$ there is a $\delta > 0$ such that for all $z \in I \setminus \{x\}$ with $|z - x| < \delta$ we have that $\left|f(z) - L\right| < \varepsilon$. We need to prove that for each sequence $\{z_n\}_{n=1}^{\infty}$ with $z_n \in I \setminus \{x\}$ for all $n \in \mathbb{N}$ and $\lim_{n \to \infty} z_n = x$ we have that $\lim_{n \to \infty} f(z_n) = L$.

Let $\{z_n\}_{n=1}^{\infty}$ be a sequence with $z_n \in I \setminus \{x\}$ for all $n \in \mathbb{N}$ and $\lim_{n \to \infty} z_n = x$, and let $\varepsilon > 0$. Then there is a $\delta > 0$ such that for all $z \in I \setminus \{x\}$ with $|z - x| < \delta$ we have $\left|f(z) - L\right| < \varepsilon$. Moreover, for δ there is an $N \in \mathbb{N}$ so that for all $n \geq N$ we have $|z_n - x| < \delta$. But then for all $n \geq N$ we infer $\left|f(z_n) - L\right| < \varepsilon$, and hence $\lim_{n \to \infty} f(z_n) = L$. This proves that $\lim_{z \to x} f(z) = L$. ■

Standard Proof Technique 3.8 In the "$\Rightarrow$" part of the proof of Theorem 3.7, for all $\delta > 0$ there is a z with $|z - x| < \delta$ and other properties. To obtain a sequence $\{z_n\}_{n=1}^{\infty}$ that converges to x so that each z_n has the other desired properties, it is standard practice to use $\delta := \frac{1}{n}$ and then pick an appropriate element z_n with $|z_n - x| < \frac{1}{n}$. □

Exercises

3-1. Explain why after Definition 3.1 it is not necessary to prove that the limit of a function is unique.

3-2. Prove Theorem 3.2.

3-3 Let $m, b \in \mathbb{R}$. Prove that $\lim_{z \to x} mz + b = mx + b$

3-4. Let I be an open interval and let $x \in I$. Prove that if $f : I \setminus \{x\} \to \mathbb{R}$ does not converge to L at x, then there are an $\varepsilon > 0$ and a sequence $\{z_n\}_{n=1}^{\infty}$ so that $\lim_{n\to\infty} z_n = x$, $z_n \in I \setminus \{x\}$ for all $n \in \mathbb{N}$ and $\left| f(z_n) - L \right| \geq \varepsilon$ for all $n \in \mathbb{N}$.

3-5 Prove that if $m \in \mathbb{Z}$ and $\lfloor\ \rfloor$ is the floor function, then $\lim_{z \to m} \lfloor z \rfloor$ does not exist.

3-6 Prove that $\lim_{z \to 3} \frac{z-3}{z^2-9} = \frac{1}{6}$.

3-7. Prove that the **Dirichlet function** $f(x) = \begin{cases} 1; & \text{for } x \in \mathbb{Q}, \\ 0; & \text{for } x \notin \mathbb{Q}, \end{cases}$ does not converge at any $x \in \mathbb{R}$.

3-8 Explain why, with the present definition, $\lim_{z \to 0} \sqrt{z}$ is not defined. Then state what the limit should be and how we could circumvent our *purely formal* problem
Note. This problem will be resolved in Exercise 16-28.

3.2 Limit Laws

Just as for limits of sequences, we are interested in how limits of functions relate to the algebraic operations, because this should simplify the computation of limits. First note that all algebraic operations on functions are defined pointwise.

Definition 3.9 *Let $D \subseteq \mathbb{R}$ be a set and let $f, g : D \to \mathbb{R}$ be functions. For all $x \in D$ we define $(f+g)(x) := f(x) + g(x)$, $(f-g)(x) := f(x) - g(x)$, and $(f \cdot g)(x) := f(x) \cdot g(x)$. For all $x \in D$ with $g(x) \neq 0$ we define $\left(\frac{f}{g}\right)(x) := \frac{f(x)}{g(x)}$.*

Theorem 3.10 Limit laws *for functions. Let $I \subseteq \mathbb{R}$ be an open interval, let $x \in I$ and let $f, g : I \setminus \{x\} \to \mathbb{R}$ be functions such that $\lim_{z \to x} f(z)$ and $\lim_{z \to x} g(z)$ exist. Then the following hold.*

1. $\lim_{z \to x} (f+g)(z) = \lim_{z \to x} f(z) + \lim_{z \to x} g(z)$.

2. $\lim_{z \to x} (f-g)(z) = \lim_{z \to x} f(z) - \lim_{z \to x} g(z)$.

3. $\lim_{z \to x} (f \cdot g)(z) = \lim_{z \to x} f(z) \cdot \lim_{z \to x} g(z)$.

4. *If* $\lim_{z \to x} g(z) \neq 0$, *then* $\lim_{z \to x} \left(\frac{f}{g}\right)(z) = \frac{\lim_{z \to x} f(z)}{\lim_{z \to x} g(z)}$.

Each equation implicitly asserts that the limit on the left side exists (see box on p. 33). Moreover, formally, in part 4 we would need to demand also that $g(z) \neq 0$ for all $z \in I \setminus \{x\}$. But $\lim_{z \to x} g(z) \neq 0$ implies that $g(z) \neq 0$ for all $z \in I \setminus \{x\}$ that are near x. Hence, if g has zeros, we implicitly assume that g has been restricted appropriately rather than worry about zeros that do not affect the convergence behavior.

Proof. Throughout this proof let $L_f := \lim_{z\to x} f(z)$ and let $L_g := \lim_{z\to x} g(z)$.

In this proof we will use Definition 3.1 as well as Theorem 3.7. Although it will turn out that Definition 3.1 in conjunction with the limit laws for sequences is more effective for this proof, it will also be instructive to see how ε-δ proofs are constructed. The reader will compare the two approaches by proving each part with the respective other approach in Exercises 3-9 and 3-10.

To prove part 1, we use Theorem 3.7. We need to prove that for all $\varepsilon > 0$ there is a $\delta > 0$ so that for all $z \in I\setminus\{x\}$ with $|z-x| < \delta$ we have $\left|(f+g)(z) - (L_f + L_g)\right| < \varepsilon$.

Let $\varepsilon > 0$. Then there are $\delta_f > 0$ and $\delta_g > 0$ so that for all $z \in I \setminus \{x\}$ with $|z-x| < \delta_f$ we have $\left|f(z) - L_f\right| < \frac{\varepsilon}{2}$ and for all $z \in I \setminus \{x\}$ with $|z-x| < \delta_g$ we have $\left|g(z) - L_g\right| < \frac{\varepsilon}{2}$. Let $\delta := \min\{\delta_f, \delta_g\}$ (compare with Standard Proof Technique 2.6). Then for all $z \in I \setminus \{x\}$ with $|z-x| < \delta$ we obtain via the triangular inequality that

$$\left|(f(z)+g(z)) - (L_f + L_g)\right| \le \left|f(z) - L_f\right| + \left|g(z) - L_g\right| < \frac{\varepsilon}{2} + \frac{\varepsilon}{2} = \varepsilon.$$

This means we have proved that for all $\varepsilon > 0$ there is a $\delta > 0$ so that for all $z \in I \setminus \{x\}$ with $|z-x| < \delta$ we have $\left|(f+g)(z) - (L_f + L_g)\right| < \varepsilon$. Consequently, $\lim_{z\to x}(f+g)(z) = L_f + L_g$.

To prove part 2 using Definition 3.1, we need to prove that for all sequences $\{z_n\}_{n=1}^{\infty}$ with $z_n \in I \setminus \{x\}$ for all $n \in \mathbb{N}$ and $\lim_{n\to\infty} z_n = x$ we have $\lim_{n\to\infty} f(z_n) = L$.

Let $\{z_n\}_{n=1}^{\infty}$ be a sequence in $I \setminus \{x\}$ with $\lim_{n\to\infty} z_n = x$. By Theorem 2.14 we infer

$$\lim_{n\to\infty} f(z_n) - g(z_n) = \lim_{n\to\infty} f(z_n) - \lim_{n\to\infty} g(z_n) = L_f - L_g.$$

Because $\{z_n\}_{n=1}^{\infty}$ was arbitrary this implies $\lim_{z\to x}(f-g)(z) = L_f - L_g$.

The proofs of parts 1 and 2 show that to prove the limit laws, Definition 3.1 is more effective than Theorem 3.7. Nonetheless, both ways are actually equally complex overall. If we compare the proof of part 1 with the proof of part 1 of Theorem 2.14 we see striking similarities in the arguments. This means that the complexity of a proof using Definition 3.1 is simply delegated to the proof of an earlier result (Theorem 2.14). The reader will have the chance to analyze the similarities and the differences in the proofs for part 3 in Exercise 3-9. The similarity between the proofs here and the proofs for Theorem 2.14 can be used to translate the proof for part 4 into a complete proof of part 4 of Theorem 2.14. The rather complicated choices in the header were of course made after the final inequalities had been analyzed carefully.

The proof of part 3 is left to the reader as Exercise 3-9.

For part 4, we use Theorem 3.7. So we need to prove that for all $\varepsilon > 0$ there is a $\delta > 0$ so that for all $z \in I \setminus \{x\}$ with $|z-x| < \delta$ we have $\left|\frac{f}{g}(z) - \frac{L_f}{L_g}\right| < \varepsilon$.

Let $\varepsilon > 0$. Then there is a $\delta_f > 0$ such that for all $z \in I \setminus \{x\}$ with $|z-x| < \delta_f$ we have $\left|f(z) - L_f\right| < \frac{\varepsilon|L_g|}{4}$. Moreover, there is a $\delta_g > 0$ such that for all $z \in I \setminus \{x\}$

with $|z-x|<\delta_g$ we have $\left|g(z)-L_g\right| < \frac{\varepsilon\left|L_g\right|^2}{2\left(2\left|L_f\right|+1\right)}$. Finally, there is a $\nu>0$ so that for all $z\in I\setminus\{x\}$ with $|z-x|<\nu$ we have $\left|g(z)-L_g\right|<\frac{\left|L_g\right|}{2}$, which by the reverse triangular inequality means $\left|L_g\right|-\left|g(z)\right|\leq\left|\left|L_g\right|-\left|g(z)\right|\right|\leq\left|L_g-g(z)\right|<\frac{\left|L_g\right|}{2}$, and hence $\left|g(z)\right|>\frac{\left|L_g\right|}{2}$. Let $\delta := \min\{\delta_f,\delta_g,\nu\}$. Then for all $z\in I\setminus\{x\}$ with $|z-x|<\delta$ we obtain

$$\begin{aligned}
\left|\frac{f(z)}{g(z)}-\frac{L_f}{L_g}\right| &= \left|\frac{f(z)L_g-L_fg(z)}{g(z)L_g}\right| \\
&\leq \frac{\left|f(z)L_g-L_fL_g\right|+\left|L_fL_g-L_fg(z)\right|}{\left|g(z)\right|\left|L_g\right|} \\
&\leq \frac{\left|f(z)-L_f\right|\left|L_g\right|+\left|L_f\right|\left|L_g-g(z)\right|}{\frac{\left|L_g\right|}{2}\left|L_g\right|} \\
&= \left|f(z)-L_f\right|\frac{2}{\left|L_g\right|}+\frac{2\left|L_f\right|}{\left|L_g\right|^2}\left|L_g-g(z)\right| \\
&< \frac{\varepsilon\left|L_g\right|}{4}\frac{2}{\left|L_g\right|}+\frac{2\left|L_f\right|}{\left|L_g\right|^2}\frac{\varepsilon\left|L_g\right|^2}{2\left(2\left|L_f\right|+1\right)} < \frac{\varepsilon}{2}+\frac{\varepsilon}{2}=\varepsilon.
\end{aligned}$$

We have proved that for all $\varepsilon>0$ there is a $\delta>0$ so that for all $z\in I\setminus\{x\}$ with $|z-x|<\delta$ we have $\left|\frac{f}{g}(z)-\frac{L_f}{L_g}\right|<\varepsilon$. Therefore the limit of the quotient exists and $\lim_{z\to x}\left(\frac{f}{g}\right)(z)=\frac{\lim_{z\to x}f(z)}{\lim_{z\to x}g(z)}$. ■

Just like limits of sequences, limits of functions preserve inequalities and there also is a Squeeze Theorem.

Definition 3.11 *Let $D\subseteq\mathbb{R}$ be a set and let $f,g : D\to\mathbb{R}$ be functions. We say $f\leq g$ iff f is pointwise less than or equal to g, that is, iff for all $x\in D$ the inequality $f(x)\leq g(x)$ holds.*

Theorem 3.12 *Let $I\subseteq\mathbb{R}$ be an open interval, let $x\in I$ and let $f,g : I\setminus\{x\}\to\mathbb{R}$ be functions. If $f\leq g$ on $I\setminus\{x\}$ and f and g converge at x, then $\lim_{z\to x}f(z)\leq\lim_{z\to x}g(z)$.*

Proof. Exercise 3-11. ■

Theorem 3.13 *The* **Squeeze Theorem** *for functions. Let $I\subseteq\mathbb{R}$ be an open interval, let $x\in I$ and let $f,g,h : I\setminus\{x\}\to\mathbb{R}$ be functions. If $f\leq g\leq h$ on $I\setminus\{x\}$ and f and h converge at x with $\lim_{z\to x}f(z)=\lim_{z\to x}h(z)$, then g converges at x and $\lim_{z\to x}g(z)=\lim_{z\to x}f(z)=\lim_{z\to x}h(z)$.*

Proof. Exercise 3-12. ■

Finally, convergence is also preserved by the composition of functions.

Theorem 3.14 *Let $I, J \subseteq \mathbb{R}$ be open intervals, let $x \in I$, let $g : I \setminus \{x\} \to \mathbb{R}$ and $f : J \to \mathbb{R}$ be functions with $g[I \setminus \{x\}] \subseteq J$, and let $\lim_{z\to x} g(z) = L \in J$. Assume that $\lim_{y\to L} f(y)$ exists and that $g[I \setminus \{x\}] \subseteq J \setminus \{g(x)\}$, or, in case $g(x) \in g[I \setminus \{x\}]$, that $\lim_{y\to L} f(y) = f(L)$. Then $f \circ g$ converges at x and $\lim_{z\to x} f \circ g(x) = \lim_{y\to L} f(y)$.*

Proof. Let $M := \lim_{y\to L} f(y)$ and let $\{z_n\}_{n=1}^{\infty}$ be a sequence in the set $I \setminus \{x\}$ so that $\lim_{n\to\infty} z_n = x$. Then $\lim_{n\to\infty} g(z_n) = L$. If no z_n satisfies $g(z_n) = g(x)$, we obtain $\lim_{n\to\infty} f(g(z_n)) = M$, while if some $g(z_n)$ are equal to $g(x)$, then $M = f(L)$ and we can infer $\lim_{n\to\infty} f(g(z_n)) = f(L) = M$ in this case also. Because the sequence $\{z_n\}_{n=1}^{\infty}$ was arbitrary the result is established. ■

Exercises

3-9 Completing the proof of Theorem 3 10 Let $x \in I$ and let $f, g : I \setminus \{x\} \to \mathbb{R}$ be functions such that $\lim_{z\to x} f(z)$ and $\lim_{z\to x} g(z)$ exist

(a) Prove part 3 using Definition 3 1 (b) Prove part 3 using Theorem 3 7.

3-10 Alternative proofs for the proved parts of Theorem 3.10 Let $x \in I$ and let $f, g : I \setminus \{x\} \to \mathbb{R}$ be functions such that $\lim_{z\to x} f(z)$ and $\lim_{z\to x} g(z)$ exist

(a) Prove that $\lim_{z\to x} (f+g)(z) = \lim_{z\to x} f(z) + \lim_{z\to x} g(z)$ (part 1) using Definition 3 1

(b) Prove that $\lim_{z\to x} (f-g)(z) = \lim_{z\to x} f(z) - \lim_{z\to x} g(z)$ (part 2) using Theorem 3.7.

(c) Prove that if $\lim_{z\to x} g(z) \neq 0$, then $\lim_{z\to x} \left(\frac{f}{g}\right)(z) = \frac{\lim_{z\to x} f(z)}{\lim_{z\to x} g(z)}$ (part 4) using Definition 3 1

3-11 Prove Theorem 3 12

(a) Using Definition 3 1 (b) Using Theorem 3.7

3-12 Prove Theorem 3 13

(a) Using Definition 3 1 (b) Using Theorem 3 7

3-13 Explain the similarities between the proof of part 1 of Theorem 3 10 as presented and the proof of part 1 of Theorem 2.14.

3-14 Computation of limits

(a) Let I be an open interval, let $x \in I$ and let $f : I\setminus\{x\} \to \mathbb{R}$ be a function Prove that $\lim_{z\to x} f(z)$ exists iff $\lim_{h\to 0} f(x+h)$ exists and that in this case $\lim_{z\to x} f(z) = \lim_{h\to 0} f(x+h)$

(b) Compute each of the following limits

i $\lim_{x\to 3} \frac{x^2 - 5x + 6}{x^2 - x - 6}$

ii. $\lim_{x\to 2} \frac{3x^3 - x^2 - 12x + 4}{x^2 - 4}$

iii $\lim_{x\to 9} \frac{x - 9}{x^2 - (9 + \sqrt{3})x + \sqrt{243}}$

3.3 One-Sided Limits and Infinite Limits

Section 3.5 will show that it is advantageous to consider continuous functions (see Definition 3.23) on closed intervals. But that means we also need a notion of convergence at the endpoints of a closed interval. One-sided limits provide just that.

Definition 3.15 Sequence formulation of the left limit of a function. *Let $a < b$. The number $L \in \mathbb{R}$ is called the* **left limit** *of the function $f : [a, b) \to \mathbb{R}$ at b iff for all sequences $\{z_n\}_{n=1}^{\infty}$ in $[a, b)$ with $\lim\limits_{n\to\infty} z_n = b$ we have $\lim\limits_{n\to\infty} f(z_n) = L$. In this case, we denote $\lim\limits_{z\to b^-} f(z) := L$ and we say f* **converges** *(to L) at b from the left.*

The **right limit** *at a for a function $f : (a, b] \to \mathbb{R}$ is defined similarly. It is denoted $\lim\limits_{z\to a^+} f(z)$, and we say f* **converges** *at a from the right.*

If $f : D \to \mathbb{R}$ is a function and the domain D contains an interval $[a, b)$ with $a < b$, we define $\lim\limits_{z\to b^-} f(z) := \lim\limits_{z\to b^-} f|_{[a,b)}(z)$ if it exists. (Exercise 3-15 shows that these left limits are well defined.) Right limits are defined similarly.

Similar to limits, we prove most results for functions defined on half-open intervals. These results are also valid for functions with larger domains. We simply apply them to the appropriate restrictions.

Example 3.16 *Let $m \in \mathbb{Z}$. For the floor and ceiling functions of Definition 1.33, we have $\lim\limits_{z\to m^-} \lfloor z \rfloor = m - 1$, $\lim\limits_{z\to m^+} \lfloor z \rfloor = m$, $\lim\limits_{z\to m^-} \lceil z \rceil = m$, and $\lim\limits_{z\to m^+} \lceil z \rceil = m + 1$.* □

Definitions 3.15 and 3.1 differ in only one way. For a one-sided limit, the sequences must all stay on one side of the number, while in Definition 3.1 the sequences can have values on either side. To emphasize the ability to approach from either side, limits of functions are sometimes called **two-sided limits**. With such strong similarity in the definitions, it is only natural that the theorems that govern one-sided limits are similar to the theorems that govern (two-sided) limits.

Theorem 3.17 Limit laws *for one-sided limits. Let $f, g : [a, b) \to \mathbb{R}$ be functions such that $\lim\limits_{z\to b^-} f(z)$ and $\lim\limits_{z\to b^-} g(z)$ exist. Then the following hold.*

1. $\lim\limits_{z\to b^-} (f + g)(z) = \lim\limits_{z\to b^-} f(z) + \lim\limits_{z\to b^-} g(z)$.

2. $\lim\limits_{z\to b^-} (f - g)(z) = \lim\limits_{z\to b^-} f(z) - \lim\limits_{z\to b^-} g(z)$.

3. $\lim\limits_{z\to b^-} (f \cdot g)(z) = \lim\limits_{z\to b^-} f(z) \cdot \lim\limits_{z\to b^-} g(z)$.

4. *If $\lim\limits_{z\to b^-} g(z) \neq 0$, then* $\lim\limits_{z\to b^-} \left(\frac{f}{g}\right)(z) = \dfrac{\lim_{z\to b^-} f(z)}{\lim_{z\to b^-} g(z)}$.

Each equation implicitly asserts that the limit on the left side exists. (See box on page 33.) Moreover, because $\lim\limits_{z\to b^-} g(z) \neq 0$ in part 4 implies that $g(z) \neq 0$ for z near b (where it matters), we did not demand $g(z) \neq 0$ for all $z \in [a, b)$. Similar limit laws hold for right limits.

Proof. Exercise 3-16. ■

Theorem 3.18 **ε-δ formulation of the left limit of a function.** *The number $L \in \mathbb{R}$ is the left limit of the function $f : [a, b) \to \mathbb{R}$ at b iff for all $\varepsilon > 0$ there is a $\delta > 0$ such that for all $z \in [a, b)$ with $|z - b| < \delta$ we have $\left|f(z) - L\right| < \varepsilon$.*

Proof. Exercise 3-17. ■

Theorem 3.19 connects one-sided limits to (two-sided) limits. Note that to make the proof efficient, we use the sequence formulation of the limit for one direction and the ε-δ formulation for the other direction.

Theorem 3.19 *Let $I \subseteq \mathbb{R}$ be an open interval, let $x \in I$ and let $f : I \setminus \{x\} \to \mathbb{R}$ be a function. Then $\lim_{z\to x} f(z)$ exists iff $\lim_{z\to x^-} f(z)$ and $\lim_{z\to x^+} f(z)$ both exist and are equal. In this case the limit is equal to the left and the right limit.*

Proof. For "$\Rightarrow$," let $L := \lim_{z\to x} f(x)$. Using Definition 3.15 we must prove that for all sequences $\{z_n\}_{n=1}^{\infty}$ in I with $z_n < x$ for all $n \in \mathbb{N}$ and $\lim_{n\to\infty} z_n = x$ we have $\lim_{n\to\infty} f(z_n) = L$ and we must prove the same result for all sequences $\{z_n\}_{n=1}^{\infty}$ in I with $z_n > x$ for all $n \in \mathbb{N}$ and $\lim_{n\to\infty} z_n = x$.

Let $\{z_n\}_{n=1}^{\infty}$ be a sequence in I with $z_n < x$ for all $n \in \mathbb{N}$ and $\lim_{n\to\infty} z_n = x$. By Definition 3.1 we have $\lim_{n\to\infty} f(z_n) = L$, which was to be proved. Sequences with $z_n > x$ for all $n \in \mathbb{N}$ are treated similarly. Hence, $\lim_{z\to x^-} f(z) = \lim_{z\to x^+} f(z) = L$.

For "$\Leftarrow$," let $\lim_{z\to x^-} f(z) = \lim_{z\to x^+} f(z) =: L$. By Theorem 3.7, we must prove that for all $\varepsilon > 0$ there is a $\delta > 0$ so that for all $z \in I \setminus \{x\}$ with $|z - x| < \delta$ we have $\left|f(z) - L\right| < \varepsilon$.

Let $\varepsilon > 0$. By Theorem 3.18 there is a $\delta_l > 0$ so that for all $z \in I \setminus \{x\}$ with $z < x$ and $|z - x| < \delta_l$ we have $\left|f(z) - L\right| < \varepsilon$. By the corresponding result for right limits, there is a $\delta_r > 0$ so that for all $z \in I \setminus \{x\}$ with $z > x$ and $|z - x| < \delta_r$ we have $\left|f(z) - L\right| < \varepsilon$. Let $\delta := \min\{\delta_l, \delta_r\}$. Then for all $z \in I \setminus \{x\}$ with $|z - x| < \delta$ we infer $\left|f(z) - L\right| < \varepsilon$. By Theorem 3.7, this proves $\lim_{z\to x} f(z) = L$. ■

Standard Proof Technique 3.20 To prove that a function has a limit at $x \in \mathbb{R}$ one often proves that the left and the right limits exist and that they are equal. □

Infinity and negative infinity are not numbers, so formally they do not qualify as limits. But knowing that a function grows beyond all bounds near a point gives more information than a statement that the limit does not exist. Hence, we extend the language to allow infinite limits.

Definition 3.21 *Let $f : [a, b) \to \mathbb{R}$ be a function. Then the left limit of f at b is said to be* **infinity** *iff for all sequences $\{z_n\}_{n=1}^{\infty}$ in $[a, b)$ with $\lim_{n\to\infty} z_n = b$ we have $\lim_{n\to\infty} f(z_n) = \infty$. We denote $\lim_{z\to b^-} f(z) := \infty$.*

Infinite right limits at a for a function $f : (a, b] \to \mathbb{R}$ and infinite (two-sided) limits of a function $f : I \setminus \{x\} \to \mathbb{R}$ at $x \in I$, where I is an open interval, are defined similarly. Limits equal to **negative infinity** *are also defined similarly and they are denoted by $-\infty$.*

Finally, as before, infinite one-sided and two-sided limits of functions with larger domains are defined via the limits of appropriate restrictions.

We chose to put one-sided infinite limits in the spotlight in Definition 3.21, because functions often have different behavior to the left and to the right of a point. For example, $\lim_{z\to 0^+} \frac{1}{z} = \infty$ and $\lim_{z\to 0^-} \frac{1}{z} = -\infty$.

It is not surprising that there is a formulation of infinite limits that is similar to the ε-δ formulation of finite limits.

Theorem 3.22 ***M-δ* formulation of infinite left limits.** *The left limit of the function $f : [a, b) \to \mathbb{R}$ at b is infinite iff for all $M \in \mathbb{R}$ there is a $\delta > 0$ such that for all $z \in [a, b)$ with $|z - b| < \delta$ we have $f(z) > M$.*

Proof. Exercise 3-18. ■

Of course, similar results also hold for right-sided and two-sided limits. Limit laws for infinite limits and a version of Theorem 3.19 are given in Exercises 3-22 and 3-23.

Exercises

3-15 Let $f, g : [a, b) \to \mathbb{R}$ be functions. Prove that if there is a number $\delta > 0$ so that f and g are equal on the subset $\{ z \in [a, b) : |z - b| < \delta \}$ of $[a, b)$, then the left limit of f at b exists iff the left limit of g at b exists and in this case we have $\lim_{z\to b^-} f(z) = \lim_{z\to b^-} g(z)$.

3-16 Prove Theorem 3.17. That is, let $f, g : [a, b) \to \mathbb{R}$ be functions such that $\lim_{z\to b^-} f(z)$ and $\lim_{z\to b^-} g(z)$ exist and prove each of the following.

(a) $\lim_{z\to b^-} (f + g)(z) = \lim_{z\to b^-} f(z) + \lim_{z\to b^-} g(z)$.

(b) $\lim_{z\to b^-} (f - g)(z) = \lim_{z\to b^-} f(z) - \lim_{z\to b^-} g(z)$.

(c) $\lim_{z\to b^-} (f \cdot g)(z) = \lim_{z\to b^-} f(z) \cdot \lim_{z\to b^-} g(z)$

(d) If $\lim_{z\to b^-} g(z) \neq 0$, then $\lim_{z\to b^-} \left(\frac{f}{g}\right)(z) = \dfrac{\lim_{z\to b^-} f(z)}{\lim_{z\to b^-} g(z)}$.

3-17. Prove Theorem 3.18.

3-18 Prove Theorem 3.22.

3-19. Prove that $f(x) = \begin{cases} x + 3; & \text{for } x < 1, \\ (x + 1)^2; & \text{for } x > 1, \\ 0; & \text{for } x = 1, \end{cases}$ has a limit at $x = 1$ and state its value

3-20 Prove that $\lim_{z\to 0^+} \sqrt{z} = 0$ Explain why this is a satisfactory resolution of the formal problem in Exercise 3-8 or why it is not

3-21. A function f is called **nondecreasing** on $I \subseteq \mathbb{R}$ iff for all $x_1 < x_2$ in I we have $f(x_1) \leq f(x_2)$ Let $f : [a, b] \to \mathbb{R}$ be a nondecreasing function

(a) Prove that for every $x \in (a, b]$ we have $\lim_{z \to x^-} f(z) = \sup \{ f(z) : z < x \}$.

Hint. Mimic the proof of the Monotone Sequence Theorem.

(b) Prove that for every $x \in [a, b)$ the right limit $\lim_{z \to x^+} f(z)$ exists and state its value.

3-22 **Limit laws** involving infinite limits of functions. Let $f, g : [a, b) \to \mathbb{R}$ be functions, and let $\lim_{z \to b^-} f(z) = \infty$ Prove the following

(a) If g is **bounded** (that is, there is an $M > 0$ so that for all $z \in [a, b)$ we have $|g(z)| < M$), then $\lim_{z \to b^-} f(z) + g(z) = \infty$ and if all $f(z)$ are nonzero, then $\lim_{z \to b^-} \frac{g(z)}{f(z)} = 0$

(b) If $\lim_{z \to b^-} g(z) = \infty$, then $\lim_{z \to b^-} f(z) + g(z) = \infty$ and $\lim_{z \to b^-} f(z)g(z) = \infty$

(c) If $c > 0$ is a real number, then $\lim_{z \to b^-} cf(z) = \infty$.

(d) If $c < 0$ is a real number, then $\lim_{z \to b^-} cf(z) = -\infty$.

3-23. Let $I \subseteq \mathbb{R}$ be an open interval, let $x \in I$ and let $f : I \setminus \{x\} \to \mathbb{R}$ be a function. Prove that $\lim_{z \to x} f(z) = \infty$ iff $\lim_{z \to x^-} f(z) = \lim_{z \to x^+} f(z) = \infty$

3.4 Continuity

Continuous functions are usually defined as functions for which the limit is computed by substituting the value. Because we may need to take care of endpoints, the formalization of the elementary definition from calculus requires an extra item (see number 4 below).

Definition 3.23 *Let $D \subseteq \mathbb{R}$ be an interval of nonzero length from which at most finitely many points have been removed and let $f : D \to \mathbb{R}$ be a function. Then f is called* **continuous at** x *iff*

1. *$f(x)$ is defined, that is, $x \in D$, and*

2. *$\lim_{z \to x} f(z)$ exists, and*

3. *$\lim_{z \to x} f(z) = f(x)$, and*

4. *If x is an endpoint of D, use left or right limits in 2 and 3, as appropriate.*

f is called **continuous** *(on D) iff f is continuous at every $x \in D$.*

We could also define continuity for functions defined on sets for which every point of the domain is contained in an interval of nonzero length. Exercise 3-32 shows that with the present definition this idea is a bit too simple to produce a sensible result. This is not a problem, because in the early part of the text the only functions whose domains are not intervals are rational functions. For these functions the pathology of Exercise 3-32 is not an issue. Therefore we relegate all concerns regarding more complicated domains to Section 16.3.

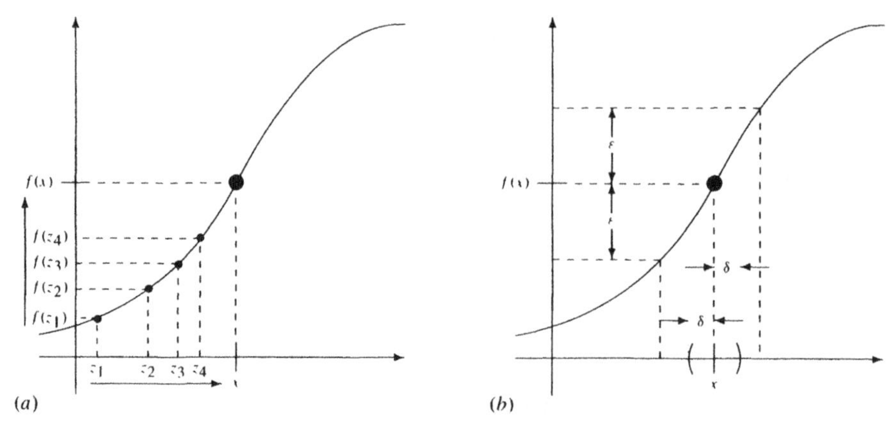

Figure 6: Two ways to view continuity. Continuity at x means that for all input sequences that converge to x, the output sequence converges to $f(x)$ ((a) and part 2 of Theorem 3.25). Continuity at x also means that for every tolerance $\varepsilon > 0$ around the output value $f(x)$ there is a threshold $\delta > 0$ so that if the input is closer than δ to x, then the output is closer than ε to $f(x)$ ((b) and part 3 of Theorem 3.25).

For theorems, we will usually work with functions that are defined on intervals, because if D is an interval from which at most finitely many points were removed, then $f : D \to \mathbb{R}$ is continuous at $x \in D$ iff $f|_I$ is continuous at x, where I is a maximum-sized (with respect to containment) interval contained in D that contains x.

Example 3.24

1. *Constant functions are continuous at every $x \in \mathbb{R}$.*

2. *The function $f(x) = x$ is continuous at every $x \in \mathbb{R}$.*

3. *The function $f(x) = |x|$ is continuous at every $x \in \mathbb{R}$.*

Parts 1 and 2 are trivial and part 3 follows from part 2 of Example 3.5. □

It is useful to incorporate the definition of limits directly into a characterization of continuity. With such a characterization it is not necessary to resolve the multiple parts of the original definition whenever we work with continuity. Figure 6 shows graphical interpretations of the conditions.

Theorem 3.25 *Let $I \subseteq \mathbb{R}$ be an interval, let $x \in I$ and let $f : I \to \mathbb{R}$ be a function. Then the following are equivalent.*

1. *f is continuous at x.*

2. *For every sequence $\{z_n\}_{n=1}^{\infty}$ with $z_n \in I$ for all $n \in \mathbb{N}$ and $\lim_{n\to\infty} z_n = x$, we have $\lim_{n\to\infty} f(z_n) = f(x)$.*

3. *For every $\varepsilon > 0$, there is a $\delta > 0$ such that for all $z \in I$ with $|z - x| < \delta$ we have $\left|f(z) - f(x)\right| < \varepsilon$.*

Proof. *We will prove "1⇒2," "2⇒3" and "3⇒1." The remaining implications follow because logical implications are* **transitive**. *That is, the implications "1⇒2" and "2⇒3" imply "1⇒3," the implications "2⇒3" and "3⇒1" imply "2⇒1," and so on.*

We will assume throughout that x is not an endpoint of the domain. The arguments are easily modified (by using appropriate one-sided limits and theorems) for the case that x is an endpoint.

For "1⇒2," we need to prove that for every sequence $\{z_n\}_{n=1}^{\infty}$ with $z_n \in I$ for all $n \in \mathbb{N}$ and $\lim_{n\to\infty} z_n = x$ we have $\lim_{n\to\infty} f(z_n) = f(x)$.

Because f is continuous at x, we have $\lim_{z\to x} f(z) = f(x)$. By Definition 3.1, this means that for all sequences $\{z_n\}_{n=1}^{\infty}$ with $z_n \in I \setminus \{x\}$ and $\lim_{n\to\infty} z_n = x$ we have that $\lim_{n\to\infty} f(z_n) = f(x)$. Now let $\{z_n\}_{n=1}^{\infty}$ be a sequence with $z_n \in I$ and $\lim_{n\to\infty} z_n = x$. If there is an $N \in \mathbb{N}$ so that $z_n = x$ for all $n \geq N$, then there is nothing to prove. Otherwise let $\{z_{n_k}\}_{k=1}^{\infty}$ be the subsequence of all elements with $z_{n_k} \in I \setminus \{x\}$. Then $\lim_{k\to\infty} z_{n_k} = x$, and hence $\lim_{k\to\infty} f(z_{n_k}) = f(x)$. Now let $\varepsilon > 0$. Then there is a $K \in \mathbb{N}$ so that for all $k \geq K$ we have $\left|f(z_{n_k}) - f(x)\right| < \varepsilon$. Then for all $n \geq n_K$ either $z_n = x$ and $\left|f(z_n) - f(x)\right| = 0 < \varepsilon$ or there is a $k \geq K$ with $n = n_k$ and $\left|f(z_{n_k}) - f(x)\right| < \varepsilon$. This means $\lim_{n\to\infty} f(z_n) = f(x)$ for every sequence $\{z_n\}_{n=1}^{\infty}$ with $z_n \in I$ for all $n \in \mathbb{N}$ and $\lim_{n\to\infty} z_n = x$, which establishes this part of the proof.

The proof of "2⇒3" is similar to the proof of "⇒" in Theorem 3.7 (Exercise 3-24).

For "3⇒1," note that, because f is defined on I and $x \in I$, f is defined at x. The condition in part 3 implies by Theorem 3.7 that $\lim_{z\to x} f(z) = f(x)$, which completes the proof. ■

Because of the formal problems with limits of functions outlined at the end of Section 16.3.1, in general settings one of the conditions in Theorem 3.25 is normally used to define continuity.

Standard Proof Technique 3.26 To prove the equivalence of several conditions, it is standard practice to prove that the first implies the second, the second implies the third, and so on, and finally that the last condition implies the first. All other implications follow from the **transitivity of logical implications**. □

The proofs of parts 5 and 6 of Theorem 3.27 serve as good examples how the conditions in Theorem 3.25 can be used in conjunction.

Theorem 3.27 *Let $I \subseteq \mathbb{R}$ be an interval, let $x \in I$ and let $f, g : I \to \mathbb{R}$ be continuous at $x \in I$. Then the following hold.*

1. *$f + g$ is continuous at x.*

2. *$f - g$ is continuous at x.*

3. *$f \cdot g$ is continuous at x.*

4. *If $g(x) \neq 0$ for all $x \in I$, then $\frac{f}{g}$ is continuous at x.*

5. $\max\{f, g\}$ *is continuous at* x. *(The* **maximum** *is defined pointwise.)*

6. $\min\{f, g\}$ *is continuous at* x. *(The* **minimum** *is defined pointwise.)*

Proof. The first four parts are direct consequences of the corresponding limit laws.

For part 5, let $x \in I$. We will establish continuity of $\max\{f, g\}$ at x by proving that for all sequences $\{z_n\}_{n=1}^{\infty}$ with $z_n \in I$ for all $n \in \mathbb{N}$ and $\lim_{n\to\infty} z_n = x$ we have $\lim_{n\to\infty} \max\{f(z_n), g(z_n)\} = \max\{f(x), g(x)\}$.

Let $\{z_n\}_{n=1}^{\infty}$ be a sequence in I with $\lim_{n\to\infty} z_n = x$. In case $f(x) \neq g(x)$, assume without loss of generality that $f(x) > g(x)$. To prove that the limit of the image sequence is $\lim_{n\to\infty} \max\{f, g\}(z_n) = \max\{f, g\}(x) = f(x)$, let $\varepsilon > 0$. Because f and g are continuous at x, there is a $\delta > 0$ so that for all $z \in I$ with $|z - x| < \delta$ we have $\left|f(z) - f(x)\right| < \min\left\{\varepsilon, \frac{f(x) - g(x)}{2}\right\}$ and $\left|g(z) - g(x)\right| < \frac{f(x) - g(x)}{2}$. In particular, for all $z \in I$ with $|z - x| < \delta$ we obtain

$$f(z) > f(x) - \frac{f(x) - g(x)}{2} = \frac{f(x) + g(x)}{2} = g(x) + \frac{f(x) - g(x)}{2} > g(z),$$

and hence $\max\{f, g\}(z) = f(z)$. For δ, find an $N \in \mathbb{N}$ so that for all $n \geq N$ we have $|z_n - x| < \delta$. Then for all $n \geq N$ we infer

$$\left|\max\{f, g\}(z_n) - \max\{f, g\}(x)\right| = \left|f(z_n) - f(x)\right| < \varepsilon,$$

and hence $\max\{f, g\}$ is continuous at x when $f(x) > g(x)$.

This leaves the case $f(x) = g(x)$. Let $\varepsilon > 0$. Because f and g are continuous at x, there is a $\delta > 0$ so that for all $z \in I$ with $|z - x| < \delta$ we have that $\left|f(z) - f(x)\right| < \varepsilon$ and $\left|g(z) - g(x)\right| < \varepsilon$. For δ, find an $N \in \mathbb{N}$ so that for all $n \geq N$ we have that $|z_n - x| < \delta$. For all $n \geq N$, the maximum $\max\{f, g\}(z_n)$ is equal to $f(z_n)$ or $g(z_n)$. Because $|z_n - x| < \delta$, if $\max\{f, g\}(z_n) = f(z_n)$ we infer $\left|\max\{f, g\}(z_n) - \max\{f, g\}(x)\right| = \left|f(z_n) - f(x)\right| < \varepsilon$ and if $\max\{f, g\}(z_n) = g(z_n)$ we infer $\left|\max\{f, g\}(z_n) - \max\{f, g\}(x)\right| = \left|g(z_n) - g(x)\right| < \varepsilon$. Thus the maximum $\max\{f, g\}$ is continuous at x when $f(x) = g(x)$.

The proof of part 6 is left as Exercise 3-25b. ■

There is a faster proof of part 5 that relies on Theorem 3.30 below and on an algebraic representation of the maximum (see Exercise 3-26a). Because our main focus is on standard techniques in analysis, the longer proof was presented here.

Theorem 3.27 gives access to two standard examples of continuous functions.

Example 3.28 A **polynomial** is a function $p : \mathbb{R} \to \mathbb{R}$ for which there are an $n \in \mathbb{N}$ and $a_0, \ldots, a_n \in \mathbb{R}$ so that $a_n \neq 0$ and for all $x \in \mathbb{R}$ we have $p(x) = \sum_{j=0}^{n} a_j x^j$. The number n is called the **degree** of the polynomial. The constant function $p(x) = 0$ is also considered to be a polynomial. Its degree is defined to be $-\infty$. Every polynomial is continuous on $\mathbb{R}$. (See Exercise 3-28.) □

Example 3.29 A **rational function** is a function r for which there are two polynomials $p, q : \mathbb{R} \to \mathbb{R}$ so that for all $x \in \mathbb{R}$ for which $q(x) \neq 0$ we have $r(x) = \frac{p(x)}{q(x)}$. By Theorem 3.27 and Example 3.28, every rational function is continuous on its domain $\{x \in \mathbb{R} : q(x) \neq 0\}$. (We implicitly use here that every polynomial has at most finitely many zeroes.) □

Continuity is also preserved by compositions.

Theorem 3.30 *Let $I, J \subseteq \mathbb{R}$ be intervals, let $g : I \to \mathbb{R}$ be continuous at $x \in I$, let $g[I] \subseteq J$ and let $f : J \to \mathbb{R}$ be continuous at $g(x)$. Then $f \circ g : I \to \mathbb{R}$ is continuous at x.*

Proof. We will prove that for every $\varepsilon > 0$ there is a $\delta > 0$ so that for all $z \in I$ with $|z - x| < \delta$ we have $\left|f(g(z)) - f(g(x))\right| < \varepsilon$.

Let $\varepsilon > 0$. Because f is continuous at $g(x)$, there is a $\delta_f > 0$ so that for all $y \in J$ with $\left|y - g(x)\right| < \delta_f$ we have $\left|f(y) - f(g(x))\right| < \varepsilon$. Because g is continuous at x, for δ_f there is a $\delta > 0$ so that for all $z \in I$ with $|z - x| < \delta$ we have $\left|g(z) - g(x)\right| < \delta_f$. But then because $\left|g(z) - g(x)\right| < \delta_f$ we infer $\left|f(g(z)) - f(g(x))\right| < \varepsilon$.

Therefore we have proved that for every $\varepsilon > 0$ there is a $\delta > 0$ so that for all $z \in I$ with $|z - x| < \delta$ we have $\left|f(g(z)) - f(g(x))\right| < \varepsilon$, which means that $f \circ g$ is continuous at x. ■

We conclude this section by characterizing discontinuities.

Definition 3.31 *Let D be an interval of nonzero length from which at most finitely many points $x_1, \ldots, x_n$ have been removed. If the function $f : D \to \mathbb{R}$ is not continuous at $x \in D \cup \{x_1, \ldots, x_n\}$, we speak of a* **discontinuity** *at x. There are several types of discontinuities. (For visualizations, see Figure 7, for examples, see Exercise 3-31.)*

1. *If $\lim_{z \to x} f(z)$ exists, or if x is an endpoint of $D \cup \{x_1, \ldots, x_n\}$ and the appropriate one-sided limit exists, but the limit is not equal to $f(x)$; or if x is not an endpoint of $D \cup \{x_1, \ldots, x_n\}$ and f is not defined at x, we speak of a* **removable discontinuity**.

2. *If $\lim_{z \to x^-} f(z)$ and $\lim_{z \to x^+} f(z)$ exist, but they are not equal, we speak of a* **jump discontinuity**.

3. *If (at least) one of $\lim_{z \to x^-} f(z)$ and $\lim_{z \to x^+} f(z)$ does not exist and if there is a sequence $\{z_n\}_{n=1}^{\infty}$ in D that converges to x and $\lim_{n \to \infty} \left|f(z_n)\right| = \infty$, we speak of an* **infinite discontinuity**.

4. *If (at least) one of $\lim_{z \to x^-} f(z)$ and $\lim_{z \to x^+} f(z)$ does not exist, if there is a $\delta > 0$ so that f is bounded in $\{z \in D : |z - x| < \delta\}$ and if there are two sequences $\{z_n\}_{n=1}^{\infty}$ and $\{w_n\}_{n=1}^{\infty}$ that converge to x such that for all $n \in \mathbb{N}$ we have $z_n, w_n < x$ (or $z_n, w_n > x$) and such that both $\lim_{n \to \infty} f(z_n)$ and $\lim_{n \to \infty} f(w_n)$ exist, but they are not equal, we speak of a* **discontinuity by oscillation**.

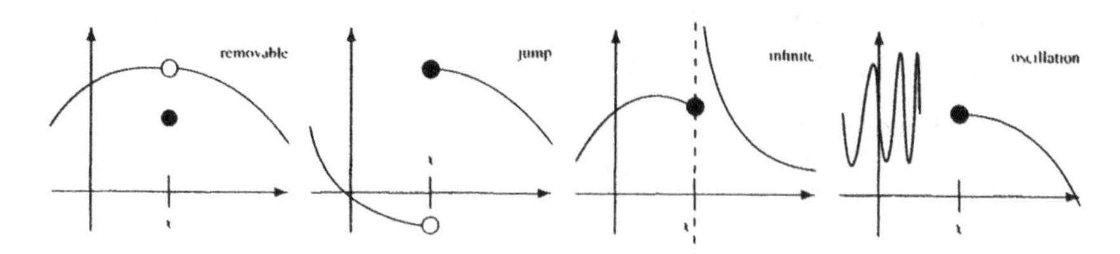

Figure 7: Visualization of the possible discontinuities of a function.

Theorem 3.32 *Let $D \subseteq \mathbb{R}$ be an interval of nonzero length from which at most finitely many points $x_1, \ldots, x_n$ have been removed and let $f : D \to \mathbb{R}$ be a function. Then every discontinuity $x \in D \cup \{x_1, \ldots, x_n\}$ of f is of one of the four types listed in Definition 3.31.*

Proof. Let $x \in D$ or let x be one of the finitely many elements that were removed and assume that f is not continuous at x. Let the discontinuity at x be neither a removable discontinuity, nor a jump discontinuity, nor an infinite discontinuity. We will prove that it must be a discontinuity by oscillation. If $\lim_{z \to x^-} f(x)$ and $\lim_{z \to x^+} f(x)$ both existed, then they would either be equal and the discontinuity would be removable, or not, in which case the discontinuity would be a jump discontinuity. Hence, one of the one-sided limits does not exist at x. If x is the supremum or the infimum of D, then f is defined at x and one of the two one-sided limits does not exist by default. In this case, the respective other one-sided limit also must not exist, because otherwise there would be a removable discontinuity at x. By symmetry, we can assume without loss of generality that $\lim_{z \to x^-} f(x)$ does not exist and f is defined on some interval $[x - \delta, x)$.

Because the discontinuity at x is not an infinite discontinuity, there is a $\nu > 0$ so that f is bounded on $[x-\nu, x)$. Let $z_n := x - \frac{\nu}{n}$. Then $\lim_{n \to \infty} z_n = x$. First consider the case that $\lim_{n \to \infty} f(z_n) =: L$ exists. Because $\lim_{z \to x^-} f(z)$ does not exist, there must be an $\varepsilon > 0$ and a sequence $\{v_k\}_{k=1}^{\infty}$ so that $\lim_{k \to \infty} v_k = x$, and for all $k \in \mathbb{N}$ we have $x - \nu \leq v_k < x$ and $\left|f(v_k) - L\right| \geq \varepsilon$. Because f is bounded on $[x - \nu, x)$, the sequence $\left\{f(v_k)\right\}_{k=1}^{\infty}$ is bounded. By the Bolzano-Weierstrass Theorem, it has a convergent subsequence $\left\{f(v_{k_n})\right\}_{n=1}^{\infty}$. Then $\lim_{n \to \infty} f(v_{k_n}) \neq L$, and hence $\{w_n\}_{n=1}^{\infty} := \left\{v_{k_n}\right\}_{n=1}^{\infty}$ and $\{z_n\}_{n=1}^{\infty}$ are sequences as required in the definition of a discontinuity by oscillation.

If $\left\{f(z_n)\right\}_{n=1}^{\infty}$ does not converge, note that it is bounded, and hence it has a convergent subsequence $\left\{f(z_{n_k})\right\}_{k=1}^{\infty}$. Now f has a discontinuity by oscillation at x, because we can replace $\{z_n\}_{n=1}^{\infty}$ with $\left\{z_{n_k}\right\}_{k=1}^{\infty}$ in the above argument and then repeat it. ■

Standard Proof Technique 3.33 When an argument requires a convergent sequence and all that is guaranteed for a given sequence $\{z_n\}_{n=1}^{\infty}$ is a convergent *sub*sequence, then one often assumes without loss of generality that $\{z_n\}_{n=1}^{\infty}$ converges. This is because, just like at the end of the proof of Theorem 3.32, the given sequence $\{z_n\}_{n=1}^{\infty}$ can be replaced with a **convergent subsequence** that (usually) has all the properties of $\{z_n\}_{n=1}^{\infty}$, plus it converges. □

Although Theorem 3.32 characterizes all possible discontinuities for functions from $\mathbb{R}$ to $\mathbb{R}$, other types of discontinuities exist for functions with infinite dimensional range (see Exercise 16-36).

Exercises

3-24. Prove part "2⇒3" of Theorem 3.25.

3-25 Completing the proof of Theorem 3.27

(a) Give all details of the proofs of parts 1-4.

(b) Prove part 6.

3-26. Alternative proofs of parts 5 and 6 of Theorem 3 27. Let I be an interval and let $f, g : I \to \mathbb{R}$ be functions

(a) Use parts 1 and 2 and Theorem 3.30 to prove that if f and g are continuous at $x \in I$, then $\max\{f, g\}$ is continuous at x.

Hint. First prove that for all $a, b \in \mathbb{R}$ the equality $\max\{a, b\} = \frac{1}{2}\left(a + b + |a - b|\right)$ holds

(b) Give a similar proof of part 6 of Theorem 3.27

3-27. Prove that for all $n \in \mathbb{N}$ the function $f(x) = x^n$ is continuous on $\mathbb{R}$ Then prove that for all $m \in \mathbb{N}$ the function $f(x) = x^{-m}$ is continuous on $\mathbb{R} \setminus \{0\}$.

3-28. Prove that all polynomials are continuous on $\mathbb{R}$.
Hint. Induction on the degree.

3-29. Alternative proofs of Theorem 3.30.

(a) Prove Theorem 3 30 using Definition 3 23 and Theorem 3.14.

(b) Prove Theorem 3.30 using part 2 of Theorem 3.25.

3-30. Explain why we demand that the interval in Definition 3.23 must have nonzero length.

3-31. Examples of discontinuities.

(a) Prove that $f(x) = \frac{x^2}{x}$ has a removable discontinuity at 0.

(b) Prove that $f(x) = \lceil x \rceil$ has a jump discontinuity at 0.

(c) Prove that $f(x) = \frac{1}{x}$ has an infinite discontinuity at 0.

(d) Let $g(x) = \begin{cases} 2x, & \text{for } 0 \le x \le \frac{1}{2}, \\ -2x + 2; & \text{for } \frac{1}{2} \le x \le 1. \end{cases}$

Prove that $f(x) := \begin{cases} g\left(2^n\left(x - \frac{1}{2^n}\right)\right); & \text{for } \frac{1}{2^n} \le x \le \frac{1}{2^{n-1}} \text{ and } n \in \mathbb{N}, \\ 0; & \text{otherwise}, \end{cases}$

has a discontinuity by oscillation at 0.

3-32 Let $U := \bigcup_{n=1}^{\infty} \left[\frac{1}{n} - \frac{1}{10n}, \frac{1}{n}\right] := \left\{x \in \mathbb{R} : \left(\exists n \in \mathbb{N} \; x \in \left[\frac{1}{n} - \frac{1}{10n}, \frac{1}{n}\right]\right)\right\}$ and let the function $f : [-1, 0] \cup U \to \mathbb{R}$ be defined by $f(x) := \begin{cases} 0; & \text{for } x \in [-1, 0], \\ 1; & \text{for } x \in U. \end{cases}$

(a) Prove that f is continuous on every interval I that is contained in its domain $D := [-1, 0] \cup U$.

(b) Prove that every point in D is contained in an interval of nonzero length

(c) Explain why the function still should not be considered to be continuous on D.

(d) Suggest a generalization of the definition of continuity that would allow domains such as D and that would make f discontinuous at 0.

This function will be revisited in Exercise 16-29.

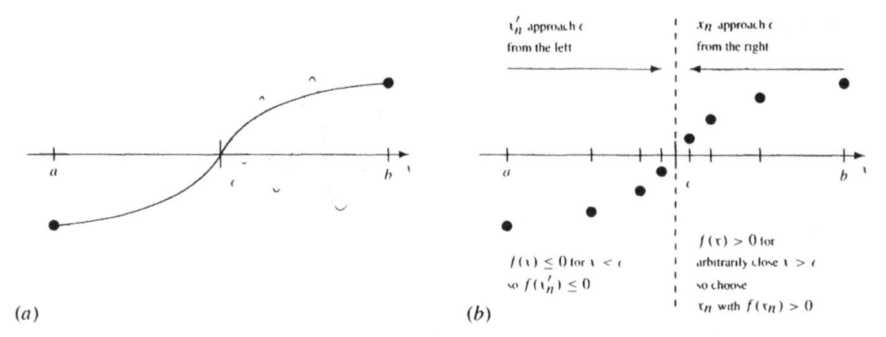

Figure 8: Visualizing the Intermediate Value Theorem. Intuitively (a) it is clear that an unbroken graph that goes from a point below the x-axis to a point above the x-axis must cross the x-axis at least once (solid graph) and that it could even cross the x-axis multiple times (dotted continuation past the first intercept). Part (b) gives a visualization of the proof. Because the sequences $\{x_n\}_{n=1}^{\infty}$ and $\{x_n'\}_{n=1}^{\infty}$ meet in the middle at $(c, f(c))$, we conclude $f(c) = 0$.

3.5 Properties of Continuous Functions

Aside from their obvious connection to limits, continuous functions are interesting because they have several useful properties.

Theorem 3.34 Intermediate Value Theorem. *Let $a < b$ and let $f : [a, b] \to \mathbb{R}$ be a continuous function. If $f(a) < 0$ and $f(b) > 0$ (or vice versa) then there is a $c \in (a, b)$ such that $f(c) = 0$. (Also see Figure 8(a).)*

Proof. Assume without loss of generality that $f(a) < 0$ and $f(b) > 0$. The proof for the other case is similar.

The set $G := \{x \in [a, b] : f(x) > 0\}$ contains b and it is bounded below by a. Let $c := \inf(G)$. We will show that c is as claimed in the theorem.

First, we show that $c \notin \{a, b\}$. For a contradiction suppose that $c = a$. Then by the version of Proposition 1.21 for infima, for each $n \in \mathbb{N}$ there would be an $x_n \in \left(a, a + \frac{1}{n}\right)$ with $f(x_n) > 0$. *(Note that we are using Standard Proof Technique 3.8 here.)* But then $\lim_{n\to\infty} x_n = a$, and by continuity of f we could infer that $0 > f(a) = \lim_{n\to\infty} f(x_n) \geq 0$, a contradiction. The inequality $c \neq b$ is proved similarly (see Exercise 3-33).

Because $c \neq b$, again by the version of Proposition 1.21 for infima, for each $n \in \mathbb{N}$ there is an $x_n \in \left(c, c + \frac{1}{n}\right)$ with $f(x_n) > 0$. In particular, $\lim_{n\to\infty} x_n = c$. Because $c \neq a$, there is an $N \in \mathbb{N}$ so that for all $n \geq N$ the number $x_n' := c - \frac{1}{n}$ is greater than or equal to a. Hence, the sequence $\{x_n'\}_{n=N}^{\infty}$ satisfies $\lim_{n\to\infty} x_n' = c$ and $f(x_n') \leq 0$ for all $n \in \mathbb{N}$. For a visualization of the sequences, consider Figure 8.

Because f is continuous, we infer $\lim_{n\to\infty} f(x_n) = \lim_{n\to\infty} f\left(x_n'\right) = f(c)$. But the inequalities for $f(x_n)$ and $f\left(x_n'\right)$ show that $\lim_{n\to\infty} f(x_n) \geq 0$ and $\lim_{n\to\infty} f\left(x_n'\right) \leq 0$. Thus $f(c)$ must be greater than or equal to zero and less than or equal to zero, which implies $f(c) = 0$. ■

The Intermediate Value Theorem immediately implies that continuous images of intervals are intervals, too.

Definition 3.35 *Let A, B be sets, let $f : A \to B$ be a function and let $C \subseteq A$. Then the* **image** *of C under f is defined to be $f[C] := \{f(c) : c \in C\}$.*

Theorem 3.36 *Let $I \subseteq \mathbb{R}$ be an interval and let $f : I \to \mathbb{R}$ be continuous. Then $f[I]$ is an interval.*

Proof. Let $l, u \in f[I]$ with $l < u$ and let $m \in (l, u)$. Then there are $a, b \in I$ with $f(a) = l$ and $f(b) = u$. Without loss of generality assume $a < b$. The function $g(x) := f(x) - m$ is continuous on $[a, b]$. By the Intermediate Value Theorem there is a $c \in (a, b)$ so that $g(c) = 0$. But then $f(c) = g(c) + m = m$. Because m, l, u were arbitrary we have proved that for any two elements $l < u$ of $f[I]$ and any element m between them we have that $m \in f[I]$. Therefore $f[I]$ is an interval. ■

We now turn to inverse functions.

Definition 3.37 *Let A, B be sets and let $f : A \to B$ be a bijective function. Then the* **inverse function** *of f is the unique (see Exercise 3-34) function $f^{-1} : B \to A$ that maps each $b \in B$ to the unique $a \in A$ so that $f(a) = b$.*

Theorem 3.38 *Let $I \subseteq \mathbb{R}$ be an interval and let $f : I \to \mathbb{R}$ be a continuous injective function. Then the inverse function $f^{-1} : f[I] \to I$ is also continuous.*

Proof. Clearly, f maps I bijectively onto $f[I]$, and hence f has an inverse function f^{-1} that is defined on $f[I]$. By Theorem 3.36, $f[I]$ is an interval. All that is left is to show that if a sequence $\{y_n\}_{n=1}^{\infty}$ in $f[I]$ converges to $y \in f[I]$, then $\left\{f^{-1}(y_n)\right\}_{n=1}^{\infty}$ converges to $f^{-1}(y)$.

Let $\{y_n\}_{n=1}^{\infty}$ be a sequence in $f[I]$ with $\lim_{n\to\infty} y_n = y$. For $n \in \mathbb{N}$, let $x_n := f^{-1}(y_n)$ and let $x := f^{-1}(y)$. For $\varepsilon > 0$, let $J := \{z \in I : |z - x| < \varepsilon\}$, $J_l := \{z \in J : z \leq x\}$ and $J_u := \{z \in J : z \geq x\}$. Then $f[J]$, $f[J_l]$ and $f[J_u]$ are all intervals that contain $y = f(x)$. Because f is injective, the only point $f[J_l]$ and $f[J_u]$ have in common is y. Therefore y is the maximum of one of the intervals $f[J_l]$ and $f[J_u]$ and it is the minimum of the other.

If both $f[J_l]$ and $f[J_u]$ have more than one element, then there is a $\delta > 0$ so that $(y - \delta, y + \delta) \subseteq f[J]$. In this case, because $\{y_n\}_{n=1}^{\infty}$ converges to y, there is an $N \in \mathbb{N}$ so that for all $n \geq N$ we have $y_n \in (y - \delta, y + \delta) \subseteq f[J]$. Consequently, for all $n \geq N$ the point x_n is in J, which means $|x_n - x| < \varepsilon$.

If $f[J_u]$ has exactly one element, then J_u has exactly one element and x is the largest point of I. Therefore $f[J_l]$ has more than one element and there is a $\delta > 0$

so that $(y-\delta, y] \subseteq f[J_l]$ or $[y, y+\delta) \subseteq f[J_l]$. Without loss of generality assume $(y-\delta, y] \subseteq f[J_l]$. We claim that then y is the largest point of $f[I]$. Suppose for a contradiction that there was a $y' > y$ in $f[I]$. Then $a := f^{-1}(y') < x$ and there would also be a $b < x$ with $f(b) < y$. By Theorem 3.36 there would be a $c \neq x$ between a and b so that $f(c) = y = f(x)$, contradicting the injectivity of f. Thus $y = \sup f[I]$. Because $\{y_n\}_{n=1}^{\infty}$ converges to y there is an $N \in \mathbb{N}$ so that for all $n \geq N$ we have $y_n \in (y-\delta, y] \subseteq f[J_l]$. Thus for all $n \geq N$ we infer $x_n \in J_l$, which means $|x_n - x| < \varepsilon$.

The case in which $f[J_l]$ has exactly one element is handled similarly.

We have shown that in each of the above cases $f^{-1}\left(\lim_{n\to\infty} y_n\right) = \lim_{n\to\infty} f^{-1}(y_n)$, which implies that f^{-1} is continuous. ■

Now that we know that inverses of continuous functions are continuous, we can establish limit laws for powers with rational exponents.

Definition 3.39 *A number $n \in \mathbb{Z}$ is called* **even** *iff there is a $k \in \mathbb{Z}$ so that $n = 2k$ and it is called* **odd** *iff there is a $k \in \mathbb{Z}$ so that $n = 2k+1$.*

Corollary 3.40 *Let $d \in \mathbb{N}$. Then $f(x) = x^{\frac{1}{d}}$ is continuous on $[0, \infty)$ if d is even and it is continuous on $\mathbb{R}$ if d is odd, where for $x < 0$ and q odd we define $x^{\frac{1}{q}} := -|x|^{\frac{1}{q}}$.*

Proof. Use Theorem 3.38 (Exercise 3-35). ■

Corollary 3.41 *Let $r \in \mathbb{Q}$ be positive. Then $f(x) = x^r$ is continuous on $[0, \infty)$, and if r can be represented as a fraction with odd denominator, f is continuous on $\mathbb{R}$. For negative $r \in \mathbb{Q}$, $f(x) = x^r$ is continuous on $(0, \infty)$, and if r can be represented as a fraction with odd denominator, f is continuous on $\mathbb{R} \setminus \{0\}$.*

Proof. Use Exercise 3-27 and Corollary 3.40 (Exercise 3-36). ■

Theorem 3.42 *Let $\{a_n\}_{n=1}^{\infty}$ be a convergent sequence of nonnegative numbers. Then for all $r \in \mathbb{Q}$ we have $\lim_{n\to\infty} a_n^r = \left(\lim_{n\to\infty} a_n\right)^r$.*

Proof. Exercise 3-37. ■

We conclude this section by showing that on closed and bounded intervals continuous functions assume an absolute minimum and an absolute maximum.

Definition 3.43 *Let $I \subseteq \mathbb{R}$ be an interval and let $f : I \to \mathbb{R}$ be a real valued function. The number y_m is called the* **absolute minimum value** *of f (in I) if and only if there is an $x_m \in I$ with $f(x_m) = y_m$ and for all $x \in I$ we have $f(x) \geq f(x_m)$. The number y_M is called the* **absolute maximum value** *of f (in I) if and only if there is an $x_M \in I$ with $f(x_M) = y_M$ and for all $x \in I$ we have $f(x) \leq f(x_M)$ A value that is the absolute maximum or the absolute minimum is also called an* **absolute extremum**.

Theorem 3.44 *Let $f : [a, b] \to \mathbb{R}$ be continuous. Then there is an $x \in [a, b]$ such that for all $z \in [a, b]$ we have $f(x) \geq f(z)$*

Proof. First we show that f is bounded above on $[a, b]$. For a contradiction, suppose that for every $n \in \mathbb{N}$ there is an $x_n \in [a, b]$ such that $f(x_n) \geq n$. Then by the Bolzano-Weierstrass Theorem there is a convergent subsequence $\left\{x_{n_k}\right\}_{k=1}^{\infty}$ with limit $x \in [a, b]$. But then for all k we have $f\left(x_{n_k}\right) \geq n_k$ while at the same time $\lim_{k\to\infty} f\left(x_{n_k}\right) = f(x) < \infty$, which is not possible.

Thus f is bounded above. Let $M := \sup\left\{f(z) : z \in [a, b]\right\}$. For each $n \in \mathbb{N}$ find $x_n \in [a, b]$ with $f(x_n) \geq M - \frac{1}{n}$. Again by the Bolzano-Weierstrass Theorem there is a convergent subsequence $\left\{x_{n_k}\right\}_{k=1}^{\infty}$ with limit $x \in [a, b]$. For all k we have $M - \frac{1}{n_k} \leq f(x_{n_k}) \leq M$, which means by the Squeeze Theorem that

$$f(x) = \lim_{k\to\infty} f(x_{n_k}) = M = \sup\left\{f(z) : z \in [a, b]\right\}.$$

■

Exercises

3-33. Prove that $c \neq b$ in the Intermediate Value Theorem.

3-34. Prove that if $f : A \to B$ is bijective and $g, h : B \to A$ are inverses of f, then $g(b) = h(b)$ for all $b \in B$.

3-35. Prove Corollary 3.40.

3-36 Prove Corollary 3.41.

3-37. Prove Theorem 3 42

3-38. Let $f : [a, b] \to \mathbb{R}$ be continuous Prove that there is an $x \in [a, b]$ such that for all $z \in [a, b]$ the inequality $f(x) \leq f(z)$ holds.

3-39 Let $I \subseteq \mathbb{R}$ be an interval Give an example that shows that even if $f : I \to \mathbb{R}$ is continuous, $f[I]$ need not be bounded. Then explain why this example does not contradict Theorem 3.44.

3-40. Let I be an interval and let $f : I \to \mathbb{R}$ be continuous and injective. Prove that f is either **increasing** (that is, for all $x_1 < x_2$ we have $f(x_1) < f(x_2)$) or **decreasing** (that is, for all $x_1 < x_2$ we have $f(x_1) > f(x_2)$).

3-41 Although the contrapositive is often used to clarify a mathematical statement, some contrapositives can be quite confusing Determine if the statement "If $f(x) \neq 0$ for all $c \in [a, b]$, then $f(a) \geq 0$ or $f(b) \leq 0$ or f is not continuous on $[a, b]$." is true or false by analyzing its (much simpler) contrapositive.

3.6 Limits at Infinity

For functions defined on an interval (t, ∞) it is sensible to investigate the behavior as the argument x gets large. The resulting notion of a limit at infinity is set up in exactly the same way as the other limits discussed in this chapter. Thus it is not surprising that similar laws hold.

Definition 3.45 *Let L be a real number and let $f : (t, \infty) \to \mathbb{R}$ be a function. We say f* **converges** *to the* **limit** *L at ∞ and write $\lim_{z\to\infty} f(z) = L$ iff for every sequence $\{z_n\}_{n=1}^{\infty}$ in (t, ∞) such that $\lim_{n\to\infty} z_n = \infty$ we have $\lim_{n\to\infty} f(z_n) = L$.*

Limits at $-\infty$ are defined similarly. For functions with domains that contain intervals (t, ∞) or $(-\infty, t)$ the limits at $\pm\infty$ are defined as the limits of the appropriate restriction.

Theorem 3.46 Limit laws *for limits of functions at ∞. Let $f, g : (t, \infty) \to \mathbb{R}$ be functions so that $\lim_{z\to\infty} f(z)$ and $\lim_{z\to\infty} g(z)$ exist. Then the following hold.*

1. $\lim\limits_{z\to\infty}(f+g)(z) = \lim\limits_{z\to\infty} f(z) + \lim\limits_{z\to\infty} g(z)$.

2. $\lim\limits_{z\to\infty}(f-g)(z) = \lim\limits_{z\to\infty} f(z) - \lim\limits_{z\to\infty} g(z)$.

3. $\lim\limits_{z\to\infty}(f\cdot g)(z) = \lim\limits_{z\to\infty} f(z) \cdot \lim\limits_{z\to\infty} g(z)$.

4. *If* $\lim\limits_{z\to\infty} g(z) \neq 0$, *then* $\lim\limits_{z\to\infty}\left(\frac{f}{g}\right)(z) = \frac{\lim_{z\to\infty} f(z)}{\lim_{z\to\infty} g(z)}$.

Each equation implicitly asserts that if the right side exists, so does the left side and in this case the equality holds.

Proof. Exercise 3-42. ■

Theorem 3.47 ε**-**M **formulation for the limit of a function at infinity.** *The function $f : (t, \infty) \to \mathbb{R}$ converges to the limit L at infinity if and only if for every $\varepsilon > 0$ there is an M such that for all $z \geq M$ we have $\left|f(z) - L\right| < \varepsilon$.*

Proof. Exercise 3-43. ■

Similar results hold for limits at $-\infty$. Moreover, infinite limits at $\pm\infty$ can be defined similar to infinite limits at a finite number and similar limit laws hold. By now, the reader is sufficiently familiar with the underlying ideas to formulate these definitions independently (see Exercise 3-44).

Exercises

3-42. Prove Theorem 3 46

(a) Prove part 1 of Theorem 3 46.

(b) Prove part 2 of Theorem 3.46.

(c) Prove part 3 of Theorem 3 46

(d) Prove part 4 of Theorem 3.46

3-43. Prove Theorem 3.47

3-44 Infinite limits at infinity.

(a) State the definition of $\lim\limits_{z\to\infty} f(z) = \infty$.

(b) State limit laws for infinite limits at ∞

(c) State and prove a result similar to Theorem 3.47 for infinite limits at infinity

3-45. Explain why we do not define left and right limits at infinity.

Chapter 4

Differentiable Functions

From here on, we will no longer consistently restate what we need to prove at the start of the proof and we will also dispense with closing statements that we have indeed proved what we set out to prove. The restatements were included earlier to reenforce the fundamental definitions and the habit of mentally clarifying what needs to be proved before starting a proof. The reader can now be expected to mentally continue this practice.

Geometrically speaking, differentiable functions have unbroken graphs without corners. This "smoothness" of differentiable functions is useful in applications. In the present chapter differentiability is introduced in Section 4.1, the relation between differentiability and the common operations on functions is considered in Section 4.2, and some geometric consequences of differentiability are provided in Section 4.3.

4.1 Differentiability

The derivative provides the slope of the tangent line, if the function has a tangent line at the point. Definition 4.1 encodes this idea by demanding that as we fix one point and move the other one closer to this "base point," the slopes of the secant lines through these points approach a limit. The left part of Figure 9 shows that this convergence of the slopes means that the secant lines tilt towards a "limiting line," the tangent line. It should be noted that differentiability typically is considered on open intervals so that secant lines "in both directions" can be used to obtain the limit.

Definition 4.1 *A function* $f : (a, b) \to \mathbb{R}$ *is* **differentiable** *at* $x \in (a, b)$ *iff the limit* $\lim_{z \to x} \frac{f(z) - f(x)}{z - x}$ *exists. In this case we set* $f'(x) := \lim_{z \to x} \frac{f(z) - f(x)}{z - x}$ *and call it the* **derivative** *of* f *at* x. *Other notations for the derivative at* x *are* $\frac{df}{dx}(x)$ *and* $Df(x)$.

Similar to limits of functions, if $D \subseteq R$, *a function* $f : D \to \mathbb{R}$ *is called* **differentiable** *at* $x \in D$ *iff there is an open interval* $(a, b) \subseteq D$ *so that* $x \in (a, b)$ *and so that*

$f|_{(a,b)}$ is differentiable at x. Moreover, similar to what we did so far, we will mostly work with functions defined on open intervals, trusting that the reader can make the jump to larger domains.

Example 4.2 *The following are verified with routine computations (Exercise 4-1).*

1. *At every $x \in \mathbb{R}$ the function $f(x) = x$ is differentiable with $f'(x) = 1$.*

2. *The function $f(x) = |x|$ is not differentiable at $x = 0$.* □

As with continuity, the local property of differentiability can be demanded at every point of the domain to obtain a global definition.

Definition 4.3 *Let $D \subseteq \mathbb{R}$. A function $f : D \to \mathbb{R}$ is* **differentiable**, *or* **differentiable on** *D, iff it is differentiable at every $x \in D$.*

Continuity at x means that $\lim_{z \to x} f(z) - f(x) = 0$. Theorem 4.4 affirms that convergence of the quotient $\frac{f(z) - f(x)}{z - x}$ to a number is a stronger condition.

Theorem 4.4 *Let $f : (a, b) \to \mathbb{R}$ be differentiable at x. Then f is continuous at x. Hence, every differentiable function is continuous. However, not every continuous function is differentiable.*

Proof. Let f be differentiable at x. Then $\lim_{z \to x} \frac{f(z) - f(x)}{z - x}$ exists. This implies

$$0 = \lim_{z \to x}(z - x) \lim_{z \to x} \frac{f(z) - f(x)}{z - x} = \lim_{z \to x}(z - x)\frac{f(z) - f(x)}{z - x} = \lim_{z \to x} f(z) - f(x),$$

that is, $\lim_{z \to x} f(z) = \lim_{z \to x} \big(f(z) - f(x)\big) + \lim_{z \to x} f(x) = 0 + f(x) = f(x)$, and f is continuous at x.

To see that not every continuous function is differentiable, consider $f(x) = |x|$, which is continuous, but it is not differentiable at $x = 0$. ■

Ultimately, we will generalize differentiability to higher dimensions. In higher dimensions, division is not possible, but it is possible to define entities that are similar to tangent lines. Theorem 4.5 shows the difference between differentiability at x and continuity at x without using division. A differentiable function f can be approximated by a straight line $g(z) = f(x) + f'(x)(z - x)$ through $\big(x, f(x)\big)$ in such a way that the difference between f and g goes to zero faster than $|z - x|$. Geometrically, this means (see Exercise 4-2a and Figure 9(b)) that, no matter how small the width, near x the differentiable function f will enter all "cones" which are centered at $\big(x, f(x)\big)$ and symmetric about the line g. This idea ultimately leads to the definition of differentiability in higher dimensional spaces (see Definition 17.24). For a continuous function f, the difference between f and any straight line through $\big(x, f(x)\big)$ goes to zero (see Exercise 4-2c), but the function need not enter arbitrarily narrow cones about the line.

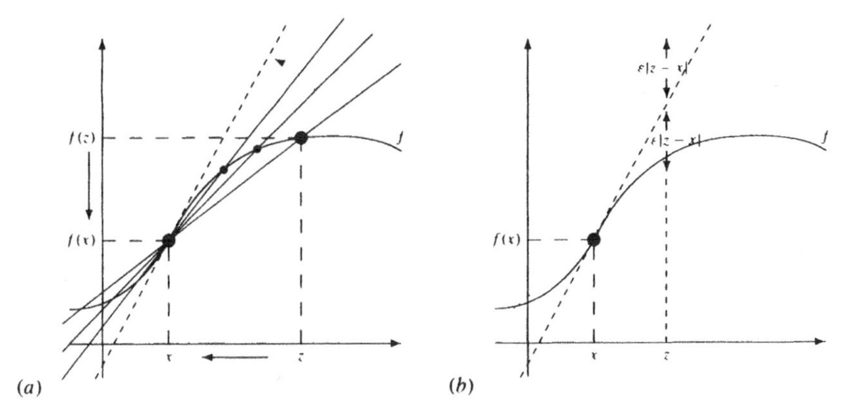

Figure 9: Two ways to view differentiability. In (a), as in Definition 4.1, the slopes of the secant lines through $(x, f(x))$ and $(z, f(z))$ approach a number, which is the slope of the tangent line. In (b), as in Theorem 4.5 and Exercise 4-2a, the function and a certain straight line, the tangent line, are such that for any width $\varepsilon > 0$, near x the function will ultimately enter the "cone of width ε" about the tangent line.

Theorem 4.5 *Let $f : (a, b) \to \mathbb{R}$ be a function and let $x \in (a, b)$. Then f is differentiable at x iff there is an $L \in \mathbb{R}$ so that for every $\varepsilon > 0$ there is a $\delta > 0$ such that for all $z \neq x$ with $|z - x| < \delta$ we have $\left|f(z) - f(x) - L(z - x)\right| < \varepsilon|z - x|$. Moreover, in this case $f'(x) = L$.*

Proof. For "$\Rightarrow$," let $x \in (a, b)$ and let $L := f'(x)$. Because f is differentiable at x, for all $\varepsilon > 0$ there is a $\delta > 0$ so that for all $z \neq x$ with $|z - x| < \delta$ we have that $\left|\frac{f(z) - f(x)}{z - x} - L\right| < \varepsilon$. Multiplying both sides by $|z - x|$ turns this inequality into the desired inequality $\left|f(z) - f(x) - L(z - x)\right| < \varepsilon|z - x|$.

For "$\Leftarrow$," let $x \in (a, b)$ be such that there is an $L \in \mathbb{R}$ so that for all $\varepsilon > 0$ there is a $\delta > 0$ so that for all $z \neq x$ with $|z-x| < \delta$ we have $\left|f(z)-f(x)-L(z-x)\right| < \varepsilon|z-x|$. Dividing the inequality by $|z - x|$ shows that for all $\varepsilon > 0$ there is a $\delta > 0$ so that for all $z \neq x$ with $|z - x| < \delta$ we have that $\left|\frac{f(z) - f(x)}{z - x} - L\right| < \varepsilon$. This means that $L = \lim_{z \to x} \frac{f(z) - f(x)}{z - x}$ and f is differentiable at x with $f'(x) = L$. ∎

Exercises

4-1. (a) Prove that the function $f(x) = x$ is differentiable at every $x \in \mathbb{R}$ and that $f'(x) = 1$.

(b) Prove that the function $f(x) = |x|$ is not differentiable at $x = 0$.

4-2. Differentiability versus continuity.

(a) Let $f : (a, b) \to \mathbb{R}$. Prove that f is differentiable at x iff there is an $L \in \mathbb{R}$ so that for every $\varepsilon > 0$ there is a $\delta > 0$ so that for all $z \in \mathbb{R}$ with $|z - x| < \delta$ we have

$$\big(f(x) + L(z - x) \big) - \varepsilon|z - x| \leq f(z) \leq \big(f(x) + L(z - x) \big) + \varepsilon|z - x|.$$

Also prove that in this case $f'(x) = L$. Use this result to explain the right part of Figure 9.

(b) Prove that for any $m \in \mathbb{R}$ we have $\lim_{z\to 0} |z| - mz = 0$, even though $f(x) = |x|$ is not differentiable at $x = 0$.

(c) Let $f : (a, b) \to \mathbb{R}$ be a function that is continuous at $x \in \mathbb{R}$ and let $m \in \mathbb{R}$. Prove that $\lim_{z\to x} \left| f(z) - [f(x) + m(z - x)] \right| = 0$.

4-3. For $f : (a, b) \to \mathbb{R}$ and $x \in (a, b)$ we define the **left-sided derivative** of f at x via the left limit $D^l f(x) := \lim_{z\to x^-} \frac{f(z) - f(x)}{z - x}$ if the limit exists and we define the **right-sided derivative** of f at x via the right limit $D^r f(x) := \lim_{z\to x^+} \frac{f(z) - f(x)}{z - x}$ if the limit exists.

(a) Prove that f is differentiable at x iff the left-sided and right-sided derivatives exist at x and $D^r f(x) = D^l f(x)$.

(Compare with Standard Proof Technique 3.20.)

(b) Prove that for $f(x) = |x|$ we have that $D^r f(0) = 1$ and $D^l f(0) = -1$.

4.2 Differentiation Rules

The results in this section show how differentiability relates to the algebraic operations for functions as well as to composition. So far, whenever limits were concerned, we worked with estimates that ultimately made certain differences smaller than ε. However, we can also use the available limit theorems and we will do so in this section. The following computations reemphasize the fact that *facility in symbolic computation remains important in abstract mathematics.*

Theorem 4.6 *Let the functions $f : (a, b) \to \mathbb{R}$ and $g : (a, b) \to \mathbb{R}$ both be differentiable at $x \in (a, b)$ and let c be a real number. Then the functions $f + g$, $f - g$ and cf are all differentiable at x and*

$$\begin{aligned} (f+g)'(x) &= f'(x) + g'(x), \\ (f-g)'(x) &= f'(x) - g'(x), \\ (cf)'(x) &= cf'(x). \end{aligned}$$

Proof. We use the definition of the derivative with the goal of extracting the difference quotients for f and g, respectively.

$$(f+g)'(x) = \lim_{z\to x} \frac{(f+g)(z) - (f+g)(x)}{z - x} = \lim_{z\to x} \frac{f(z) + g(z) - f(x) - g(x)}{z - x}$$

The key in computing derivatives with the definition is the recognition of difference quotients for the functions involved. To see more clearly where we can find terms of the form $\frac{f(z) - f(x)}{z - x}$ and $\frac{g(z) - g(x)}{z - x}$ in the present computation, we switch the two middle terms.

$$= \lim_{z\to x} \frac{f(z) - f(x) + g(z) - g(x)}{z - x}$$

$$= \lim_{z\to x}\frac{f(z)-f(x)}{z-x}+\lim_{z\to x}\frac{g(z)-g(x)}{z-x}$$
$$= f'(x)+g'(x)$$

The remainder of the proof is left to the reader as Exercises 4-4 and 4-5. ■

Derivatives are also well-behaved when products, quotients, powers and compositions are involved. We first prove the quotient rule, which is a bit harder than the product rule and we leave the proof of the product rule as an exercise.

Theorem 4.7 Quotient Rule. *Let the functions $f, g : (a,b) \to \mathbb{R}$ be differentiable at $x \in (a,b)$ and let $g(x) \neq 0$. Then the quotient $\frac{f}{g}$ is differentiable at x with* $\left(\frac{f}{g}\right)'(x) = \frac{f'(x)g(x)-g'(x)f(x)}{\left(g(x)\right)^2}$.

Proof. Similar to the proof of Theorem 4.6 we compute the limit of the difference quotients. The computations are a bit more involved than before.

$$\lim_{z\to x}\frac{\frac{f}{g}(z)-\frac{f}{g}(x)}{z-x}$$

The first step in a computation involving complicated quotients is almost always to simplify and find common denominators.

$$= \lim_{z\to x}\frac{\frac{f(z)}{g(z)}-\frac{f(x)}{g(x)}}{z-x} = \lim_{z\to x}\frac{\frac{f(z)g(x)-f(x)g(z)}{g(z)g(x)}}{z-x}$$
$$= \lim_{z\to x}\frac{1}{g(z)g(x)}\frac{f(z)g(x)-f(x)g(z)}{z-x}$$

We cannot compute the limit here because nothing can be factored out to obtain difference quotients for f or g. This is remedied by introducing two terms of the form $f(x)g(x)$ and pairing them up with the terms we already have (see Standard Proof Technique 2.5).

$$= \lim_{z\to x}\frac{1}{g(z)g(x)}\frac{f(z)g(x)-f(x)g(x)+f(x)g(x)-f(x)g(z)}{z-x}$$

The remainder of the proof relies on the right factorizations, the differentiability of f and g and the limit laws.

$$= \lim_{z\to x}\frac{1}{g(z)g(x)}\frac{f(z)g(x)-f(x)g(x)-\big(f(x)g(z)-f(x)g(x)\big)}{z-x}$$
$$= \lim_{z\to x}\frac{1}{g(z)g(x)}\left(\frac{f(z)g(x)-f(x)g(x)}{z-x}-\frac{f(x)g(z)-f(x)g(x)}{z-x}\right)$$
$$= \lim_{z\to x}\frac{1}{g(z)g(x)}\left(\frac{f(z)-f(x)}{z-x}g(x)-\frac{g(z)-g(x)}{z-x}f(x)\right)$$

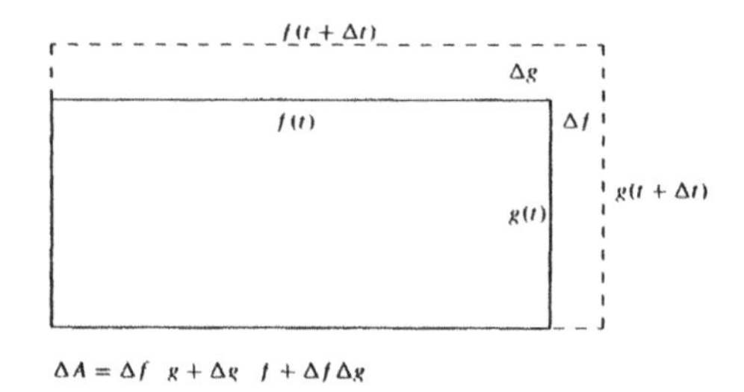

Figure 10: Visualization of the Product Rule. The growth rate of a rectangle with side lengths f and g can be obtained from the picture above by dividing the formula for ΔA by Δt and letting Δt go to zero. It is $A' = f'g + g'f$. The term $\dfrac{\Delta f \Delta g}{\Delta t}$ does not contribute to the rate because in its numerator *two* quantities are going to zero. The proof of the product rule (Exercise 4-6) makes this idea more precise.

$$
\begin{aligned}
&= \frac{1}{\left(g(x)\right)^2}\left(f'(x)g(x) - g'(x)f(x)\right) \\
&= \frac{f'(x)g(x) - g'(x)f(x)}{\left(g(x)\right)^2}.
\end{aligned}
$$

■

Theorem 4.8 Product Rule. *Let $f, g : (a,b) \to \mathbb{R}$ be differentiable at $x \in (a,b)$. Then fg is differentiable at x with $(fg)'(x) = f'(x)g(x) + g'(x)f(x)$. (For a visualization, consider Figure 10.)*

Proof. Exercise 4-6. ■

Now that products and quotients are taken care of, we can consider powers. The Power Rule could have been proved earlier in a more direct fashion, but the present proof is faster.

Theorem 4.9 Power Rule. *For every integer $n \neq 0$, the function $f(x) = x^n$ is differentiable with $\dfrac{d}{dx}x^n = nx^{n-1}$ at every $x \in \mathbb{R}$ for which the right side is defined.*

Proof. For $n > 0$, we use induction on the exponent n. For the base step with $n = 1$, note that $\dfrac{d}{dx}x^1 = \lim\limits_{z \to x} \dfrac{z - x}{z - x} = 1$ for all $x \in \mathbb{R}$.

For the induction step $n \to (n+1)$, we need to prove $\dfrac{d}{dx}x^{n+1} = (n+1)x^n$ for all $x \in \mathbb{R}$ and we can use the induction hypothesis that $\dfrac{d}{dx}x^n = nx^{n-1}$. Via the Product Rule we obtain $\dfrac{d}{dx}x^{n+1} = \dfrac{d}{dx}\left(x \cdot x^n\right) = 1 \cdot x^n + nx^{n-1} \cdot x = (n+1)x^n$.

This establishes the power rule for positive integer exponents n. For any $m \in \mathbb{N}$, we have $-m < 0$ and we can differentiate x^{-m} as follows.

$$
\frac{d}{dx}x^{-m} = \frac{d}{dx}\frac{1}{x^m} = \frac{0 \cdot x^m - mx^{m-1} \cdot 1}{x^{2m}} = -mx^{-m-1}.
$$

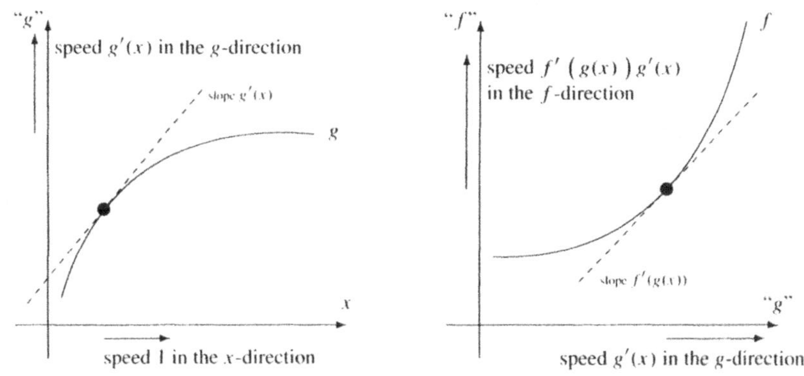

The output of g becomes the input of f. Position and speed are preserved.

Figure 11: Visualization of the chain rule. The derivative can also be understood as a magnification factor for speeds. If a particle at x moves at unit speed along the horizontal axis, then its image particle under g moves at speed $g'(x)$ along the vertical axis. Now, if a particle at $g(x)$ moves at speed $g'(x)$ along the horizontal axis then its image particle moves at speed $f'(g(x))g'(x)$ along the vertical axis.

The above shows that the power rule holds for all nonzero integers. *Now that we have the power rule, the last computation can only be viewed as a clumsy way to compute the derivative of powers with negative exponents. However, in this proof we had no choice, because we were proving the very rule that abbreviates this computation.* ■

We will revisit the Power Rule in Theorems 4.22 and 12.10. With algebraic operations taken care of, we turn our attention to composition. Figure 11 shows a kinematic way to explain the Chain Rule.

Theorem 4.10 Chain Rule. *Let $g : (a, b) \to \mathbb{R}$ and $f : (c, d) \to \mathbb{R}$ be functions with $g\big[(a, b)\big] \subseteq (c, d)$ and let $x \in (a, b)$ be such that g is differentiable at x and f is differentiable at $g(x)$. Then $f \circ g : (a, b) \to \mathbb{R}$ is differentiable at x and the derivative is $(f \circ g)'(x) = f'\big(g(x)\big)g'(x)$.*

Proof. First, consider the case that there is no sequence $\{z_n\}_{n=1}^{\infty}$ with $\lim_{n\to\infty} z_n = x$ and $z_n \neq x$ and $g(z_n) = g(x)$ for all $n \in \mathbb{N}$. In this case, we proceed as follows.

$$\lim_{z\to x} \frac{f \circ g(z) - f \circ g(x)}{z - x}$$

> There must be a $\delta > 0$ so that for $z \neq x$ and $|z - x| < \delta$ we have $g(z) \neq g(x)$. By Theorem 3.2 we can assume for this computation that $g(z) \neq g(x)$ overall. Thus we can multiply and divide by $g(z) - g(x)$. (This is the first time that we use the multiplicative version of Standard Proof Technique 2.5).

$$= \lim_{z\to x} \frac{f\big(g(z)\big) - f\big(g(x)\big)}{z - x} \cdot \frac{g(z) - g(x)}{g(z) - g(x)}$$

$$\begin{aligned}
&= \lim_{z\to x} \frac{f(g(z)) - f(g(x))}{g(z)-g(x)} \cdot \frac{g(z)-g(x)}{z-x} \\
&= \lim_{z\to x} \frac{f(g(z)) - f(g(x))}{g(z)-g(x)} \cdot \lim_{z\to x} \frac{g(z)-g(x)}{z-x}
\end{aligned}$$

Apply Theorem 3.14 to the first limit.

$$\begin{aligned}
&= \lim_{u\to g(x)} \frac{f(u) - f(g(x))}{u-g(x)} \cdot \lim_{z\to x} \frac{g(z)-g(x)}{z-x} \\
&= f'(g(x))g'(x).
\end{aligned}$$

This leaves the case in which there is a sequence $\{z_n\}_{n=1}^{\infty}$ with $\lim_{n\to\infty} z_n = x$ and $z_n \neq x$ and $g(z_n) = g(x)$ for all $n \in \mathbb{N}$. In this case,

$$g'(x) = \lim_{z\to x} \frac{g(z)-g(x)}{z-x} = \lim_{n\to\infty} \frac{g(z_n)-g(x)}{z_n - x} = 0.$$

Because f is differentiable at $g(x)$, there is a $\nu > 0$ so that for all $u \neq g(x)$ with $|u - g(x)| < \nu$ we have $\left| \frac{f(u) - f(g(x))}{u-g(x)} - f'(g(x)) \right| < 1$, and hence (by the reverse triangular inequality) $\left| \frac{f(u) - f(g(x))}{u-g(x)} \right| < |f'(g(x))| + 1$. Moreover, for all $\varepsilon > 0$ there is a $\delta > 0$ so that for all $z \neq x$ with $|z - x| < \delta$ we have that $\left| \frac{g(z)-g(x)}{z-x} \right| < \frac{\varepsilon}{|f'(g(x))| + 1}$ (because $g'(x) = 0$) and $|g(z) - g(x)| < \nu$ (by Theorem 4.4). *Formally, we would need to find a δ so that the first inequality holds and another so that the second one holds and then use the minimum of the two. We used a simple modification of Standard Proof Technique 2.6 to abbreviate this step.* Therefore for all $z \neq x$ with $|z - x| < \delta$ we obtain the following. In case $g(z) = g(x)$, we have $\frac{f\circ g(z) - f\circ g(x)}{z-x} = 0$. In case $g(z) \neq g(x)$, we have $|g(z) - g(x)| < \nu$, and hence

$$\begin{aligned}
\left| \frac{f\circ g(z) - f\circ g(x)}{z-x} \right| &= \left| \frac{f(g(z)) - f(g(x))}{z-x} \right| \left| \frac{g(z)-g(x)}{g(z)-g(x)} \right| \\
&= \left| \frac{f(g(z)) - f(g(x))}{g(z)-g(x)} \right| \left| \frac{g(z)-g(x)}{z-x} \right| \\
&< \left(|f'(g(x))| + 1 \right) \frac{\varepsilon}{|f'(g(x))| + 1} = \varepsilon.
\end{aligned}$$

Because ε was arbitrary we conclude

$$\lim_{z\to x} \frac{f\circ g(z) - f\circ g(x)}{z-x} = 0 = f'(g(x)) \cdot 0 = f'(g(x))g'(x)$$

and the proof is complete. ■

With compositions taken care of, it would be natural to also consider inverse functions. Because we need Rolle's Theorem to dispose of a technicality, consideration of inverse functions is postponed to Theorem 4.21.

Example 4.11 *Maxima and minima do not preserve differentiability.*
Even though $f(x) = x$ and $g(x) = -x$ are both differentiable, the functions $\max\{f, g\}(x) = |x|$ and $\min\{f, g\}(x) = -|x|$ are not differentiable at 0. Thus, while differentiability is compatible with the natural algebraic operations for functions, it is not compatible with the natural order-theoretical operations for functions. □

If a function $f : (a, b) \to \mathbb{R}$ is differentiable on (a, b), then the derivative f' is a function in its own right. Hence, we can consider continuity and differentiability for f'.

Definition 4.12 *Let $D \subseteq \mathbb{R}$ and let $f : D \to \mathbb{R}$ be a function. Then f is called* **continuously differentiable** *iff it is differentiable on D and the derivative $f' : D \to \mathbb{R}$ is continuous. The function f is also often considered to be its own* "**zeroth derivative**" *$f^{(0)} := f$. The function f is called n* **times differentiable** *iff it is $n-1$ times differentiable and its $(n-1)^{\text{st}}$ derivative $f^{(n-1)} : D \to \mathbb{R}$ is differentiable. The nth* **derivative** *of f is $f^{(n)}(x) := \frac{d}{dx} f^{(n-1)}(x)$, also denoted $\frac{d^n}{dx^n} f$. The function f is called n* **times continuously differentiable** *iff it is n times differentiable and its nth derivative $f^{(n)} : D \to \mathbb{R}$ is continuous. Finally, f is called* **infinitely differentiable** *iff for all $n \in \mathbb{N}$ it is n times differentiable.*

The differentiation rules we have derived here immediately show the following.

Example 4.13 *Polynomials and rational functions are infinitely differentiable on their domains.* □

Example 4.14 *For every $n \in \mathbb{N}$, the function $f(x) = \begin{cases} x^{n+1}; & \text{for } x \geq 0, \\ -x^{n+1}; & \text{for } x < 0, \end{cases}$ is n times continuously differentiable, but it is not $(n+1)$-times differentiable. (Exercise 4-7.)* □

Exercises

4-4. Prove that if $f, g : (a, b) \to \mathbb{R}$ are both differentiable at $x \in (a, b)$, then $f - g$ is differentiable at x and $(f - g)'(x) = f'(x) - g'(x)$.

4-5 Prove that if $f : (a, b) \to \mathbb{R}$ is differentiable at $x \in (a, b)$ and $c \in \mathbb{R}$, then cf is differentiable at x and $(cf)'(x) = cf'(x)$.

4-6 Prove the product rule.
Hint. It's similar to the proof of the quotient rule, but simpler.

4-7 Prove the claim in Example 4.14.
Hint. Use induction and at $x = 0$ use Exercise 4-3.

4-8. Prove that $f : (a, b) \to \mathbb{R}$ is differentiable at $x \in (a, b)$ iff the limit $\lim_{h \to 0} \frac{f(x+h) - f(x)}{h}$ exists and that in this case $f'(x) = \lim_{h \to 0} \frac{f(x+h) - f(x)}{h}$

4-9. Let $n \in \mathbb{N}$. Use the Binomial Theorem and Exercise 4-8 to prove the Power Rule for $f(x) = x^n$ without using induction.

4-10. Prove that the derivative of $f(x) = \sqrt{x}$ is $f'(x) = \frac{1}{2\sqrt{x}}$.

4-11. Compute the derivative of each of the following functions

(a) $f(x) = \frac{3x^3 - 2x}{7 - 5x^2}$

(b) $f(x) = \left(x^2 + 5\right)^4 \sqrt{3x^2 - 2}$ (You may use Exercise 4-10.)

(c) $f(x) = \left[\frac{\left(x^2 + 1\right)^3 \left(x^4 - 1\right)^5}{(3x + 2)^2 \left(2x^3 + x\right)^4}\right]^7$

4-12 Use induction to prove that $\frac{d^n}{dx^n}\left(x^{-1}\right) = (-1)^n n! x^{-n-1}$ for all $x \neq 0$.

4-13. Prove that if $f, g : (a, b) \to \mathbb{R}$ are both n times differentiable, then the product fg is n times differentiable and $(fg)^{(n)} = \sum_{k=0}^{n} \binom{n}{k} f^{(k)} g^{(n-k)}$.

Hint. Mimic the proof of the Binomial Theorem.

4.3 Rolle's Theorem and the Mean Value Theorem

One of the main applications of derivatives is to use the sign of the derivative to compute relative extrema of a function and intervals where a function is increasing or decreasing. The formal justification follows from Rolle's Theorem and the Mean Value Theorem.

Definition 4.15 *Let $f : (a, b) \to \mathbb{R}$ be a function. Then f is said to have a* **relative (or local) minimum** *at x_m iff there is a $\delta > 0$ such that $f(x_m) \leq f(x)$ for all $x \in (x_m - \delta, x_m + \delta)$. f is said to have a* **relative (or local) maximum** *at x_M if and only if there is a $\delta > 0$ such that $f(x_M) \geq f(x)$ for all $x \in (x_M - \delta, x_M + \delta)$. If f has a local maximum or a local minimum at the point c we also say that f has a* **relative (or local) extremum** *at c.*

Intuitively, relative extrema are the locally highest or lowest points of the graph. (Note, however, that stagnation also is possible, see Exercise 4-14.) At the location of a relative maximum there cannot be an incline in any direction. Hence, the derivative should be zero at a relative maximum.

Theorem 4.16 *Let $f : (a, b) \to \mathbb{R}$ be a function and let $m \in (a, b)$. If f is differentiable at m and f has a relative maximum at m, then $f'(m) = 0$.*

Proof. Because f has a relative maximum at m there is a positive number δ so that $f(z) - f(m) \leq 0$ for all z with $|z - m| < \delta$. We infer that $\lim_{z \to m^+} \frac{f(z) - f(m)}{z - m} \leq 0$ and $\lim_{z \to m^-} \frac{f(z) - f(m)}{z - m} \geq 0$. Because f is differentiable at m, these two limits must

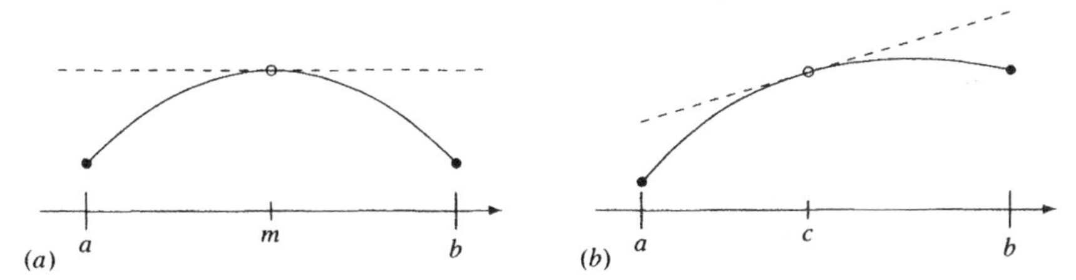

Figure 12: Rolle's Theorem states that if a differentiable function starts and ends at the same height, then it must have a flat tangent in between (a). The Mean Value Theorem states that some tangent must be parallel to the secant through the starting point and the ending point (b).

be equal to $f'(m)$. Therefore $f'(m)$ is greater than or equal to 0 *and* smaller than or equal to 0. This implies $f'(m) = 0$. ■

Rolle's Theorem states that if a function's values are equal at the endpoints of an interval, then the function must have a horizontal tangent line in the interval (see Figure 12). Note that the proof is a collection of direct proofs (modus ponens), which is possible because we can rely on strong results that we proved earlier.

Theorem 4.17 Rolle's Theorem. *Let $f : [a, b] \to \mathbb{R}$ be differentiable on the open interval (a, b) and continuous on the closed interval $[a, b]$. If $f(a) = f(b)$, then there is an $m \in (a, b)$ with $f'(m) = 0$.*

Proof. Because the result is trivial if f is constant, we can assume f is not constant.

By Theorem 3.44, f assumes an absolute maximum and an absolute minimum on $[a, b]$. Because f is not constant, one of these values is not equal to $f(a)$. Assume without loss of generality that the absolute maximum value is greater than $f(a)$. Let $m \in [a, b]$ be so that $f(m) \geq f(x)$ for all $x \in [a, b]$.

Because $f(m) > f(a) = f(b)$ we infer $m \in (a, b)$. Hence, f is differentiable at m. Because $f(m) \geq f(x)$ for all $x \in (a, b)$, f has a relative maximum at m. Now by Theorem 4.16 we conclude $f'(m) = 0$. ■

The Mean Value Theorem generalizes Rolle's Theorem by no longer demanding that the values at the endpoints are equal. It guarantees that there is a point in the interval so that the tangent line at that point is parallel to the secant line through the endpoints. In the proof, we subtract the line $l(x) = (x - a)\frac{f(b) - f(a)}{b - a}$ from the function f. This reduces the proof to an application of Rolle's Theorem.

Theorem 4.18 Mean Value Theorem. *Let $f : [a, b] \to \mathbb{R}$ be differentiable on the open interval (a, b) and continuous on the closed interval $[a, b]$. Then there is a number $c \in (a, b)$ so that $\frac{f(b) - f(a)}{b - a} = f'(c)$.*

Proof. The function $g(x) := f(x) - (x - a)\frac{f(b) - f(a)}{b - a}$ is continuous on $[a, b]$, differentiable on (a, b) and $g(b) = f(a) = g(a)$. By Rolle's Theorem there is a c in the

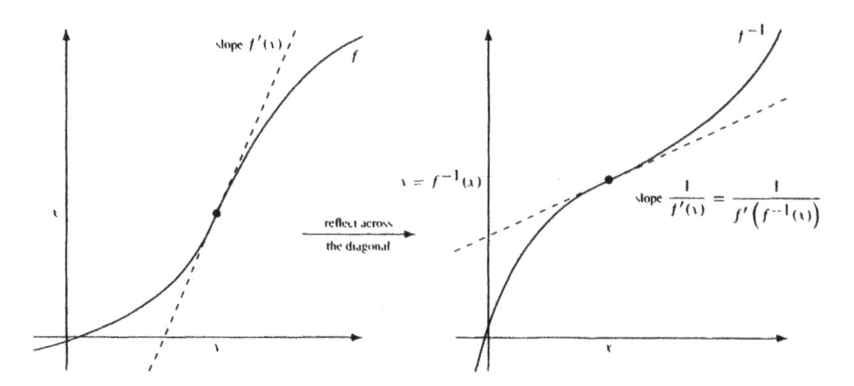

Figure 13: Visualization of Theorem 4.21. Reflection across the diagonal produces the graph of the inverse function and the slope of the reflected line is the multiplicative inverse of the slope of the original line.

interval (a, b) such that $g'(c) = 0$. But that means $f'(c) - \dfrac{f(b) - f(a)}{b - a} = g'(c) = 0$, that is, $f'(c) = \dfrac{f(b) - f(a)}{b - a}$. ■

With the Mean Value Theorem available we can now prove a sufficient criterion for when a function is increasing or decreasing.

Definition 4.19 *Let $I \subseteq \mathbb{R}$ be an interval and let $f : I \to \mathbb{R}$ be a function. Then f is called (strictly)* **increasing** *on I if and only if for all $x_1 < x_2$ in I we have $f(x_1) < f(x_2)$. f is called (strictly)* **decreasing** *on I if and only if for all $x_1 < x_2$ in I we have $f(x_1) > f(x_2)$.*

Moreover, f is called **nondecreasing** *on I if and only if for all $x_1 < x_2$ in I we have $f(x_1) \leq f(x_2)$ and f is called* **nonincreasing** *on I if and only if for all $x_1 < x_2$ in I we have $f(x_1) \geq f(x_2)$.*

Theorem 4.20 *Let $f : [a, b] \to \mathbb{R}$ be differentiable on the open interval (a, b) and continuous on the closed interval $[a, b]$. If $f'(x) > 0$ for all $x \in (a, b)$, then f is increasing on $[a, b]$.*

Proof. Let $f'(x) > 0$ for all x in (a, b). Suppose, for a contradiction, that there are points x_1, x_2 in $[a, b]$ such that $x_1 < x_2$ and $f(x_1) \geq f(x_2)$. Then by the Mean Value Theorem there is a c in the interval (x_1, x_2) (and therefore in (a, b)) such that $f'(c) = \dfrac{f(x_2) - f(x_1)}{x_2 - x_1} \leq 0$, which is a contradiction. ■

Exercise 4-21 provides a similar criterion for nonincreasing and nondecreasing functions. Rolle's Theorem also enables us to prove that the inverses of differentiable functions are differentiable. For a visualization, consider Figure 13.

Theorem 4.21 *Let $f : (a,b) \to \mathbb{R}$ be a continuous injective function, let f^{-1} be its inverse and let $x \in f\big[(a,b)\big]$ be such that f is differentiable at $f^{-1}(x)$ with $f'\left(f^{-1}(x)\right) \neq 0$. Then f^{-1} is differentiable at x and $\left(f^{-1}\right)'(x) = \dfrac{1}{f'\left(f^{-1}(x)\right)}$.*

Proof. By Theorem 3.36 $f\big[(a,b)\big]$ is an interval and by Theorem 4.16 x is neither its maximum nor its minimum. Thus there is a $\delta > 0$ so that the containment $(x-\delta, x+\delta) \subseteq f\big[(a,b)\big]$ holds. Hence, we can talk about differentiability of the function f^{-1} at x.

Let $\{z_n\}_{n=1}^{\infty}$ be a sequence in $f\big[(a,b)\big] \setminus \{x\}$ so that $\lim_{n\to\infty} z_n = x$. By Definition 3.1, we are done if we can show that $\lim_{n\to\infty} \dfrac{f^{-1}(z_n) - f^{-1}(x)}{z_n - x} = \dfrac{1}{f'\left(f^{-1}(x)\right)}$. For each $n \in \mathbb{N}$ let $y_n := f^{-1}(z_n)$ and let $y = f^{-1}(x)$. Then $y_n = f^{-1}(z_n) \neq f^{-1}(x) = y$ for all $n \in \mathbb{N}$. By assumption f is continuous, and hence by Theorem 3.38 so is f^{-1}. This means that $\lim_{n\to\infty} y_n = \lim_{n\to\infty} f^{-1}(z_n) = f^{-1}\left(\lim_{n\to\infty} z_n\right) = f(x) = y$, and therefore

$$\begin{aligned}
\lim_{n\to\infty} \frac{f^{-1}(z_n) - f^{-1}(x)}{z_n - x} &= \lim_{n\to\infty} \frac{f^{-1}\big(f(y_n)\big) - f^{-1}\big(f(y)\big)}{f(y_n) - f(y)} \\
&= \lim_{n\to\infty} \frac{y_n - y}{f(y_n) - f(y)} = \lim_{n\to\infty} \frac{1}{\frac{f(y_n)-f(y)}{y_n - y}} \\
&= \frac{1}{f'(y)} = \frac{1}{f'\left(f^{-1}(x)\right)}.
\end{aligned}$$

■

Theorem 4.21 allows us to extend the Power Rule to rational exponents.

Theorem 4.22 Power Rule. *Let $r \in \mathbb{Q} \setminus \{0\}$. Then $f(x) = x^r$ is differentiable with $\dfrac{d}{dx} x^r = r x^{r-1}$ at every x for which the right side is defined.*

Proof. Let $m \in \mathbb{N}$ and consider $f(x) = x^{\frac{1}{m}}$. Then f is the inverse function of $g(x) = x^m$. By Theorem 4.21 for all nonzero $x \in \mathbb{R}$ for which $x^{\frac{1}{m}}$ is defined, we obtain $f'(x) = \left(g^{-1}\right)'(x) = \dfrac{1}{g'\left(g^{-1}(x)\right)} = \dfrac{1}{m\left(x^{\frac{1}{m}}\right)^{m-1}} = \dfrac{1}{m} x^{\frac{1}{m}-1}$.

Now let $n \in \mathbb{Z}$ and $m \in \mathbb{N}$. By the Chain Rule, for all nonzero x for which $x^{\frac{n}{m}}$ is defined we conclude $\dfrac{d}{dx} x^{\frac{n}{m}} = \dfrac{d}{dx}\left(x^{\frac{1}{m}}\right)^n = n\left(x^{\frac{1}{m}}\right)^{n-1} \dfrac{1}{m} x^{\frac{1}{m}-1} = \dfrac{n}{m} x^{\frac{n}{m}-1}$. ■

Exercises

4-14. Geometry versus utility in the definition of relative extrema.

(a) Explain why by Definition 4.15 every point in the interval $(0, 1)$ is a relative maximum and a relative minimum of the function $f(x) = \begin{cases} x^3, & \text{for } x \leq 0, \\ 0, & \text{for } x \in [0,1], \\ (x-1)^3, & \text{for } x \geq 1. \end{cases}$

(b) Sketch the graph of the function and comment whether intuition agrees that f has a relative maximum and a relative minimum at each $x \in (0, 1)$

(c) Use the proof of Rolle's Theorem to explain why the definition of relative maxima as in Definition 4.15 is preferable to a definition of relative maxima that requires the value at a relative maximum to be *strictly* larger than the values of the function near the relative maximum.

4-15. Prove that if $f : (a, b) \to \mathbb{R}$ is differentiable and has a relative minimum at $m \in (a, b)$, then $f'(m) = 0$.

4-16. Let $f : [a, b] \to \mathbb{R}$ be differentiable on the open interval (a, b) and continuous on the closed interval $[a, b]$. Prove that if $f'(x) < 0$ for all $x \in (a, b)$, then f is decreasing on $[a, b]$.

4-17 Let $f : [a, b] \to \mathbb{R}$ be differentiable on the open interval (a, b) and continuous on the closed interval $[a, b]$. Prove that if $f'(x) = 0$ for all $x \in (a, b)$, then f is constant on $[a, b]$.

4-18. Give a direct proof of Theorem 4.20. That is, give a proof that does not argue via contradiction.

4-19 Let $f(x) := ax^2 + bx + c$ be a quadratic function defined on $\mathbb{R}$ Prove that if $a > 0$ then f is decreasing on $\left(-\infty, -\frac{b}{2a}\right)$ and increasing on $\left(-\frac{b}{2a}, \infty\right)$.
State and prove a similar result for $a < 0$

4-20. Let $f(x) := ax^3 + bx^2 + cx + d$ be a cubic function defined on $\mathbb{R}$ with $a \neq 0$. Prove that if $4b^2 - 12ac > 0$, then f has two relative extrema. For $a > 0$ and for $a < 0$ separately describe where f is increasing and where it is decreasing.

4-21. Let $f : [a, b] \to \mathbb{R}$ be differentiable on (a, b) and continuous on $[a, b]$

(a) Prove that f is nondecreasing iff for all $x \in (a, b)$ we have $f'(x) \geq 0$. (Note that this is an equivalence.)

(b) Prove that f is nonincreasing iff for all $x \in (a, b)$ we have $f'(x) \leq 0$ (Note that this is an equivalence.)

(c) Give an example that shows that the condition in Theorem 4 20 is not equivalent to f being strictly increasing.

4-22. *Use Theorem 4 21* to prove that the derivative of $f(x) = \sqrt{x}$ is $f'(x) = \frac{1}{2\sqrt{x}}$.

4-23. Explain with the following is *not* a proof for Theorem 4.21.

"Proof." We know that $x = f\left(f^{-1}(x)\right)$. Differentiating both sides with respect to x gives
$1 = \frac{d}{dx}x = \frac{d}{dx}f\left(f^{-1}(x)\right) = f'\left(f^{-1}(x)\right)\left(f^{-1}\right)'(x)$, so $\left(f^{-1}\right)'(x) = \frac{1}{f'\left(f^{-1}(x)\right)}$. ■??

4-24. Let $f : (a, b) \to \mathbb{R}$ be continuous and let $x \in (a, b)$ Prove that if $f'(z)$ exists for all $z \in (a, b) \setminus \{x\}$ and $\lim_{z \to x} f'(z)$ exists, then f is differentiable at x with $f'(x) = \lim_{z \to x} f'(z)$.

4-25 Let $f : (a, b) \to \mathbb{R}$ be differentiable. Prove that f' (which need not be continuous) has the **intermediate value property**. That is, prove that for all $c < d$ in (a, b) and all v between $f'(c)$ and $f'(d)$ there is an $m \in (c, d)$ so that $f'(m) = v$

Hint. Show that $g(x) := \begin{cases} \frac{f(x)-f(c)}{x-c}; & \text{for } x > c, \\ f'(c); & \text{for } x = c, \end{cases}$ and $h(x) := \begin{cases} \frac{f(x)-f(d)}{x-d}; & \text{for } x < d, \\ f'(d); & \text{for } x = d, \end{cases}$ are continuous on $[c, d]$, that $g(d) = h(c)$, apply the Intermediate Value Theorem to one of them, then apply the Mean Value Theorem to f.

4-26. Prove that for $n \geq 2$ the nth derivative of $f(x) = \sqrt{x}$ is
$$f^{(n)}(x) = (-1)^{n+1}\frac{(2n-2)!}{2^{n-1}(n-1)!2^n}x^{-\frac{2n-1}{2}} = (-1)^{n+1}\frac{1 \cdot 3 \cdot 5 \cdots (2n-3)}{2^n}x^{-\frac{2n-1}{2}}.$$

4-27 For each function, find an expression for $f^{(n)}(x)$ and prove that it is the nth derivative of f.

(a) $f(x) = \frac{1}{\sqrt{x}}$ (b) $f(x) = \sqrt[3]{x}$

Chapter 5

The Riemann Integral I

From here on, references to standard proof techniques will be sporadic and the writing style in proofs will gradually become more condensed.

The geometric goal of integration is to compute the area under a graph. In Riemann integration this is done by approximating the area with rectangles. This chapter presents the idea behind Riemann integration and some integration criteria, examples and theorems. Although intuitively the Riemann integral seems to be the right idea, by the end of the chapter we will need more machinery to fully characterize Riemann integrability and we will also have exposed some key weaknesses. An equivalent criterion for Riemann integrability will be presented in Theorem 8.12. The observed weaknesses of the Riemann integral will be addressed in Chapter 9.

5.1 Riemann Sums and the Integral

To define the area of rectangles under the graph of a function, we first need to determine the base for each rectangle. This is done with a partition of the interval.

Definition 5.1 *Let $[a, b]$ be a closed interval. Then any finite set $P \subseteq [a, b]$ such that $a, b \in P$ will be called a* **partition** *of $[a, b]$. Because the order of the points will be important, we also write $P = \{a = x_0 < x_1 < \cdots < x_n = b\}$ when working with a partition P.*

With the partition giving the bases of the rectangles, we still need to determine the heights. Each height will be a value that the function assumes within the respective interval of the partition (see Figure 14(a)).

Definition 5.2 *Let $[a, b]$ be a closed interval and let $f : [a, b] \to \mathbb{R}$ be bounded. For any partition $P = \{a = x_0 < x_1 < \cdots < x_n = b\}$ a set $T = \{t_1, \ldots, t_n\}$ such that for all $i \in \{1, \ldots, n\}$ we have that $t_i \in [x_{i-1}, x_i]$ will be called an* **evaluation set**. *We*

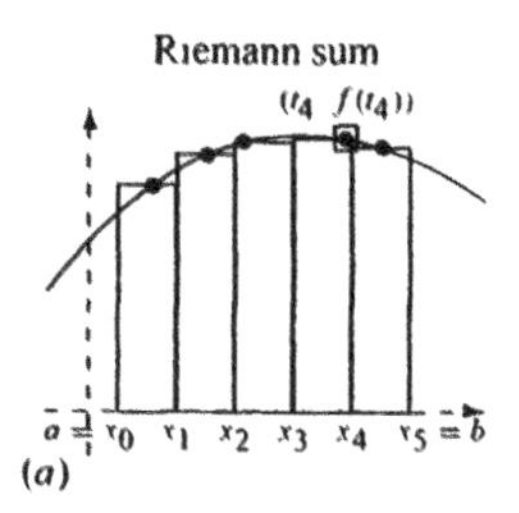

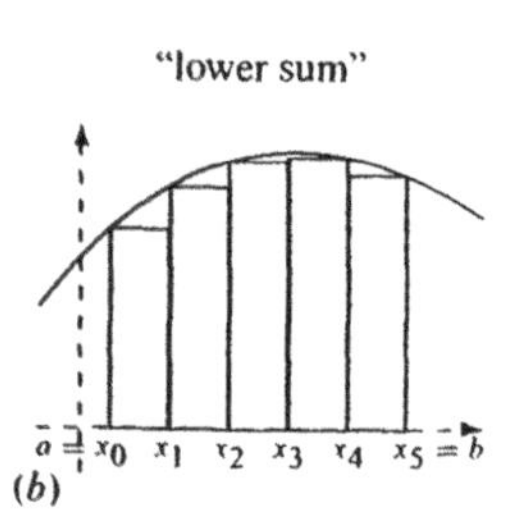

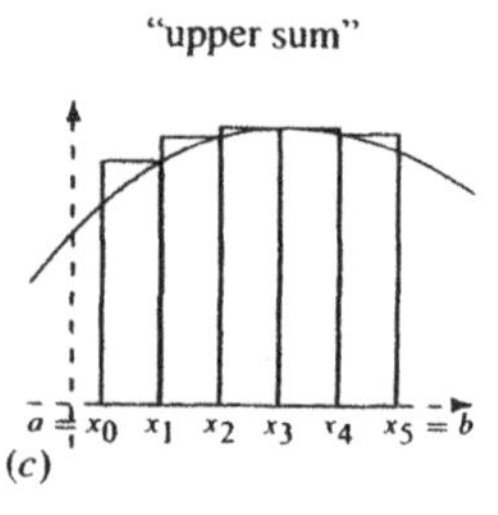

Figure 14: The Riemann integral approximates the area under the graph of a function with the areas of rectangles. Definition 5.2 demands that the height of each rectangle is the value of the function at some point in the base interval (one point marked in (a)). Of particular importance are the lower sum (see Definition 5.13) of the areas of the largest rectangles that can be fit under the graph of the function (b) and the upper sum (see Definition 5.13) of the areas of the smallest rectangles that contain the graph of the function (c).

define the **Riemann sum** *of f with respect to the partition P and evaluation set T to be* $R(f, P, T) := \sum_{i=1}^{n} f(t_i)(x_i - x_{i-1})$. *We will also use the notation* $\Delta x_i := x_i - x_{i-1}$.

Clearly, a Riemann sum can only accidentally be equal to the area under the graph. However, the narrower we make the rectangles, the closer the Riemann sums should be to the actual area. The norm of a partition gives a uniform measure of how narrow the rectangles in a partition are.

Definition 5.3 *Let $[a, b]$ be a closed interval in the real numbers. For a partition $P = \{a = x_0 < x_1 < \cdots < x_n = b\}$, we define the* **norm** *of the partition P to be* $\|P\| := \max\left\{(x_i - x_{i-1}) : i = 1, \ldots, n\right\}$.

We now say that a function is Riemann integrable iff all Riemann sums get close to one value, the integral, as the norm of the partitions is made small.

Definition 5.4 *Let D be a set. A function $f : D \to \mathbb{R}$ is called* **bounded** *iff there is an $M \in \mathbb{R}$ so that $\left|f(x)\right| \leq M$ for all $x \in D$.*

Definition 5.5 *The function $f : [a, b] \to \mathbb{R}$ is called* **Riemann integrable** *(on $[a, b]$) iff f is bounded and there is a number I such that for all $\varepsilon > 0$ there is a $\delta > 0$ so that for all partitions P with $\|P\| < \delta$ and all evaluation sets T the inequality* $\left|\sum_{i=1}^{n} f(t_i)\Delta x_i - I\right| = \left|R(f, P, T) - I\right| < \varepsilon$ *holds. The number I will be called the* **Riemann integral** *of f and it will also be denoted* $\int_a^b f(x)\,dx := I$. *The function f is also called the* **integrand** *and a and b are called the* **bounds** *of the integral, with a being the* **lower bound** *and b being the* **upper bound**

The definition of the Riemann integral allows us to compute the value of the integral, if it exists, as a limit of a sequence of Riemann sums. Note that in Lemma 5.6 below the only condition on the partitions is that their norm goes to zero and that we are free to choose the evaluation sets any way we want to.

Lemma 5.6 *Let $f : [a,b] \to \mathbb{R}$ be Riemann integrable. Then for any sequence $\{P_k\}_{k=1}^{\infty}$ of partitions of $[a,b]$ with $\lim_{k\to\infty} \|P_k\| = 0$ and any associated sequence of evaluation sets $\{T_k\}_{k=1}^{\infty}$ we have $\lim_{k\to\infty} R(f, P_k, T_k) = \int_a^b f(x)\,dx$.*

Proof. Let $\{P_k\}_{k=1}^{\infty}$ be a sequence of partitions of $[a,b]$ with $\lim_{k\to\infty} \|P_k\| = 0$, let $\{T_k\}_{k=1}^{\infty}$ be an associated sequence of evaluation sets and let $\varepsilon > 0$. There is a number $\delta > 0$ so that for all partitions P of $[a,b]$ with $\|P\| < \delta$ and any associated evaluation set T we have $\left| R(f, P, T) - \int_a^b f(x)\,dx \right| < \varepsilon$. Because $\lim_{k\to\infty} \|P_k\| = 0$ there is a $K \in \mathbb{N}$ so that for all $k \geq K$ we have $\|P_k\| < \delta$. Hence, for all $k \geq K$ we infer that $\left| R(f, P_k, T_k) - \int_a^b f(x)\,dx \right| < \varepsilon$. ■

The idea in Lemma 5.6 is very useful for numerical integration (see Exercise 13-25) and for the proof of the Fundamental Theorem of Calculus (see Theorem 5.23). To demonstrate how Lemma 5.6 is applied, consider the following example.

Example 5.7 If $f(x) = x$ *is Riemann integrable on* $[0,1]$, *then* $\int_0^1 x\,dx = \frac{1}{2}$.

If f is Riemann integrable, then by Lemma 5.6 for any sequence $\{P_n\}_{n=1}^{\infty}$ of partitions with evaluation sets T_n we have $\lim_{n\to\infty} R(f, P_n, T_n) = \int_a^b f(x)\,dx$. We choose $P_n := \left\{ \frac{j}{n} : j = 0, \ldots, n \right\}$ and $T_n := \left\{ \frac{j}{n} : j = 1, \ldots, n \right\}$. Then

$$\begin{aligned}
\lim_{n\to\infty} R(f, P_n, T_n) &= \lim_{n\to\infty} \sum_{j=1}^{n} \left(\frac{j}{n}\right)\left[\frac{j}{n} - \frac{j-1}{n}\right] = \lim_{n\to\infty} \sum_{j=1}^{n} \left(\frac{j}{n}\right)\left[\frac{1}{n}\right] \\
&= \lim_{n\to\infty} \frac{1}{n^2} \sum_{j=1}^{n} j \qquad \text{Now use the summation formula from Example 1.41.} \\
&= \lim_{n\to\infty} \frac{1}{n^2} \frac{n}{2}(n+1) = \lim_{n\to\infty} \frac{n^2+n}{2n^2} = \frac{1}{2}.
\end{aligned}$$

□

Example 5.7 exhibits a key problem that we will ultimately resolve in Theorem 8.12. Although we may be able to compute a value that *should be* the Riemann integral, it may not be clear if the function actually *is* Riemann integrable.

The following results show that it is reasonably simple to prove Riemann integrability if the value of the integral is known. Note that the strong similarity between Theorem 5.8 and the appropriate parts of Theorems 2.14 and 3.10 is not accidental. The definition of the Riemann integral is similar to the definitions of the limits of sequences and functions. Hence, similar "limit laws" hold and they are proved with similar methods. Unfortunately the similarity does not apply to integrals of products and quotients. (Products are addressed in Exercises 5-21 and 5-22 and quotients are typically treated as products in integration.)

Theorem 5.8 *Let $f, g : [a, b] \to \mathbb{R}$ be Riemann integrable and let $c \in \mathbb{R}$. Then $f + g$ and cf are Riemann integrable and the following equations hold.*

1. $\displaystyle\int_a^b (f+g)(x)\,dx = \int_a^b f(x)\,dx + \int_a^b g(x)\,dx.$

2. $\displaystyle\int_a^b cf(x)\,dx = c\int_a^b f(x)\,dx.$

Proof. We will only prove that $f + g$ is Riemann integrable and that equation 1 holds, leaving the rest to Exercise 5-3. Let $\varepsilon > 0$. Then there is a $\delta > 0$ so that for all partitions P of $[a, b]$ with $\|P\| < \delta$ and all associated evaluation sets T we have $\left|\sum_{i=1}^n f(t_i)\Delta x_i - \int_a^b f(x)\,dx\right| < \frac{\varepsilon}{2}$ and $\left|\sum_{i=1}^n g(t_i)\Delta x_i - \int_a^b g(x)\,dx\right| < \frac{\varepsilon}{2}$. Therefore for all partitions P of $[a, b]$ with $\|P\| < \delta$ and all associated evaluation sets T we obtain

$$\begin{aligned}
&\left|\sum_{i=1}^n (f+g)(t_i)\Delta x_i - \left(\int_a^b f(x)\,dx + \int_a^b g(x)\,dx\right)\right| \\
&= \left|\sum_{i=1}^n f(t_i)\Delta x_i + \sum_{i=1}^n g(t_i)\Delta x_i - \left(\int_a^b f(x)\,dx + \int_a^b g(x)\,dx\right)\right| \\
&\leq \left|\sum_{i=1}^n f(t_i)\Delta x_i - \int_a^b f(x)\,dx\right| + \left|\sum_{i=1}^n g(t_i)\Delta x_i - \int_a^b g(x)\,dx\right| \\
&< \frac{\varepsilon}{2} + \frac{\varepsilon}{2} = \varepsilon.
\end{aligned}$$

By Definition 5.5, this implies that $f + g$ is Riemann integrable and the equation $\displaystyle\int_a^b (f+g)(x)\,dx = \int_a^b f(x)\,dx + \int_a^b g(x)\,dx$ holds. ∎

The prototypes with which we approximate the area under functions are rectangles, that is, areas under step functions. Therefore it is only natural that step functions should be Riemann integrable and that their integrals should be the sum of the areas of the rectangles involved.

Definition 5.9 *Let M be a set and let $S \subseteq M$ be a subset. Then the* **indicator function** *of S is the function $\mathbf{1}_S(x) := \begin{cases} 1; & \text{for } x \in S, \\ 0; & \text{for } x \notin S. \end{cases}$*

Proposition 5.10 *Let $a < b$ and let $c, d \in [a, b]$ with $c < d$. Then $\mathbf{1}_{[c,d)}$ is Riemann integrable over $[a, b]$ and $\int_a^b \mathbf{1}_{[c,d)}\, dx = d - c$.*

Proof. Let $\varepsilon > 0$, let $\delta := \min\left\{\frac{\varepsilon}{2}, \frac{d-c}{3}\right\}$, let $P = \{a = x_0 < \cdots < x_n = b\}$ be a partition with $\|P\| < \delta$ and let $T = \{t_1, \ldots, t_n\}$ be an associated evaluation set. Because $\delta \le \frac{d-c}{3}$, there are $j, k \in \{1, \ldots, n\}$ with $j \le k$ so that $\mathbf{1}_{[c,d)}(t_i) = 1$ iff $i \in \{j, j+1, \ldots, k\}$ and $\mathbf{1}_{[c,d)}(t_i) = 0$ otherwise. Then $c \in (t_{j-1}, t_j] \subseteq [x_{j-2}, x_j]$, or, if $j = 1$, $c \in [x_0, t_1] \subseteq [x_{j-1}, x_j]$. Similarly, $d \in [x_{k-1}, x_{k+1}]$, or, if $k = n$, $d \in [x_{n-1}, x_n]$. Either way, $|x_{j-1} - c| < \frac{\varepsilon}{2}$ and $|x_k - d| < \frac{\varepsilon}{2}$. Hence,

$$\begin{aligned}
&\left|\sum_{i=1}^{n} \mathbf{1}_{[c,d)}(t_i)(x_i - x_{i-1}) - (d-c)\right| \\
&= \left|\sum_{i=j}^{k}(x_i - x_{i-1}) - (d-c)\right| = \left|\sum_{i=j}^{k} x_i - \sum_{i=j-1}^{k-1} x_i - (d-c)\right| \\
&= \left|(x_k - x_{j-1}) - (d-c)\right| \le |x_k - d| + \left|x_{j-1} - c\right| < \frac{\varepsilon}{2} + \frac{\varepsilon}{2} = \varepsilon.
\end{aligned}$$

By Definition 5.5, $\mathbf{1}_{[c,d)}$ is Riemann integrable and the integral is $(d - c)$. ■

Standard Proof Technique 5.11 A sum of differences as in the proof of Proposition 5.10, for which the negative part of one term cancels the positive part of the previous or the next term, is also called a **telescoping sum**. Telescoping sums are typically used to collapse sums of many terms into shorter sums (see the proof of Theorem 5.23), or, conversely (see, for example, the proofs of Theorems 13.14 and 17.33) to make a connection between two terms using differences that are easier to work with. □

Proposition 5.12 *Let $Q := \{a = z_0 < z_1 < \cdots < z_m = b\}$ be a partition of $[a, b]$ and for $k \in \{1, \ldots, m\}$ let $a_k \in \mathbb{R}$. Then $f := \sum_{k=1}^{m} a_k \mathbf{1}_{[z_{k-1}, z_k)}$ is Riemann integrable and $\int_a^b \sum_{k=1}^{m} a_k \mathbf{1}_{[z_{k-1}, z_k)}\, dx = \sum_{k=1}^{m} a_k (z_k - z_{k-1})$.*

Proof. Combine Theorem 5.8 and Proposition 5.10 (Exercise 5-5). ■

Exercises

5-1 Prove that Lemma 5.6 can be turned into a biconditional. That is, let $f : [a, b] \to \mathbb{R}$ be a bounded function and prove that f is Riemann integrable iff there is an $I \in \mathbb{R}$ so that for every sequence $\{P_k\}_{k=1}^{\infty}$ of partitions of $[a, b]$ with $\lim_{k\to\infty} \|P_k\| = 0$ and any associated sequence of evaluation sets $\{T_k\}_{k=1}^{\infty}$ we have $\lim_{k\to\infty} R(f, P_k, T_k) = I$.

5-2. Prove each of the following

(a) If the function $f(x) = x^2$ is Riemann integrable on $[0, 1]$, then $\int_0^1 x^2\,dx = \frac{1}{3}$.

(b) If the function $f(x) = x^3$ is Riemann integrable on $[0, 1]$, then $\int_0^1 x^3\,dx = \frac{1}{4}$.

(c) If the function $f(x) = x^4$ is Riemann integrable on $[0, 1]$, then $\int_0^1 x^4\,dx = \frac{1}{5}$.

Hint Use summation formulas from Exercise 1-33.

5-3. Prove part 2 of Theorem 5.8.

5-4. Prove that if $f_1, \ldots, f_m : [a, b] \to \mathbb{R}$ are Riemann integrable on $[a, b]$, then $\sum_{k=1}^{m} f_k$ is Riemann integrable on $[a, b]$ and $\int_a^b \sum_{k=1}^{m} f_k(x)\,dx = \sum_{k=1}^{m} \int_a^b f_k(x)\,dx$.

5-5 Prove Proposition 5.12. You may use the result of Exercise 5-4

5-6 Let f, g $[a, b] \to \mathbb{R}$ be Riemann integrable. Prove that $f - g$ is Riemann integrable and $\int_a^b (f - g)(x)\,dx = \int_a^b f(x)\,dx - \int_a^b g(x)\,dx$.

5-7 Let $a < b$ and let $c, d \in [a, b]$ with $c < d$ Prove that $\mathbf{1}_{[c,d]}$ is Riemann integrable over $[a, b]$ and $\int_a^b \mathbf{1}_{[c,d]}\,dx = d - c$

5-8. Prove that if c is a constant, then $f_c(x) := c$ is Riemann integrable and $\int_a^b f_c(x)\,dx = c(b - a)$.

5-9. Let $f \cdot [a, b] \to \mathbb{R}$ be Riemann integrable and let $m, M \in \mathbb{R}$ be such that for all $x \in [a, b]$ we have $m \leq f(x) \leq M$ Prove that $m(b - a) \leq \int_a^b f(x)\,dx \leq M(b - a)$.

5-10 In the definitions of the Riemann integral and of Riemann sums we demand that the function is bounded. We could define Riemann sums for unbounded functions also, but this exercise shows that the **Riemann integral cannot be defined for unbounded functions**

Let $f : [a, b] \to \mathbb{R}$ be a function Prove that if there is a number I such that for all $\varepsilon > 0$ there is a $\delta > 0$ so that for all partitions P with $\|P\| < \delta$ and all associated evaluation sets T we have $\left| R(f, P, T) - I \right| < \varepsilon$, then f must be bounded.

Hint. Assume the function is not bounded and choose the right partitions and evaluation sets to obtain a contradiction to Riemann integrability

5-11 **Cauchy Criterion** for Riemann integrability. Let $f : [a, b] \to \mathbb{R}$ be a function. Prove that if for all $\varepsilon > 0$ there is a $\delta > 0$ so that for all partitions P, Q with $\|P\|, \|Q\| < \delta$ and all associated evaluation sets T, U we have $\left| R(f, P, T) - R(f, Q, U) \right| < \varepsilon$, then f must be Riemann integrable.

5-12. **Riemann-Stieltjes integrals** Let $[a, b]$ be a closed interval, let $g : [a, b] \to \mathbb{R}$ be nondecreasing and let $f : [a, b] \to \mathbb{R}$ be bounded For any partition $P = \{a = x_0 < x_1 < \cdots < x_n = b\}$ and associated evaluation set $T = \{t_1, \ldots, t_n\}$, define the **Riemann-Stieltjes sum** of f with respect to the partition P, evaluation set T and **integrator** g as $S_g(f, P, T) := \sum_{i=1}^{n} f(t_i)\left(g(x_i) - g(x_{i-1})\right)$.

We will also use the notation $\Delta g_i = g(x_i) - g(x_{i-1})$.

The **integrand** f is called **Riemann-Stieltjes integrable** on $[a, b]$ with respect to the **integrator** g iff f is bounded and there is a number I such that for all $\varepsilon > 0$ there is a $\delta > 0$ so that for all partitions P with $\|P\| < \delta$ and all associated evaluation sets T we have $\left| S_g(f, P, T) - I \right| < \varepsilon$.

The number I will be called the **Riemann-Stieltjes integral** of f, denoted $\int_a^b f\,dg := I$.

(a) Prove that if f_1, f_2 are Riemann-Stieltjes integrable with respect to g, then f_1+f_2 is Riemann-Stieltjes integrable with respect to g and $\int_a^b (f_1 + f_2)\, dg = \int_a^b f_1\, dg + \int_a^b f_2\, dg$.

(b) Prove that if f is Riemann-Stieltjes integrable with respect to g and $c \in \mathbb{R}$, then cf is Riemann-Stieltjes integrable with respect to g and $\int_a^b cf\, dg = c\int_a^b f\, dg$

(c) Prove that if $g : [0, 1] \to \mathbb{R}$ is the indicator function $\mathbf{1}_{\left[\frac{1}{2}, 1\right]}$, then the function $f = \mathbf{1}_{\left[0, \frac{1}{2}\right)}$ is *not* Riemann-Stieltjes integrable with respect to g.

5.2 Uniform Continuity and Integrability of Continuous Functions

Unlike continuity and differentiability, Riemann integrability is not easily checked. Therefore, we need to invest some effort into finding criteria for Riemann integrability that are simpler than the definition. The first such criterion is that continuity implies Riemann integrability.

The fundamental idea behind the Squeeze Theorem (see Theorem 2.21) is that if a sequence is trapped between two other sequences that converge to the same limit, then the sequence in the middle must converge to that limit, too. To use this idea in our investigation of Riemann integrability, we define lower and upper bounds for all Riemann sums associated with a given partition.

Definition 5.13 *Let $f : [a, b] \to \mathbb{R}$ be bounded, let $P = \{a = x_0 < \ldots < x_n = b\}$ be a partition of $[a, b]$ and for each $i \in \{1, \ldots, n\}$ let*

$$m_i := \inf\left\{f(x) : x \in [x_{i-1}, x_i]\right\} \quad \text{and} \quad M_i := \sup\left\{f(x) : x \in [x_{i-1}, x_i]\right\}.$$

We define the **lower sum** *of f with respect to P to be $L(f, P) := \sum_{i=1}^{n} m_i \Delta x_i$ and we define the* **upper sum** *of f with respect to P to be $U(f, P) := \sum_{i=1}^{n} M_i \Delta x_i$.*

Because the supremum and the infimum in Definition 5.13 need not be assumed by f, lower and upper sums need not be Riemann sums of f. However, the next result shows that the Riemann sums are "trapped" between appropriate upper and lower sums.

Lemma 5.14 *Let $f : [a, b] \to \mathbb{R}$ be bounded and let $P = \{a = x_0, \ldots, x_n = b\}$ be a partition of $[a, b]$. Then for all associated evaluation sets $T = \{t_1, \ldots, t_n\}$ the inequalities $L(f, P) \leq R(f, P, T) \leq U(f, P)$ hold.*

Proof. Because $t_i \in [x_{i-1}, x_i]$ for all $i \in \{1, \ldots, n\}$ we have $m_i \leq f(t_i) \leq M_i$ for all $i \in \{1, \ldots, n\}$. Hence, $\sum_{i=1}^{n} m_i \Delta x_i \leq \sum_{i=1}^{n} f(t_i)\Delta x_i \leq \sum_{i=1}^{n} M_i \Delta x_i$, as claimed. ■

To prove that a function is Riemann integrable, we would need to show that lower and upper sums get closer to a certain limit as the norm of the partition tends to zero. The following monotonicity properties are a first step in this direction.

Definition 5.15 *Let P, Q be partitions of $[a,b]$. Then Q is called a* **refinement** *of P iff $P \subseteq Q$.*

Lemma 5.16 *Let $f : [a,b] \to \mathbb{R}$ be bounded and let P, Q be partitions of $[a,b]$ so that Q is a refinement of P. Then $L(f,P) \leq L(f,Q) \leq U(f,Q) \leq U(f,P)$.*

Proof. From Lemma 5.14, we know that $L(f,Q) \leq U(f,Q)$. Therefore we can focus on the other two inequalities. Let $\{z_1, \ldots, z_k\} := Q \setminus P$. Then with $Q_0 := P$ and $Q_j := Q_{j-1} \cup \{z_j\}$ for $j = 1, \ldots, k$ we have that each Q_j is a refinement of Q_{j-1} and $Q_k = Q$. Thus we are done if we can prove the outer inequalities for $Q = P \cup \{z\}$ for some $z \in [a,b] \setminus P$. We will only prove $L(f,P) \leq L(f,Q)$, leaving the other inequality to Exercise 5-13. Let $P = \{a = x_0 < x_1 < \cdots < x_n = b\}$ and let $j \in \{1, \ldots, n\}$ be such that $z \in [x_{j-1}, x_j]$. Then

$$L(f,P) = \sum_{i=1}^{n} m_i \Delta x_i = \left(\sum_{i=1}^{j-1} m_i \Delta x_i \right) + m_j \Delta x_j + \left(\sum_{i=j+1}^{n} m_i \Delta x_i \right)$$

$m_j = \inf\{f(x) : x \in [x_{j-1}, x_j]\}$, so by Exercise 1-18b we have $m_j \leq \inf\{f(x) : x \in [x_{j-1}, z]\}$ and $m_j \leq \inf\{f(x) : x \in [z, x_j]\}$.

$$\leq \left(\sum_{i=1}^{j-1} m_i \Delta x_i \right) + \inf\{f(x) : x \in [x_{j-1}, z]\}(z - x_{j-1})$$

$$+ \inf\{f(x) : x \in [z, x_j]\}(x_j - z) + \left(\sum_{i=j+1}^{n} m_i \Delta x_i \right)$$

$$= L(f,Q).$$

■

Lemma 5.17 *Let $f : [a,b] \to \mathbb{R}$ be bounded and let P, Q be partitions of $[a,b]$. Then $L(f,Q) \leq U(f,P)$.*

Proof. Because $P \cup Q$ is a refinement of P and of Q, the proof follows from Lemma 5.16: $L(f,Q) \leq L(f, P \cup Q) \leq U(f, P \cup Q) \leq U(f,P)$. ■

Turning to continuous functions, we note that continuity guarantees that for every point x and every $\varepsilon > 0$, there is an interval $(x - \delta, x + \delta)$ so that the inequality $\sup\{f(z) : z \in (x - \delta, x + \delta)\} - \inf\{f(z) : z \in (x - \delta, x + \delta)\} < \varepsilon$ holds. With this difference expected to be small, the difference between upper and lower sums of continuous functions should also become small as the partitions become finer. Unfortunately, while the norm of a partition is given for the whole interval, the δ may vary from point to point. To prove that continuous functions are Riemann integrable, we need to show that for any $\varepsilon > 0$ we can choose a δ as above that works at *every* point of the closed interval $[a,b]$. This is the idea behind uniform continuity.

Definition 5.18 *Let $J \subseteq \mathbb{R}$ be an interval. The function $f : J \to \mathbb{R}$ is called* **uniformly continuous** *iff for every $\varepsilon > 0$ there is a $\delta > 0$ such that for all $u, v \in J$ with $|u - v| < \delta$ we have $|f(u) - f(v)| < \varepsilon$.*

Lemma 5.19 *Let $f : [a, b] \to \mathbb{R}$ be continuous. Then f is uniformly continuous.*

Proof. Suppose for a contradiction that f is not uniformly continuous. Then there is an $\varepsilon > 0$ such that for all $\delta > 0$ there are $u, v \in [a, b]$ such that $|u - v| < \delta$ and $\left|f(u) - f(v)\right| \geq \varepsilon$. Therefore, for all $k \in \mathbb{N}$ with $\delta_k := \frac{1}{k}$ we can find numbers $u_k, v_k \in [a, b]$ so that $|u_k - v_k| < \frac{1}{k}$ and $\left|f(u_k) - f(v_k)\right| \geq \varepsilon$. By the Bolzano-Weierstrass Theorem, the bounded sequence $\{u_k\}_{k=1}^{\infty}$ has a subsequence $\{u_{k_m}\}_{m=1}^{\infty}$ that converges to a $t \in [a, b]$. Because for all $m \in \mathbb{N}$ the inequality $\left|u_{k_m} - v_{k_m}\right| < \frac{1}{k_m}$ holds, we conclude that $\lim_{m\to\infty} v_{k_m} = \lim_{m\to\infty} u_{k_m} = t$. But then, because f is continuous, $\lim_{m\to\infty} f\left(v_{k_m}\right) = f(t) = \lim_{m\to\infty} f\left(u_{k_m}\right)$, which means $\lim_{m\to\infty} \left|f\left(v_{k_m}\right) - f\left(u_{k_m}\right)\right| = 0$, a contradiction to $\left|f(u_k) - f(v_k)\right| \geq \varepsilon$ for all natural numbers k. ■

Uniform continuity was the last piece in the puzzle to establish that continuous functions on closed and bounded intervals are Riemann integrable.

Theorem 5.20 *Let $f : [a, b] \to \mathbb{R}$ be continuous. Then f is Riemann integrable.*

Proof. By Theorem 3.44, there is a real number M such that $f(x) \leq M$ for all $x \in [a, b]$. Then for all partitions P of $[a, b]$ we have that $L(f, P) \leq M(b - a)$. Set $I := \sup\left\{L(f, P) : P \text{ is a partition of } [a, b]\right\}$. Because by Lemma 5.17 for any partitions P, Q of $[a, b]$ we have $L(f, Q) \leq U(f, P)$, we infer $L(f, P) \leq I \leq U(f, P)$ for all partitions P of $[a, b]$. By Lemma 5.14 for all evaluation sets T, we have $L(f, P) \leq R(f, P, T) \leq U(f, P)$, so by Exercise 1-13, for all partitions P of $[a, b]$ the inequality $\left|R(f, P, T) - I\right| \leq U(f, P) - L(f, P)$ holds.

Now let $\varepsilon > 0$. By Lemma 5.19, there is a $\delta > 0$ such that for all $x, y \in [a, b]$ with $|x - y| < \delta$ we have $\left|f(x) - f(y)\right| < \frac{\varepsilon}{b - a}$. Let $P = \{a = x_0 < x_1 < \ldots < x_n = b\}$ be any partition of $[a, b]$ with $\|P\| < \delta$. Because f is continuous on $[x_{i-1}, x_i]$, by Theorem 3.44 and Exercise 3-38 there are $t_i^L, t_i^U \in [x_{i-1}, x_i]$ so that $f\left(t_i^L\right) = m_i$ and $f\left(t_i^U\right) = M_i$. Therefore

$$\begin{aligned}
\left|R(f, P, T) - I\right| &\leq U(f, P) - L(f, P) = \sum_{i=1}^{n} f\left(t_i^U\right) \Delta x_i - \sum_{i=1}^{n} f\left(t_i^L\right) \Delta x_i \\
&= \sum_{i=1}^{n} \left[f\left(t_i^U\right) - f\left(t_i^L\right)\right] \Delta x_i < \sum_{i=1}^{n} \frac{\varepsilon}{b - a} \Delta x_i = \varepsilon.
\end{aligned}$$

By Definition 5.5 f is Riemann integrable with $\int_a^b f(x)\, dx = I$. ■

We conclude this section by proving that for continuous functions the "right" choice of an evaluation point will produce the value of the integral. Proposition 5.21 is a lemma needed to establish the result, which is stated in Theorem 5.22.

Proposition 5.21 *Let $f, g : [a, b] \to \mathbb{R}$ be Riemann integrable. If $f(x) \leq g(x)$ for all $x \in [a, b]$, then $\int_a^b f(x)\,dx \leq \int_a^b g(x)\,dx$.*

Proof. Let $\varepsilon > 0$ and let $\delta > 0$ be such that for all partitions $P = \{a = x_0 < x_1 < \cdots < x_n = b\}$ with $\|P\| < \delta$ and all evaluation sets $T = \{t_1, \ldots, t_n\}$ we have $\left|\int_a^b f(x)\,dx - \sum_{i=1}^n f(t_i)\Delta x_i\right| < \frac{\varepsilon}{2}$ and $\left|\int_a^b g(x)\,dx - \sum_{i=1}^n g(t_i)\Delta x_i\right| < \frac{\varepsilon}{2}$. Then with P being a partition with $\|P\| < \delta$ and T being an evaluation set we obtain

$$\begin{aligned}\int_a^b g(x)\,dx &\geq \sum_{i=1}^n g(t_i)\Delta x_i - \frac{\varepsilon}{2} \geq \sum_{i=1}^n f(t_i)\Delta x_i - \frac{\varepsilon}{2} \\ &\geq \left(\int_a^b f(x)\,dx - \frac{\varepsilon}{2}\right) - \frac{\varepsilon}{2} = \int_a^b f(x)\,dx - \varepsilon,\end{aligned}$$

which proves the result (see Standard Proof Technique 2.7). ∎

Theorem 5.22 Mean Value Theorem for the Integral. *Let $f : [a, b] \to \mathbb{R}$ be continuous. Then there is a $c \in (a, b)$ so that $\int_a^b f(x)\,dx = f(c)(b - a)$.*

Proof. Let $M = \max\{f(x) : x \in [a, b]\}$, $m := \min\{f(x) : x \in [a, b]\}$ and let $x_m, x_M \in [a, b]$ be such that $f(x_m) = m$ and $f(x_M) = M$. Then the constant functions $k_m(x) := m$ and $k_M(x) := M$ are continuous, and hence Riemann integrable over $[a, b]$. Moreover, $k_m(x) \leq f(x) \leq k_M(x)$ for all $x \in [a, b]$ and it is easy to show (see Exercise 5-8) that if k is a constant, then $\int_a^b k\,dx = k(b - a)$.

Therefore by Proposition 5.21 we obtain $m(b - a) \leq \int_a^b f(x)\,dx \leq M(b - a)$. But then $f(x_m) = m \leq \frac{\int_a^b f(x)\,dx}{b - a} \leq M = f(x_M)$ and by the Intermediate Value Theorem there must be a c between x_m and x_M so that $f(c) = \frac{\int_a^b f(x)\,dx}{b - a}$. ∎

Exercises

5-13. Finish the proof of Lemma 5.16 by proving that if $f : [a, b] \to \mathbb{R}$ is bounded, P is a partition of $[a, b]$ and $Q = P \cup \{z\}$ for some $z \in [a, b] \setminus P$, then $U(f, Q) \leq U(f, P)$.

5-14. Prove that $f : (0, 1] \to \mathbb{R}$ defined by $f(x) = \frac{1}{x}$ is continuous but not uniformly continuous. Then explain why this example is not a contradiction to Lemma 5.19.

5-15. A function $f : [a, b] \to \mathbb{R}$ is called **Lipschitz continuous** iff there is an $L > 0$ such that for all $x, y \in [a, b]$ the inequality $\left|f(x) - f(y)\right| \leq L|x - y|$ holds.

(a) Prove that if $f : [a, b] \to \mathbb{R}$ is Lipschitz continuous, then f is uniformly continuous.

(b) Prove that if $f : (c, d) \to \mathbb{R}$ is differentiable and f' is bounded on $[a, b] \subset (c, d)$, then f is Lipschitz continuous on $[a, b]$.

(c) Prove that $f(x) := \sqrt{x}$ is uniformly continuous on $[0, 1]$, but not Lipschitz continuous

5-16. Prove that if $f : [a, b] \to [0, \infty)$ is a continuous nonnegative function with $\int_a^b f(x)\,dx = 0$, then $f(x) = 0$ for all $x \in [a, b]$.

5-17. Let $f : [a, b] \to \mathbb{R}$ be Riemann integrable and let $\{P_k\}_{k=1}^{\infty}$ be a sequence of partitions with $\lim_{k\to\infty} \|P_k\| = 0$ Prove that $\lim_{k\to\infty} L(f, P_k) = \int_a^b f(x)\,dx$.

5-18. Prove that if $f : [a, b] \to [0, \infty)$ is a nonnegative Riemann integrable function with $\int_a^b f(x)\,dx > 0$, then there are an $\varepsilon > 0$ and $c < d$ so that $f(x) \geq \varepsilon$ for all $x \in [c, d]$.
Hint. For a contradiction, suppose the opposite Use lower sums to show the integral would be zero.

5-19 Let $f : [a, b] \to \mathbb{R}$ be bounded and let $g : [a, b] \to \mathbb{R}$ be nondecreasing With m_i and M_i as in Definition 5 13, let $L_g(f, P) := \sum_{i=1}^{n} m_i \Delta g_i$ be the **lower sum** and let $U_g(f, P) := \sum_{i=1}^{n} M_i \Delta g_i$ be the **upper sum** of f with respect to g and P.

(a) Let P, Q be partitions of $[a, b]$ so that Q is a refinement of P.
Prove that $L_g(f, P) \leq L_g(f, Q) \leq U_g(f, Q) \leq U_g(f, P)$.

(b) Prove that if f is continuous, then f is Riemann-Stieltjes integrable with respect to g

(c) Let $f_1, f_2 : [a, b] \to \mathbb{R}$ be Riemann-Stieltjes integrable with respect to g. Prove that if $f_1(x) \leq f_2(x)$ for all $x \in [a, b]$, then $\int_a^b f_1\,dg \leq \int_a^b f_2\,dg$

(d) **Mean Value Theorem for Riemann Stieltjes Integrals.** Let $f : [a, b] \to \mathbb{R}$ be continuous. Prove that there is a $c \in (a, b)$ so that $\int_a^b f\,dg = f(c)\big(g(b) - g(a)\big)$.

(e) Let $c \in (a, b)$ and let $g := \mathbf{1}_{[c,b]}$.

i Prove that if $f : [a, b] \to \mathbb{R}$ is continuous at c, then f is Riemann-Stieltjes integrable with respect to g and $\int_a^b f\,d\mathbf{1}_{[c,b]} = f(c)$.

ii Prove that if $f : [a, b] \to \mathbb{R}$ is not continuous at c, then f is not Riemann-Stieltjes integrable with respect to g.

(f) Prove that if f is continuous and g is continuously differentiable, then, with the integral on the right being a Riemann integral, we have $\int_a^b f\,dg = \int_a^b f(x)g'(x)\,dx$.
Note. Integrals against dg are often used to abbreviate Riemann integrals as on the right side.

5.3 The Fundamental Theorem of Calculus

The Fundamental Theorem of Calculus connects differential and integral calculus by stating that differentiation and integration are essentially inverses of each other.

Theorem 5.23 Fundamental Theorem of Calculus, Antiderivative Form. *Let $[a, b]$ be contained in (c, d) and let $F : (c, d) \to \mathbb{R}$ be a differentiable function whose derivative f is Riemann integrable on $[a, b]$. Then* $\int_a^b f(x)\,dx = F(b) - F(a) =: F(x)\Big|_a^b$.

Proof. Because f is Riemann integrable, the integral exists. Hence, we can use Lemma 5.6 to prove the equation. For $k \in \mathbb{N}$, let $P_k = \left\{a = x_0^{(k)} < \cdots < x_{n_k}^{(k)} = b\right\}$ be a partition with $\|P_k\| = \frac{1}{k}$. By the Mean Value Theorem (see Theorem 4.18), for each $k \in \mathbb{N}$ and each $i \in \{1, \ldots, n_k\}$ there is a point $t_i^{(k)} \in \left[x_{i-1}^{(k)}, x_i^{(k)}\right]$ such that the equation $f\left(t_i^{(k)}\right) = F'\left(t_i^{(k)}\right) = \dfrac{F\left(x_i^{(k)}\right) - F\left(x_{i-1}^{(k)}\right)}{x_i^{(k)} - x_{i-1}^{(k)}}$ holds. Therefore we obtain $f\left(t_i^{(k)}\right) \Delta x_i^{(k)} = F\left(x_i^{(k)}\right) - F\left(x_{i-1}^{(k)}\right)$. Let $T_k := \left\{t_i^{(k)} : i = 1, \ldots, n_k\right\}$. Then T_k is an evaluation set for P_k and by Lemma 5.6 we conclude via a telescoping sum

$$\begin{aligned}
\int_a^b f(x)\,dx &= \lim_{k\to\infty} R(f, P_k, T_k) = \lim_{k\to\infty} \sum_{i=1}^{n_k} f\left(t_i^{(k)}\right) \Delta x_i^{(k)} \\
&= \lim_{k\to\infty} \sum_{i=1}^{n_k} F\left(x_i^{(k)}\right) - F\left(x_{i-1}^{(k)}\right) = \lim_{k\to\infty} \sum_{i=1}^{n_k} F\left(x_i^{(k)}\right) - \sum_{i=0}^{n_k-1} F\left(x_i^{(k)}\right) \\
&= \lim_{k\to\infty} F\left(x_{n_k}^{(k)}\right) - F\left(x_0^{(k)}\right) = F(b) - F(a),
\end{aligned}$$

which proves the result. ■

Definition 5.24 *Because of the Fundamental Theorem of Calculus, if F and f are functions with $F' = f$, then F will also be called an* **indefinite integral** *of f, denoted $F = \int f(x)\,dx$. One way to compute Riemann integrals is to evaluate an indefinite integral at the upper and lower bound and to compute the difference.*

The hypothesis that the derivative of F is Riemann integrable feels quite artificial. Nonetheless, this hypothesis is best possible. Exercise 12-24 will exhibit a differentiable function whose derivative is bounded, but not Riemann integrable. Although this example is a bit pathological, it points out a weakness of the Riemann integral that motivates the development of the Lebesgue integral. The Antiderivative Form of the Fundamental Theorem of Calculus for the Lebesgue integral, which has no artificial looking hypotheses, will be proved in Exercise 23-8. The Derivative Form of the Fundamental Theorem of Calculus is proved in Theorem 8.17 for the Riemann integral and in Exercise 18-6 for the Lebesgue integral.

Exercises

5-20 **Power Rule** for integration Let $r \in \mathbb{Q} \setminus \{-1\}$ and let $a < b$ In case $r < 0$, let a and b either both be positive or both be negative. Prove that $\int_a^b x^r\,dx = \frac{1}{r+1} b^{r+1} - \frac{1}{r+1} a^{r+1}$ Then explain why we needed to require a and b to be both positive or both negative for $r < 0$

5-21. **Integration by Parts**. Let $[a, b] \subset (c, d)$ and let $F, g : (c, d) \to \mathbb{R}$ be continuously differentiable with derivatives f and g' Prove that $\int_a^b f(x)g(x)\,dx = F(b)g(b) - F(a)g(a) - \int_a^b F(x)g'(x)\,dx$.

5-22. **Integration by Substitution**. Let $[a, b] \subset (c, d)$, let $g : (c, d) \to \mathbb{R}$ be continuously differentiable with derivative g' and let F be continuously differentiable with derivative f such that the domain of F contains $g\left[[a, b]\right]$. Prove that $\int_a^b f\left(g(x)\right)g'(x)\,dx = F\left(g(b)\right) - F\left(g(a)\right)$

5-23. What would we need to prove so that the hypothesis that the derivatives are continuous in Exercises 5-21 and 5-22 can be replaced with the hypothesis that the derivatives are Riemann integrable?

5.4 The Darboux Integral

Lemma 5.6 is an efficient tool to establish properties of the Riemann integral, provided all functions involved are Riemann integrable. It is now time to look for a similarly efficient criterion to prove that a function is Riemann integrable.

Riemann's Condition below is inspired by the idea of trapping the Riemann sums between lower and upper sums, which was already used in the proof that continuous functions are Riemann integrable. Riemann's Condition is simpler to verify than Definition 5.5 of Riemann integrability, because, for $\varepsilon > 0$, instead of working with *all* partitions of sufficiently small norm and *all* evaluation sets, we only need to find *one* partition so that the upper and lower sums are closer together than ε. The price is paid in the proof, where for the "$\Leftarrow$" direction, our only tool is one partition and we must prove something for all partitions of sufficiently small norm.

Theorem 5.25 Riemann's Condition. *Let $f : [a, b] \to \mathbb{R}$ be a bounded function. Then f is Riemann integrable on $[a, b]$ iff for all $\varepsilon > 0$ there is a partition P of $[a, b]$ such that $U(f, P) - L(f, P) < \varepsilon$.*

Proof. For "$\Leftarrow$," let $f : [a, b] \to \mathbb{R}$ be bounded and such that for all $\varepsilon > 0$ there is a partition P of $[a, b]$ such that $U(f, P) - L(f, P) < \varepsilon$. By Lemma 5.17 the set $B := \left\{L(f, P) : P \text{ is a partition of } [a, b]\right\}$ is bounded above. Let $\mathcal{L} := \sup B$ and let $\varepsilon > 0$. Let $P = \{a = x_0 < x_1 < \ldots < x_n = b\}$ be a partition of $[a, b]$ such that $U(f, P) - L(f, P) < \frac{\varepsilon}{2}$. Then for all refinements Q of P and all evaluation sets T for Q, Lemmas 5.14 and 5.16 imply

$$\left|R(f, Q, T) - \mathcal{L}\right| \leq U(f, Q) - L(f, Q) \leq U(f, P) - L(f, P) < \frac{\varepsilon}{2}.$$

To show that f is Riemann integrable, let $M := \sup\left\{|f(x)| : x \in [a, b]\right\}$ and let $\delta := \min\left\{\frac{\varepsilon}{4n(M+1)}, \frac{\Delta x_1}{3}, \ldots, \frac{\Delta x_n}{3}\right\}$. We will now show that for all partitions S with $\|S\| < \delta$ and all associated evaluation sets T_S we have $\left|R(f, S, T_S) - \mathcal{L}\right| < \varepsilon$.

Let $S = \left\{a = x_0^S < x_1^S < \cdots < x_{n_S}^S = b\right\}$ be any partition of $[a, b]$ with $\|S\| < \delta$. Then $Q : S \cup P$ is a refinement of P. Therefore, for any evaluation set T for Q we have $\left|R(f, Q, T) - \mathcal{L}\right| < \frac{\varepsilon}{2}$. Moreover, by choice of δ, every interval $\left[x_{j-1}^S, x_j^S\right]$ contains at most one point of P that is not in S and any two intervals $\left[x_{j-1}^S, x_j^S\right]$ that contain such a point do not intersect. Let T_S be any evaluation set for S and let T be an evaluation set for Q that contains T_S. Then $T \setminus T_S = \{t_{i_1}, \ldots, t_{i_k}\}$ with $k < n$, because the addition of at most $n-1$ points to S that are all in distinct intervals $\left[x_{j-1}^S, x_j^S\right]$ adds at most $n-1$ intervals that do not already have an evaluation point assigned. Therefore

$$
\begin{aligned}
&\left|R(f,Q,T)-R(f,S,T_S)\right| \\
&= \left|\sum_{t_m\in T} f(t_m)\,\Delta x_m - \sum_{t_l^S\in T_S} f\left(t_l^S\right)\Delta x_l^S\right| \\
&= \left|\sum_{t_m\in T_S} f(t_m)\,\Delta x_m + \sum_{j=1}^{k} f\left(t_{i_j}\right)\Delta x_{i_j} - \sum_{t_l^S\in T_S} f\left(t_l^S\right)\Delta x_l^S\right|
\end{aligned}
$$

For each $t_m \in T$ that is in T_S, we have $t_m = t_{l(m)}^S$ for some $l(m)$. Moreover, because each interval $\left[x_{m-1}^S, x_m^S\right]$ contains at most one point of $T\setminus T_S$ there is a function $t : \{1,\ldots,k\}\to T_S$ so that if $t_m = t(j)$ then $\Delta x_m - \Delta x_{l(m)}^S = -\Delta x_{i_j}$ and if $t_m \notin t[\{1,\ldots,k\}]$ then $\Delta x_m = \Delta x_{l(m)}^S$.

$$
\begin{aligned}
&= \left|\sum_{j=1}^{k} f\left(t_{i_j}\right)\Delta x_{i_j} - \sum_{j=1}^{k} f(t(j))\Delta x_{i_j}\right| \le \sum_{j=1}^{k} 2M\Delta x_{i_j} \le \sum_{j=1}^{k} 2M\|S\| \\
&< \sum_{j=1}^{k} 2M\frac{\varepsilon}{4n(M+1)} = \sum_{j=1}^{k}\frac{\varepsilon}{2n} \le \frac{\varepsilon}{2},
\end{aligned}
$$

and hence

$$
\begin{aligned}
\left|R(f,S,T_S)-\mathcal{L}\right| &\le \left|R(f,S,T_S)-R(f,Q,T)\right| + \left|R(f,Q,T)-\mathcal{L}\right| \\
&< \frac{\varepsilon}{2}+\frac{\varepsilon}{2}=\varepsilon.
\end{aligned}
$$

Thus f is Riemann integrable and its Riemann integral is $\mathcal{L}$.

For "$\Rightarrow$," let $f:[a,b]\to\mathbb{R}$ be Riemann integrable and let $\varepsilon>0$. Then there is a number I such that for all $\varepsilon>0$ there is a $\delta>0$ such that for all partitions P with $\|P\|<\delta$ and all evaluation sets T we have that $\left|R(f,P,T)-I\right|<\frac{\varepsilon}{4}$. Let $P=\{a=x_0<x_1<\ldots<x_n=b\}$ be such a partition and for each $i\in\{1,\ldots,n\}$ find $t_i^L, t_i^U\in[x_{i-1},x_i]$ such that $f\left(t_i^L\right)\le m_i+\frac{\varepsilon}{4n\Delta x_i}$ and $f\left(t_i^U\right)\ge M_i-\frac{\varepsilon}{4n\Delta x_i}$. Let $T^L:=\left\{t_i^L : i=1,\ldots,n\right\}$ and $T^U:=\left\{t_i^U : i=1,\ldots,n\right\}$. Then

$$
\begin{aligned}
&U(f,P)-L(f,P) \\
&\le \left[U(f,P)-R\left(f,P,T^U\right)\right]+\left|R\left(f,P,T^U\right)-I\right| \\
&\quad +\left|I-R\left(f,P,T^L\right)\right|+\left[R\left(f,P,T^L\right)-L(f,P)\right]
\end{aligned}
$$

$$< \sum_{i=1}^{n}\left(M_i - f\left(t_i^U\right)\right)\Delta x_i + \frac{\varepsilon}{4} + \frac{\varepsilon}{4} + \sum_{i=1}^{n}\left(f\left(t_i^L\right) - m_i\right)\Delta x_i$$

$$< \sum_{i=1}^{n}\frac{\varepsilon}{4n\Delta x_i}\Delta x_i + \frac{\varepsilon}{2} + \sum_{i=1}^{n}\frac{\varepsilon}{4n\Delta x_i}\Delta x_i = \frac{\varepsilon}{4} + \frac{\varepsilon}{2} + \frac{\varepsilon}{4} = \varepsilon,$$

which was to be proved. ■

Riemann's Condition shows that if the lower and upper sums of a function f can get arbitrarily close to each other, then f is Riemann integrable. The only way this cannot happen is if the function oscillates too much, that is, if the function is highly discontinuous.

Example 5.26 *The* **Dirichlet function** $f(x) = \begin{cases} 0; & \textit{for } x \in [0,1]\setminus\mathbb{Q}, \\ 1; & \textit{for } x \in [0,1]\cap\mathbb{Q}, \end{cases}$ *is not Riemann integrable on* $[0, 1]$.

All lower sums of the Dirichlet function are zero and all upper sums are 1. By Theorem 5.25 the Dirichlet function is not Riemann integrable. □

It is natural to ask now how much oscillation or discontinuity there can be without losing Riemann integrability. Moreover, there are further important results about Riemann integrals that could be presented here. Theorem 5.25 could be used to prove these results, but the proofs would involve substantial work with Riemann sums. We will avoid this work by first using Theorem 5.25 to prove the Lebesgue criterion for Riemann integrability in Theorem 8.12. This criterion makes the proofs of further results about Riemann integrals, presented in Section 8.3, more effective. Moreover, the Lebesgue criterion will answer how discontinuous a Riemann integrable function can be.

To formulate the Lebesgue criterion, we need to introduce the Lebesgue measure of a set, which is fundamental for more advanced analysis. To define Lebesgue measure we need "infinite summations" (introduced in Chapter 6) and some set theoretical notions of "size" for sets (introduced in Chapter 7). In Chapter 8, we will continue where we leave off here.

To conclude this chapter, we note that the approximation of the area under a function with lower and upper sums is also known as Darboux integration. To acquaint the reader with the language involved, we formally introduce Darboux integration, which is equivalent to Riemann integration.

Definition 5.27 *Let* $f : [a,b] \to \mathbb{R}$ *be a bounded function. We define*

$$\begin{aligned} \mathcal{L}_f &:= \sup\left\{L(f,P) : P \text{ is a partition of } [a,b]\right\}, \\ \mathcal{U}_f &:= \inf\left\{U(f,P) : P \text{ is a partition of } [a,b]\right\}. \end{aligned}$$

$\mathcal{L}_f$ *is also called the* **lower integral** *of* f *and* $\mathcal{U}_f$ *is also called the* **upper integral** *of* f. *We will say that* f *is* **Darboux integrable** *on* $[a,b]$ *iff* $\mathcal{L}_f = \mathcal{U}_f$. *In this case* $\mathcal{L}_f = \mathcal{U}_f$ *is also called the* **Darboux integral** *of* f.

Proposition 5.28 *Let* $f : [a,b] \to \mathbb{R}$ *be a bounded function. Then* $\mathcal{L}_f \leq \mathcal{U}_f$.

Proof. Easy consequence of Lemma 5.17 (Exercise 5-24). ■

Theorem 5.29 *Let $f : [a,b] \to \mathbb{R}$ be a bounded function. Then f is Darboux integrable on $[a,b]$ iff f is Riemann integrable on $[a,b]$. Moreover, the Darboux and Riemann integrals are equal in this case, that is,* $\int_a^b f(x)\,dx = \mathcal{L} = \mathcal{U}$.

Proof. For "$\Rightarrow$," let $f : [a,b] \to \mathbb{R}$ be Darboux integrable and let $\varepsilon > 0$. Then $\mathcal{U} = \mathcal{L}$ and there are partitions Q and R of $[a,b]$ such that $\mathcal{L} - \frac{\varepsilon}{2} < L(f,Q)$ and $U(f,R) < \mathcal{U} + \frac{\varepsilon}{2}$. Now $P := Q \cup R$ is a refinement of Q and R and by Lemma 5.16 we infer

$$\mathcal{L} - \frac{\varepsilon}{2} < L(f,Q) \le L(f,P) \le \mathcal{L} = \mathcal{U} \le U(f,P) \le U(f,R) < \mathcal{U} + \frac{\varepsilon}{2} = \mathcal{L} + \frac{\varepsilon}{2},$$

and hence $U(f,P) - L(f,P) < \varepsilon$. By Theorem 5.25 f is Riemann integrable.

For "$\Leftarrow$," let $f : [a,b] \to \mathbb{R}$ be Riemann integrable. Then by Theorem 5.25 for all $\varepsilon > 0$ there is a partition P of $[a,b]$ such that $U(f,P) - L(f,P) < \varepsilon$. Hence, for any $\varepsilon > 0$ we infer $\mathcal{U} - \mathcal{L} \le U(f,P) - L(f,P) < \varepsilon$, which means $\mathcal{U} = \mathcal{L}$.

The "moreover" part follows upon examination of the "$\Leftarrow$" part of the proof of Theorem 5.25, which also shows that the Riemann integral of f is equal to $\mathcal{L}$. ■

Exercises

5-24. Prove Proposition 5.28.

5-25. Let $f, g : [a,b] \to \mathbb{R}$ be Darboux integrable. Prove that then $f+g$ also is Darboux integrable and $\mathcal{L}_{f+g} = \mathcal{L}_f + \mathcal{L}_g$.

5-26. Let $f : [a,b] \to \mathbb{R}$ be Riemann integrable on $[a,b]$.

(a) For each $n \in \mathbb{N}$, let $P_n = \left\{a = x_0^{(n)} < \cdots < x_{k_n}^{(n)} = b\right\}$ be a partition of the interval $[a,b]$ with $\|P_n\| < \frac{1}{n}$ and let $s_n := \left(\sum_{i=1}^{k_n - 1} m_i^{(n)} \mathbf{1}_{\left[x_{i-1}^{(n)}, x_i^{(n)}\right)}\right) + m_{k_n}^{(n)} \mathbf{1}_{\left[x_{k_n-1}^{(n)}, x_{k_n}^{(n)}\right]}$, where $m_i^{(n)} = \inf\left\{f(x) : x \in \left[x_{i-1}^{(n)}, x_i^{(n)}\right]\right\}$

i. Prove that for all $n \in \mathbb{N}$ and all $x \in [a,b]$ we have $s_n(x) \le f(x)$

ii. Prove that if f is continuous at $x \in [a,b]$, then $\left\{s_n(x)\right\}_{n=1}^{\infty}$ converges to $f(x)$.

iii. Prove that $\lim_{n\to\infty} \int_a^b |f - s_n|\,dx = 0$.

(b) Prove that there is a sequence $\{c_n\}_{n=1}^{\infty}$ of continuous functions on $[a,b]$ such that for all $n \in \mathbb{N}$ we have $|c_n| \le |f|$ and so that $\lim_{n\to\infty} \int_a^b |f - c_n|\,dx = 0$.

5-27. Prove **Riemann's Condition** for Riemann-Stieltjes integrals. That is, let $f : [a,b] \to \mathbb{R}$ be bounded, let $g : [a,b] \to \mathbb{R}$ be nondecreasing and prove that f is Riemann-Stieltjes integrable on $[a,b]$ with respect to g iff for all $\varepsilon > 0$ there is a partition P of $[a,b]$ such that $U_g(f,P) - L_g(f,P) < \varepsilon$.

Readers interested in how the theory of integrals and derivatives is used to obtain numerical approximations could read Chapter 13 before continuing. The only missing prerequisites would be the existence of the transcendental functions and their derivatives (for the examples) and geometric series (for the proof of Theorem 13.14). This should not be a problem, because these results should be sufficiently well known from calculus.

Chapter 6

Series of Real Numbers I

Series facilitate the computation of "sums with infinitely many terms." This first introduction mostly showcases the results needed to define outer Lebesgue measure (see Definition 8.1). Further results about series will be presented in Chapter 10.

6.1 Series as a Vehicle To Define Infinite Sums

Of course it is impossible to add infinitely many numbers. But we can consider the convergence behavior of a sequence of finite sums.

Definition 6.1 *Let $\{a_j\}_{j=1}^{\infty}$ be a sequence of real numbers. The* **partial sums** *of the* **series** $\sum_{j=1}^{\infty} a_j$ *of real numbers are defined to be $s_n := \sum_{j=1}^{n} a_j$. The series is said to* **converge** *iff the sequence of partial sums $\left\{\sum_{j=1}^{n} a_j\right\}_{n=1}^{\infty}$ converges and it is said to* **diverge** *otherwise. For a convergent series, the limit is usually denoted $\sum_{j=1}^{\infty} a_j$.*

Series that start at numbers other than 1 are defined similarly (Exercise 6-1). This section is devoted to introductory examples and some fundamental results.

Geometric series are the prototypical examples of convergent series. Because there is a simple formula for the partial sums, geometric series nicely fit the mold of Definition 6.1. Figure 15 gives an indication why geometric series are called "geometric" and Exercise 6-2 gives some examples of computations.

Theorem 6.2 Geometric sums *and* **geometric series**. *Let $a, q \in \mathbb{R}$ and let $q \neq 1$. Then for all $n \in \mathbb{N}$ the summation formula $\sum_{j=1}^{n} aq^j = aq\frac{1-q^n}{1-q}$ holds. Moreover, for*

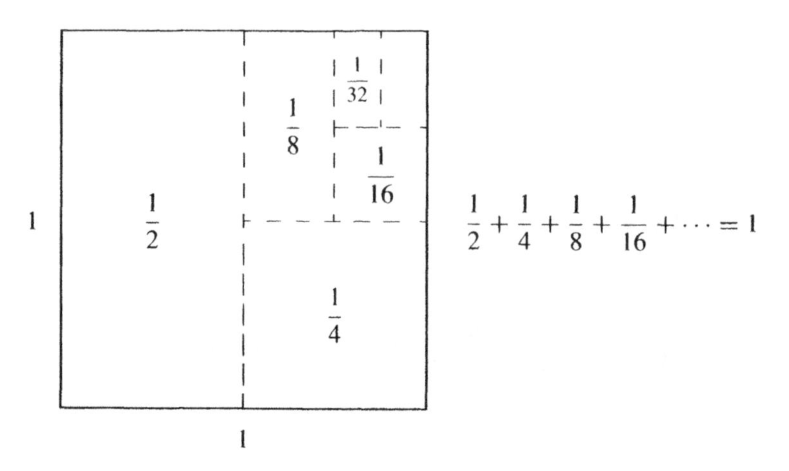

Figure 15: The "sum" of infinitely many terms can be finite. Each rectangle inside the square above is one-half of the size of the previous rectangle and the sum of all their areas should be the area of the square.

all real numbers $|q| < 1$ *the series* $\sum_{j=1}^{\infty} aq^j$ *converges and* $\sum_{j=1}^{\infty} aq^j = \frac{aq}{1-q}$.

Proof. The first statement follows from the following telescoping sum argument.

$$\begin{aligned}(1-q)\sum_{j=1}^{n} aq^j &= \sum_{j=1}^{n} aq^j - \sum_{j=1}^{n} aq^{j+1} = \sum_{j=1}^{n} aq^j - \sum_{j=2}^{n+1} aq^j \\ &= aq - aq^{n+1}.\end{aligned}$$

For $|q| < 1$, note that $\sum_{j=1}^{\infty} aq^j = \lim_{n\to\infty} \sum_{j=1}^{n} aq^j = \lim_{n\to\infty} aq\frac{1-q^n}{1-q} = \frac{aq}{1-q}$. ■

Series with nonnegative terms play an important role in analysis. Their partial sums are nondecreasing. Therefore, by the Monotone Sequence Theorem, whenever the terms of a series are nonnegative and the partial sums are bounded, the series converges.

Lemma 6.3 *For all* $j \in \mathbb{N}$, *let* $a_j \geq 0$. *Then the series* $\sum_{j=1}^{\infty} a_j$ *converges iff the sequence of the partial sums* $\left\{\sum_{j=1}^{n} a_j\right\}_{n=1}^{\infty}$ *is bounded above.*

Proof. For "⇒," recall that by Proposition 2.34 any convergent sequence is bounded, and hence, in particular, it is bounded above.

For "$\Leftarrow$," let $\sum_{j=1}^{\infty} a_j$ be a series such that all a_j are nonnegative and such that the sequence $\left\{\sum_{j=1}^{n} a_j\right\}_{n=1}^{\infty}$ of its partial sums is bounded above. Because for all $n \in \mathbb{N}$ we have $\sum_{j=1}^{n+1} a_j - \sum_{j=1}^{n} a_j = a_{n+1} \geq 0$, the sequence of the partial sums is nondecreasing. Thus by the Monotone Sequence Theorem $\left\{\sum_{j=1}^{n} a_j\right\}_{n=1}^{\infty}$ must converge, and hence the series converges. ■

Series can easily be added, subtracted, and multiplied by numbers. Multiplication of series is more complicated and is therefore deferred to Theorem 10.14.

Theorem 6.4 *Let $\sum_{j=1}^{\infty} a_j$ and $\sum_{j=1}^{\infty} b_j$ be convergent series of real numbers and let c be a real number. Then the following hold.*

1. *The series $\sum_{j=1}^{\infty} a_j + b_j$ converges and $\sum_{j=1}^{\infty} a_j + b_j = \sum_{j=1}^{\infty} a_j + \sum_{j=1}^{\infty} b_j$.*

2. *The series $\sum_{j=1}^{\infty} a_j - b_j$ converges and $\sum_{j=1}^{\infty} a_j - b_j = \sum_{j=1}^{\infty} a_j - \sum_{j=1}^{\infty} b_j$.*

3. *The series $\sum_{j=1}^{\infty} ca_j$ converges and $\sum_{j=1}^{\infty} ca_j = c\sum_{j=1}^{\infty} a_j$.*

Proof. To prove part 1, note that

$$\begin{aligned}\sum_{j=1}^{\infty} a_j + \sum_{j=1}^{\infty} b_j &= \lim_{n\to\infty} \sum_{j=1}^{n} a_j + \lim_{n\to\infty} \sum_{j=1}^{n} b_j = \lim_{n\to\infty} \sum_{j=1}^{n} a_j + \sum_{j=1}^{n} b_j \\ &= \lim_{n\to\infty} \sum_{j=1}^{n} a_j + b_j = \sum_{j=1}^{\infty} a_j + b_j,\end{aligned}$$

where in each step the existence of the quantity on the left side of the equal sign implies the existence of the quantity on the right side. Hence, if $\sum_{j=1}^{\infty} a_j$ and $\sum_{j=1}^{\infty} b_j$ converge, then $\sum_{j=1}^{\infty} a_j + b_j$ converges and we have $\sum_{j=1}^{\infty} a_j + b_j = \sum_{j=1}^{\infty} a_j + \sum_{j=1}^{\infty} b_j$.

The remaining parts are left to the reader as Exercises 6-3a and 6-3b. ■

Similar to the algebraic operations, comparabilities can be moved through the summations.

Theorem 6.5 *Let* $\sum_{j=1}^{\infty} a_j$ *and* $\sum_{j=1}^{\infty} b_j$ *be convergent series so that* $a_j \leq b_j$ *for all* $j \in \mathbb{N}$. *Then* $\sum_{j=1}^{\infty} a_j \leq \sum_{j=1}^{\infty} b_j$.

Proof. Exercise 6-4. ■

The most concrete example of series in our daily experience is so fundamental, it is often not even recognized as a series. Geometric series are the foundation for the decimal expansion of real numbers. Formally, decimal expansions are an identification of numbers with sequences of integers. The connection is made via series with nonnegative terms as shown below.

Proposition 6.6 *Let D be the set of all sequences $\{d_j\}_{j=1}^{\infty}$ of integers in the set (of digits) $\{0, 1, 2, 3, 4, 5, 6, 7, 8, 9\}$ such that there is a $k \in \mathbb{N}$ with $d_k \neq 0$ and so that for every $n \in \mathbb{N}$ there is an $m \geq n$ so that $d_m \neq 9$. Then for each $\{d_j\}_{j=1}^{\infty} \in D$ the series* $r\left(\{d_j\}_{j=1}^{\infty}\right) := \sum_{j=1}^{\infty} \frac{d_j}{10^j}$ *converges and the function $r : D \to (0, 1)$ is bijective.*

For every $x \in (0, 1)$, we call $r^{-1}(x)$ the **decimal expansion** *of x.*

Proof. Let $\{d_j\}_{j=1}^{\infty} \in D$. For all $n \in \mathbb{N}$, we infer the inequalities

$$0 \leq \sum_{j=1}^{n} \frac{d_j}{10^j} \leq \sum_{j=1}^{n} \frac{9}{10^j} = 9\frac{1}{10}\frac{1-\left(\frac{1}{10}\right)^n}{1-\frac{1}{10}} \leq \frac{9}{10}\frac{1}{\frac{9}{10}} = 1.$$

Because the terms $\frac{d_j}{10^j}$ are nonnegative, Lemma 6.3 guarantees that $\sum_{j=1}^{\infty} \frac{d_j}{10^j}$ converges. Moreover, because at least one d_k is positive and not all d_k are equal to nine, we infer that $0 < \sum_{j=1}^{\infty} \frac{d_j}{10^j} < 1$ for all $\{d_j\}_{j=1}^{\infty} \in D$.

To prove that r is injective, let $\{c_j\}_{j=1}^{\infty} \neq \{d_j\}_{j=1}^{\infty}$ be sequences in D. Let $n \in \mathbb{N}$ be the smallest natural number so that $c_n \neq d_n$ and assume without loss of generality that $c_n > d_n$. Then

$$\begin{aligned} r\left(\{c_j\}_{j=1}^{\infty}\right) - r\left(\{d_j\}_{j=1}^{\infty}\right) &= \sum_{j=n}^{\infty} \frac{c_j - d_j}{10^j} \geq \frac{1}{10^n} + \sum_{j=n+1}^{\infty} \frac{c_j - d_j}{10^j} \\ &> \frac{1}{10^n} - \sum_{j=n+1}^{\infty} \frac{9}{10^j} = \frac{1}{10^n} - \sum_{j=1}^{\infty} \frac{9}{10^n}\frac{1}{10^j} \\ &= \frac{1}{10^n} - \frac{9}{10^n}\frac{\frac{1}{10}}{1-\frac{1}{10}} = \frac{1}{10^n} - \frac{1}{10^n} = 0, \end{aligned}$$

which implies that $r\left(\{c_j\}_{j=1}^{\infty}\right) \neq r\left(\{d_j\}_{j=1}^{\infty}\right)$.

For surjectivity, let $x \in (0, 1)$ and construct $\{d_j\}_{j=1}^{\infty}$ recursively as follows. Set the digit $d_0 := 0$. Once d_n has been defined so that $0 \leq x - \sum_{j=1}^{n} \frac{d_j}{10^j} < \frac{1}{10^n}$, let d_{n+1} be the largest integer k so that $0 \leq x - \sum_{j=1}^{n} \frac{d_j}{10^j} - \frac{k}{10^{n+1}}$. Then d_{n+1} is at most 9 and the inequalities $0 \leq x - \sum_{j=1}^{n+1} \frac{d_j}{10^j} < \frac{1}{10^{n+1}}$ hold.

The series $\sum_{j=1}^{\infty} \frac{d_j}{10^j}$ converges to x, because for all $n \in \mathbb{N}$ we have by construction that $0 \leq x - \sum_{j=1}^{n} \frac{d_j}{10^j} < \frac{1}{10^n}$. Finally, for each $n \in \mathbb{N}$ there must be an $m > n$ with $d_m \neq 9$, because otherwise we would have

$$x - \sum_{j=1}^{n} \frac{d_j}{10^j} = \sum_{j=1}^{\infty} \frac{d_j}{10^j} - \sum_{j=1}^{n} \frac{d_j}{10^j} = \sum_{j=n+1}^{\infty} \frac{9}{10^j} = \frac{1}{10^n} \frac{9 \cdot \frac{1}{10}}{1 - \frac{1}{10}} = \frac{1}{10^n},$$

which cannot be. ■

In particular, the representation in Proposition 6.6 can be used to convert infinite repeating decimals into fractions (see Exercise 6-5). Of course not every series converges. For example, if all terms are equal to 1, the sum is infinite. The limit test below is a criterion for divergence, because it says (Exercise 6-6) that if the terms do not go to zero, the series diverges.

Theorem 6.7 Limit Test. *If the series* $\sum_{j=1}^{\infty} a_j$ *converges, then* $\lim_{j\to\infty} a_j = 0$.

Proof. Note that if $\sum_{j=1}^{\infty} a_j$ converges, then

$$\lim_{n\to\infty} a_n = \lim_{n\to\infty} \sum_{j=1}^{n} a_j - \sum_{j=1}^{n-1} a_j = \sum_{j=1}^{\infty} a_j - \sum_{j=1}^{\infty} a_j = 0.$$

■

If Theorem 6.7 was a biconditional, the theory of series would be very easy indeed. The next example shows that this is *not* the case.

Example 6.8 *The* **harmonic series** $\sum_{j=1}^{\infty} \frac{1}{j}$ *diverges, even though the terms converge to zero.*

The partial sums $s_n = \sum_{j=1}^{n} \frac{1}{j}$ form an increasing sequence. If we can show that a subsequence $\left\{s_{n_i}\right\}_{i=1}^{\infty}$ goes to infinity, then the sequence $\{s_n\}_{n=1}^{\infty}$ diverges. We choose $n_i = 2^i$.

$$\begin{aligned} s_{2^i} &= \sum_{j=1}^{2^i} \frac{1}{j} = 1 + \sum_{j=2}^{2^i} \frac{1}{j} = 1 + \sum_{k=0}^{i-1} \sum_{j=2^k+1}^{2^{k+1}} \frac{1}{j} \geq 1 + \sum_{k=0}^{i-1} \sum_{j=2^k+1}^{2^{k+1}} \frac{1}{2^{k+1}} \\ &\geq 1 + \sum_{k=0}^{i-1} 2^k \frac{1}{2^{k+1}} = 1 + \sum_{k=0}^{i-1} \frac{1}{2} = 1 + \frac{i}{2}. \end{aligned}$$

Because the s_{2^i} go to infinity, the sequence of partial sums diverges, which means that $\sum_{j=1}^{\infty} \frac{1}{j}$ diverges. □

We will be confronted with "infinite sums" like the harmonic series throughout measure theory. Therefore it is useful to formally define infinite sums.

Definition 6.9 *Let $\sum_{j=1}^{\infty} a_j$ be a series. We write $\sum_{j=1}^{\infty} a_j = \infty$ and call it* **infinite** *or an* **infinite sum** *iff $\lim_{n\to\infty} \sum_{j=1}^{n} a_j = \infty$. We write $\sum_{j=1}^{\infty} a_j = -\infty$ iff $\lim_{n\to\infty} \sum_{j=1}^{n} a_j = -\infty$.*

Exercises

6-1. Let $k \in \mathbb{Z}$ and for each $j \in \mathbb{Z}$ let $a_j \in \mathbb{R}$. Define the series $\sum_{j=k}^{\infty} a_j$ and $\sum_{j=-\infty}^{k} a_j$.

6-2. Compute the value of each of the series below.

(a) $\sum_{j=1}^{\infty} \left(\frac{2}{3}\right)^j$ (b) $\sum_{j=1}^{\infty} \frac{1}{4^j}$ (c) $\sum_{j=1}^{\infty} \frac{(-3)^j}{7^{j+1}}$ (d) $\sum_{j=4}^{\infty} \frac{2^{j-2}}{5^{j+4}}$

6-3. More on the arithmetic of series (Theorem 6.4)

(a) Prove part 2 of Theorem 6 4

(b) Prove part 3 of Theorem 6 4.

(c) Give an example of two series $\sum_{j=1}^{\infty} a_j$ and $\sum_{j=1}^{\infty} b_j$ such that $\sum_{j=1}^{\infty} a_j + b_j$ converges, but neither $\sum_{j=1}^{\infty} a_j$ nor $\sum_{j=1}^{\infty} b_j$ converges.

(d) Is it possible to find two series $\sum_{j=1}^{\infty} a_j$ and $\sum_{j=1}^{\infty} b_j$ such that $\sum_{j=1}^{\infty} a_j + b_j$ converges, and exactly one of $\sum_{j=1}^{\infty} a_j$ and $\sum_{j=1}^{\infty} b_j$ diverges?

(e) Is there a series $\sum_{j=1}^{\infty} a_j$ and a $c \in \mathbb{R}$ such that $\sum_{j=1}^{\infty} ca_j$ converges, and $\sum_{j=1}^{\infty} a_j$ diverges?

6-4 Prove Theorem 6.5.

6-5 Convert each of the following infinite repeating decimals below into a fraction. A bar over a set of digits means that these digits repeat indefinitely.

(a) 0.25 (b) $0.\overline{25}$ (c) $0.\overline{9462}$ (d) $0.1\overline{473}$ (e) $12.004\overline{95}$

6-6 Why does the limit test say that if $\lim_{j\to\infty} a_j \neq 0$, then the series $\sum_{j=1}^{\infty} a_j$ diverges?

6-7. Another proof for the summation formula in Theorem 6.2. Let $q \neq 1$ and a be real numbers. Prove *by induction* that $\sum_{j=1}^{n} aq^j = aq\frac{1-q^n}{1-q}$.

6-8. 2^k **test** Let $\{a_j\}_{j=1}^{\infty}$ be a nonincreasing sequence with nonnegative terms. Prove that $\sum_{j=1}^{\infty} a_j$ converges iff $\sum_{k=1}^{\infty} 2^k a_{2^k}$ converges.
Hint. Use Example 6.8 as guidance.

6-9. Prove that $\sum_{j=1}^{\infty} (-1)^{j+1}\frac{1}{j}$ converges by showing that the partial sums form a Cauchy sequence.

Hint. Show that $\left|\sum_{j=m}^{n} (-1)^{j+1}\frac{1}{j}\right| \leq \frac{1}{m} + \frac{1}{n}$. Be careful to distinguish all possible combinations of n, m being even or odd.

6-10 Translating between sequences and series. Let $\{a_n\}_{n=1}^{\infty}$ be a sequence of real numbers. Prove that $\{a_n\}_{n=1}^{\infty}$ converges iff the series $\sum_{j=1}^{\infty} (a_{j+1} - a_j)$ converges and that in this case we obtain the limit as $\lim_{n\to\infty} a_n = a_1 + \sum_{j=1}^{\infty} (a_{j+1} - a_j)$.

6-11 Use Lemma 6.3 to prove the Monotone Sequence Theorem.

6-12 Let $\sum_{j=1}^{\infty} a_j$ be a convergent series, let $\{j_k\}_{k=0}^{\infty}$ be a strictly increasing sequence of natural numbers with $j_0 = 1$ and for all $k \in \mathbb{N}$ let $A_k := \sum_{j=j_{k-1}}^{j_k - 1} a_j$. Prove that $\sum_{k=1}^{\infty} A_k$ converges and $\sum_{k=1}^{\infty} A_k = \sum_{j=1}^{\infty} a_j$.

6.2 Absolute Convergence and Unconditional Convergence

For series, there is no simple condition that is equivalent to convergence. Therefore we need criteria that imply convergence. This section presents the criteria that are needed to define and work with outer Lebesgue measure. More criteria will be presented in Section 10.2.

Cauchy sequences play an important role throughout analysis, so it is natural to relate the convergence of series to Cauchy sequences.

Proposition 6.10 Cauchy Criterion. *A series* $\sum_{j=1}^{\infty} a_j$ *converges iff for all* $\varepsilon > 0$ *there is an* $N \in \mathbb{N}$ *so that for all* $n \geq m \geq N$ *we have* $\left|\sum_{j=m}^{n} a_j\right| < \varepsilon$.

Proof. The series $\sum_{j=1}^{\infty} a_j$ converges iff the sequence $\{s_n\}_{n=1}^{\infty}$ of its partial sums converges, which by Theorem 2.27 is the case iff it is a Cauchy sequence. This is the case iff for all $\varepsilon > 0$ there is an $N \in \mathbb{N}$ so that for all $n \geq m \geq N$ the inequality $|s_n - s_{m-1}| < \varepsilon$ holds. Since $\left|\sum_{j=m}^{n} a_j\right| = |s_n - s_{m-1}|$ we have proved the result. ■

The Alternating Series Test shows that there are indeed many convergent series.

Theorem 6.11 Alternating Series Test. *Let* $\{b_j\}_{j=1}^{\infty}$ *be a nonincreasing nonnegative sequence such that* $\lim_{j\to\infty} b_j = 0$. *Then* $\sum_{j=1}^{\infty}(-1)^{j+1}b_j$ *converges.*

Proof. Let $\varepsilon > 0$ be arbitrary. There is an $N \in \mathbb{N}$ so that for all $n \geq N$ we have $b_n < \frac{\varepsilon}{2}$. Then for all $m > n \geq N$ we obtain

$$\begin{aligned}\left|\sum_{j=n}^{m}(-1)^{j+1}b_j\right| &= \sum_{i=0}^{m-n}(-1)^i b_{n+i} \\ &\leq b_{m+1} + \underbrace{\sum_{i=0}^{\lfloor\frac{m-n}{2}\rfloor}\left(b_{n+2i} - b_{n+2i+1}\right)}_{\text{use } b_{n+2i+1}\geq b_{n+2(i+1)} \text{ and a telescoping sum}} \\ &\leq b_{m+1} + b_n - b_{n+2\lfloor\frac{m-n}{2}\rfloor+1} \\ &\leq b_{m+1} + b_n < \frac{\varepsilon}{2} + \frac{\varepsilon}{2} = \varepsilon,\end{aligned}$$

If $m - n$ is odd, the b_{m+1} on the next line is simply an extra term.

and by the Cauchy Criterion we conclude that the series converges. ■

In many situations we will work with series of nonnegative numbers. If negative summands occur, it is natural to take absolute values and hope the sum still converges. This is not always the case, but the idea of absolute convergence is fundamental.

Definition 6.12 *A series* $\sum_{j=1}^{\infty} a_j$ **converges absolutely** *iff the series* $\sum_{j=1}^{\infty} |a_j|$ *converges.*

Absolute convergence is a strictly stronger condition than convergence. That is, absolutely convergent series converge, but the converse is not true.

Proposition 6.13 *If the series* $\sum_{j=1}^{\infty} a_j$ *converges absolutely, then it converges. Moreover, the* **triangular inequality** $\left|\sum_{j=1}^{\infty} a_j\right| \leq \sum_{j=1}^{\infty} |a_j|$ *holds.*

Proof. Let $\varepsilon > 0$. Because $\sum_{j=1}^{\infty} a_j$ converges absolutely, $\sum_{j=1}^{\infty} |a_j|$ converges. Thus there is an $N \in \mathbb{N}$ so that for all $n \geq m \geq N$ we have that $\sum_{j=m}^{n} |a_j| < \varepsilon$. But then for all $n \geq m \geq N$ we obtain $\left|\sum_{j=m}^{n} a_j\right| \leq \sum_{j=m}^{n} |a_j| < \varepsilon$. Therefore by the Cauchy Criterion the series $\sum_{j=1}^{\infty} a_j$ converges.

Moreover, for all $n \in \mathbb{N}$ we have $\left|\sum_{j=1}^{n} a_j\right| \leq \sum_{j=1}^{n} |a_j| \leq \sum_{j=1}^{\infty} |a_j|$, which implies the triangular inequality. ■

Example 6.14 By the Alternating Series Test, the series $\sum_{j=1}^{\infty} (-1)^{j+1} \frac{1}{j}$ converges. However, $\sum_{j=1}^{\infty} \left|(-1)^{j+1} \frac{1}{j}\right| = \sum_{j=1}^{\infty} \frac{1}{j} = \infty$, so it does not converge absolutely. □

Absolute convergence of a series is often established with the Comparison Test.

Theorem 6.15 Comparison Test. *Let* $\sum_{j=1}^{\infty} a_j$ *and* $\sum_{j=1}^{\infty} b_j$ *be series with* $0 \leq a_j \leq b_j$ *for all* $j \in \mathbb{N}$. *If* $\sum_{j=1}^{\infty} b_j$ *converges, then* $\sum_{j=1}^{\infty} a_j$ *converges, too.*

Proof. Note that $\left\{\sum_{j=1}^{n} a_j\right\}_{n=1}^{\infty}$ is a nondecreasing sequence and for all n we have $\sum_{j=1}^{n} a_j \leq \sum_{j=1}^{n} b_j \leq \sum_{j=1}^{\infty} b_j < \infty$. This means that $\left\{\sum_{j=1}^{n} a_j\right\}_{n=1}^{\infty}$ is bounded above by $\sum_{j=1}^{\infty} b_j$. By the Monotone Sequence Theorem, the series $\sum_{j=1}^{\infty} a_j$ converges. ■

The Comparison Test allows us to establish convergence and divergence for series that can be compared to series with known convergence behavior. For example, $\sum_{j=1}^{\infty} \frac{1}{2^{j^2}}$ converges because $\frac{1}{2^{j^2}} \leq \left(\frac{1}{2}\right)^j$. Similarly, $\sum_{j=1}^{\infty} \frac{1}{\sqrt{j}}$ diverges because $\frac{1}{\sqrt{j}} \geq \frac{1}{j}$. We will use the Comparison Test extensively in Section 10.2. Moreover, it should be noted that the Comparison Test can also be applied when there is a $k \in \mathbb{N}$ so that the comparability $0 \leq a_j \leq b_j$ holds only for $j \geq k$ (Exercise 6-16).

We conclude this section with the subtle, but nonetheless necessary, notion of unconditional convergence. Assume that the terms of a series are provided in no specific order. (This is the case in the definition of outer Lebesgue measure.) Then we could add the terms in any order we choose. It would be catastrophic if the value of this summation depended on the order in which we sum the terms. Unfortunately, for arbitrary series this can actually happen. For example, $\sum_{j=1}^{\infty}(-1)^{j+1}\frac{1}{j}$ converges. However, a rearrangement can destroy the convergence. Consider that $\sum_{j=1}^{\infty} \frac{1}{2j+1} = \infty$ and $\sum_{j=1}^{\infty}(-1)\frac{1}{2j} = -\infty$ (Exercise 6-13). Suppose we first sum enough odd numbered terms to get a number > 1, say, $\frac{1}{1} + \frac{1}{3} > 1$, then add the first even numbered term, so we get $\frac{1}{1} + \frac{1}{3} + \left(-\frac{1}{2}\right)$, then add enough odd numbered terms to get a number > 2, say, $\frac{1}{1} + \frac{1}{3} + \left(-\frac{1}{2}\right) + \sum_{j=2}^{20} \frac{1}{2j+1} > 2$, then add the second even numbered term, so we get $\frac{1}{1} + \frac{1}{3} + \left(-\frac{1}{2}\right) + \sum_{j=2}^{20} \frac{1}{2j+1} + \left(-\frac{1}{4}\right)$, then add enough odd numbered terms to get a number > 3, say $\frac{1}{1} + \frac{1}{3} + \left(-\frac{1}{2}\right) + \sum_{j=2}^{20} \frac{1}{2j+1} + \left(-\frac{1}{4}\right) + \sum_{k=21}^{254} \frac{1}{2k+1} > 3$, then add the third even numbered term, so we get

$$\frac{1}{1}+\frac{1}{3}+\left(-\frac{1}{2}\right)+\sum_{j=2}^{20}\frac{1}{2j+1}+\left(-\frac{1}{4}\right)+\sum_{k=21}^{254}\frac{1}{2k+1}+\left(-\frac{1}{6}\right),$$

and proceed in this fashion indefinitely. This process uses all the available numbers and it produces a sequence of sums that goes to ∞. Of course we will need to use more and more odd numbered terms to make up for the one new even numbered term and the increment of 1, but nonetheless, in one arrangement the series sums to a finite number, while in another the sum is infinite. We will make this idea more precise in the proof of Theorem 6.18 and we will push it to its full extent in Exercise 6-23.

Series in which we can rearrange the terms in any way without affecting the "sum" are called unconditionally convergent.

Definition 6.16 *The series* $\sum_{j=1}^{\infty} a_j$ **converges unconditionally** *iff for all bijective functions* $\sigma : \mathbb{N} \to \mathbb{N}$ *the series* $\sum_{i=1}^{\infty} a_{\sigma(i)}$ *converges and* $\sum_{i=1}^{\infty} a_{\sigma(i)} = \sum_{j=1}^{\infty} a_j$. *If a series converges, but not unconditionally, we will say it* **converges conditionally**.

For series of real numbers, Theorem 6.18 below shows that unconditional convergence and absolute convergence are equivalent. This is very helpful, because it is a lot easier to check whether *one* series of absolute values converges than to check if *all* rearranged series converge to the same limit.

Definition 6.17 *Let* $\sum_{j=1}^{\infty} a_j$ *be a series and let* $B \subseteq \mathbb{N}$. *We define the series* $\sum_{j\in B} a_j$ *to be the series* $\sum_{j=1}^{\infty} b_j$, *where* $b_j := \begin{cases} a_j; & \text{for } j \in B, \\ 0; & \text{for } j \notin B. \end{cases}$

Theorem 6.18 *The series* $\sum_{j=1}^{\infty} a_j$ *converges absolutely iff it converges unconditionally.*

Proof. For "$\Rightarrow$," let $\varepsilon > 0$ and let $N \in \mathbb{N}$ be such that for all $n \geq m \geq N$ we have $\sum_{j=m}^{n} |a_j| < \frac{\varepsilon}{2}$. Then $\sum_{j=N}^{\infty} |a_j| \leq \frac{\varepsilon}{2} < \varepsilon$.

Now let $\sigma : \mathbb{N} \to \mathbb{N}$ be an arbitrary bijection. Then there is an $I \in \mathbb{N}$ so that $\{1, \ldots, N-1\} \subseteq \sigma\big[\{1, \ldots, I\}\big]$. Hence, for all $m \geq I$ we obtain

$$\left|\sum_{i=1}^{m} a_{\sigma(i)} - \sum_{j=1}^{\infty} a_j\right| = \left|\sum_{j\notin\sigma[\{1,\ldots,m\}]} a_j\right| \leq \sum_{j\notin\sigma[\{1,\ldots,m\}]} |a_j| \leq \sum_{j=N}^{\infty} |a_j| < \varepsilon,$$

which means $\sum_{i=1}^{\infty} a_{\sigma(i)} = \sum_{j=1}^{\infty} a_j$.

For "$\Leftarrow$," we prove that if $\sum_{j=1}^{\infty} a_j$ does not converge absolutely, then it does not converge unconditionally. There is nothing to prove if $\sum_{j=1}^{\infty} a_j$ does not converge, so we can assume $\sum_{j=1}^{\infty} a_j$ converges, but not absolutely.

For $j \in \mathbb{N}$, let $a_j^+ := \max\{a_j, 0\}$ and let $a_j^- := -\min\{a_j, 0\}$. We first claim that $\sum_{j=1}^{\infty} a_j^+ = \sum_{j=1}^{\infty} a_j^- = \infty$. Suppose for a contradiction the series $\sum_{j=1}^{\infty} a_j^+$ did converge. Then the series $\sum_{j=1}^{\infty} a_j^- = \sum_{n=1}^{\infty} -\left(a_j - a_j^+\right)$ would converge, so $\sum_{j=1}^{\infty} |a_j| = \sum_{j=1}^{\infty} a_j^+ + a_j^-$ would converge, and hence $\sum_{j=1}^{\infty} a_j$ would converge absolutely, a contradiction.

We now recursively construct a bijection $\sigma : \mathbb{N} \to \mathbb{N}$ so that $\sum_{i=1}^{\infty} a_{\sigma(i)}$ diverges. Because $\sum_{j=1}^{\infty} a_j^+ = \infty$ there are $n_1 \in \mathbb{N}$ and $\sigma(1) < \sigma(2) < \cdots < \sigma(n_1)$ so that for all $i \in \{1, \ldots, n_1\}$ we have $a_{\sigma(i)} > 0$, so that for all $j \leq \sigma(n_1)$ with $a_j > 0$ we have $j = \sigma(i)$ for some $i \in \{1, \ldots, n_1\}$ and so that $\sum_{i=1}^{n_1 - 1} a_{\sigma(i)} \leq 1 < \sum_{i=1}^{n_1} a_{\sigma(i)}$. Because $\sum_{j=1}^{\infty} a_j^- = \infty$ there are $m_1 \in \mathbb{N}$ and $\sigma(n_1+1) < \sigma(n_1+2) < \cdots < \sigma(n_1+m_1)$ so that for all $i \in \{n_1 + 1, \ldots, n_1 + m_1\}$ we have $a_{\sigma(i)} \leq 0$, so that for all $j \leq \sigma(n_1 + m_1)$ with $a_j \leq 0$ we have $j = \sigma(i)$ for some $i \in \{n_1 + 1, \ldots, n_1 + m_1\}$ and so that $\sum_{i=1}^{n_1+m_1-1} a_{\sigma(i)} > 1 \geq \sum_{i=1}^{n_1+m_1} a_{\sigma(i)}$.

For the recursive step, let $n_0 := m_0 := 0$ and assume that $n_1 < \cdots < n_k$, $m_1, \ldots, m_k \in \mathbb{N}$ and distinct $\sigma(1), \ldots, \sigma(n_k + m_k) \in \mathbb{N}$ have been chosen as follows. For all $i \in P_k := \bigcup_{p=1}^{k} \{n_{p-1} + m_{p-1} + 1, \ldots, n_p\}$ we have $a_{\sigma(i)} > 0$, for all $j \leq \sigma(n_k)$ with $a_j > 0$ we have $j = \sigma(i)$ for some $i \in P_k$, and $\sum_{i=1}^{n_k - 1} a_{\sigma(i)} \leq k < \sum_{i=1}^{n_k} a_{\sigma(i)}$.

Moreover, for all $i \in N_k := \bigcup_{p=1}^{k} \{n_p + 1, \ldots, n_p + m_p\}$ we have $a_{\sigma(i)} \leq 0$, for all $j \leq \sigma(n_k + m_k)$ with $a_j \leq 0$ we have $j = \sigma(i)$ for some $i \in N_k$ and the sums satisfy the inequalities $\sum_{i=1}^{n_k+m_k-1} a_{\sigma(i)} > k \geq \sum_{i=1}^{n_k+m_k} a_{\sigma(i)}$.

Choose natural numbers $\sigma(n_k + m_k + 1) < \cdots < \sigma(n_{k+1})$ greater than $\sigma(n_k)$ so that for all $i \in P_{k+1} := \bigcup_{p=1}^{k+1} \{n_{p-1} + m_{p-1} + 1, \ldots, n_p\}$ we have $a_{\sigma(i)} > 0$, for all $j \leq \sigma(n_{k+1})$ with $a_j > 0$ we have $j = \sigma(i)$ for some $i \in P_{k+1}$ and the inequalities $\sum_{i=1}^{n_{k+1}-1} a_{\sigma(i)} \leq k + 1 < \sum_{i=1}^{n_{k+1}} a_{\sigma(i)}$ hold. (This can be done because $\sum_{j=1}^{\infty} a_j^+ = \infty$.)

Because $k \geq \sum_{i=1}^{n_k+m_k} a_{\sigma(i)}$ we infer $n_{k+1} > n_k + m_k$. Choose the natural numbers $\sigma(n_{k+1}+1) < \cdots < \sigma(n_{k+1}+m_{k+1})$ greater than $\sigma(n_k+m_k)$ such that for all indices $i \in N_{k+1} := \bigcup_{p=1}^{k+1} \{n_p+1, \ldots, n_p+m_p\}$ we have $a_{\sigma(i)} \leq 0$, for all $j \leq \sigma(n_{k+1}+m_{k+1})$ with $a_j \leq 0$ we have $j = \sigma(i)$ for some $i \in N_{k+1}$ and the sums satisfy the inequalities $\sum_{i=1}^{n_{k+1}+m_{k+1}-1} a_{\sigma(i)} > k + 1 \geq \sum_{i=1}^{n_{k+1}+m_{k+1}} a_{\sigma(i)}$. (This can be done because $\sum_{j=1}^{\infty} a_j^- = \infty$.)

Because $k + 1 < \sum_{i=1}^{n_{k+1}} a_{\sigma(i)}$ we infer $m_{k+1} \geq 1$.

Then $\sigma : \mathbb{N} \to \mathbb{N}$ as constructed is injective by construction, surjective because in every recursive step at least one more positive term and at least one more nonpositive term of $\sum_{j=1}^{\infty} a_i$ are added to the set $\{a_{\sigma(1)}, \ldots, a_{\sigma(n_k+m_k)}\}$, and the construction shows that $\left\{\sum_{i=1}^{n} a_{\sigma(i)}\right\}_{n=1}^{\infty}$ is unbounded, and hence divergent. ■

Standard Proof Technique 6.19 In the proof of "⇒" of Theorem 6.18, our estimates only yielded a nonstrict inequality $\sum_{j=N}^{\infty} |a_j| \leq \ldots$. By making the sum less than or equal to $\frac{\varepsilon}{2}$ we made it strictly less than ε. Because ε can be replaced with $\frac{\varepsilon}{2}$ in any argument, in analysis it is usually not a problem if an estimate only yields a nonstrict inequality $\leq \varepsilon$ rather than a strict inequality $< \varepsilon$. □

Sometimes we need to partition an infinite sum into infinitely many "chunks" (see the proof of part 3 of Theorem 8.6). For this type of rearrangement, we use double series. For brevity's sake, we limit ourselves to double series of nonnegative terms in Proposition 6.22.

Definition 6.20 *For $i, j \in \mathbb{N}$, we define the* **ordered pair** $(i, j) := \{i, \{i, j\}\}$ *and we define* $\mathbb{N}\times\mathbb{N} = \{(i, j) : i, j \in \mathbb{N}\}$. *A* **(doubly indexed) family** *of numbers* $\{a_{(i,j)}\}_{i,j=1}^{\infty}$ *is a function from* $\mathbb{N} \times \mathbb{N}$ *to* $\mathbb{R}$.

The definition of ordered pairs guarantees that the order in which the numbers are listed matters (Exercise 6-14). Definition 6.20 also indicates that it is time to investigate set theory in more detail. We will do so in the next chapter.

Definition 6.21 *Let* $\{a_{(i,j)}\}_{i,j=1}^{\infty}$ *be a doubly indexed family of numbers. The* **double series** $\sum_{i=1}^{\infty}\sum_{j=1}^{\infty} a_{(i,j)}$ *is called* **convergent** *iff for all* $i \in \mathbb{N}$ *the series* $\sum_{j=1}^{\infty} a_{(i,j)}$ *converges and furthermore the series* $\sum_{i=1}^{\infty}\left(\sum_{j=1}^{\infty} a_{(i,j)}\right)$ *also converges. We will also denote the double series as well as its limit by* $\sum_{i=1}^{\infty}\sum_{j=1}^{\infty} a_{i,j}$ *instead of* $\sum_{i=1}^{\infty}\sum_{j=1}^{\infty} a_{(i,j)}$.

Proposition 6.22 shows that for double series of nonnegative numbers the order of summation is immaterial.

Proposition 6.22 *Let* $\{a_{(i,j)}\}_{i,j=1}^{\infty}$ *be a family of nonnegative numbers. Then the double series* $\sum_{i=1}^{\infty}\sum_{j=1}^{\infty} a_{(i,j)}$ *converges iff for all bijections* $\sigma : \mathbb{N} \times \mathbb{N} \to \mathbb{N} \times \mathbb{N}$ *the sum* $\sum_{i=1}^{\infty}\sum_{j=1}^{\infty} a_{\sigma(i,j)}$ *converges. Furthermore, in this case the values are equal.*

Proof. The direction "$\Leftarrow$" is trivial, because we can choose $\sigma(i, j) := (i, j)$. To prove "$\Rightarrow$," let $\sum_{i=1}^{\infty}\sum_{j=1}^{\infty} a_{(i,j)}$ be convergent and let $\sigma : \mathbb{N} \times \mathbb{N} \to \mathbb{N} \times \mathbb{N}$ be a bijection.

Let $i \in \mathbb{N}$. Then for any $m \in \mathbb{N}$ we have $\sum_{j=1}^{m} a_{\sigma(i,j)} \leq \sum_{k=1}^{\infty}\sum_{j=1}^{\infty} a_{(k,j)}$. Hence, by Lemma 6.3 each series $\sum_{j=1}^{\infty} a_{\sigma(i,j)}$ converges.

For the convergence of the rearranged double series to $\sum_{i=1}^{\infty}\sum_{j=1}^{\infty} a_{(i,j)}$, let $\varepsilon > 0$ be arbitrary but fixed, and let $m \in \mathbb{N}$. Then for all $i \in \{1, \ldots, m\}$ there is an N_i so that

$\sum_{j=N_i+1}^{\infty} a_{\sigma(i,j)} < \frac{\varepsilon}{m}$. Let $N := \max\{N_i : i \in \{1, \ldots, m\}\}$. Then

$$\sum_{i=1}^{m}\sum_{j=1}^{\infty} a_{\sigma(i,j)} \leq \sum_{i=1}^{m}\left(\sum_{j=1}^{N} a_{\sigma(i,j)} + \frac{\varepsilon}{m}\right) = \varepsilon + \sum_{i=1}^{m}\sum_{j=1}^{N} a_{\sigma(i,j)}$$
$$\leq \varepsilon + \sum_{i=1}^{\infty}\sum_{j=1}^{\infty} a_{(i,j)}.$$

Because ε was arbitrary, for all $m \in \mathbb{N}$ the inequality $\sum_{i=1}^{m}\sum_{j=1}^{\infty} a_{\sigma(i,j)} \leq \sum_{i=1}^{\infty}\sum_{j=1}^{\infty} a_{(i,j)}$ holds. In particular, this means that the double series $\sum_{i=1}^{\infty}\sum_{j=1}^{\infty} a_{\sigma(i,j)}$ converges and $\sum_{i=1}^{\infty}\sum_{j=1}^{\infty} a_{\sigma(i,j)} \leq \sum_{i=1}^{\infty}\sum_{j=1}^{\infty} a_{(i,j)}$, which proves "$\Rightarrow$."

To show that the inequality actually is an equation, note that the roles of the double series can be reversed in the above argument to arrive at the reversed inequality (see Exercise 6-24).

Therefore, we conclude that $\sum_{i=1}^{\infty}\sum_{j=1}^{\infty} a_{\sigma(i,j)} = \sum_{i=1}^{\infty}\sum_{j=1}^{\infty} a_{(i,j)}$, which completes the proof. ■

We will revisit the summation of double series that also have negative terms with more sophisticated tools in Exercise 14-50.

Exercises

6-13. Prove that $\sum_{j=1}^{\infty} \frac{1}{2j+1} = \infty$ and $\sum_{j=1}^{\infty} (-1)\frac{1}{2j} = -\infty$.

6-14. Let $i, j, m, n \in \mathbb{N}$ Prove that $(i, j) = (m, n)$ iff $i = m$ and $j = n$.

6-15 **Cauchy Criterion** for absolute convergence Prove that a series $\sum_{j=1}^{\infty} a_j$ converges absolutely iff for all $\varepsilon > 0$ there is an $N \in \mathbb{N}$ so that for all $n \geq m \geq N$ we have $\sum_{j=m}^{n} |a_j| < \varepsilon$

6-16. **Comparison Test** revisited. Let $k \in \mathbb{N}$ and let $\sum_{j=1}^{\infty} a_j$ and $\sum_{j=1}^{\infty} b_j$ be series with $0 \leq a_j \leq b_j$ for all $j \geq k$. Prove that if $\sum_{j=1}^{\infty} b_j$ converges, then $\sum_{j=1}^{\infty} a_j$ converges, too.

6-17 Prove that a series $\sum_{j=1}^{\infty} a_j$ converges absolutely iff the sequence $\left\{\sum_{j=1}^{n} |a_j|\right\}_{n=1}^{\infty}$ is bounded

6-18. Prove that the sum of two absolutely convergent series also converges absolutely.

6-19. Give an example of an absolutely convergent series $\sum_{j=1}^{\infty} a_j$ so that $\sum_{j=1}^{\infty} a_j \neq \sum_{j=1}^{\infty} |a_j|$.

6-20. Let $\sum_{j=1}^{\infty} a_j$ be an absolutely convergent series and let $\{b_j\}_{j=1}^{\infty}$ be a bounded sequence. Prove that $\sum_{j=1}^{\infty} a_j b_j$ converges absolutely.

6-21. Determine which of the following series converges. If it converges, determine if it converges absolutely.

(a) $\sum_{j=1}^{\infty} \frac{1}{j!}$ (b) $\sum_{j=1}^{\infty} \frac{(-1)^j}{j+\sqrt{j}}$ (c) $\sum_{j=1}^{\infty} \frac{(-1)^j}{4^j}$

6-22. **Limit Comparison Test** for series. Let $\sum_{j=1}^{\infty} a_j$ and $\sum_{j=1}^{\infty} b_j$ be series with positive terms. Prove that if $\lim_{j\to\infty} \frac{a_j}{b_j} = c > 0$, then either both series converge or both series diverge.
Hint. For $\varepsilon > 0$ there is an $N \in \mathbb{N}$ so that for all $j \geq N$ we have $(c-\varepsilon)b_j \leq a_j \leq (c+\varepsilon)b_j$.

6-23. Let $\sum_{j=1}^{\infty} a_j$ be a conditionally convergent series and let $z \in \mathbb{R}$. Prove that there is a bijective function $\sigma : \mathbb{N} \to \mathbb{N}$ such that $\sum_{j=1}^{\infty} a_{\sigma(j)} = z$.
Hint. Mimic the proof of the "$\Leftarrow$" part of Theorem 6.18. Oscillate about z rather than increasing with k. Use that the limit of the terms is zero and that the distance of the partial sum $\sum_{j=1}^{n} a_{\sigma(j)}$ to z ultimately is always bounded by some $|a_m|$ with m being large.

6-24. Reverse the roles of $\sum_{i=1}^{\infty}\sum_{j=1}^{\infty} a_{\sigma(i,j)}$ and $\sum_{i=1}^{\infty}\sum_{j=1}^{\infty} a_{(i,j)}$ as indicated in the proof of Proposition 6.22 to show that $\sum_{i=1}^{\infty}\sum_{j=1}^{\infty} a_{\sigma(i,j)} \geq \sum_{i=1}^{\infty}\sum_{j=1}^{\infty} a_{(i,j)}$.

6-25. Let $\{a_{(i,j)}\}_{i,j=1}^{\infty}$ be a family of real numbers so that the double series $\sum_{i=1}^{\infty}\sum_{j=1}^{\infty} |a_{(i,j)}|$ converges. Prove that $\sum_{i=1}^{\infty}\sum_{j=1}^{\infty} a_{(i,j)}$ converges to a number L and for all bijections $\sigma : \mathbb{N}\times\mathbb{N} \to \mathbb{N}\times\mathbb{N}$ the sum $\sum_{i=1}^{\infty}\sum_{j=1}^{\infty} a_{\sigma(i,j)}$ converges to the same number L.

Readers interested in further results on series could read Chapter 10 before proceeding.

Chapter 7

Some Set Theory

Up to this point our development of analysis had minimal need for details of set theory. However, to define the outer Lebesgue measure of a set (see Definition 8.1), we will cover the set with an infinite family of intervals and sum the lengths. To be able to compute the sum as a series, we can use at most as many intervals as there are natural numbers. Surprisingly enough this is not trivial, because infinite sets can come in different sizes. This chapter presents the fundamentals on arbitrary families of sets and on the notion of size for sets. These ideas will be needed immediately for outer Lebesgue measure and they will be useful throughout our work with measures. All sophisticated set theory needed in this text is presented in this chapter.

7.1 The Algebra of Sets

Recall from the introduction to Chapter 1 that the terms "set" and "element" remain undefined. To be able to work with arbitrary families of sets, we need to formally define them and show some fundamental properties. Because every family of sets must consist of subsets of another set, we start with the power set of a set.

Definition 7.1 *Let S be a set. The* **power set** $\mathcal{P}(S)$ *of S is the set of all subsets of S.*

Definition 7.2 *Let S be a set. A* **family** *of sets $\mathcal{C}$ is simply a set of subsets of S, that is, $\mathcal{C} \subseteq \mathcal{P}(S)$. An* **indexed family** *of sets is denoted $\{C_\iota\}_{\iota\in I}$, where it is understood that there is a function from the* **index set** *I to $\mathcal{P}(S)$ that maps each $i \in I$ to C_i.*

So while families of sets do not introduce any new set theoretical ideas, we call them "families of sets" rather than "sets of sets" to indicate that we are now considering objects whose elements are set theoretically more complicated than just elements that cannot be decomposed any further. A family of sets can always be turned into an indexed family of sets by making each set its own index. Conversely, an indexed family can be turned into a nonindexed family by considering the set $\mathcal{C} := \{C_\iota : i \in I\}$, but this process will collapse sets $C_i = C_j$ with $i \neq j$ into one set in $\mathcal{C}$. In our work, we may need repeated sets, so most of the families of sets we consider will be indexed.

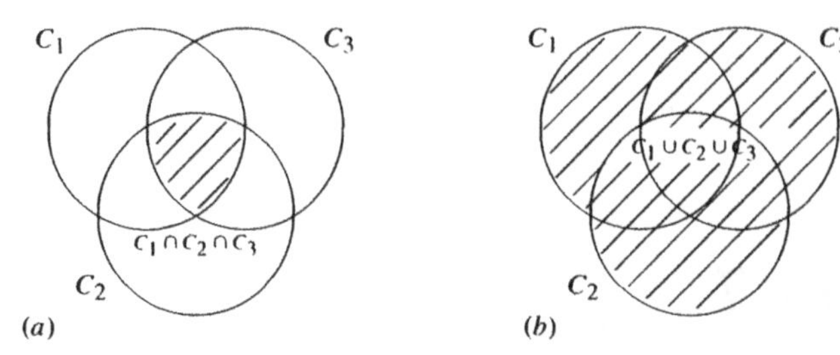

Figure 16: Venn diagrams of the intersection (a) and the union (b) of three sets.

Nonetheless, proofs for both kinds of families are very similar and we will present both notations in this section. The most elementary operations for families of sets are unions and intersections (also see the **Venn diagrams** in Figure 16).

Definition 7.3 *Let $\mathcal{C}$ be a family of sets in S. Then we define the* **union** *of $\mathcal{C}$ as the set $\bigcup \mathcal{C} := \{x \in S : (\exists C \in \mathcal{C} : x \in C)\}$. If $\{C_\iota\}_{\iota \in I}$ is an indexed family, the union is denoted $\bigcup_{\iota \in I} C_\iota := \{x \in S : (\exists i \in I : x \in C_\iota)\}$.*

Definition 7.4 *Let $\mathcal{C}$ be a family of sets in S. Then we define the* **intersection** *of $\mathcal{C}$ as the set $\bigcap \mathcal{C} := \{x \in S : (\forall C \in \mathcal{C} : x \in C)\}$. If $\{C_\iota\}_{\iota \in I}$ is an indexed family, the intersection is $\bigcap_{i \in I} C_\iota := \{x \in S : (\forall i \in I : x \in C_\iota)\}$.*

The relation between complements (see page 1) and unions/intersections is described by DeMorgan's Laws.

Theorem 7.5 DeMorgan's Laws. *Let $\mathcal{C}$ and $\{C_\iota\}_{\iota \in I}$ be families of sets and let X be another set. Then*

1. $X \setminus \bigcup_{\iota \in I} C_\iota = \bigcap_{\iota \in I} X \setminus C_\iota$ *and* $X \setminus \bigcup \mathcal{C} = \bigcap \{X \setminus C : C \in \mathcal{C}\}$.

2. $X \setminus \bigcap_{\iota \in I} C_\iota = \bigcup_{\iota \in I} X \setminus C_\iota$ *and* $X \setminus \bigcap \mathcal{C} = \bigcup \{X \setminus C : C \in \mathcal{C}\}$.

Proof. *Proofs for indexed and nonindexed families are very similar. Hence, we will only present one proof for each part.*

For part 1, we will prove the containments $X \setminus \bigcup \mathcal{C} \subseteq \bigcap \{X \setminus C : C \in \mathcal{C}\}$ and $X \setminus \bigcup \mathcal{C} \supseteq \bigcap \{X \setminus C : C \in \mathcal{C}\}$, which (see introduction to Chapter 1) shows the two sets are equal.

For the containment relation $X \setminus \bigcup \mathcal{C} \subseteq \bigcap \{X \setminus C : C \in \mathcal{C}\}$, consider an arbitrary $x \in X \setminus \bigcup \mathcal{C}$. Then $x \notin \bigcup \mathcal{C}$, which means that $x \notin C$ for all $C \in \mathcal{C}$. But then $x \in X \setminus C$ for all $C \in \mathcal{C}$, which means $x \in \bigcap \{X \setminus C : C \in \mathcal{C}\}$. This proves that $X \setminus \bigcup \mathcal{C}$ is contained in $\bigcap \{X \setminus C : C \in \mathcal{C}\}$.

For the reversed containment relation $X \setminus \bigcup \mathcal{C} \supseteq \bigcap \{X \setminus C : C \in \mathcal{C}\}$, consider an arbitrary $x \in \bigcap \{X \setminus C : C \in \mathcal{C}\}$. Then for all $C \in \mathcal{C}$ we have $x \in X \setminus C$, that is, $x \notin C$. But then $x \notin \bigcup \mathcal{C}$, which means $x \in X \setminus \bigcup \mathcal{C}$. This proves that $X \setminus \bigcup \mathcal{C}$ contains $\bigcap \{X \setminus C : C \in \mathcal{C}\}$.

Because each set is contained in the respective other set, the sets must be equal. It is also possible to prove part 1 with a chain of equalities, like part 2 below (Exercise 7-1a). The proof of part 1 for indexed families of sets is similar (Exercise 7-1b.).

We prove part 2 for indexed families.

$$\begin{aligned} X \setminus \bigcap_{i \in I} C_i &= \left\{ x \in X : x \notin \bigcap_{i \in I} C_i \right\} = \{x \in X : (\exists i \in I : x \notin C_i)\} \\ &= \{x \in X : (\exists i \in I : x \in X \setminus C_i)\} = \bigcup_{i \in I} X \setminus C_i . \end{aligned}$$

It is also possible to prove part 2 with a mutual containment argument as done for part 1 (Exercise 7-1c). The proof of part 2 for (nonindexed) families of sets is similar (Exercise 7-1d). ■

Standard Proof Technique 7.6 To prove equality of two sets A and B it is common to prove **mutual containment**, that is, $A \subseteq B$ and $B \subseteq A$. This is similar to proving that two numbers are equal by proving that one number is both greater than or equal to and less than or equal to the other one. □

Ordered pairs and products also have formal set theoretical definitions.

Definition 7.7 *Let A and B be sets and let $a \in A$ and $b \in B$. Then we define the* **ordered pair** *(a, b) to be $(a, b) := \{a, \{a, b\}\}$. If $A_1, \ldots, A_n$ are sets and $a_i \in A_i$ for $i = 1, \ldots, n$, we define $(a_1, \ldots, a_n) := \{(a_1, \ldots, a_{n-1}), \{(a_1, \ldots, a_{n-1}), a_n\}\}$ and call it an* **ordered n-tuple**.

Definition 7.8 *Let $A_1, \ldots, A_n$ be sets. Then the* **product** *$A_1 \times \cdots \times A_n$ of these sets is defined to be the set of all ordered n-tuples $(a_1, \ldots, a_n)$ so that for all $i = 1, \ldots, n$ we have $a_i \in A_i$. The product is also denoted $\prod_{i=1}^{n} A_i$. Moreover, for $j \in \{1, \ldots, n\}$ we define the natural* **projection** *$\pi_{A_j} : \prod_{i=1}^{n} A_i \to A_j$ onto A_j by $\pi_{A_j}(a_1, \ldots, a_n) := a_j$. We will also abbreviate the natural projection onto the jth factor as π_j.*

For a visualization of the product of two sets, see Figure 17. The definition of a product is fundamental for the formal definition of functions and relations. Because we will not need this level of detail in this text, these definitions have been relegated to Appendix B.2.

Theorem 7.9 *Let $\{C_i\}_{i \in I}$ be a family of sets and let A be a set. Then the equalities $A \times \bigcup_{i \in I} C_i = \bigcup_{i \in I} A \times C_i$ and $A \times \bigcap_{i \in I} C_i = \bigcap_{i \in I} A \times C_i$ hold.*

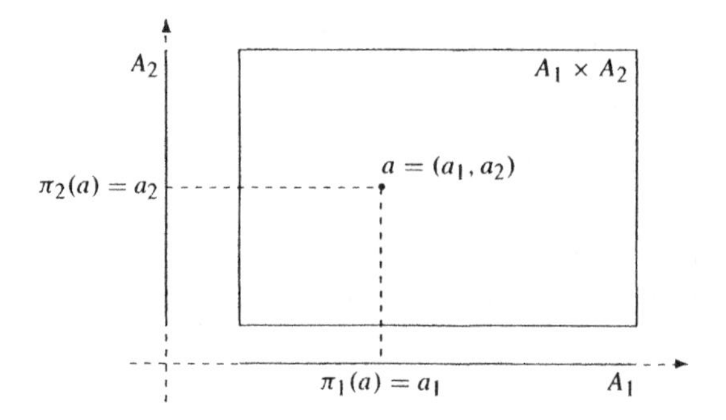

Figure 17: The product of two sets is a "rectangle" with the natural projections π_i providing the "coordinates" of an element.

Proof. Left to Exercise 7-2. ■

The distributive laws express the relationship between unions and intersections.

Theorem 7.10 Distributive laws. *Let $\{A_i\}_{i\in I}$ and $\{B_j\}_{j\in J}$ be indexed families of sets. Then* $\bigcup_{i\in I} A_i \cap \bigcup_{j\in J} B_j = \bigcup_{(i,j)\in I\times J} A_i \cap B_j$ *and* $\bigcap_{i\in I} A_i \cup \bigcap_{j\in J} B_j = \bigcap_{(i,j)\in I\times J} A_i \cup B_j$.

Proof. For the first equation let $x \in \bigcup_{i\in I} A_i \cap \bigcup_{j\in J} B_j$. Then there are an $i \in I$ so that $x \in A_i$ and a $j \in J$ so that $x \in B_j$, which means $x \in A_i \cap B_j \subseteq \bigcup_{(i,j)\in I\times J} A_i \cap B_j$. For the reverse inclusion, let $x \in \bigcup_{(i,j)\in I\times J} A_i \cap B_j$. Then there is an $(i, j) \in I \times J$ so that $x \in A_i \cap B_j \subseteq \bigcup_{i\in I} A_i \cap \bigcup_{j\in J} B_j$.

For the second equation, first note that if $x \in \bigcap_{i\in I} A_i \cup \bigcap_{j\in J} B_j$, then $x \in A_i \cup B_j$ for all $(i, j) \in I \times J$. For the reverse inclusion, let $x \in \bigcap_{(i,j)\in I\times J} A_i \cup B_j$. If $x \in A_i$ for all $i \in I$, then $x \in \bigcap_{i\in I} A_i \subseteq \bigcap_{i\in I} A_i \cup \bigcap_{j\in J} B_j$ and if $x \notin A_{i_0}$ for some $i_0 \in I$, then $x \in A_{i_0} \cup B_j$ for all $j \in J$, and hence $x \in \bigcap_{j\in J} B_j \subseteq \bigcap_{i\in I} A_i \cup \bigcap_{j\in J} B_j$. ■

Exercises

7-1 Proving DeMorgan's Laws.

(a) Prove part 1 of Theorem 7.5 for families of sets using a chain of equalities.

(b) Prove part 1 of Theorem 7.5 for indexed families of sets.

(c) Prove part 2 of Theorem 7.5 for indexed families of sets using a mutual containment argument

(d) Prove part 2 of Theorem 7.5 for nonindexed families of sets.

7-2. Prove Theorem 7.9. That is, let $\{C_\iota\}_{\iota\in I}$ be a family of sets and let A be a set.

(a) Prove that $A \times \bigcup_{\iota\in I} C_\iota = \bigcup_{\iota\in I} A \times C_\iota$.

(b) Prove that $A \times \bigcap_{\iota\in I} C_\iota = \bigcap_{\iota\in I} A \times C_\iota$.

7-3 Let A, B, C be sets. Prove each of the following.

(a) $C \setminus (A \cap B) = (C \setminus A) \cup (C \setminus B)$

(b) $C \setminus (A \cup B) = (C \setminus A) \cap (C \setminus B)$

(c) $C \cup (A \cap B) = (C \cup A) \cap (C \cup B)$

(d) $C \cap (A \cup B) = (C \cap A) \cup (C \cap B)$

(e) $(A \setminus B) \cap C = (A \cap C) \setminus (B \cap C)$

(f) If $A, B \subseteq C$, then $A \setminus B = A \cap (C \setminus B)$

7-4. Give an example of a family of sets with $\bigcap \mathcal{C} = \emptyset$ so that any two sets in the family intersect.

7-5. Let X be a set. An **algebra** is a set of sets $\mathcal{A} \subseteq \mathcal{P}(X)$ such that $\emptyset \in \mathcal{A}$, if $A \in \mathcal{A}$, then $X \setminus A \in \mathcal{A}$, and if $A_j \in \mathcal{A}$ for all $j = 1, \ldots, n$, then $\bigcup_{j=1}^{n} A_j \in \mathcal{A}$

(a) Prove that if $A_j \in \mathcal{A}$ for all $j = 1, \ldots, n$, then $\bigcap_{j=1}^{n} A_j \in \mathcal{A}$.

(b) Let X be a set Prove that the power set of X is an algebra

(c) Let X be a set. Prove that $\mathcal{A} := \{A \subseteq X \quad A$ or $X \setminus A$ is finite $\}$ is an algebra.

(d) Prove that an algebra need not contain countable unions of its elements

7-6. Let $A \subseteq X$ and $B \subseteq Y$ be sets. Prove that
$(X \times Y) \setminus (A \times B) = \big[(X \setminus A) \times (Y \setminus B)\big] \cup \big[(X \setminus A) \times B\big] \cup \big[A \times (Y \setminus B)\big]$.

7-7. Let A, B, C, D be sets Prove that $(A \times B) \cap (C \times D) = (A \cap C) \times (B \cap D)$.

7-8 Let f be a function whose domain is contained in X and whose range is contained in Y.

(a) Prove that $\pi_X \big[\{(x, y) \in X \times Y : y = f(x)\}\big]$ is the domain of f

(b) Prove that $\pi_Y \big[\{(x, y) \in X \times Y : y = f(x)\}\big]$ is the range of f

7-9. Unions and intersections versus functions. Let X, Y be sets, let $f : X \to Y$ be a function, let $\{X_\iota\}_{\iota\in I}$ be a family of subsets of X and let $\{Y_\iota\}_{\iota\in I}$ be a family of subsets of Y. Prove each of the following.

(a) $f^{-1}\left[\bigcup_{\iota\in I} Y_\iota\right] = \bigcup_{\iota\in I} f^{-1}[Y_\iota]$.

(b) $f^{-1}\left[\bigcap_{\iota\in I} Y_\iota\right] = \bigcap_{\iota\in I} f^{-1}[Y_\iota]$.

(c) $f\left[\bigcup_{\iota\in I} X_\iota\right] = \bigcup_{\iota\in I} f[X_\iota]$.

(d) $f\left[\bigcap_{\iota\in I} X_\iota\right] \subseteq \bigcap_{\iota\in I} f[X_\iota]$.

(e) For $f(x) := x^2$ there are sets A, B with $f[A \cap B] \neq f[A] \cap f[B]$.

(f) $f\left[\bigcap_{\iota\in I} X_\iota\right] = \bigcap_{\iota\in I} f[X_\iota]$ holds if f is injective

7.2 Countable Sets

Two sets are considered to be of the same size iff their elements can be matched one-by-one without any leftovers on either side. Recall that bijective functions were defined in Definition 2.24.

Definition 7.11 *Two sets A and B are called* **equivalent** *iff there is a bijective function $f : A \to B$.*

With this language, Definition 2.25 says that a nonempty set is finite iff it is equivalent to a subset $\{1, \ldots, n\}$ of $\mathbb{N}$. In Figure 3 on page 36, only the two sets on the right are equivalent. For analysis, the most important size distinction between infinite sets is if they are countable or uncountable. Countable sets are, roughly speaking, sets for which it is possible to "count" the elements, where the counting process may stop or it may not.

Definition 7.12 *A set C is called* **countably infinite** *iff there is a bijective function $f : \mathbb{N} \to C$. A set C is called* **countable** *iff C is finite or countably infinite.*

Subsets of a set cannot be larger than the set, so it is natural that subsets of countable sets are countable.

Theorem 7.13 *If C is countable and $S \subseteq C$, then S is countable.*

Proof. If S is finite, there is nothing to prove. Hence, we can assume that S is infinite. Let $f : \mathbb{N} \to C$ be a bijection and let $n_1 := \min f^{-1}[S]$. For $k \in \mathbb{N}$, we define n_{k+1} recursively. Once $n_1, \ldots, n_k$ are chosen, let $n_{k+1} := \min \left(f^{-1}[S] \setminus \{n_1, \ldots, n_k\} \right)$. Define $g : \mathbb{N} \to S$ by $g(k) := f(n_k)$. Because all n_k are in $f^{-1}[S]$, g maps $\mathbb{N}$ into S. Because no two n_k are equal and f is injective, g is injective.

Finally, suppose for a contradiction that g is not surjective. Let b be the smallest element of $f^{-1}\big[S \setminus g[\mathbb{N}]\big]$ and let the number of elements of $f^{-1}[S] \cap \{1, \ldots, b-1\}$ be k. Then $f^{-1}[S] \cap \{1, \ldots, b-1\} = \{1, \ldots, n_k\}$ (with $n_0 = 0$ in case $k = 0$) and $b = \min \left(f^{-1}[S] \setminus \{n_1, \ldots, n_k\} \right)$, which means $b = n_{k+1}$, a contradiction. ■

Theorem 7.13 says that to prove that a set is countable, it is enough to embed a copy of it into a countable set. In the language of analysis, it is good enough to find an "upper bound" that is countable. We will use this idea repeatedly in the remainder of this section. Interestingly enough, even sets that look uncountable may be countable.

Lemma 7.14 *The set $\mathbb{N} \times \mathbb{N}$ is countable.*

Proof. The function $f : \mathbb{N} \times \mathbb{N} \to \mathbb{N}$ defined by $f(m, n) := 2^m 3^n$ is injective. Thus the set $\mathbb{N} \times \mathbb{N}$ is equivalent to a subset of $\mathbb{N}$, and hence by Theorem 7.13 $\mathbb{N} \times \mathbb{N}$ is countable. ■

Note how Theorem 7.13 allowed us to avoid the detailed construction of a bijective function between $\mathbb{N}$ and $\mathbb{N} \times \mathbb{N}$. Defining such a function is not trivial. Exercise 7-10 explicitly presents a bijective function between $\mathbb{N} \times \mathbb{N}$ and $\mathbb{N}$.

Definition 7.15 *Two sets A and B are called* **disjoint** *iff $A \cap B = \emptyset$. A family $\{C_i\}_{i \in I}$ is called* **pairwise disjoint** *iff for all $i \neq j$ we have $C_i \cap C_j = \emptyset$.*

Theorem 7.16 *Countable unions of countable sets are countable.*

Proof. Let $\{C_n\}_{n=1}^{\alpha}$ with $\alpha \in \mathbb{N} \cup \{\infty\}$ be a countable family of countable sets. For each C_n, let $B_n := C_n \setminus \bigcup_{j=1}^{n-1} C_j$. We claim $\bigcup_{n=1}^{\alpha} B_n = \bigcup_{n=1}^{\alpha} C_n$. The containment "$\subseteq$" follows from $B_n \subseteq C_n$ for all $n \in \mathbb{N}$. For "$\supseteq$" let $x \in \bigcup_{n=1}^{\alpha} C_n$. Let $n \in \mathbb{N}$ be the smallest natural number so that $x \in C_n$. Then $x \in B_n$, which proves "$\supseteq$." Moreover, the B_n are pairwise disjoint, because if $m < n$, then $B_m = C_m \setminus \bigcup_{j=1}^{m-1} C_j \subseteq C_m$ and $B_n = C_n \setminus \bigcup_{j=1}^{n-1} C_j \subseteq C_n \setminus C_m$. Now for each n let $B_n = \left\{ b_n^k : k \in I_n \right\}$, where I_n is $\mathbb{N}$ or a set of the form $\{1, \ldots, m_n\}$. Then $f(n, k) := b_n^k$ is a bijective function between $\left\{ (n, k) \in \mathbb{N} \times \mathbb{N} : n \leq \alpha, k \in I_n \right\} \subseteq \mathbb{N} \times \mathbb{N}$ and $\bigcup_{n=1}^{\alpha} B_n = \bigcup_{n=1}^{\alpha} C_n$. Thus $\bigcup_{n=1}^{\infty} C_n$ is countable. ■

Maybe the most surprising fact about countability is that the rational numbers are countable. The reason is that the order in which the rational numbers are counted has nothing to do with their natural ordering.

Theorem 7.17 *The rational numbers $\mathbb{Q}$ are countable.*

Proof. We have $\{r \in \mathbb{Q} : r > 0\} = \bigcup_{d=1}^{\infty} \left\{ \frac{n}{d} : n \in \mathbb{N} \right\}$, that is, the positive rational numbers are a countable union of countable sets, and hence by Theorem 7.16 they are countable. Similarly, $\{r \in \mathbb{Q} : r < 0\}$ is countable and of course $\{0\}$ is finite. Now $\mathbb{Q} = \{r \in \mathbb{Q} : r > 0\} \cup \{r \in \mathbb{Q} : r < 0\} \cup \{0\}$ is countable by Theorem 7.16. ■

Exercises

7-10. Prove that the function $f : \mathbb{N} \times \mathbb{N} \to \mathbb{N}$ defined by $f(m, n) := \frac{1}{2}(m + n - 1)(m + n - 2) + n$ is bijective. (For a visualization, consider the middle of Figure 18.)

7-11. Prove that the set of integers $\mathbb{Z}$ is countable.

7-12. Prove that the set of integers $\mathbb{Z}$ is countable by constructing a bijective function $f : \mathbb{N} \to \mathbb{Z}$.
Hint. Figure 18(a).

7-13. Prove that the set of dyadic rational numbers is countable.

7-14. Use Theorem 7.16 to prove Lemma 7.14.

7-15. Give a *direct* proof that the union of *two* countable sets is countable.

7-16. Prove that if $C_1, \ldots, C_n$ are countable, then $C_1 \times C_2 \times \cdots \times C_n$ is countable.

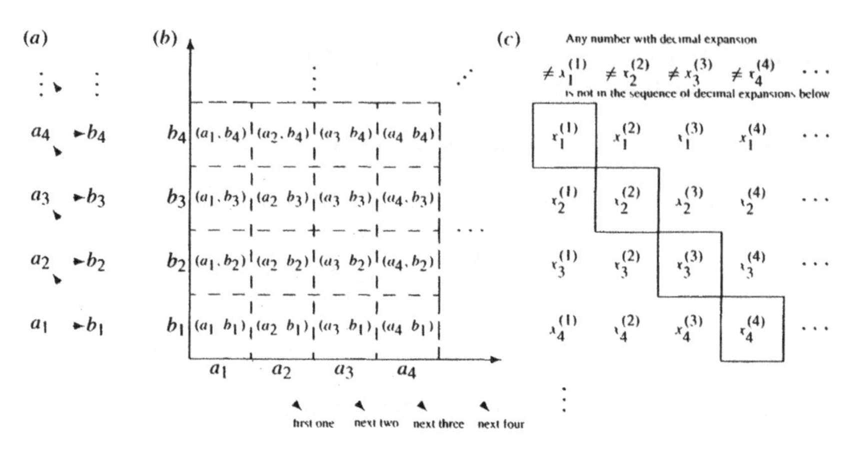

Figure 18: Some standard visualizations for countability arguments. Part (a) shows the construction for a direct proof that the union of two countable sets is countable. Part (b) shows an explicit bijective function between $\mathbb{N}$ and the product of two countable sets. Part (c) shows the idea behind the proof that $(0, 1)$ is not countable.

7-17. Let $\{a_{(i,j)}\}_{i,j=1}^{\infty}$ be a family of nonnegative numbers. Then the double series $\sum_{i=1}^{\infty}\sum_{j=1}^{\infty} a_{(i,j)}$ converges iff for all bijections $\sigma : \mathbb{N} \to \mathbb{N} \times \mathbb{N}$ the sum $\sum_{i=1}^{\infty} a_{\sigma(i)}$ converges. Furthermore, in this case the values are equal. *Hint.* This is similar to the proof of Proposition 6.22.

7.3 Uncountable Sets

Not all sets are countable. In fact, (see Exercise 7-18) there is an infinite hierarchy of sizes for infinite sets, because any time we form a power set, we obtain a set that is strictly larger than the set we started with.

Theorem 7.18 *If X is a set, then X is not equivalent to its power set $\mathcal{P}(X)$.*

Proof. Suppose for a contradiction that $f : X \to \mathcal{P}(X)$ is a bijection. Define $B := \{x \in X : x \notin f(x)\}$. Because f is surjective, there is a $b \in X$ with $B = f(b)$. Now $b \in B$ would imply $b \in B = f(b)$ and by definition of B this would mean $b \notin f(b) = B$. Thus we infer $b \notin B$. But then $b \notin B = f(b)$, which by definition of B forces $b \in B$, a contradiction. ∎

For analysis, we typically do not need the full hierarchy. Instead we only need to distinguish sets that are countable from those that are not.

Definition 7.19 *A set U is called* **uncountable** *iff it is not countable.*

The real numbers, which are fundamental for analysis, are not countable.

Theorem 7.20 *The interval $(0, 1)$ is uncountable.*

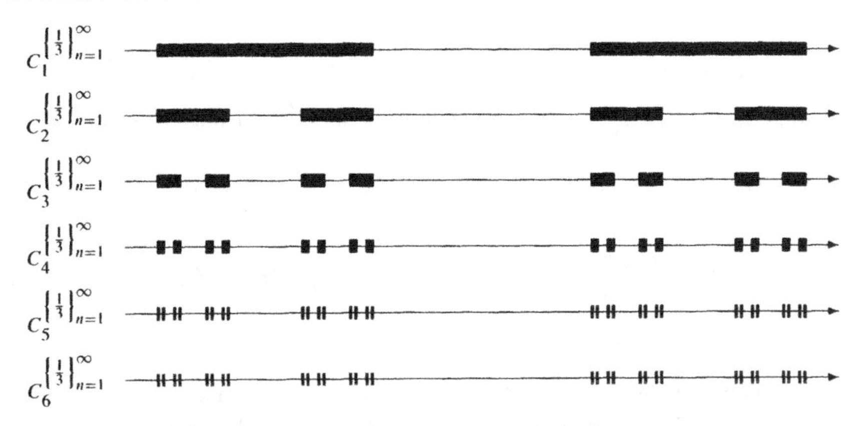

Figure 19: Cantor sets are the intersection of a sequence of unions of closed intervals where at each step only the left and right segments of each interval are kept.

Proof. Suppose for a contradiction that $(0, 1)$ was countable. Then there is a sequence $\{x_n\}_{n=1}^{\infty}$ such that for every $x \in (0, 1)$ there is an $n \in \mathbb{N}$ so that $x = x_n$. For each $n \in \mathbb{N}$, let $\left\{x_n^{(k)}\right\}_{k=1}^{\infty}$ be the decimal expansion of x_n as in Proposition 6.6. For each $n \in \mathbb{N}$, let y_n be a number in the set $\{1, 2, 3, 4, 5, 6, 7, 8\} \setminus \left\{x_n^{(n)}\right\}$. Then $\{y_n\}_{n=1}^{\infty}$ is a decimal expansion of a number $y \in (0, 1)$. However, for all $n \in \mathbb{N}$ we have that $y_n \neq x_n^{(n)}$, and hence $y \neq x_n$, contradiction. ■

The remainder of the proof that $\mathbb{R}$ is uncountable is left to Exercise 7-19b. The uncountability of the real numbers shows that, even though in real life we work mostly with rational numbers, there are many more irrational numbers than there are rational numbers.

Theorem 7.21 *The set $\mathbb{R} \setminus \mathbb{Q}$ of irrational numbers is uncountable.*

Proof. By Exercise 7-19b, the real numbers are uncountable. Now for a contradiction suppose that $\mathbb{R} \setminus \mathbb{Q}$ was countable. Then $\mathbb{R} = (\mathbb{R} \setminus \mathbb{Q}) \cup \mathbb{Q}$ would be countable by Theorem 7.16, contradiction. ■

We conclude this section by defining Cantor sets. These sets are very useful to construct counterexamples which show that certain hypotheses in analytical theorems cannot be dispensed with. Because these counterexamples can be considered a bit pathological, we defer their construction to the exercises and we will only refer to Cantor sets when necessary. Cantor sets are constructed from a sequence of unions of closed intervals so that in each step we remove the middle of each interval. Figure 19 shows the first six stages in the construction of the **ternary Cantor set**, which is constructed by successively removing the middle third of the intervals at each stage.

Definition 7.22 *Let $[a, b]$ be an interval and let $0 < q < \frac{1}{2}$. Then we define the left part of $[a, b]$ to be the interval $L_q[a, b] := \left[a, a + q(b - a)\right]$ and the right part to be the interval $R_q[a, b] := \left[b - q(b - a), b\right]$.*

For a sequence $Q := \{q_n\}_{n=1}^{\infty}$ with $0 < q_n < \frac{1}{2}$ for all $n \in \mathbb{N}$ define C_n^Q recursively as follows. Let $C_0^Q := I_{1,0}^Q := [0,1]$ and once the set C_n^Q is defined as a union of pairwise disjoint closed intervals $\bigcup_{i=1}^{2^n} I_{i,n}^Q$, for $i = 1, \ldots, 2^n$ let $I_{2i-1,n+1}^Q := L_{q_{n+1}}\left[I_{i,n}^Q\right]$, let $I_{2i,n+1}^Q := R_{q_{n+1}}\left[I_{i,n}^Q\right]$ and let $C_{n+1}^Q := \bigcup_{i=1}^{2^{n+1}} I_{i,n+1}^Q$. Then $C^Q := \bigcap_{n=1}^{\infty} C_n^Q$ is called the **Cantor set** *associated with the sequence Q.*

Even though the construction looks like it should only leave the boundary points of the intervals, Cantor sets are in fact uncountable. The details can be explored in Exercise 7-25.

Exercises

7-18 Construct a sequence $\{P_n\}_{n=1}^{\infty}$ of infinite sets so that no two sets P_n and P_{n+1} are equivalent, but for each $n \in \mathbb{N}$ there is an injective function $f_n : P_n \to P_{n+1}$.

7-19. Containment and uncountable sets

(a) Let U, V be sets. Prove that if U is uncountable and $U \subseteq V$, then V is uncountable.

(b) Prove that $\mathbb{R}$ is not a countable set.

7-20. Prove that every uncountable set contains a countably infinite subset

7-21 Prove that the set of all functions from $\mathbb{N}$ to $\{0, 1\}$ is uncountable

7-22. Let $\{a_i\}_{i\in I}$ be an uncountable family of positive numbers. Prove that there are an $\varepsilon > 0$ and a countable subfamily $\{a_{i_n}\}_{n\in\mathbb{N}}$ so that $a_{i_n} > \varepsilon$ for all $n \in \mathbb{N}$

7-23 Let $F : [a, b] \to \mathbb{R}$ be a **nondecreasing** bounded function Prove that F can have at most countably many discontinuities.

Hint. Suppose the set of discontinuities is uncountable and use Exercise 3-21 to conclude that there must be an ε so that the set of all x with $\lim_{z\to x^+} F(z) - \lim_{z\to x^-} F(z) > \varepsilon$ is uncountable.

7-24. Prove that for every countable subset $A \subseteq \mathbb{R}$ there is a **nondecreasing** function $f : \mathbb{R} \to [0, 1]$ that is continuous on $\mathbb{R} \setminus A$ and discontinuous at every $a \in A$

Hint. With $A = \{a_n : n \in \mathbb{N}\}$ set $f(x) = \sum_{n: a_n < x} \frac{1}{2^n}$

7-25 Cantor sets.

(a) Prove that for any sequence $Q = \{q_n\}_{n=1}^{\infty}$ of numbers $q_n \in \left(0, \frac{1}{2}\right)$ there is a bijective function from C^Q to the set of all sequences of zeroes and ones

Hint. For $x \in C^Q$ and $n \geq 1$, set $a_n(x) := 0$ iff $x \in I_{j,n}^Q = L_{q_n}\left(I_{k,n-1}^Q\right)$, that is, iff we have to turn "left" at the nth stage of the construction to keep x in the interval.

(b) Prove that C^Q is uncountable. *Hint.* Exercise 7-21.

(c) Prove that the set of endpoints of the intervals $I_{j,n}^Q$ that make up the C_n^Q is countable.

(d) Prove that every $x \in C^Q$ is the limit of a sequence of endpoints of intervals $I_{j,n}^Q$

Chapter 8

The Riemann Integral II

The Dirichlet function in Example 5.26 shows that functions with too many discontinuities may not be Riemann integrable. This is because for a discontinuity d there is an $\varepsilon > 0$ so that, independent of δ, the numbers $\inf\{f(x) : x \in (d-\delta, d+\delta)\}$ and $\sup\{f(x) : x \in (d-\delta, d+\delta)\}$ will always be at least ε apart. By Theorem 5.25, too many such discontinuities will cause the function to not be Riemann integrable. At the same time, Proposition 5.12 shows that a function can have some discontinuities and still be Riemann integrable. To determine when a function is Riemann integrable, we need to determine "how many" discontinuities are acceptable. For a graphical motivation, consider Figure 20 on page 133. Section 8.1 introduces outer Lebesgue measure, which is the tool to measure "how many" discontinuities a function has. Section 8.2 introduces Lebesgue's integrability criterion and Section 8.3 shows how this criterion allows us to easily obtain new results about the Riemann integral. We conclude in Section 8.4 with improper integrals.

8.1 Outer Lebesgue Measure

Outer Lebesgue measure covers a set with open intervals and assigns the infimum of the sums of the lengths of these intervals as the measure ("size") of the set. With respect to Riemann integrability, we should note that if all discontinuities of our function are trapped in a union of intervals, then outside of this union of intervals the function is continuous and Riemann integrability should not be a problem there.

Definition 8.1 *For an open interval $I = (a, b)$ in $\mathbb{R}$, we define $|I| := b - a$. For any set $S \subseteq \mathbb{R}$, we define the* **outer Lebesgue measure** *of S to be*

$$\lambda(S) := \inf\left\{ \sum_{j=1}^{\infty} |I_j| : S \subseteq \bigcup_{j=1}^{\infty} I_j,\ \text{each } I_j \text{ is an open interval} \right\},$$

where we set $\lambda(S) = \infty$ if none of the series in the set on the right converge.

The proof of Proposition 8.5 will show why we use open intervals in the definition of outer Lebesgue measure. We first turn our attention to sets with outer Lebesgue measure zero. These sets, and properties that hold on the complement of such a set, will be of particular importance for the Riemann integral.

Definition 8.2 *A set of outer Lebesgue measure zero is called a set of* **measure zero** *or a* **null set**. *A property $P(x)$ such that $\lambda(\{x \in D : P(x) \text{ is not true }\}) = 0$ is said to hold* **almost everywhere** *in D. Almost everywhere is also abbreviated as* **a.e.**

Countable sets are considered "small" in set theory and they are also "small" with respect to outer Lebesgue measure.

Proposition 8.3 *Countable subsets of $\mathbb{R}$ have outer Lebesgue measure* 0.

Proof. Let $C = \{c_1, c_2, \ldots\}$ be a countable subset of $\mathbb{R}$ and let $\varepsilon > 0$. For $j \in \mathbb{N}$ let $I_j := \left(c_j - \frac{\varepsilon}{2}\frac{1}{2^j}, c_j + \frac{\varepsilon}{2}\frac{1}{2^j}\right)$. Then $C \subseteq \bigcup_{j=1}^{\infty} I_j$ and $\sum_{j=1}^{\infty} |I_j| = \sum_{j=1}^{\infty} \varepsilon \frac{1}{2^j} = \varepsilon$. Thus $\lambda(C) = 0$. ■

Although the proof of Proposition 8.3 is quite simple, it takes a little to get used to the result. Recall that $\mathbb{Q}$ is countable, which means it is a null set! Exercise 8-3d will show that null sets can be uncountable, too. Of course, not all subsets of $\mathbb{R}$ are null sets. To prove that outer Lebesgue measure provides the "right" measure for intervals, we need the Heine-Borel Theorem. The conclusion of the Heine-Borel Theorem is the inspiration for the topological definition of compactness (see Theorem 16.72). Until we formally introduce compactness in Section 16.5 we will rely on Standard Proof Technique 2.28 as used in the proof of the Heine-Borel Theorem.

Theorem 8.4 Heine-Borel Theorem. *Let $[a, b] \subset \mathbb{R}$ and let $\mathcal{I}$ be a family of open intervals with $[a, b] \subseteq \bigcup_{I \in \mathcal{I}} I$. Then there are finitely many intervals $I_1, I_2, \ldots, I_n \in \mathcal{I}$ so that $[a, b] \subseteq \bigcup_{j=1}^{n} I_j$.*

Proof. Suppose for a contradiction that

$$C := \left\{ x \in [a, b] : \left(\forall I_1, I_2, \ldots, I_n \in \mathcal{I} : [a, x] \not\subseteq \bigcup_{j=1}^{n} I_j \right) \right\} \neq \emptyset.$$

Let $c := \inf C$. Because $a \in I$ for some $I \in \mathcal{I}$ we infer $c > a$. Let $J \in \mathcal{I}$ be an open interval with $c \in J$. If $c < b$ there is a $\delta > 0$ with $(c - \delta, c + \delta) \subseteq J \cap [a, b]$. If $c = b$ there is a $\delta > 0$ with $(c - \delta, b] \subseteq J \cap [a, b]$. By definition of c, there are finitely many $I_1, I_2, \ldots, I_n \in \mathcal{I}$ so that $[a, c - \delta] \subseteq \bigcup_{j=1}^{n} I_j$. But then if $c < b$ we obtain $[a, c+\delta] \subseteq J \cup \bigcup_{j=1}^{n} I_j$, contradicting the definition of c. Hence, $c = b$ and we obtain $[a, b] \subseteq J \cup \bigcup_{j=1}^{n} I_j$, implying $C = \emptyset$, a contradiction. ■

Proposition 8.5 *Let $a, b \in \mathbb{R}$ with $a < b$. Then $\lambda([a, b]) = b - a$.*

Proof. Let $\varepsilon > 0$. Because $[a, b] \subseteq \left(a - \frac{\varepsilon}{4}, b + \frac{\varepsilon}{4}\right) \cup \bigcup_{n=2}^{\infty} \left(-\frac{\varepsilon}{2 \cdot 2^n}, \frac{\varepsilon}{2 \cdot 2^n}\right)$ we obtain $\lambda([a, b]) < (b - a) + \varepsilon$. Because $\varepsilon > 0$ was arbitrary, $\lambda([a, b]) \leq b - a$.

To show the reverse inequality, let $\{I_j\}_{j=1}^{\infty}$ be a countable family of open intervals so that $[a, b] \subseteq \bigcup_{j=1}^{\infty} I_j$. By the Heine-Borel Theorem, there is a finite number of intervals $I_{j_1}, \ldots, I_{j_n}$ so that $[a, b] \subseteq \bigcup_{k=1}^{n} I_{j_k}$. For $k = 1, \ldots, n$ let $I_{j_k} = (a_k, b_k)$. Without loss of generality assume that no interval I_{j_k} is contained in another. Reorder the intervals so that for $k = 2, \ldots, n$ we have $b_{k-1} \leq b_k$. Then for all $k = 2, \ldots, n$ we have $a_{k-1} \leq a_k$. Moreover, $b_n > b$, $a_1 < a$ and for all $k = 2, \ldots, n$ we infer $a_k < b_{k-1}$. Hence, $\sum_{k=1}^{n} b_k - a_k > b_1 - a_1 + \sum_{k=2}^{n} b_k - b_{k-1} = b_1 - a_1 + b_n - b_1 \geq b - a$, which implies $\sum_{j=1}^{\infty} |I_j| > b - a$, and hence $\lambda([a, b]) \geq b - a$. ■

Aside from its use related to Riemann integration, outer Lebesgue measure also is the foundation for Lebesgue integration. We conclude this section with some of the properties of outer Lebesgue measure, which will also be helpful for some exercises.

Theorem 8.6 *The* **properties of outer Lebesgue measure** *λ. With ∞ defined to be greater than all real numbers and the sum of a divergent series of nonnegative numbers being ∞ we have the following.*

1. *$\lambda(\emptyset) = 0$.*
2. *If $A \subseteq B$, then $\lambda(A) \leq \lambda(B)$.*
3. *Outer Lebesgue measure is* **countably subadditive**. *That is, for all sequences $\{A_n\}_{n=1}^{\infty}$ of subsets $A_n \subseteq \mathbb{R}$ the inequality $\lambda\left(\bigcup_{n=1}^{\infty} A_n\right) \leq \sum_{n=1}^{\infty} \lambda(A_n)$ holds.*

Proof. For part 2, let $A \subseteq B$. Then we obtain the following.

$$\begin{aligned}\lambda(A) &= \inf\left\{\sum_{j=1}^{\infty}|I_j| : A\subseteq\bigcup_{j=1}^{\infty}I_j,\ \text{each } I_j \text{ is an open interval}\right\}\end{aligned}$$

Because $A \subseteq B$, the set above contains the set below. Exercise 1-18b provides the inequality for the infima.

$$\begin{aligned}&\leq \inf\left\{\sum_{j=1}^{\infty}|I_j| : B\subseteq\bigcup_{j=1}^{\infty}I_j,\ \text{each } I_j \text{ is an open interval}\right\}\\ &= \lambda(B).\end{aligned}$$

For part 1, note that for all $n \in \mathbb{N}$ we have $\emptyset \subseteq \left[-\frac{1}{n}, \frac{1}{n}\right]$ and thus $\lambda(\emptyset) \leq \frac{2}{n}$, which via part 2 implies $\lambda(\emptyset) = 0$.

For part 3, first note that there is nothing to prove if the right side is infinite. So assume the right side is finite and let $\varepsilon > 0$. For each $n \in \mathbb{N}$, find a countable family $\left\{I_j^n\right\}_{j=1}^{\infty}$ of open intervals such that $A_n \subseteq \bigcup_{j=1}^{\infty} I_j^n$ and $\sum_{j=1}^{\infty}\left|I_j^n\right| \leq \lambda(A_n) + \frac{\varepsilon}{2^n}$. Then by Lemma 7.14 the family $\left\{I_j^n\right\}_{j,n=1}^{\infty}$ is a countable family of open intervals. The union of this family is such that $\bigcup_{n=1}^{\infty} A_n \subseteq \bigcup_{n=1}^{\infty}\bigcup_{j=1}^{\infty} I_j^n = \bigcup_{j,n=1}^{\infty} I_j^n$. By Proposition 6.22, the convergence behavior and value of a doubly infinite sum of nonnegative numbers do not depend on the order of summation and by Exercise 7-17 it does not matter if we represent the sum as a double sum or a single sum. Thus we can conclude

$$\lambda\left(\bigcup_{n=1}^{\infty}A_n\right) \leq \sum_{j,n=1}^{\infty}\left|I_j^n\right| = \sum_{n=1}^{\infty}\sum_{j=1}^{\infty}\left|I_j^n\right| \leq \sum_{n=1}^{\infty}\left(\lambda(A_n)+\frac{\varepsilon}{2^n}\right) = \sum_{n=1}^{\infty}\lambda(A_n)+\varepsilon.$$

Because ε was arbitrary, this proves part 3. ■

Exercises

8-1. Let $\{A_n\}_{n=1}^{\infty}$ be a countable family of null sets. Prove that $\lambda\left(\bigcup_{n=1}^{\infty} A_n\right) = 0$.

8-2. Let $a, b \in \mathbb{R}$ with $a < b$. Prove that $\lambda\left((a,b)\right) = b - a$.
Hint. For "$\geq$" approximate the open interval "from the inside" with closed intervals.

8-3. Let $Q = \{q_n\}_{n=1}^{\infty}$ be a sequence of numbers $q_n \in \left(0, \frac{1}{2}\right)$ and let C^Q be the associated **Cantor set** as in Definition 7.22. We will use the notation of Definition 7.22 throughout this exercise.

(a) Prove that $\lambda\left(C_n^Q\right) = \prod_{j=1}^{n} 2q_j$.

Hint. Prove that for a finite union of pairwise disjoint intervals the outer Lebesgue measure is the sum of the lengths of the intervals. This requires repeated use of the argument in the proof of Proposition 8.5.

(b) Prove that $\left\{\prod_{j=1}^{n} 2q_j\right\}$ converges.

(c) Prove that $\lambda\left(C^Q\right) = \lim_{n\to\infty} \prod_{j=1}^{n} 2q_j$.

Hint. For "$\geq$," first consider a family $\mathcal{I}$ of open intervals so that $C^Q \subseteq \bigcup \mathcal{I}$. Prove that $C = \left\{x \in [0,1] : \left(\forall I_1, I_2, \ldots, I_n \in \mathcal{I} : C^Q \cap [0,x] \not\subseteq \bigcup_{j=1}^{n} I_j\right)\right\} = \emptyset$, by assuming it is not empty and showing that $\inf C \in C^Q$, which leads to a contradiction similar to the proof of the Heine-Borel Theorem. (It also helps to look ahead to the proof of Lemma 8.11 for a similar argument.) Then show that for any countable family $\{I_j\}_{j=1}^{\infty}$ with $C^Q \subseteq \bigcup_{j=1}^{\infty} I_j$ there are intervals $I_{j_1}, \ldots, I_{j_k}$ with $C^Q \subseteq \bigcup_{k=1}^{m} I_{j_k}$. Conclude that there must be an $n \in \mathbb{N}$ so that $C_n^Q \subseteq \bigcup_{k=1}^{m} I_{j_k}$.

(d) Prove that for any $q \in \left(0, \frac{1}{2}\right)$ the constant sequence $Q = \{q\}_{n=1}^{\infty}$ yields a Cantor set C^Q of measure zero.

Note. By Exercise 7-25b in Section 7.3 Cantor sets are uncountable. This means there are uncountable sets of measure zero.

(e) Use $Q = \left\{\frac{2^{n+1}-1}{2^{n+2}}\right\}_{n=1}^{\infty}$ to prove that there are Cantor sets that are not of measure zero.

Hint. Prove that for all $n \in \mathbb{N}$ we have $\prod_{j=1}^{n} 2q_j \geq 1 - \sum_{j=2}^{n+1} \frac{1}{2^j}$.

(f) Prove that there are Cantor sets whose Lebesgue measure is arbitrarily close to 1.

Hint. For $k \in \mathbb{N}$ fixed, use $n+k$ instead of n in Exercise 8-3e.

8-4. Use the Heine-Borel Theorem and the axioms for $\mathbb{R}$ except for Axiom 1.19 to prove the Bolzano-Weierstrass Theorem.

Hint. We will ultimately do this in a more abstract setting in Theorem 16.72.

8-5. Prove that if $f : [a,b] \to \mathbb{R}$ is continuous and $\lambda\left(\{x \in [a,b] : f(x) \neq 0\}\right) = 0$, then $f(x) = 0$ for all $x \in [a,b]$.

8-6. Prove that if $f, g : [a,b] \to \mathbb{R}$ are continuous almost everywhere, then $f+g$ continuous almost everywhere.

8.2 Lebesgue's Criterion for Riemann Integrability

The oscillation of a function is a quantitative measure "how discontinuous" the function is. It is the last tool we need to characterize Riemann integrable functions.

Definition 8.7 *Let $f : [a, b] \to \mathbb{R}$ be a bounded function and let $x \in [a, b]$. For any interval I let $\omega_f(I) := \sup\left\{\left|f(y) - f(z)\right| : y, z \in I \cap [a, b]\right\}$ be the* **oscillation** *of f over the interval I. Define the* **oscillation** *of f at the point $x \in [a, b]$ as the infimum $\omega_f(x) := \inf\left\{\omega_f(J) : x \in J, J \text{ is an open interval }\right\}$.*

Exercise 8-8 shows that the oscillation measures the height of "jumps" in the function and Exercise 12-23 shows that the oscillation is also a measure of the size of oscillations. Regarding the details of the definition, Exercise 8-10 shows that it is important that we use *open* intervals in the definition of the oscillation $\omega_f(x)$ at a point.

Theorem 8.8 *The bounded function $f : [a, b] \to \mathbb{R}$ is continuous at $x \in [a, b]$ iff $\omega_f(x) = 0$.*

Proof. For "$\Rightarrow$," let f be continuous at $x \in [a, b]$ and let $\varepsilon > 0$. Then there is a $\delta > 0$ such that for all $y \in [a, b]$ with $|y - x| < \delta$ we have $\left|f(y) - f(x)\right| < \frac{\varepsilon}{2}$. Then for all $y, z \in (x - \delta, x + \delta) \cap [a, b]$ we obtain

$$\left|f(y) - f(z)\right| \leq \left|f(y) - f(x)\right| + \left|f(x) - f(z)\right| < \frac{\varepsilon}{2} + \frac{\varepsilon}{2} = \varepsilon.$$

Hence, $\omega_f(x) \leq \omega_f\left((x - \delta, x + \delta)\right) \leq \varepsilon$ and because ε was arbitrary we conclude that $\omega_f(x) = 0$.

Conversely, for "$\Leftarrow$," let $x \in [a, b]$ with $\omega_f(x) = 0$ and let $\varepsilon > 0$. Then there is an interval (c, d) with $x \in (c, d)$ such that $\omega_f\left((c, d)\right) < \varepsilon$. Let $\delta := \min\{x - c, d - x\}$ unless $x = a$, in which case we set $\delta := d - x$, or $x = b$, in which case we set $\delta := x - c$. Then for all $y \in [a, b]$ with $|x - y| < \delta$ we have $y \in (c, d)$, and hence $\left|f(x) - f(y)\right| < \varepsilon$. Thus f is continuous at x. ∎

The next two lemmas show that, when it comes to Riemann integrability, small oscillation means that lower and upper sums can get close to each other, while nonzero oscillation on a set of positive outer Lebesgue measure prevents Riemann integrability (also see Figure 20).

Lemma 8.9 *Let $f : [a, b] \to \mathbb{R}$ be bounded. If $\omega_f(x) < \varepsilon$ for all $x \in [a, b]$, then there is a partition P of $[a, b]$ so that $U(f, P) - L(f, p) < \varepsilon|b - a|$.*

Proof. For every $z \in [a, b]$, there is an open interval $I_z = (z - \delta_z, z + \delta_z)$ so that $\omega_f(I_z) < \varepsilon$. By the Heine-Borel Theorem, there are finitely many $z_1, \ldots, z_m \in [a, b]$ so that $[a, b] \subseteq \bigcup_{j=1}^{m}\left(z_j - \frac{\delta_{z_j}}{2}, z_j + \frac{\delta_{z_j}}{2}\right)$. Let $P := \{a = x_0 < x_1 < \cdots < x_n = b\}$ be the set comprised of a, b and the endpoints $z_j \pm \frac{\delta_{z_j}}{2}$ that are in $[a, b]$. Then each interval $[x_{i-1}, x_i]$ is contained in an interval I_{z_j} and consequently, with notation as in Definition 5.13, we infer $M_i - m_i < \varepsilon$. But then

$$\begin{aligned} U(f, P) - L(f, p) &= \sum_{i=1}^{n} M_i \Delta x_i - \sum_{i=1}^{n} m_i \Delta x_i = \sum_{i=1}^{n} (M_i - m_i) \Delta x_i \\ &< \sum_{i=1}^{n} \varepsilon \Delta x_i = \varepsilon|b - a|, \end{aligned}$$

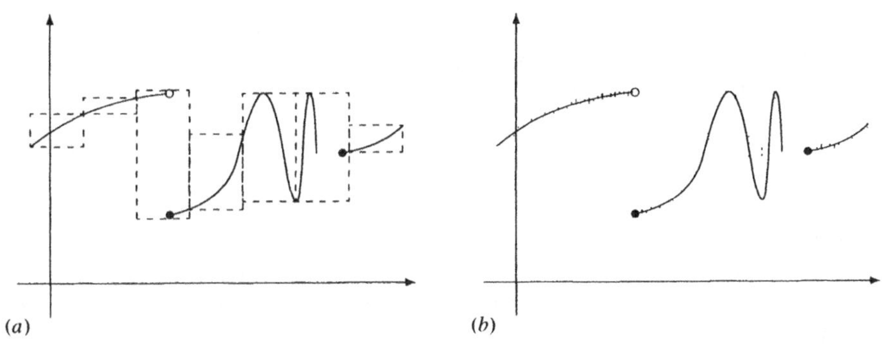

Figure 20: The area of the boxes indicates the difference between the lower and the upper sum of the function for the given partition. Boxes are tall where the function has large slopes or discontinuities. Comparison of (a) and (b) shows that as the norm of the partition goes to zero, the height of the boxes goes to zero where the function is continuous. Where the function is discontinuous the height remains bounded away from zero (consider the two discontinuities). Thus f can only be Riemann integrable if the discontinuities (unavoidable tall boxes) can be trapped inside a set of intervals whose total length is small. This is the idea for the Lebesgue criterion.

which finishes the proof. ■

Lemma 8.10 *Let $f : [a, b] \to \mathbb{R}$ be bounded. If $\lambda\left(\left\{x \in [a, b] : \omega_f(x) > 0\right\}\right) > 0$, then f is not Riemann integrable.*

Proof. Because $\left\{x \in [a, b] : \omega_f(x) > 0\right\} = \bigcup_{j=1}^{\infty} \left\{x \in [a, b] : \omega_f(x) > \frac{1}{j}\right\}$ there is an $\varepsilon > 0$ so that $L := \lambda\left(\left\{x \in [a, b] : \omega_f(x) > \varepsilon\right\}\right) > 0$.

Let $P = \{a = x_0 < x_1 < \cdots < x_n = b\}$ be any partition of $[a, b]$. Define the set $D := \left\{x \in [a, b] \setminus \{x_0, \ldots, x_n\} : \omega_f(x) > \varepsilon\right\}$. By Theorem 8.6, we obtain

$$\begin{aligned} L &\geq \lambda(D) = \lambda(D) + \sum_{j=1}^{n} \lambda\left(\{x_j\}\right) \geq \lambda\left(D \cup \bigcup_{j=1}^{n} \{x_j\}\right) \\ &\geq \lambda\left(\left\{x \in [a, b] : \omega_f(x) > \varepsilon\right\}\right) = L, \end{aligned}$$

which means $\lambda(D) = L$.

With $i_1, \ldots, i_k \in \{1, \ldots, n\}$ being the indices of the intervals (x_{i-1}, x_i) that intersect D we obtain $D \subseteq \bigcup_{j=1}^{k} (x_{i_j-1}, x_{i_j})$ and $\sum_{j=1}^{k} \Delta x_{i_j} \geq \lambda(D) = L$. But then in each interval (x_{i_j-1}, x_{i_j}) there is an s_j with $\omega_f(s_j) > \varepsilon$, and hence $\omega_f(x_{i_j-1}, x_{i_j}) > \varepsilon$. For each $j = 1, \ldots, k$ we infer $M_{i_j} - m_{i_j} > \varepsilon$ and thus

$$U(f, P) - L(f, P) = \sum_{i=1}^{n} (M_i - m_i)\Delta x_i \geq \sum_{j=1}^{k} (M_{i_j} - m_{i_j})\Delta x_{i_j} > \varepsilon L > 0.$$

Because P was arbitrary we have shown that for any partition P of $[a, b]$ we have $U(f, P) - L(f, P) > \varepsilon L > 0$, where ε and L are fixed. By Theorem 5.25 this means that f is not Riemann integrable. ■

Lemma 8.10 shows that only functions that are continuous almost everywhere have a chance to be Riemann integrable. Lemma 8.9 shows that if we can "trap" the discontinuities in a small enough set, a function should be Riemann integrable. The only obstacle left is that outer Lebesgue measure works with countably many open intervals. To obtain a complement that is made up of closed intervals, it would be nice if we could use finitely many open intervals to cover the set of discontinuities. The next lemma shows that this is possible.

Lemma 8.11 *Let $f : [a, b] \to \mathbb{R}$ be a bounded function and for each $p > 0$ define $B_p := \{x \in [a, b] : \omega_f(x) \geq p\}$. If $\mathcal{I}$ is a family of open intervals such that $B_p \subseteq \bigcup_{I\in\mathcal{I}} I$, then there are finitely many $I_1, I_2, \ldots, I_n \in \mathcal{I}$ so that $B_p \subseteq \bigcup_{j=1}^{n} I_j$.*

Proof. Suppose for a contradiction that

$$C := \left\{ x \in [a, b] : \left(\forall I_1, I_2, \ldots, I_n \in \mathcal{I} : B_p \cap [a, x] \not\subseteq \bigcup_{j=1}^{n} I_j \right) \right\} \neq \emptyset.$$

Then $c := \inf C \leq b$.

We first claim $c \in B_p$. There is a sequence $\{a_n\}_{n=1}^{\infty}$ of elements of B_p so that $a_n < c$ and $\lim_{n\to\infty} a_n = c$. Let $\varepsilon > 0$. Then there is an $n \in \mathbb{N}$ so that $|a_n - c| < \frac{\varepsilon}{2}$. Moreover, because $a_n \in B_p$ we have that $\omega_f\left(\left(a_n - \frac{\varepsilon}{2}, a_n + \frac{\varepsilon}{2}\right)\right) \geq p$. Because of the containment $\left(a_n - \frac{\varepsilon}{2}, a_n + \frac{\varepsilon}{2}\right) \subseteq (c - \varepsilon, c + \varepsilon)$ we infer $\omega_f\big((c - \varepsilon, c + \varepsilon)\big) \geq p$. Because for all $l < c < r$ there is an $\varepsilon > 0$ with $(c - \varepsilon, c + \varepsilon) \subseteq (l, r)$ we conclude that $\omega_f(c) \geq p$.

Because $c \in B_p$, there is an open interval $I \in \mathcal{I}$ that contains c. Clearly, $c \neq a$. But then $x := \max\{a, \inf(I)\} \in [a, b]$ and because $x \notin C$, there are intervals $I_1, \ldots, I_n \in \mathcal{I}$ so that $B_p \cap [a, x] \subseteq \bigcup_{j=1}^{n} I_j$. But then for all $y \in I$ with $y \geq c$ we obtain $B_p \cap [a, y] \subseteq I \cup \bigcup_{j=1}^{n} I_j$. If $c < b$ this means $\inf C \geq \sup(I) > c$ and if $c = b$ this means $C = \emptyset$, a contradiction either way. This proves the result. ■

Now we can characterize Riemann integrability as follows.

Theorem 8.12 Lebesgue's criterion for Riemann integrability. *The bounded function $f : [a, b] \to \mathbb{R}$ is Riemann integrable on $[a, b]$ iff f is continuous a.e. on $[a, b]$.*

Proof. The part "$\Rightarrow$" is the contrapositive of Lemma 8.10.

For "$\Leftarrow$," let f be continuous a.e. on $[a, b]$ and let $\varepsilon > 0$. Choose $m \in \mathbb{N}$ so that $\frac{b-a}{m} < \frac{\varepsilon}{2}$. Let $X_m := \left\{ x \in [a, b] : \omega_f(x) \geq \frac{1}{m} \right\}$. Then $\lambda(X_m) = 0$, and there is a sequence $\{I_j\}_{j=1}^{\infty}$ of open intervals with $X_m \subseteq \bigcup_{j=1}^{\infty} I_j$ and $\sum_{j=1}^{\infty} |I_j| < \frac{\varepsilon}{2\left(\omega_f([a,b]) + 1\right)}$. By Lemma 8.11, there are finitely many open intervals $I_{j_1}, \ldots, I_{j_k}$ so that $X_m \subseteq \bigcup_{l=1}^{k} I_{j_l}$ and without loss of generality we may assume the I_{j_l} are pairwise disjoint. For each I_{j_l} let $C_{j_l} := I_{j_l} \cup \left\{ \sup(I_{j_l}), \inf(I_{j_l}) \right\}$. The set $[a, b] \setminus \bigcup_{l=1}^{k} I_{j_l}$ is a finite union of closed intervals $J_1, \ldots, J_p$ and singleton sets $\{s_1\}, \ldots, \{s_q\}$. By Lemma 8.9 for each J_i, there is a partition P_i so that $U(f, P_i) - L(f, P_i) < \frac{|J_i|}{m}$. Set $P := \bigcup_{i=1}^{p} P_i \cup \{a, b\}$. Then

$$U(f,P) - L(f,P) \leq \sum_{i=1}^{p} \left[U(f, P_i) - L(f, P_i) \right] + \sum_{l=1}^{k} \omega_f(C_{j_l}) |I_{j_l}|$$

> **Note how in the first sum the function is continuous and upper and lower sums are close to each other (flat boxes in Figure 20) and how the second sum traps the discontinuities in narrow intervals (tall, narrow boxes in Figure 20).**

$$\leq \frac{1}{m} \sum_{i=1}^{p} |J_i| + \omega_f\left([a,b]\right) \sum_{l=1}^{k} |I_{j_l}|$$

$$< \frac{b-a}{m} + \omega_f\left([a,b]\right) \frac{\varepsilon}{2\left(\omega_f([a,b]) + 1\right)} < \frac{\varepsilon}{2} + \frac{\varepsilon}{2} = \varepsilon.$$

By Theorem 5.25 f is Riemann integrable on $[a, b]$. ■

Lebesgue's integrability criterion shows some of our earlier examples in a different light. Clearly, the Lebesgue criterion implies that continuous functions are Riemann integrable, so it supersedes Theorem 5.20. Similarly, any function with finitely many discontinuities is Riemann integrable, so Proposition 5.12 also is an easy consequence of the Lebesgue criterion. Regarding the Dirichlet function we can say the following.

Example 8.13 *(Example 5.26 revisited.) The Dirichlet function is not Riemann integrable on* $[0, 1]$.

In light of Theorem 8.12, the function $f(x) = \begin{cases} 0; & \text{for } x \in [0,1] \setminus \mathbb{Q}, \\ 1; & \text{for } x \in [0,1] \cap \mathbb{Q}, \end{cases}$ is not Riemann integrable, because it is discontinuous at every point of $[0, 1]$ and by Proposition 8.5 we have $\lambda\left([0, 1]\right) = 1 > 0$. □

Exercises

8-7. Let $f : [a, b] \to \mathbb{R}$ be bounded and let I, J be intervals with $I \subseteq J$. Prove that $\omega_f(I) \leq \omega_f(J)$.

8-8. The oscillation as a measure of "jumps" in the function.

(a) Let $f(x) := \mathbf{1}_{[0,\infty)}$. Prove that $\omega_f(0) = 1$.

(b) Let $f : [a, b] \to \mathbb{R}$ be bounded and let $x \in (a, b)$ be so that $\lim_{z\to x^+} f(z)$ and $\lim_{z\to x^-} f(z)$ both exist. Prove that $\omega_f(x) \geq \left| \lim_{z\to x^+} f(z) - \lim_{z\to x^-} f(z) \right|$.

(c) State when the inequality in Exercise 8-8b actually is an equality and prove your claim.

8-9. Let $f : [a, b] \to \mathbb{R}$ be bounded and let $x \in [a, b]$. Prove that $\omega_f(x) = \lim_{n\to\infty} \omega_f\left(x - \frac{1}{n}, x + \frac{1}{n}\right)$.

8-10. Prove that for $f(x) := \begin{cases} 0, & \text{for } x \geq 0, \\ 1; & \text{for } x < 0, \end{cases}$ we have $\inf\left\{ \omega_f(J) : 0 \in J, J \text{ is an interval } \right\} = 0$, even though f is discontinuous at 0. Then explain why this does not contradict Exercise 8-8b.

8-11. Prove that any **nondecreasing** function $F : [a, b] \to \mathbb{R}$ is Riemann integrable. *Hint.* Exercise 7-23.

8-12. Let $a < b$. A function $f : [a, b] \to \mathbb{R}$ is said to be of **bounded variation** iff

$$V_a^b f := \sup\left\{ \sum_{i=1}^{n} \left| f(a_i) - f(a_{i-1}) \right| : a = a_0 < a_1 < \cdots < a_n = b \right\} < \infty.$$

(a) Prove that every function of bounded variation is the difference of two nondecreasing functions. *Hint.* Use $V_a^x(f) := \sup\left\{ \sum_{i=1}^{n} \left| f(a_i) - f(a_{i-1}) \right| : a = a_0 < a_1 < \cdots < a_n = x \right\}$

(b) Prove that every function of bounded variation has at most countably many discontinuities. *Hint.* Exercise 7-23.

(c) Prove that every function of bounded variation on $[a, b]$ is Riemann integrable on $[a, b]$.

(d) Prove that if f and g are of bounded variation and $\alpha \in \mathbb{R}$, then $f + g$ and αf are of bounded variation, too.

8-13. A function $f : [a, b] \to \mathbb{R}$ is called **piecewise continuous** iff there are $a = z_0 < z_1 < \cdots < z_n = b$ so that for all $i = 1, \ldots, n$ the restriction $f|_{(z_{i-1}, z_i)}$ is continuous. Prove that piecewise continuous bounded functions are Riemann integrable.

8-14. Prove that if $A, N \subseteq \mathbb{R}$ and N is a null set, then $\lambda(A \setminus N) = \lambda(A)$.

8-15. Prove that if C^Q is a **Cantor set** with nonzero measure (see Exercise 8-3e), then $\mathbf{1}_{C^Q}$ is not Riemann integrable. *Hint.* Prove that $\mathbf{1}_{C^Q}$ is discontinuous at every $x \in C^Q$.

8.3 More Integral Theorems

The most tedious part in any proof involving Riemann integrable functions is to prove Riemann integrability. Lebesgue's integrability criterion is a powerful tool to address this issue. Without it, the proofs of Theorems 8.14 and 8.16 would be a lot longer.

Theorem 8.14 *Let $f, g : [a, b] \to \mathbb{R}$ be Riemann integrable. Then*

1. *The product fg is Riemann integrable on $[a, b]$,*

2. *If there is an* $\varepsilon > 0$ *so that* $|g(x)| \geq \varepsilon$ *for all* $x \in [a, b]$, *then the quotient* $\frac{f}{g}$ *is Riemann integrable on* $[a, b]$,

3. *The absolute value* $|f|$ *is Riemann integrable and the* **triangular inequality** $\left|\int_a^b f(x)\,dx\right| \leq \int_a^b |f(x)|\,dx$ *holds,*

Proof. For part 1, note that if f and g are continuous a.e., then the set of discontinuities $\{x : \omega_{fg}(x) \neq 0\}$ is contained in the union $\{x : \omega_f(x) \neq 0\} \cup \{x : \omega_g(x) \neq 0\}$, which has measure zero. Thus fg is continuous a.e., and hence Riemann integrable.

The remaining parts of this proof are left as Exercise 8-16. ■

It is now easy to see that once $f : [a, b] \to \mathbb{R}$ is Riemann integrable, then it is also Riemann integrable over any closed subinterval. Formally, we define the following.

Definition 8.15 *Let* $f : [a, b] \to \mathbb{R}$ *and let* $c, d \in [a, b]$ *with* $c < d$. *Then* f *is called* **Riemann integrable** *over* $[c, d]$ *iff the restriction* $f|_{[c,d]}$ *is Riemann integrable. In this case we set* $\int_c^d f(x)\,dx := \int_c^d f|_{[c,d]}(x)\,dx$.

Theorem 8.16 *Let* $f : [a, b] \to \mathbb{R}$ *and let* $m \in (a, b)$. *Then* f *is Riemann integrable over* $[a, b]$ *iff* f *is Riemann integrable over* $[a, m]$ *and over* $[m, b]$. *In this case, the integrals satisfy the equation* $\int_a^b f(x)\,dx = \int_a^m f(x)\,dx + \int_m^b f(x)\,dx$.

Proof. Exercise 8-17. ■

We can now consider integrals in which the upper bound is an independent variable. This idea provides another connection between derivatives and integrals. For any continuous function f, definite integrals produce a function G with $G' = f$.

Theorem 8.17 Fundamental Theorem of Calculus, Derivative Form. *Let* f *be a Riemann integrable function on* $[a, b]$. *Then the function* $G(x) := \int_a^x f(t)\,dt$ *is uniformly continuous on* $[a, b]$ *and if* f *is continuous at* $x \in (a, b)$, *then* G *is differentiable at* x *with* $G'(x) = \frac{d}{dx}\left(\int_a^x f(t)\,dt\right) = f(x)$.

Proof. Let B be an upper bound so that $|f(x)| < B$ for all $x \in [a, b]$. To see that G is uniformly continuous on $[a, b]$, let $\varepsilon > 0$. Set $\delta := \frac{\varepsilon}{B}$. Then for all $x, z \in [a, b]$ with $|x - z| < \delta$ we obtain (assuming without loss of generality that $x < z$)

$$\begin{aligned} |G(z) - G(x)| &= \left|\int_x^z f(t)\,dt\right| \leq \int_x^z |f(t)|\,dt \leq \int_x^z B\,dt = B|z - x| \\ &< B\frac{\varepsilon}{B} = \varepsilon, \end{aligned}$$

which means that G is uniformly continuous.

Now let $x \in (a,b)$, let f be continuous at x and let $\varepsilon > 0$. Then there is a $\delta > 0$ such that for all $z \in [a,b]$ with $|z-x| < \delta$ we have $\left|f(z) - f(x)\right| < \varepsilon$. Hence, for all $z > x$ with $z \in [a,b]$ and $|z-x| < \delta$ we obtain

$$\begin{aligned} \left|\frac{G(z)-G(x)}{z-x} - f(x)\right| &= \left|\frac{\int_a^z f(t)\,dt - \int_a^x f(t)\,dt}{z-x} - \frac{f(x)(z-x)}{z-x}\right| \\ &= \left|\frac{\int_x^z f(t)\,dt}{z-x} - \frac{\int_x^z f(x)\,dt}{z-x}\right| \\ &\leq \frac{\int_x^z \left|f(t)-f(x)\right|\,dt}{z-x} < \frac{\int_x^z \varepsilon\,dt}{z-x} = \varepsilon. \end{aligned}$$

For $z < x$, the proof is similar, and hence $\lim_{z\to x} \frac{G(z)-G(x)}{z-x} = f(x)$. ■

Functions as mentioned in Theorem 8.17 sometimes are also defined with an integral such that $a > x$. If we went through a partition from right to left rather than left to right, then all the bases of the rectangles in a Riemann sum would have negative length. Thus it makes sense to define the following.

Definition 8.18 *Let $f : [a,b] \to \mathbb{R}$ be Riemann integrable. Then we define the integral with the reversed bounds to be* $\int_b^a f(x)\,dx := -\int_a^b f(x)\,dx$.

Corollary 8.19 *Let $f : [a,b] \to \mathbb{R}$ be Riemann integrable and let $x_0 \in [a,b]$. Then the function $G(x) := \int_{x_0}^x f(t)\,dt$ is uniformly continuous on $[a,b]$ and if f is continuous at $x \in (a,b)$, then* $\frac{d}{dx}\left(\int_{x_0}^x f(t)\,dt\right) = f(x)$.

Proof. Exercise 8-18. ■

Exercises

8-16. Proving Theorem 8 14

(a) Prove part 2 of Theorem 8 14

(b) Prove part 3 of Theorem 8.14

Hint. Use Lemma 5.6 to prove the inequality

(c) To see how valuable the Lebesgue criterion is, use Riemann's Condition (Theorem 5.25) to prove that $|f|$ is Riemann integrable over $[a,b]$ Then compare this proof with the proof using the Lebesgue criterion and state which proof is simpler

Hint. Use a partition P with $U(f,P) - L(f,P) < \varepsilon$

8-17 Proving Theorem 8 16.

(a) Prove Theorem 8 16

Hint In each direction of the proof use the Lebesgue criterion to establish Riemann integrability. Use Lemma 5 6 for the equations, making sure that the point m is an element of each partition P_k.

(b) To see how valuable the Lebesgue criterion is, use Riemann's Condition (Theorem 5.25) to prove that if f is Riemann integrable over $[a, b]$, then f is Riemann integrable over $[a, m]$. Then compare this proof with the proof using the Lebesgue criterion and state which proof is simpler

Hint. Use a partition P with $U(f, P) - L(f, P) < \varepsilon$.

8-18. Prove Corollary 8 19.

8-19 Use Riemann's Condition (Theorem 5.25) to prove that if f and g are Riemann integrable over $[a, b]$, then fg is Riemann integrable over $[a, b]$. Then compare this proof with the proof using the Lebesgue criterion and state which proof is simpler.

8-20. Determine which result from this section that was used in the proof of Theorem 8.17 was not available in Section 5 3 (This prevented us from placing Theorem 8.17 right after the Antiderivative Form of the Fundamental Theorem of Calculus.)

8-21. **Mean Value Theorem for the Integral**. Let the function $f : [a, b] \to \mathbb{R}$ be continuous and let the function $g : [a, b] \to [0, \infty)$ be nonnegative and Riemann integrable. Prove that there is a $c \in [a, b]$ so that the integral satisfies $\int_a^b f(x)g(x)\, dx = f(c) \int_a^b g(x)\, dx$.

8-22 Let $g : [a, b] \to [0, \infty)$ be Riemann integrable and let $f : [a, b] \to \mathbb{R}$ be nondecreasing. Prove that there is a $c \in [a, b]$ so that $\int_a^b f(x)g(x)\, dx = f(a) \int_a^c g(x)\, dx + f(b) \int_c^b g(x)\, dx$. You may use that nondecreasing functions are Riemann integrable (see Exercise 8-11).

8-23 **Integration by Parts**. Let $[a, b] \subset (c, d)$ and let $F, g : (c, d) \to \mathbb{R}$ be differentiable functions with derivatives f and g' that are Riemann integrable on $[a, b]$. Prove that the integral satisfies $\int_a^b f(x)g(x)\, dx = F(b)g(b) - F(a)g(a) - \int_a^b F(x)g'(x)\, dx$.

8-24. **Integration by Substitution**. Let $[a, b] \subset (c, d)$, let $g : (c, d) \to \mathbb{R}$ be differentiable, let its derivative g' be Riemann integrable on $[a, b]$, let $(u, v) \supseteq g\left[[a, b]\right]$ and let $F : (u, v) \to \mathbb{R}$ be differentiable with continuous derivative f. Prove that $\int_a^b f\left(g(x)\right) g'(x)\, dx = F\left(g(b)\right) - F\left(g(a)\right)$.

8-25 Let $a > 0$. A function $f : [-a, a] \to \mathbb{R}$ is called **even** iff $f(x) = f(-x)$ for all $x \in [-a, a]$. A function $f . [-a, a] \to \mathbb{R}$ is called **odd** iff $f(x) = -f(-x)$ for all $x \in [-a, a]$.

(a) Prove that if f is even and Riemann integrable, then $\int_{-a}^a f(x)\, dx = 2 \int_0^a f(x)\, dx$.

(b) Prove that if f is odd and Riemann integrable, then $\int_{-a}^a f(x)\, dx = 0$.

(c) Prove that any function $f : [-a, a] \to \mathbb{R}$ is the sum of an even and an odd function.

Hint $\frac{f(x) + f(-x)}{2}$ is even.

8-26. Compute the derivative (Should the integrands be unknown, simply note that they are continuous.)

(a) $\frac{d}{dx} \int_0^x e^{t^2}\, dt$ (b) $\frac{d}{dx} \int_x^1 \sin\left(\frac{1}{t}\right) dt$ (c) $\frac{d}{dx} \int_0^{x^2} e^{\sqrt{t}}\, dt$

8-27. Let $f : [a, b] \to \mathbb{R}$ be continuous and let $l, u : (c, d) \to [a, b]$ be differentiable. Prove that $\frac{d}{dx}\left(\int_{l(x)}^{u(x)} f(t)\, dt\right) = f\left(u(x)\right) u'(x) - f\left(l(x)\right) l'(x)$.

8-28. Construct a function $f \cdot [0, 1] \to \mathbb{R}$ so that $|f|$ is Riemann integrable and f is not Riemann integrable.

8-29. A function $f \cdot [a, b] \to \mathbb{R}$ is called **absolutely continuous** iff for every $\varepsilon > 0$ there is a $\delta > 0$ so that for all sequences $(a_1, b_1), \ldots, (a_n, b_n)$ of pairwise disjoint open intervals the inequality $\sum_{i=1}^n (b_i - a_i) < \delta$, implies $\sum_{i=1}^n \left| f(b_i) - f(a_i) \right| < \varepsilon$.

(a) Let $f : [a, b] \to \mathbb{R}$ be a Riemann integrable function. Prove that the function $G : [a, b] \to \mathbb{R}$ defined by $G(x) := \int_a^x f(t)\, dt$ is absolutely continuous on $[a, b]$.

(b) Prove that every absolutely continuous function is uniformly continuous

(c) Prove that $f(x) = \frac{1}{x}$ is continuous, but not absolutely continuous on $(0, 1]$

8-30 Results for Riemann-Stieltjes integrals. Let $g : [a, b] \to \mathbb{R}$ be nondecreasing, and let the functions $f, h : [a, b] \to \mathbb{R}$ be bounded and Riemann-Stieltjes integrable on $[a, b]$ with respect to g.

(a) Prove that $|f|$ is Riemann-Stieltjes integrable on $[a, b]$ with respect to g and that the **triangular inequality** $\left| \int_a^b f\, dg \right| \leq \int_a^b |f|\, dg$ holds

(b) Prove that for all $m \in [a, b]$ the function f is Riemann-Stieltjes integrable with respect to g over $[a, m]$ and over $[m, b]$ and that $\int_a^b f\, dg = \int_a^m f\, dg + \int_m^b f\, dg$.

(c) Prove that the product fh is Riemann-Stieltjes integrable on $[a, b]$ with respect to g

Hint (for all) Use the Riemann Condition for Riemann-Stieltjes integrals (see Exercise 5-27)

8.4 Improper Riemann Integrals

The Riemann integral allows us to compute the "area" under bounded functions defined on closed and bounded intervals. Sometimes we are interested in the area under functions that are defined on infinite intervals or that are unbounded. These areas can be approximated with Riemann integrals.

Definition 8.20 *Let $a \in \mathbb{R}$ and let $f : [a, \infty) \to \mathbb{R}$ be such that for all $c > a$ the restriction $f|_{[a,c]}$ is Riemann integrable. If the limit $\lim_{t\to\infty} \int_a^t f(x)\, dx$ exists, it is called the* **improper Riemann integral** *of f over $[a, \infty)$ and it is denoted $\int_a^\infty f(x)\, dx$. Improper Riemann integrals for functions $f : (-\infty, b] \to \mathbb{R}$ are defined similarly (Exercise 8-31).*

Example 8.21 *The* **p-integral test** *for integrals over infinite intervals. Let $p > 0$ be rational. Then $f(x) = \frac{1}{x^p}$ is improperly Riemann integrable over $[1, \infty)$ iff $p > 1$.*

For $p > 1$, we have $\lim_{t\to\infty} \int_1^t \frac{1}{x^p}\, dx = \lim_{t\to\infty} \frac{1}{(1-p)x^{p-1}} \bigg|_1^t = \frac{1}{p-1}$, while for $p < 1$ we have $\lim_{t\to\infty} \int_1^t \frac{1}{x^p}\, dx = \lim_{t\to\infty} \frac{x^{1-p}}{1-p} \bigg|_1^t = \infty$.

Finally, we need to show that the improper integral does not exist for $p = 1$. To do this, note that $f(x) = \frac{1}{x}$ is greater than $\frac{1}{m+1}$ on each interval $[m, m+1)$. Therefore, for all $n \in \mathbb{N}$ we infer $f \geq \sum_{k=1}^{n} \frac{1}{k+1} \mathbf{1}_{[k,k+1)}$, and hence

$$\int_1^{n+1} \frac{1}{x}\, dx \geq \int_1^{n+1} \sum_{k=1}^{n} \frac{1}{k+1} \mathbf{1}_{[k,k+1)}\, dx \geq \sum_{k=1}^{n} \frac{1}{k+1}.$$

The latter sum is unbounded, so $\lim_{t\to\infty} \int_1^t \frac{1}{x}\, dx$ does not exist. □

Note that because we have not yet defined logarithms, the last argument in Example 8.21 is unavoidable. Similarly, we had to restrict ourselves to rational powers, because powers with arbitrary real exponents have not been defined yet. Of course, the p-integral test ultimately holds for real exponents p and we will not restate it once real powers are defined.

Improper integrals can also be defined for (potentially) unbounded functions.

Definition 8.22 *Let $a, b \in \mathbb{R}$ with $a < b$ and let $f : [a, b) \to \mathbb{R}$ be such that for all $c \in (a, b)$ the restriction $f|_{[a,c]}$ is Riemann integrable. If the limit $\lim_{t\to b^-} \int_a^t f(x)\, dx$ exists, it is called the* **improper Riemann integral** *of f over $[a, b)$ and it is denoted $\int_a^b f(x)\, dx$. Improper Riemann integrals for functions $f : (a, b] \to \mathbb{R}$ are defined similarly (Exercise 8-32).*

For Riemann integrable functions on $[a, b]$, the improper Riemann integral over $[a, b)$ agrees with the Riemann integral over $[a, b]$.

Proposition 8.23 *Let $a, b \in \mathbb{R}$ with $a < b$ and let $f : [a, b] \to \mathbb{R}$ be Riemann integrable. Then the improper Riemann integral of f over $[a, b)$ exists and it is equal to the Riemann integral of f over $[a, b]$.*

Proof. Let $B > 0$ be an upper bound of $|f|$ and let $\varepsilon > 0$. Then for $\delta := \frac{\varepsilon}{B}$ and all $t \in [a, b)$ with $|t - b| < \delta$ we obtain

$$\left| \int_a^t f(x)\, dx - \int_a^b f(x)\, dx \right| \leq \int_t^b |f(x)|\, dx < B\frac{\varepsilon}{B} = \varepsilon\,. \qquad \blacksquare$$

Riemann integrable functions are not the only functions that are improperly Riemann integrable. Example 8.24 shows that there are unbounded functions for which the improper Riemann integral exists.

Example 8.24 *The* **p-integral test** *for improper integrals over $(0, 1]$. Let $p > 0$ be rational. Then $f(x) = \frac{1}{x^p}$ is improperly Riemann integrable over $(0, 1]$ iff $p < 1$.*

Mimic the proof for Example 8.21. (Exercise 8-33.) □

Note that the improper integrals $\int_0^1 \frac{1}{x^p}\,dx$ converge for $p < 1$, while the integrals $\int_1^\infty \frac{1}{x^p}\,dx$ converge for $p > 1$. Irrespective of this important difference, for both types of improper integrals similar laws hold and there also is a Comparison Test that is similar to the Comparison Test for series.

Theorem 8.25 *Let $f, g : [a, b) \to \mathbb{R}$ (where b could be ∞) be improperly Riemann integrable over $[a, b)$ and let $c \in \mathbb{R}$. Then*

1. *$f + g$ is improperly Riemann integrable over $[a, b)$ and* $\int_a^b (f+g)(x)\,dx = \int_a^b f(x)\,dx + \int_a^b g(x)\,dx.$
2. *cf is improperly Riemann integrable over $[a, b)$ and* $\int_a^b cf(x)\,dx = c\int_a^b f(x)\,dx.$

Proof. Similar to the proof of Theorem 6.4. (Exercise 8-34.) ■

Theorem 8.26 *Let $f : [a, b) \to \mathbb{R}$ (where b could be ∞) be Riemann integrable over all intervals $[a, c] \subseteq [a, b)$. Then f is improperly Riemann integrable over $[a, b)$ if and only if for all $c \in [a, b)$ the function f is improperly Riemann integrable over $[c, b)$ and in this case* $\int_a^b f(x)\,dx = \int_a^c f(x)\,dx + \int_c^b f(x)\,dx.$

Proof. Exercise 8-35. ■

Theorem 8.27 Comparison Test *for improper integrals. Let $f, g : [a, b) \to \mathbb{R}$ (where b could be ∞) be such that $0 \le f \le g$, f is Riemann integrable over every closed interval in $[a, b)$ and g is improperly Riemann integrable over $[a, b)$. Then f is improperly Riemann integrable over $[a, b)$ and* $\int_a^b f(x)\,dx \le \int_a^b g(x)\,dx.$

Proof. The function $F(t) := \int_a^t f(x)\,dx$ is continuous and nondecreasing on $[a, b)$ and it is bounded by $\int_a^b g(x)\,dx$. The reader will show in Exercise 8-36 that $\lim_{t\to b^-} F(t) = \sup\left\{F(t) : t \in [a, b)\right\} \le \int_a^b g(x)\,dx$ to complete the proof. ■

Theorem 8.28 *Let $f : [a, b) \to \mathbb{R}$ (where b could be ∞) be Riemann integrable over every closed interval in $[a, b)$. If $|f|$ is improperly Riemann integrable over $[a, b)$, then f is improperly Riemann integrable over $[a, b)$ and the* **triangular inequality** $\left|\int_a^b f(x)\,dx\right| \le \int_a^b |f(x)|\,dx$ *holds.*

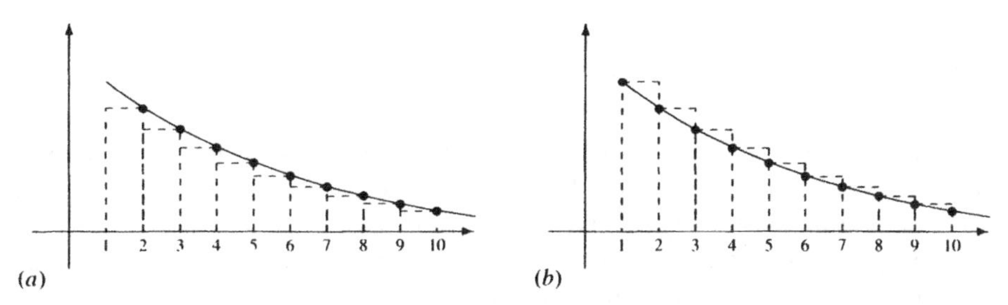

Figure 21: In the integral test, the improper integral of a nonincreasing function is related to the series that give the Riemann sums with left and right endpoint evaluations for the partition with step length 1. The integral test says that for an improperly integrable function the series obtained by right endpoint evaluation cannot be infinite (a), while for a function that is not improperly integrable the series obtained by left endpoint evaluation cannot be finite (b).

Proof. Exercise 8-37. ■

Finally, we should note that the occurrence of series in Example 8.21 is not an accident. The Integral Test connects the convergence of certain series to the convergence of improper integrals over infinite intervals.

Theorem 8.29 Integral Test. *Let $f : [1, \infty) \to [0, \infty)$ be a bounded nonincreasing nonnegative function. Then the series $\sum_{j=1}^{\infty} f(j)$ converges iff the improper integral $\int_1^{\infty} f(x)\,dx$ converges (also see Figure 21).*

Proof. Throughout the proof let $g(x) = \sum_{j=1}^{\infty} f(j)\mathbf{1}_{[j,j+1)}(x)$. (For every $x \geq 1$, this sum has at most one nonzero term.)

For "$\Rightarrow$," let $\sum_{j=1}^{\infty} f(j)$ be convergent. Then g as defined above is improperly Riemann integrable over $[1, \infty)$ with $\int_1^{\infty} g(x)\,dx = \sum_{j=1}^{\infty} f(j) < \infty$. Because $0 \leq f \leq g$, the Comparison Test for improper integrals implies that $\int_1^{\infty} f(x)\,dx$ converges.

Conversely, for "$\Leftarrow$," let $\int_1^{\infty} f(x)\,dx$ be convergent. Then the improper integral $\int_2^{\infty} f(x-1)\,dx = \int_1^{\infty} f(x)\,dx$ converges. Because $g(x) \leq f(x-1)$ for all $x \geq 2$, by Comparison Test for improper integrals the integral $\int_2^{\infty} g(x)\,dx$ converges, which means the series $\sum_{j=2}^{\infty} f(j)$ converges. ■

The connections and similarities between integrals over infinite intervals and series need to be considered with caution. For example, Exercise 8-38 shows that there is no Limit Test for improper integrals over infinite intervals.

Finally, improper integrals can also be defined for functions on the whole real line and for functions with multiple singularities.

Definition 8.30 *Let $a = a_0 < a_1 < a_2 < \cdots < a_{n-1} < a_n = b$ (where a could be $-\infty$ and b could be ∞) and let $f : (a,b)\setminus\{a_1, \ldots, a_{n-1}\} \to \mathbb{R}$ be Riemann integrable over all closed subintervals of its domain. We define the* **improper Riemann integral** *of f as follows. Let $r_1 < r_2 < \cdots < r_n$ be so that $a_{j-1} < r_j < a_j$ and define $\int_a^b f(x)\,dx := \sum_{j=1}^{n} \left[\int_{a_{j-1}}^{r_j} f(x)\,dx + \int_{r_j}^{a_j} f(x)\,dx \right]$ if all the summands exist.*

Exercises

8-31. Let $b \in \mathbb{R}$ and let $f : (-\infty, b] \to \mathbb{R}$ be such that for all $a < b$ the restriction $f|_{[a,b]}$ is Riemann integrable. Define the improper Riemann integral $\int_{-\infty}^{b} f(x)\,dx$ of f over $(-\infty, b]$.

8-32 Let $a, b \in \mathbb{R}$ with $a < b$ and let $f\ (a, b] \to \mathbb{R}$ be such that for all $c \in (a, b)$ the restriction $f|_{[c,b]}$ is Riemann integrable Define the improper Riemann integral $\int_a^b f(x)\,dx$ of f over $[a, b]$

8-33. Prove the claim in Example 8.24

8-34. Prove Theorem 8.25

8-35 Prove Theorem 8 26

8-36. Finish the proof of Theorem 8 27 by proving that $\lim_{t\to b^-} F(t) = \sup\left\{ F(t) : t \in [a, b) \right\}$.

8-37. Prove Theorem 8.28. *Hint.* $0 \leq f + |f| \leq 2|f|$

8-38 Construct a function $f : [1, \infty) \to [0, 1]$ that is improperly Riemann integrable, but does not converge to zero as $x \to \infty$ *Hint.* Triangles of height 1 and area $\frac{1}{2^n}$.

8-39 **Cauchy Criterion** for improper Riemann integrability Let $f\ .\ [a, b) \to \mathbb{R}$ (where b could be ∞) be Riemann integrable over all intervals $[a, c] \subseteq [a, b)$. Prove that f is improperly Riemann integrable over $[a, b)$ iff for all $\varepsilon > 0$ there is an $M \in [a, b)$, so that for all $c, d \in (M, b)$ we have $\left| \int_c^d f(x)\,dx \right| < \varepsilon$.

8-40. **Limit Comparison Test** for improper integrals. Let $f, g : [a, b) \to [0, \infty)$ (where b could be ∞) be Riemann integrable over all intervals $[a, c] \subseteq [a, b)$ Prove that if $\lim_{x\to b^-} \frac{f(x)}{g(x)} = K > 0$, then $\int_a^b f(x)\,dx$ converges iff $\int_a^b g(x)\,dx$ converges
Hint. Close to b we have $g(x)(K - \varepsilon) \leq f(x) \leq g(x)(K + \varepsilon)$

8-41 Prove that the function $f \cdot (a, b] \to \mathbb{R}$ is improperly Riemann integrable over $(a, b]$ iff the function $g(x) := \frac{1}{x^2} f\left(a + \frac{1}{x}\right)$ is improperly Riemann integrable over $\left[\frac{1}{b-a}, \infty\right)$ and that in this case the integrals are equal

Chapter 9

The Lebesgue Integral

If so desired, the Lebesgue integral can be delayed until after Chapter 13.

The geometric idea behind integration is to approximate the area under the graph of a function with areas that are easier to compute. In the Riemann integral, we partition the x-axis and erect a rectangle over each partition interval to approximate the area under the graph. However, Lebesgue's criterion for Riemann integrability shows that geometric rectangles will not approximate the area well if the function "oscillates" too much. That is, if the differences between the possible choices for the heights of the rectangles do not shrink to zero, then we cannot uniquely identify the area. Equivalently, in the Darboux formulation (see Section 5.4) excessive oscillations force the upper and lower approximations of the area with rectangles to stay a finite distance apart.

If we change our point of view and partition the y-axis instead of the x-axis, the problem with different choices for the height goes away (see Figure 22). For a set $S = \{x \in [a, b] : y_{i-1} \leq f(x) < y_i\}$, all sensible values for the height of a generalized rectangle with base S are between y_{i-1} and y_i. Because the difference between these values can be made small, oscillations are not an issue. However, this approach requires that the bases of our generalized rectangles are no longer intervals. The area of such a generalized rectangle will be the Lebesgue measure of the base set times the height of the rectangle. In this fashion, we retain all benefits of the geometric motivation for integration, while being able to integrate many more functions.[1]

This chapter introduces the fundamentals of Lebesgue integration. These fundamentals will be revisited in Chapter 14 when we generalize our work to arbitrary measure spaces. Our presentation is designed to readily translate to the more abstract setting of Chapter 14.

Before we start, it is time to extend our arithmetic from the real numbers to the

[1]The Lebesgue integral also remedies a more abstract problem with the Riemann integral. Spaces of Riemann integrable functions are usually not complete (see Exercise 16-15d), while spaces of Lebesgue integrable functions are (see Theorem 16.19). Completeness is such a fundamental abstract property that this may be the main reason why the Lebesgue integral is preferred.

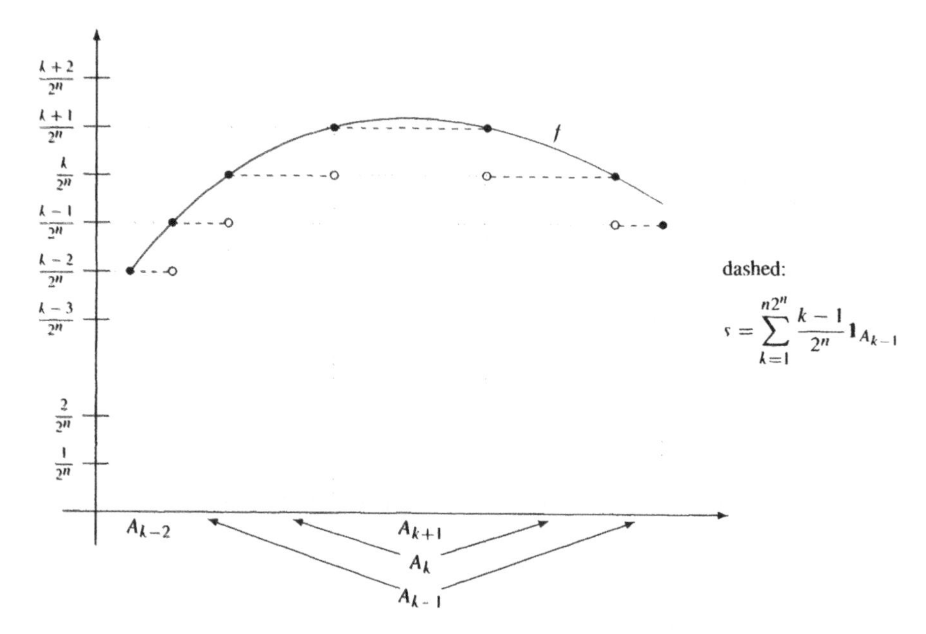

Figure 22: The idea behind the Lebesgue integral is to partition the y-axis instead of the x-axis and to approximate the area with simple functions. This figure shows that this can lead to "scattered" base sets on the x-axis, which we will treat first in this chapter. The proof of Theorem 9.19 uses the partition of the y-axis into intervals with dyadic rational endpoints.

real numbers with ∞ and $-\infty$ included. This is sometimes called the **extended real number system**. In the extended real number system, every set has a supremum. This will make our definitions of measures and of the Lebesgue integral simpler, because we will not need to explicitly distinguish between bounded and unbounded sets.

Proposition 9.1 *In the extended real number system* $[-\infty, \infty] := \mathbb{R} \cup \{\infty, -\infty\}$, ∞ *is the greatest element and* $-\infty$ *is the smallest element. Consequently,* every *set has an infimum and a supremum.*

Proof. Let $A \subseteq [-\infty, \infty]$. The supremum of $\{-\infty\}$ is $-\infty$, and the supremum of $\emptyset$ also is $-\infty$, so we can assume that $A \neq \emptyset, \{-\infty\}$. If A is bounded above by a real number, then $A \cap \mathbb{R} \neq \emptyset$ has a supremum in the real numbers, which is also the supremum of A in $[-\infty, \infty]$. If A is not bounded above by a real number, then $\sup(A) = \infty$. Infima are treated similarly. ■

Proposition 9.1 assures that from here on, we can take suprema and infima of sets of numbers fairly indiscriminately, as long as we know how to handle infinite values algebraically. Arithmetic on $[-\infty, \infty]$ is the same as on $\mathbb{R}$ with the additional conventions of Definition 9.2. These conventions are inspired by the corresponding limit laws in Theorems 2.44 and 2.46.

Definition 9.2 Arithmetic involving ∞. *Let $c \in \mathbb{R}$. Then*

$$c + \infty = \infty, \qquad c - \infty = -\infty,$$

$$c \cdot \infty = \begin{cases} \infty; & \text{if } c > 0, \\ -\infty; & \text{if } c < 0, \\ \text{undefined}; & \text{if } c = 0, \end{cases} \qquad \frac{c}{\infty} = 0,$$

$$c \cdot (-\infty) = \begin{cases} -\infty; & \text{if } c > 0, \\ \infty; & \text{if } c < 0, \\ \text{undefined}; & \text{if } c = 0, \end{cases} \qquad \frac{c}{-\infty} = 0,$$

$$\infty \cdot \infty = \infty, \qquad \infty \cdot (-\infty) = -\infty,$$

$$(-\infty) \cdot (-\infty) = \infty.$$

All other attempts at "arithmetic with infinity" lead to what is called *indeterminate forms*. For these, the result can be any number, and hence rules of arithmetic cannot be stated. We will consider indeterminate forms in Section 12.3.

9.1 Lebesgue Measurable Sets

We will work with generalized rectangles whose bases are no longer intervals. Therefore we need to measure the size of sets that are more complicated than rectangles. Outer Lebesgue measure is a reliable upper bound for the (one-dimensional) "volume" of a set. For all examples we have seen, it gives the right "volume." Hence, we can (and do) consider it the right way to measure the "outer volume" of a set.

Unfortunately, there are complications. It is possible to split a set T into two sets so that the outer Lebesgue measures of the two sets add up to more than the outer Lebesgue measure of T. If we were to involve such pathological sets in a definition of integration, the integral would not even be a reliable measure of the area of generalized rectangles. It would make no sense if the total "length" of the base would depend on how we split the base. The definition of Lebesgue measurable sets is designed to safeguard exactly against this problem (see Theorem 9.11). Because all our sets are subsets of $\mathbb{R}$ (and later of a measure space M), we introduce abbreviated notation for the complement.

Notation 9.3 *When there is one underlying set X that contains all sets that we currently investigate (as is the case in measure theory) and $S \subseteq X$, then we denote the* **complement** *of S in X also as $S' := X \setminus S$.* □

Definition 9.4 *A subset $S \subseteq \mathbb{R}$ is called* **Lebesgue measurable** *iff for all $T \subseteq \mathbb{R}$ the equality $\lambda(T) = \lambda(S \cap T) + \lambda\left(S' \cap T\right)$ holds. We will also call the set T a* **test set**. *We denote the set of Lebesgue measurable subsets of $\mathbb{R}$ by Σ_λ.*

The existence of non-Lebesgue measurable sets, which would cause the above-mentioned problems, is equivalent to the Axiom of Choice. That is, whether or not non-Lebesgue measurable sets exist depends on what axiomatic system of set theory is used.

In practical terms it means that non-Lebesgue measurable sets do not occur in physical phenomena. Hence, from an applied point-of-view we need not be too concerned with nonmeasurable sets and we will not consider them any further in this text. Exercise 9-7 illustrates the problems mentioned above with a simpler measure of "size," the Jordan content, for which even simple sets can behave badly. The construction is more complicated for Lebesgue measure and the interested reader can find such constructions in [14] and [29]. For the remainder of this section, we explore the properties of the set of Lebesgue measurable subsets of $\mathbb{R}$.

Definition 9.5 *Let $S \subseteq \mathbb{R}$ be a Lebesgue measurable set. Then the outer Lebesgue measure $\lambda(S)$ of S is also called the* **Lebesgue measure** *of S.*

We first note that "half" of the definition of measurability is always satisfied.

Corollary 9.6 *For all subsets $S, T \subseteq \mathbb{R}$, we have $\lambda(T) \leq \lambda(S \cap T) + \lambda\left(S' \cap T\right)$.*

Proof. This follows from part 3 of Theorem 8.6 with $A_1 := S \cap T$, $A_2 := S' \cap T$ and $A_n := \emptyset$ for $n \geq 3$. ∎

It is now easy to see that null sets are Lebesgue measurable.

Proposition 9.7 *If $\lambda(S) = 0$, then S is Lebesgue measurable.*

Proof. Let S be a null set. By part 2 of Theorem 8.6 for all sets $T \subseteq \mathbb{R}$ we obtain $0 \leq \lambda(S \cap T) \leq \lambda(S) = 0$. Hence, for all subsets $T \subseteq \mathbb{R}$ we have the inequality $\lambda(S \cap T) + \lambda\left(S' \cap T\right) \leq \lambda\left(S' \cap T\right) \leq \lambda(T)$, which by Corollary 9.6 is all we need to establish Lebesgue measurability of S. ∎

In the following, we establish that certain set theoretical operations preserve measurability. Theorem 9.10 summarizes the most important facts. Although there are more results, the properties listed in Theorem 9.10 suffice for our purposes. Set systems that satisfy these properties are called σ-algebras (see Definition 14.1). These set systems are fundamental for measure theory.

Lemma 9.8 *If A and B are Lebesgue measurable sets, then the intersection $A \cap B$ is Lebesgue measurable.*

Proof. *Proofs of Lebesgue measurability typically involve the appropriate rewriting of terms and the use of the right test sets.* To show that $A \cap B$ is Lebesgue measurable, let $T \subseteq \mathbb{R}$ be any subset of $\mathbb{R}$. Then

$$\begin{aligned}
&\lambda((A \cap B) \cap T) + \lambda\left((A \cap B)' \cap T\right) \\
&= \lambda(A \cap B \cap T) + \lambda\left(\left(A' \cup B'\right) \cap T\right)
\end{aligned}$$

To break up the second term, use that A is Lebesgue measurable and apply it to the test set $\left(A' \cup B'\right) \cap T$.

$$\begin{aligned}
&= \lambda(A \cap B \cap T) + \lambda\left(A \cap \left(A' \cup B'\right) \cap T\right) + \lambda\left(A' \cap \left(A' \cup B'\right) \cap T\right) \\
&= \lambda(B \cap A \cap T) + \lambda\left(B' \cap A \cap T\right) + \lambda\left(A' \cap T\right)
\end{aligned}$$

B is Lebesgue measurable, so combine the first two terms.

$$= \lambda(A \cap T) + \lambda\left(A' \cap T\right)$$

Use that A is Lebesgue measurable to combine terms.

$$= \lambda(T).$$

Because T was an arbitrary subset of $\mathbb{R}$ we have proved that $A \cap B$ is Lebesgue measurable. ■

The next lemma will be useful in two ways. It is a step toward proving that countable unions of Lebesgue measurable sets are Lebesgue measurable. It also is a step toward proving that if a Lebesgue measurable set consists of countably many pairwise disjoint Lebesgue measurable pieces, then the Lebesgue measure of the set is the sum of the Lebesgue measures of the pieces.

Lemma 9.9 *Let $\{A_n\}_{n=1}^{\infty}$ be a sequence of pairwise disjoint Lebesgue measurable sets. Then the union $\bigcup_{n=1}^{\infty} A_n$ is Lebesgue measurable and for all $T \subseteq \mathbb{R}$ we have*

$$\lambda(T) = \sum_{n=1}^{\infty} \lambda(A_n \cap T) + \lambda\left(\left(\bigcup_{n=1}^{\infty} A_n\right)' \cap T\right).$$

Proof. Let $T \subseteq \mathbb{R}$. We first prove by induction that for all $k \in \mathbb{N}$ we have that $\lambda(T) = \sum_{n=1}^{k} \lambda(A_n \cap T) + \lambda\left(\left(\bigcup_{n=1}^{k} A_n\right)' \cap T\right)$. The base step with $k = 1$ follows from the Lebesgue measurability of A_1.

For the induction step $k \to (k+1)$, we can assume that the induction hypothesis $\lambda(T) = \sum_{n=1}^{k} \lambda(A_n \cap T) + \lambda\left(\left(\bigcup_{n=1}^{k} A_n\right)' \cap T\right)$ is true, and we need to prove that $\lambda(T) = \sum_{n=1}^{k+1} \lambda(A_n \cap T) + \lambda\left(\left(\bigcup_{n=1}^{k+1} A_n\right)' \cap T\right)$. We start the induction step with the induction hypothesis.

$$\begin{aligned}
\lambda(T) &= \sum_{n=1}^{k} \lambda(A_n \cap T) + \lambda\left(\left(\bigcup_{n=1}^{k} A_n\right)' \cap T\right) \\
&= \sum_{n=1}^{k} \lambda(A_n \cap T) + \lambda\left(\bigcap_{n=1}^{k} A_n' \cap T\right)
\end{aligned}$$

To break up the second term, use that A_{k+1} is Lebesgue measurable.

$$= \sum_{n=1}^{k} \lambda(A_n \cap T) + \lambda\left(A_{k+1} \cap \bigcap_{n=1}^{k} A_n' \cap T\right) + \lambda\left(A_{k+1}' \cap \bigcap_{n=1}^{k} A_n' \cap T\right)$$

For the middle term, recall that the A_k are pairwise disjoint, which means that $A_{k+1} \cap \bigcap_{n=1}^{k} A'_n = A_{k+1}$.

$$= \sum_{n=1}^{k} \lambda(A_n \cap T) + \lambda(A_{k+1} \cap T) + \lambda\left(\bigcap_{n=1}^{k+1} A'_n \cap T\right)$$

$$= \sum_{n=1}^{k+1} \lambda(A_n \cap T) + \lambda\left(\left(\bigcup_{n=1}^{k+1} A_n\right)' \cap T\right),$$

which finishes the induction step.

Thus for all $k \in \mathbb{N}$ we have $\lambda(T) \geq \sum_{n=1}^{k} \lambda(A_n \cap T) + \lambda\left(\left(\bigcup_{n=1}^{\infty} A_n\right)' \cap T\right)$ and by letting k go to infinity we obtain the following.

$$\lambda(T) \geq \sum_{n=1}^{\infty} \lambda(A_n \cap T) + \lambda\left(\left(\bigcup_{n=1}^{\infty} A_n\right)' \cap T\right)$$

Use the countable subadditivity of λ.

$$\geq \lambda\left(\left(\bigcup_{n=1}^{\infty} A_n\right) \cap T\right) + \lambda\left(\left(\bigcup_{n=1}^{\infty} A_n\right)' \cap T\right)$$

Use Corollary 9.6.

$$\geq \lambda(T).$$

The above establishes measurability of the union as well as the desired equality. ■

Theorem 9.10 *The set Σ_λ of Lebesgue measurable subsets of $\mathbb{R}$ has the following properties.*

1. $\emptyset \in \Sigma_\lambda$.
2. *If $S \in \Sigma_\lambda$, then $S' \in \Sigma_\lambda$.*
3. *If $A_n \in \Sigma_\lambda$ for all $n \in \mathbb{N}$, then $\bigcup_{n=1}^{\infty} A_n \in \Sigma_\lambda$.*

Proof. Parts 1 and 2 are left to the reader as Exercises 9-1a and 9-1b. For part 3, let $A_n \in \Sigma_\lambda$ for all $n \in \mathbb{N}$ and let $T \subseteq \mathbb{R}$. Define $B_1 := A_1$ and then inductively for $n \in \mathbb{N}$ set $B_{n+1} := A_{n+1} \cap \left(\bigcup_{i=1}^{n} B_i\right)' = A_{n+1} \cap \bigcap_{i=1}^{n} B'_i$. An easy induction shows that for all $n \in \mathbb{N}$ the set B_n is Lebesgue measurable (use part 2 and Lemma 9.8), that $B_1, \ldots, B_n$ are pairwise disjoint and that $\bigcup_{i=1}^{n} A_i = \bigcup_{i=1}^{n} B_i$ (see Exercise 9-1c). But then $\bigcup_{i=1}^{\infty} A_i = \bigcup_{i=1}^{\infty} B_i$ (see Exercise 9-1d) and the latter set is Lebesgue measurable by Lemma 9.9. ■

Now that we have established that countable unions preserve Lebesgue measurability, it is reassuring to note that Lebesgue measure is additive for countable unions of pairwise disjoint Lebesgue measurable sets. This is exactly what we expect from a sensible measure. The size of a whole set is the sum of the sizes of its pairwise disjoint parts.

Theorem 9.11 *Let $\{A_n\}_{n=1}^{\infty}$ be a sequence of pairwise disjoint Lebesgue measurable sets. Then $\bigcup_{n=1}^{\infty} A_n$ is Lebesgue measurable and $\lambda\left(\bigcup_{n=1}^{\infty} A_n\right) = \sum_{n=1}^{\infty} \lambda(A_n)$.*

Proof. The union is Lebesgue measurable by Lemma 9.9 and if we apply Lemma 9.9 to $T := \bigcup_{n=1}^{\infty} A_n$ we obtain

$$\begin{aligned}\lambda\left(\bigcup_{n=1}^{\infty} A_n\right) &= \sum_{n=1}^{\infty} \lambda\left(A_n \cap \bigcup_{n=1}^{\infty} A_n\right) + \lambda\left(\left(\bigcup_{n=1}^{\infty} A_n\right)' \cap \bigcup_{n=1}^{\infty} A_n\right) \\ &= \sum_{n=1}^{\infty} \lambda(A_n) + 0 = \sum_{n=1}^{\infty} \lambda(A_n),\end{aligned}$$

which proves the result. ■

The next result is a nice application of the countable additivity we just proved and it will also be needed when we show that sums of Lebesgue integrable functions are Lebesgue integrable.

Theorem 9.12 *Let $\{A_n\}_{n=1}^{\infty}$ be a sequence of Lebesgue measurable subsets of $\mathbb{R}$ so that $A_n \subseteq A_{n+1}$ for all $n \in \mathbb{N}$. Then $\lambda\left(\bigcup_{n=1}^{\infty} A_n\right) = \lim_{n\to\infty} \lambda(A_n)$.*

Proof. Set $B_1 := A_1$ and for all $n \in \mathbb{N}$ define $B_{n+1} := A_{n+1} \setminus A_n$. Then for all $N \in \mathbb{N}$ the equality $\bigcup_{n=1}^{N} B_n = \bigcup_{n=1}^{N} A_n = A_N$ holds, which means $\bigcup_{n=1}^{\infty} B_n = \bigcup_{n=1}^{\infty} A_n$. Now $\lambda\left(\bigcup_{n=1}^{\infty} A_n\right) = \lambda\left(\bigcup_{n=1}^{\infty} B_n\right) = \sum_{n=1}^{\infty} \lambda(B_n) = \lim_{N\to\infty} \sum_{n=1}^{N} \lambda(B_n) = \lim_{N\to\infty} \lambda(A_N)$. ■

The whole idea of Lebesgue measurability is only useful if it indeed allows us to extend the idea of Riemann integrability. For that to happen, intervals must be Lebesgue measurable. We have delayed this result to the end of the section, because now we can use some of the machinery built so far.

Proposition 9.13 *Intervals are Lebesgue measurable and the Lebesgue measure of an interval is its length.*

Proof. Because singletons are null sets and thus Lebesgue measurable, and because countable unions of Lebesgue measurable sets are Lebesgue measurable, intervals are shown to be Lebesgue measurable if we can prove that open intervals of finite length are Lebesgue measurable. So let $A = (a, b)$ be an open interval of finite length and let $T \subseteq \mathbb{R}$. By Corollary 9.6, we only need to prove $\lambda(T) \geq \lambda(A \cap T) + \lambda\left(A' \cap T\right)$. The inequality is trivial if $\lambda(T) = \infty$, so we only need to prove the inequality if $\lambda(T) < \infty$. Let $T \subseteq \mathbb{R}$ be a set of finite outer Lebesgue measure, let $\varepsilon > 0$ and let $\{I_j\}_{j=1}^{\infty}$ be a family of open intervals with $T \subseteq \bigcup_{j=1}^{\infty} I_j$ and $\sum_{j=1}^{\infty} |I_j| \leq \lambda(T) + \frac{\varepsilon}{2}$. Then $\mathcal{I} := \left\{I_j \cap A : j \in \mathbb{N}\right\}$ is a countable family of open intervals whose union contains $A \cap T$. Moreover,

$$\begin{aligned} \mathcal{O} &:= \left\{I_j \setminus [a, \infty) : j \in \mathbb{N}\right\} \cup \left\{I_j \setminus (-\infty, b] : j \in \mathbb{N}\right\} \\ &\quad \cup \left\{\left(a - \frac{\varepsilon}{8}, a + \frac{\varepsilon}{8}\right), \left(b - \frac{\varepsilon}{8}, b + \frac{\varepsilon}{8}\right)\right\} \end{aligned}$$

is a countable family of open intervals whose union contains $A' \cap T$. Thus

$$\begin{aligned} &\lambda(A \cap T) + \lambda\left(A' \cap T\right) \\ &\leq \sum_{I \in \mathcal{I}} |I| + \sum_{I \in \mathcal{O}} |I| \\ &= \left|\left(a - \frac{\varepsilon}{8}, a + \frac{\varepsilon}{8}\right)\right| + \left|\left(b - \frac{\varepsilon}{8}, b + \frac{\varepsilon}{8}\right)\right| \\ &\quad + \sum_{j=1}^{\infty} \left[\left|I_j \cap (a, b)\right| + \left|I_j \setminus [a, \infty)\right| + \left|I_j \setminus (-\infty, b]\right|\right] \\ &\overset{(*)}{\leq} \frac{\varepsilon}{4} + \frac{\varepsilon}{4} + \sum_{j=1}^{\infty} |I_j| \leq \frac{\varepsilon}{4} + \frac{\varepsilon}{4} + \lambda(T) + \frac{\varepsilon}{2} \\ &= \lambda(T) + \varepsilon \end{aligned}$$

Because ε was arbitrary this proves that $\lambda(T) \geq \lambda(A \cap T) + \lambda\left(A' \cap T\right)$ and we have proved that A is Lebesgue measurable.

Regarding the Lebesgue measure of intervals, by Proposition 8.5 for closed intervals in $\mathbb{R}$ we have $\lambda\big([a, b]\big) = b - a$. For closed, unbounded intervals of the form $[a, \infty)$, we infer for all $b > a$ that $\lambda\big([a, \infty)\big) \geq \lambda\big([a, b]\big) = b - a$, and hence $\lambda\big([a, \infty)\big) = \infty$. Intervals of the form $(-\infty, b]$ are handled similarly. For open and half-open intervals, note that the singleton sets consisting of the endpoints have measure zero. This means (see Exercise 9-2) that adding or removing these points does not affect the Lebesgue measurability of the set or its Lebesgue measure. ■

Standard Proof Technique 9.14 Note that the inequality marked with $(*)$ in the proof of Proposition 9.13 can actually be shown to be an equality. This is not necessary because we only need the inequality. In complicated estimates, it can happen that an inequality sign is put between quantities that are actually equal. Usually, this happens when the equality would not have helped in the proof (as in the example just mentioned) and when the writer did not want (the reader) to spend extra effort to think about why the quantities may be equal. □

Exercises

9-1. Finish the proof of Theorem 9.10. That is,

(a) Prove that $\emptyset \in \Sigma_\lambda$.

(b) Prove that if $S \in \Sigma_\lambda$, then $S' \in \Sigma_\lambda$.

(c) Perform the induction mentioned in the proof of part 3.

(d) Prove that if $\{A_i\}_{i=1}^\infty$ and $\{B_i\}_{i=1}^\infty$ are countable families of sets so that for all $n \in \mathbb{N}$ we have $\bigcup_{i=1}^n A_i = \bigcup_{i=1}^n B_i$, then $\bigcup_{i=1}^\infty A_i = \bigcup_{i=1}^\infty B_i$.

9-2. Let A be a Lebesgue measurable set and let N be a null set. Prove that $A \setminus N$ is Lebesgue measurable and that $\lambda(A \setminus N) = \lambda(A)$.
Hint. Use Theorem 9.10.

9-3. Let $A, B \subseteq \mathbb{R}$ be Lebesgue measurable sets. Prove that $A \setminus B$ is Lebesgue measurable.

9-4. Let $\{A_n\}_{n=1}^N$ be a finite sequence of pairwise disjoint Lebesgue measurable sets. Prove that the union $\bigcup_{n=1}^N A_n$ is Lebesgue measurable with $\lambda\left(\bigcup_{n=1}^N A_n\right) = \sum_{n=1}^N \lambda(A_n)$.

9-5. Let $\{A_n\}_{n=1}^\infty$ be a sequence of Lebesgue measurable sets. Prove that $\bigcap_{n=1}^\infty A_n$ is Lebesgue measurable and that for all $k \in \mathbb{N}$ we have $\lambda\left(\bigcap_{n=1}^\infty A_n\right) \leq \lambda(A_k)$.

9-6. Let C^Q be a **Cantor set**. Prove that C^Q is Lebesgue measurable.
Hint. Use Exercise 9-5.

9-7. For all $S \subseteq [0,1]$, let $J(S) = \inf\left\{\sum_{j=1}^n |I_j| : S \subseteq \bigcup_{j=1}^n I_j, \text{ each } I_j \text{ is an open interval}\right\}$ be the **Jordan content** of S. Prove that $J\left([0,1] \cap \mathbb{Q}\right) = 1$ and $J\left([0,1] \setminus \mathbb{Q}\right) = 1$.

9-8. Let $A \subseteq B \subseteq \mathbb{R}$ and let A (but not necessarily B) be Lebesgue measurable. Prove that $\lambda(B) = \lambda(A) + \lambda(B \setminus A)$.

9.2 Lebesgue Measurable Functions

We now introduce the functions for which the Lebesgue integral can be defined. By first defining what (potentially) integrable functions should look like, we avoid the Riemann integral's conceptual complications that are characterized in Lebesgue's criterion (see Theorem 8.12). Existence or nonexistence of the Lebesgue integral, defined in Section 9.3, will then merely be a question of whether there is too much area under the graph of the function. We will revisit the original motivation of partitioning the y-axis after the proof of Theorem 9.19. To approximate areas, in Lebesgue integration indicator functions take the place of rectangles. Recall that by Definition 5.9 the **indicator function** of a set $S \subseteq \mathbb{R}$ is $\mathbf{1}_S(x) := \begin{cases} 1; & \text{for } x \in S, \\ 0; & \text{for } x \notin S. \end{cases}$

Just as Riemann integrable functions can be approximated a.e. with step functions (see Exercise 5-26), Lebesgue integrable functions will be approximated with functions that are constant on measurable sets and which only assume finitely many values.

Definition 9.15 *A function $s : \mathbb{R} \to \mathbb{R}$ is called a* **simple Lebesgue measurable function**, *or, a* **simple function**, *iff there are numbers $a_1, \ldots, a_n \in \mathbb{R}$ and pairwise disjoint Lebesgue measurable sets $A_1, \ldots, A_n \subseteq \mathbb{R}$ so that $s = \sum_{k=1}^{n} a_k \mathbf{1}_{A_k}$.*

For functions f that assume more than finitely many values, we consider the positive and negative parts of f separately.

Definition 9.16 *For $f : \mathbb{R} \to [-\infty, \infty]$, we define $f^+(x) := \max\{f(x), 0\}$ and $f^-(x) := -\min\{f(x), 0\}$ for all $x \in \mathbb{R}$.*

Because we will successively approximate measurable functions from below we want to speak of sequences of functions.

Definition 9.17 *A family $\{f_n\}_{n\in\mathbb{N}}$ of functions will also be called a* **sequence** *of functions, denoted $\{f_n\}_{n=1}^{\infty}$.*

Definition 9.18 *A function $f : \mathbb{R} \to [0, \infty]$ is called* **Lebesgue measurable** *iff there is a sequence $\{s_n\}_{n=1}^{\infty}$ of simple functions $s_n : \mathbb{R} \to [0, \infty)$ such that for all $x \in \mathbb{R}$ the sequence $\{s_n(x)\}_{n=1}^{\infty}$ is nondecreasing and $\lim_{n\to\infty} s_n(x) = f(x)$.*

A function $f : \mathbb{R} \to [-\infty, \infty]$ is called **Lebesgue measurable** *iff f^+ and f^- are both Lebesgue measurable.*

The key problem in Riemann integration is that for some functions the approximations from above and below will not "meet." Definition 9.18 does not simply circumvent this problem by only focusing on approximations from below. Exercise 9-9 shows that a bounded function (only bounded functions are considered in Riemann integration) is Lebesgue measurable iff it can be approximated from above with simple functions. That is, Definition 9.18 may look biased, but for bounded functions the concept of Lebesgue measurability could also be defined with approximations from above. Because we are also interested in unbounded functions, we choose to work with approximations from below throughout.

Because Lebesgue measurability is a key concept, it is useful to have several equivalent formulations available.

Theorem 9.19 *Let $f : \mathbb{R} \to [-\infty, \infty]$ be a function. Then the following are equivalent.*

1. *The function f is Lebesgue measurable.*

2. *For all $a \in \mathbb{R}$, the set $\{x \in \mathbb{R} : f(x) > a\}$ is Lebesgue measurable.*

3. *For all $a \in \mathbb{R}$, the set $\{x \in \mathbb{R} : f(x) \leq a\}$ is Lebesgue measurable.*

4. *For all $a \in \mathbb{R}$, the set $\{x \in \mathbb{R} : f(x) < a\}$ is Lebesgue measurable.*

5. *For all $a \in \mathbb{R}$, the set $\{x \in \mathbb{R} : f(x) \geq a\}$ is Lebesgue measurable.*

Proof. We first prove the result for a function $f : \mathbb{R} \to [0, \infty]$.

For the implication "1⇒2," let $a \in \mathbb{R}$ and let $\{s_n\}_{n=1}^{\infty}$ be a sequence of simple functions such that for all $x \in \mathbb{R}$ the sequence $\{s_n(x)\}_{n=1}^{\infty}$ is nondecreasing and converges to $f(x)$. For all $x \in \mathbb{R}$, if $f(x) > a$, then for some $n \in \mathbb{N}$ the inequality $s_n(x) > a$ holds. Conversely, if for some $n \in \mathbb{N}$ we have that $s_n(x) > a$, then because $\{s_n(x)\}_{n=1}^{\infty}$ is nondecreasing we must have $f(x) > a$. This means that $\{x \in \mathbb{R} : f(x) > a\} = \bigcup_{n=1}^{\infty} \{x \in \mathbb{R} : s_n(x) > a\}$. But each set $\{x \in \mathbb{R} : s_n(x) > a\}$ is a union of finitely many Lebesgue measurable sets, which means it is Lebesgue measurable. Therefore, as a countable union of Lebesgue measurable sets, the set $\{x : f(x) > a\}$ is Lebesgue measurable.

For "2⇒3," let $a \in \mathbb{R}$. Then $\{x \in \mathbb{R} : f(x) \leq a\} = \mathbb{R} \setminus \{x \in \mathbb{R} : f(x) > a\}$, which is the complement of a Lebesgue measurable set.

For "3⇒4," note that for all real numbers a the set $\{x \in \mathbb{R} : f(x) < a\}$ is equal to the union $\{x \in \mathbb{R} : f(x) < a\} = \bigcup_{k=1}^{\infty} \left\{x \in \mathbb{R} : f(x) \leq a - \frac{1}{k}\right\}$ and the latter set is a countable union of Lebesgue measurable sets.

For "4⇒5," let $a \in \mathbb{R}$. Then $\{x \in \mathbb{R} : f(x) \geq a\} = \mathbb{R} \setminus \{x \in \mathbb{R} : f(x) < a\}$, which is the complement of a Lebesgue measurable set.

For "5⇒1," by Exercise 9-10 the set $\{x \in \mathbb{R} : f(x) = \infty\}$ is Lebesgue measurable. For each $n \in \mathbb{N}$, define the simple function

$$s_n := \sum_{k=1}^{n2^n} \frac{k-1}{2^n} \left(\mathbf{1}_{\left\{x \in \mathbb{R} : f(x) \geq \frac{k-1}{2^n}\right\}} - \mathbf{1}_{\left\{x \in \mathbb{R} : f(x) \geq \frac{k}{2^n}\right\}} \right) + n\mathbf{1}_{\{x \in \mathbb{R} : f(x) = \infty\}}.$$

Then each s_n is nonnegative, because for all $k, n \in \mathbb{N}$ we have

$$\mathbf{1}_{\left\{x \in \mathbb{R} : f(x) \geq \frac{k-1}{2^n}\right\}} - \mathbf{1}_{\left\{x \in \mathbb{R} : f(x) \geq \frac{k}{2^n}\right\}} = \mathbf{1}_{\left\{x \in \mathbb{R} : f(x) \geq \frac{k-1}{2^n}\right\} \setminus \left\{x \in \mathbb{R} : f(x) \geq \frac{k}{2^n}\right\}}.$$

To see that each $\{s_n(x)\}_{n=1}^{\infty}$ is nondecreasing, let $x \in \mathbb{R}$ and let $n \in \mathbb{N}$. The claim is trivial if $f(x) = \infty$, so we can assume $f(x) < \infty$. If $f(x) \geq n$, then $s_n(x) = 0 \leq s_{n+1}(x)$. If $f(x) < n$, find $k \in \{1, \ldots, n2^n\}$ so that $s_n(x) = \frac{k-1}{2^n}$. Then $\frac{k-1}{2^n} \leq f(x) < \frac{k}{2^n}$, and hence $s_{n+1}(x) \geq \frac{2(k-1)}{2^{n+1}} = \frac{k-1}{2^n} = s_n(x)$. Finally, each $\{s_n(x)\}_{n=1}^{\infty}$ converges to $f(x)$ because, if $f(x) \in [N-1, N)$, then for all $n \geq N$ the inequality $|f(x) - s_n(x)| < \frac{1}{2^n}$ holds, and if $f(x) = \infty$, then $f(x) = n$.

Finally, consider $f : \mathbb{R} \to [-\infty, \infty]$. For "1⇒2" let f be Lebesgue measurable. Then f^- and f^+ are Lebesgue measurable. But then for $a \geq 0$ we have that $\{x \in \mathbb{R} : f(x) > a\} = \{x \in \mathbb{R} : f^+(x) > a\}$, which is Lebesgue measurable, and for $a < 0$ we have $\{x \in \mathbb{R} : f(x) > a\} = \{x \in \mathbb{R} : f^-(x) < -a\}$, which is also Lebesgue measurable. Parts "2⇒3," "3⇒4," and "4⇒5" are similar to what was done for nonnegative functions. To prove part "5⇒1," first note that for all $a > 0$ we have $\{x \in \mathbb{R} : f^+(x) \geq a\} = \{x \in \mathbb{R} : f(x) \geq a\}$, which is Lebesgue measurable and for all $a \leq 0$ we have $\{x \in \mathbb{R} : f^+(x) \geq a\} = \mathbb{R}$, which is also Lebesgue measurable. Hence, f^+ is Lebesgue measurable. Considering the negative part f^-, for all $a \geq 0$ we have that $\{x \in \mathbb{R} : f^-(x) \leq a\} = \{x \in \mathbb{R} : f(x) \geq -a\}$, which is Lebesgue measurable and for $a < 0$ we have $\{x \in \mathbb{R} : f^-(x) \leq a\} = \emptyset$, which is also Lebesgue measurable. Hence, f^- is Lebesgue measurable and because we already proved that f^+ is Lebesgue measurable we have proved that f is Lebesgue measurable. ∎

The underlying idea of the proof of part "5⇒1" for nonnegative functions f is to partition the interval $[0, n)$ on the y-axis into intervals of length $\frac{1}{2^n}$. The proof shows that the area under the functions that are used to approximate f should approximate the area under f, which means the idea of partitioning the y-axis can lead to a sensible notion of integration.

Once measurable functions are characterized, it is helpful to determine how measurability relates to common algebraic operations.

Theorem 9.20 *Let $f, g : \mathbb{R} \to [-\infty, \infty]$ be Lebesgue measurable functions. If $f + g$ is defined everywhere, then $f + g$ is Lebesgue measurable. Similarly, if $f - g$ or $f \cdot g$ is defined everywhere, then it is Lebesgue measurable. Moreover, f^+, f^- and $|f|$ are Lebesgue measurable.*

Proof. To see that $f + g$ is Lebesgue measurable, let $a \in \mathbb{R}$. We will use Theorem 1.36 to show that the set $\{x \in \mathbb{R} : (f + g)(x) < a\}$ is a countable union of Lebesgue measurable sets. If $(f + g)(x) < a$, then there is an $\varepsilon > 0$ so that $(f + g)(x) + 2\varepsilon < a$. By Theorem 1.36, there are rational numbers r and s so that $f(x) < r < f(x) + \varepsilon$ and $g(x) < s < g(x) + \varepsilon$. This means that $f(x) + g(x) < r + s < (f + g)(x) + 2\varepsilon < a$, which proves the containment "$\subseteq$" in the equation below. The containment "$\supseteq$" is trivial.

$$\{x \in \mathbb{R} : (f + g)(x) < a\} = \bigcup_{r,s \in \mathbb{Q}, r+s<a} \{x \in \mathbb{R} : f(x) < r\} \cap \{x \in \mathbb{R} : g(x) < s\}.$$

Because the latter set is a countable union of Lebesgue measurable sets, the set $\{x \in \mathbb{R} : (f + g)(x) < a\}$ is Lebesgue measurable. Because $a \in \mathbb{R}$ was arbitrary this means that $f + g$ is a Lebesgue measurable function.

The proofs that $f - g$ and $f \cdot g$ are Lebesgue measurable functions are similar (see Exercise 9-11). The functions f^+ and f^- are Lebesgue measurable by Definition 9.18 and the Lebesgue measurability of $|f| = f^+ + f^-$ follows from the Lebesgue measurability of sums of Lebesgue measurable functions. ∎

Exercises

9-9. Prove that a bounded function $f : \mathbb{R} \to [0, \infty)$ is Lebesgue measurable iff there is a sequence $\{s_n\}_{n=1}^{\infty}$ of simple functions $s_n : \mathbb{R} \to \mathbb{R}$ such that for all $x \in \mathbb{R}$ the sequence $\{ s_n(x) \}_{n=1}^{\infty}$ is nonincreasing and $\lim_{n\to\infty} s_n(x) = f(x)$.

Hint. Mimic part "5⇒1" of the proof of Theorem 9.19 for nonnegative functions.

9-10. Let $f : \mathbb{R} \to [-\infty, \infty]$ be a Lebesgue measurable function. Use part 4 of Theorem 9.19 to prove that $\{ x \in \mathbb{R} : f(x) = \infty \}$ is Lebesgue measurable.

9-11. Finish the proof of Theorem 9.20. That is,

(a) Prove that if $f, g : \mathbb{R} \to [-\infty, \infty]$ are Lebesgue measurable and $f - g$ is defined everywhere, then $f - g$ is Lebesgue measurable.

(b) Prove that if $f, g : \mathbb{R} \to [-\infty, \infty]$ are Lebesgue measurable and $f \cdot g$ is defined everywhere, then $f \cdot g$ is Lebesgue measurable

Hint. This one is complicated because of negative signs. Prove the result first for $f, g \geq 0$, then use $f = f^+ - f^-$ and $g = g^+ - g^-$.

9-12 Prove that the sum of two simple functions is again a simple function.

9-13. Prove that $f : \mathbb{R} \to [-\infty, \infty]$ is Lebesgue measurable iff for any two numbers $a < b$ in $\mathbb{R}$ the set $\{ x \in \mathbb{R} : f(x) \in [a, b) \}$ is Lebesgue measurable

9-14 Use Definition 9 18 to prove that if $f, g : \mathbb{R} \to [0, \infty]$ are Lebesgue measurable functions, then $f + g$ is Lebesgue measurable.

9-15. Let $f, h : \mathbb{R} \to [-\infty, \infty]$ and let f be Lebesgue measurable. Prove that if $f = h$ a.e., then h is Lebesgue measurable.

9-16. Let $f, g : \mathbb{R} \to [-\infty, \infty]$ be Lebesgue measurable functions.

(a) Prove that if $f, g : \mathbb{R} \to [-\infty, \infty]$ are Lebesgue measurable and $f(x) + g(x)$ is defined almost everywhere, then $(f+g)(x) := \begin{cases} f(x) + g(x); & \text{if } f(x) + g(x) \text{ is defined,} \\ 0; & \text{otherwise ,} \end{cases}$ is Lebesgue measurable.

Hint Apply Exercise 9-15 to the right auxiliary functions and then use Theorem 9.20.

(b) Prove that if $f, g : \mathbb{R} \to [-\infty, \infty]$ are Lebesgue measurable and $f(x) - g(x)$ is defined almost everywhere, then $(f-g)(x) := \begin{cases} f(x) - g(x); & \text{if } f(x) - g(x) \text{ is defined,} \\ 0; & \text{otherwise ,} \end{cases}$ is Lebesgue measurable.

(c) Prove that if $f, g : \mathbb{R} \to [-\infty, \infty]$ are Lebesgue measurable and $f(x)g(x)$ is defined almost everywhere, then $(fg)(x) = \begin{cases} f(x)g(x); & \text{if } f(x)g(x) \text{ is defined,} \\ 0; & \text{otherwise ,} \end{cases}$ is Lebesgue measurable

9-17. Let $f, g : \mathbb{R} \to [-\infty, \infty]$ be Lebesgue measurable functions.

(a) Prove that $\{ x \in \mathbb{R} : f(x) = g(x) \}$ is Lebesgue measurable.

(b) Prove that $\{ x \in \mathbb{R} : f(x) \leq g(x) \}$ is Lebesgue measurable.

(c) Prove that $\{ x \in \mathbb{R} : f(x) < g(x) \}$ is Lebesgue measurable.

9-18. Let $f, g : \mathbb{R} \to [-\infty, \infty]$ be Lebesgue measurable functions.

(a) Prove that $\max\{f, g\}$ (defined pointwise) is Lebesgue measurable.

(b) Prove that $\min\{f, g\}$ (defined pointwise) is Lebesgue measurable.

9-19. Let $f : \mathbb{R} \to \mathbb{R}$ be a nondecreasing function. Prove that f is Lebesgue measurable.

9.3 Lebesgue Integration

Independent of whether the base is an interval or a potentially more scattered Lebesgue measurable set, the area of a "rectangle" should be the measure of the base times the height. This idea is behind the Lebesgue integral of simple functions.

Definition 9.21 *Let $A_1, \ldots, A_n \subseteq \mathbb{R}$ be pairwise disjoint Lebesgue measurable sets, let $a_1, \ldots, a_n \in [0, \infty)$ be nonnegative numbers and let $s = \sum_{k=1}^{n} a_k \mathbf{1}_{A_k}$ be a simple function. We define the* **Lebesgue integral** *of s by $\int_{\mathbb{R}} \sum_{k=1}^{n} a_k \mathbf{1}_{A_k} \, d\lambda := \sum_{k=1}^{n} a_k \lambda(A_k)$.*

By Exercise 9-20a for any given simple function s the value $\sum_{k=1}^{n} a_k \lambda(A_k)$ does not depend on the representation $s = \sum_{k=1}^{n} a_k \mathbf{1}_{A_k}$ that was chosen for s. Hence, Definition 9.21 is sensible and we can proceed to more general functions. For a more general function, the Lebesgue integral is defined by approximating the area under the function from below with the area under simple functions.

Definition 9.22 *Let $f : \mathbb{R} \to [0, \infty]$ be a Lebesgue measurable function. We define the* **Lebesgue integral** *of f to be*

$$\int_{\mathbb{R}} f \, d\lambda := \sup \left\{ \int_{\mathbb{R}} s \, d\lambda : s \text{ is a simple function with } 0 \leq s \leq f \right\}$$

and we will call f **Lebesgue integrable** *iff the supremum is finite.*

A function $g : \mathbb{R} \to \mathbb{R}$ will be called **Lebesgue integrable** *iff g^+ and g^- are both Lebesgue integrable. We set $\int_{\mathbb{R}} g \, d\lambda := \int_{\mathbb{R}} g^+ \, d\lambda - \int_{\mathbb{R}} g^- \, d\lambda$ and call it the* **Lebesgue integral** *of g.*

Continuing our comparison with the Riemann integral, Exercise 9-21 guarantees that for bounded functions that differ from zero only on a set of finite Lebesgue measure (the Riemann integral is defined for bounded functions on bounded intervals) there also is an approximation from above that will give the value from Definition 9.22. That is, unlike the Riemann/Darboux integral (see Definition 5.27 and Theorem 5.29), for bounded functions that differ from zero on closed and bounded intervals the Lebesgue integral does not have any problems with an upper and a lower approximation not being equal.

As noted after the definition of Lebesgue measurable sets (see Definition 9.4), non-measurable sets are quite hard to come by. Similarly, although we will always need to prove measurability for functions that we want to integrate, nonmeasurable functions are not expected to arise in practical applications. This means that the only possible problem in the definition of the Lebesgue integral is the potential for infinite area under the graph. This is not a problem, because functions whose graphs enclose an infinite area cannot have a finite integral, independent of what notion of integration is used.

Now that we have a sensible notion of integration that (as it ultimately turns out) does not have the weaknesses of the Riemann integral, we can establish some theorems about Lebesgue integrals.

Theorem 9.23 *Let $f, g : \mathbb{R} \to [-\infty, \infty]$ be Lebesgue measurable functions.*

1. *If $0 \leq f \leq g$ a.e. and g is Lebesgue integrable, then f is also Lebesgue integrable and $\int_{\mathbb{R}} f\, d\lambda \leq \int_{\mathbb{R}} g\, d\lambda$.*
2. *f is Lebesgue integrable iff $|f|$ is Lebesgue integrable and in this case the* **triangular inequality** $\left|\int_{\mathbb{R}} f\, d\lambda\right| \leq \int_{\mathbb{R}} |f|\, d\lambda$ *holds.*
3. *If $f \geq 0$, then $\int_{\mathbb{R}} f\, d\lambda = 0$ iff $f = 0$ a.e..*

Proof. For part 1, let $N := \{x \in \mathbb{R} : f(x) \geq g(x)\}$. By hypothesis, N is a null set. Let $s : \mathbb{R} \to [0, \infty)$ be a simple function with $0 \leq s \leq f$. Then $s\mathbf{1}_{\mathbb{R}\setminus N} \leq g$ is a simple function also and $\int_{\mathbb{R}} s\, d\lambda = \int_{\mathbb{R}} s\mathbf{1}_{\mathbb{R}\setminus N}\, d\lambda$. Hence,

$$\begin{aligned}
&\sup\left\{\int_{\mathbb{R}} s\, d\lambda : s \text{ is a simple function with } 0 \leq s \leq f\right\} \\
&= \sup\left\{\int_{\mathbb{R}} s\mathbf{1}_{\mathbb{R}\setminus N}\, d\lambda : s \text{ is a simple function with } 0 \leq s \leq f\right\} \\
&\leq \sup\left\{\int_{\mathbb{R}} s\, d\lambda : s \text{ is a simple function with } 0 \leq s \leq g\right\}.
\end{aligned}$$

Because g is Lebesgue integrable, the latter supremum is finite, and hence f is Lebesgue integrable. Because the suprema are equal to the respective Lebesgue integrals, we conclude that $\int_{\mathbb{R}} f\, d\lambda \leq \int_{\mathbb{R}} g\, d\lambda$.

For part 2, first note that by Theorem 9.20 the function $|f|$ is Lebesgue measurable. The direction "$\Leftarrow$" of the claim follows straight from part 1, because if $|f|$ is Lebesgue integrable, then $0 \leq f^+ \leq |f|$ and $0 \leq f^- \leq |f|$ imply that f^+ and f^- are both Lebesgue integrable, which means by Definition 9.22 that f is Lebesgue integrable.

For the direction "$\Rightarrow$," let $f : \mathbb{R} \to [-\infty, \infty]$ be Lebesgue integrable. Then f^+ and $-f^-$ are Lebesgue integrable. To prove that $|f| = f^+ - f^-$ is Lebesgue integrable let $s = \sum_{k=1}^{n} a_k \mathbf{1}_{A_k}$ be a simple function with $0 \leq s \leq |f|$. Let $P := \{x \in \mathbb{R} : f(x) \geq 0\}$ and $N := \{x \in \mathbb{R} : f(x) < 0\}$. Then $P \cup N = \mathbb{R}$ and P and N are disjoint. Therefore with $s_+ = \sum_{k=1}^{n} a_k \mathbf{1}_{A_k \cap P}$ and $s_- = \sum_{k=1}^{n} a_k \mathbf{1}_{A_k \cap N}$ we obtain $s = s_+ + s_-$, $0 \leq s_+ \leq f^+$

and $0 \leq s_- \leq f^-$. But this means that

$$\begin{aligned} \int_{\mathbb{R}} s\, d\lambda &= \sum_{k=1}^{n} a_k \lambda(A_k) = \sum_{k=1}^{n} a_k \lambda(A_k \cap P) + \sum_{k=1}^{n} a_k \lambda(A_k \cap N) \\ &\leq \int_{\mathbb{R}} f^+\, d\lambda + \int_{\mathbb{R}} f^-\, d\lambda < \infty. \end{aligned}$$

Therefore we conclude that $\int_{\mathbb{R}} |f|\, d\lambda \leq \int_{\mathbb{R}} f^+\, d\lambda + \int_{\mathbb{R}} f^-\, d\lambda < \infty$ and thus $|f|$ is Lebesgue integrable.

Finally, because $f^-, f^+ \leq |f|$ we conclude by part 1 that

$$\begin{aligned} \left| \int_{\mathbb{R}} f\, d\lambda \right| &= \left| \int_{\mathbb{R}} f^+\, d\lambda - \int_{\mathbb{R}} f^-\, d\lambda \right| \leq \max \left\{ \int_{\mathbb{R}} f^+\, d\lambda, \int_{\mathbb{R}} f^-\, d\lambda \right\} \\ &\leq \int_{\mathbb{R}} |f|\, d\lambda. \end{aligned}$$

For part 3, let $f \geq 0$. First, consider the direction "$\Leftarrow$." Because $f = 0$ a.e., we obtain $0 \leq f \leq 0$ a.e. and by part 1 we conclude that $\int_{\mathbb{R}} f\, d\lambda \leq \int_{\mathbb{R}} 0\, d\lambda = 0$, which means $\int_{\mathbb{R}} f\, d\lambda = 0$.

Conversely, for the direction "$\Rightarrow$" let $\int_{\mathbb{R}} f\, d\lambda = 0$ and suppose for a contradiction that $\lambda(\{x \in \mathbb{R} : f(x) > 0\}) > 0$. Then, because the countable union of null sets is again a null set and $\{x \in \mathbb{R} : f(x) > 0\} = \bigcup_{n=1}^{\infty} \left\{ x \in \mathbb{R} : f(x) > \frac{1}{n} \right\}$, there must be an $n \in \mathbb{N}$ so that $\lambda\left(\left\{ x \in \mathbb{R} : f(x) > \frac{1}{n} \right\}\right) > 0$. But then with $A := \left\{ x \in \mathbb{R} : f(x) > \frac{1}{n} \right\}$ and $s := \frac{1}{n} \mathbf{1}_A$ we obtain $\int_{\mathbb{R}} s\, d\lambda = \frac{1}{n} \lambda(A) > 0$ and $0 \leq s \leq f$, which is not possible by part 1. Therefore we conclude that $f = 0$ a.e.. ■

Exercise 9-15 shows that if a function is equal a.e. to a Lebesgue measurable function, then it must be Lebesgue measurable, too. Part 1 of Theorem 9.23 can be used to show that if two Lebesgue measurable functions are equal a.e., then their Lebesgue integrals must be equal, too (see Exercise 9-26). Basically this means that for integration null sets are insignificant. The following definition is therefore sensible because independent of how we extend the function f to all of $\mathbb{R}$, either all extensions are Lebesgue measurable or none of them are (because for all $a > 0$ the set $\{x \in \mathbb{R} : g(x) < a\}$ with g as in Definition 9.24 below differs from $\{x \in \mathbb{R} : f(x)$ exists and $f(x) < a\}$ at most by a null set) and either all extensions are Lebesgue integrable with the same Lebesgue integral or none of them are Lebesgue integrable (by part 1 of Theorem 9.23).

Definition 9.24 *If the function $f : \mathbb{R} \to [-\infty, \infty]$ is defined a.e. and the function $g(x) := \begin{cases} f(x); & \text{if } f(x) \text{ is defined,} \\ 0; & \text{if } f(x) \text{ is not defined,} \end{cases}$ is Lebesgue measurable, then we will call f*

Lebesgue integrable *iff g is Lebesgue integrable and we define the* **Lebesgue integral** *of f to be* $\int_{\mathbb{R}} f\, d\lambda := \int_{\mathbb{R}} g\, d\lambda$.

Theorem 9.25 shows that the Lebesgue integral is well-behaved with respect to the linear operations of multiplying with a real number and addition. Because of Exercise 9-16a and Definition 9.24 and because the set where the sum $f+g$ is undefined is a null set (see Exercise 9-27), we do not need to place any additional hypotheses on the functions in part 2 of Theorem 9.25.

Theorem 9.25 *Let $f, g : \mathbb{R} \to [-\infty, \infty]$ be Lebesgue integrable functions and let $\alpha \in \mathbb{R}$. The the following are true:*

1. *αf is Lebesgue integrable and* $\int_{\mathbb{R}} \alpha f\, d\lambda = \alpha \int_{\mathbb{R}} f\, d\lambda$.

2. *$f+g$ is Lebesgue integrable and* $\int_{\mathbb{R}} f+g\, d\lambda = \int_{\mathbb{R}} f\, d\lambda + \int_{\mathbb{R}} g\, d\lambda$.

Proof. For part 1, note that by Theorem 9.20 with $g(x) = \alpha$ the function αf is Lebesgue measurable. If $f \geq 0$ and $\alpha \geq 0$, then αf is Lebesgue integrable because

$$\begin{aligned} \int_{\mathbb{R}} \alpha f\, d\lambda &= \sup\left\{ \int_{\mathbb{R}} s\, d\lambda : s \text{ is a simple function with } 0 \leq s \leq \alpha f \right\} \\ &= \sup\left\{ \int_{\mathbb{R}} \alpha s\, d\lambda : s \text{ is a simple function with } 0 \leq \alpha s \leq \alpha f \right\} \\ &= \alpha \sup\left\{ \int_{\mathbb{R}} s\, d\lambda : s \text{ is a simple function with } 0 \leq s \leq f \right\} \\ &= \alpha \int_{\mathbb{R}} f\, d\lambda < \infty. \end{aligned}$$

But this means that for any Lebesgue integrable function f and any $\alpha \geq 0$ the functions $(\alpha f)^+ = \alpha f^+$ and $(\alpha f)^- = \alpha f^-$ are Lebesgue integrable and

$$\begin{aligned} \int_{\mathbb{R}} \alpha f\, d\lambda &= \int_{\mathbb{R}} (\alpha f)^+\, d\lambda - \int_{\mathbb{R}} (\alpha f)^-\, d\lambda = \alpha \int_{\mathbb{R}} f^+\, d\lambda - \alpha \int_{\mathbb{R}} f^-\, d\lambda \\ &= \alpha \int_{\mathbb{R}} f\, d\lambda. \end{aligned}$$

Finally, for any Lebesgue integrable function f and any $\alpha < 0$ the functions $(\alpha f)^+ = -\alpha f^-$ and $(\alpha f)^- = -\alpha f^+$ are Lebesgue integrable and

$$\begin{aligned} \int_{\mathbb{R}} \alpha f\, d\lambda &= \int_{\mathbb{R}} (\alpha f)^+\, d\lambda - \int_{\mathbb{R}} (\alpha f)^-\, d\lambda = \int_{\mathbb{R}} -\alpha f^-\, d\lambda - \int_{\mathbb{R}} -\alpha f^+\, d\lambda \\ &= -\alpha \int_{\mathbb{R}} f^-\, d\lambda + \alpha \int_{\mathbb{R}} f^+\, d\lambda = \alpha \int_{\mathbb{R}} f\, d\lambda. \end{aligned}$$

For part 2 first note that by Exercise 9-16a, Definition 9.24 and the preceding discussion, we can assume that $f+g$ is defined and finite everywhere and that $f+g$ is

Lebesgue measurable. (Simply set f and g equal to zero where the sum is not defined.) Also note that part 2 is easily proved for simple functions (Exercise 9-20b), so we can use the additivity of the Lebesgue integral for simple functions in the following.

We will first prove the result for nonnegative f and g. To see that $f+g$ is Lebesgue integrable, suppose for a contradiction that $f+g$ is not Lebesgue integrable. Then for each $n \in \mathbb{N}$ there is a simple function s_n with $0 \leq s_n \leq f+g$ and $\int_{\mathbb{R}} s_n \, d\lambda > n$. Let
$$F := \left\{ x \in \mathbb{R} : f(x) \geq \frac{f(x)+g(x)}{2} \right\} \quad \text{and} \quad G := \left\{ x \in \mathbb{R} : g(x) \geq \frac{f(x)+g(x)}{2} \right\}.$$
For each $n \in \mathbb{N}$ one of the inequalities $\int_{\mathbb{R}} \frac{1}{2} s_n \mathbf{1}_F \, d\lambda > \frac{n}{4}$ or $\int_{\mathbb{R}} \frac{1}{2} s_n \mathbf{1}_G \, d\lambda > \frac{n}{4}$ holds. Without loss of generality, assume that $\int_{\mathbb{R}} \frac{1}{2} s_n \mathbf{1}_F \, d\lambda > \frac{n}{4}$ holds for all $n \in \mathbb{N}$. Because $0 \leq \frac{1}{2} s_n \mathbf{1}_F \leq \frac{f+g}{2} \mathbf{1}_F \leq f$ this implies that f is not Lebesgue integrable, which is a contradiction. Hence, $f+g$ must be Lebesgue integrable.

Now if s_1 is a simple function with $0 \leq s_1 \leq f$ and s_2 is a simple function with $0 \leq s_2 \leq g$, then $s_1 + s_2$ is a simple function with $0 \leq s_1 + s_2 \leq f+g$. This implies $\int_{\mathbb{R}} s_1 \, d\lambda + \int_{\mathbb{R}} s_2 \, d\lambda = \int_{\mathbb{R}} s_1 + s_2 \, d\lambda \leq \int_{\mathbb{R}} f + g \, d\lambda$ and because s_1, s_2 were arbitrary, we obtain $\int_{\mathbb{R}} f \, d\lambda + \int_{\mathbb{R}} g \, d\lambda \leq \int_{\mathbb{R}} f + g \, d\lambda$.

For the reversed inequality, let $C_n := \left\{ x \in \mathbb{R} : \frac{1}{n} \leq f(x) + g(x) \leq n \right\}$ for each $n \in \mathbb{N}$. We first prove that $\lim_{n \to \infty} \int_{\mathbb{R}} (f+g) \mathbf{1}_{C_n} \, d\lambda = \int_{\mathbb{R}} f + g \, d\lambda$. Let $\varepsilon > 0$ and let s be a simple function so that $0 \leq s \leq f+g$ and $\int_{\mathbb{R}} s \, d\lambda > \int_{\mathbb{R}} f + g \, d\lambda - \frac{\varepsilon}{2}$. Note that $\bigcup_{n=1}^{\infty} C_n = \{ x \in \mathbb{R} : f(x) + g(x) > 0 \}$. Hence, by Theorem 9.12 for all measurable subsets $A \subseteq \{ x \in \mathbb{R} : f(x) + g(x) > 0 \}$ we obtain $\lim_{n \to \infty} \lambda(A \cap C_n) = \lambda(A)$. With $s = \sum_{j=1}^{m} a_j \mathbf{1}_{A_j}$ and all $a_j > 0$ this implies

$$\lim_{n \to \infty} \int_{\mathbb{R}} s \mathbf{1}_{C_n} \, d\lambda = \lim_{n \to \infty} \sum_{j=1}^{m} a_j \lambda(A_j \cap C_n) = \sum_{j=1}^{m} a_j \lambda(A_j) = \int_{\mathbb{R}} s \, d\lambda.$$

Therefore there is an $n \in \mathbb{N}$ so that the inequality $\int_{\mathbb{R}} s \mathbf{1}_{C_n} \, d\lambda > \int_{\mathbb{R}} s \, d\lambda - \frac{\varepsilon}{2}$ holds. But then $0 \leq s \mathbf{1}_{C_n} \leq (f+g) \mathbf{1}_{C_n}$, and hence

$$\int_{\mathbb{R}} (f+g) \mathbf{1}_{C_n} \, d\lambda \geq \int_{\mathbb{R}} s \mathbf{1}_{C_n} \, d\lambda > \int_{\mathbb{R}} s \, d\lambda - \frac{\varepsilon}{2} > \int_{\mathbb{R}} f + g \, d\lambda - \varepsilon.$$

Because f and g are both nonnegative, the sequence $\left\{\int_{\mathbb{R}} (f+g)\mathbf{1}_{C_n}\,d\lambda\right\}_{n=1}^{\infty}$ is nondecreasing and we conclude that $\lim_{n\to\infty}\int_{\mathbb{R}} (f+g)\mathbf{1}_{C_n}\,d\lambda = \int_{\mathbb{R}} f+g\,d\lambda$.

Now let $\varepsilon > 0$ and let $n \in \mathbb{N}$ be so that $\int_{\mathbb{R}} (f+g)\mathbf{1}_{C_n}\,d\lambda > \int_{\mathbb{R}} f+g\,d\lambda - \frac{\varepsilon}{2}$. Let $\nu := \min\left\{\frac{1}{n}, \frac{\varepsilon}{4(\lambda(C_n)+1)}\right\}$. Because f and g are bounded by n on C_n, the proof of part "5⇒1" of Theorem 9.19 shows that there are simple functions s_f and s_g so that $0 \leq s_f \leq f\mathbf{1}_{C_n}$, $0 \leq s_g \leq g\mathbf{1}_{C_n}$ and for all $x \in \mathbb{R}$ the inequalities $f(x)\mathbf{1}_{C_n}(x) - s_f(x) < \nu$ and $g(x)\mathbf{1}_{C_n}(x) - s_g(x) < \nu$ hold. Therefore,

$$\int_{\mathbb{R}} f\,d\lambda + \int_{\mathbb{R}} g\,d\lambda$$

$$\geq \int_{\mathbb{R}} s_f\,d\lambda + \int_{\mathbb{R}} s_g\,d\lambda = \int_{\mathbb{R}} s_f + s_g\,d\lambda \geq \int_{\mathbb{R}} (f+g)\mathbf{1}_{C_n} - 2\nu\mathbf{1}_{C_n}\,d\lambda$$

We can move the difference out of the integral because $(f+g)\mathbf{1}_{C_n} - 2\nu\mathbf{1}_{C_n} \geq 0$ and because the integral of $(f+g)\mathbf{1}_{C_n}$ is the supremum of the integrals of all step functions s with $\nu\mathbf{1}_{C_n} \leq s \leq (f+g)\mathbf{1}_{C_n}$ (also see Exercise 9-23).

$$\geq \int_{\mathbb{R}} (f+g)\mathbf{1}_{C_n}\,d\lambda - \int_{\mathbb{R}} \frac{\varepsilon}{2(\lambda(C_n)+1)}\mathbf{1}_{C_n}\,d\lambda = \int_{\mathbb{R}} (f+g)\mathbf{1}_{C_n}\,d\lambda - \frac{\varepsilon}{2}$$

$$> \int_{\mathbb{R}} f+g\,d\lambda - \varepsilon.$$

Because $\varepsilon > 0$ was arbitrary, this proves the additivity of the Lebesgue integral for nonnegative functions.

For not necessarily nonnegative functions, note that because $|f+g| \leq |f| + |g|$, the above and part 2 of Theorem 9.23 show that $f+g$ is Lebesgue integrable when f and g are Lebesgue integrable.

For the equality of the integrals, first notice that if f_1, f_2, g_1, g_2 are nonnegative, integrable and satisfy $f_1 - f_2 = g_1 - g_2$, then via $\int_{\mathbb{R}} f_1 + g_2\,d\lambda = \int_{\mathbb{R}} f_2 + g_1\,d\lambda$ we can conclude that $\int_{\mathbb{R}} f_1\,d\lambda - \int_{\mathbb{R}} f_2\,d\lambda = \int_{\mathbb{R}} g_1\,d\lambda - \int_{\mathbb{R}} g_2\,d\lambda$. Therefore with $(f+g)^+ - (f+g)^- = f+g = (f^+ + g^+) - (f^- + g^-)$ we obtain

$$\int_{\mathbb{R}} f+g\,d\lambda = \int_{\mathbb{R}} (f+g)^+\,d\lambda - \int_{\mathbb{R}} (f+g)^-\,d\lambda$$

Note that for this step we need the above argument. Pointwise, we usually have $(f+g)^+ \neq f^+ + g^+$ and $(f+g)^- \neq f^- + g^-$ (see Exercise 9-24).

$$= \int_{\mathbb{R}} (f^+ + g^+)\,d\lambda - \int_{\mathbb{R}} (f^- + g^-)\,d\lambda$$

$$= \int_{\mathbb{R}} f^+ \, d\lambda - \int_{\mathbb{R}} f^- \, d\lambda + \int_{\mathbb{R}} g^+ \, d\lambda - \int_{\mathbb{R}} g^- \, d\lambda$$
$$= \int_{\mathbb{R}} f \, d\lambda + \int_{\mathbb{R}} g \, d\lambda,$$

which completes the proof. ■

The approximation of the integral of $f+g$ with integrals of functions $(f+g)\mathbf{1}_{C_n}$ in part 2 of the proof of Theorem 9.23, shows that sequences of functions should be powerful tools. Convergence of sequences of functions is discussed in Chapter 11 and the fundamental limit theorems for (Lebesgue) integration are introduced in Section 14.5. Moreover, Exercise 14-33 gives a more efficient proof of part 2 using limit theorems for integrals.

Exercises

9-20. Integration of simple functions

(a) Let s be a simple nonnegative function, let $y_1, \ldots, y_m \in (0, \infty)$ be the nonzero values that s assumes, let $a_1, \ldots, a_n \in [0, \infty)$, and let $A_1, \ldots, A_n \subseteq \mathbb{R}$ be pairwise disjoint Lebesgue measurable sets so that $s = \sum_{k=1}^{n} a_k \mathbf{1}_{A_k}$. Prove that $\sum_{k=1}^{n} a_k \lambda(A_k) = \sum_{j=1}^{m} y_j \lambda\left(s^{-1}(y_j)\right)$.

(This proves that the integral in Definition 9.21 does not depend on the representation of s.)

Hint. Each of the a_k must be a y_j or zero. Group the first sum so that equal values a_k are contiguous and then prove that the union of the corresponding sets A_k is the inverse image of the appropriate y_j.

(b) Prove that if s_1 and s_2 are simple functions, then $\int_{\mathbb{R}} s_1 + s_2 \, d\lambda = \int_{\mathbb{R}} s_1 \, d\lambda + \int_{\mathbb{R}} s_2 \, d\lambda$.

Hint. Find Lebesgue measurable sets $A_1, \ldots, A_n$ so that s_1 and s_2 are constant on each A_j.

9-21 Let $f : \mathbb{R} \to [0, \infty)$ be bounded and Lebesgue measurable so that $\lambda\left(\left\{x \in \mathbb{R} : f(x) > 0\right\}\right)$ is finite. Prove that f is Lebesgue integrable and

$$\int_{\mathbb{R}} f \, d\lambda = \inf\left\{\int_{\mathbb{R}} s \, d\lambda : s \text{ is a simple function with } f \leq s\right\}$$

9-22. Let $f : \mathbb{R} \to [0, \infty]$ be a Lebesgue measurable function. Prove that f is Lebesgue integrable iff the supremum $S := \sup\left\{\int_{\mathbb{R}} \min(f, n)\mathbf{1}_{[-n,n]} \, d\lambda : n \in \mathbb{N}\right\}$ is finite and that in this case S is the Lebesgue integral of f

9-23 Let $f : \mathbb{R} \to [0, \infty]$ be Lebesgue integrable, let $a > 0$ and let $A \subseteq \mathbb{R}$ be Lebesgue measurable and so that $f - a\mathbf{1}_A \geq 0$. Prove that $\int_{\mathbb{R}} f - a\mathbf{1}_A \, d\lambda = \int_{\mathbb{R}} f \, d\lambda - a\lambda(A)$.

9-24. Construct Lebesgue integrable functions $f : \mathbb{R} \to \mathbb{R}$ and $g : \mathbb{R} \to \mathbb{R}$ so that $(f+g)^+ \neq f^+ + g^+$ and $(f+g)^- \neq f^- + g^-$

9-25. Let $a_1, \ldots, a_n \in \mathbb{R}$, let $A_1, \ldots, A_n \subseteq \mathbb{R}$ be (not necessarily pairwise disjoint) Lebesgue measurable sets and let $f = \sum_{k=1}^{n} a_k \mathbf{1}_{A_k}$ be a simple function Prove that $\int_{\mathbb{R}} f \, d\lambda = \sum_{k=1}^{n} a_k \lambda(A_k)$ Then explain how this result differs from the result in Exercise 9-20a and why we could not have used it to prove Exercise 9-20a.

Hint For the proof of the equation, you may use Theorem 9.25.

9-26. Let $f, g : \mathbb{R} \to [-\infty, \infty]$ be Lebesgue measurable functions so that $f = g$ a.e. Use only Theorem 9.23 to prove that f is Lebesgue integrable iff g is Lebesgue integrable and that in this case we have $\int_{\mathbb{R}} f \, d\lambda = \int_{\mathbb{R}} g \, d\lambda$.

9-27. Let $f : \mathbb{R} \to [-\infty, \infty]$ be Lebesgue integrable Prove that $\{ x \in \mathbb{R} : f(x) = \infty \}$ is a null set.

9-28. Let $f, g : \mathbb{R} \to [-\infty, \infty]$ be Lebesgue integrable functions.

(a) Prove that $\max\{f, g\}$ (defined pointwise) is Lebesgue integrable

(b) Prove that $\min\{f, g\}$ (defined pointwise) is Lebesgue integrable.

(c) Prove that $f - g$ is Lebesgue integrable and $\int_{\mathbb{R}} f - g \, d\lambda = \int_{\mathbb{R}} f \, d\lambda - \int_{\mathbb{R}} g \, d\lambda$.

9-29 Let C^Q be a **Cantor set**. Prove that $\mathbf{1}_{C^Q}$ is Lebesgue integrable.
Hint. Exercise 9-6.

9-30 Prove that the Dirichlet function $f(x) = \mathbf{1}_{\mathbb{Q}\cap[0,1]}$ is Lebesgue integrable.

9.4 Lebesgue Integrals versus Riemann Integrals

We conclude this chapter by establishing the relationship between the Lebesgue integral and regular as well as improper Riemann integrals. The Lebesgue integral truly is an extension of the Riemann integral of bounded functions on closed and bounded intervals. To see this, we first show that Riemann integrable functions are Lebesgue integrable and that for these functions the Lebesgue integral equals the Riemann integral. Theorem 9.26 also shows how to overcome the small nuisance of Riemann integrals being defined on sets $[a, b]$, while the Lebesgue integral is defined on $\mathbb{R}$.

Theorem 9.26 *If f is Riemann integrable over $[a, b]$, then $f_{\mathbb{R}} : \mathbb{R} \to \mathbb{R}$ defined by $f_{\mathbb{R}}(x) := \begin{cases} f(x); & \text{for } x \in [a, b], \\ 0; & \text{otherwise,} \end{cases}$ is Lebesgue integrable and the Riemann integral of f is equal to the Lebesgue integral of $f_{\mathbb{R}}$. That is, $\int_{\mathbb{R}} f_{\mathbb{R}} \, d\lambda = \int_a^b f \, dx$.*

Proof. First, let $f \geq 0$. Let $P = \{a = x_0 < \cdots < x_n = b\}$ be a partition of $[a, b]$. With m_i and M_i as in Definition 5.13, let $s_L^P := \sum_{k=1}^{n} m_k \mathbf{1}_{[x_{k-1}, x_k)} + m_n \mathbf{1}_{\{x_n\}}$ and $s_U^P := \sum_{k=1}^{n} M_k \mathbf{1}_{[x_{k-1}, x_k)} + M_n \mathbf{1}_{\{x_n\}}$. Then $0 \leq s_L^P \leq f \leq s_U^P$.

For all $n \in \mathbb{N}$, consider the partition $P_n := \left\{ a + k\frac{b-a}{2^n} : k = 0, \ldots, 2^n \right\}$. Then $s_L^{P_n} \leq s_L^{P_{n+1}}$ for all $n \in \mathbb{N}$ and by Lebesgue's criterion for Riemann integrability (Theorem 8.12) we infer that $\lim_{n\to\infty} s_L^{P_n}(x) = f(x)$ a.e. Because $f_{\mathbb{R}}$ and the $s_L^{P_n}$ are zero outside $[a, b]$, by Exercise 9-15 this means that $f_{\mathbb{R}}$ is Lebesgue measurable. Because $f_{\mathbb{R}} \leq M\mathbf{1}_{[a,b]}$ for some $M > 0$, we conclude that $f_{\mathbb{R}}$ is Lebesgue integrable. Moreover, for all $n \in \mathbb{N}$ we have $L(f, P_n) = \int_{\mathbb{R}} s_L^{P_n} \, d\lambda \leq \int_{\mathbb{R}} f_{\mathbb{R}} \, d\lambda \leq \int_{\mathbb{R}} s_U^{P_n} \, d\lambda = U(f, P_n)$.

For each $n \in \mathbb{N}$, there is an evaluation set $T_n = \left\{t_1^{(n)}, \ldots, t_{2^n}^{(n)}\right\}$ so that for all integers $k = 1, \ldots, 2^n$ the inequality $f\left(t_k^{(n)}\right) - m_k < \frac{1}{n(b-a)}$ holds. But then we infer $R(f, P_n, T_n) - L(f, P_n) < \frac{1}{n}$. Because $\lim_{n\to\infty} \|P_n\| = 0$ we obtain

$$\lim_{n\to\infty} L(f, P_n) = \lim_{n\to\infty} R(f, P_n, T_n) = \int_a^b f(x)\,dx.$$

Similarly, we can prove $\lim_{n\to\infty} U(f, P_n) = \int_a^b f(x)\,dx$. But this means

$$\int_a^b f(x)\,dx = \lim_{n\to\infty} L(f, P_n) \le \int_{\mathbb{R}} f_{\mathbb{R}}\,d\lambda \le \lim_{n\to\infty} U(f, P_n) = \int_a^b f(x)\,dx,$$

which establishes the desired equality.

For f being an arbitrary Riemann integrable function on $[a, b]$, note that because f is bounded there is a real number B so that $h := f + B\mathbf{1}_{[a,b]} \ge 0$. Then $h_{\mathbb{R}}$ is Lebesgue integrable, so $f_{\mathbb{R}} = h_{\mathbb{R}} - B\mathbf{1}_{[a,b]}$ is Lebesgue integrable and

$$\begin{aligned}\int_{\mathbb{R}} f_{\mathbb{R}}\,d\lambda &= \int_{\mathbb{R}} h_{\mathbb{R}} - B\mathbf{1}_{[a,b]}\,d\lambda = \int_a^b h\,dx - B(b-a)\\ &= \int_a^b h - B\,dx = \int_a^b f\,dx.\end{aligned}$$

■

The Lebesgue integral also incorporates parts of the improper Riemann integral as the next result shows. For improperly Riemann integrable functions as in Theorem 9.27, we also often say that the improper Riemann integral **converges absolutely**.

Theorem 9.27 *If f is Riemann integrable over every closed subinterval of $[a, b)$ and $|f|$ is improperly Riemann integrable over $[a, b)$ (where b is either a number or infinity), then $f_{\mathbb{R}} : \mathbb{R} \to \mathbb{R}$ defined pointwise by $f_{\mathbb{R}}(x) := \begin{cases} f(x); & \text{for } x \in [a, b),\\ 0; & \text{otherwise,}\end{cases}$ is Lebesgue integrable and the integrals are equal, that is, $\int_{\mathbb{R}} f_{\mathbb{R}}\,d\lambda = \int_a^b f\,dx$.*

A similar result holds for functions that are improperly Riemann integrable over an interval $(a, b]$ (where a is either a number or negative infinity) or over a set of the form $(a, b) \setminus \{a_1, \ldots, a_n\}$.

Proof. We will prove the result for $b \in (a, \infty)$ and leave the case $b = \infty$ to Exercise 9-31b. For Lebesgue measurability, first consider the case that $f \ge 0$. For all $n \in \mathbb{N}$, let $P_n := \left\{a + k\frac{b-a}{2^n} : k = 0, \ldots, 2^n\right\}$ and with m_i and M_i as in Definition 5.13, let $s_L^n := \sum_{k=1}^{2^n-1} m_k \mathbf{1}_{\left[a+(k-1)\frac{b-a}{2^n}, a+k\frac{b-a}{2^n}\right)}$. Then for all $n \in \mathbb{N}$ we have

$s_L^n \leq s_L^{n+1} \leq f$ and by Lebesgue's criterion for Riemann integrability (Theorem 8.12) for almost all $x \in [a, b)$ we infer that $\lim_{n\to\infty} s_L^n(x) = f(x)$. Because $f_{\mathbb{R}}$ and the s_L^n are zero outside $[a, b)$, by Exercise 9-15 this means that if $f \geq 0$, then $f_{\mathbb{R}}$ is Lebesgue measurable. Applying this result to f^+ and f^- separately implies that f is Lebesgue measurable regardless of what sign it takes.

Now for all simple functions s with $0 \leq s \leq |f|$ and all $n \in \mathbb{N}$ the inequalities $\int_{\mathbb{R}} s\mathbf{1}_{\left[a,b-\frac{1}{n}\right]}\,d\lambda \leq \int_a^{b-\frac{1}{n}} |f(x)|\,dx \leq \int_a^b |f(x)|\,dx < \infty$ hold. Because $n \in \mathbb{N}$ was arbitrary we can conclude that for all simple functions s with $0 \leq s \leq |f|$ we have the inequality $\int_{\mathbb{R}} s\,d\lambda \leq \int_a^b |f(x)|\,dx < \infty$ (Exercise 9-31a). Hence, $|f_{\mathbb{R}}|$ is Lebesgue integrable, so $f_{\mathbb{R}}$ is Lebesgue integrable and $\int_{\mathbb{R}} |f_{\mathbb{R}}|\,d\lambda \leq \int_a^b |f(x)|\,dx < \infty$.

To prove that the two integrals are equal, let $\varepsilon > 0$ and find a $\delta > 0$ so that $\int_{b-\delta}^b |f(x)|\,dx < \frac{\varepsilon}{2}$. Then

$$\left|\int_{\mathbb{R}} f_{\mathbb{R}}\,d\lambda - \int_a^b f(x)\,dx\right|$$

$$\leq \left|\int_{\mathbb{R}} f_{\mathbb{R}}\,d\lambda - \int_{\mathbb{R}} f_{\mathbb{R}}\mathbf{1}_{[a,b-\delta]}\,d\lambda\right| + \left|\int_{\mathbb{R}} f_{\mathbb{R}}\mathbf{1}_{[a,b-\delta]}\,d\lambda - \int_a^{b-\delta} f(x)\,dx\right|$$
$$+ \left|\int_a^{b-\delta} f(x)\,dx - \int_a^b f(x)\,dx\right|$$

The middle term is zero by Theorem 9.26.

$$\leq \int_{\mathbb{R}} \left|f_{\mathbb{R}}\mathbf{1}_{[b-\delta,b)}\right|\,d\lambda + \int_{b-\delta}^b |f(x)|\,dx$$

$\int_{\mathbb{R}} \left|f_{\mathbb{R}}\mathbf{1}_{[b-\delta,b)}\right|\,d\lambda \leq \int_{b-\delta}^b |f(x)|\,dx$ by what we have shown above.

$$\leq 2\int_{b-\delta}^b |f(x)|\,dx < \varepsilon.$$

Because ε was arbitrary the two integrals must be equal. ■

Note that not every improperly Riemann integrable function is Lebesgue integrable (Exercise 9-34). Moreover, not every Lebesgue integrable function can be integrated with a regular or improper Riemann integral (Exercises 9-29, 9-30, 12-24g). Hence, neither integral can formally be replaced by the other. In fact, because cancellations as in Exercise 9-34 are sometimes desired in Lebesgue integration, **improper Lebesgue integrals** can be defined similar to improper Riemann integrals (Exercise 9-35).

If we consider the earlier development of the Riemann integral in Chapter 5, it would be natural to target the Fundamental Theorem of Calculus as the next big result after the fundamental properties and examples that were presented in this chapter. Unfortunately, although it is quite beautiful, the Fundamental Theorem of Calculus for the

Lebesgue integral has a rather technical proof. Exercise 9-36 is the first lemma for this result. Further lemmas are presented in Exercises 10-7, 11-20, 14-36 and 14-38 before the result itself is proved in Exercises 18-6 and 23-8.

Exercises

9-31. Finish the proof of Theorem 9.27.

(a) Prove that if $a, b \in \mathbb{R}$, $a < b$, $M \geq 0$ and s is a nonnegative simple function so that for all $n \in \mathbb{N}$ we have that $\int_{\mathbb{R}} s\mathbf{1}_{\left[a, b-\frac{1}{n}\right]} \, d\lambda \leq M$, then $\int_{\mathbb{R}} s \, d\lambda \leq M$.

Hint. Recall that simple functions are bounded.

(b) Prove Theorem 9.27 for $b = \infty$.

9-32. Let $f(x) = \frac{1}{x}\mathbf{1}_{(0,1]}$. Prove that $\sqrt{f}$ is Lebesgue integrable while f is not.

9-33. Let $f(x) = \frac{1}{x}\mathbf{1}_{[1,\infty)}$. Prove that f^2 is Lebesgue integrable while f is not.

9-34. Let $f(x) := \sum_{n=1}^{\infty}(-1)^{n+1}\frac{1}{n}\mathbf{1}_{[n,n+1)}$, where the sum is taken pointwise. Prove that the improper Riemann integral of f over $[1, \infty)$ exists and that f is not Lebesgue integrable.

Hint. Harmonic series and the Alternating Series Test.

9-35. Define the improper Lebesgue integral of a function $f : [a, b) \to \mathbb{R}$, where $b \in \mathbb{R} \cup \{\infty\}$.

9-36. Use the following steps to prove that if $E \subseteq [a, b]$ and $\mathcal{I}$ is a family of closed subintervals of $[a, b]$ with $E \subseteq \bigcup \mathcal{I}$, then there is a finite, pairwise disjoint subfamily $\mathcal{F} \subseteq \mathcal{I}$ so that $\sum_{I \in \mathcal{F}} |I| \geq \frac{1}{6}\lambda(E)$.

(a) For any closed interval I, let I^* be the closed interval with the same midpoint and $|I^*| = 5|I|$. Prove that if $I, J \in \mathcal{I}$, $I \cap J \neq 0$ and $|I| < 2|J|$, then $I \subseteq J^*$.

(b) Let $\delta_0 := \sup\left\{|I| : I \in \mathcal{I}\right\}$, let $I_1 \in \mathcal{I}$ be so that $|I_1| > \frac{\delta_0}{2}$. Assume pairwise disjoint $I_1, \ldots, I_n$ are chosen so that, with $A_n := \bigcup_{j=1}^{n} I_j$, for each $I \in \mathcal{I}$ we have $I \cap A_n = \emptyset$ or $I \subseteq \bigcup_{j=1}^{n} I_j^*$. Prove that if there is an $I \in \mathcal{I}$ with $I \cap A_n = \emptyset$, we can continue the construction.

Hint. Use $\delta_{n+1} := \sup\left\{|I| : I \in \mathcal{I}, I \cap A_n = \emptyset\right\}$.

(c) Prove that, independent of whether the construction in part 9-36b terminates in finitely many steps or not, if $\mathcal{J}$ is the family of intervals constructed in part 9-36b, then $\lambda(E) < 5\sum_{I \in \mathcal{J}} |I|$.

(d) Prove that part 9-36c yields the requisite intervals.

It is possible to continue with Chapter 14 as long as the reader is willing to pick up the necessary results on the limit superior and the limit inferior and the language on pointwise convergence when they are needed.

Chapter 10

Series of Real Numbers II

The essentials on series were introduced in Chapter 6 to facilitate the "summation of infinitely many numbers." This chapter presents further aspects of the theory of series. Particular attention is given to power series, which are needed to define the transcendental functions in Chapter 12.

10.1 Limits Superior and Inferior

The terms of a series are usually easier to handle than the partial sums. Hence, it would be useful to have convergence criteria based on properties of the terms. The first obstacle for devising such criteria is that sequences obtained from the terms of a series need not converge. The limits superior and inferior defined in this section can be computed for any sequence and they describe the sequence's limiting behavior.

Proposition 10.1 *Let $\{a_n\}_{n=1}^{\infty}$ be a sequence of real numbers that is bounded above. Then the sequence $\left\{ \sup\{a_j : j \geq n\}\right\}_{n=1}^{\infty}$ converges to a real number or to $-\infty$.*

Proof. The sequence of suprema is nonincreasing. If it is bounded below, then it converges to a real number. If not, then it converges to $-\infty$. ■

Proposition 10.1 and a simple modification (see Exercise 10-1) show that the following definition is sensible.

Definition 10.2 *Let $\{a_n\}_{n=1}^{\infty}$ be a sequence of real numbers that is bounded above. The* **limit superior** *of $\{a_n\}_{n=1}^{\infty}$ is defined to be $\limsup_{n\to\infty} a_n := \lim_{n\to\infty} \sup\{a_j : j \geq n\}$. For sequences that are not bounded above, we say that the limit superior is ∞.*

If $\{a_n\}_{n=1}^{\infty}$ is a sequence of real numbers that is bounded below, then the **limit inferior** *of $\{a_n\}_{n=1}^{\infty}$ is defined to be $\liminf_{n\to\infty} a_n := \lim_{n\to\infty} \inf\{a_j : j \geq n\}$. For sequences that are not bounded below, we say that the limit inferior is $-\infty$.*

The relationship between limit superior and limit inferior is the obvious one.

Proposition 10.3 *Let $\{a_n\}_{n=1}^{\infty}$ be a sequence in $\mathbb{R}$. Then* $\liminf\limits_{n\to\infty} a_n \leq \limsup\limits_{n\to\infty} a_n$.

Proof. Clearly, for all $n \in \mathbb{N}$ we have $\inf\{a_j : j \geq n\} \leq \sup\{a_j : j \geq n\}$, which implies $\liminf\limits_{n\to\infty} a_n \leq \limsup\limits_{n\to\infty} a_n$. ■

Exercise 10-3 shows that precise arithmetic with limits inferior and superior is impossible for most operations. At least for negative signs we have a simple "reversal."

Proposition 10.4 *For any sequence of real numbers $\{a_n\}_{n=1}^{\infty}$, we have the equations* $\liminf\limits_{n\to\infty} -a_n = -\limsup\limits_{n\to\infty} a_n$ *and* $\limsup\limits_{n\to\infty} -a_n = -\liminf\limits_{n\to\infty} a_n$.

Proof. Recalling Exercise 1-19 we note that

$$\liminf_{n\to\infty} -a_n = \lim_{n\to\infty} \inf\{-a_j : j \geq n\} = \lim_{n\to\infty} -\sup\{a_j : j \geq n\} = -\limsup_{n\to\infty} a_n.$$

The other equation is proved similarly. ■

Finally, we should establish the relationship between the limit (if it exists) and the limits superior and inferior.

Lemma 10.5 *Let $\{a_n\}_{n=1}^{\infty}$ be a sequence of real numbers and let $\{b_n\}_{n=1}^{\infty}$ be a convergent sequence of real numbers so that for all $n \in \mathbb{N}$ the inequality $b_n \leq a_n$ holds. Then* $\lim\limits_{n\to\infty} b_n \leq \liminf\limits_{n\to\infty} a_n$.

Proof. Let $\varepsilon > 0$ and let $L = \lim\limits_{n\to\infty} b_n$. Then there is an $N \in \mathbb{N}$ so that for all $n \geq N$ we have $|b_n - L| < \varepsilon$, and hence $a_n \geq b_n > L - \varepsilon$. This means that for all $n \geq N$ we have $\inf\{a_j : j \geq n\} \geq L - \varepsilon$, and hence $\liminf\limits_{n\to\infty} a_n \geq L - \varepsilon$. Because $\varepsilon > 0$ was arbitrary we conclude $\liminf\limits_{n\to\infty} a_n \geq L$, which was to be proved. ■

Theorem 10.6 *A sequence of real numbers $\{a_n\}_{n=1}^{\infty}$ converges iff its limits superior and inferior are equal and real, that is,* $\limsup\limits_{n\to\infty} a_n = \liminf\limits_{n\to\infty} a_n \in \mathbb{R}$. *In this case, we have the equalities* $\lim\limits_{n\to\infty} a_n = \limsup\limits_{n\to\infty} a_n = \liminf\limits_{n\to\infty} a_n$.

Proof. For "$\Rightarrow$," let $\{a_n\}_{n=1}^{\infty}$ be a convergent sequence. Then by Lemma 10.5 and Exercise 10-4a we obtain $\limsup\limits_{n\to\infty} a_n \leq \lim\limits_{n\to\infty} a_n \leq \liminf\limits_{n\to\infty} a_n \in \mathbb{R}$. By Proposition 10.3, we also have $\liminf\limits_{n\to\infty} a_n \leq \limsup\limits_{n\to\infty} a_n$, which implies the two must be equal to each other and to the limit of the sequence.

For "$\Leftarrow$," let $\limsup\limits_{n\to\infty} a_n = \liminf\limits_{n\to\infty} a_n =: L \in \mathbb{R}$ and let $\varepsilon > 0$. There is an $N \in \mathbb{N}$ so that for all $n \geq N$ we have $\inf\{a_j : j \geq n\} > L - \varepsilon$ and $\sup\{a_j : j \geq n\} < L + \varepsilon$. In particular, for all $n \geq N$ we infer $|a_n - L| < \varepsilon$, so $\{a_n\}_{n=1}^{\infty}$ converges to L.

The equation at the end of the theorem is proved in either of the above parts. ■

Exercises

10-1. State and prove a version of Proposition 10.1 that can be used to justify the definition of the limit inferior.

10-2. Limit superior, limit inferior, and subsequences. Let $\{a_n\}_{n=1}^{\infty}$ be a sequence of real numbers.

(a) Prove that the limit superior of $\{a_n\}_{n=1}^{\infty}$ is the largest $S \in [-\infty, \infty]$ such that there is a subsequence $\{a_{n_k}\}_{k=1}^{\infty}$ with $\lim_{k\to\infty} a_{n_k} = S$.

(b) Prove that the limit inferior of $\{a_n\}_{n=1}^{\infty}$ is the smallest $S \in [-\infty, \infty]$ such that there is a subsequence $\{a_{n_k}\}_{k=1}^{\infty}$ with $\lim_{k\to\infty} a_{n_k} = S$.

10-3. Let $\{a_n\}_{n=1}^{\infty}$ and $\{b_n\}_{n=1}^{\infty}$ be bounded sequences of real numbers.

(a) Prove that $\liminf_{n\to\infty} a_n + b_n \geq \liminf_{n\to\infty} a_n + \liminf_{n\to\infty} b_n$.

(b) Give an example that shows that the inequality can be strict.

(c) Give examples that show that the limit inferior of a product of sequences can be greater or smaller than the product of the individual limit inferiors.

10-4. Let $\{a_n\}_{n=1}^{\infty}$ and $\{b_n\}_{n=1}^{\infty}$ be sequences of real numbers so that $a_n \leq b_n$ for all $n \in \mathbb{N}$.

(a) Prove that if $\{b_n\}_{n=1}^{\infty}$ converges, then $\limsup_{n\to\infty} a_n \leq \lim_{n\to\infty} b_n$.

(b) Prove that in general it is possible to have $\limsup_{n\to\infty} a_n > \liminf_{n\to\infty} b_n$.

10-5. Let $\{a_n\}_{n=1}^{\infty}$ be a bounded sequence of positive numbers so that the sequence of the reciprocals is bounded, too. Prove each of the following.

(a) $\liminf_{n\to\infty} \frac{1}{a_n} = \frac{1}{\limsup_{n\to\infty} a_n}$. (b) $\limsup_{n\to\infty} \frac{1}{a_n} = \frac{1}{\liminf_{n\to\infty} a_n}$.

10-6. Let $\{a_n\}_{n=1}^{\infty}$ be a sequence of positive numbers and let $\{b_n\}_{n=1}^{\infty}$ be a convergent sequence of positive numbers with nonzero limit. Prove each of the following.

(a) $\limsup_{n\to\infty} a_n b_n = \limsup_{n\to\infty} a_n \lim_{n\to\infty} b_n$ (b) $\limsup_{n\to\infty} a_n^{b_n} = \left(\limsup_{n\to\infty} a_n\right)^{\lim_{n\to\infty} b_n}$

10-7. **Lebesgue's Differentiation Theorem** states that for any function $f : [a, b] \to \mathbb{R}$ of bounded variation the derivative $f'(x)$ exists a.e. We will prove this result using the following steps.

(a) For any function $f : (a, b) \to \mathbb{R}$ and $x \in (a, b)$, define $D^+ f(x) := \limsup_{h\to 0^+} \frac{f(x+h) - f(x)}{h}$,

$D_+ f(x) := \liminf_{h\to 0^+} \frac{f(x+h) - f(x)}{h}$, $D^- f(x) := \limsup_{h\to 0^-} \frac{f(x+h) - f(x)}{h}$, and

$D_- f(x) := \liminf_{h\to 0^-} \frac{f(x+h) - f(x)}{h}$, where "$h \to 0^+$" means that h approaches zero form the right ($h > 0$) and "$h \to 0^-$" means that h approaches zero form the left ($h < 0$). Prove that f is differentiable at x iff $D^+ f(x) = D_+ f(x) = D^- f(x) = D_- f(x)$ and that in this case these values are equal to $f'(x)$.

Note. The four values above are also known as the **Dini derivatives** of f at x.

(b) Let $f : [a, b] \to \mathbb{R}$ be of bounded variation, let $A := \left\{ x \in (a, b) : D_- f(x) < D^+ f(x) \right\}$, let $B := \left\{ x \in (a, b) : D^- f(x) > D_+ f(x) \right\}$, and let $C := \left\{ x \in (a, b) : \left| D_+ f(x) \right| = \infty \right\}$. Prove that f is differentiable at every $x \in (a, b) \setminus (A \cup B \cup C)$.

Note. In particular, this means we have proved Lebesgue's Theorem if we can prove that A, B and C are null sets. The remaining parts of this exercise are devoted to this task.

(c) Suppose for a contradiction that $\lambda(A) > 0$ and obtain a contradiction as follows.

i. Prove that $\lambda\left(\left\{x \in A : D_- f(x) < q - \frac{1}{n} < q < q + \frac{1}{n} < D^+ f(x)\right\}\right) > 0$ for some $q \in \mathbb{Q}$ and $n \in \mathbb{N}$

ii. With q, n as above let $E := \left\{x \in A : D_- f(x) < q - \frac{1}{n} < q < q + \frac{1}{n} < D^+ f(x)\right\}$, let $g(x) := f(x) - qx$, let $\varepsilon := \frac{1}{n}$ and let $a = x_0 < x_1 < \cdots < x_n = b$ be so that $\sum_{k=1}^{n} \left| g(x_k) - g(x_{k-1}) \right| > V_a^b g - \frac{\varepsilon}{6}\lambda(E)$. Prove that for all $x \in E \cap (x_{k-1}, x_k)$ there are α_x and β_x with $x_{k-1} < \alpha_x < \beta_x < x_k$ and $\frac{g(\beta_x) - g(\alpha_x)}{\beta_x - \alpha_x} < -\varepsilon$ if $g(x_{k-1}) \leq g(x_k)$ or $\frac{g(\beta_x) - g(\alpha_x)}{\beta_x - \alpha_x} > \varepsilon$ if $g(x_{k-1}) \geq g(x_k)$.

iii. Prove that for every finite disjoint family $\mathcal{I}_k$ of intervals $[\alpha_x, \beta_x] \subset (x_{k-1}, x_k)$ as above we have $V_{x_{k-1}}^{x_k} g > \left| g(x_k) - g(x_{k-1}) \right| + 2\varepsilon \sum_{I \in \mathcal{I}_k} |I|$.

iv. Use Exercise 9-36 to obtain a finite pairwise disjoint family $\mathcal{I}$ of intervals $[\alpha_x, \beta_x]$ with $\sum_{I \in \mathcal{I}} |I| > \frac{1}{6}\lambda(E)$ and use this family to obtain the contradiction $\sum_{k=1}^{n} V_{x_{k-1}}^{x_k} g > V_a^b g$

(d) Use $D^- f(x) = -D_-(-f)(x)$ and $D_+ f(x) = -D^+(-f)(x)$ to prove that $\lambda(B) = 0$

(e) Lead the assumption $\lambda(C) > 0$ to a contradiction by letting $M := \frac{6V_a^b f}{\lambda(C)}$ and finding for each $x \in C$ an h_x with $\left| f(x + h_x) - f(x) \right| > Mh_x$.
Use Exercise 9-36 as in part 10-7(c)iv and arrive at a contradiction in a similar way.

10.2 The Root Test and the Ratio Test

When it works, the Ratio Test for the convergence of a series is computationally convenient, because it only involves a division. Note that by using the limit superior and the limit inferior we avoid any issues with convergence of the sequence of quotients.

Theorem 10.7 Ratio Test. *Let $\sum_{j=1}^{\infty} a_j$ be a series with $a_j \neq 0$ for all $j \in \mathbb{N}$.*

1. *If $\limsup_{j\to\infty} \left|\frac{a_{j+1}}{a_j}\right| < 1$ then $\sum_{j=1}^{\infty} a_j$ converges absolutely.*

2. *If $\liminf_{j\to\infty} \left|\frac{a_{j+1}}{a_j}\right| > 1$ then $\sum_{j=1}^{\infty} a_j$ diverges.*

If neither of the above conditions holds, then the series might converge or diverge. We also say that the Ratio Test failed in this situation.

Proof. For part 1, let $q := \frac{1}{2}\left(\limsup_{j\to\infty} \left|\frac{a_{j+1}}{a_j}\right| + 1\right) < 1$. Then there is a $J \in \mathbb{N}$ such that for all $j \geq J$ we have $|a_{j+1}| < q|a_j|$, and hence $|a_j| < q^{j-J}|a_J|$. Let

$M := \frac{|a_J|}{q^J}$. Then for all $j \geq J$ we obtain $0 \leq |a_j| < q^j \frac{|a_J|}{q^J} = Mq^j$. By Comparison Test (Theorem 6.15 and Exercise 6-16), $\sum_{j=1}^{\infty} a_j$ converges absolutely.

For part 2, note that if $\liminf_{j\to\infty} \left|\frac{a_{j+1}}{a_j}\right| > 1$, then there is a $J \in \mathbb{N}$ so that for all $j \geq J$ we have $|a_{j+1}| > |a_j|$. Hence, for all $j > J$ we obtain $|a_j| > |a_J| > 0$ and the terms of the series do not converge to zero. Thus the series diverges.

The last statement will be illuminated in Example 10.10. ■

The limit superior and the limit inferior remove any analytic concerns about convergence of the quotient from the Ratio Test. But algebraic concerns remain regarding the possible division by zero. The Ratio Test cannot be applied directly to any series for which a subsequence of the terms is zero. The Root Test does not involve divisions. Thus it can be applied directly to any series.

Theorem 10.8 Root Test. *Let $\sum_{j=1}^{\infty} a_j$ be a series.*

1. *If $\limsup_{j\to\infty} \sqrt[j]{|a_j|} < 1$, then $\sum_{j=1}^{\infty} a_j$ converges absolutely.*

2. *If $\limsup_{j\to\infty} \sqrt[j]{|a_j|} > 1$, then $\sum_{j=1}^{\infty} a_j$ diverges.*

If $\limsup_{j\to\infty} \sqrt[j]{|a_j|} = 1$, then the series might converge or diverge. We also say that the Root Test failed in this situation.

Proof. For part 1, let $q := \frac{1}{2}\left(\limsup_{j\to\infty} \sqrt[j]{|a_j|} + 1\right) < 1$. There is a $J \in \mathbb{N}$ such that $|a_j| < q^j$ for all $j \geq J$. By Comparison Test, $\sum_{j=1}^{\infty} a_j$ converges absolutely.

For part 2, note that if $\limsup_{j\to\infty} \sqrt[j]{|a_j|} > 1$, then for any $n \in \mathbb{N}$ there is a $j > n$ with $|a_j| > 1$. Hence, the terms do not converge to zero, and so the series diverges.

The last statement will be illuminated in Example 10.10. ■

In theoretical investigations, the Root Test usually is preferred over the Ratio Test, because we do not need to worry about terms that are equal to zero. Exercise 10-8 shows that for any series for which convergence can be proved with the Ratio Test, convergence can (in theory) also be proved with the Root Test. This is another indication that the Root Test is preferable when developing a theory.

The p-series test below provides examples of convergent and divergent series for which the Root Test and the Ratio Test both fail.

Theorem 10.9 **p-series test.** *Let $p \in \mathbb{Q}$. The sum $\sum_{j=1}^{\infty} \frac{1}{j^p}$ converges iff $p > 1$.*

Proof. If $p \leq 1$, then for all $j \geq 1$ we have $\frac{1}{j^p} \geq \frac{1}{j}$. Because the harmonic series $\sum_{j=1}^{\infty} \frac{1}{j}$ diverges (Example 6.8), by Comparison Test $\sum_{j=1}^{\infty} \frac{1}{j^p}$ diverges for $p \leq 1$.

This leaves the case $p > 1$. Note that the sequence of partial sums $s_k = \sum_{j=1}^{k} \frac{1}{j^p}$ is increasing. Hence, we are done if we can show that it is bounded. To do this we use a chunking argument similar to the proof in Example 6.8.

$$\begin{aligned} s_{2^m} &= \sum_{j=1}^{2^m} \frac{1}{j^p} = 1 + \sum_{l=0}^{m-1} \sum_{j=2^l+1}^{2^{l+1}} \frac{1}{j^p} = 1 + \sum_{l=0}^{m-1} \sum_{j=2^l+1}^{2^{l+1}} \frac{1}{j} \frac{1}{j^{p-1}} \\ &\leq 1 + \sum_{l=0}^{m-1} \frac{1}{\left(2^l\right)^{p-1}} \sum_{j=2^l+1}^{2^{l+1}} \frac{1}{j} \leq 1 + \sum_{l=0}^{m-1} \frac{1}{\left(2^{p-1}\right)^l} \leq 1 + \sum_{l=0}^{\infty} \left(\frac{1}{2^{p-1}}\right)^l \\ &= 1 + \frac{1}{1 - \frac{1}{2^{p-1}}} = 1 + \frac{2^{p-1}}{2^{p-1} - 1} < \infty. \end{aligned}$$

Because the right side of this inequality does not depend on m, the s_k are bounded and the series converges for $p > 1$. ■

Example 10.10 The Root Test and the Ratio Test both fail for the harmonic series $\sum_{j=1}^{\infty} \frac{1}{j}$, which diverges, and for the series $\sum_{j=1}^{\infty} \frac{1}{j^2}$, which converges. For the Ratio Test, this is a simple computation. For the Root Test, we need to use that $\lim_{j \to \infty} \sqrt[j]{j} = 1$. We postpone the verification to Exercise 10-16c. □

Exercises

10-8. The Root Test versus the Ratio Test Let $\sum_{j=1}^{\infty} a_j$ be a series with $a_j \neq 0$ for all j.

(a) Prove that if $\limsup_{j \to \infty} \left| \frac{a_{j+1}}{a_j} \right| < 1$ then $\limsup_{j \to \infty} \sqrt[j]{|a_j|} < 1$

(b) Explain why part 10-8a shows that if the Ratio Test proves that a series converges, then the Root Test also proves that the series converges.

(c) Prove that if $\liminf_{j \to \infty} \left| \frac{a_{j+1}}{a_j} \right| > 1$ then $\limsup_{j \to \infty} \sqrt[j]{|a_j|} \geq 1$

10-9 Construct a convergent series $\sum_{j=1}^{\infty} a_j$ with $\limsup_{j\to\infty} \left| \frac{a_{j+1}}{a_j} \right| > 1$.

10-10. Prove the p-series test (Theorem 10.9) using the Integral Test (Theorem 8.29).

10-11. Use the Ratio Test or the Root Test to determine which of the following series converge. If a series converges, determine if it converges absolutely.

(a) $\sum_{j=1}^{\infty} \frac{1}{j!}$ (b) $\sum_{j=1}^{\infty} \frac{(-1)^j}{j!}$ (c) $\sum_{j=1}^{\infty} \frac{j^{100}}{j!}$ (d) $\sum_{j=1}^{\infty} \frac{(-1)^j j^j}{j!}$

10-12. Use any of the convergence tests introduced so far (including those from Chapter 6) to determine which of the following series converge. If a series converges, determine if it converges absolutely.

(a) $\sum_{j=1}^{\infty} \frac{j^2}{2^j}$ (b) $\sum_{j=1}^{\infty} \frac{(-1)^j j}{\sqrt{j}+20}$ (c) $\sum_{j=1}^{\infty} \frac{(-1)^j j}{j^{\frac{3}{2}}+1}$ (d) $\sum_{j=1}^{\infty} \frac{(-1)^j j}{j^3-7}$

10.3 Power Series

With series giving access to "sums with infinitely many terms" it is natural to consider what happens when we let the sum in the definition of polynomials have infinitely many terms. The resulting notion of a power series is very versatile. In particular, it will enable us to introduce the transcendental functions in Chapter 12.

Definition 10.11 *For a sequence $\{c_k\}_{k=0}^{\infty}$, the expression $\sum_{k=0}^{\infty} c_k(x-a)^k$ is called a* **power series** *centered at a (or a power series about a) with coefficients c_k. The power series is called* **convergent** *at $x \in \mathbb{R}$ if and only if $\lim_{N\to\infty} \sum_{k=0}^{N} c_k(x-a)^k$ exists. Otherwise it is called* **divergent** *at x.*

Theorem 10.12 shows that power series converge on a symmetric open interval $(a-R, a+R)$ about the center point (where R could be infinite), and they diverge outside $[a-R, a+R]$. So only if R is finite are there two points, $a+R$ and $a-R$ for which the convergence behavior is unknown. Note how in Theorem 10.12 R is defined in terms of the Root Test. Exercise 10-13 provides an equally useful conceptual characterization of R.

Theorem 10.12 *Let $\sum_{k=0}^{\infty} c_k(x-a)^k$ be a power series. If $\limsup_{k\to\infty} \sqrt[k]{|c_k|} \in (0,\infty)$ let $R := \frac{1}{\limsup_{k\to\infty} \sqrt[k]{|c_k|}}$, if $\limsup_{k\to\infty} \sqrt[k]{|c_k|} = \infty$ let $R := 0$ and if $\limsup_{k\to\infty} \sqrt[k]{|c_k|} = 0$ let $R := \infty$. Then $\sum_{k=1}^{\infty} c_k(x-a)^k$ converges absolutely for all $x \in \mathbb{R}$ with $|x-a| < R$ and it diverges for all $x \in \mathbb{R}$ with $|x-a| > R$.*

Proof. We apply the Root Test. Let the formal quotient $\frac{1}{0}$ be ∞. For $x \neq a$, the limit superior $\limsup_{k\to\infty} \sqrt[k]{\left|c_k(x-a)^k\right|} = |x-a| \limsup_{k\to\infty} \sqrt[k]{|c_k|}$ is less than 1 iff $|x-a| < \frac{1}{\limsup_{k\to\infty} \sqrt[k]{|c_k|}}$ and it is greater than 1 iff $|x-a| > \frac{1}{\limsup_{k\to\infty} \sqrt[k]{|c_k|}}$. This proves the result. ∎

Definition 10.13 *Let $\sum_{k=0}^{\infty} c_k(x-a)^k$ be a power series. Then, with $\frac{1}{0} := \infty$ this once, $R := \frac{1}{\limsup_{k\to\infty} \sqrt[k]{|c_k|}}$ is called the* **radius of convergence** *of $\sum_{k=0}^{\infty} c_k(x-a)^k$.*

Theorem 10.12 shows that power series define functions on intervals. These functions are limits of polynomials of increasing degree. To find out more about the analytic properties of these functions we need to investigate sequences of functions, which is done in Chapter 11. We conclude the present section with some remarks on the arithmetic of power series. Sums, differences and constant multiples of power series are straightforward (see Exercise 10-14).

The multiplication of power series is similar to that of polynomials. We want to multiply each term of the first series with each term of the second and then collect terms with equal exponents. The product of a term with exponent i in the first power series with a term with exponent $k-i$ in the second power series gives a term with exponent k for the product. We use this observation to first define a product for series.

Theorem 10.14 *The* **Cauchy Product** *of two series. Let $\sum_{k=0}^{\infty} a_k$ and $\sum_{k=0}^{\infty} b_k$ be absolutely convergent series and for each $k \geq 0$ define $p_k := \sum_{i=0}^{k} a_i b_{k-i}$. Then $\sum_{k=0}^{\infty} p_k$ converges absolutely and $\sum_{k=0}^{\infty} p_k = \sum_{k=0}^{\infty} a_k \sum_{k=0}^{\infty} b_k$.*

Proof. First note that $\sum_{k=0}^{n} p_k = \sum_{k=0}^{n} \sum_{i=0}^{k} a_i b_{k-i} = \sum_{0\leq j+l\leq n} a_j b_l = \sum_{k=0}^{n} \sum_{i=0}^{n-k} a_k b_i$ for every $n \in \mathbb{N}$. To see that $\sum_{k=0}^{\infty} p_k$ converges absolutely, note that for all $n \in \mathbb{N}$ the inequalities

$$\sum_{k=0}^{n} |p_k| \leq \sum_{k=0}^{n} \sum_{i=0}^{n-k} |a_k||b_i| \leq \sum_{k=0}^{n} \sum_{i=0}^{n} |a_k||b_i| \leq \sum_{k=0}^{\infty} |a_k| \sum_{i=0}^{\infty} |b_i| < \infty$$

hold. Now let $\varepsilon > 0$ and assume without loss of generality that each series has at least one nonzero term. Then there is an even $N \in \mathbb{N}$ so that $\sum_{k=\frac{N}{2}}^{\infty} |a_k| < \frac{\varepsilon}{3\sum_{k=0}^{\infty}|b_k|}$ and $\sum_{k=\frac{N}{2}}^{\infty} |b_k| < \frac{\varepsilon}{3\sum_{k=0}^{\infty}|a_k|}$. Therefore for all $n \geq N$ we obtain

$$\begin{aligned}
&\left|\sum_{k=0}^{n} p_k - \sum_{k=0}^{\infty} a_k \sum_{k=0}^{\infty} b_k\right| \\
&= \left|\sum_{k=0}^{n}\sum_{i=0}^{n-k} a_k b_i - \sum_{k=0}^{\infty} a_k \sum_{k=0}^{\infty} b_k\right| \\
&\leq \left|\sum_{k=0}^{n} a_k \sum_{i=0}^{n-k} b_i - \sum_{k=0}^{n} a_k \sum_{i=0}^{\infty} b_i\right| + \left|\sum_{k=0}^{n} a_k \sum_{k=0}^{\infty} b_k - \sum_{k=0}^{\infty} a_k \sum_{k=0}^{\infty} b_k\right| \\
&\leq \sum_{k=0}^{n} |a_k| \sum_{i=n-k+1}^{\infty} |b_i| + \sum_{k=n+1}^{\infty} |a_k| \sum_{k=0}^{\infty} |b_k| \\
&\leq \sum_{k=0}^{\lfloor\frac{n}{2}\rfloor} |a_k| \sum_{i=\lfloor\frac{n}{2}\rfloor+1}^{\infty} |b_i| + \sum_{k=\lfloor\frac{n}{2}\rfloor}^{n} |a_k| \sum_{i=0}^{\infty} |b_i| + \sum_{k=n+1}^{\infty} |a_k| \sum_{k=0}^{\infty} |b_k| \\
&< \sum_{k=0}^{\infty} |a_k| \frac{\varepsilon}{3\sum_{k=0}^{\infty}|a_k|} + \frac{\varepsilon}{3\sum_{k=0}^{\infty}|b_k|} \sum_{i=0}^{\infty} |b_i| + \frac{\varepsilon}{3\sum_{k=0}^{\infty}|b_k|} \sum_{k=0}^{\infty} |b_k| \\
&= \frac{\varepsilon}{3} + \frac{\varepsilon}{3} + \frac{\varepsilon}{3} = \varepsilon.
\end{aligned}$$

Hence, $\sum_{k=0}^{\infty} p_k = \sum_{k=0}^{\infty} a_k \sum_{k=0}^{\infty} b_k$, which completes the proof. ■

It can be shown that in Theorem 10.14 it is enough to demand that one series converges (not necessarily absolutely) and the other converges absolutely. Because the proof is much more involved, we demanded here that both series converge absolutely.

Corollary 10.15 *Let* $\sum_{k=0}^{\infty} c_k(x-a)^k$ *and* $\sum_{k=0}^{\infty} d_k(x-a)^k$ *be power series with radii of convergence* R_c *and* R_d*, respectively. With* $p_k := \sum_{i=0}^{k} c_i d_{k-i}$ *for* $k \geq 0$*, the radius of convergence of* $\sum_{k=0}^{\infty} p_k(x-a)^k$ *is at least* $R_p := \min\{R_c, R_d\}$ *and for all* x *with* $|x-a| < R_p$ *the product of the power series is the power series with coefficients* p_k*, that is,* $\sum_{k=0}^{\infty} p_k(x-a)^k = \left(\sum_{k=0}^{\infty} c_k(x-a)^k\right)\left(\sum_{k=0}^{\infty} d_k(x-a)^k\right)$.

Proof. Exercise 10-15. ■

Exercises

10-13 Prove that the radius of convergence of a power series $\sum_{k=0}^{\infty} a_k(x-a)^k$ is the largest $R \in [0, \infty]$ so that $\sum_{k=0}^{\infty} a_k(x-a)^k$ converges absolutely for all x with $|x-a| < R$.

Note. This conceptual formulation is often more helpful than the formula in Definition 10.13.

10-14 Let $\sum_{k=0}^{\infty} c_k(x-a)^k$ and $\sum_{k=0}^{\infty} d_k(x-a)^k$ be power series with radii of convergence R_c and R_d.

(a) Prove that the radius of convergence of $\sum_{k=0}^{\infty}(c_k+d_k)(x-a)^k$ is at least $\min\{R_c, R_d\}$ and that $\sum_{k=0}^{\infty}(c_k+d_k)(x-a)^k = \sum_{k=0}^{\infty} c_k(x-a)^k + \sum_{k=0}^{\infty} d_k(x-a)^k$ for all x for which both series converge.

Hint. Use the result from Exercise 10-13.

(b) State and prove a result similar to part 10-14a for the difference of the two power series.

(c) Prove that if $b \in \mathbb{R} \setminus \{0\}$, then the radius of convergence of $\sum_{k=0}^{\infty} bc_k(x-a)^k$ is R_c and for all x for which $\sum_{k=0}^{\infty} c_k(x-a)^k$ converges, we have $\sum_{k=0}^{\infty} bc_k(x-a)^k = b\sum_{k=0}^{\infty} c_k(x-a)^k$.

10-15 Prove Corollary 10.15.

10-16 Consider the power series $\sum_{k=1}^{\infty} \frac{x^k}{k}$.

(a) Use the Ratio Test to show that the radius of convergence is 1.

(b) Use Theorem 10.12 to show that $\lim_{n\to\infty} \sqrt[n]{n} = 1$.

(c) Prove that the Root Test fails for $\sum_{j=1}^{\infty} \frac{1}{j}$ and $\sum_{j=1}^{\infty} \frac{1}{j^2}$.

10-17. Use the formula from Definition 10.13 to compute the radius of convergence of the given power series. You may use the result from Exercise 10-16b.

(a) $\sum_{k=1}^{\infty} \frac{x^k}{k^2}$ (b) $\sum_{k=1}^{\infty} \frac{x^k}{k^k}$ (c) $\sum_{k=1}^{\infty} \frac{k!x^k}{k^k}$ (d) $\sum_{k=1}^{\infty} \frac{x^{2k}}{k}$

Chapter 11

Sequences of Functions

In Section 10.3, we considered power series as series of numbers for fixed $x \in \mathbb{R}$. Another way to look at them is as sequences of polynomials. This change in point-of-view is the first step toward function spaces, which are very important in analysis. This chapter describes ways in which sequences of functions can converge. Formally, a sequence of functions is a map from the natural numbers into a function space. Until we encounter function spaces in Example 15.3, *a sequence of functions will simply be, as in Definition 9.17, a sequence whose "values" are functions, not numbers.*

11.1 Notions of Convergence

The most obvious way in which a sequence of functions can converge is at every point.

Definition 11.1 *Let S be a set and for every $n \in \mathbb{N}$ let $f_n : S \to \mathbb{R}$ be a function. The sequence of functions $\{f_n\}_{n=1}^{\infty}$ is called* **pointwise convergent** *to the function $f : S \to \mathbb{R}$ iff for all $x \in S$ we have $\lim_{n\to\infty} f_n(x) = f(x)$.*

Example 11.2 *For $n \in \mathbb{N}$, let $f_n(x) := x^{\frac{1}{n}}$. On the set $(0, 1)$ the sequence $\{f_n\}_{n=1}^{\infty}$ converges pointwise to the constant function $f(x) = 1$.*

By Exercise 2-46, for all $x \in (0, 1)$ the sequence $\left\{x^{\frac{1}{n}}\right\}_{n=1}^{\infty}$ converges to 1. □

Example 11.3 *By Definition 9.18 every real valued Lebesgue measurable function f is the pointwise limit of a sequence of simple functions.*

This is because if $\{s_n\}_{n=1}^{\infty}$ is a sequence of simple functions that converges pointwise to f^+ and $\{t_n\}_{n=1}^{\infty}$ is a sequence of simple functions that converges pointwise to f^-, then the sequence $\{s_n - t_n\}_{n=1}^{\infty}$ converges pointwise to $f = f^+ - f^-$. □

Pointwise convergence means the sequence of functions converges at every point, but the "speed" of convergence can vary from point to point. Uniform convergence requires that convergence happens at a uniform minimum speed (see Figure 23).

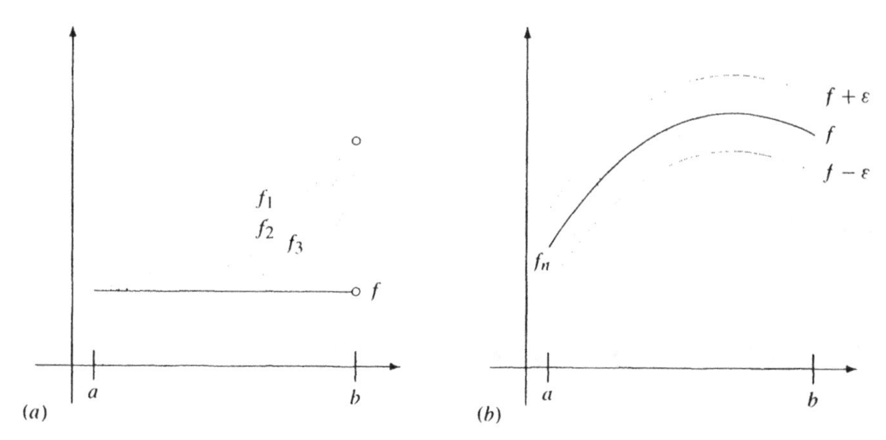

Figure 23: Pointwise convergence allows the sequence of functions to converge at different "speeds" at different points (a). In uniform convergence, for every $\varepsilon > 0$ there is an $N \in \mathbb{N}$ so that for all $n \geq N$ the functions f_n are trapped between $f - \varepsilon$ and $f + \varepsilon$ (b). In this fashion, convergence occurs at a uniformly given "minimum speed."

Definition 11.4 *Let S be a set and for every $n \in \mathbb{N}$ let $f_n : S \to \mathbb{R}$ be a function. The sequence of functions $\{f_n\}_{n=1}^{\infty}$ is called* **uniformly convergent** *to the function $f : S \to \mathbb{R}$ iff for all $\varepsilon > 0$ there is an $N \in \mathbb{N}$ so that for all $n \geq N$ and all $x \in S$ we have $\left|f_n(x) - f(x)\right| < \varepsilon$.*

If a sequence of functions only converges (pointwise or uniformly) on a part T of the common domain S, we also say that the sequence converges (pointwise or uniformly) on T. It is easy to see that uniform convergence implies pointwise convergence (Exercise 11-1). The converse is false, as the next example shows.

Example 11.5 *For $n \in \mathbb{N}$, let $f_n(x) := x^{\frac{1}{n}}$. Then for any $\delta > 0$ on the interval $(\delta, 1)$ the sequence $\{f_n\}_{n=1}^{\infty}$ converges uniformly to the constant function $f(x) = 1$. However, on the set $(0, 1)$ the sequence $\{f_n\}_{n=1}^{\infty}$ does not converge uniformly to the constant function $f(x) = 1$.*

For the first claim, let $\delta \in (0, 1)$. Then for all points $x \in (\delta, 1)$ we have that $\delta^{\frac{1}{n}} < x^{\frac{1}{n}} < 1$. If $\varepsilon > 0$ is given, there is an $N \in \mathbb{N}$ so that for all $n \geq N$ we have $1 - \varepsilon < \delta^{\frac{1}{n}} < 1$. Consequently, for all points $x \in (\delta, 1)$ and all $n \geq N$ we infer $1 - \varepsilon < \delta^{\frac{1}{n}} < x^{\frac{1}{n}} < 1$. Therefore on the interval $(\delta, 1)$ the sequence $\{f_n\}_{n=1}^{\infty}$ converges uniformly to the constant function $f(x) = 1$.

To prove that the sequence $\{f_n\}_{n=1}^{\infty}$ does not converge uniformly to $f(x) = 1$ on $(0, 1)$, suppose for a contradiction that it does. Then for every $\varepsilon \in (0, 1)$ there is an $N \in \mathbb{N}$ so that for all $n \geq N$ and all $x \in (0, 1)$ we have $1 - \varepsilon < x^{\frac{1}{n}} < 1$. But $y := (1 - \varepsilon)^N$ is in $(0, 1)$ and $f_N\left((1 - \varepsilon)^N\right) = \left((1 - \varepsilon)^N\right)^{\frac{1}{N}} = 1 - \varepsilon \not> 1 - \varepsilon$, a contradiction. Therefore, on the set $(0, 1)$ the sequence $\{f_n\}_{n=1}^{\infty}$ does not converge uniformly to the constant function $f(x) = 1$. □

Power series are important examples of uniformly convergent series.

Theorem 11.6 *Let* $\sum_{k=0}^{\infty} c_k(x-a)^k$ *be a power series with radius of convergence* R. *Then* $\sum_{k=0}^{\infty} c_k(x-a)^k$ *converges uniformly on any closed subinterval of* $(a-R, a+R)$.

Proof. Any closed subinterval of $(a-R, a+R)$ is contained in a symmetric closed subinterval $[a-r, a+r]$ with $0 < r < R$. Thus we are done if we can prove the result for symmetric closed subintervals $[a-r, a+r] \subseteq (a-R, a+R)$. Let $r \in (0, R)$. For all $x \in [a-r, a+r]$ and all $k \in \mathbb{N} \cup \{0\}$, we have that $\left|c_k(x-a)^k\right| \leq |c_k| r^k$. For each $x \in [a-r, a+r]$, let $P(x) := \sum_{k=0}^{\infty} c_k(x-a)^k$ be the limit of the power series. By Theorem 10.12, $\sum_{k=0}^{\infty} c_k\big((a+r)-a\big)^k$ converges absolutely. Thus for any $\varepsilon > 0$ there is an $N \in \mathbb{N}$ so that for all $n \geq N$ we have $\sum_{k=n}^{\infty} |c_k| r^k < \varepsilon$. Therefore for all $x \in [a-r, a+r]$ and all $n \geq N$ we obtain

$$\begin{aligned}\left|P(x) - \sum_{k=0}^{n} c_k(x-a)^k\right| &= \left|\sum_{k=n+1}^{\infty} c_k(x-a)^k\right| \leq \sum_{k=n+1}^{\infty} \left|c_k(x-a)^k\right| \\ &\leq \sum_{k=n+1}^{\infty} |c_k| r^k < \varepsilon,\end{aligned}$$

and hence the power series converges uniformly on $[a-r, a+r]$. ■

More notions of convergence for sequences of functions are discussed in Section 14.6.

Exercises

11-1 Prove that every uniformly convergent sequence of functions is pointwise convergent.

11-2 In Example 11.5, consider the proof that the $f_n(x) = x^{\frac{1}{n}}$ do not converge uniformly to the constant function $f(x) = 1$ on $(0, 1)$. Could we have used $\varepsilon = \frac{1}{2}$ in this part of the proof? Explain.

11-3 Prove that for each $\delta > 0$ the sequence $\{f_n\}_{n=1}^{\infty}$ with $f_n(x) = x^n$ converges uniformly on $(0, 1-\delta)$, but it does not converge uniformly on $(0, 1)$.

11-4 Prove that a bounded function $f : \mathbb{R} \to \mathbb{R}$ is Lebesgue measurable iff it is a uniform limit of simple functions. *Hint.* Part "5⇒1" of Theorem 9.19.

11-5 Give an example of a sequence $\{f_n\}_{n=1}^{\infty}$ of bounded functions on an interval I that converges pointwise to an unbounded function $f : I \to \mathbb{R}$.

11-6 A sequence of functions $\{f_n\}_{n=1}^{\infty}$ defined on a set S is called **pointwise Cauchy** on the set S iff for all $x \in S$ the sequence $\left\{f_n(x)\right\}_{n=1}^{\infty}$ is a Cauchy sequence. Prove that every pointwise Cauchy sequence of functions is pointwise convergent to a function $f : S \to \mathbb{R}$.

11-7. Approximating continuous functions with differentiable functions. Let $[a, b]$ be an interval.

(a) For continuous $f : [a, b] \to \mathbb{R}$ and all $x \in \mathbb{R}$, let $\tilde{f}(x) := \begin{cases} f(x), & \text{if } x \in [a, b], \\ f(a); & \text{if } x < a, \\ f(b), & \text{if } x > b. \end{cases}$

For $n \in \mathbb{N}$ and all $x \in \mathbb{R}$, let $f_n(x) := n \int_x^{x+\frac{1}{n}} \tilde{f}(z)\, dz$. Prove that f_n is differentiable on $\mathbb{R}$

(b) Prove that for every continuous function on $[a, b]$ there is a sequence of functions that converges uniformly to f and so that every function in the sequence is continuous on $[a, b]$ and differentiable on (a, b)

11-8 A sequence of functions $\{f_n\}_{n=1}^{\infty}$ defined on a set S is called **uniformly Cauchy** on the set S iff for all $\varepsilon > 0$ there is an $N \in \mathbb{N}$ so that for all $n, m \geq N$ and all $x \in S$ we have $\left| f_n(x) - f_m(x) \right| < \varepsilon$ Prove that every uniform Cauchy sequence of functions is uniformly convergent to a function $f : S \to \mathbb{R}$.

11-9. For each $j \in \mathbb{N}$, let $f_j : [a, b] \to \mathbb{R}$ be so that $a_j := \sup \left\{ \left| f_j(x) \right| : x \in [a, b] \right\} < \infty$. Prove that if $\sum_{j=1}^{\infty} a_j$ converges, then $\sum_{j=1}^{\infty} f_j$ converges absolutely at every $x \in [a, b]$ and uniformly on $[a, b]$

11-10. Let $f : (a, b) \to \mathbb{R}$ be continuous, let $[c, d] \subset (a, b)$ be a closed subinterval and let $N := \left\lceil \frac{1}{b-d} \right\rceil$.

Prove that $\{g_n\}_{n=N}^{\infty}$ defined by $g_n(x) := f\left(x + \frac{1}{n} \right)$ converges uniformly to f on $[c, d]$.

Hint Uniform continuity

11-11 Let $f : (a, b) \to \mathbb{R}$ be continuously differentiable, let $[c, d] \subset (a, b)$ be a closed subinterval and let $N := \left\lceil \frac{1}{b-d} \right\rceil$ Prove that the sequence $\{g_n\}_{n=N}^{\infty}$ defined by $g_n(x) := \frac{f\left(x + \frac{1}{n}\right) - f(x)}{\frac{1}{n}}$ converges uniformly to f' on $[c, d]$. *Hint* Mean Value Theorem

11.2 Uniform Convergence

If the function f is the (pointwise or uniform) limit of a sequence of functions $\{f_n\}_{n=1}^{\infty}$, it is natural to ask which properties of the functions in the sequence are inherited by the limit f. It turns out that pointwise convergence preserves neither continuity (see Exercise 11-12) nor Riemann integrability (see Exercise 11-13). Uniform convergence on the other hand preserves continuity (see Theorem 11.7), as well as Riemann integrability (see Theorem 11.9).

Theorem 11.7 *Let $\{f_n\}_{n=1}^{\infty}$ be a sequence of functions on $[a, b]$ that converges uniformly to $f : [a, b] \to \mathbb{R}$ and let all f_n be continuous at $x \in [a, b]$. Then f is continuous at x.*

Proof. Let $\varepsilon > 0$. Because $\{f_n\}_{n=1}^{\infty}$ converges uniformly to f on $[a, b]$, there is an $N \in \mathbb{N}$ so that for all $z \in [a, b]$ we have $\left| f_N(z) - f(z) \right| < \frac{\varepsilon}{3}$. Moreover, because f_N is continuous at x, there is a $\delta > 0$ so that for all $z \in [a, b]$ with $|z - x| < \delta$ we have $\left| f_N(z) - f_N(x) \right| < \frac{\varepsilon}{3}$. Then for all $z \in [a, b]$ with $|z - x| < \delta$ we conclude

$$\begin{aligned} \left| f(z) - f(x) \right| &\leq \left| f(z) - f_N(z) \right| + \left| f_N(z) - f_N(x) \right| + \left| f_N(x) - f(x) \right| \\ &< \frac{\varepsilon}{3} + \frac{\varepsilon}{3} + \frac{\varepsilon}{3} = \varepsilon, \end{aligned}$$

and hence f is continuous at x. ■

Corollary 11.8 *Let $\{f_n\}_{n=1}^{\infty}$ be a sequence of functions on $[a,b]$ that converges uniformly to $f:[a,b]\to\mathbb{R}$ and let all f_n be continuous on $[a,b]$. Then f is continuous on $[a,b]$.*

Proof. Apply Theorem 11.7 at every point in $[a,b]$. ■

Theorem 11.9 *Let $\{f_n\}_{n\in\mathbb{N}}$ be a sequence of Riemann integrable functions on the interval $[a,b]$ that converges uniformly to the function $f:[a,b]\to\mathbb{R}$. Then f is Riemann integrable and $\displaystyle\int_a^b f(x)\,dx=\lim_{n\to\infty}\int_a^b f_n(x)\,dx$.*

Proof. For each $n\in\mathbb{N}$, the function f_n is Riemann integrable. Therefore by Theorem 8.12 each set $B_n:=\left\{x\in[a,b]:\omega_f(x)>0\right\}$ is a null set. Hence, by countable subadditivity of outer Lebesgue measure (part 3 of Theorem 8.6), the set $B:=\displaystyle\bigcup_{n=1}^{\infty}B_n$ is a null set. Now for every $x\in[a,b]\setminus B$ all functions f_n are continuous at x, and by Theorem 11.7 we infer that f is continuous at x. Because B is a null set, by Theorem 8.12 f is Riemann integrable.

For the integrals, let $\varepsilon>0$ and find an $N\in\mathbb{N}$ so that for all $n\geq N$ and all $x\in[a,b]$ we have $\left|f_n(x)-f(x)\right|<\dfrac{\varepsilon}{b-a}$. Then for all $n\geq N$ we obtain

$$\left|\int_a^b f(x)\,dx-\int_a^b f_n(x)\,dx\right|\leq\int_a^b\left|f(x)-f_n(x)\right|dx<\int_a^b\frac{\varepsilon}{b-a}\,dx=\varepsilon,$$

which establishes the equality. ■

Lebesgue's integrability criterion for Riemann integrals reveals that being Riemann integrable is a substantially weaker property than continuity. It would be nice if this weaker property would be preserved by a weaker notion of convergence, such as pointwise convergence. However, Exercise 11-13 shows that the pointwise limit of Riemann integrable functions need not be Riemann integrable. This is another situation in which the Lebesgue integral has an advantage over the Riemann integral. For pointwise limits of (Lebesgue) integrable functions, please consider Section 14.5.

With continuity and integrability investigated, we turn to differentiability. For sequences of differentiable functions, uniform convergence does not guarantee the differentiability of the limit.

Example 11.10 *Uniform limits of differentiable functions need not be differentiable. For every $n\in\mathbb{N}$, the function $f_n(x):=x^{\frac{2n+2}{2n+1}}$ is differentiable on $(-1,1)$, but the uniform limit of $\{f_n\}_{n=1}^{\infty}$ is $f(x)=|x|$, which is not differentiable (also see Figure 24).*

It is clear that the f_n are all differentiable. To see the uniform convergence to the absolute value function, let $\varepsilon>0$. For all $x\in(-1,1)$ we have

$$\left|x^{\frac{2n+2}{2n+1}}-|x|\right|=\left||x|^{\frac{2n+2}{2n+1}}-|x|\right|=|x|\left||x|^{\frac{1}{2n+1}}-1\right|,$$

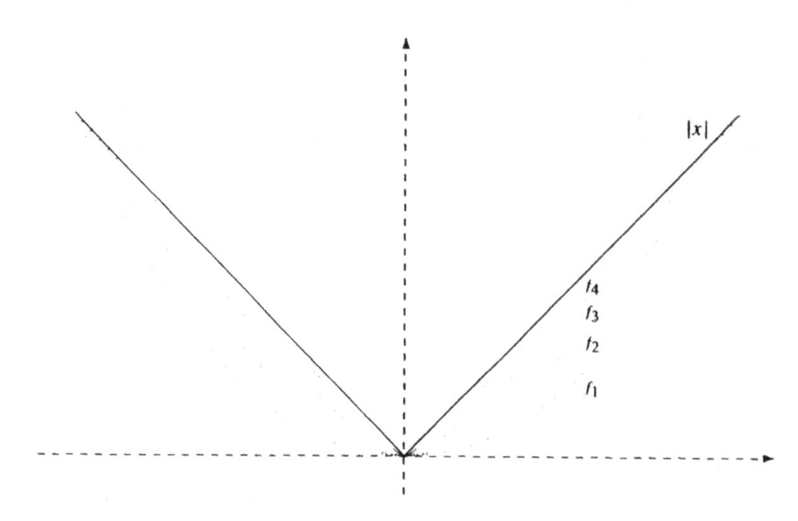

Figure 24: A sequence of differentiable functions that converges uniformly to $|x|$. For an indication how pathological the situation can become, consider that by Exercise 11-7 every continuous function is the uniform limit of differentiable functions, while by Exercise 11-21 there are continuous functions that are not differentiable at *any* $x \in \mathbb{R}$.

as well as $\left||x|^{\frac{1}{2n+1}} - 1\right| \leq 1$ and $|x| < 1$. Moreover, $\lim_{n\to\infty} r^{\frac{1}{2n+1}} = 1$ for every positive real number r. Let $\varepsilon > 0$ and let $N \in \mathbb{N}$ be such that for all $n \geq N$ we have $\left|\varepsilon^{\frac{1}{2n+1}} - 1\right| < \varepsilon$. Then for all $x \in (-\varepsilon, \varepsilon)$ and all $n \in \mathbb{N}$ we obtain

$$\left|x^{\frac{2n+2}{2n+1}} - |x|\right| = \left||x|^{\frac{2n+2}{2n+1}} - |x|\right| = |x|\left||x|^{\frac{1}{2n+1}} - 1\right| < \varepsilon \cdot 1 = \varepsilon,$$

while for all $x \in (-1, 1) \setminus (-\varepsilon, \varepsilon)$ and all $n \geq N$ we obtain

$$\left|x^{\frac{2n+2}{2n+1}} - |x|\right| = |x|\left||x|^{\frac{1}{2n+1}} - 1\right| \leq 1 \cdot \left|\varepsilon^{\frac{1}{2n+1}} - 1\right| < \varepsilon.$$

We have proved that for all $\varepsilon > 0$ there is an $N \in \mathbb{N}$ so that for all $n \geq N$ and for all $x \in (-1, 1)$ we have $\left|f_n(x) - f(x)\right| < \varepsilon$, which means that the sequence $\{f_n\}_{n=1}^{\infty}$ converges uniformly on $(-1, 1)$ to $f(x) = |x|$. □

To obtain a differentiable limit function, we require continuous derivatives and we impose a uniform convergence criterion on the derivatives.

Theorem 11.11 *Let $\{f_n\}_{n=1}^{\infty}$ be a sequence of differentiable functions on (a, b) so that for all $n \in \mathbb{N}$ the derivative f_n' is continuous on (a, b). If $\{f_n\}_{n=1}^{\infty}$ converges pointwise to $f : (a, b) \to \mathbb{R}$ and the sequence $\left\{f_n'\right\}_{n=1}^{\infty}$ converges uniformly to the function $g : (a, b) \to \mathbb{R}$, then f is differentiable and $f' = g$.*

Proof. By Theorem 11.7, the function g is continuous, because (on every closed subinterval of (a, b)) it is the uniform limit of continuous functions. To show that

$f' = g$, let $x \in (a,b)$ and let $\varepsilon > 0$. Let $\delta > 0$ be such that for all $y \in (a,b)$ with $|y-x| < \delta$ we have $\left|g(y) - g(x)\right| < \frac{\varepsilon}{3}$. Let $z \in (a,b) \setminus \{x\}$ with $|z-x| < \delta$ be fixed. Because $\left\{f_n'\right\}_{n=1}^{\infty}$ converges uniformly to g, there is an $N_g \in \mathbb{N}$ so that for all $n \geq N_g$ and for all $y \in (a,b)$ we have $\left|f_n'(y) - g(y)\right| < \frac{\varepsilon}{3}$. Because $\{f_n\}_{n=1}^{\infty}$ converges pointwise to f, there is an $N_f \in \mathbb{N}$ so that for all $n \geq N_f$ the inequality $\left|\frac{f(z)-f(x)}{z-x} - \frac{f_n(z)-f_n(x)}{z-x}\right| < \frac{\varepsilon}{3}$ holds. Let $n \geq \max\{N_f, N_g\}$. By the Mean Value Theorem there is a c between z and x so that $\frac{f_n(z)-f_n(x)}{z-x} = f_n'(c)$. Therefore we conclude

$$\begin{aligned}
&\left|\frac{f(z)-f(x)}{z-x} - g(x)\right| \\
&\quad\leq \left|\frac{f(z)-f(x)}{z-x} - \frac{f_n(z)-f_n(x)}{z-x}\right| + \left|\frac{f_n(z)-f_n(x)}{z-x} - f_n'(c)\right| \\
&\qquad + \left|f_n'(c) - g(c)\right| + \left|g(c) - g(x)\right| \\
&\quad< \frac{\varepsilon}{3} + 0 + \frac{\varepsilon}{3} + \frac{\varepsilon}{3} = \varepsilon.
\end{aligned}$$

We have proved that for any $\varepsilon > 0$ there is a $\delta > 0$ so that for all $z \in (a,b) \setminus \{x\}$ with $|z-x| < \delta$ we have $\left|\frac{f(z)-f(x)}{z-x} - g(x)\right| < \varepsilon$. Hence, f is differentiable at x and $f'(x) = g(x)$. Because $x \in (a,b)$ was arbitrary, the result is established. ■

Example 11.10 shows that demanding uniform convergence of the derivatives is crucial in Theorem 11.11, because the sequence of functions in Example 11.10 satisfies all other hypotheses. Theorem 11.11 allows us to show that power series, considered as functions, are differentiable, which will help our investigation of transcendental functions in Chapter 12.

Corollary 11.12 *Let* $\sum_{k=0}^{\infty} c_k(x-a)^k$ *be a power series with radius of convergence* $R = \frac{1}{\limsup_{k\to\infty} \sqrt[k]{|c_k|}}$. *Then* $\sum_{k=0}^{\infty} c_k(x-a)^k$ *is differentiable on* $(a-R, a+R)$ *and the derivative is* $\sum_{k=1}^{\infty} kc_k(x-a)^{k-1}$.

Proof. First, note that (by Exercises 10-6 and 10-16b)

$$\begin{aligned}
\frac{1}{\limsup_{k\to\infty} \sqrt[k-1]{|kc_k|}} &= \frac{1}{\limsup_{k\to\infty} \left(\sqrt[k]{k}\sqrt[k]{|c_k|}\right)^{\frac{k}{k-1}}} \\
&= \frac{1}{\limsup_{k\to\infty} \sqrt[k]{|c_k|}} = R,
\end{aligned}$$

which means that the termwise derivative $\sum_{k=1}^{\infty} kc_k(x-a)^{k-1}$ has the same radius of convergence as $\sum_{k=0}^{\infty} c_k(x-a)^k$. Let $r \in (0, R)$. Then for every $n \in \mathbb{N}$ and all $x \in (a-r, a+r)$ we have $\frac{d}{dx}\left(\sum_{k=0}^{n} c_k(x-a)^k\right) = \sum_{k=1}^{n} kc_k(x-a)^{k-1}$, and the derivative is continuous on $(a-r, a+r)$. Moreover, $\left\{\sum_{k=0}^{n} c_k(x-a)^k\right\}_{n=1}^{\infty}$ converges pointwise to $\sum_{k=0}^{\infty} c_k(x-a)^k$ on $(a-r, a+r)$ and the sequence of the derivatives $\left\{\sum_{k=1}^{n} kc_k(x-a)^{k-1}\right\}_{n=1}^{\infty}$ converges uniformly to $\sum_{k=1}^{\infty} kc_k(x-a)^{k-1}$ on $(a-r, a+r)$. Therefore by Theorem 11.11 the function $\sum_{k=0}^{\infty} c_k(x-a)^k$ is differentiable on the interval $(a-r, a+r)$ and the derivative is $\sum_{k=1}^{\infty} kc_k(x-a)^{k-1}$. Because $r \in (0, R)$ was arbitrary, $\sum_{k=0}^{\infty} c_k(x-a)^k$ is differentiable at every point $x \in (a-R, a+R)$ and the derivative at x is $\sum_{k=1}^{\infty} kc_k(x-a)^{k-1}$. ■

In particular, Corollary 11.12 shows that the derivative of a power series is again a power series. Therefore Corollary 11.12 can be applied to derivatives and higher derivatives of power series and we obtain the following.

Example 11.13 Any power series $f(x) = \sum_{k=0}^{\infty} c_k(x-a)^k$ with radius of convergence $R > 0$ is infinitely differentiable on the interval $(a-R, a+R)$. Moreover, the kth derivative at $x = a$ is $f^{(k)}(a) = k!c_k$. □

In light of Example 11.13, it is natural to ask if every infinitely differentiable function is a power series. Lemma 18.8 will show that this is not the case.

Exercises

11-12 Let $n \in \mathbb{N}$. Prove that $f_n(x) := \begin{cases} 0, & \text{for } 0 \le x \le 1 - \frac{1}{n}, \\ n\left[x - \left(1 - \frac{1}{n}\right)\right]; & \text{for } 1 - \frac{1}{n} \le x \le 1, \\ 1; & \text{for } 1 \le x \le 2, \end{cases}$

defines a continuous function on $[0, 2]$ and that $f(x) = \begin{cases} 0; & \text{for } 0 \le x < 1, \\ 1; & \text{for } 1 \le x \le 2, \end{cases}$ is not continuous on

$[0, 2]$. Then prove that $\{f_n\}_{n=1}^{\infty}$ converges pointwise to f

11-13. Let $\{q_j\}_{j=1}^{\infty}$ be an enumeration of all rational numbers in $[0, 1]$ and let $n \in \mathbb{N}$. Prove that each function $f_n(x) := \begin{cases} 0, & \text{for } x \in [0,1] \setminus \{q_1, \ldots, q_n\}, \\ 1; & \text{for } x \in \{q_1, \ldots, q_n\}, \end{cases}$ is Riemann integrable on $[0, 1]$. Then prove that the sequence $\{f_n\}_{n=1}^{\infty}$ converges pointwise to the Dirichlet function $f(x) = \begin{cases} 0; & \text{for } x \in [0,1] \setminus \mathbb{Q}, \\ 1; & \text{for } x \in [0,1] \cap \mathbb{Q}, \end{cases}$ which is not Riemann integrable on $[0, 1]$

11-14. Prove Theorem 11.11 as follows. Fix $x_0 \in (a, b)$. Prove that g is Riemann integrable from x_0 to any $x \in (a, b)$. Then use the Fundamental Theorem of Calculus to prove that for all $x \in (a, b)$ we have $\int_{x_0}^{x} g(t)\,dt = \lim_{n\to\infty} f_n(x) - f_n(x_0)$. From this finding, conclude that $f' = g$

11-15 Explain why in the proof of Theorem 11.11 the hypothesis that all f_n' are continuous can be replaced with the hypothesis that g is continuous

11-16 A sequence $\{f_n\}_{n=1}^{\infty}$ of functions $f_n : [a, b] \to \mathbb{R}$ is called **equicontinuous** at x iff for all $\varepsilon > 0$ there is a $\delta > 0$ so that for all $n \in \mathbb{N}$ and all $z \in [a, b]$ with $|z - x| < \delta$ we have $\left| f_n(z) - f_n(x) \right| < \varepsilon$.

(a) Prove that if $\{f_n\}_{n=1}^{\infty}$ is equicontinuous at x and converges pointwise to $f : [a, b] \to \mathbb{R}$, then f is continuous at x.

(b) Prove that if $\{f_n\}_{n=1}^{\infty}$ is a sequence of continuous functions that converges uniformly to the continuous function f, then $\{f_n\}_{n=1}^{\infty}$ is equicontinuous at every $x \in [a, b]$.

11-17. Give an example of a sequence of functions $\{f_n\}_{n=1}^{\infty}$ on $[0, 1]$ that converges pointwise on $[0, 1]$ to a Riemann integrable function $f : [0, 1] \to \mathbb{R}$ and $\int_0^1 f(x)\,dx \neq \lim_{n\to\infty} \int_0^1 f_n(x)\,dx$

11-18. Prove that if $\{f_n\}_{n=1}^{\infty}$ is a sequence of bounded functions on an interval I that converges uniformly to the function $f : I \to \mathbb{R}$, then f is bounded also

11-19 Let $\sum_{k=0}^{\infty} c_k(x-a)^k$ and $\sum_{k=0}^{\infty} d_k(x-a)^k$ be two power series with positive radius of convergence and let $\varepsilon > 0$ be so that $\sum_{k=0}^{\infty} c_k(x-a)^k = \sum_{k=0}^{\infty} d_k(x-a)^k$ for all $x \in \mathbb{R}$ with $|x - a| < \varepsilon$ Prove that then for all $k \in \mathbb{N}$ we have $c_k = d_k$. *Hint.* Derivatives at a

11-20 **Fubini's Differentiation Theorem** states that if $\{f_n\}_{n=1}^{\infty}$ is a sequence of nondecreasing functions on $[a, b]$ so that $f_n \le f_{n+1}$ for all $n \in \mathbb{N}$, $f_{n+1} - f_n$ is nondecreasing for all $n \in \mathbb{N}$ and so that $\{f_n\}_{n=1}^{\infty}$ converges pointwise to $f : [a, b] \to \mathbb{R}$, then $f'(x) = \lim_{n\to\infty} f_n'(x)$ for almost all $x \in [a, b]$. We will prove this result using the steps below.

(a) Prove that f is nondecreasing.

(b) Prove that there is a set $A \subseteq [a, b]$ so that $\lambda\left([a, b] \setminus A\right) = 0$ and all the derivatives involved exist at every $x \in A$. Conclude that we only need to prove the equality for almost all $x \in A$.
Hint Lebesgue's Differentiation Theorem (Exercise 10-7).

(c) Prove that for all $n \in \mathbb{N}$ and all $x \in A$ we have $f_n'(x) \le f_{n+1}'(x)$.
Hint. Consider that $f_{n+1} = f_n + (f_{n+1} - f_n)$

(d) Prove that for all $n \in \mathbb{N}$ the function $f - f_n$ is nondecreasing.

(e) Prove that for all $x \in A$ we have $f_n'(x) \le f'(x)$.

(f) Construct a subsequence $\{f_{n_k}\}_{k=1}^{\infty}$ so that $\sum_{k=1}^{\infty} \left[f(b) - f_{n_k}(b)\right]$ converges and prove that $\sum_{k=1}^{\infty} \left[f(x) - f_{n_k}(x)\right]$ converges for all $x \in [a, b]$

(g) Apply what we have proved so far to $g(x) := \sum_{k=1}^{\infty} \left[f(x) - f_{n_k}(x) \right]$ to prove that the series $\sum_{k=1}^{\infty} \left[f'(x) - f'_{n_k}(x) \right]$ converges at almost every $x \in A$ with limit less than or equal to $g'(x)$ Conclude that $\left\{ f'(x) - f'_{n_k}(x) \right\}_{k=1}^{\infty}$ converges to zero at almost every $x \in A$.

(h) Prove that $\lim_{n\to\infty} f'_n(x) = f'(x)$ at almost every $x \in A$, which establishes the result.

11-21. **A continuous nowhere differentiable function**. Let
$$t(x) := \begin{cases} x, & \text{for } 0 < x < \frac{1}{2}, \\ 1 - x, & \text{for } \frac{1}{2} \le x < 1, \\ 0; & \text{for } x \notin (0, 1), \end{cases}$$
and let $s(x) := t(x) + \sum_{j=1}^{\infty} \left(t(x-j) + t(x+j) \right)$. For all $n \in \mathbb{N}$ let $f_n(x) := \frac{1}{100^n} s\left(1000^n x\right)$ and define $f := \sum_{n=1}^{\infty} f_n$.

(a) Prove that f is continuous because it is the uniform limit of continuous functions.

(b) Prove that for every $x \in \mathbb{R}$ and $n \in \mathbb{N}$ there is a $z \in \mathbb{R}$ with $|z - x| = \frac{1}{4 \cdot 1000^n}$ so that $\left| \frac{f(z) - f(x)}{z - x} \right| > \frac{10^n}{2}$ and conclude that f is not differentiable at any $x \in \mathbb{R}$.

Hint. The nth sawtooth function f_n is made up of straight line segments with slopes $\pm 10^n$ on intervals of length $\frac{1}{2 \cdot 1000^n}$

(c) Prove that for every continuous $f : [a, b] \to \mathbb{R}$ that is differentiable on (a, b) there is a sequence $\{f_n\}_{n=1}^{\infty}$ of continuous functions that are not differentiable at *any* point in (a, b), which converges uniformly to f

11-22 By Exercise 4-17, if the derivative of the function $f : (a, b) \to \mathbb{R}$ is zero for all $x \in (a, b)$, then f is constant This exercise shows that if the derivative is only zero almost everywhere, then f need not be constant even if it is continuous The function ψ below is called **Lebesgue's singular function**. Let $Q = \left\{ \frac{1}{3} \right\}_{n=1}^{\infty}$, let C^Q be the ternary Cantor set and use the notation of Definition 7 22

(a) Let $n \in \mathbb{N}$ Prove that for all $i \in \left\{1, \ldots, 2^n\right\}$ we have $\left| I_{i,n}^Q \right| = \frac{1}{3^n}$.

(b) Let $n \in \mathbb{N}$ For every $x \in [0, 1] \setminus I_{1,n}^Q$ let $i_x \in \left\{1, \ldots, 2^n\right\}$ be the largest number in $\left\{1, \ldots, 2^n\right\}$ so that x is an upper bound of $I_{i,n}^Q$ and $x \notin I_{i,n}^Q$. For $x \in I_{1,n}^Q$ let $i_x := 0$. Define
$$\psi_n(x) := \begin{cases} \sum_{j=1}^{i_x} \frac{1}{2^n}, & \text{for } x \notin \bigcup_{i=1}^{2^n} I_{i,n}^Q, \\ \left(\sum_{j=1}^{i_x} \frac{1}{2^n} \right) + \frac{3^n}{2^n} \left(x - \min\left(I_{i,n}^Q \right) \right); & \text{for } x \in I_{i,n}^Q. \end{cases}$$
Prove that $\psi_n : [0, 1] \to [0, 1]$ is continuous, surjective and $\psi_n \left[\bigcup_{i=1}^{2^n} I_{i,n}^Q \right] = [0, 1]$.

(c) Prove that $\{\psi_n\}_{n=1}^{\infty}$ converges uniformly to a continuous, nondecreasing function ψ.

(d) Prove that $\psi\left[C^Q\right] = [0, 1]$ and that ψ is constant on every subinterval of $[0, 1] \setminus C^Q$, which means $\psi' = 0$ a e.

Chapter 12

Transcendental Functions

In this chapter, we use the machinery built so far to define the transcendental functions and to investigate their properties. It would have been possible to define the transcendental functions earlier, but the proofs would have been harder.

12.1 The Exponential Function

The question "Is there a nonzero function that is equal to its own derivative?" can be tackled with Corollary 11.12. If there is a power series with this property, then for all $k \in \mathbb{N}$ we infer $kc_k = c_{k-1}$. That is, $c_k = \frac{1}{k}c_{k-1} = \cdots = \frac{1}{k!}c_0$. The most natural nonzero value for c_0 is 1. Thus we are interested in the power series $\sum_{k=0}^{\infty} \frac{x^k}{k!}$.

Definition 12.1 *For* $x \in \mathbb{R}$, *we define* $\exp(x) := \sum_{k=0}^{\infty} \frac{x^k}{k!}$.

Theorem 12.2 *The series that defines* $\exp(\cdot)$ *has infinite radius of convergence. Therefore* $\exp(\cdot)$ *is differentiable on* $\mathbb{R}$ *with* $\frac{d}{dx}\exp(x) = \exp(x)$ *for all* $x \in \mathbb{R}$.

Proof. Because the last $\left\lfloor \frac{k}{2} \right\rfloor$ terms of $k!$ exceed $\left\lfloor \frac{k}{2} \right\rfloor$, we obtain for the radius of convergence

$$R = \lim_{k\to\infty} \sqrt[k]{k!} \geq \lim_{k\to\infty} \sqrt[k]{\left\lfloor \frac{k}{2} \right\rfloor^{\left\lfloor \frac{k}{2} \right\rfloor}} \geq \lim_{k\to\infty} \left\lfloor \frac{k}{2} \right\rfloor^{\left\lfloor \frac{k}{2} \right\rfloor \frac{1}{k}} \geq \lim_{k\to\infty} \left\lfloor \frac{k}{2} \right\rfloor^{\frac{1}{3}} = \infty,$$

which proves that the radius of convergence is infinite. The differentiability of $\exp(\cdot)$ on $\mathbb{R}$ follows from Corollary 11.12. To see that $\frac{d}{dx}\exp(x) = \exp(x)$ we also use Corollary 11.12 and obtain the following.

$$\frac{d}{dx}\exp(x)=\sum_{k=1}^{\infty}k\frac{x^{k-1}}{k!}=\sum_{k=1}^{\infty}\frac{x^{k-1}}{(k-1)!}=\sum_{n=0}^{\infty}\frac{x^n}{n!}=\exp(x).$$ ■

Aside from being equal to its own derivative, the function $\exp(\cdot)$ has interesting algebraic properties.

Theorem 12.3 *For all $x, y \in \mathbb{R}$, we have* $\exp(x)\exp(y)=\exp(x+y)$.

Proof. Let $x, y \in \mathbb{R}$. Then via Theorem 10.14 (or Corollary 10.15) and the Binomial Theorem we obtain

$$\begin{aligned}\exp(x)\exp(y) &= \sum_{k=0}^{\infty}\frac{x^k}{k!}\sum_{k=0}^{\infty}\frac{y^k}{k!}=\sum_{k=0}^{\infty}\sum_{j=0}^{k}\frac{x^j}{j!}\frac{y^{k-j}}{(k-j)!}\\ &= \sum_{k=0}^{\infty}\frac{1}{k!}\sum_{j=0}^{k}\frac{k!}{j!(k-j)!}x^jy^{k-j}=\sum_{k=0}^{\infty}\frac{1}{k!}(x+y)^k=\exp(x+y).\end{aligned}$$

■

Theorem 12.4 *For all rational numbers r, we have* $\exp(r)=\left(\exp(1)\right)^r$.

Proof. With an easy induction it can be proved that for all $n \in \mathbb{N}$ and $x \in \mathbb{R}$ we have $\exp(nx)=\left(\exp(x)\right)^n$ (Exercise 12-1). It is also easy to see that $\exp(0)=1$. Moreover, for all $x \in \mathbb{R}$ we have $\exp(x)\exp(-x)=\exp(x-x)=\exp(0)=1$, which means that $\exp(-x)=\dfrac{1}{\exp(x)}$. In particular, for all $n \in \mathbb{Z}$ we infer $\exp(nx)=\left(\exp(x)\right)^n$.

Hence, for all rational numbers $\frac{p}{q}$ with $p \in \mathbb{Z}$ and $q \in \mathbb{N}$ we obtain

$$\left[\exp\left(\frac{p}{q}\right)\right]^q=\exp\left(\frac{p}{q}q\right)=\exp(p),$$

which implies $\exp\left(\dfrac{p}{q}\right)=\left[\exp(p)\right]^{\frac{1}{q}}=\left[(\exp(1))^p\right]^{\frac{1}{q}}=\left(\exp(1)\right)^{\frac{p}{q}}$. ■

Theorem 12.4 shows that on $\mathbb{Q}$ the function $\exp(\cdot)$ is an exponential function. Exponentials with irrational exponents have remained undefined so far. Because $\exp(\cdot)$ is differentiable, it is continuous. The most sensible way to extend an exponential function defined on $\mathbb{Q}$ to all of $\mathbb{R}$ is so that the resulting function is continuous. Thus the following definition is reasonable.

Definition 12.5 **Euler's number** *is defined to be $e := \exp(1)$. The function $\exp(\cdot)$ is called the* **natural exponential function** *and for all $x \in \mathbb{R}$ we set $e^x := \exp(x)$.*

By Theorem 12.4 for $x \in \mathbb{Q}$, the equality $\exp(x)=e^x$ holds with the power on the right side being a power as in Definition 1.50. Next, we want to define powers with real exponents for all positive real bases. To do so, we first need to prove that e^x has an inverse function.

Proposition 12.6 *The natural exponential function is a strictly increasing, bijective function from $\mathbb{R}$ to $(0, \infty)$.*

Proof. Let $x < y$. By Definition 12.1 for any $z > 0$, we have $e^z > e^0 = 1$. But this means $e^y = e^{y-x+x} = e^{y-x}e^x > e^x$, so the natural exponential function is strictly increasing and, in particular, injective. Moreover, its values are greater than or equal to 1 on $[0, \infty)$.

For $z < 0$ note that $e^z e^{-z} = e^{z-z} = 1$, which means that $e^z = \dfrac{1}{e^{-z}}$ must be greater than 0. Therefore, e^x maps $\mathbb{R}$ into $(0, \infty)$.

Now note that for $x > 0$ we have $e^x > 1 + x$, and hence $\lim_{x\to\infty} e^x = \infty$. Consequently, $\lim_{x\to-\infty} e^x = \lim_{x\to-\infty} \dfrac{1}{e^{-x}} = 0$. By Theorem 12.2, the natural exponential function is differentiable, and hence continuous. The above limits and the Intermediate Value Theorem show that the natural exponential function is surjective onto $(0, \infty)$. *(Mentally fill in the details.)* ■

Definition 12.7 *We define the* **natural logarithm function** $\ln : (0, \infty) \to \mathbb{R}$ *to be the inverse function of* $\exp : \mathbb{R} \to (0, \infty)$.

The natural logarithm allows us to define powers of positive bases with arbitrary real exponents. With an argument similar to the proof of Theorem 12.4 we can show that this definition agrees with Definition 1.50 for rational exponents. (Exercise 12-2.)

Definition 12.8 *Let $a > 0$ be a positive real number and let $r \in \mathbb{R}$. Then we define the* r^{th} **power** *of a to be* $a^r := \exp\big(r \ln(a)\big) = e^{r \ln(a)}$.

Theorem 12.9 *For all positive numbers a and b, and all real numbers x and y the following hold.*

$$a^x a^y = a^{x+y}, \qquad (ab)^x = a^x b^x, \qquad \frac{a^x}{a^y} = a^{x-y} = \frac{1}{a^{y-x}},$$

$$\left(\frac{a}{b}\right)^x = \frac{a^x}{b^x}, \qquad \left(a^x\right)^y = a^{xy}, \qquad a^x > 0.$$

Proof. All properties follow directly from corresponding properties of the natural exponential function. The first property is proved as follows.

$$a^x a^y = e^{x\ln(a)} e^{y\ln(a)} = e^{x\ln(a)+y\ln(a)} = e^{(x+y)\ln(a)} = a^{x+y}.$$

Working rows left to right we obtain the following for the second property.

$$\begin{aligned}(ab)^x &= \left(e^{\ln(a)}e^{\ln(b)}\right)^x = \left(e^{\ln(a)+\ln(b)}\right)^x = e^{\ln\left(e^{\ln(a)+\ln(b)}\right)x} \\ &= e^{\left(\ln(a)+\ln(b)\right)x} = e^{\ln(a)x+\ln(b)x} = e^{\ln(a)x}e^{\ln(b)x} = a^x b^x.\end{aligned}$$

The third property follows from $a^{x-y}a^{y-x} = 1$ and $a^{x-y}a^y = a^x$. The remaining properties are left as Exercise 12-3. ■

Of course, the properties of the natural logarithm function are also of interest.

Theorem 12.10 *The natural logarithm function is differentiable on* $(0, \infty)$ *with derivative* $\frac{d}{dx}\ln(x) = \frac{1}{x}$. *Consequently, all power functions* $f(x) := x^r$ *with* $r \in \mathbb{R}\setminus\mathbb{Q}$ *are differentiable on* $(0, \infty)$ *and the* **Power Rule** $\frac{d}{dx}x^r = rx^{r-1}$ *holds.*

Proof. By Theorem 4.21, the natural logarithm function is differentiable at every $x \in (0, \infty)$ and $\frac{d}{dx}\ln(x) = \frac{1}{e^{\ln(x)}} = \frac{1}{x}$. For the derivative of powers, the Chain Rule implies $\frac{d}{dx}x^r = \frac{d}{dx}e^{r\ln(x)} = e^{r\ln(x)}r\frac{1}{x} = rx^{r-1}$. ■

Because the natural logarithm function is differentiable, it is continuous, which allows us to extend the limit law for powers to arbitrary exponents.

Corollary 12.11 *Let* $\{a_n\}_{n=1}^{\infty}$ *be a convergent sequence of nonnegative numbers and let* $r \in \mathbb{R}$. *Then* $\lim_{n\to\infty} a_n^r = \left(\lim_{n\to\infty} a_n\right)^r$ *unless* $r < 0$ *and* $\lim_{n\to\infty} a_n = 0$.

Proof. Exercise 12-4b. ■

Further properties of the natural logarithm function are exhibited in Exercises 12-5 and 12-6.

Exercises

12-1. Prove by induction that for all $n \in \mathbb{N}$ we have $\exp(n) = \left(\exp(1)\right)^n$ as claimed in the proof of Theorem 12 4.

12-2. Prove that for all $x > 0$ and all $r \in \mathbb{Q}$ the definitions of the power x^r in Definition 1.50 and Definition 12 8 agree.

12-3 Finish the proof of Theorem 12 9 That is, prove the following for all $a, b > 0$ and $x, y \in \mathbb{R}$

(a) $\left(\frac{a}{b}\right)^x = \frac{a^x}{b^x}$ (b) $\left(a^x\right)^y = a^{xy}$ (c) $a^x > 0$

12-4. The limit law for powers

(a) Prove that $\lim_{x\to-\infty} e^x = 0$.

(b) Prove Corollary 12 11 Be careful with sequences that go to zero

12-5. Prove that the natural logarithm function is a strictly increasing bijective function from $(0, \infty)$ to $\mathbb{R}$ with $\lim_{x\to\infty}\ln(x) = \infty$ and $\lim_{x\to 0^+}\ln(x) = -\infty$

12-6. Let $u, v > 0$. Prove each of the following

(a) $\ln(1) = 0$ (b) $\ln(e) = 1$ (c) $\ln(uv) = \ln(u) + \ln(v)$

(d) $\ln\left(\frac{u}{v}\right) = \ln(u) - \ln(v)$ (e) $\ln\left(u^v\right) = v\ln(u)$

12-7 Limits of nth roots.

(a) Prove that if $\lim_{n\to\infty} a_n = a > 0$, then $\lim_{n\to\infty}\sqrt[n]{a_1\cdots a_n} = a$

Hint. Exercises 2-16 and 12-6.

(b) Prove that $\lim_{n\to\infty} \sqrt[n]{n} = 1$.

12-8. Compute the integral $\int_0^1 x^3 e^{x^2}\,dx$.

12-9. **Gronwall's Inequality**. Let $u, v : [a, b] \to [0, \infty)$ be continuous functions and let $c \geq 0$ be so that for all $x \in [a, b]$ we have $u(x) \leq c + \int_a^x u(t)v(t)\,dt$. Prove that for all $x \in [a, b]$ we have $u(x) \leq ce^{\int_a^x v(t)\,dt}$.

Hint. Divide by the right side, multiply by $v(x)$ and integrate.

12-10 The function $\Gamma(\alpha) := \int_0^\infty x^{\alpha-1}e^{-x}\,dx$ defined for $\alpha > 0$ is called the **Gamma function**.

(a) Prove that the improper integral $\int_0^\infty x^{\alpha-1}e^{-x}\,dx$ converges for $\alpha \geq 1$.

(b) Prove that the improper integral also converges for $0 < \alpha < 1$.

(c) Prove that the improper integral diverges for $\alpha \leq 0$.

(d) Prove that $\Gamma(1) = 1$.

(e) Prove that for all $\alpha \geq 1$ we have $\Gamma(\alpha + 1) = \alpha\Gamma(\alpha)$.

(f) Prove by induction that for every natural number n we have $\Gamma(n) = (n - 1)!$.

For this reason, the Gamma function is also referred to as the **generalized factorial function**.

12-11. Compute the following parameter dependent indefinite integrals. These integrals are useful for the integration of rational functions after a **partial fraction decomposition**

(a) $\int \frac{1}{x-c}\,dx$

(b) $\int \frac{1}{(x-c)^n}\,dx \qquad (n > 1)$

(c) $\int \frac{x}{x^2+b^2}\,dx$

(d) $\int \frac{x}{(x^2+b^2)^n}\,dx \qquad (n > 1)$

12.2 Sine and Cosine

Similar to the natural exponential function, the sine and cosine functions are defined via power series that have the right derivatives at the origin. Of course, this is a bit of "reverse engineering," because to obtain these derivatives we would need to quote arguments that rely on the geometric definition of these functions.

Definition 12.12 *For $x \in \mathbb{R}$, we define* $\sin(x) := \sum_{k=0}^{\infty} \frac{(-1)^k x^{2k+1}}{(2k+1)!}$, *which is called the* **sine function**, *and* $\cos(x) := \sum_{k=0}^{\infty} \frac{(-1)^k x^{2k}}{(2k)!}$, *which is called the* **cosine function**.

Theorem 12.13 *The power series that define* $\sin(\cdot)$ *and* $\cos(\cdot)$ *have infinite radius of convergence. Therefore, both* $\sin(\cdot)$ *and* $\cos(\cdot)$ *are differentiable and moreover* $\frac{d}{dx}\sin(x) = \cos(x)$ *and* $\frac{d}{dx}\cos(x) = -\sin(x)$ *for all* $x \in \mathbb{R}$.

Proof. We prove the result for $\sin(\cdot)$, leaving $\cos(\cdot)$ to the reader in Exercise 12-12. For the sine function, note that for every $x \in \mathbb{R}$ we have

$$\lim_{k\to\infty}\left|\frac{\frac{(-1)^{k+1}x^{2(k+1)+1}}{(2(k+1)+1)!}}{\frac{(-1)^k x^{2k+1}}{(2k+1)!}}\right| = \lim_{k\to\infty}\left|\frac{x^{2k+3}}{(2k+3)!}\frac{(2k+1)!}{x^{2k+1}}\right| = \lim_{k\to\infty}\frac{x^2}{(2k+2)(2k+3)}$$
$$= 0.$$

Hence, by the Ratio Test, the power series converges for all $x \in \mathbb{R}$, so its radius of convergence is infinite. By Corollary 11.12, the sine function is differentiable on $\mathbb{R}$ and

$$\begin{aligned}\frac{d}{dx}\sin(x) &= \frac{d}{dx}\sum_{k=0}^{\infty}\frac{(-1)^k x^{2k+1}}{(2k+1)!} = \sum_{k=0}^{\infty}\frac{(-1)^k(2k+1)x^{2k}}{(2k+1)!}\\ &= \sum_{k=0}^{\infty}\frac{(-1)^k x^{2k}}{(2k)!} = \cos(x).\end{aligned}$$

■

The following identities are useful when working with sine and cosine.

Theorem 12.14 *For all $x, y \in \mathbb{R}$ the following identities hold.*

1. $\sin^2(x) + \cos^2(x) = 1$ *(trigonometric law of* **Pythagoras***)*
2. $\sin(x + y) = \sin(x)\cos(y) + \cos(x)\sin(y)$
3. $\cos(x + y) = \cos(x)\cos(y) - \sin(x)\sin(y)$

Proof. To prove the first identity, we proceed as follows.

$$\begin{aligned}&\sin^2(x) + \cos^2(x)\\ &= \left(\sum_{k=0}^{\infty}\frac{(-1)^k x^{2k+1}}{(2k+1)!}\right)^2 + \left(\sum_{k=0}^{\infty}\frac{(-1)^k x^{2k}}{(2k)!}\right)^2\\ &= \sum_{n=1}^{\infty}\left(\sum_{2j+1+2k+1=2n}\frac{(-1)^j}{(2j+1)!}\frac{(-1)^k}{(2k+1)!}\right)x^{2n}\\ &\quad+\sum_{n=0}^{\infty}\left(\sum_{2j+2k=2n}\frac{(-1)^j}{(2j)!}\frac{(-1)^k}{(2k)!}\right)x^{2n}\\ &= 1+\sum_{n=1}^{\infty}\left[\sum_{j+k=n-1}\frac{(-1)^j}{(2j+1)!}\frac{(-1)^k}{(2k+1)!}+\sum_{j+k=n}\frac{(-1)^j}{(2j)!}\frac{(-1)^k}{(2k)!}\right]x^{2n}\\ &= 1+\sum_{n=1}^{\infty}(-1)^n\left[\sum_{k=0}^{n-1}\frac{-1}{(2n-2k-1)!(2k+1)!}+\sum_{k=0}^{n}\frac{1}{(2n-2k)!(2k)!}\right]x^{2n}\end{aligned}$$

$$= 1+\sum_{n=1}^{\infty}\frac{(-1)^n}{(2n)!}\left[\sum_{k=0}^{n-1}\frac{(2n!)}{(2n-(2k+1))!(2k+1)!}(-1)^{2k+1}1^{2n-(2k+1)}\right.$$
$$\left.+\sum_{k=0}^{n}\frac{(2n)!}{(2n-2k)!(2k)!}(-1)^{2k}1^{2n-2k}\right]x^{2n}$$

Note that the two sums in square brackets above are over the odd and the even terms, respectively, of the sum in square brackets below.

$$= 1+\sum_{n=1}^{\infty}\frac{(-1)^n}{(2n)!}\left[\sum_{j=0}^{2n}\frac{(2n!)}{(2n-j)!j!}(-1)^j1^{2n-j}\right]x^{2n}$$
$$= 1+\sum_{n=1}^{\infty}\frac{(-1)^n}{(2n)!}\left[(-1+1)^{2n}\right]x^{2n}=1.$$

The remaining two identities are left for Exercise 12-13. ■

The smallest positive zero of the sine function also has a special place in mathematics. Of course, we first must show that the sine function *has* a positive zero.

Proposition 12.15 *The function* $\sin(x)$ *is positive on* $\left(0,\sqrt{6}\right)$ *and it is negative at* 4.

Proof. For all $x\in\left(0,\sqrt{6}\right)$, we have

$$\sin(x)=\sum_{k=0}^{\infty}(-1)^k\frac{x^{2k+1}}{(2k+1)!}=\sum_{k=0}^{\infty}\frac{x^{4k+1}}{(4k+1)!}\left(1-\frac{x^2}{(4k+2)(4k+3)}\right)>0.$$

On the other hand, for $x=4$ we obtain

$$\begin{aligned}\sin(4) &= \sum_{k=0}^{\infty}(-1)^k\frac{4^{2k+1}}{(2k+1)!}\\ &= \sum_{k=0}^{4}(-1)^k\frac{4^{2k+1}}{(2k+1)!}-\sum_{k=2}^{\infty}\frac{4^{4k+3}}{(4k+3)!}\left(1-\frac{4^2}{(4k+4)(4k+5)}\right)<0,\end{aligned}$$

where the first term is negative because

$$\begin{aligned}&\sum_{k=0}^{4}(-1)^k\frac{4^{2k+1}}{(2k+1)!}\\ &= 4-\frac{4^3}{1\cdot2\cdot3}+\frac{4^5}{1\cdot2\cdot3\cdot4\cdot5}-\frac{4^7}{1\cdot2\cdot3\cdot4\cdot5\cdot6\cdot7}+\frac{4^9}{1\cdot2\cdot3\cdot4\cdot5\cdot6\cdot7\cdot8\cdot9}\\ &= 4-\frac{5\cdot4^3}{1\cdot2\cdot3\cdot5}+\frac{4\cdot4^3}{1\cdot2\cdot3\cdot5}-\frac{9\cdot4^6}{1\cdot2\cdot3\cdot5\cdot6\cdot7\cdot9}+\frac{2\cdot4^6}{1\cdot2\cdot3\cdot5\cdot6\cdot7\cdot9}\\ &= 4-\frac{6\cdot9\cdot4^3}{1\cdot2\cdot3\cdot5\cdot6\cdot9}-\frac{7\cdot4^3\cdot4^3}{1\cdot2\cdot3\cdot5\cdot6\cdot7\cdot9}\end{aligned}$$

$$= \quad 4 - \frac{118 \cdot 4^3}{1 \cdot 2 \cdot 3 \cdot 5 \cdot 6 \cdot 9} = 4 - 4 \cdot \frac{472}{405} < 0.$$

■

Definition 12.16 *The smallest positive zero of the function* $\sin(x)$ *is called* π.

We conclude by proving that the sine and cosine functions are periodic.

Definition 12.17 *Let* $p > 0$ *be a real number. A function* $f : \mathbb{R} \to \mathbb{R}$ *is called* **periodic** *with period p iff for all* $x \in \mathbb{R}$ *we have* $f(x + p) = f(x)$.

Theorem 12.18 *Both the sine and the cosine function have period* 2π.

Proof. Note that $\sin(2\pi) = \sin(\pi + \pi) = \sin(\pi)\cos(\pi) + \cos(\pi)\sin(\pi) = 0$. This implies $\cos^2(2\pi) = 1 - \sin^2(2\pi) = 1$ and because

$$\cos(2\pi) = \cos(\pi + \pi) = \cos(\pi)\cos(\pi) - \sin(\pi)\sin(\pi) = \cos^2(\pi) > 0,$$

we infer that $\cos(2\pi) = 1$. Hence, for all $x \in \mathbb{R}$ we obtain

$$\begin{aligned} \sin(x + 2\pi) &= \sin(x)\cos(2\pi) + \cos(x)\sin(2\pi) = \sin(x) \\ \cos(x + 2\pi) &= \cos(x)\cos(2\pi) - \sin(x)\sin(2\pi) = \cos(x). \end{aligned}$$

■

Exercises

12-12. Prove that the power series that defines the cosine function has infinite radius of convergence and that $\frac{d}{dx}\cos(x) = -\sin(x)$.

12-13 Finish the proof of Theorem 12.14

(a) Use power series to prove $\sin(x + y) = \sin(x)\cos(y) + \cos(x)\sin(y)$ for all $x, y \in \mathbb{R}$

(b) Use the above and the law of Pythagoras to prove $\cos\left(\frac{\pi}{2}\right) = 0$ and $\sin\left(\frac{\pi}{2}\right) = 1$.

(c) Use the power series to prove $\sin(-x) = -\sin(x)$ and $\cos(-x) = \cos(x)$ for all $x \in \mathbb{R}$.

(d) Prove that $\sin\left(\frac{\pi}{2} - x\right) = \cos(x)$ for all $x \in \mathbb{R}$. (Use parts 12-13a, 12-13b and 12-13c)

(e) Prove that $\cos\left(\frac{\pi}{2} - x\right) = \sin(x)$ for all $x \in \mathbb{R}$.

(f) Prove that $\cos(x + y) = \cos(x)\cos(y) - \sin(x)\sin(y)$ for all $x, y \in \mathbb{R}$

12-14 Prove that $\pi > 3$

12-15 A function $f : [0, 2\pi] \to \mathbb{R}$ of the form $f(x) = a_0 + \sum_{j=1}^{n} \left(a_j \cos(jx) + b_j \sin(jx) \right)$ is called a **trigonometric polynomial**

(a) Prove the following **product-to-sum formulas**.

i. $\cos(x)\cos(y) = \frac{1}{2}\left[\cos(x + y) + \cos(x - y) \right]$

ii. $\sin(x)\sin(y) = \frac{1}{2}\left[\cos(x - y) - \cos(x + y) \right]$

iii. $\sin(x)\cos(y) = \frac{1}{2}\left[\sin(x+y) + \sin(x-y)\right]$

(b) Prove that the product of two trigonometric polynomials is a trigonometric polynomial.

12-16. **Inverse trigonometric functions**

(a) Prove that the sine function is injective on $\left[-\frac{\pi}{2}, \frac{\pi}{2}\right]$.

(b) The inverse of the sine function restricted to $\left[-\frac{\pi}{2}, \frac{\pi}{2}\right]$ is called the **arcsine** $\arcsin(\cdot)$ Prove that $\frac{d}{dx}\arcsin(x) = \frac{1}{\sqrt{1-x^2}}$ for all $x \in \left(-\frac{\pi}{2}, \frac{\pi}{2}\right)$.

(c) Prove that the cosine function is injective on $[0, \pi]$.

(d) The inverse of the cosine function restricted to $[0, \pi]$ is called the **arccosine** $\arccos(\ )$. Prove that $\frac{d}{dx}\arccos(x) = -\frac{1}{\sqrt{1-x^2}}$ for all $x \in (0, \pi)$.

12-17 The **tangent function** is defined to be $\tan(x) := \frac{\sin(x)}{\cos(x)}$.

(a) Prove that the tangent function is differentiable on its domain and $\frac{d}{dx}\tan(x) = \frac{1}{\cos^2(x)}$.

(b) Prove that the tangent function is injective on $\left(-\frac{\pi}{2}, \frac{\pi}{2}\right)$.

(c) The inverse of the tangent function restricted to $\left(-\frac{\pi}{2}, \frac{\pi}{2}\right)$ is called the **arctangent** $\arctan(\ )$. Prove that $\frac{d}{dx}\arctan(x) = \frac{1}{1+x^2}$

(d) (Another integral for integration with **partial fraction decompositions**.) Compute the integral $\int \frac{1}{x^2+b^2}\,dx$.

12-18 (The last integral needed for integration with **partial fraction decompositions**) Proceed as follows to prove that for all natural numbers $n > 1$ we have

$$\int \frac{1}{\left(x^2+b^2\right)^n}\,dx = \frac{x}{2(n-1)b^2\left(x^2+b^2\right)^{n-1}} + \frac{2n-3}{2(n-1)b^2}\int \frac{1}{\left(x^2+b^2\right)^{n-1}}\,dx.$$

(a) Use integration by parts on $\int \frac{1}{\left(x^2+b^2\right)^n}\,dx = \int 1 \cdot \frac{1}{\left(x^2+b^2\right)^n}\,dx$.

(b) The resulting equation contains an integral $\int \frac{x^2}{\left(x^2+b^2\right)^{n+1}}\,dx$ Expand the numerator with $\pm b^2$ and cancel what can be canceled

(c) Solve the equation for $\int \frac{1}{\left(x^2+b^2\right)^{n+1}}\,dx$.

12-19. Compute each of the integrals below.

(a) $\int e^{-x}\sin(x)\,dx$.

(b) $\int xe^{x^2}\sin\left(x^2\right)\,dx$.

12-20. A representation of π.

(a) Prove that $\int_a^b \sin^n(x)\,dx = -\frac{1}{n}\sin^{n-1}(x)\cos(x)\Big|_a^b + \frac{n-1}{n}\int_a^b \sin^{n-2}(x)\,dx$ for all natural numbers $n \in \mathbb{N}$ and all real numbers $a < b$.

Hint. Use $\sin^n(x) = \sin^{n-2}(x)\left(1-\cos^2(x)\right)$ and use integration by parts for the summand $\left(\sin^{n-2}(x)\cos(x)\right)\cos(x)$.

(b) Prove by induction that $\int_0^{\frac{\pi}{2}} \sin^{2n}(x)\,dx = \frac{2n-1}{2n} \cdot \frac{2n-3}{2n-2} \cdot \frac{1}{2}\frac{\pi}{2}$ for all $n \in \mathbb{N}$

(c) Prove that $\int_0^{\frac{\pi}{2}} \sin^{2n+1}(x)\,dx = \frac{2n}{2n+1} \cdot \frac{2n-2}{2n-1} \cdot \frac{2}{3}$ for all $n \in \mathbb{N}$

(d) Prove that $\frac{\pi}{2} = \lim_{n\to\infty} \frac{1}{2n} \prod_{k=1}^{n} \frac{(2k)^2}{(2k-1)^2}$. This is called **Wallis' Product Formula**

Hint Prove $\int_0^{\frac{\pi}{2}} \sin^{2n-1}(x)\,dx \geq \int_0^{\frac{\pi}{2}} \sin^{2n}(x)\,dx \geq \int_0^{\frac{\pi}{2}} \sin^{2n+1}(x)\,dx$ for all $n \in \mathbb{N}$, substitute the above expressions and divide by the expression in front of the $\frac{\pi}{2}$

12-21 Prove that if $f : [0, n] \to \mathbb{R}$ is Riemann integrable and has a Riemann integrable derivative, then $\sum_{k=0}^{n} f(k) = \int_0^n f(x)\,dx + \frac{f(0)+f(n)}{2} + \int_0^n \left(x - \frac{1}{2} - \lfloor x \rfloor\right) f'(x)\,dx$ This formula is called **Euler's Summation Formula**

Hint. Note that $\int_0^1 \left(x - \frac{1}{2}\right) f'(x) = \frac{1}{2}\left[f(0) + f(1)\right] - \int_0^1 f(x)\,dx$ and that the last integral on the right side of Euler's Summation Formula can be turned into a sum of integrals like these by integrating from one integer to the next and applying the appropriate shift.

12-22. An **asymptotic expression for $n!$** To see the idea for the first step, note that $\ln(n!) = \sum_{k=1}^{n} \ln(k)$.

(a) Apply Euler's Summation Formula to $f(x) = \ln(x+1)$ to prove that for all $n \in \mathbb{N}$ we have $\sum_{k=1}^{n} \ln(k) = n\ln(n) - n + \frac{1}{2}\ln(n) + \int_0^{n-1} \left(x - \frac{1}{2} - \lfloor x \rfloor\right) \frac{1}{x+1}\,dx$

(b) Prove that $\left\{\int_0^n \left(x - \frac{1}{2} - \lfloor x \rfloor\right) \frac{1}{x+1}\,dx\right\}_{n=1}^{\infty}$ converges.

Hint $\int_0^1 \left(x - \frac{1}{2}\right) \frac{1}{x+k}\,dx = \frac{1}{2}\left(x - \frac{1}{2}\right)^2 \frac{1}{x+k}\Big|_0^1 + \frac{1}{2}\int_0^1 \left(x - \frac{1}{2}\right)^2 \frac{1}{(x+k)^2}\,dx.$

(c) Let $b_n = \frac{n!e^n}{n^n\sqrt{n}}$ Prove that $\{b_n\}_{n=1}^{\infty}$ converges to a limit $b \in \mathbb{R}$

(d) Use Wallis' Product Formula to show that $\frac{1}{b} = \lim_{n\to\infty} \frac{b_{2n}}{b_n^2} = \frac{1}{\sqrt{2\pi}}$

(e) Prove that $\lim_{n\to\infty} \frac{n!e^n}{n^n\sqrt{2\pi n}} = 1$

This result is called **Stirling's Formula** It is often written as $n! \sim \frac{n^n\sqrt{2\pi n}}{e^n}$ where "$\sim$" is read "is asymptotically equal to"

12-23 Let $f(x) = \begin{cases} \sin\left(\frac{1}{x}\right); & \text{for } x \neq 0, \\ 0; & \text{for } x = 0. \end{cases}$ Prove that $\omega_f(0) = 2$.

12-24. A **differentiable function with bounded, but not Riemann integrable, derivative** (see Figure 25).

(a) Prove that for all $\delta > 0$ there is an $x_\delta \in (0, \delta)$ so that $2x\cos\left(\frac{1}{x}\right) + \sin\left(\frac{1}{x}\right) = 0$

(b) Prove that $f_\delta(x) := \begin{cases} 0, & \text{for } x \leq 0, \\ x^2\cos\left(\frac{1}{x}\right), & \text{for } x \in (0, x_\delta), \\ x_\delta^2\cos\left(\frac{1}{x_\delta}\right), & \text{for } x \geq x_\delta, \end{cases}$ where x_δ is a positive number as in part 12-24a, is differentiable on $\mathbb{R}$, but f_δ' is not continuous at $x = 0$

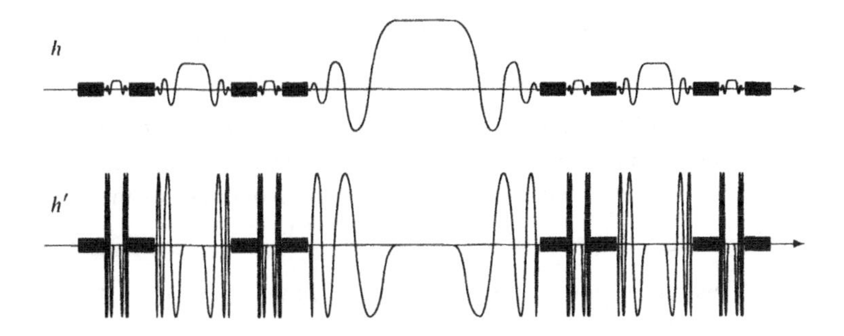

Figure 25: The differentiable function h in Exercise 12-24 oscillates so that the derivative h' is discontinuous on a Cantor set (marked by blocks).

(c) Prove that $g_\delta(x) := \begin{cases} f_{\frac{\delta}{2}}(x), & \text{for } x \leq \frac{\delta}{2}, \\ f_{\frac{\delta}{2}}(\delta - x); & \text{for } x > \frac{\delta}{2}, \end{cases}$ is differentiable on $\mathbb{R}$, g'_δ is discontinuous at 0 and at δ and $\{ x : g_\delta(x) \neq 0 \} \subseteq (0, \delta)$.

(d) For any open interval (a, b), let $g_{(a,b)}(x) := g_{|b-a|}(x - a)$. Prove that $g_{(a,b)}$ is differentiable, $g'_{(a,b)}$ is discontinuous at a and at b and $\{x \in \mathbb{R} : g_{(a,b)}(x) \neq 0\} \subseteq (a, b)$.

(e) Let $C^Q = \bigcap_{n=1}^{\infty} C_n^Q$ be a **Cantor set** of nonzero Lebesgue measure, which exists by Exercise 8-3e. For each $n \in \mathbb{N}$ let D_n^i, $i = 1, \ldots, 2^n - 1$ be a sequence of pairwise disjoint open intervals so that $[0, 1] \setminus C_n^Q = \bigcup_{i=1}^{2^n-1} D_n^i$. Let $h_n := \sum_{i=1}^{2^n-1} g_{D_n^i}$. Prove that $h := \lim_{n\to\infty} h_n$ (taken pointwise) is differentiable on $\mathbb{R}$ and h' is discontinuous on C^Q.

(f) Prove that h' is bounded, but not Riemann integrable on $[0, 1]$.

(g) Prove that h' is Lebesgue measurable, and hence Lebesgue integrable on $[0, 1]$.

Hint. Represent h' as a sum of measurable functions and obtain $(h')^{-1}\left[(a, \infty)\right]$ as a union or intersection of preimages of (a, ∞) under these functions

12.3 L'Hôpital's Rule

Limits of functions involving transcendental functions can be hard to compute. The algebra either involves power series, or it might even be so complicated that it is virtually impossible. L'Hôpital's Rule is a way to replace the limit of a quotient with the limit of the quotient of the derivatives, which may be easier to compute. The idea behind L'Hôpital's Rule is easily explained with power series. If $f(x) = \sum_{k=0}^{\infty} f_k(x-a)^k$ and $g(x) = \sum_{k=0}^{\infty} g_k(x-a)^k$ are power series with $f(a) = g(a) = 0$ and if $g'(a) \neq 0$, then

$$\lim_{x\to a} \frac{f(x)}{g(x)} = \lim_{x\to a} \frac{\sum_{k=1}^{\infty} f_k(x-a)^k}{\sum_{k=1}^{\infty} g_k(x-a)^k} = \lim_{x\to a} \frac{(x-a)\sum_{k=1}^{\infty} f_k(x-a)^{k-1}}{(x-a)\sum_{k=1}^{\infty} g_k(x-a)^{k-1}}$$

$$= \lim_{x\to a} \frac{\sum_{k=1}^{\infty} f_k(x-a)^{k-1}}{\sum_{k=1}^{\infty} g_k(x-a)^{k-1}} = \frac{f_1}{g_1} = \frac{f'(a)}{g'(a)}.$$

That is, the limit of the quotient of the functions is the quotient of the derivatives. Because not every function is a power series and because derivatives cannot be defined at infinity, in general we expect the *limit* of the quotient of the derivatives to be on the right side. To prove $\lim_{x\to a} \frac{f(x)}{g(x)} = \lim_{x\to a} \frac{f'(x)}{g'(x)}$, it is tempting to apply the Mean Value Theorem to the quotient, using that $f(a) = g(a) = 0$. But arguing $\lim_{x\to a} \frac{f(x)}{g(x)} = \lim_{x\to a} \frac{\frac{f(x)-f(a)}{x-a}}{\frac{g(x)-g(a)}{x-a}} = \lim_{x\to a} \frac{f'(c_f)}{g'(c_g)}$ is problematic. If c_f and c_g approach a at different rates, then the limit need not be $\lim_{x\to a} \frac{f'(x)}{g'(x)}$. To get the right limit, we need $c_f = c_g$, which can be achieved with a stronger form of the Mean Value Theorem.

Theorem 12.19 Generalized Mean Value Theorem. *Let f, g be functions that are continuous on $[a, b]$ and differentiable on (a, b) so that $g(a) \neq g(b)$ and $g'(x) \neq 0$ for all $x \in (a, b)$ Then there is a number $c \in (a, b)$ such that* $\frac{f(b)-f(a)}{g(b)-g(a)} = \frac{f'(c)}{g'(c)}$.

Proof. Let $h(x) := [f(b) - f(a)]g(x) - [g(b) - g(a)]f(x)$ and apply Rolle's Theorem. (See Exercise 12-25.) ■

Now we are ready to prove L'Hôpital's Rule.

Theorem 12.20 L'Hôpital's Rule *Let $a \in [-\infty, \infty]$ and let f, g be differentiable functions defined on an interval (z, ∞) (if $a = \infty$), or $(-\infty, z)$ (if $a = -\infty$), or $(a-\delta, a+\delta)\setminus\{a\}$ (if $a \in \mathbb{R}$). If the limits of f and g satisfy $\lim_{x\to a} f(x) = \lim_{x\to a} g(x) = 0$ or both limits are in $\{\pm\infty\}$ and $\lim_{x\to a} \frac{f'(x)}{g'(x)}$ exists as a number or is infinite, then* $\lim_{x\to a} \frac{f(x)}{g(x)} = \lim_{x\to a} \frac{f'(x)}{g'(x)}$. *This rule also applies to one-sided limits.*

Proof. Let $L := \lim_{x\to a} \frac{f'(x)}{g'(x)}$. First consider $a \in (-\infty, \infty]$ and $L \neq -\infty$. We claim that for all $y_0 < L$ there is an $x_0 < a$ so that for all $x \in (x_0, a)$ we have $\frac{f(x)}{g(x)} > y_0$.

Let $y_0 < L$. Then there is an $x_1 < a$ such that for all $x \in (x_1, a)$ we have $\frac{f'(x)}{g'(x)} > y_0$.

In case $\lim_{x\to a} f(x) = \lim_{x\to a} g(x) = 0$ and $a \in \mathbb{R}$, set $f(a) := g(a) := 0$. Then by the Generalized Mean Value Theorem for all $x \in (x_1, a)$ there is a $c \in (x, a)$ so that $\frac{f(x)}{g(x)} = \frac{f(x)-f(a)}{g(x)-g(a)} = \frac{f'(c)}{g'(c)} > y_0$. Thus the claim holds with $x_0 := x_1$. For $a = \infty$ the argument is similar to the next case (in fact, it is a little simpler).

In case $\lim_{x\to a} f(x) = \lim_{x\to a} g(x) = \infty$, we can assume without loss of generality that f and g are positive on $[x_1, a)$. Let $\varepsilon > 0$ be such that $\frac{y_0}{(1-\varepsilon)^2} < L$. Find an

$x_2 \in (x_1, a)$ so that for all $c \in (x_2, a)$ we have that $\frac{f'(c)}{g'(c)} > \frac{y_0}{(1-\varepsilon)^2}$. Then find an $x_0 \in (x_2, a)$ so that for all $x \in (x_0, a)$ we have that $\frac{f(x)}{f(x) - f(x_2)} > 1 - \varepsilon$ and $\frac{g(x) - g(x_2)}{g(x)} > 1 - \varepsilon$. Then for all $x \in (x_0, a)$ by the Generalized Mean Value Theorem there is a $c \in (x_2, x)$ so that $\frac{f(x) - f(x_2)}{g(x) - g(x_2)} = \frac{f'(c)}{g'(c)} > \frac{y_0}{(1-\varepsilon)^2}$, and hence for all $x \in (x_0, a)$ we obtain

$$\frac{f(x)}{g(x)} = \frac{f(x) - f(x_2)}{g(x) - g(x_2)} \frac{f(x)}{f(x) - f(x_2)} \frac{g(x) - g(x_2)}{g(x)} > \frac{f'(c)}{g'(c)}(1-\varepsilon)^2 > y_0,$$

which proves the claim in this case.

The other cases for the limits of f and g being infinite are proved similarly, so the claim is proved. The claim proves the result for $L = \infty$ for left limits at $a \in \mathbb{R}$ and for limits at infinity.

Similar to the above we can prove that for $L \neq \infty$ for all $y_0 > L$ there is an $x_0 < a$ so that for all $x \in (x_0, a)$ we have $\frac{f(x)}{g(x)} < y_0$. This proves the result for $L = -\infty$ for left limits at $a \in \mathbb{R}$ and for limits at infinity.

Putting the two results above together, for $L \in \mathbb{R}$ and every $\varepsilon > 0$ there is an $x_0 < a$ so that for all $x \in (x_0, a)$ we have $L - \varepsilon < \frac{f(x)}{g(x)} < L + \varepsilon$, which proves the result for $L \in \mathbb{R}$ for left limits at $a \in \mathbb{R}$ and for limits at ∞.

Repeating the above process for $a \in [-\infty, \infty)$ to the right of a establishes the result for $a = -\infty$, for right limits at $a \in \mathbb{R}$ and for two-sided limits at $a \in \mathbb{R}$. ■

Exercise 12-28 shows that L'Hôpital's Rule is not a one-for-one swap. The limit $\lim_{x \to a} \frac{f(x)}{g(x)}$ can exist even if $\lim_{x \to a} \frac{f'(x)}{g'(x)}$ fails to exist.

Aside from the obvious applications to quotients, L'Hôpital's Rule also allows us to derive a well-known representation for the **exponential function**.

Theorem 12.21 *For all $x \in \mathbb{R}$ we have $e^x = \lim_{n \to \infty} \left(1 + \frac{x}{n}\right)^n$.*

Proof. For all $x \in \mathbb{R}$, we obtain

$$\begin{aligned} \lim_{n \to \infty} n \ln\left(1 + \frac{x}{n}\right) &= \lim_{n \to \infty} \frac{\ln\left(1 + \frac{x}{n}\right)}{\frac{1}{n}} = \lim_{n \to \infty} \frac{\frac{1}{1+\frac{x}{n}}\left(-\frac{x}{n^2}\right)}{-\frac{1}{n^2}} \\ &= \lim_{n \to \infty} x \frac{1}{1 + \frac{x}{n}} = x. \end{aligned}$$

Because the natural exponential function is continuous, the result follows. ■

Exercises

12-25 Prove Theorem 12.19 Remember to prove that Rolle's Theorem can be applied

12-26. Compute each of the limits below.

(a) $\lim_{x\to 4} \frac{2x^4 - 4x^3 + x^2 + x - 276}{x^2 - 16}$ (b) $\lim_{x\to 0^+} x\ln(x)$ (c) $\lim_{x\to\infty} \frac{\ln(x)}{x}$

(d) $\lim_{x\to\infty} x^2 e^{-x}$ (e) $\lim_{x\to 0+} x^x$ (f) $\lim_{x\to 0+} x^{\sin(x)}$

(g) $\lim_{x\to 0+} \frac{1}{\sin(x)} - \frac{1}{\cos(x) - 1}$

12-27 Prove by induction that for all $n \in \mathbb{N}$ we have $\lim_{x\to\infty} x^n e^{-x} = 0$

12-28. Prove that for $f(x) = 2x + \sin(x)$, $g(x) = 2x - \sin(x)$ and $a = \infty$ we have $\lim_{x\to\infty} \frac{f(x)}{g(x)} = 1$, but $\lim_{x\to\infty} \frac{f'(x)}{g'(x)}$ does not exist.

12-29 Let $f : (0, \infty) \to \mathbb{R}$ be differentiable Does $\lim_{x\to\infty} f'(x) = 0$ imply that $\lim_{x\to\infty} f(x)$ exists? Justify your answer.

12-30 Is there a differentiable function $f : (0, \infty) \to \mathbb{R}$ with $\lim_{x\to\infty} f'(x) = 0$ so that for every real number L there is a sequence $\{x_n\}_{n=1}^{\infty}$ that goes to infinity and so that $\lim_{n\to\infty} f(x_n) = L$? Justify your answer

12-31 Creating **summation formulas** This exercise shows one way in which summation formulas for powers of integers (see Exercise 1-33) can be discovered. Let $f(x) := \sum_{k=1}^{n} e^{kx}$

(a) Prove that for all $p \in \mathbb{N}$ we have $\sum_{k=1}^{n} k^p = f^{(p)}(0)$

(b) Prove that for all $x \in \mathbb{R}$ we have $\sum_{k=1}^{n} e^{kx} = \frac{e^x - e^{(n+1)x}}{1 - e^x}$.

(c) Prove that for all $p \in \mathbb{N}$ we have $f^{(p)}(0) = \lim_{x\to 0} \frac{d^p}{dx^p} \frac{e^x - e^{(n+1)x}}{1 - e^x}$.

(d) Use l'Hôpital's Rule to verify that $\sum_{k=1}^{n} k = \frac{n}{2}(n + 1)$ for all $n \in \mathbb{N}$

(e) Use a computer algebra system to generate a closed formula for $\sum_{k=1}^{n} k^5$.

12-32 Use Cauchy's Limit Theorem (see Exercise 2-51) to give an alternative proof of l'Hôpital's Rule

Chapter 13

Numerical Methods

Readers familiar with geometric series and the transcendental functions could read this chapter after Chapter 5.

Many problems cannot be solved exactly. Therefore it is natural to consider computational approaches to mathematics. Numerical analysis is a wide field. For any problem that can be solved exactly (under good circumstances), there is at least one numerical method to provide an approximate solution in case exact methods fail. Usually, a numerical method contains a parameter, call it n, that indicates the computational effort required to obtain the approximation. With enough computational effort, a numerical method should provide approximations close to the exact solution. More formally, this means that as n goes to infinity the limit of our approximations should be the exact solution. But just having approximations that converge to the exact answers usually is not enough. We want to obtain good approximations with as little computational effort as possible. This means we not only need to assure that, given enough computational effort, the approximations converge to the correct result. We also must analyze *how fast* the approximations converge.

In the language that we have developed, this means that just showing that for every $\varepsilon > 0$ there is an $N \in \mathbb{N}$ so that for all $n \geq N$ the nth approximation is within ε of the exact solution is not enough. We also want our estimates to be sharp enough so that when, say, $N = 10$ guarantees a desired accuracy, we do not use a larger N and waste computational effort. So, where in proofs so far we were satisfied with the fact that N exists, in numerical analysis we want to know what N is. Where in proofs so far we were satisfied that estimates ultimately showed that a certain difference is smaller than ε, in numerical analysis we want to perform the estimate with an N that is as small as possible.

For this reason, this chapter will emphasize error analysis. We present numerical approaches for three typical tasks: The representation of functions in Section 13.1, the solution of equations in Section 13.2 and the computation of integrals in Section 13.3.

13.1 Approximation with Taylor Polynomials

It seems mundane, but the most fundamental numerical task is the computation of the values of functions such as exponential and trigonometric functions. The exact values of these functions can only be computed for certain special input values x. For all other input values, we need to use approximation techniques. The most fundamental of these techniques is the approximation with Taylor polynomials. There are several ways to motivate the use of polynomials. Most importantly, polynomials are easy to compute, which is a paramount concern in numerical analysis. Moreover, for each of the many functions defined as power series there is a sequence of polynomials that converges to it.

Geometrically, we can argue that the tangent line of a differentiable function at a point a has the same value and the same first derivative as f at a and, locally, it approximates f rather well. We have reason to hope that, by increasing the number of derivatives that agree with the derivatives of the function at a, we can enlarge the interval on which we have a good approximation for f. To increase the number of derivatives that agree at a, we need to use polynomials of degree greater than 1.

Theorem 13.1 *Let the function f be n times differentiable at a. Then the polynomial* $T_n(x) := \sum_{j=0}^{n} \frac{f^{(j)}(a)}{j!}(x-a)^j$ *is such that the first n derivatives of T_n at a are equal to the first n derivatives of f at a. That is, $T_n^{(k)}(a) = f^{(k)}(a)$ for $k = 0, \ldots, n$.*

Proof. Prove by induction that for $1 \leq k \leq n$ the kth derivative of $T_n(x)$ is $T_n^{(k)}(x) = \sum_{j=k}^{n} \frac{f^{(j)}(a)}{j!} j \cdot (j-1) \cdots (j-k+1)(x-a)^{j-k}$. (Exercise 13-1.) ■

Theorem 13.1 motivates the definition of Taylor polynomials and Taylor series.

Definition 13.2 *Let the function f be n times differentiable at a. Then the polynomial* $T_n(x) := \sum_{j=0}^{n} \frac{f^{(j)}(a)}{j!}(x-a)^j$ *is called the nth* **Taylor polynomial** *of f at a. If f is infinitely differentiable at a, the series* $T(x) := \sum_{j=0}^{\infty} \frac{f^{(j)}(a)}{j!}(x-a)^j$ *is called the* **Taylor series** *of f at a.*

The definitions of the exponential and the trigonometric functions guarantee that the Taylor polynomials at $a = 0$ ultimately provide good approximations for these functions (see Exercise 13-2). However, as mentioned in the introduction, for numerical purposes it is not sufficient to just know that for *some* degree n the nth Taylor polynomial of f is close to f. We need to know *how close* T_n is to f.

Theorem 13.3 Taylor's Formula. *If f is $(n+1)$ times continuously differentiable on $(a-R, a+R)$, then for all $x \in \mathbb{R}$ with $|x-a| < R$ we have*

$$f(x) = T_n(x) + (x-a)^{n+1} \int_0^1 \frac{(1-u)^n}{n!} f^{(n+1)}\big(a + u(x-a)\big)\, du.$$

In particular, if $|x-a|<R$ and M is such that for all $x\in\mathbb{R}$ with $|x-a|<R$ we have $\left|f^{(n+1)}(x)\right|\leq M$, then $|f(x)-T_n(x)|\leq\dfrac{M}{(n+1)!}|x-a|^{n+1}$.

Proof. This proof is an induction on n. For the base step $n=0$ note that for all $x\in(a-R,a+R)$ we have

$$\begin{aligned} & f(a)+(x-a)\int_0^1 f'\big(a+u(x-a)\big)\,du \\ = \;& f(a)+\big[f\big(a+u(x-a)\big)\big]_0^1 = f(a)+f(x)-f(a)=f(x). \end{aligned}$$

For the induction step, we use integration by parts on the induction hypothesis. Let $x\in(a-R,a+R)$ and let f be $n+2$ times continuously differentiable on $(a-R,a+R)$. Then

$$\begin{aligned} f(x) &= T_n(x)+(x-a)^{n+1}\int_0^1\frac{(1-u)^n}{n!}f^{(n+1)}\big(a+u(x-a)\big)\,du \\ &= T_n(x)+(x-a)^{n+1}\Bigg[-\frac{(1-u)^{n+1}}{(n+1)!}f^{(n+1)}\big(a+u(x-a)\big)\Bigg|_0^1 \\ &\qquad -\int_0^1-\frac{(1-u)^{n+1}}{(n+1)!}f^{(n+2)}\big(a+u(x-a)\big)(x-a)\,du\Bigg] \\ &= T_n(x)+(x-a)^{n+1}\frac{f^{(n+1)}(a)}{(n+1)!} \\ &\qquad +(x-a)^{n+2}\int_0^1\frac{(1-u)^{n+1}}{(n+1)!}f^{(n+2)}\big(a+u(x-a)\big)\,du \\ &= T_{n+1}(x)+(x-a)^{n+2}\int_0^1\frac{(1-u)^{n+1}}{(n+1)!}f^{(n+2)}\big(a+u(x-a)\big)\,du. \end{aligned}$$

The remainder of the proof is left to Exercise 13-3. ■

There are two ways to use error estimates like Taylor's formula. In an **a posteriori** or **after the fact** estimate, the polynomial and the interval are given and we estimate the error. This is straight substitution into the formula (see Exercise 13-4). In an **a priori** or **before the fact** estimate, we have a desired accuracy for an interval and need to find an n that guarantees this accuracy.

Example 13.4 *Determine the degree n so that the nth Taylor polynomial of $f(x)=e^x$ about $a=0$ is within 10^{-4} of e^x for all $x\in[-2,2]$.*

To bound the error by the maximum acceptable error of 10^{-4}, we make the *upper bound* of the error from Theorem 13.3 less than the specified acceptable error. This makes the actual error less than the acceptable error. In the notation of Theorem 13.3, we have $a=0$, $|x-a|\leq 2$, and n is to be determined. All derivatives of e^x are again e^x and e^x is increasing. Hence, the upper bound M for the $(n+1)^{\text{st}}$ derivative on $[-2,2]$

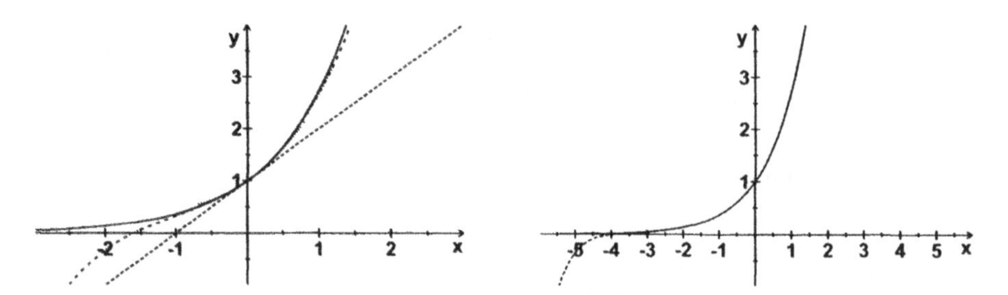

Figure 26: Taylor polynomials are local approximations to the function. The left graph shows the exponential function and its first (dashed), second (dotted) and third (dash-dotted) Taylor polynomials. Note how the approximation gets better as the degree of the polynomials increases. The right graph shows the exponential function and its eleventh Taylor polynomial as demanded in Example 13.4.

is e^2. We set $\left|f(x) - T_n(x)\right| \leq \frac{e^2}{(n+1)!}2^{n+1} \overset{!}{\leq} 10^{-4}$, where the inequality we need to solve is marked with an exclamation sign.

The above inequality cannot be solved algebraically for n. However, $n!$ grows much faster than 2^n. Thus by substituting values for n and checking if the inequality is satisfied we find that $n = 11$ is large enough to guarantee the desired accuracy. For a visualization of the Taylor polynomials, consider Figure 26. □

Remark 13.5 Not every function can be approximated well with Taylor polynomials. For example, Lemma 18.8 exhibits a function that is infinitely differentiable, not identical to zero, and yet all its Taylor polynomials at $a = 0$ are identical to zero. □

Remark 13.6 Early operating systems, such as the one on the Commodore 64 in the 1980s, used Taylor polynomials of sufficiently high degree to compute many functions. Nowadays, computational schemes that are faster than Taylor polynomials, but also more memory intensive, are used to compute functions. The reason is that memory is not as much of an issue as it was in the early days of computing, while speed remains a crucial concern. □

Remark 13.7 Taylor polynomials are also used in physics to obtain low-order approximations of complicated functions f. Typically, if f is to be evaluated at $x + \Delta x$, where Δx is small, the exact expression $f(x + \Delta x)$ is replaced with the approximately equal expression $f(x) + f'(x)\Delta x$. This is feasible because, as Taylor's Formula shows, the difference is often bounded by $C(\Delta x)^2$, where C is a constant. If Δx is small, terms of the order $(\Delta x)^2$ are usually negligible. The determination what is small and what is negligible is made based on practical, nonmathematical considerations. A posteriori, if the approximate formula correctly predicts an experiment, then the approximation must have been permissible. A priori, one could say that if other effects influence the quantity given by f by, say, 0.1% (of the underlying base unit), and Δx is at most 1%, then $(\Delta x)^2$ is less than 0.01%, so it can be ignored because other effects will have greater influence. If a first order approximation as indicated does not work, higher-order Taylor polynomials can be used in more sophisticated models. □

Functions can also be approximated with trigonometric polynomials. This idea is motivated by problems as described in Section 21.3. We will present the corresponding series, called Fourier series, in Section 20.2. The powerful tools available by then allow for a more efficient presentation than what would be possible now.

Exercises

13-1 Prove Theorem 13.1

13-2 Prove that if $f(x) = \sum_{k=0}^{\infty} c_k x^k$ is a power series with nonzero radius of convergence, then the Taylor series of f about $a = 0$ is $\sum_{k=0}^{\infty} c_k x^k$.

13-3. Prove that if f is $(n+1)$ times continuously differentiable on $(a - R, a + R)$, then for all $x \in \mathbb{R}$ with $|x - a| < R$ we have that $\left| f(x) - T_n(x) \right| \leq \frac{M}{(n+1)!}|x - a|^{n+1}$, where M is such that for all $|x - a| < R$ we have $\left| f^{(n+1)}(x) \right| < M$.

13-4 Find an upper bound for the error incurred when approximating f on $[l, r]$ with its nth degree Taylor polynomial at a.

(a) $f(x) = e^x, a = 0, n = 10, [l, r] = [-5, 5]$

(b) $f(x) = \sin(x), a = 0, n = 7, [l, r] = \left[\frac{-\pi}{2}, \frac{\pi}{2}\right]$

(c) $f(x) = \arctan(x), a = 0, n = 5, [l, r] = \left[-\frac{9}{10}, \frac{9}{10}\right]$

13-5 Use induction to prove that the given expression is the nth derivative of the given function

(a) $f(x) = \ln|x|, f^{(n)}(x) = \frac{(-1)^{n+1}(n-1)!}{x^n}$ for $n \geq 1$

(b) $f(x) = 2^x, f^{(n)}(x) = 2^x \left(\ln(2)\right)^n$

(c) $f(x) = xe^x, f^{(n)}(x) = xe^x + ne^x$

(d) $f(x) = xe^{ax}, f^{(n)}(x) = a^n xe^{ax} + \frac{1 - a^n}{1 - a}e^{ax}$

13-6. Determine the smallest n so that $\left| T_n(x) - f(x) \right| < \varepsilon$ for all $x \in [l, r]$, where T_n is the nth Taylor polynomial of f about a.

(a) $f(x) = e^x, a = 0, [l, r] = [-10, 10], \varepsilon = 10^{-5}$

(b) $f(x) = \cos(x), a = 0, [l, r] = [-\pi, \pi], \varepsilon = 10^{-12}$

(c) $f(x) = \sqrt{x}, a = 1, [l, r] = [.5, 1.5], \varepsilon = 10^{-10}$

(d) $f(x) = \ln(x), a = 2, [l, r] = [1, 3], \varepsilon = 10^{-8}$

13-7. Prove that for any $a > 0$, the Taylor polynomials of $f(x) = \ln|x|$ at a converge for $x \in (0, 2a)$ and they diverge for $|x - a| > a$.

13-8. **Second Derivative Test** Let $f : (a, b) \to \mathbb{R}$ be twice continuously differentiable and let $x \in (a, b)$ be so that $f'(x) = 0$.

(a) Prove that if $f''(x) > 0$, then there is an $\varepsilon > 0$ so that for all $z \neq x$ with $|z - x| < \varepsilon$ we have that $f(x) < f(z)$.

(b) Prove that if $f''(x) < 0$, then there is an $\varepsilon > 0$ so that for all $z \neq x$ with $|z - x| < \varepsilon$ we have that $f(x) > f(z)$.

(c) State and prove a similar result for $f : (a, b) \to \mathbb{R}$ being n times continuously differentiable with $f'(x) = f''(x) = \cdots = f^{(n-1)}(x) = 0$. (Distinguish even and odd n.)

13-9. **Efficient evaluation of polynomials** Let $p(x) = \sum_{j=0}^{n} a_j x^j$ be a polynomial.

(a) Prove that $p(x) = a_0 + x(a_1 + x(a_2 + \cdots + x(a_{n-1} + x(a_n)) \cdots))$.

(b) Count the number of operations in the evaluation of the sum $\sum_{j=0}^{n} a_j x^j$ and in the evaluation in part 13-9a to prove that evaluation as in part 13-9a takes fewer operations (and is thus more efficient) than evaluation of the original sum.

Hint. Evaluating $a_j x^j$ takes j floating point multiplications and floating point multiplications take much more time than floating point additions.

(c) State an n step recursive procedure that evaluates polynomials as in part 13-9a

Hint. Start with $H_n := a_n$ and define $H_{n-1}, \ldots, H_1$ in such a way that $H_1 = p(x)$.

13.2 Newton's Method

Solving equations is a common numerical task. The Intermediate Value Theorem guarantees that for equations $f(x) = 0$ the issue usually is not *if* we can find solutions, but rather *how fast* we can compute them.

Example 13.8 The **bisection method**. Let $f : [a, b] \to \mathbb{R}$ be continuous with $f(a)f(b) < 0$. Then by the Intermediate Value Theorem f has a zero in $[a, b]$. To simplify the presentation, assume without loss of generality that f has a unique zero z in $[a, b]$. We will recursively construct a sequence $\{x_n\}_{n=0}^{\infty}$ that converges to z.

Let $x_0 := a$, $x_1 := b$, and $j(1) := 0$. For the recursive construction, let $x_0, \ldots, x_n$ and $j(n) \in \{0, \ldots, n-1\}$ be so that $f(x_n)f(x_{j(n)}) < 0$ and $\left|x_n - x_{j(n)}\right| = \frac{b-a}{2^{n-1}}$. Define $x_{n+1} := \frac{x_n + x_{j(n)}}{2}$. If $f(x_{n+1}) = 0$, stop the recursion because x_{n+1} is the desired zero. Otherwise, if $f(x_{n+1})f(x_{j(n)}) < 0$ set $j(n+1) := j(n)$, and if $f(x_{n+1})f(x_n) < 0$ set $j(n+1) := n$. In either case, $\left|x_{n+1} - x_{j(n+1)}\right| = \frac{b-a}{2^n}$, and we can continue the recursion.

By the Intermediate Value Theorem and the uniqueness of z, for each $n \in \mathbb{N}$ we obtain $|x_n - z| \leq \frac{b-a}{2^{n-1}}$, and hence $\lim_{n\to\infty} x_n = z$.

If f has multiple zeroes in $[a, b]$ the argument becomes a bit messier, but the sequence will converge to one of the zeroes of f. □

Aside from establishing convergence, the estimate that in the nth step we are within $\frac{b-a}{2^{n-1}}$ of a zero of f tells us how many steps we need to execute to obtain an estimate of a given quality. For example, if $b - a = 1$ and we want a number that approximates the first 15 decimal places of the zero z, we would need to execute 51 steps of the bisection method $\left(2^{-51+1} = 2^{-50} = 1{,}024^{-5} < 10^{-15}\right)$.

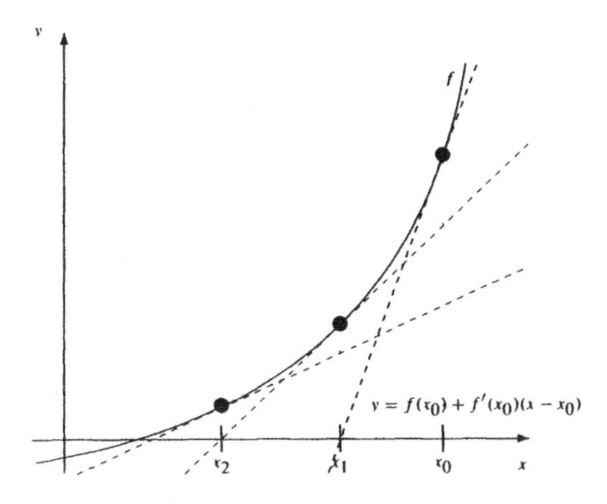

Figure 27: An illustration of Newton's method that will be analyzed precisely in Exercise 13-12.

This is a large number of steps to carry out by hand. Nonetheless, to solve a single equation, a program would find the desired approximation rather quickly. However, when these methods were first designed, computers had not been invented, so a faster computation was desirable. For today's applications, speed remains an issue, because in large scale computations we may need to solve billions of equations. This means every step we save would be magnified by a large factor, leading to a more efficient use of computational resources.

Newton's method reduces the (hard) computation of the zero of a differentiable function to the (easier) computation of the zero of the tangent line at a point. Say we are currently at a point x_n. Then the tangent line of f at x_n is the unique straight line that goes through the point $\big(x_n, f(x_n)\big)$ and has slope $f'(x_n)$. Setting its equation $y = f(x_n) + f'(x_n)(x - x_n)$ equal to zero and solving for x gives the expression $x = x_n - \dfrac{f(x_n)}{f'(x_n)}$ for the zero of the tangent line. Of course, the zero of the tangent line is not the zero of the function, but we would hope that it is closer to the zero of f than x_n was. Hence, we set $x_{n+1} := x_n - \dfrac{f(x_n)}{f'(x_n)}$ and repeat the above with x_{n+1} in place of x_n (see Figure 27).

Definition 13.9 *Let f be a differentiable function and let x_0 be a point in its domain. Then* **Newton's method** *started at x_0 generates a sequence defined recursively by $x_{n+1} := x_n - \dfrac{f(x_n)}{f'(x_n)}$ for $n \geq 0$.*

If the sequence generated by Newton's method converges, then under mild hypotheses the limit is a solution of the equation $f(x) = 0$.

Theorem 13.10 *Let f be a continuously differentiable function. If the sequence generated by Newton's method converges to a limit L in the domain of f and $f'(L) \neq 0$, then $f(L) = 0$.*

Proof. Because the sequence $\{x_n\}_{n=0}^{\infty}$ converges, we can take limits on both sides of the recurrence relation. $\lim_{n\to\infty} x_{n+1} = \lim_{n\to\infty} \left(x_n - \frac{f(x_n)}{f'(x_n)} \right)$ leads to the equation $L = L - \frac{f(L)}{f'(L)}$, which clearly implies $f(L) = 0$. ■

The reader will prove in Exercise 13-10 that L is also a zero of f if $f'(L) = 0$.

Another way of looking at the proof of Theorem 13.10 is to say that the limit L is a fixed point of the function $F(x) = x - \frac{f(x)}{f'(x)}$. Fixed points are a standard tool in numerical and abstract mathematics. We will encounter fixed points again in Theorems 17.64, 17.65, and 22.6.

Theorem 13.10 says that *if* Newton's method converges, the limit is a zero of the function. It does not guarantee *that* the sequence from Newton's method converges. However, there are mild conditions that guarantee convergence.

Lemma 13.11 *Let $q \in [0, 1)$, let $F : (a, b) \to \mathbb{R}$ be a differentiable function so that $|F'(x)| < q$ for all $x \in (a, b)$ and let $p \in (a, b)$ be so that $F(p) = p$ (p is also called a* **fixed point** *of F). Then p is the only fixed point of F in (a, b) and for all $x_0 \in (a, b)$ the sequence generated by the recursive equation $x_{n+1} := F(x_n)$ converges to p.*

Proof. To prove that there is no $\widetilde{p} \in (a, b) \setminus \{p\}$ with $\widetilde{p} = F(\widetilde{p})$ note that if $\widetilde{p} = F(\widetilde{p})$, then by the Mean Value Theorem there would be a c between $\widetilde{p}$ and p so that $|\widetilde{p} - p| = |F(\widetilde{p}) - F(p)| = |F'(c)|\,|\widetilde{p} - p| < |\widetilde{p} - p|$, which is a contradiction. Hence, p is the only fixed point of F in (a, b).

Now let $x_0 \in (a, b)$. Because the powers q^n converge to zero, the result is proved if we can establish the inequality $|x_n - p| \le q^n |x_0 - p|$ for all $n \in \mathbb{N}$. The inequality is proved by induction on n.

For the base step $n = 1$ note that by Mean Value Theorem between x_0 and p there must be a c so that $|x_1 - p| = |F(x_0) - F(p)| = |F'(c)|\,|x_0 - p| \le q|x_0 - p|$.

For the induction step note that there must be a c between x_n and p so that $|x_{n+1} - p| = |F(x_n) - F(p)| = |F'(c)|\,|x_n - p| \le q\left(q^n|x_0 - p|\right) = q^{n+1}|x_0 - p|$. ■

Theorem 13.12 *Let $f : (a, b) \to \mathbb{R}$ be twice continuously differentiable. Then for each zero z of f in (a, b) with $f'(z) \neq 0$ there is a $\delta_z > 0$ so that for every starting point $x_0 \in (z - \delta_z, z + \delta_z)$ the sequence generated by Newton's method converges to z.*

Proof. We apply Lemma 13.11 to $F(x) := x - \frac{f(x)}{f'(x)}$. First note that

$$F'(x) = \frac{d}{dx}\left(x - \frac{f(x)}{f'(x)} \right) = 1 - \frac{f'(x)f'(x) - f''(x)f(x)}{\left(f'(x)\right)^2} = \frac{f''(x)f(x)}{\left(f'(x)\right)^2}.$$

In particular this means that $F'(z) = 0$, and hence there is a $\delta_z > 0$ so that $|F'(x)| < \frac{1}{2} < 1$ for all $x \in (z - \delta_z, z + \delta_z)$. But then by Lemma 13.11 every sequence generated by Newton's method started at any $x_0 \in (z - \delta_z, z + \delta_z)$ converges to z. ■

From Theorem 13.12 and its proof we infer that the closer x_0 is to a zero of f, the faster Newton's method will converge. But when Newton's method converges, the numbers x_n will get ever closer to a zero of f. Hence, as Newton's method is executed, the speed of convergence should accelerate. Theorem 13.14 below makes this statement more precise.

Lemma 13.13 *Let $f : (a,b) \to \mathbb{R}$ be continuously differentiable and let $\gamma > 0$ be so that for all $x, z \in (a,b)$ we have that $\left|f'(z) - f'(x)\right| \leq \gamma|z - x|$. Then for all $x, z \in (a,b)$ the inequality $\left|f(z) - f(x) - f'(x)(z - x)\right| \leq \frac{\gamma}{2}|z - x|^2$ holds.*

Proof. Without loss of generality assume that $x < z$. Then

$$\begin{aligned}
\left|f(z) - f(x) - f'(x)(z - x)\right| &= \left|\int_x^z f'(t)\,dt - \int_x^z f'(x)\,dt\right| \\
&\leq \int_x^z \left|f'(t) - f'(x)\right|\,dt \leq \int_x^z \gamma|t - x|\,dt \\
&= \frac{\gamma}{2}(t - x)^2\Big|_x^z = \frac{\gamma}{2}(z - x)^2.
\end{aligned}$$

■

Theorem 13.14 *Let $f : (a,b) \to \mathbb{R}$ be a continuously differentiable function so that $f'(x) \neq 0$ for all $x \in (a,b)$ Assume there are $x_0 \in (a,b)$ and $\alpha, \beta, \gamma > 0$ so that $\left|\frac{f(x_0)}{f'(x_0)}\right| \leq \alpha$, so that for all $x \in (a,b)$ we have $\left|\frac{1}{f'(x)}\right| \leq \beta$, so that for all $x, z \in (a,b)$ we have $\left|f'(z) - f'(x)\right| \leq \gamma|z - x|$, so that $h := \frac{\alpha\beta\gamma}{2} < 1$ and so that with $r := \frac{\alpha}{1 - h}$ we have $[x_0 - r, x_0 + r] \subseteq (a,b)$ Then*

1 *Each recursively defined point $x_{n+1} := x_n - \frac{f(x_n)}{f'(x_n)}$ is in $(x_0 - r, x_0 + r)$.*

2 *The sequence $\{x_n\}_{n=0}^{\infty}$ converges to a point $u \in [x_0 - r, x_0 + r]$ with $f(u) = 0$*

3 *For all $n \geq 0$ we have $|u - x_n| \leq \alpha\frac{h^{2^n - 1}}{1 - h^{2^n}}$.*

Proof. We first prove by induction that for all $n \in \mathbb{N}$ the point x_n is well defined, $|x_n - x_{n-1}| \leq \alpha h^{2^{n-1}-1}$, and $|x_n - x_0| < r$. For $n = 1$ the above is trivial. For the induction step, $n \to n+1$ first note that because $|x_n - x_0| < r$ the number x_{n+1} is well defined. The definition of x_n implies that $f(x_{n-1}) + f'(x_{n-1})(x_n - x_{n-1}) = 0$, which implies

$$\begin{aligned}
|x_{n+1} - x_n| &= \left|-\frac{f(x_n)}{f'(x_n)}\right| \leq \beta\left|f(x_n)\right| \\
&= \beta\left|f(x_n) - f(x_{n-1}) - f'(x_{n-1})(x_n - x_{n-1})\right| \\
&\quad \boxed{\text{Apply Lemma 13.13.}} \\
&\leq \beta\frac{\gamma}{2}|x_n - x_{n-1}|^2 \leq \frac{\beta\gamma}{2}\alpha^2 h^{2^n - 2} = \alpha h^{2^n - 1}.
\end{aligned}$$

But then, using a telescoping sum, we obtain the following.

$$|x_{n+1} - x_0| \leq \sum_{k=1}^{n+1} |x_k - x_{k-1}| \leq \alpha \sum_{k=1}^{n+1} h^{2^{k-1}-1} < \alpha \frac{1}{1-h} = r,$$

which finishes the induction.

To prove that $\{x_n\}_{n=1}^{\infty}$ converges, let $m \geq n$. Then

$$\begin{aligned} |x_m - x_n| &\leq \sum_{k=n+1}^{m} |x_k - x_{k-1}| \leq \alpha \sum_{k=n+1}^{m} h^{2^{k-1}-1} = \alpha \sum_{j=0}^{m-n-1} h^{2^n-1} h^{2^n(2^j-1)} \\ &< \alpha h^{2^n-1} \sum_{k=0}^{\infty} \left(h^{2^n}\right)^j = \alpha h^{2^n-1} \frac{1}{1-h^{2^n}}. \end{aligned}$$

Because $0 < h < 1$ the bound $\alpha h^{2^n-1} \frac{1}{1-h^{2^n}}$ goes to zero as n goes to infinity, so $\{x_n\}_{n=0}^{\infty}$ is a Cauchy sequence. *(Mentally fill in the argument.)* In particular, $\{x_n\}_{n=0}^{\infty}$ converges to a number u, which by Theorem 13.10 must satisfy $f(u) = 0$. Letting m go to infinity in the above estimate also shows that $|u - x_n| \leq \alpha h^{2^n-1} \frac{1}{1-h^{2^n}}$, which finishes the proof. ■

Theorem 13.14 shows that once Newton's method is "close enough" to a zero of a function, it converges quite rapidly. Indeed, near a zero u of a continuously differentiable function f, the hypotheses of Theorem 13.14 can be satisfied if $f'(u) \neq 0$, because we can make α and h small by starting near u. For comparison with the bisection method, suppose, for argument's sake, $\alpha \leq 1$ and $h \leq \frac{1}{2}$. Then $|u - x_n| \leq \frac{1}{2^{2^n-1} - \frac{1}{2}}$ and because $\frac{1}{2^{2^6-1} - \frac{1}{2}} = \frac{1}{2^{63} - \frac{1}{2}} < 1{,}024^{-6} < 10^{-15}$ we only need to iterate Newton's method six times to obtain an approximation that gives the first 15 digits behind the decimal point of u.

Another nice feature of Theorem 13.14 is that it can be generalized to several variables (see Exercise 17-44).

Finally, note that even though Newton's method is only applicable to differentiable functions, Exercise 13-13 shows that it can be modified to provide a method that is applicable to all functions.

Exercises

13-10 Let f be a continuously differentiable function. Prove that if the sequence generated by Newton's method converges to a limit L in the domain of f and $f'(L) = 0$, then $f(L) = 0$

13-11 Let $q > 1$, let F $(a\ b) \to \mathbb{R}$ be a differentiable function so that $\left|F'(x)\right| > q$ for all $x \in (a, b)$ and let $p \in (a, b)$ be so that $F(p) = p$ Prove that p is the only fixed point of F in (a, b) and that for all $x_0 \in (a, b) \setminus \{p\}$ the sequence generated by the recursive equation $x_{n+1} := F(x_n)$ terminates after finitely many steps with a value that is not in $(a\ b)$

13-12 Let $f: (a, b) \to \mathbb{R}$ be twice differentiable

(a) Let $z \in (a, b)$ be so that $f(z) = 0$, $f'(x) > 0$ for all $x \in (z, b)$ (that is, f is increasing on (z, b)), and so that $f''(x) > 0$ for all $x \in (z, b)$ (that is, f is concave up on (z, b)) Prove that the sequence generated by Newton's method started at any $x_0 \in (z, b)$ converges to z.
Hint. Prove that $z \leq x_{n+1} \leq x_n$ for all $n \in \mathbb{N}$
Note. Figure 27 provides a geometric visualization of the claim in this exercise.
Second note. In particular, it is allowed that $f'(z) = 0$ in this exercise.

(b) Prove that if f'' is continuous, then for each $z \in (a, b)$ with $f(z) = f'(z) = 0$ and $f''(z) \neq 0$ there is a $\delta_z > 0$ so that for every starting point $x_0 \in (z - \delta_z, z + \delta_z)$ the sequence generated by Newton's method converges to z.

13-13 Let f be a function, let $x_0, x_1 \in \mathbb{R}$ with $f(x_0) \neq f(x_1)$ and consider the recursively defined sequence $x_{n+1} = x_n - \dfrac{x_n - x_{n-1}}{f(x_n) - f(x_{n-1})} f(x_n)$.

(a) Show that the recursive formula is obtained by taking the equation of the secant line of f through $\big(x_{n-1}, f(x_{n-1})\big)$ and $\big(x_n, f(x_n)\big)$ and computing its unique zero.

(b) Prove that if the sequence generated with this method converges to L and $\dfrac{f(z) - f(x)}{z - x}$ is bounded for x, z near L, then $f(L) = 0$.

(c) Prove that if z is a zero of f, $z < x_1 < x_0$, f is twice differentiable on $(z - \delta, x_0 + \delta)$ for some $\delta > 0$, and f is increasing and concave up on $(z - \delta, x_0 + \delta)$ (that is, $f''(x) > 0$ for all $x \in (z - \delta, x_0 + \delta)$), then the sequence generated by this method converges to z
Hint Prove that $z \leq x_{n+1} \leq x_n$ for all $n \in \mathbb{N}$.

(d) Prove that in a situation as in part 13-13c the sequence $\{x_n\}_{n=0}^{\infty}$ converges at least as fast to z as the sequence $\left\{x_n^N\right\}_{n=0}^{\infty}$ generated with Newton's method and $x_0^N = x_0$.
Hint. Prove that $z \leq x_n \leq x_n^N$ for all n.

13-14. Explain why $x_0 = 1$ is not a useful starting point for finding the zeroes of $f(x) = x^3 - 3x + 4$ with Newton's method.

13-15. Explain why $x_0 = 1$ is not a useful starting point for using Newton's method to find the zeroes of the function $f(x) = x^6 - 4x^2 - x + 1$. Sketch a rough graph of f and of the tangent lines used to compute x_1 and x_2 to illustrate your point

13-16. For the function $f(x) = x^3 - 5x - 5$, execute Newton's method started with $x_0 = -2$. Use a calculator or a computer.

(a) Find x_1, x_2, x_3, x_4.

(b) Find the first n such that your computer shows $x_n = x_{n+1}$. Explain why n is so large.

13-17. Apply Newton's method to $f(x) = 6x^4 - 18x^2 - 6x + 1$ with $x_0 = -1$. Explain why the limit is not the zero of f that is closest to the starting point.

13-18. The limit L computed with Newton's method does not always give $f(L) \approx 0$.

(a) Let $f(x) = \left(10^{100} + 10^{50}\right)x - 10^{100}$. Apply Newton's method with $x_0 = 1$. Use a computer and call the *apparent* limit L.

(b) Compute $f(L)$. Is $f(L) \approx 0$?

(c) Compute the zero of f exactly and compare it to L. Is L close to the zero?

(d) Explain why Newton's method cannot produce a better approximation to the actual zero of f.

13-19. Square roots.

(a) Use Newton's method and the fact that $\sqrt{a}$ is the solution of the equation $x^2 - a = 0$ to devise a recursive method to approximate square roots
Note. This is one algorithm that is used in computers to approximate square roots.

(b) Prove that for any $x_0 > 0$ the sequence generated in part 13-19a converges to $\sqrt{a}$ and for any $x_0 < 0$ the sequence generated in part 13-19a converges to $-\sqrt{a}$.

13.3 Numerical Integration

We conclude our introduction to numerics with numerical integration. Although the Darboux and Lebesgue integrals are useful, even essential, for the development of integration theory, the Riemann integral and the Fundamental Theorem of Calculus are the best tools for numerical considerations. The key to numerical integration is to replace the function we want to integrate with a function that is close to it and which can easily be integrated. The simplest geometric idea is to choose points on the function and let a polynomial go through these points. This section will focus solely on this idea. For other approaches to numerical integration, consider [28].

Definition 13.15 *Let $x_0, \ldots, x_n \in \mathbb{R}$ with $x_i \neq x_k$ for $i \neq k$ and define the polynomial $L_i(x) := \prod_{k \neq i} \frac{x - x_k}{x_i - x_k}$. Then L_i is called a* **Lagrange polynomial.**

Theorem 13.16 Lagrange's Interpolation Formula. *Let $n \in \mathbb{N}$, $x_0, \ldots, x_n \in \mathbb{R}$ with $x_i \neq x_k$ for $i \neq k$, and let $f_0, \ldots, f_n \in \mathbb{R}$. Then the polynomial defined by $P_n(x) := \sum_{i=0}^{n} f_i L_i(x) = \sum_{i=0}^{n} f_i \prod_{k \neq i} \frac{x - x_k}{x_i - x_k}$ is the unique polynomial of degree $\leq n$ so that for all $i = 0, \ldots, n$ we have $P_n(x_i) = f_i$.*

Proof. The equation follows easily from the fact that for all $j \in \{0, \ldots, n\}$ we have $L_i(x_j) := \prod_{k \neq i} \frac{x_j - x_k}{x_i - x_k} = \begin{cases} 0, & \text{if } j \neq i, \\ 1, & \text{if } j = i. \end{cases}$ Regarding uniqueness, suppose that Q was another polynomial of degree $\leq n$ with $Q(x_j) = f_j$ for all $j = 0, \ldots, n$. Then $P - Q$ is a polynomial of degree $\leq n$ with $n + 1$ zeroes. By the Fundamental Theorem of Algebra (see Exercise 16-74 for a proof), this means that $P - Q = 0$ or $P = Q$. ■

With the polynomial P_n going through the points (x_i, f_i) the most natural way to approximate a function f is to set $f_i = f(x_i)$. We first note that if we start with a polynomial f of sufficiently low degree, then we simply get f back.

Corollary 13.17 *Let f be a polynomial of degree $\leq n$ and let P_n be a polynomial computed as in Theorem 13.16 with numbers $f_i = f(x_i)$ for distinct numbers $x_i \in \mathbb{R}$, $i = 0, \ldots, n$. Then $P_n(x) = f(x)$ for all $x \in \mathbb{R}$.*

Proof. Both P_n and f are polynomials of degree $\leq n$ that go through the points $\big(x_i, f(x_i)\big)$. The equality follows from the uniqueness of P_n. ■

To obtain a convenient integration formula, we need to express the integral of P_n in terms of the $f_i = f(x_i)$. Interestingly enough, as long as the x_i are equidistant, the coefficients in the formula do not depend on the interval over which we integrate.

Proposition 13.18 Newton-Cotes formulas. *Let $n \in \mathbb{N}$ and let $[c, d]$ be an interval. Define $h := \frac{d - c}{n}$, for $i = 0, \ldots, n$ let $x_i := c + ih$, let $f_i \in \mathbb{R}$ and*

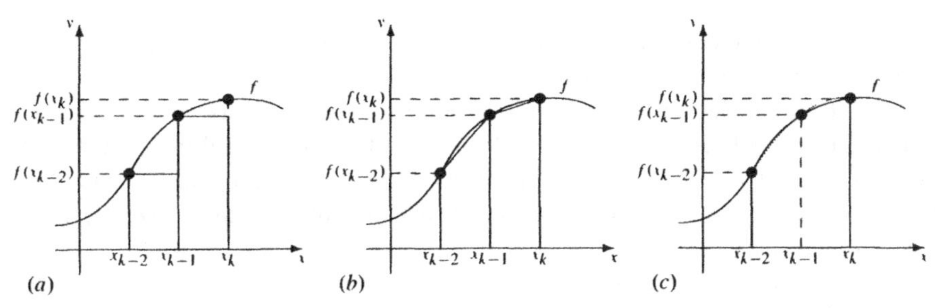

Figure 28: Visualization of the approximation of an integral over two subintervals with Riemann sums with evaluation at left endpoints (a), trapezoidal sums (b) and Simpson's Rule (c).

set $P_n(x) := \sum_{i=0}^{n} f_i L_i(x) = \sum_{i=0}^{n} f_i \prod_{k\neq i} \frac{x - x_k}{x_i - x_k}$. *Then with* $\alpha_i := \int_0^n \prod_{k\neq i} \frac{t-k}{i-k}\, dt$ *we have that* $\int_c^d P_n(x)\, dx = h \sum_{i=0}^{n} f_i \alpha_i$.

Proof. The key to the proof is the substitution $t := \frac{x-c}{h}$.

$$\begin{aligned}
\int_c^d \prod_{k\neq i} \frac{x - x_k}{x_i - x_k}\, dx &= \int_c^d \prod_{k\neq i} \frac{x - c - kh}{(i-k)h}\, dx = \int_c^d \prod_{k\neq i} \frac{\frac{x-c}{h} - k}{i-k}\, dx \\
&= h \int_0^n \prod_{k\neq i} \frac{t-k}{i-k}\, dt.
\end{aligned}$$

Now $\int_c^d P_n(x)\, dx = \sum_{i=0}^{n} f_i \int_c^d \prod_{k\neq i} \frac{x - x_k}{x_i - x_k}\, dx = h \sum_{i=0}^{n} f_i \alpha_i$. ■

The most obvious way to approximate an integral is to approximate a function $f : [a, b] \to \mathbb{R}$ with a polynomial of sufficiently large degree and then use the appropriate Newton-Cotes formula. However, for $n \geq 7$ some of the α_i become negative (see Exercise 13-20), which leads to problems with cancellations of digits. Moreover, the α_i are hard to compute for large n. Therefore an interval $[a, b]$ is usually first partitioned into shorter subintervals, the formula from Proposition 13.18 is applied on each subinterval and then the results are added. This type of integration formula is called a **composite integration formula**.

For example, for $n = 1$, we obtain $\alpha_0 = \frac{1}{2}$ and $\alpha_1 = \frac{1}{2}$. If we now partition the interval $[a, b]$ into N subintervals of length $\Delta x := \frac{b-a}{N}$ and apply Proposition 13.18 on each subinterval we obtain the approximation

$$\int_a^b f(x)\, dx \approx \frac{\Delta x}{2} \left(f(a) + f(b) + 2 \sum_{k=1}^{N-1} f(a + k\Delta x) \right),$$

which is called the **trapezoidal rule**, also shown in Figure 28(b). In the trapezoidal rule, we need to evaluate the function $N+1$ times.

The effort in numerical integration is proportional to the number of times the function f is evaluated. To compare the performance of two numerical integration formulas, it is thus important that both formulas evaluate the function equally many times. For this reason, we will construct composite Newton-Cotes formulas so that f is evaluated $N+1$ times.

For $n=2$, note that each additional subinterval of length Δx requires two additional evaluations of the function. Thus for $n=2$ we must demand that N is even. With $\alpha_0=\frac{1}{3}$, $\alpha_1=\frac{4}{3}$ and $\alpha_2=\frac{1}{3}$ we obtain the approximation formula

$$\int_a^b f(x)\,dx \approx \frac{\Delta x}{3}\left(f(a)+f(b)+\sum_{k=1}^{N-1}\left(3+(-1)^{k+1}\right)f(a+k\Delta x)\right),$$

which is called **Simpson's Rule**, also shown in Figure 28(c). In Exercise 13-21, the reader will state composite integration formulas based on Newton-Cotes formulas for $n=3$–6.

We now turn to the error analysis once more. Peano's error representation (Theorem 13.19) shows that a more "abstract" point-of-view can have benefits for concrete tasks, which makes it a perfect conclusion for Part I of this text and a lead-in for the more abstract Part II. Consider that the integral as well as the Newton-Cotes formulas are functions themselves. They map functions on an interval to real numbers. Moreover, the integral as well as the Newton-Cotes formulas are linear. That is, sums and constant factors can be moved through the integral (see Theorems 5.8 and 9.25) and also through the Newton-Cotes formulas (easy computation). Linear functions on vector spaces (like vector spaces of integrable functions) will be important in abstract analysis (see Chapter 17). Because we adopt this more abstract point-of-view, the error representation actually is valid for any numerical approximation of integrals that is linear, that gives exact results for low-degree polynomials and that can be "moved into the integral" as described in the hypothesis of Theorem 13.19 (see [28] for examples beyond the Newton-Cotes formulas).

By Corollary 13.17 the nth Newton-Cotes formula is exact for polynomials up to degree n. Because it only evaluates the function at select points x_k, multiplies the values with numbers, and adds the results, it is linear and it can be moved into the integral. Hence, Peano's error representation applies to the nth Newton-Cotes formula. Although the hypotheses for Peano's error representation look complicated, the proof shows that they are just what is needed to get an estimate.

Finally, Peano's error representation also shows that we need to establish more abstract results to fully justify more concrete results, like error estimates. In the proof of Peano's error representation, we work with double integrals and we reverse their order. Formally, we have not proved yet that this is possible. Fubini's Theorem (see Theorem 14.66) will show that this reversal is indeed allowed. For the specific case of double Lebesgue integrals, Fubini's Theorem is stated in Exercise 16-80. Because Theorem 13.19 is not used to prove these results, we will reverse the order of integration in the proof of Theorem 13.19, anticipating that this can be justified.

Theorem 13.19 Peano's error representation. *Let $n \in \mathbb{N}$, $c < d$ and let $F(\cdot)$ be an integration formula which is linear, that is, $F(\alpha f + \beta g) = \alpha F(f) + \beta F(g)$ for all $\alpha, \beta \in \mathbb{R}$ and $f, g : [c, d] \to \mathbb{R}$, which for polynomials p of degree at most n gives $F(p) = \int_c^d p(x)\,dx$, and which for all continuous functions $g : [c, d] \to \mathbb{R}$ satisfies $F\left(\int_c^d g(t)(x-t)^n \mathbf{1}_{x-t\geq 0}\,dt\right) = \int_c^d g(t) F\left((x-t)^n \mathbf{1}_{x-t\geq 0}\right)\,dt$. For every Riemann integrable function $f : [c, d] \to \mathbb{R}$, let $R(f) := F(f) - \int_c^d f(x)\,dx$ be the error of the numerical integration and let $K(t) := \frac{1}{n!} R\left((\cdot - t)^n \mathbf{1}_{\cdot - t\geq 0}\right)$. Then for every function f that for some $\delta > 0$ is $n+1$ times continuously differentiable on the interval $(c-\delta, d+\delta)$ we have $R(f) = \int_c^d f^{(n+1)}(t) K(t)\,dt$. The function K is also called the* **Peano kernel**.

Proof. By Taylor's Formula, we obtain the following for all $x \in [c, d]$.

$$f(x) = T_n(x) + (x-c)^{n+1} \int_0^1 f^{(n+1)}\big(c + u(x-c)\big) \frac{1}{n!}(1-u)^n\,du$$

Substitute $t := c + u(x-c)$ and use that $1 - u = \frac{x-t}{x-c}$.

$$= T_n(x) + \int_c^x f^{(n+1)}(t) \frac{1}{n!}(x-t)^n\,dt$$

$$= T_n(x) + \int_c^d f^{(n+1)}(t) \frac{1}{n!}(x-t)^n \mathbf{1}_{x-t\geq 0}\,dt,$$

where $T_n(x)$ is the nth Taylor polynomial of f at c. Because the integration formula is linear, we infer for all functions g, h that

$$\begin{aligned} R(g+h) &= F(g+h) - \int_c^d g + h\,dx \\ &= F(g) - \int_c^d g\,dx + F(h) - \int_c^d h\,dx \\ &= R(g) + R(h). \end{aligned}$$

Now because the integration formula gives exact results for polynomials of degree $\leq n$ we know that $R(T_n) = 0$. Because the integral in Taylor's formula above is a function of x we obtain the following.

$$\begin{aligned} R(f) &= R(T_n) + R\left(\int_c^d f^{(n+1)}(t) \frac{1}{n!}(x-t)^n \mathbf{1}_{x-t\geq 0}\,dt\right) \\ &= F\left(\int_c^d f^{(n+1)}(t) \frac{1}{n!}(x-t)^n \mathbf{1}_{x-t\geq 0}\,dt\right) \end{aligned}$$

$$-\int_c^d \int_c^d f^{(n+1)}(t)\frac{1}{n!}(x-t)^n \mathbf{1}_{x-t\geq 0}\,dt\,dx$$

Use the hypothesis for the first integral and switch the order of integration in the second one.

$$= \int_c^d f^{(n+1)}(t)F\left(\frac{1}{n!}(x-t)^n \mathbf{1}_{x-t\geq 0}\right)dt$$

$$-\int_c^d f^{(n+1)}(t)\int_c^d \frac{1}{n!}(x-t)^n \mathbf{1}_{x-t\geq 0}\,dx\,dt$$

$$= \int_c^d f^{(n+1)}(t)\frac{1}{n!}\left[F\left((x-t)^n \mathbf{1}_{x-t\geq 0}\right) - \int_c^d (x-t)^n \mathbf{1}_{x-t\geq 0}\,dx\right]dt$$

$$= \int_c^d f^{(n+1)}(t)K(t)\,dt.$$

■

Although Peano's error representation is very versatile, we need something more concrete for applications. One key weakness of Peano's error representation is that it depends in a rather complicated way on the interval of integration. The next result makes this dependency more manageable for the Newton-Cotes formulas by reducing the error to a power of half the length of the interval, times a constant, times a value of an appropriate derivative.

Theorem 13.20 *Let $n \in \mathbb{N}$ and let $f : [c,d] \to \mathbb{R}$ be $n+1$ times continuously differentiable if n is odd and $n+2$ times continuously differentiable if n is even. Let $R_n^{[c,d]}(f)$ be the error when approximating the integral $\int_c^d f(t)\,dt$ with the nth Newton-Cotes formula $F_n(f)$, that is, $R_n^{[c,d]}(f) = F_n(f) - \int_c^d f(t)\,dt$. For odd n, there is a $\xi \in (c,d)$ so that $R_n^{[c,d]}(f) = \left(\frac{d-c}{2}\right)^{n+2} \frac{R_n^{[-1,1]}\left(x^{n+1}\right)}{(n+1)!} f^{(n+1)}(\xi)$ and for even n there is a $\xi \in (c,d)$ so that $R_n^{[c,d]}(f) = \left(\frac{d-c}{2}\right)^{n+3} \frac{R_n^{[-1,1]}\left(x^{n+2}\right)}{(n+2)!} f^{(n+2)}(\xi)$.*

Proof. For odd n, by its definition the nth Newton-Cotes formula gives exact results for polynomials of degree n. For even n, the nth Newton-Cotes formula even gives exact results for polynomials of degree $n+1$ (see Exercise 13-22). Because the proofs for even and odd n are very similar, we will assume throughout that n is fixed and odd. The idea for the proof is a substitution that changes the domain of the integral from $[c,d]$ to $[-1,1]$.

For all real numbers $c < d$, let $K_n^{[c,d]}$ denote the Peano kernel for the nth Newton-Cotes formula for the interval $[c,d]$. This Peano kernel can be reduced to the Peano kernel $K_n^{[-1,1]}$ on $[-1,1]$ as follows. A simple substitution (similar to the computation below, see Exercise 13-23) shows that $K_n^{[c,d]}\left(t + \frac{c+d}{2}\right) = K_n^{\left[-\frac{d-c}{2}, \frac{d-c}{2}\right]}(t)$ for all

$t \in \left[-\frac{d-c}{2}, \frac{d-c}{2}\right]$. Moreover, for all $x \in [-1, 1]$ we have

$$
\begin{aligned}
& n!K_n^{\left[-\frac{d-c}{2}, \frac{d-c}{2}\right]}\left(\frac{d-c}{2}x\right) \\
= \; & F_n^{\left[-\frac{d-c}{2}, \frac{d-c}{2}\right]}\left(\left(\cdot - \frac{d-c}{2}x\right)^n \mathbf{1}_{\cdot - \frac{d-c}{2}x \geq 0}\right) \\
& - \int_{-\frac{d-c}{2}}^{\frac{d-c}{2}} \left(z - \frac{d-c}{2}x\right)^n \mathbf{1}_{z - \frac{d-c}{2}x \geq 0}\, dz \\
= \; & \frac{d-c}{2} F_n^{[-1,1]}\left(\left(\frac{d-c}{2}(\cdot) - \frac{d-c}{2}x\right)^n \mathbf{1}_{\frac{d-c}{2}(\cdot) - \frac{d-c}{2}x \geq 0}\right) \\
& - \frac{d-c}{2}\int_{-1}^{1} \left(\frac{d-c}{2}u - \frac{d-c}{2}x\right)^n \mathbf{1}_{\frac{d-c}{2}u - \frac{d-c}{2}x \geq 0}\, du \\
= \; & \left(\frac{d-c}{2}\right)^{n+1}\left[F_n^{[-1,1]}\left((\cdot - x)^n \mathbf{1}_{\cdot - x \geq 0}\right) - \int_{-1}^{1} (u-x)^n \mathbf{1}_{u-x \geq 0}\, du\right] \\
= \; & \left(\frac{d-c}{2}\right)^{n+1} n!K_n^{[-1,1]}(x).
\end{aligned}
$$

To complete the error estimate, we need to use that the Peano kernels $K_n^{[c,d]}(t)$ for the Newton-Cotes formulas on $[c, d]$ are of constant sign. The proof of this fact is geometrically obvious for $n = 1$ (see Exercise 13-24b) and still manageable for $n = 2$ (see Exercise 13-24c), but the algebra becomes increasingly tedious. Because we will focus only on $n = 1$ and $n = 2$ (trapezoidal and Simpson's Rule) in the examples, the presentation remains (somewhat) self-contained. For a general argument that also holds for other integration formulas, consider [10], which uses matrix methods to prove that the Peano kernel does not change its sign.

Because the Peano kernel for the Newton-Cotes formulas does not change sign, by the Mean Value Theorem for the Integral as in Exercise 8-21 there is a $\xi \in (c, d)$ so that

$$
\begin{aligned}
R_n^{[c,d]}(f) &= f^{(n+1)}(\xi)\int_c^d K_n^{[c,d]}(t)\, dt \\
&= f^{(n+1)}(\xi)\int_{-\frac{d-c}{2}}^{\frac{d-c}{2}} K_n^{[c,d]}\left(u + \frac{c+d}{2}\right) du
\end{aligned}
$$

Easy substitution, see Exercise 13-23.

$$
\begin{aligned}
&= f^{(n+1)}(\xi)\int_{-\frac{d-c}{2}}^{\frac{d-c}{2}} K_n^{\left[-\frac{d-c}{2}, \frac{d-c}{2}\right]}(u)\, du \\
&= f^{(n+1)}(\xi)\int_{-1}^{1} \frac{d-c}{2} K_n^{\left[-\frac{d-c}{2}, \frac{d-c}{2}\right]}\left(\frac{d-c}{2}x\right) dx
\end{aligned}
$$

Tedious substitution, see computation above.

$$
\begin{aligned}
&= f^{(n+1)}(\xi)\int_{-1}^{1}\left(\frac{d-c}{2}\right)^{n+2} K_n^{[-1,1]}(x)\,dx \\
&= \frac{f^{(n+1)}(\xi)}{(n+1)!}\left(\frac{d-c}{2}\right)^{n+2}\int_{-1}^{1}\left(\frac{d^{n+1}}{dx^{n+1}}x^{n+1}\right) K_n^{[-1,1]}(x)\,dx
\end{aligned}
$$

Now use Peano's error representation for $f(x) = x^{n+1}$.

$$
= \frac{f^{(n+1)}(\xi)}{(n+1)!}\left(\frac{d-c}{2}\right)^{n+2} R_n^{[-1,1]}\left(x^{n+1}\right).
$$

■

Note that the proof of Theorem 13.20 (and hence its conclusion) will work for any linear integration formula which can be moved into the integral, which gives exact results for polynomials up to degree n and for which the Peano kernel does not change sign. The reader will verify this for the Midpoint Rule in Exercise 13-25.

With Theorem 13.20 providing a bound for the error for individual Newton-Cotes formulas, bounds for the error for composite Newton-Cotes formulas are now an easy consequence.

Corollary 13.21 *Let $n \in \mathbb{N}$, $a < b$, $\delta > 0$ and let $f : (a-\delta, b+\delta) \to \mathbb{R}$ be a function. If n is odd let f be $n+1$ times continuously differentiable and let $M := \max\left\{\left|f^{(n+1)}(x)\right| : x \in [a,b]\right\}$. If n is even let f be $n+2$ times continuously differentiable and let $M := \max\left\{\left|f^{(n+2)}(x)\right| : x \in [a,b]\right\}$. Let $C_n^{[a,b]}(f)$ be the absolute value of the error when approximating the integral $\int_a^b f(t)\,dt$ with the nth* composite *Newton-Cotes formula with $N = nj$ intervals, that is, $N+1 = nj+1$ evaluations, where $j \in \mathbb{N}$. Then if n is odd we have $C_n^{[a,b]}(f) \le M\frac{(b-a)^{n+2}}{N^{n+1}}\frac{n^{n+1}R_n^{[-1,1]}\left(x^{n+1}\right)}{(n+1)!2^{n+2}}$, while if n is even we have $C_n^{[a,b]}(f) \le M\frac{(b-a)^{n+3}}{N^{n+2}}\frac{n^{n+2}R_n^{[-1,1]}\left(x^{n+2}\right)}{(n+2)!2^{n+3}}$.*

Proof. If n is odd, by Theorem 13.20 we obtain

$$
\begin{aligned}
C_n^{[a,b]}(f) &\le j\left(\frac{\frac{b-a}{j}}{2}\right)^{n+2}\frac{R_n^{[-1,1]}\left(x^{n+1}\right)}{(n+1)!}M = M\frac{(b-a)^{n+2}}{j^{n+1}2^{n+2}}\frac{R_n^{[-1,1]}\left(x^{n+1}\right)}{(n+1)!} \\
&= M\frac{(b-a)^{n+2}}{N^{n+1}}\frac{n^{n+1}R_n^{[-1,1]}\left(x^{n+1}\right)}{(n+1)!2^{n+2}}
\end{aligned}
$$

and a similar computation gives the result for even n. ■

Note that in the error bounds in Corollary 13.21 only M depends on the function f, only $(b-a)$ depends on the interval and only N depends on the number of evaluations. The remainder, although it looks complicated, is merely a constant. Therefore, as $N \to \infty$ the numerical approximation of an integral of an $n+1$ times (if n is odd) or

$n+2$ times (if n is even) continuously differentiable function with the nth composite Newton-Cotes formula will converge to the actual integral. Of course, more concrete formulas will be better for given fixed n. For the trapezoidal rule and Simpson's Rule, we obtain the error bounds indicated below.

Corollary 13.22 *Let $f:(a-\delta, b+\delta)\to\mathbb{R}$ be a function.*

1. *If f is twice continuously differentiable, we have $C_1^{[a,b]}(f)\leq K\dfrac{(b-a)^3}{12N^2}$, where $K=\max\left\{\left|f''(x)\right| : x\in[a,b]\right\}$. This is the error formula for the* **trapezoidal rule** *with N intervals.*

2. *If f is four times continuously differentiable, we have $C_2^{[a,b]}(f)\leq C\dfrac{(b-a)^5}{180N^4}$, where $C=\max\left\{\left|f^{(iv)}(x)\right| : x\in[a,b]\right\}$. This is the error formula for* **Simpson's Rule** *with N intervals.*

Proof. For $n=1$, we obtain

$$C_1^{[a,b]}(f)\leq K\frac{(b-a)^3}{N^2}\frac{1^2R_1^{[-1,1]}\left(x^2\right)}{2!2^3}=K\frac{(b-a)^3}{16N^2}R_1^{[-1,1]}\left(x^2\right)\quad\text{and}$$

$$R_1^{[-1,1]}\left(x^2\right)=\frac{2}{2}\left((-1)^2+1^2\right)-\int_{-1}^{1}x^2\,dx=2-\frac{2}{3}=\frac{4}{3},$$

which proves $C_1^{[a,b]}(f)\leq K\dfrac{(b-a)^3}{12N^2}$. The computation for Simpson's Rule is similar and left to the reader as Exercise 13-26. ■

If the function, the interval and N are given, the *a posteriori* error estimate is simply a substitution into the formula (see Exercise 13-27). If the function, the interval and a desired accuracy are given, the *a priori* estimate is obtained by demanding the error bound is less than the desired accuracy and solving for N.

Example 13.23 *Find the number of intervals needed to estimate $\int_{-1}^{1}e^{-\frac{x^2}{2}}\,dx$ with the trapezoidal rule so that the error is less than 10^{-4}.*

First, note that with $f(x)=e^{-\frac{x^2}{2}}$ we have the derivatives $f'(x)=-xe^{-\frac{x^2}{2}}$ and $f''(x)=\left(x^2-1\right)e^{-\frac{x^2}{2}}$. The maximum of $\left|f''\right|$ is assumed at $x=0$, so $K=1$. Now $K\dfrac{(b-a)^3}{12N^2}<10^{-4}$ means $1\dfrac{2^3}{12N^2}<10^{-4}$ or $N^2>\dfrac{2}{3}10^4$. Hence, for N we obtain $N>100\sqrt{\dfrac{2}{3}}\approx 81.65$ and so $N=82$ would work. Note that in the last step, we always need to round up. □

Exercises

13-20. Compute the coefficients α_i in the Newton-Cotes formulas for $n = 7$ Use a computer

13-21 For each given n compute $\alpha_0, \ldots, \alpha_n$ for the Newton-Cotes formula Then state the corresponding composite integration formula for N divisible by n. Finally, compute the error for the Newton-Cotes formula and for the composite integration formula For the computation of the α_i, a computer is recommended.

(a) $n = 3$ This is called the $\frac{3}{8}$**-rule**

(b) $n = 4$. This is called **Milne's Rule**.

(c) $n = 5$. This rule does not have a specific name.

(d) $n = 6$. This is called **Weddle's Rule**.

13-22. Prove as follows that for even n the nth Newton-Cotes formula gives accurate results for polynomials of degree $n + 1$

(a) Use the linearity property to prove that if the nth Newton-Cotes formula gives the exact integral for one polynomial of degree $n + 1$, then it gives the exact integral for all of them

(b) Prove that the coefficients α_i of the nth Newton-Cotes formula satisfy $\alpha_i = \alpha_{n-i}$ for all $i \in \{0, \ldots, n\}$.

(c) Prove that if n is even, the nth Newton-Cotes formula gives the exact integral for the polynomial $p(x) = \left(x - \frac{c+d}{2}\right)^{n+1}$ and conclude that the nth Newton-cotes formula gives the exact integral for all polynomials of degree $n + 1$.

(d) To illustrate that the result does not work for odd n, prove that the first Newton-Cotes formula does not give the integral of $f(x) = x^2$ on $[-1, 1]$.

13-23 Prove that for all $t \in \left[-\frac{d-c}{2}, \frac{d-c}{2}\right]$ we have $K_n^{[c,d]}\left(t + \frac{c+d}{2}\right) = K_n^{\left[-\frac{d-c}{2}, \frac{d-c}{2}\right]}(t)$

13-24 The sign of the Peano kernel for the Newton-Cotes formulas.

(a) Explain why proving that the Peano kernel $K_n^{[-1,1]}$ for the nth Newton-Cotes formula on $[-1, 1]$ does not change its sign on $[-1, 1]$ proves that all Peano kernels $K_n^{[c,d]}$ for the nth Newton-Cotes formula on $[c, d]$ do not change signs on $[c, d]$.

(b) Prove that for $n = 1$ the Peano kernel $K_n^{[-1,1]}$ is nonnegative on $[-1, 1]$.

Hint Direct computation of the integrals, using that the approximating polynomial is a straight line

(c) Prove that for $n = 2$ the Peano kernel $K_n^{[-1,1]}$ is nonnegative on $[-1, 1]$.

Hint This computation is more tedious. Make sure you use $(x - t)^3 \mathbf{1}_{x-t\geq 0}$ and do separate computations for $t > 0$ and $t \leq 0$

(d) In a computer algebra system implement a short program that graphs the Peano kernel $K_n^{[-1,1]}$ on $[-1, 1]$ for arbitrary n

Note. While these graphs do not formally prove that the Peano kernel does not change its sign, the graphs and the implementation are instructive.

13-25. Let $f : [a, b] \to \mathbb{R}$ be twice continuously differentiable. Prove that when approximating $\int_a^b f(x)\,dx$ with the **midpoint rule**, which uses Riemann sums with evaluation in the middle of the interval to approximate the integral, the error is bounded by $K\frac{(b-a)^3}{24n^2}$, where $K = \max\left\{|f''(x)| \; a \leq x \leq b\right\}$.

Hint. Use Peano's representation of the error and use the fact that the midpoint rule is accurate for polynomials of degree $n \leq 1$ Then emulate the rest of the proof for the Newton-Cotes formulas, including the proof that the Peano kernel does not change its sign

13-26. Prove part 2 of Corollary 13 22.

13-27. In each part, give an upper bound for the error of the approximation of the given integral with the given rule and the given number of intervals.

(a) $\int_{-1}^{1} e^{-\frac{x^2}{2}}\, dx$, trapezoidal rule, $N = 20$

(b) $\int_{-1}^{1} e^{-\frac{x^2}{2}}\, dx$, Simpson's Rule, $N = 20$

(c) $\int_{0}^{5} \sin\left(x^2\right)\, dx$, trapezoidal rule, $N = 50$

(d) $\int_{0}^{5} \sin\left(x^2\right)\, dx$, Simpson's Rule, $N = 50$

13-28. Compute the number N of intervals needed to approximate the integral with the indicated rule so that the error is at most the given ν

(a) $\int_{-1}^{1} e^{-\frac{x^2}{2}}\, dx$, Simpson's Rule, $\nu \leq 10^{-4}$

(b) $\int_{0}^{2} \sqrt{1+x^3}\, dx$, trapezoidal rule, $\nu \leq 10^{-7}$

(c) $\int_{2}^{4} \ln(x)\sqrt{x}\, dx$, Simpson's Rule, $\nu \leq 10^{-8}$

13-29. Approximate the integral with the indicated rule so that the error is at most the given ν.

Hint Use the error bounds in Corollary 13.22 to compute the number of intervals. Then compute the requisite sum with a computer

(a) $\int_{-1}^{1} \frac{1}{\sqrt{2\pi}} e^{-\frac{x^2}{2}}\, dx$, trapezoidal rule, $\nu \leq 10^{-8}$

(b) $\int_{-1}^{1} \frac{1}{\sqrt{2\pi}} e^{-\frac{x^2}{2}}\, dx$, Simpson's Rule, $\nu \leq 10^{-8}$

(c) $\int_{-2}^{2} \frac{1}{\sqrt{2\pi}} e^{-\frac{x^2}{2}}\, dx$, Simpson's Rule, $\nu \leq 10^{-8}$

(d) $\int_{-3}^{3} \frac{1}{\sqrt{2\pi}} e^{-\frac{x^2}{2}}\, dx$, Simpson's Rule, $\nu \leq 10^{-8}$

(e) $\int_{\frac{1}{10}}^{1} \sin\left(\frac{1}{x}\right)\, dx$, trapezoidal rule, $\nu \leq 10^{-2}$

(f) $\int_{1}^{5} \sqrt{x}e^{x}\, dx$, Simpson's Rule, $\nu \leq 10^{-6}$

13-30. Let the function $f : [a, b] \to \mathbb{R}$ be continuously differentiable. Use Euler's Summation Formula (see Exercise 12-21) and an appropriate substitution to prove that the approximations of the integral $\int_a^b f(x)\, dx$ with the trapezoidal rule with N trapezoids converge to $\int_a^b f(x)\, dx$ as $N \to \infty$

Part II

Analysis in Abstract Spaces

Chapter 14

Integration on Measure Spaces

Throughout this second part of the text, we need to integrate multivariable functions. Therefore we start our investigation of the more abstract realms of analysis with integration. Recall that the fundamental idea behind Lebesgue integration was the partition of the range (see Figure 22). Moreover, when we worked with Lebesgue measure, we were concerned with properties stated in terms of sets. We rarely used the fact that these sets were subsets of the real numbers. Therefore integration of functions from a more general space to $[-\infty, \infty]$ should be similar to the theory developed in Chapter 9. Indeed, Sections 14.1–14.4 basically recast and sort the results of Chapter 9 to show how these ideas generalize to arbitrary measure spaces. In particular, this generalization makes it possible to talk about Lebesgue integration in $\mathbb{R}^d$. Sections 14.5 and 14.6 provide fundamental results on sequences of integrable functions. These results are the cornerstone for the proofs of many important facts about integrable functions. Finally, Sections 14.7 and 14.8 show how measures on products (such as $\mathbb{R}^2 = \mathbb{R} \times \mathbb{R}$) are related to measures on the factors.

In this chapter, we experience the full power of abstraction for the first time. Once the abstract core of concrete results for the real numbers is identified, the familiar results from the real numbers can be established in much more general contexts, sometimes even with the same proof. It is important to realize, however, that this generalization comes at a price. We must very carefully check that we did not use any specific properties of the real line in the proof of the concrete result. The most important property of the real line that is lost in the general setting is the linear ordering. Proofs and results that depend on this ordering do not generalize easily. This is why the linear ordering of the real line was used sparingly in the first part of the text.

14.1 Measure Spaces

To be most widely applicable, abstract integration is defined on sets equipped with a structure that makes them a "measure space." In this fashion, we do not need to worry about details regarding the shape and dimension of the domain. The fundamental idea for integration in arbitrary spaces is the same as for Lebesgue integration. Partition the

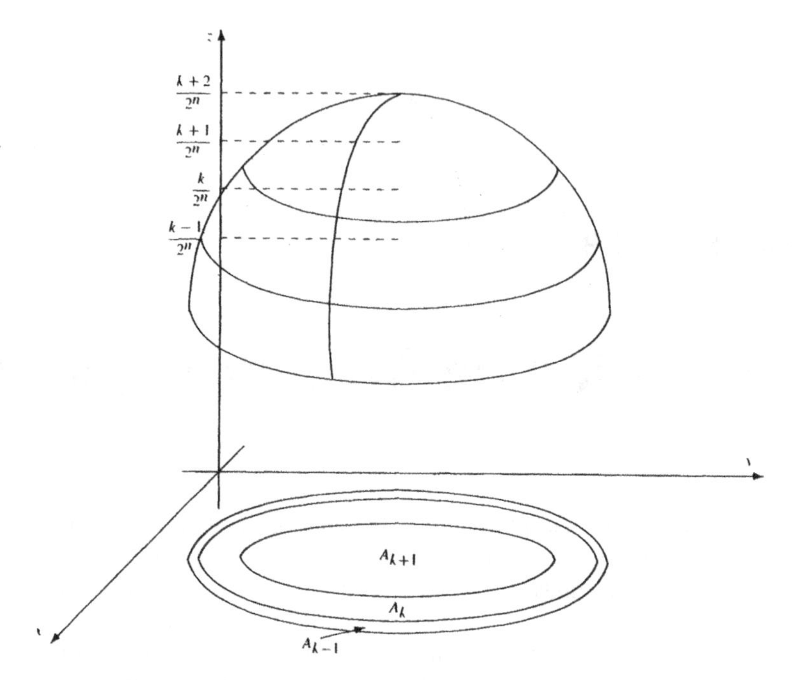

Figure 29: For integration in more complicated spaces than the real line, we retain the main idea from Lebesgue integration. We partition the range (here the z-axis) and measure the size of the inverse images A_k of intervals in the range (the z-axis). The sum of these sizes times the corresponding heights $\left(\text{here } \frac{k}{2^n}\right)$ should approximate the "volume" under the graph of the function.

range and approximate the "area" or "volume" under the function with "generalized rectangles" or "generalized boxes" whose bases are sets for which we can determine the "measure" (see Figure 29). It was noted after Definition 9.4 that even on the real line there are sets for which we cannot determine a sensible "measure" using outer Lebesgue measure. Thus, it is not surprising that on a general set M we need to consider the subset of the power set $\mathcal{P}(M)$ that contains all subsets for which we can determine the "measure." The properties that define these subsets are directly inspired by properties of Lebesgue measurable sets (see Theorem 9.10).

Definition 14.1 *Let M be a set. A subset $\Sigma \subseteq \mathcal{P}(M)$ is called a* **sigma algebra** *or* **σ-algebra** *iff*

1. $\emptyset \in \Sigma$,

2. *If $S \in \Sigma$, then $S' \in \Sigma$,*

3. *If $A_n \in \Sigma$ for all $n \in \mathbb{N}$, then $\bigcup_{n=1}^{\infty} A_n \in \Sigma$.*

Our most important examples of σ-algebras so far are the power set itself and the set of Lebesgue measurable subsets of $\mathbb{R}$.

Example 14.2

1. *Let M be a set. Then the power set $\mathcal{P}(M)$ of M is a σ-algebra.*

2. *The set Σ_λ of Lebesgue measurable subsets of $\mathbb{R}$ is a σ-algebra.*

Part 1 is trivial and part 2 is Theorem 9.10. □

More examples of σ-algebras are given in Exercise 14-1 and Theorem 14.25. Because σ-algebras are newly introduced entities, we need to prove that the properties we know from Lebesgue measurable sets also hold in this general context.

Proposition 14.3 *Let M be a set and let $\Sigma \subseteq \mathcal{P}(M)$ be a σ-algebra. If for every $n \in \mathbb{N}$ we have $A_n \in \Sigma$, then $\bigcap_{n=1}^{\infty} A_n \in \Sigma$.*

Proof. For each $n \in \mathbb{N}$ by part 2 of Definition 14.1, we infer $M \setminus A_n \in \Sigma$. By part 3 of Definition 14.1, we obtain $\bigcup_{n=1}^{\infty} M \setminus A_n \in \Sigma$. By DeMorgan's Laws, this means that $M \setminus \bigcap_{n=1}^{\infty} A_n \in \Sigma$. Therefore by part 2 of Definition 14.1 $\bigcap_{n=1}^{\infty} A_n \in \Sigma$, which concludes the proof. ■

Proposition 14.4 *Let M be a set, let $\Sigma \subseteq \mathcal{P}(M)$ be a σ-algebra, let $N \in \mathbb{N}$ and let $A_1, \ldots, A_N \in \Sigma$. Then $\bigcup_{n=1}^{N} A_n \in \Sigma$.*

Proof. For $n \geq N+1$, let $A_n := \emptyset$. Then $\bigcup_{n=1}^{N} A_n = \bigcup_{n=1}^{\infty} A_n \in \Sigma$. ■

Further properties of σ-algebras are presented in Exercise 14-2. Recall that for the definition of Lebesgue measurable functions we never referred to the measure itself. Thus for some purposes it will be sufficient to work with a set and its measurable subsets. This is the idea behind a measurable space.

Definition 14.5 *A* **measurable space** *is a pair (M, Σ) consisting of a set X and a σ-algebra $\Sigma \subseteq \mathcal{P}(M)$. The sets in Σ will also be called* **Σ-measurable sets**.

Finally, a measure is a function that assigns each measurable set its "measure." The only conditions for a sensible "measure" function are that the empty set has no volume and that the volumes of pairwise disjoint sets are added to obtain the volume of their union.

Definition 14.6 *Let (M, Σ) be a measurable space. Then $\mu : \Sigma \to [0, \infty]$ is called a* **measure** *iff*

1. $\mu(\emptyset) = 0$, *and*

2. μ *is* **countably additive**, *that is, if $\{A_n\}_{n=1}^{\infty}$ is a sequence of pairwise disjoint sets $A_n \in \Sigma$, then* $\mu\left(\bigcup_{n=1}^{\infty} A_n\right) = \sum_{n=1}^{\infty} \mu(A_n)$.

The definition of a measure space connects the "measurable sets" with a function that assigns the measure.

Definition 14.7 *A* **measure space** *is a triple (M, Σ, μ) consisting of a set M, a sigma algebra $\Sigma \subseteq \mathcal{P}(M)$ and a measure $\mu : \Sigma \to [0, \infty]$.*

Example 14.8 *With λ denoting outer Lebesgue measure and Σ_λ denoting the sigma algebra of Lebesgue measurable sets, $(\mathbb{R}, \Sigma_\lambda, \lambda)$ is a measure space.*

We have already noted that Σ_λ is a σ-algebra. Trivially $\lambda(\emptyset) = 0$, and the countable additivity of Lebesgue measure is given by Theorem 9.11. □

Example 14.9 *Let M be a set. For $A \subseteq M$, we define the* **counting measure** *$\gamma_M(A)$ to be the number of elements of A if A is finite and ∞ if A is infinite. Then $\big(M, \mathcal{P}(M), \gamma_M\big)$ is a measure space.*

The reader will prove this in Exercise 14-3. Note that counting measure will allow us to model absolutely convergent series as integrals in Examples 14.36 and 14.37 below. □

In Chapter 9, we defined Lebesgue integration over $\mathbb{R}$, but we never formally defined Lebesgue integration over subsets of $\mathbb{R}$ such as intervals. The formalism of measure spaces allows us to easily fill this gap. Every measurable subset of a measure space can be equipped with the structure of a measure space.

Example 14.10 *Let (M, Σ, μ) be a measure space and let $\Omega \in \Sigma$ be measurable. Let $\Sigma^{\Omega} := \{S \in \Sigma : S \subseteq \Omega\}$ and let $\mu_\Omega := \mu|_{\Sigma^\Omega}$. Then $\big(\Omega, \Sigma^{\Omega}, \mu_\Omega\big)$ is a measure space.*

The reader will prove this in Exercise 14-4. □

Because measure spaces are newly introduced, we must prove their properties "from scratch." Specifically, we cannot use familiar properties of Lebesgue measure without first proving them for measure spaces. Nonetheless, the next three propositions should be familiar from Lebesgue measure.

Proposition 14.11 *Let (M, Σ, μ) be a measure space and let $A_1, \ldots, A_N$ be pairwise disjoint sets in Σ. Then* $\mu\left(\bigcup_{n=1}^{N} A_n\right) = \sum_{n=1}^{N} \mu(A_n)$.

Proof. By Proposition 14.4, we have that $\bigcup_{n=1}^{N} A_n \in \Sigma$. For $n \geq N+1$, let $A_n := \emptyset$.

Then $\mu\left(\bigcup_{n=1}^{N} A_n\right) = \mu\left(\bigcup_{n=1}^{\infty} A_n\right) = \sum_{n=1}^{\infty} \mu(A_n) = \sum_{n=1}^{N} \mu(A_n)$. ■

Proposition 14.12 *Let (M, Σ, μ) be a measure space and let $A, B \in \Sigma$ with $A \subseteq B$. Then $\mu(A) \leq \mu(B)$.*

Proof. By Exercise 14-2b we have $B \setminus A \in \Sigma$. Therefore

$$\mu(A) \leq \mu(A) + \mu(B \setminus A) = \mu\big(A \cup (B \setminus A)\big) = \mu(B).$$ ■

Definition 14.13 *Let (M, Σ, μ) be a measure space. Then $A \in \Sigma$ is called a set of* **measure zero** *or a* **null set** *iff $\mu(A) = 0$. A property is said to hold* **almost everywhere** *in M iff the subset of M for which the property does not hold is of measure zero. Almost everywhere is also abbreviated as* **a.e.**

Proposition 14.14 *Let (M, Σ, μ) be a measure space and for all $n \in \mathbb{N}$ let the set $A_n \in \Sigma$ be a null set. Then $\bigcup_{n=1}^{\infty} A_n$ also is a null set.*

Proof. By Proposition 14.12, subsets of null sets are null sets also. For $n \in \mathbb{N}$, define $B_n := A_n \setminus \bigcup_{j=1}^{n-1} A_j$. Then each B_n is a null set and $\bigcup_{n=1}^{\infty} B_n = \bigcup_{n=1}^{\infty} A_n$. But $\mu\left(\bigcup_{n=1}^{\infty} B_n\right) = \sum_{n=1}^{\infty} \mu(B_n) = 0$, which establishes the result. ■

We conclude this section with a result that shows that the measure of the union of a nested sequence of sets is equal to the limit of the nondecreasing sequence of the measures of the sets

Theorem 14.15 *Let (M, Σ, μ) be a measure space and let $\{A_n\}_{n=1}^{\infty}$ be a sequence of sets in Σ so that $A_n \subseteq A_{n+1}$ for all $n \in \mathbb{N}$. Then $\mu\left(\bigcup_{n=1}^{\infty} A_n\right) = \lim_{n\to\infty} \mu(A_n)$.*

Proof. Mimic the proof of Theorem 9.12. (Exercise 14-5.) ■

Exercises

14-1 Let M be a set. Prove that the set of all subsets $S \subseteq M$ so that S is countable or $M \setminus S$ is countable is a σ-algebra.

14-2. Let M be a set and let $\Sigma \subseteq \mathcal{P}(M)$ be a σ-algebra.

(a) Prove that if $A_1, \ldots, A_N \in \Sigma$, then $\bigcap_{n=1}^{N} A_n \in \Sigma$

(b) Prove that if $A, B \in \Sigma$, then $A \setminus B \in \Sigma$.

14-3 Counting measure. Let M be a set and let $\gamma_M : \mathcal{P}(M) \to [0, \infty]$ be counting measure on M

(a) Prove that $\gamma_M(A) = 0$ iff $A = \emptyset$

(b) Prove that γ_M is a measure

14-4. Measures on subsets. Let (M, Σ, μ) be a measure space, let $\Omega \in \Sigma$, let $\Sigma^{\Omega} := \{S \in \Sigma : S \subseteq \Omega\}$ and let $\mu_{\Omega} := \mu|_{\Sigma^{\Omega}}$

(a) Prove that Σ^{Ω} is a σ-algebra

(b) Prove that μ_{Ω} is a measure

14-5 Prove Theorem 14.15

14-6 Let (M, Σ, μ) be a measure space and let $\{A_n\}_{n=1}^{\infty}$ be a sequence of sets in Σ Prove the inequality

$$\mu\left(\bigcup_{n=1}^{\infty} A_n\right) \le \sum_{n=1}^{\infty} \mu(A_n)$$

14-7. Let M be a set. Prove that a subset $\Sigma \subseteq \mathcal{P}(M)$ is a σ-algebra iff $\emptyset \in \Sigma$; if $S \in \Sigma$, then $S' \in \Sigma$; and if $A_n \in \Sigma$ for all $n \in \mathbb{N}$, then $\bigcap_{n=1}^{\infty} A_n \in \Sigma$.

14-8. Let (M, Σ, μ) be a measure space and let $A, B \in \Sigma$ Prove $\mu(A) - \mu(B) = \mu(A \setminus B) - \mu(B \setminus A)$

14-9 The measure of nested intersections

(a) Let (M, Σ, μ) be a measure space and let $\{A_n\}_{n=1}^{\infty}$ be a sequence of sets in Σ such that for all $n \in \mathbb{N}$ we have $A_n \supseteq A_{n+1}$ and such that for some $m \in \mathbb{N}$ we have $\mu(A_m) < \infty$. Prove that $\mu\left(\bigcap_{n=1}^{\infty} A_n\right) = \lim_{n\to\infty} \mu(A_n)$

(b) Show that the condition $\mu(A_m) < \infty$ for some $m \in \mathbb{N}$ cannot be dropped.
Hint. Let $\gamma_{\mathbb{N}}$ be counting measure on $\mathbb{N}$ and let $A_n := \{i \in \mathbb{N} : i \ge n\}$

14-10 Let (M, Σ, μ) be a measure space

(a) Prove that $\Sigma_{\mu} := \left\{ A \subseteq M : \left(\exists E, F \in \Sigma . E \subseteq A \subseteq F, \mu(F \setminus E) = 0\right) \right\}$ is a σ-algebra that contains Σ.

(b) Prove that for all $A \in \Sigma_{\mu}$ and all $E, F \in \Sigma$ with $E \subseteq A \subseteq F$ and $\mu(F \setminus E) = 0$ we have $\mu(E) = \mu(F)$.

(c) For all $A \in \Sigma_{\mu}$, let $E, F \in \Sigma$ with $E \subseteq A \subseteq F$ and $\mu(F \setminus E) = 0$ and define $\overline{\mu}(A) := \mu(F)$ Prove that $\overline{\mu} : \Sigma_{\mu} \to [0, \infty]$ is a measure (You also need to show that $\overline{\mu}$ is well-defined.)

(d) Prove that for all $B \in \Sigma$ we have $\overline{\mu}(B) = \mu(B)$.

(e) Prove that the measure space $(M, \Sigma_{\mu}, \overline{\mu})$ is **complete**. That is, prove that if $N \in \Sigma_{\mu}$ is so that $\mu(N) = 0$ and $S \subseteq N$, then $S \in \Sigma_{\mu}$

The σ-algebra Σ_{μ} is also called the **completion** of Σ and $\overline{\mu}$ is also called the **completion** of μ.

14.2 Outer Measures

Although not all subsets of the real numbers are Lebesgue measurable, outer Lebesgue measure is defined for all subsets of $\mathbb{R}$. The idea of an outer measure can be transplanted to an abstract space. This section shows that outer measures produce a measure space similar to how outer Lebesgue measure produced a measure space. In particular, we will obtain a Lebesgue measure on d-dimensional space.

Definition 14.16 *Let M be a set. Then $\mu : \mathcal{P}(M) \to [0, \infty]$ is called an* **outer measure** *iff*

1. $\mu(\emptyset) = 0$,
2. *If $A \subseteq B$, then $\mu(A) \leq \mu(B)$,*
3. *The function μ is* **countably additive**, *that is, for all sequences $\{A_n\}_{n=1}^{\infty}$ of sets in M we have $\mu\left(\bigcup_{n=1}^{\infty} A_n\right) \leq \sum_{n=1}^{\infty} \mu(A_n)$.*

By Theorem 8.6, outer Lebesgue measure on $\mathbb{R}$ is an outer measure. To integrate in $\mathbb{R}^d$ (and via Example 14.10 on subsets $\Omega \subseteq \mathbb{R}^d$) we need to define an outer measure on $\mathbb{R}^d$ that is similar to outer Lebesgue measure on $\mathbb{R}$. The definition is very much the same as for the real line, except that instead of open intervals we use d-dimensional boxes. The thus defined outer measure is also called outer Lebesgue measure and it is also denoted with λ.

Definition 14.17 *Let $d \geq 1$ and for $i = 1, \ldots, d$ let $a_i < b_i$. Then a set of the form $B := \prod_{i=1}^{d}(a_i, b_i)$ is called an* **open box** *in $\mathbb{R}^d$. We define $|B| := \prod_{i=1}^{d}(b_i - a_i)$. For a set $S \subseteq \mathbb{R}^d$ we define the* **outer Lebesgue measure** *of S to be*

$$\lambda(S) := \inf\left\{\sum_{j=1}^{\infty}|B_j| : S \subseteq \bigcup_{j=1}^{\infty} B_j, \text{ each } B_j \text{ is an open box in } \mathbb{R}^d\right\},$$

where we set $\lambda(S) = \infty$ if none of the series in the set on the right converge.

It would be nice to show here, similar to Proposition 8.5, that the outer Lebesgue measure of a box is exactly the volume of the box. We postpone this result to Theorem 16.81, where we can use compactness for an efficient proof.

The argument that outer Lebesgue measure on $\mathbb{R}^d$ defines a measure on its measurable sets is exactly the same as for $\mathbb{R}$. Thus the following results and the definition of measurable sets can be seen as a recap of the appropriate parts of Section 9.1.

Theorem 14.18 *Outer Lebesgue measure on $\mathbb{R}^d$ is an outer measure.*

Proof. Mimic the proof of Theorem 8.6. (Exercise 14-11.) ∎

The motivation and definition of measurable sets in the abstract setting are the same as for Lebesgue measure (see Definition 9.4). We must prevent that the sum of the measures of the pairwise disjoint measurable parts of a set is different from the measure of the whole measurable set.

Definition 14.19 *Let M be a set and let $\mu : \mathcal{P}(M) \to [0, \infty]$ be an outer measure. A subset $S \subseteq M$ is called μ-***measurable** *iff for all $T \subseteq M$ we have that $\mu(T) = \mu(S \cap T) + \mu\left(S' \cap T\right)$. The set of μ-measurable subsets of M is denoted Σ_μ.*

As in Definition 9.5 the Lebesgue measure is obtained by restricting outer Lebesgue measure to the set Σ_λ of Lebesgue measurable sets.

Definition 14.20 *Let $S \subseteq \mathbb{R}^d$ be a Lebesgue measurable set. Then the outer Lebesgue measure $\lambda(S)$ of S is also called the* **Lebesgue measure** *of S.*

As before, "half" of the equality for measurability is always satisfied. Moreover, the proof that outer measures induce measure spaces runs along the same lines as the proofs of the corresponding results in Section 9.1.

Proposition 14.21 *Let M be a set and let $\mu : \mathcal{P}(M) \to [0, \infty]$ be an outer measure. For all subsets $S, T \subseteq M$, we have $\mu(T) \leq \mu(S \cap T) + \mu\left(S' \cap T\right)$.*

Proof. Mimic the proof of Corollary 9.6. (Exercise 14-12.) ■

Proposition 14.22 *Let M be a set and let $\mu : \mathcal{P}(M) \to [0, \infty]$ be an outer measure. If $\mu(S) = 0$, then S is μ-measurable.*

Proof. Mimic the proof of Proposition 9.7. (Exercise 14-13.) ■

Lemma 14.23 *Let M be a set and let $\mu : \mathcal{P}(M) \to [0, \infty]$ be an outer measure. If A and B are μ-measurable, then the intersection $A \cap B$ is μ-measurable.*

Proof. Mimic the proof of Lemma 9.8. (Exercise 14-14.) ■

Lemma 14.24 *Let M be a set, let $\mu : \mathcal{P}(M) \to [0, \infty]$ be an outer measure and let $\{A_n\}_{n=1}^{\infty}$ be a sequence of pairwise disjoint μ-measurable sets. Then $\bigcup_{n=1}^{\infty} A_n$ is μ-measurable and $\mu(T) = \sum_{n=1}^{\infty} \mu\left(A_n \cap T\right) + \mu\left(\left(\bigcup_{n=1}^{\infty} A_n\right)' \cap T\right)$ for all $T \subseteq M$.*

Proof. Mimic the proof of Lemma 9.9. (Exercise 14-15.) ■

Theorem 14.25 *Let M be a set, let $\mu : \mathcal{P}(M) \to [0, \infty]$ be an outer measure and let Σ_μ be the set of μ-measurable sets. Then $\left(M, \Sigma_\mu, \mu\right)$ is a measure space.*

Proof. Mimic the proof of Theorem 9.10 to prove that Σ_μ is a σ-algebra and then mimic the proof of Theorem 9.11 to prove that μ is countably additive on Σ_μ. (Exercise 14-16.) ■

In particular, Theorem 14.25 shows that the triple $\left(\mathbb{R}^d, \Sigma_\lambda, \lambda\right)$ is a measure space. Lebesgue measure is the standard measure for integration in d-dimensional space. Thus, as we proceed to define measurable and integrable functions, we are constructing a theory that allows us to integrate on $\mathbb{R}^d$ and its subsets.

Exercises

14-11 Prove Theorem 14.18

14-12 Prove Proposition 14.21

14-13 Prove Proposition 14.22.

14-14. Prove Lemma 14.23.

14-15. Prove Lemma 14.24.

14-16. Prove Theorem 14.25.

14-17. Let $d \geq 1$ and for $i = 1, \ldots, d$ let the numbers $a_i < b_i$ be dyadic rational numbers Then a set of the form $D := \prod_{i=1}^{d}(a_i, b_i)$ is called a **dyadic open box** in $\mathbb{R}^d$ Prove that for any set $S \subseteq \mathbb{R}^d$ we have $\lambda(S) = \inf\left\{\sum_{j=1}^{\infty}|D_j| : S \subseteq \bigcup_{j=1}^{\infty} D_j,\text{ each } D_j \text{ is a dyadic open box in } \mathbb{R}^d\right\}$.

14-18. Prove that if $A, B \subseteq \mathbb{R}$ and $\lambda(A) = 0$ (where λ is Lebesgue measure on $\mathbb{R}$), then $\lambda(A \times B) = 0$ (where λ is Lebesgue measure on $\mathbb{R}^2$).

14-19 Looking at Riemann integrals from a measure theoretic point-of-view. Let $[a, b]$ be an interval. For a set $S \subseteq [a, b]$ we define $J(S) := \inf\left\{\sum_{j=1}^{n}|I_j| : S \subseteq \bigcup_{j=1}^{n} I_j,\text{ each } I_j \text{ is an open interval}\right\}$ and call it the **Jordan content** of S. Note that the only difference between the Jordan content and outer Lebesgue measure is that the sums over which the infimum is taken are finite.

(a) Prove that for any closed interval $[c, d] \subseteq [a, b]$ we have $J\left([c, d]\right) = d - c$.

(b) Prove that the Jordan content is *not* an outer measure.
Hint: $\mathbb{Q}$.

(c) Let $J_i(S) := \sup\left\{\sum_{j=1}^{n}|b_j - a_j| : S \supseteq \bigcup_{j=1}^{n}[a_j, b_j], a_1 \leq b_1 \leq a_2 \leq b_2 \leq \cdots \leq a_n \leq b_n\right\}$, and call it the **inner Jordan content** of $S \subseteq [a, b]$. Prove that for any closed interval $[c, d] \subseteq [a, b]$ we have $J_i\left([c, d]\right) = d - c$.

(d) Prove that the inner Jordan content is *not* an outer measure.

(e) Call a set $S \subseteq \mathbb{R}$ **Jordan measurable** iff $J(S) = J_i(S)$. An **algebra** of sets satisfies the first two properties of a σ-algebra, but it is only closed under finite unions rather than infinite unions. Prove that the set $\mathcal{J}_{[a,b]}$ of Jordan measurable subsets of $[a, b]$ forms an algebra

(f) Prove that the Jordan content $J : \mathcal{J}_{[a,b]} \to [0, \infty)$ is a **finitely additive measure**. That is, $J(\emptyset) = 0$ and if $\{A_n\}_{n=1}^{N}$ is a finite sequence of pairwise disjoint sets $A_n \in \mathcal{J}_{[a,b]}$, then $J\left(\bigcup_{n=1}^{N} A_n\right) = \sum_{n=1}^{N} J(A_n)$

(g) Prove that a set $S \subseteq [a, b]$ is Jordan measurable iff its indicator function $\mathbf{1}_S$ is Riemann integrable.

Note. This exercise shows it is fair to say that the problem with the Riemann integral is that its associated notion of a content is only finitely additive.

14-20. Let $a < b$ be real numbers and let $g : [a, b] \to \mathbb{R}$ be nondecreasing. For numbers $c, d \in (a, b)$ so that $c < d$ define $\left|(c, d)\right|_g := \lim_{x \to d^-} g(x) - \lim_{x \to c^+} g(x)$, define $\left|[a, d)\right|_g := \lim_{x \to d^-} g(x) - g(a)$, and define $\left|(c, b]\right|_g := g(b) - \lim_{x \to c^+} g(x)$. Set $\left|[a, b]\right|_g := g(b) - g(a)$. Let $\mathcal{I}$ be the set of all subintervals of $[a, b]$ that are either open, closed at a and open on the right, closed at b and open on the left or equal to $[a, b]$ For any set $S \subset [a, b]$, we define the **outer Lebesgue-Stieltjes measure** of S to be $\lambda_g(S) := \inf\left\{\sum_{j=1}^{\infty}|I_j|_g : S \subseteq \bigcup_{j=1}^{\infty} I_j,\text{ each } I_j \text{ is in } \mathcal{I}\right\}$

(a) Prove that the outer Lebesgue-Stieltjes measure really is an outer measure

(b) Prove that open intervals $(c, d) \subseteq [a, b]$ are λ_g-measurable.
Hint. Compare with Proposition 9.13.

(c) Prove that for $c < d$ both in $[a, b]$ we have $\lambda_g\big([c, d]\big) = \lim_{x\to d^+} g(x) - \lim_{x\to c^-} g(x)$, where the one-sided limit is understood to be the value of g if $c = a$ or $d = b$.

(d) Let $g(x) := \begin{cases} 0, & \text{for } x < c, \\ 1, & \text{for } x \geq c. \end{cases}$ Prove that $\lambda_g(S) := \begin{cases} 0; & \text{if } c \notin S, \\ 1; & \text{if } c \in S, \end{cases}$ for any set $S \subseteq [a, b]$

(e) Construct a nondecreasing function $g : [a, b] \to \mathbb{R}$ and an interval $[c, d] \subseteq [a, b]$ so that $\lambda_g\big([c, d]\big) > g(d) - g(c)$

(f) Prove that the function $f : [a, b] \to \mathbb{R}$ is **Riemann-Stieltjes integrable** with respect to g iff $\lambda_g\big(\{x : f \text{ is discontinuous at } x\}\big) = 0$

14.3 Measurable Functions

As we did on the real line, we first define the functions for which there is a chance that the integral exists and then we define the integral. Indicator functions (see Definition 5.9) are once more our "rectangles" and simple functions are composed of finitely many disjoint "rectangles." From here on, **algebraic operations on functions will always be understood to be pointwise**, that is, for example, the sum $f + g$ of two functions f and g with the same domain is defined as $(f + g)(x) := f(x) + g(x)$ for all elements x of the domain.

Definition 14.26 *Let (M, Σ) be a measurable space. A* **simple measurable function** *is a function $s : M \to \mathbb{R}$ such that there are $n \in \mathbb{N}$, $a_1, \ldots, a_n \in \mathbb{R}$ and pairwise disjoint sets $A_1, \ldots, A_n \in \Sigma$ so that $s = \sum_{k=1}^{n} a_k \mathbf{1}_{A_k}$. We will also call these functions* **simple functions**.

For measurable functions, we consider the positive and negative parts separately.

Definition 14.27 *Let M be a set. For $f : M \to [-\infty, \infty]$ we define the* **positive part** *$f^+(x) := \max\big\{f(x), 0\big\}$ and the* **negative part** *$f^-(x) := -\min\big\{f(x), 0\big\}$.*

Definition 14.28 *Let (M, Σ) be a measurable space. The nonnegative function $f : M \to [0, \infty]$ is called* **measurable** *iff there is a sequence $\{s_n\}_{n=1}^{\infty}$ of simple functions $s_n : M \to [0, \infty)$ such that for all $x \in M$ the sequence $\big\{s_n(x)\big\}_{n=1}^{\infty}$ is nondecreasing and $\lim_{n\to\infty} s_n(x) = f(x)$.*

A function $f : M \to [-\infty, \infty]$ is called **measurable** *iff f^+ and f^- are both measurable. If it is necessary to distinguish between several σ-algebras, we will also call these functions* **Σ-measurable**.

Once again, measurable functions have many characterizations.

Theorem 14.29 *Let (M, Σ) be a measurable space and let $f : M \to [-\infty, \infty]$ be a function. Then the following are equivalent.*

1. *f is measurable,*
2. *For all $a \in \mathbb{R}$, we have $\{x \in M : f(x) > a\} \in \Sigma$,*
3. *For all $a \in \mathbb{R}$, we have $\{x \in M : f(x) \leq a\} \in \Sigma$,*
4. *For all $a \in \mathbb{R}$, we have $\{x \in M : f(x) < a\} \in \Sigma$,*
5. *For all $a \in \mathbb{R}$, we have $\{x \in M : f(x) \geq a\} \in \Sigma$.*

Proof. Mimic the proof of Theorem 9.19 (Exercise 14-21). ■

The characterizations of measurable functions can be used to prove that certain operations preserve measurability.

Theorem 14.30 *Let (M, Σ) be a measurable space and let $f, g : M \to [-\infty, \infty]$ be measurable functions. If $f + g$ is defined everywhere, then $f + g$ is measurable. If $f - g$ or $f \cdot g$ is defined everywhere, then it is measurable. Finally, f^+, f^- and $|f|$ are measurable.*

Proof. Mimic the proof of Theorem 9.20. (Exercise 14-22.) ■

Exercises

14-21. Prove Theorem 14.29

14-22. Prove Theorem 14 30.

14-23. Let (M, Σ) be a measurable space. Prove that a bounded function $f : M \to [0, \infty)$ is measurable iff there is a sequence $\{s_n\}_{n=1}^{\infty}$ of simple functions $s_n : M \to [0, \infty)$ such that for all $x \in M$ the sequence $\{ s_n(x) \}_{n=1}^{\infty}$ is nonincreasing and $\lim_{n\to\infty} s_n(x) = f(x)$.
Hint. Mimic part "5⇒1" of the proof of Theorem 14.29 for nonnegative functions.

14-24. Let (M, Σ) be a measurable space and let $f, g : M \to [-\infty, \infty]$ be measurable functions

(a) Prove that $\{ x \in M : f(x) = g(x) \} \in \Sigma$

(b) Prove that $\{ x \in M : f(x) \leq g(x) \} \in \Sigma$

(c) Prove that $\{ x \in M : f(x) < g(x) \} \in \Sigma$.

14.4 Integration of Measurable Functions

Once measurable functions are defined, the definition of the integral also is similar to that of the Lebesgue integral.

Definition 14.31 *Let (M, Σ, μ) be a measure space, let $n \in \mathbb{N}$, let $A_1, \ldots, A_n \in \Sigma$ be pairwise disjoint, let $a_1, \ldots, a_n \in [0, \infty)$ and let $s = \sum_{k=1}^{n} a_k \mathbf{1}_{A_k}$ be a simple function. We define $\int_M \sum_{k=1}^{n} a_k \mathbf{1}_{A_k} \, d\mu := \sum_{k=1}^{n} a_k \mu(A_k)$.*

The fact that the integral in Definition 14.31 does not depend on the representation of s can be proved just as for the Lebesgue integral in Exercise 9-20a.

Definition 14.32 *Let (M, Σ, μ) be a measure space. Let $f : M \to [0, \infty]$ be a measurable function. We define the* **integral** *of f to be*

$$\int_M f \, d\mu := \sup\left\{\int_M s \, d\mu : s \text{ is a simple function with } 0 \leq s \leq f\right\}$$

and we call f **integrable** *iff the supremum is finite.*

A function $g : M \to [-\infty, \infty]$ is called **integrable** *iff g^+ and g^- are both integrable. Its* **integral** *is defined to be $\int_M g \, d\mu := \int_M g^+ \, d\mu - \int_M g^- \, d\mu$.*

The integral on measure spaces is very versatile. Examples 14.33 and 14.34 show that integration on the real line as well as on d-dimensional spaces are special cases. Examples 14.36 and 14.37 show that absolutely convergent series and absolutely convergent double series are in one-to-one correspondence with integrable functions on the right measure spaces.

Example 14.33 With $M = \mathbb{R}$, $\Sigma = \Sigma_\lambda$ and $\mu = \lambda$, Definition 14.32 gives the Lebesgue integral on the real line. □

Example 14.34 With $M = \mathbb{R}^d$, $\Sigma = \Sigma_\lambda$ and $\mu = \lambda$, Definition 14.32 gives the Lebesgue integral in d-dimensional space. We will investigate in Section 14.8 how the Lebesgue integrals in various dimensions are related to each other. □

For the mentioned examples on series, we first need to establish that a function is integrable iff its absolute value is integrable. As was (by now) to be expected, the properties of the abstract integral are proved in exactly the same way as for the Lebesgue integral.

Theorem 14.35 *Let (M, Σ, μ) be a measure space and let $f, g : M \to [-\infty, \infty]$ be measurable functions.*

1. *If $0 \leq f \leq g$ a.e. and g is integrable, then so is f and $\int_M f \, d\mu \leq \int_M g \, d\mu$.*
2. *f is integrable iff $|f|$ is integrable and in that case the* **triangular inequality** *$\left|\int_M f \, d\mu\right| \leq \int_M |f| \, d\mu$ holds.*
3. *If $f \geq 0$, then $\int_M f \, d\mu = 0$ iff $f = 0$ a.e..*

Proof. Mimic the proof of Theorem 9.23. (Exercise 14-25.) ■

Example 14.36 With $M = \mathbb{N}$, $\Sigma = \mathcal{P}(\mathbb{N})$ and $\mu = \gamma_{\mathbb{N}}$ being counting measure on $\mathbb{N}$, a series $\sum_{j=1}^{\infty} a_j$ converges absolutely iff the function $f : \mathbb{N} \to [-\infty, \infty]$ defined by $f(j) = a_j$ is integrable over the measure space $\left(\mathbb{N}, \mathcal{P}(\mathbb{N}), \gamma_{\mathbb{N}}\right)$. Moreover, for every integrable function f on this measure space we have $\int_{\mathbb{N}} f \, d\gamma_{\mathbb{N}} = \sum_{j=1}^{\infty} f(j)$. □

Example 14.37 Let $\{a_{(i,j)}\}_{i,j=1}^{\infty}$ be a doubly indexed countable family of numbers. We say the **double series** $\sum_{i=1}^{\infty} \sum_{j=1}^{\infty} a_{(i,j)}$ **converges absolutely** iff the double series $\sum_{i=1}^{\infty} \sum_{j=1}^{\infty} |a_{(i,j)}|$ converges.

With $M = \mathbb{N} \times \mathbb{N}$, $\Sigma = \mathcal{P}(\mathbb{N} \times \mathbb{N})$ and $\mu = \gamma_{\mathbb{N}\times\mathbb{N}}$ being counting measure on $\mathbb{N} \times \mathbb{N}$, a double series $\sum_{i=1}^{\infty} \sum_{j=1}^{\infty} a_{(i,j)}$ converges absolutely iff the function $f(i, j) := a_{(i,j)}$ is integrable on the measure space $\left(\mathbb{N} \times \mathbb{N}, \mathcal{P}(\mathbb{N} \times \mathbb{N}), \gamma_{\mathbb{N}\times\mathbb{N}}\right)$. For every integrable function f on this measure space, we have $\int_{\mathbb{N}\times\mathbb{N}} f \, d\gamma_{\mathbb{N}\times\mathbb{N}} = \sum_{i=1}^{\infty} \sum_{j=1}^{\infty} f(i, j)$. □

As for Lebesgue measure, we consider null sets to be insignificant. Therefore with the same motivation as given for Definition 9.24 we define the following.

Definition 14.38 *Let (M, Σ, μ) be a measure space. If $f : M \to [-\infty, \infty]$ is defined a.e., we call it* **integrable** *iff* $g(x) := \begin{cases} f(x); & \text{if } f(x) \text{ is defined,} \\ 0; & \text{if } f(x) \text{ is not defined,} \end{cases}$ *is integrable.*

As long as a function is constructed from other integrable functions, like the sum in Theorem 14.39, measurability of the set where the function is not defined is not an issue. However, even if this was a problem and the function is undefined on a subset of a null set, we could simply switch to the completion of the measure (see Exercise 14-10) and note that subsets of null sets are also insignificant.

Overall, with the same conventions as for the Lebesgue integral we can prove that the integral over a measure space has the right linearity properties.

Theorem 14.39 *Let (M, Σ, μ) be a measure space and let $f, g : M \to [-\infty, \infty]$ be integrable functions. Then the following hold.*

1. *For all $\alpha \in \mathbb{R}$, the scalar multiple αf is integrable and* $\int_M \alpha f \, d\mu = \alpha \int_M f \, d\mu$.

2. *The sum $f + g$ is integrable and* $\int_M f + g \, d\mu = \int_M f \, d\mu + \int_M g \, d\mu$.

Proof. Mimic the proof of Theorem 9.25. (Exercises 14-26a and 14-26b.) ■

Exercises

14-25. Prove Theorem 14.35. That is, let (M, Σ, μ) be a measure space and let f, g be measurable functions.

(a) Prove that if $0 \leq f \leq g$ a.e. and g is integrable, then so is f and $\int_M f \, d\mu \leq \int_M g \, d\mu$.

(b) Prove that f is integrable iff $|f|$ is integrable and that in this case $\left| \int_M f \, d\mu \right| \leq \int_M |f| \, d\mu$

(c) Prove that if $f \geq 0$, then $\int_M f \, d\mu = 0$ iff $f = 0$ a.e.

14-26. Let (M, Σ, μ) be a measure space and let $f, g : M \to [-\infty, \infty]$ be integrable functions

(a) Prove that if $\alpha \in \mathbb{R}$, then the scalar multiple αf is integrable and $\int_M \alpha f \, d\mu = \alpha \int_M f \, d\mu$

(b) Prove that the sum $f + g$ is integrable and $\int_M f + g \, d\mu = \int_M f \, d\mu + \int_M g \, d\mu$.

Note Exercise 14-33 gives a more effective proof than mimicking the proof of Theorem 9.25.

(c) Prove that $f - g$ is integrable and $\int_M f - g \, d\mu = \int_M f \, d\mu - \int_M g \, d\mu$.

(d) Prove that $\max\{f, g\}$ (defined pointwise) is integrable

(e) Prove that $\min\{f, g\}$ (defined pointwise) is integrable

(f) Give an example that shows that the product of f and g need not be integrable

14-27. **Markov's inequality**. Let (M, σ, μ) be a measure space and let $f : M \to [0, \infty)$ be integrable Prove that for all $c > 0$ we have $\mu\left(\left\{x \in M : f(x) > c\right\}\right) < \frac{1}{c} \int_M f \, d\mu$.

14.5 Monotone and Dominated Convergence

Sequences have been a standard tool throughout our investigation of single variable functions. Sequences play an equally important role in more abstract settings. In Exercise 11-13, we have seen that pointwise convergence of a bounded, monotone sequence of Riemann integrable functions need not produce a Riemann integrable limit function. The two fundamental results about pointwise convergence in measure theory are the Monotone Convergence Theorem, which says that the pointwise limit of a nondecreasing sequence of nonnegative integrable functions is integrable if the limit of the integrals is finite (see Theorem 14.41), and the Dominated Convergence Theorem, which says that the pointwise limit of a sequence of integrable functions is integrable provided all functions are below one integrable function that dominates them all (see Theorem 14.43). Hence, when it comes to pointwise convergence of functions, the Lebesgue-type integral has more favorable properties than the Riemann integral.

Because every subset $A \in \Sigma$ of a measure space (M, Σ, μ) can be turned into a measure space, and because by Theorem 14.35 for any integrable function g, the function $g\mathbf{1}_A$ is also integrable, we can define integrals over subsets.

Definition 14.40 *Let (M, Σ, μ) be a measure space, let $A \in \Sigma$ and let the function $f : M \to [-\infty, \infty]$ be integrable. Then we define the* **integral** *of f over the subset A*

as $\int_A g\, d\mu := \int_M g\mathbf{1}_A\, d\mu$. *For Lebesgue integrals over intervals, we will also write* $\int_a^b f(x)\, d\lambda(x) := \int_a^b f\, d\lambda := \int_{[a,b]} f\, d\lambda$.

Because the integral is defined as a supremum of integrals of simple functions, it should not be too surprising that the integral is well-behaved with respect to monotone sequences of functions.

Theorem 14.41 Monotone Convergence Theorem. *Let* (M, Σ, μ) *be a measure space and let* $\{f_n\}_{n=1}^{\infty}$ *be a sequence of nonnegative measurable functions so that* $\{f_n(x)\}_{n=1}^{\infty}$ *is nondecreasing for almost all* $x \in M$ *and* $\lim_{n\to\infty} f_n(x)$ *exists for almost all* $x \in M$. *Let* $f : M \to [0, \infty]$ *be a measurable function so that* $f(x) = \lim_{n\to\infty} f_n(x)$ *a.e. Then* $\int_M f\, d\mu = \lim_{n\to\infty} \int_M f_n\, d\mu$.

Proof. First assume that for *all* $x \in M$ we have $0 \le f_1(x) \le f_2(x) \le \cdots$ and the limit $\lim_{n\to\infty} f_n(x)$ exists.

By part 1 of Theorem 14.35, we infer $\int_M f\, d\mu \ge \lim_{n\to\infty} \int_M f_n\, d\mu$. To prove the reversed inequality, let $s = \sum_{k=1}^{m} a_k \mathbf{1}_{A_k}$ be a simple function such that $0 \le s \le f$ and let $t \in (0, 1)$. For $n \in \mathbb{N}$, define $E_n := \{x \in M : f_n(x) \ge ts(x)\}$. Then $M = \bigcup_{n=1}^{\infty} E_n$, for all $n \in \mathbb{N}$ we have the containment $E_n \subseteq E_{n+1}$, and

$$\int_M ts\, d\mu = \sum_{k=1}^{m} ta_k\mu(A_k) = \sum_{k=1}^{m} ta_k\mu\left(A_k \cap \bigcup_{n=1}^{\infty} E_n\right) = \sum_{k=1}^{m} ta_k\mu\left(\bigcup_{n=1}^{\infty} A_k \cap E_n\right)$$

By Theorem 14.15, the measure of the nested union is the limit of the individual measures.

$$= \sum_{k=1}^{m} ta_k \lim_{n\to\infty} \mu(A_k \cap E_n) = \lim_{n\to\infty} \sum_{k=1}^{m} ta_k\mu(A_k \cap E_n)$$

$$= \lim_{n\to\infty} \int_M \left(\sum_{k=1}^{m} ta_k \mathbf{1}_{A_k}\right) \mathbf{1}_{E_n}\, d\mu = \lim_{n\to\infty} \int_{E_n} ts\, d\mu.$$

Let $\varepsilon > 0$. Then there is an $N \in \mathbb{N}$ so that $\int_M ts\, d\mu - \varepsilon \le \int_{E_N} ts\, d\mu$, and hence

$$t\int_M s\, d\mu - \varepsilon = \int_M ts\, d\mu - \varepsilon \le \int_{E_N} ts\, d\mu \le \int_M f_N\, d\mu \le \lim_{n\to\infty} \int_M f_n\, d\mu.$$

Because ε was arbitrary we obtain $\int_M ts\,d\mu \leq \lim_{n\to\infty} \int_M f_n\,d\mu$, and because the number $t \in (0, 1)$ was arbitrary, we can let t approach 1 and obtain

$$\int_M s\,d\mu = \lim_{t\to 1^-} t \int_M s\,d\mu \leq \lim_{n\to\infty} \int_M f_n\,d\mu.$$

Because $s \leq f$ was an arbitrary simple function, we conclude that

$$\int_M f\,d\mu = \sup\left\{ \int_M s\,d\mu : s \text{ simple, } 0 \leq s \leq f \right\} \leq \lim_{n\to\infty} \int_M f_n\,d\mu.$$

Now assume that $0 \leq f_1(x) \leq f_2(x) \leq \cdots$ and $\lim_{n\to\infty} f_n(x) = f(x)$ for *almost all* $x \in M$. Let $N \in \Sigma$ be a null set so that $0 \leq f_1(x) \leq f_2(x) \leq \cdots$ and $\lim_{n\to\infty} f_n(x) = f(x)$ for *all* $x \in M \setminus N$. Changing a function on a null set does not change its integral. Therefore we can replace f with $f\mathbf{1}_{M\setminus N}$ and each f_n with $f_n\mathbf{1}_{M\setminus N}$ and apply what we have proved above to obtain the equation for the integrals. ■

We need the next lemma for the proof of the Dominated Convergence Theorem.

Lemma 14.42 Fatou's Lemma. *Let (M, Σ, μ) be a measure space and let $\{f_n\}_{n=1}^{\infty}$ be a sequence of nonnegative measurable functions. Then $\liminf_{n\to\infty} f_n$ (defined pointwise) is measurable and $\int_M \liminf_{n\to\infty} f_n\,d\mu \leq \liminf_{n\to\infty} \int_M f_n\,d\mu$.*

Proof. Because the limit inferior is $\liminf_{n\to\infty} f_n = \lim_{n\to\infty} \inf\{f_j : j \geq n\}$, where the infimum and the limits are taken pointwise, we first consider the sequence of functions $p_n := \inf\{f_j : j \geq n\}$.

Let $n \in \mathbb{N}$ and $a \in \mathbb{R}$. Then $\{x \in M : p_n(x) < a\} = \bigcup_{j=n}^{\infty} \{x \in M : f_j(x) < a\}$.
Because the union on the right side is measurable and a was arbitrary we conclude that p_n is measurable. Moreover, clearly we have $p_n \leq p_{n+1}$.

By Exercise 14-28a, this means that $\liminf_{n\to\infty} f_n = \lim_{n\to\infty} p_n$ is measurable and by the Monotone Convergence Theorem $\int_M \liminf_{n\to\infty} f_n\,d\mu = \lim_{n\to\infty} \int_M p_n\,d\mu$. Now for all numbers $n \in \mathbb{N}$ we have $p_n \leq f_n$, which by Lemma 10.5 means

$$\int_M \liminf_{n\to\infty} f_n\,d\mu = \lim_{n\to\infty} \int_M p_n\,d\mu \leq \liminf_{n\to\infty} \int_M f_n\,d\mu. \qquad ■$$

The inequality in Fatou's Lemma can be strict (Exercise 14-29). Finally, as long as all functions' absolute values are bounded ("dominated") by one integrable function, pointwise limits preserve integrability and the limit can be moved out of the integral.

Theorem 14.43 Dominated Convergence Theorem. *Let (M, Σ, μ) be a measure space, let $\{f_n\}_{n=1}^{\infty}$ be a sequence of measurable functions and let f be a measurable function such that $f(x) = \lim_{n\to\infty} f_n(x)$ for almost all x. Moreover, let g be an integrable* *function such that for all $n \in \mathbb{N}$ and almost all $x \in M$ we have $|f_n(x)| \leq g(x)$. Then f is integrable and $\int_M f\,d\mu = \lim_{n\to\infty} \int_M f_n\,d\mu$.*

Proof. Because $|f(x)| = \lim_{n\to\infty} |f_n(x)| \leq g(x)$ for almost all $x \in M$, by part 1 of Theorem 14.35 $|f|$ is integrable and then by part 2 of Theorem 14.35 f is integrable.

Because changing a function on a null set does not change the integral, we can assume that $|f_n(x)| \leq g(x)$ and $|f(x)| \leq g(x)$ for all $x \in M$. Now for all $n \in \mathbb{N}$ we have $f_n(x) + g(x) \geq 0$ and thus by Fatou's Lemma we obtain

$$\begin{aligned}\int_M f\,d\mu + \int_M g\,d\mu &= \int_M \lim_{n\to\infty} f_n + g\,d\mu = \int_M \liminf_{n\to\infty}(f_n + g)\,d\mu \\ &\leq \liminf_{n\to\infty} \int_M f_n + g\,d\mu = \liminf_{n\to\infty} \int_M f_n\,d\mu + \int_M g\,d\mu,\end{aligned}$$

which means that $\int_M f\,d\mu \leq \liminf_{n\to\infty} \int_M f_n\,d\mu$. The same argument applied to the functions $(-f_n)+g \geq 0$ gives $\int_M -f\,d\mu \leq \liminf_{n\to\infty} \int_M -f_n\,d\mu$, that is (recall Proposition 10.4) $\int_M f\,d\mu \geq \limsup_{n\to\infty} \int_M f_n\,d\mu$. By Proposition 10.3 and Theorem 10.6, we conclude that $\int_M f\,d\mu = \lim_{n\to\infty} \int_M f_n\,d\mu$. ■

The hypothesis that all functions f_n in the Dominated Convergence Theorem are below an integrable function g cannot be dropped. For example, on $(0, 1)$ equipped with Lebesgue measure, the sequence of functions $\left\{\mathbf{1}_{\left[\frac{1}{2^n+1}, \frac{1}{2^n}\right)} \frac{1}{x}\right\}_{n=1}^{\infty}$ converges pointwise to zero, but all integrals are equal to $\ln(2)$. Moreover, Exercise 14-28 shows that the demand that the pointwise a.e. limit is measurable cannot be dropped in general, but that it can be dropped for Lebesgue measure.

Exercises

14-28. Pointwise almost everywhere limits of measurable functions. Let (M, Σ, μ) be a measure space and let $\{f_n\}_{n=1}^{\infty}$ be a sequence of measurable functions.

(a) Prove that if $f(x) = \lim_{n\to\infty} f_n(x)$ for all $x \in M$, then f is measurable.

(b) Give an example that shows that the pointwise a.e limit of a sequence of measurable functions need not be measurable.
Hint. There is an example with $M = \{0, 1\}$, $\Sigma = \{\emptyset, M\}$ and $\mu = 0$.

(c) Prove that if (M, Σ, μ) is a complete measure space, then $f(x) = \lim_{n\to\infty} f_n(x)$ a.e. implies that f is measurable.

(d) Prove that $(\mathbb{R}^d, \Sigma_\lambda, \lambda)$ is a complete measure space.

14-29. Give an example that the inequality in Fatou's Lemma can be strict. Then explain why this is not a counterexample to the Dominated Convergence Theorem.

14-30. Let (M, Σ, μ) be a measure space and let $g : M \to [0, \infty]$ be integrable. Prove that the function $\mu_g : \Sigma \to [0, \infty)$ defined by $\mu_g(A) := \int_A g\,d\mu$ is a measure.

14-31. Let (M, Σ, μ) be a measure space and let $\{f_n\}_{n=1}^{\infty}$ be a sequence of integrable functions that converges uniformly to a measurable function f.

(a) Prove that if $\mu(M) < \infty$, then f is integrable and $\int_M f \, d\mu = \lim_{n\to\infty} \int_M f_n \, d\mu$.

(b) Give an example that shows this result need not hold for infinite measure spaces.

14-32. Let (M, Σ, μ) be a measure space and let $\{f_n\}_{n=1}^{\infty}$ be a sequence of integrable functions that converges pointwise to a measurable function f and for which there is a $B \in \mathbb{R}$ so that for all $x \in M$ we have $f_n(x) \leq B$

(a) Prove that if $\mu(M) < \infty$, then f is integrable and $\int_M f \, d\mu = \lim_{n\to\infty} \int_M f_n \, d\mu$.

(b) Give an example that shows this result need not hold for infinite measure spaces.

14-33 Let $f, g : M \to [-\infty, \infty]$ be integrable functions Give a more effective proof that $f + g$ is integrable and $\int_M f + g \, d\mu = \int_M f \, d\mu + \int_M g \, d\mu$ than was given in Theorem 9.23

Hint. Use the Monotone Convergence Theorem and two nondecreasing sequences of simple functions that converge to $|f|$ and $|g|$ respectively to prove that $|f| + |g|$ is integrable. Use a similar idea to prove the equation for the integrals. Do not use the Dominated Convergence Theorem (why?).

14-34 Explain how the hypotheses of the Dominated Convergence Theorem prevent the occurrence of examples as in Exercise 11-17 or as mentioned at the end of this section.

14-35 Prove that for every Lebesgue integrable function $f : \mathbb{R} \to \mathbb{R}$ there is a sequence of Riemann integrable functions $r_n : \mathbb{R} \to \mathbb{R}$ such that $\lim_{n\to\infty} \int_{\mathbb{R}} |f - r_n| \, d\lambda = 0$.

14.6 Convergence in Mean, in Measure, and Almost Everywhere

Theorem 15.50 will show that the "distance" between two functions can be measured with the integral of the absolute value of the difference. Because convergence with respect to this distance is fundamental in analysis, we investigate this notion of convergence and some of its consequences here.

Definition 14.44 *Let (M, Σ, μ) be a measure space, let $\{f_n\}_{n=1}^{\infty}$ be a sequence of measurable functions and let $f : M \to [-\infty, \infty]$ be a measurable function. If $\lim_{n\to\infty} \int_M |f_n - f| \, d\mu = 0$ then we say that $\{f_n\}_{n=1}^{\infty}$* **converges in mean** *to f.*

Convergence in mean is near the heart of the definition of integration.

Theorem 14.45 *Let (M, Σ, μ) be a measure space and let $f : M \to [-\infty, \infty]$ be an integrable function. Then there is a sequence $\{s_n\}_{n=1}^{\infty}$ of simple functions that converges in mean to f.*

Proof. By definition of the integral, for every $n \in \mathbb{N}$ there are simple functions $0 \leq s_n^- \leq f^-$ and $0 \leq s_n^+ \leq f^+$ so that $\int_M f^+ \, d\mu - \int_M s_n^+ \, d\mu < \frac{1}{n}$ and

$\int_M f^- \, d\mu - \int_M s_n^- \, d\mu < \frac{1}{n}$. But then $s_n := s_n^+ - s_n^-$ is simple and

$$\begin{aligned}
\int_M |f - s_n| \, d\mu &= \int_M |f^+ - s_n^+ - (f^- - s_n^-)| \, d\mu \\
&= \int_M |f^+ - s_n^+| \, d\mu + \int_M |f^- - s_n^-| \, d\mu \\
&= \int_M f^+ \, d\mu - \int_M s_n^+ \, d\mu + \int_M f^- \, d\mu - \int_M s_n^- \, d\mu \\
&< \frac{1}{n} + \frac{1}{n} = \frac{2}{n},
\end{aligned}$$

which means that $\{s_n\}_{n=1}^{\infty}$ converges in mean to f. ■

Unfortunately, convergence in mean does not even imply pointwise a.e. convergence, so the visualization of convergence in mean is a bit complicated.

Example 14.46 On the interval $[0, 1]$, for $n \in \mathbb{N}$ and $k \in \{0, \ldots, 2^n - 1\}$ define the function $f_{2^n+k} := \mathbf{1}_{\left[\frac{k}{2^n}, \frac{k+1}{2^n}\right]}$. Then the sequence $\{f_m\}_{m=1}^{\infty}$ converges in mean to 0, but it does not converge at any point in $[0, 1]$. □

We can say that convergence in mean implies that the measure of any set on which the difference $|f_n - f|$ is greater than a given number must go to zero.

Proposition 14.47 *Let (M, Σ, μ) be a measure space. If the sequence $\{f_n\}_{n=1}^{\infty}$ of measurable functions converges in mean to the measurable function f, then for all $\varepsilon > 0$ we have $\lim_{n\to\infty} \mu\left(\left\{x \in M : \left|f_n(x) - f(x)\right| > \varepsilon\right\}\right) = 0$.*

Proof. Let $\varepsilon > 0$. Then

$$\begin{aligned}
\mu\left(\left\{x \in M : \left|f_n(x) - f(x)\right| > \varepsilon\right\}\right) &= \frac{1}{\varepsilon} \int_M \varepsilon \mathbf{1}_{\left\{x \in M : \left|f_n(x) - f(x)\right| > \varepsilon\right\}} \, d\mu \\
&\leq \frac{1}{\varepsilon} \int_M |f_n - f| \, d\mu,
\end{aligned}$$

and the latter sequence goes to zero. ■

The condition derived in Proposition 14.47 is called convergence in measure. Convergence in measure implies at least that a subsequence converges a.e.

Definition 14.48 *Let (M, Σ, μ) be a measure space. The sequence $\{f_n\}_{n=1}^{\infty}$ of measurable functions* **converges in measure** *to the measurable function f iff for every $\varepsilon > 0$ we have $\lim_{n\to\infty} \mu\left(\left\{x \in M : \left|f_n(x) - f(x)\right| > \varepsilon\right\}\right) = 0$.*

Proposition 14.49 *Let (M, Σ, μ) be a measure space. If the sequence $\{f_n\}_{n=1}^{\infty}$ of measurable functions converges in measure to the measurable function f, then it has a subsequence that converges a.e. to f.*

Proof. Define $\{n_k\}_{k=1}^{\infty}$ as follows. Let $n_0 := 1$ and for $k \geq 1$ let $n_k > n_{k-1}$ be such that $\mu\left(\left\{x \in M : \left|f_{n_k}(x) - f(x)\right| > \frac{1}{k}\right\}\right) < \frac{1}{2^k}$. Suppose for a contradiction that $\left\{f_{n_k}\right\}_{k=1}^{\infty}$ does not converge a.e. to f. Then there are an $m \in \mathbb{N}$ and an $\varepsilon > 0$ so that $\mu\left(\left\{x \in M : \left(\forall K \in \mathbb{N} : \exists k \geq K : \left|f_{n_k}(x) - f(x)\right| > \frac{1}{m}\right)\right\}\right) > \varepsilon$. Let $I \in \mathbb{N}$ be so that $I > m$ and $\sum_{k=I}^{\infty} \frac{1}{2^k} < \varepsilon$. Then

$$\left\{x \in M : \left(\forall K \in \mathbb{N} : \exists k \geq K : \left|f_{n_k}(x) - f(x)\right| > \frac{1}{m}\right)\right\} \subseteq \bigcup_{k=I}^{\infty} \left\{x \in M : \left|f_{n_k}(x) - f(x)\right| > \frac{1}{k}\right\}$$

but the measure of the latter set is less than ε, a contradiction. ■

Exercises

14-36. Let (M, Σ, μ) be a measure space and let $g : M \to [0, \infty]$ be integrable

(a) Prove that $\int_M g\, d\mu = \lim_{n\to\infty} \int_M \min\{g, n\}\, d\mu$.

(b) Prove that for every $\varepsilon > 0$ there is a $\delta > 0$ so that if $\mu(A) < \delta$, then $\int_A g\, d\mu < \varepsilon$

Hint. First prove the result for bounded g

(c) Prove that with $M = [a, b]$, μ being Lebesgue measure λ on $[a, b]$ and Σ being the set Σ_λ of Lebesgue measurable subsets of $[a, b]$, the function $f(x) = \int_{[a,x]} g\, d\lambda$ is **absolutely continuous**

14-37. Pointwise convergence and convergence in measure.

(a) Give an example that shows that, in general, pointwise convergence does not imply convergence in measure.

(b) Prove that if (M, Σ, μ) is a measure space with $\mu(M) < \infty$, $\{f_n\}_{n=1}^{\infty}$ is a sequence of measurable functions that converges a.e. to the real valued measurable function f, then $\{f_n\}_{n=1}^{\infty}$ converges in measure to f

(c) Prove **Egoroff's Theorem**, which states that if (M, Σ, μ) is a measure space with $\mu(M) < \infty$, $\{f_n\}_{n=1}^{\infty}$ is a sequence of measurable functions that converges a.e. to the real valued measurable function f, then for each $\varepsilon > 0$ there is a $B \in \Sigma$ so that $\mu\left(B'\right) < \varepsilon$ and $\{f_n\}_{n=1}^{\infty}$ converges uniformly to f on B.

14-38 First results on derivatives of integrable functions.

(a) Let $f : [a, b] \to \mathbb{R}$ be a Lebesgue measurable function that is differentiable a.e. Prove that f' is Lebesgue measurable. You may use the result of Exercise 14-28c.

(b) Let $f : [a, b] \to \mathbb{R}$ be a nondecreasing function Prove that f' (which exists a e. by Exercise 10-7) is Lebesgue integrable on $[a, b]$ and $\int_a^b f'(x)\, d\lambda \leq f(b) - f(a)$.

Hint. Apply Fatou's Lemma to $f_n(x) := \frac{f\left(x+\frac{1}{n}\right) - f(x)}{\frac{1}{n}}$ (where we set $f(b+\varepsilon) := f(b)$ for all $\varepsilon > 0$) and use Exercise 7-23, Theorem 9.26 and an adaptation of the proof of the Derivative Form of the Fundamental Theorem of Calculus for Riemann integrals.

The results in Sections 14.7 and 14.8 are quite technical. The main insight is that Fubini's Theorem allows the reduction of an integral on a product space to iterated integrals on the factor spaces. Readers who are willing to accept this fact could postpone a detailed reading of the remaining two sections in this chapter and proceed to Chapter 15.

14.7 Product σ-Algebras

One motivation for introducing measure theory was the desire to define an integral on multidimensional spaces. Integration with respect to d-dimensional Lebesgue measure provides such an integral as an abstract entity. Computationally, it is standard practice to compute higher dimensional integrals as iterated lower dimensional integrals. Sections 14.7 and 14.8 show that this is also possible in measure theory. As a consequence, we not only progress towards showing that d-dimensional integrals can be computed as iterated integrals, but we also are able to establish some results on double series (see Exercise 14-50).

The relevant constructions actually start with the construction of the product space of two measure spaces. For this product space, we need a σ-algebra that contains certain sets. This section will provide the σ-algebra on the product space and Section 14.8 will provide the product measure.

As before, to safeguard against nonmeasurable sets, we must make sure that our σ-algebras do not get too large. The smallest σ-algebra that contains certain sets is generated as follows.

Definition 14.50 *Let M be a set and let $\mathcal{U} \subseteq \mathcal{P}(M)$. Then the* **$\sigma$-algebra generated by** *$\mathcal{U}$ is defined to be the intersection of all σ-algebras that contain $\mathcal{U}$.*

Proposition 14.51 *Let M be a set and let $\mathcal{U} \subseteq \mathcal{P}(M)$. Then the σ-algebra generated by $\mathcal{U}$ is a σ-algebra.*

Proof. Exercise 14-39. ∎

Exercise 14-41a shows that the σ-algebra of Lebesgue measurable sets is generated by a family of fairly simple sets. To prove the equality of two measures on the σ-algebra generated by the set $\mathcal{U}$ we want to concentrate on the sets in $\mathcal{U}$. Unfortunately, because the definition of a measure involves unions of *pairwise disjoint* sets, this is not as straightforward as it might seem. Instead of being equal on σ-algebras, measures usually agree on Dynkin systems.

Definition 14.52 *Let M be a set and let $\mathcal{D} \subseteq \mathcal{P}(M)$. Then $\mathcal{D}$ is called a* **Dynkin system** *iff*

1. *$M \in \mathcal{D}$,*
2. *If $A, B \in \mathcal{D}$ and $A \supseteq B$, then $A \setminus B \in \mathcal{D}$, and*
3. *If $\{A_n\}_{n=1}^{\infty}$ is a sequence in $\mathcal{D}$ with $A_n \subseteq A_{n+1}$ for all $n \in \mathbb{N}$, then $\bigcup_{n=1}^{\infty} A_n \in \mathcal{D}$.*

Definition 14.53 *Let (M, Σ, μ) be a measure space. Then the measure μ is called* **finite** *iff $\mu(M) < \infty$.*

Proposition 14.54 *Let (M, Σ) be a measurable space and let μ and ν be finite measures on M so that $\mu(M) = \nu(M)$. Then $\{S \in \Sigma : \mu(S) = \nu(S)\}$ is a Dynkin system.*

Proof. Exercise 14-43a (for part 3 use Theorem 14.15). ■

As with σ-algebras, the intersection of all Dynkin systems that contain a given set of sets U is again a Dynkin system, called the **Dynkin system generated by** U (see Exercise 14-44). The connection between σ-algebras and Dynkin systems is made via π-systems.

Definition 14.55 *Let M be a set and let $\mathcal{U} \subseteq \mathcal{P}(M)$. Then $\mathcal{U}$ is called a* **π-system** *iff for all $A, B \in \mathcal{U}$ the intersection satisfies $A \cap B \in \mathcal{U}$.*

Theorem 14.56 Monotone Class Theorem *or* **Dynkin's Lemma.** *Let M be a set and let $\mathcal{U}$ be a π-system that contains M. Then the σ-algebra generated by $\mathcal{U}$ equals the Dynkin system generated by $\mathcal{U}$.*

Proof. Let Σ be the σ-algebra generated by $\mathcal{U}$ and let $\mathcal{D}$ be the Dynkin system generated by $\mathcal{U}$. Because every σ-algebra also is a Dynkin system, we infer $\mathcal{D} \subseteq \Sigma$. For the reversed inclusion, first consider $\mathcal{D}_1 := \{A \in \mathcal{D} : A \cap U \in \mathcal{D} \text{ for all } U \in \mathcal{U}\}$. Because $\mathcal{U}$ contains M we infer $M \in \mathcal{D}_1$. If $A, B \in \mathcal{D}_1$ and $A \supseteq B$, then for all $U \in \mathcal{U}$ we have $(A \setminus B) \cap U = (A \cap U) \setminus (B \cap U) \in \mathcal{D}$ (see Exercise 7-3e), which means $A \setminus B \in \mathcal{D}_1$. Moreover, if $\{A_n\}_{n=1}^{\infty}$ is a sequence of sets in $\mathcal{D}_1$ with $A_n \subseteq A_{n+1}$ for all $n \in \mathbb{N}$, then for all $U \in \mathcal{U}$ we obtain $\left(\bigcup_{n=1}^{\infty} A_n\right) \cap U = \bigcup_{n=1}^{\infty} (A_n \cap U) \in \mathcal{D}$, which means that $\bigcup_{n=1}^{\infty} A_n \in \mathcal{D}_1$. Hence, $\mathcal{D}_1$ is a Dynkin system. Because $\mathcal{U}$ is a π-system, we infer $\mathcal{U} \subseteq \mathcal{D}_1$ and therefore $\mathcal{D} \subseteq \mathcal{D}_1$, which implies that $\mathcal{D} = \mathcal{D}_1$.

Now define $\mathcal{D}_2 := \{A \in \mathcal{D} : A \cap D \in \mathcal{D} \text{ for all } D \in \mathcal{D}\}$. Because $\mathcal{D}_1 = \mathcal{D}$ we infer $\mathcal{U} \subseteq \mathcal{D}_2$. Similar to the above we can prove that $\mathcal{D}_2$ is a Dynkin system, which means that $\mathcal{D} = \mathcal{D}_2$. We conclude that if $A, B \in \mathcal{D}$, then $A \cap B \in \mathcal{D}$.

Now we can prove that $\mathcal{D}$ is a σ-algebra. First, note that because $M \in \mathcal{D}$, for all sets $S \in \mathcal{D}$ the complement $S' = M \setminus S$ is in $\mathcal{D}$. In particular, we trivially obtain $\emptyset = M' \in \mathcal{D}$. Now let $\{B_n\}_{n=1}^{\infty}$ be a sequence of sets in $\mathcal{D}$. For $n \in \mathbb{N}$, define the

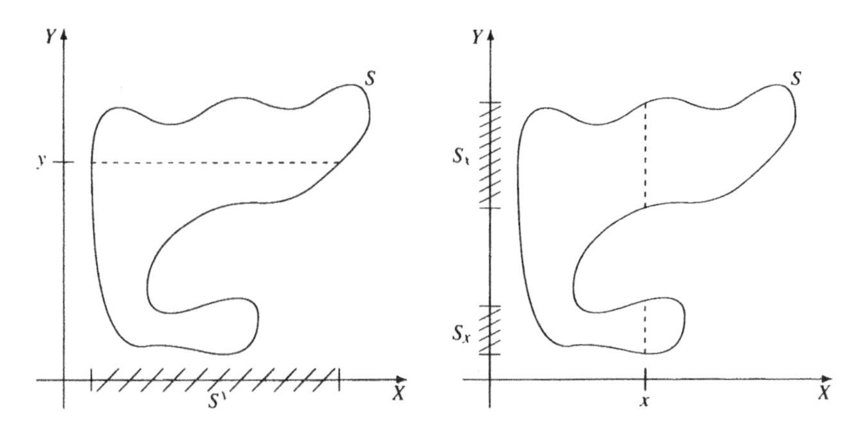

Figure 30: The "horizontal sections" S^y of a set $S \subseteq X \times Y$ are obtained by intersecting the set with "horizontal lines" $X \times \{y\}$. The "vertical sections" S_x are obtained by intersecting the set with "vertical lines" $\{x\} \times Y$.

set $A_n := \bigcup_{i=1}^{n} B_i = M \setminus \left(\bigcap_{i=1}^{n} M \setminus B_i \right) \in \mathcal{D}$. Then $A_n \subseteq A_{n+1}$ for all $n \in \mathbb{N}$ and so $\bigcup_{n=1}^{\infty} B_n = \bigcup_{n=1}^{\infty} A_n \in \mathcal{D}$ and $\mathcal{D}$ is a σ-algebra. This proves $\Sigma \subseteq \mathcal{D}$, hence $\Sigma = \mathcal{D}$. ■

The Monotone Class Theorem now allows us to prove that under certain circumstances, two measures on the same σ-algebra must be equal.

Theorem 14.57 *Let M be a set, let $\mathcal{U} \subseteq \mathcal{P}(M)$ be a π-system, let Σ be the σ-algebra generated by $\mathcal{U}$ and let $\mu, \nu : \Sigma \to [0, \infty]$ be measures with $\mu|_{\mathcal{U}} = \nu|_{\mathcal{U}}$. If there is a sequence $\{A_n\}_{n=1}^{\infty}$ in $\mathcal{U}$ so that all $\mu(A_n) = \nu(A_n)$ are finite, $A_n \subseteq A_{n+1}$ for all $n \in \mathbb{N}$ and $\bigcup_{n=1}^{\infty} A_n = M$, then $\mu = \nu$.*

Proof. First prove that each $\{A \in \Sigma : \mu(A \cap A_n) = \nu(A \cap A_n)\}$ is a Dynkin system. Then apply Theorem 14.56 to show $\mu(A \cap A_n) = \nu(A \cap A_n)$ for all $A \in \Sigma$ and $n \in \mathbb{N}$. Finally, show $\mu = \nu$ by using $A_0 := \emptyset$ and $\mu(A) = \sum_{n=1}^{\infty} \mu(A \cap (A_n \setminus A_{n-1}))$ and $\nu(A) = \sum_{n=1}^{\infty} \nu(A \cap (A_n \setminus A_{n-1}))$ for all $A \in \Sigma$. (Exercise 14-45.) ■

To define a product of two measurable spaces we now simply form all the products of sets in the respective σ-algebras and consider the σ-algebra generated by these products. We also define the intersections of sets with "horizontal lines" or "vertical lines" and the intersections of graphs of functions with "vertical planes." These intersections will allow us to reduce integrals on the product of two measure spaces to iterated inte-

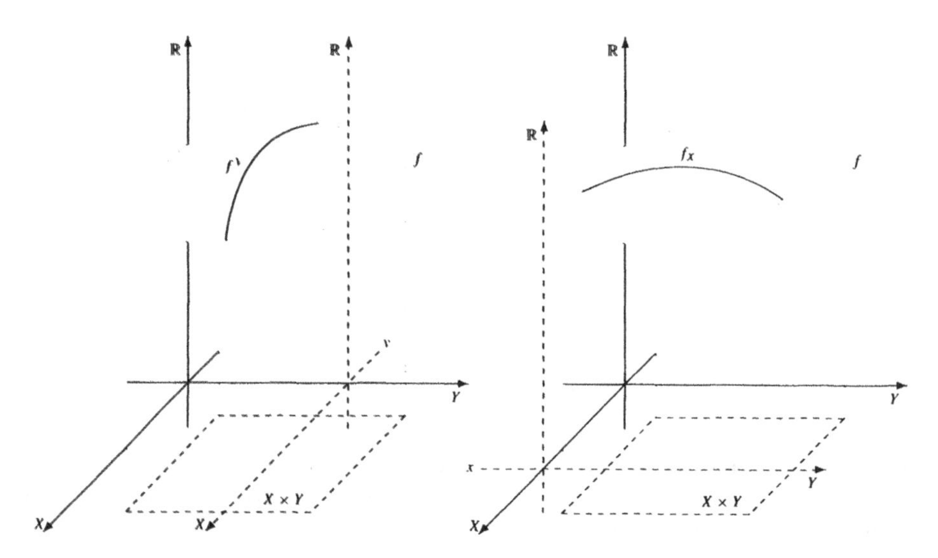

Figure 31: The "y-sections" f^y of a function $f : X \times Y \to [-\infty, \infty]$ are obtained by intersecting the graph of the function with "vertical planes $X \times \{y\} \times [-\infty, \infty]$ in the X-direction at y." The "x-sections" f_x are obtained by intersecting the graph with "vertical planes $\{x\} \times Y \times [-\infty, \infty]$ in the Y-direction at x."

grals. To emphasize the visualization as x- and y-coordinates, for the remainder of this chapter, the first factor space will be X and the second factor space will be Y.

Definition 14.58 *Let (X, Σ) and (Y, Γ) be measurable spaces. A subset S of $X \times Y$ is called a* **rectangle with measurable sides** *iff there are subsets $A \in \Sigma$ and $B \in \Gamma$ such that $S = A \times B$. The σ-algebra generated by the set of all rectangles with measurable sides is called the* **product σ-algebra** *$\Sigma \times \Gamma$ of Σ and Γ.*

Let $x \in X$ and $y \in Y$. For all sets $S \in \Sigma \times \Gamma$ in the product σ-algebra we define the sets $S_x := \pi_Y\big[\{s \in S : \pi_X(s) = x\}\big]$ and $S^y := \pi_X\big[\{s \in S : \pi_Y(s) = y\}\big]$ (see Figure 30). For all $\Sigma \times \Gamma$-measurable functions, we define $f_x(y) := f(x, y)$ and $f^y(x) := f(x, y)$ (see Figure 31).

The standard approach to prove results about product σ-algebras is to establish the result for rectangles with measurable sides and then to show that the sets for which the result holds form a σ-algebra. We start by making sure that the sections S_x, S^y, f_x and f^y of measurable sets and functions are measurable.

Proposition 14.59 *Let (X, Σ) and (Y, Γ) be measurable spaces and let $x \in X$ and $y \in Y$. Then for all sets $S \in \Sigma \times \Gamma$ the set S_x is Γ-measurable and the set S^y is Σ-measurable. Moreover, for all $\Sigma \times \Gamma$-measurable functions, the function f_x is Γ-measurable and the function f^y is Σ-measurable.*

Proof. First, we prove that for each $y \in Y$ and $S \in \Sigma \times \Gamma$ we have $S^y \in \Sigma$. To do this let $y \in Y$ and let $\mathcal{B}^y$ be the set of all $S \in \Sigma \times \Gamma$ so that $S^y \in \Sigma$. For all $A \in \Sigma$ and $B \in \Gamma$ the rectangle with measurable sides $A \times B$ is in $\mathcal{B}^y$, because if $y \in B$ we have $(A \times B)^y = A \in \Sigma$ and if $y \notin B$ we have $(A \times B)^y = \emptyset \in \Sigma$. To prove that $\mathcal{B}^y = \Sigma \times \Gamma$ we need to show that $\mathcal{B}^y$ is a σ-algebra, because then it must contain $\Sigma \times \Gamma$ (see Exercise 14-42) and must hence be equal to it. To see this, first note that $\emptyset = \emptyset \times \emptyset \in \mathcal{B}^y$. Now if $S \in \mathcal{B}^y$, then

$$\begin{aligned}(S')^y &= \pi_X\left[\left\{s \in S' : \pi_Y(s) = y\right\}\right] = \pi_X\left[\left\{s \notin S : \pi_Y(s) = y\right\}\right] \\ &= X \setminus \pi_X\left[\left\{s \in S : \pi_Y(s) = y\right\}\right] = \left(S^y\right)' \in \Sigma,\end{aligned}$$

so $S' \in \mathcal{B}^y$. Finally, if $\{S_n\}_{n=1}^{\infty}$ is a sequence in $\mathcal{B}^y$, then

$$\begin{aligned}\left(\bigcup_{n=1}^{\infty} S_n\right)^y &= \pi_X\left[\left\{s \in \bigcup_{n=1}^{\infty} S_n : \pi_Y(s) = y\right\}\right] \\ &= \bigcup_{n=1}^{\infty} \pi_X\left[\left\{s \in S_n : \pi_Y(s) = y\right\}\right] = \bigcup_{n=1}^{\infty} (S_n)^y \in \Sigma,\end{aligned}$$

so $\bigcup_{n=1}^{\infty} S_n \in \mathcal{B}^y$. Thus $\mathcal{B}^y$ is a σ-algebra that contains the rectangles with measurable sides and is contained in $\Sigma \times \Gamma$, which means that $\mathcal{B}^y = \Sigma \times \Gamma$. Hence, for all $y \in Y$ and all $S \in \Sigma \times \Gamma$ we have that $S^y \in \Sigma$. Similarly, we prove that for all $x \in X$ and all $S \in \Sigma \times \Gamma$ we have that $S_x \in \Gamma$.

Now let $f : X \times Y \to [-\infty, \infty]$ be a $\Sigma \times \Gamma$-measurable function. To prove that for every $y \in Y$, the function f^y is Σ-measurable, let $y \in Y$ and $a \in \mathbb{R}$. Then

$$\begin{aligned}\left(f^y\right)^{-1}\left[(a, \infty]\right] &= \left\{x \in X : f(x, y) > a\right\} = \pi_X\left[\left\{z \in X \times Y : f(z) > a, \pi_Y(z) = y\right\}\right] \\ &= \pi_X\left[\left\{z \in f^{-1}\left[(a, \infty]\right] : \pi_Y(z) = y\right\}\right] = \left(f^{-1}\left[(a, \infty]\right]\right)^y \in \Sigma.\end{aligned}$$

Similarly, we prove that for every $x \in X$ the function f_x is Γ-measurable. ■

We conclude this section by showing that the constructions presented here give reasonable results for Lebesgue measure.

Proposition 14.60 *Let $m, n, d \in \mathbb{N}$ be such that $m + n = d$ and let λ_m, λ_n and λ_d denote Lebesgue measure on $\mathbb{R}^m$, $\mathbb{R}^n$ and $\mathbb{R}^d$, respectively. Then $\Sigma_{\lambda_d} \supseteq \Sigma_{\lambda_m} \times \Sigma_{\lambda_n}$ and for all $A \in \Sigma_{\lambda_m}$ and $B \in \Sigma_{\lambda_n}$ we have $\lambda_d(A \times B) \le \lambda_m(A)\lambda_n(B)$.*

Proof. By Exercise 14-41b, $\Sigma_{\lambda_m} \times \Sigma_{\lambda_n}$ is generated by the sets of the form $A \times B$, where each A and B is either an open box or a null set.

The proof that products of open boxes $A \times B$ (which are of course open boxes themselves) are λ_d-measurable is similar to the proof of Proposition 9.13 (see Exercise 14-40). Now let $A \in \Sigma_{\lambda_m}$, let $B \in \Sigma_{\lambda_n}$ and let $\varepsilon > 0$. Find sequences of open boxes $\{I_j\}_{j=1}^{\infty}$ and $\{K_l\}_{l=1}^{\infty}$ so that the containments $A \subseteq \bigcup_{j=1}^{\infty} I_j$, $B \subseteq \bigcup_{l=1}^{\infty} K_l$ and the inequality $\sum_{j=1}^{\infty}|I_j|\sum_{l=1}^{\infty}|K_l| < \lambda_m(A)\lambda_n(B)+\varepsilon$ hold. Then

$$\lambda_d(A\times B)\le\sum_{j,l=1}^{\infty}|I_j||K_l|=\sum_{j=1}^{\infty}|I_j|\sum_{l=1}^{\infty}|K_l|<\lambda_m(A)\lambda_n(B)+\varepsilon,$$

and for all $A\in\Sigma_{\lambda_m}$ and $B\in\Sigma_{\lambda_n}$ we obtain $\lambda_d(A\times B)\le\lambda_m(A)\lambda_n(B)$.

In particular, this means that for $A\in\Sigma_{\lambda_m}$ and $B\in\Sigma_{\lambda_n}$ such that one of A, B is a null set we have that $\lambda_d(A\times B)=0$, which means $A\times B$ is λ_d-measurable. Together with the λ_d-measurability of products of open boxes, this means $\Sigma_{\lambda_d}\supseteq\Sigma_{\lambda_m}\times\Sigma_{\lambda_n}$, because $\Sigma_{\lambda_m}\times\Sigma_{\lambda_n}$ is generated by products of open boxes or null sets in $\mathbb{R}^m$ with open boxes or null sets in $\mathbb{R}^n$. ■

Proposition 14.60 has two apparent shortcomings. First, the σ-algebras are not shown to be equal to each other. Exercise 14-47 shows that the containment is indeed strict, so this part of the result cannot be improved. Second, we did not show that $\lambda_d(A\times B)$ and $\lambda_m(A)\lambda_n(B)$ are equal, even though they should be equal. Recall that in Proposition 8.5 we needed the Heine-Borel Theorem to prove that the Lebesgue measure of an interval is its length. To prove that $\lambda_d(A\times B)$ and $\lambda_m(A)\lambda_n(B)$ are indeed equal, we need a higher dimensional version of the Heine-Borel Theorem. This version will be provided in Theorems 16.72 and 16.80. Thus we delay the proof that $\lambda_d(A\times B)=\lambda_m(A)\lambda_n(B)$ to Theorem 16.81. *To keep notation simple, unless the distinction of different dimensions is necessary, we will denote Lebesgue measure in all dimensions by* λ.

Exercises

14-39 Prove Proposition 14.51

14-40 Prove that every d-dimensional open box is λ_d-measurable.

14-41 Generating the Lebesgue measurable sets. For $d\in\mathbb{N}$ let G_d be the set of all d-dimensional open boxes and all d-dimensional null sets

(a) Prove that Σ_{λ_d}, the σ-algebra of Lebesgue measurable sets in $\mathbb{R}^d$, is generated by G_d
Hint. $G_d\subseteq\Sigma_{\lambda_d}$ by Exercise 14-40. Prove that every $S\in\Sigma_{\lambda_d}$ with $\lambda(S)<\infty$ differs by a null set from a countable intersection of countable unions of open boxes. Then use σ-finiteness

(b) Prove that for $m, n\in\mathbb{N}$ the σ-algebra generated by $\{A\times B : A\in G_m, B\in G_n\}$ is $\Sigma_{\lambda_m}\times\Sigma_{\lambda_n}$.

(c) Prove that if $m+n=d$ and $A\in\Sigma_{\lambda_d}$, then there is a $B\in\Sigma_{\lambda_m}\times\Sigma_{\lambda_n}$ with $\lambda_d(A\setminus B\cup B\setminus A)=0$.

(d) Prove that if $m+n=d$ and the function $f:\mathbb{R}^d\to[-\infty,\infty]$ is Σ_{λ_d}-measurable, then there is a $\Sigma_{\lambda_m}\times\Sigma_{\lambda_n}$-measurable function g with $f=g$ a.e

14-42 Let M be a set and let $\mathcal{U}$ be a set of subsets of M. Prove that the σ-algebra generated by $\mathcal{U}$ is contained in all σ-algebras that contain $\mathcal{U}$

14-43. On the equality of measures.

(a) Prove Proposition 14.54

(b) Prove that the σ-algebra generated by all the singleton subsets of $\mathbb{R}$ is the set $\mathcal{C}:=\{C\subseteq\mathbb{R} : C \text{ or } \mathbb{R}\setminus C \text{ is countable}\}$.

(c) Construct two finite measures on $\mathcal{C}$ that agree on all singletons, but which are not equal.

Hint. Set $\mu\left(\{k\}\right) := \frac{1}{2^k}$ for $k \in \mathbb{N}$ and set it equal to zero for all other singletons.

(d) Construct a measurable space (X, Σ) and two finite measures μ and ν on Σ so that the equality $\mu(X) = \nu(X)$ holds, but $\left\{ S \in \Sigma : \mu(S) = \nu(S) \right\}$ is not a σ-algebra.

Hint. There is a finite example.

14-44. Prove that if $\{\mathcal{D}_i\}_{i\in I}$ is a family of Dynkin systems on the set M, then $\bigcap_{i\in I} \mathcal{D}_i$ is a Dynkin system.

14-45. Prove Theorem 14.57.

14-46. Prove that if μ is a measure defined on the σ-algebra $\mathcal{B}$ generated by the finite open intervals such that $\mu\left((a, b)\right) = b - a$ for all finite open intervals, then μ must be equal to the restriction of Lebesgue measure to $\mathcal{B}$.

14-47. The containment of the σ-algebras in Theorem 14.60 is strict. To see this, prove that for $m < d$ every subset of $\mathbb{R}^m$ (considered as a the subset $\mathbb{R}^m \times \{0\}^{d-m}$ of $\mathbb{R}^d$) can be a section of a Lebesgue measurable set in $\mathbb{R}^d$.

14.8 Product Measures and Fubini's Theorem

We are now ready to define a measure on the product σ-algebra. Because Theorems 14.56 and 14.57 are needed to construct an unambiguous product measure in Theorem 14.62, we introduce the notion of σ-finiteness.

Definition 14.61 *A measure space (M, Σ, μ) is called* σ**-finite** *iff there is a sequence of subsets $A_j \subseteq M$ of finite measure so that $M = \bigcup_{j=1}^{\infty} A_j$. We will also sometimes call the measure σ-finite.*

Clearly, Lebesgue measure is σ-finite, so we will be able to use the results from this section for d-dimensional integration.

Theorem 14.62 *Let (X, Σ, μ) and (Y, Γ, ν) be σ-finite measure spaces. Then there is a unique measure $\mu \times \nu$ on $\Sigma \times \Gamma$ so that with the convention $0 \cdot \infty = 0$ for all $A \in \Sigma$ and $B \in \Gamma$ we have $(\mu \times \nu)(A \times B) = \mu(A)\nu(B)$.*

Proof. The natural idea for the product measure of a set $S \in \Sigma \times \Gamma$ is to integrate the ν-measures of the sections S_x over X. To do this we first need to prove that for all $S \in \Sigma \times \Gamma$ the function $x \mapsto \nu\left(S_x\right)$ is Σ-measurable.

Let $\mathcal{H}$ be the set of sets $S \in \Sigma \times \Gamma$ so that the function $x \mapsto \nu\left(S_x\right)$ is Σ-measurable. We need to prove that $\mathcal{H} = \Sigma \times \Gamma$. For all $A \in \Sigma$ and $B \in \Gamma$ we have that $\nu\big((A \times B)_x\big)$ equals $\nu(B)$ when $x \in A$ and 0 when $x \notin A$. Therefore, the function $x \mapsto \nu\big((A \times B)_x\big) = \nu(B)\mathbf{1}_A(x)$ is Σ-measurable for all $A \times B \in \Sigma \times \Gamma$.

To prove $\mathcal{H}$ is a σ-algebra, first note that the rectangles with measurable sides form a π-system (see Exercise 7-7) and $X \times Y \in \mathcal{H}$. To be able to apply Theorem 14.56, we will now prove $\mathcal{H}$ is a Dynkin system when ν is finite. If $S, F \in \mathcal{H}$ and $S \subseteq F$, then for $F \setminus S$ we have $\nu\left((F \setminus S)_x\right) = \nu\left(F_x \setminus S_x\right) = \nu(F_x) - \nu(S_x)$, which implies that $F \setminus S \in \mathcal{H}$. Now let $\{A_n\}_{n=1}^{\infty}$ be a sequence in $\mathcal{H}$ with $A_n \subseteq A_{n+1}$ for all $n \in \mathbb{N}$.

Then $\nu\left(\left(\bigcup_{n=1}^{\infty} A_n\right)_x\right) = \nu\left(\bigcup_{n=1}^{\infty} (A_n)_x\right) = \lim_{n\to\infty} \nu\left((A_n)_x\right)$ by Theorem 14.15, so by Exercise 14-28a $\mathcal{H}$ is closed under the formation of increasing unions and thus $\mathcal{H}$ is a Dynkin system. Therefore, by Theorem 14.56 $\mathcal{H} = \Sigma \times \Gamma$. Hence, if ν is finite, for all $S \in \Sigma \times \Gamma$ the function $x \mapsto \nu(S_x)$ is Σ-measurable.

To prove the result for σ-finite measure spaces (Y, ν, Γ), let $\{F_n\}_{n=1}^{\infty}$ be a sequence of pairwise disjoint sets in Γ with $\nu(F_n) < \infty$ so that $\bigcup_{n=1}^{\infty} F_n = Y$. Define $\nu_n(S) := \nu(F_n \cap S)$. Then each ν_n is a finite measure and so each function $\nu_n(S_x)$ is Σ-measurable. But then $\nu(S_x) = \sum_{n=1}^{\infty} \nu_n(S_x)$ is also Σ-measurable (Exercise 14-28a), which proves that the functions $x \mapsto \nu(S_x)$ are Σ-measurable.

Now we prove that $(\mu \times \nu)(S) := \int_X \nu(S_x)\, d\mu$ defines a measure on $\Sigma \times \Gamma$. Clearly, $(\mu \times \nu)(\emptyset) = \int_X \nu(\emptyset_x)\, d\mu = \int_X 0\, d\mu = 0$. Now let $\{S_n\}_{n=1}^{\infty}$ be a sequence of pairwise disjoint sets in $\Sigma \times \Gamma$. Then for all $x \in X$ we have that $\left\{(S_n)_x\right\}_{n=1}^{\infty}$ is a sequence of pairwise disjoint sets in Γ. Therefore

$$(\mu \times \nu)\left(\bigcup_{n=1}^{\infty} S_n\right) = \int_X \nu\left(\left(\bigcup_{n=1}^{\infty} S_n\right)_x\right) d\mu = \int_X \sum_{n=1}^{\infty} \nu\left((S_n)_x\right) d\mu$$

Apply the Monotone Convergence Theorem to $\left\{\sum_{n=1}^{k} \nu\left((S_n)_x\right)\right\}_{k=1}^{\infty}$.

$$= \sum_{n=1}^{\infty} \int_X \nu\left((S_n)_x\right) d\mu = \sum_{n=1}^{\infty} (\mu \times \nu)(S_n).$$

Thus $(\mu \times \nu)$ defines a measure on $\Sigma \times \Gamma$. Moreover, for all $A \in \Sigma$ and $B \in \Gamma$ we have $(\mu \times \nu)(A \times B) = \int_X \nu\left((A \times B)_x\right) d\mu = \int_X \nu(B)\, \mathbf{1}_A\, d\mu = \mu(A)\nu(B)$, where the overall result is zero if one of A or B has measure zero. Because μ and ν are both σ-finite, any measure with the above property must be σ-finite. Thus by Theorem 14.57 $\mu \times \nu$ is the unique measure on $\Sigma \times \Gamma$ so that $(\mu \times \nu)(A \times B) = \mu(A)\nu(B)$ for all $A \in \Sigma$ and $B \in \Gamma$. ■

Definition 14.63 *The measure $\mu \times \nu$ in Theorem 14.62 is called the* **product measure** *of μ and ν.*

Theorem 14.62 not only proves the existence of the product measure, its proof also provides a representation of $\mu \times \nu$. Because the product measure is unique and we could have switched the roles of the two factors, there are two representations. These two representations are needed so that we can ultimately switch the order of integration.

Corollary 14.64 *Let (X,Σ,μ), (Y,Γ,ν) be σ-finite measure spaces and let $S\in\Sigma\times\Gamma$. Then $a(x):=\nu(S_x)$ is Σ-measurable, $b(y):=\mu(S^y)$ is Γ-measurable and the equality $(\mu\times\nu)(S)=\int_X \nu(S_x)\,d\mu=\int_Y \mu(S^y)\,d\nu$ holds.*

Proof. Similar to the proof of Theorem 14.62 we can prove that for all $S\in\Sigma\times\Gamma$ the function $y\mapsto\mu\left(S^y\right)$ is Γ-measurable and $(\mu\times\nu)(S)=\int_Y \mu\left(S^y\right)\,d\nu$. ■

We are now ready to prove that integrals with respect to product measures can be computed as iterated integrals.

Proposition 14.65 *Let (X,Σ,μ) and (Y,Γ,ν) be σ-finite measure spaces and let the function $f:X\times Y\to[0,\infty]$ be $\Sigma\times\Gamma$-measurable. Then the function $x\mapsto\int_Y f_x\,d\nu$ is Σ-measurable and the function $y\mapsto\int_X f^y\,d\mu$ is Γ-measurable. Moreover,*

$$\int_{X\times Y} f\,d(\mu\times\nu)=\int_Y\left(\int_X f^y\,d\mu\right)d\nu=\int_X\left(\int_Y f_x\,d\nu\right)d\mu.$$

Proof. We first prove the result for $f=\mathbf{1}_S$ where $S\in\Sigma\times\Gamma$. By Proposition 14.59, for all $y\in Y$ the function $f^y=(\mathbf{1}_S)^y=\mathbf{1}_{S^y}$ is Σ-measurable. Now

$$\begin{aligned}\int_Y\left(\int_X f^y\,d\mu\right)d\nu &= \int_Y\left(\int_X \mathbf{1}_{S^y}\,d\mu\right)d\nu=\int_Y \mu\left(S^y\right)d\nu\\ &= (\mu\times\nu)(S)=\int_{X\times Y}\mathbf{1}_S\,d(\mu\times\nu)\\ &= \int_{X\times Y} f\,d(\mu\times\nu),\end{aligned}$$

and similarly we can prove that $\int_X\left(\int_Y f_x\,d\nu\right)d\mu=\int_{X\times Y} f\,d(\mu\times\nu)$.

By Theorem 14.39 the equalities must hold for all simple $\Sigma\times\Gamma$-measurable functions and by the Monotone Convergence Theorem the equalities hold for all nonnegative $\Sigma\times\Gamma$-measurable functions. ■

Theorem 14.66 Fubini's Theorem. *Let (X,Σ,μ) and (Y,Γ,ν) be σ-finite measure spaces and let $f:X\times Y\to[-\infty,\infty]$ be $\mu\times\nu$-integrable. Then for μ-almost all $x\in X$ the function f_x is ν-integrable and for ν-almost all $y\in Y$ the function f^y is μ-integrable. The function $x\mapsto\begin{cases}\int_Y f_x\,d\nu; & \text{if } f_x \text{ is } \nu\text{-integrable},\\ 0; & \text{otherwise},\end{cases}$ is μ-integrable and the function $y\mapsto\begin{cases}\int_X f^y\,d\mu; & \text{if } f^y \text{ is } \mu\text{-integrable},\\ 0; & \text{otherwise},\end{cases}$ is ν-integrable. Moreover,*

$$\int_{X\times Y} f\,d(\mu\times\nu)=\int_Y\left(\int_X f^y\,d\mu\right)d\nu=\int_X\left(\int_Y f_x\,d\nu\right)d\mu.$$

Proof. By Proposition 14.65 we have that the functions $x \mapsto \int_Y (f^+)_x \, d\nu$ and $x \mapsto \int_Y (f^-)_x \, d\nu$ are μ-integrable, so f_x is ν-integrable for μ-a.e. $x \in X$. Now

$$\begin{aligned}
\int_{X\times Y} f \, d(\mu\times\nu) &= \int_{X\times Y} f^+ \, d(\mu\times\nu) - \int_{X\times Y} f^- \, d(\mu\times\nu) \\
&= \int_X \left(\int_Y (f^+)_x \, d\nu \right) d\mu - \int_X \left(\int_Y (f^-)_x \, d\nu \right) d\mu
\end{aligned}$$

Note that $(f^+)_x - (f^-)_x = (f_x)^+ - (f_x)^- = f_x$.

$$= \int_X \left(\int_Y f_x \, d\nu \right) d\mu$$

The other equation and the assertions about the f_y are proved similarly. ■

Exercises

14-48. Construct a Lebesgue integrable function $f : [0,1] \times [0,1] \to [-\infty, \infty]$ so that not all f_x and f^y are Lebesgue integrable.

14-49. Let $(\mathbb{R}, \Sigma_\lambda, \lambda)$ be the real numbers with Lebesgue measure, let $\left(\mathbb{R}, \mathcal{P}(\mathbb{R}), \gamma_{\mathbb{R}} \right)$ be the real numbers with counting measure and let $f := \mathbf{1}_{y=x}$ be the indicator function of the set $\left\{ (x,y) \in \mathbb{R}^2 : y = x \right\}$. Prove that $\int_{\mathbb{R}} \left(\int_{\mathbb{R}} f^y \, d\lambda \right) d\gamma_{\mathbb{R}} \neq \int_{\mathbb{R}} \left(\int_{\mathbb{R}} f_x \, d\gamma_{\mathbb{R}} \right) d\lambda$.

Note. This example shows that the assumption of σ-finiteness of the measure spaces is needed throughout this section.

14-50. Let $\left(\mathbb{N}, \mathcal{P}(\mathbb{N}), \gamma_{\mathbb{N}} \right)$ be the natural numbers with counting measure.

(a) Prove that $\mathcal{P}(\mathbb{N}) \times \mathcal{P}(\mathbb{N}) = \mathcal{P}(\mathbb{N}\times\mathbb{N})$, where the first product denotes the product σ-algebra of the two power sets.

(b) Prove that $\gamma_{\mathbb{N}} \times \gamma_{\mathbb{N}} = \gamma_{\mathbb{N}\times\mathbb{N}}$.

(c) Prove that if the **double series** $\sum_{i=1}^{\infty} \sum_{j=1}^{\infty} a_{(i,j)}$ converges absolutely, then we can exchange the order of summation, that is, $\sum_{i=1}^{\infty} \left(\sum_{j=1}^{\infty} a_{(i,j)} \right) = \sum_{j=1}^{\infty} \left(\sum_{i=1}^{\infty} a_{(i,j)} \right)$.

Hint. Recall Example 14.37.

(d) Prove that if the double series $\sum_{i=1}^{\infty} \sum_{j=1}^{\infty} a_{(i,j)}$ converges absolutely, then for all bijective functions $\sigma : \mathbb{N}\times\mathbb{N} \to \mathbb{N}\times\mathbb{N}$ the sum $\sum_{i=1}^{\infty} \sum_{j=1}^{\infty} a_{\sigma(i,j)}$ converges to $\sum_{i=1}^{\infty} \left(\sum_{j=1}^{\infty} a_{(i,j)} \right)$.

Hint. Dominated Convergence Theorem and Fubini.

14-51. Let (M, Σ, μ) be a σ-finite measure space, let $f : M \to [0, \infty]$ be Σ-measurable and let $(\mathbb{R}, \Sigma_\lambda, \lambda)$ be the real numbers with Lebesgue measure.

(a) Prove that $E = \left\{ (x,y) \subseteq M \times \mathbb{R} : 0 \le y < f(x) \right\}$ is in $\Sigma \times \Sigma_\lambda$.

(b) Prove that $\int_M f \, d\mu = \int_0^\infty \mu\left(\left\{ x \in M : f(x) > y \right\} \right) dy$.

Chapter 15

The Abstract Venues for Analysis

It is now time to introduce the broad spectrum of settings in which analysis is done. Applications are usually set in multidimensional Euclidean space, so we need to find a way to do analysis in $\mathbb{R}^d$. Moreover, because the solutions for many modeling processes are functions, we need to consider spaces of functions. To make the theory applicable in many situations, we analyze the abstract commonalities of the structures in question, define spaces based on these commonalities and then prove theorems in these abstract spaces. The beauty and power of the abstraction are founded in the wide applicability of the results as well as in their relative simplicity. Regarding the wide applicability, recall, for example, that the initial abstraction in Chapter 14 was but a simple step from the corresponding notions for functions of one real variable. Yet it ultimately allowed us to integrate in arbitrary dimensions. Regarding the relative simplicity, note that the notions of convergence for functions presented in Chapter 11 and in Section 14.6 involve a significant amount of detail about functions, which may be quite complicated objects themselves. The generalizations of d-dimensional space presented in this chapter will allow us to visualize the functions and their convergence behavior as if they were points and sequences in low dimensional spaces. This simplification is significant.

The subsequent chapters will provide more details on the metric, normed and inner product spaces presented in this chapter.

15.1 Abstraction I: Vector Spaces

Vector spaces describe the algebraic properties of multidimensional spaces and of spaces of functions. These properties guarantee that objects can be added to each other and multiplied with numbers.

Definition 15.1 *A* **real vector space** *or a* **vector space** *is a triple* $(X, +, \cdot)$ *of a set* X *and two binary operations,* **vector addition** $+ : X \times X \to X$ *and* **scalar multiplication** $\cdot : \mathbb{R} \times X \to X$, *so that the following hold.*

1. *Vector addition is* **associative**.
   That is, for all $x, y, z \in X$ *we have* $(x + y) + z = x + (y + z)$.

2. *Vector addition is* **commutative**.
   That is, for all $x, y \in X$ *we have* $x + y = y + x$.

3. *There is a* **neutral element** $0 \in X$ *for addition.*
   That is, there is an element $0 \in X$ *so that for all* $x \in X$ *we have* $x + 0 = x$.

4. *For every element* $x \in X$, *there is an* **additive inverse** *element.*
   That is, for every $x \in X$ *there is a* $(-x) \in X$ *so that* $x + (-x) = 0$.

5. *Scalar multiplication is* **(left) distributive** *over vector addition.*
   That is, for all $x, y \in X$ *and* $\alpha \in \mathbb{R}$ *we have* $\alpha \cdot (x + y) = \alpha \cdot x + \alpha \cdot y$.

6. *Scalar multiplication is* **(right) distributive** *over scalar addition.*
   That is, for all $x \in X$ *and* $\alpha, \beta \in \mathbb{R}$ *we have* $(\alpha + \beta) \cdot x = \alpha \cdot x + \beta \cdot x$,

7. *Scalar multiplication is "***associative***."*
   That is, for all $x \in X$ *and* $\alpha, \beta \in \mathbb{R}$ *we have* $\alpha \cdot (\beta \cdot x) = (\alpha\beta) \cdot x$,

8. *The number* $1 \in \mathbb{R}$ *is a* **neutral element** *for scalar multiplication.*
   That is, for all $x \in X$ *we have* $1 \cdot x = x$.

An element of a vector space is also called a **vector** *and a real number is also called a* **scalar** *in this context. We will usually refer to the set* X *as the vector space, implicitly assuming addition and multiplication are denoted as usual. As is customary for multiplications, the dot is usually omitted.*

The standard example and visualization for a vector space is d-dimensional space.

Example 15.2 Let $d \in \mathbb{N}$. The set $\mathbb{R}^d := \{(x_1, \ldots, x_d) : x_i \in \mathbb{R}\}$ with component-wise addition and scalar multiplication is a vector space. □

To indicate how abstraction can simplify arguments, we will not directly prove the vector space properties for d-dimensional space. Instead, we note that d-dimensional space can be interpreted as the space of functions $f : \{1, \ldots, d\} \to \mathbb{R}$. The fact that d-dimensional space is a vector space then follows from Example 15.3 below. In a way, Example 15.3 is universal and it should be thoroughly understood. All examples of spaces in this text are subspaces of spaces as in Example 15.3.

Example 15.3 *Let* D *be a set. The set* $F(D, \mathbb{R})$ *of all functions* $f : D \to \mathbb{R}$ *with addition defined pointwise by* $(f + g)(x) := f(x) + g(x)$ *and scalar multiplication defined pointwise by* $(\alpha \cdot f)(x) := \alpha f(x)$ *is a vector space.*

All properties follow from the appropriate pointwise properties for real numbers. For example, addition of functions is commutative, because for all $x \in D$ we have that $(f + g)(x) = f(x) + g(x) = g(x) + f(x) = (g + f)(x)$. The neutral element with respect to addition is the function that is equal to $0 \in \mathbb{R}$ for all $x \in X$. □

By identifying each vector $(x_1, \ldots, x_d) \in \mathbb{R}^d$ with the function that maps each index $i \in \{1, \ldots, d\}$ to x_i, we see that $\mathbb{R}^d$ is the vector space $F(\{1, \ldots, d\}, \mathbb{R})$. We conclude this section by introducing several important subspaces of spaces $F(D, \mathbb{R})$.

Definition 15.4 *A subset S of a vector space X is called a* **vector subspace** *or* **subspace** *of X iff it is a vector space when equipped with the restrictions of addition and scalar multiplication to S.*

The advantage of working with subspaces of a somewhat universal vector space like $F(D, \mathbb{R})$ is that subspaces are easily proved to be vector spaces.

Lemma 15.5 *Let X be a vector space. Then the neutral element $0 \in X$ with respect to addition is unique. Moreover, for the scalar $0 \in \mathbb{R}$ and all vectors $x \in X$ we have that $0 \cdot x = 0$.*

Proof. Suppose $0, 0' \in X$ are both neutral elements with respect to addition. Then $0 = 0+0' = 0'+0 = 0'$. Hence, the neutral element with respect to addition is unique.

Now let $x \in X$. Then for all $y \in X$ we have

$$y + 0x = y + x + 0x - x = y + (1 + 0)x - x = y + x - x = y.$$

Therefore $0x$ is neutral with respect to addition, and hence it is equal to $0 \in X$. ∎

Proposition 15.6 *A nonempty subset S of the vector space X is a subspace iff for all $x, y \in S$ we have $x + y \in S$ and for all $\alpha \in \mathbb{R}$ and $x \in S$ we have $\alpha \cdot x \in S$.*

Proof. For "$\Rightarrow$," note that if S is a subspace, then the results of operations with elements of S must again be in S.

Conversely, for "$\Leftarrow$" note that if the sums and scalar multiples of elements of S are again in S, then the properties of the restrictions of the operations to S follow from the corresponding properties for the overall operations on X. The existence of a neutral element for addition follows from the fact that if $x \in S$, then $0 = 0x \in S$. ∎

Example 15.7 *The set $l^\infty := \left\{ \{x_i\}_{i=1}^\infty : x_i \in \mathbb{R},\ \sup\left\{|x_i| : i \in \mathbb{N}\right\} < \infty \right\}$ with addition and scalar multiplication defined termwise is a vector space.*

First, note that l^∞ is a subset of $F(\mathbb{N}, \mathbb{R})$. Now if $f, g \in l^\infty$ and $\left\{|f(n)| : n \in \mathbb{N}\right\}$ is bounded by B_f and $\left\{|g(n)| : n \in \mathbb{N}\right\}$ is bounded by B_g, then for all $n \in \mathbb{N}$ we have $|(f+g)(n)| = |f(n) + g(n)| \leq |f(n)| + |g(n)| \leq B_f + B_g < \infty$, so $f + g \in l^\infty$. Similarly, if $\alpha \in \mathbb{R}$, then for all $n \in \mathbb{N}$ we have $|(\alpha f)(n)| = |\alpha||f(n)| \leq |\alpha| B_f < \infty$. Thus l^∞ is a subspace of $F(\mathbb{N}, \mathbb{R})$. □

Example 15.8 *The set $C^0([a,b], \mathbb{R})$ of continuous functions $f : [a,b] \to \mathbb{R}$ with pointwise addition and scalar multiplication is a vector space.*

Clearly, $C^0([a,b], \mathbb{R})$ is a subset of $F([a,b], \mathbb{R})$. Moreover, part 1 of Theorem 3.27 implies that sums of continuous functions are continuous. Part 3 of Theorem 3.27, applied with one function being constant, implies that scalar multiples of continuous functions are continuous. □

Notation 15.9 To simplify notation, we write $C^0[a,b]$ instead of $C^0([a,b],\mathbb{R})$. □

Example 15.8 shows how switching to more abstract situations is often just a change in point-of-view. We no longer consider individual objects, but instead we consider classes of objects. Results about individual objects are then often summarized in statements about the class. For example, the fact that sums and constant multiples of continuous functions are continuous is summarized in the statement that the continuous functions form a vector space.

Example 15.10 *For $a<b$ and $k\in\mathbb{N}$, let $C^k(a,b)$ be the set of k times* **continuously differentiable** *functions on (a,b). Then $C^k(a,b)$ with pointwise addition and scalar multiplication is a vector space. Moreover, $C^\infty(a,b):=\bigcap_{k=1}^{\infty} C^k(a,b)$, the space of infinitely differentiable functions on (a,b), with pointwise addition and scalar multiplication is a vector space.*

This is an induction using Theorem 4.6. □

Spaces of "p-integrable functions" are of central importance in analysis.

Definition 15.11 *Let (M,Σ,μ) be a measure space and let $1\le p<\infty$. We define $\mathcal{L}^p(M,\Sigma,\mu):=\left\{f\in F(M,\mathbb{R}):\int_M |f|^p\,d\mu<\infty\right\}$.*

Example 15.12 *Let (M,Σ,μ) be a measure space and let $1\le p<\infty$. Then $\mathcal{L}^p(M,\Sigma,\mu)$ is a vector space.*

Because $\mathcal{L}^p(M,\Sigma,\mu)\subseteq F(M,\mathbb{R})$ we can apply Proposition 15.6.

Let $f,g\in\mathcal{L}^p(M,\Sigma,\mu)$. Then

$$\begin{aligned}\int_M |f+g|^p\,d\mu &\le \int_M (|f|+|g|)^p\,d\mu \le \int_M (2\max\{|f|,|g|\})^p\,d\mu\\ &\le \int_M 2^p|f|^p+2^p|g|^p\,d\mu<\infty,\end{aligned}$$

so $f+g\in\mathcal{L}^p(M,\Sigma,\mu)$. Moreover, if $\alpha\in\mathbb{R}$, then

$$\int_M |\alpha\cdot f|^p\,d\mu = \int_M |\alpha|^p|f|^p\,d\mu=|\alpha|^p\int_M |f|^p\,d\mu<\infty,$$

so $\alpha\cdot f\in\mathcal{L}^p(M,\Sigma,\mu)$. Hence, $\mathcal{L}^p(M,\Sigma,\mu)$ is a vector space. □

For concrete examples, the notation for $\mathcal{L}^p$ with an abstract measure space becomes a bit cumbersome.

Definition 15.13 *When we work with real valued functions on a Lebesgue measurable subset D of $\mathbb{R}^d$, we will assume that $M=D$, $\Sigma=\Sigma_\lambda$, the σ-algebra of Lebesgue measurable subsets of D, and $\mu=\lambda$, the Lebesgue measure. In this case, we denote $\mathcal{L}^p(D):=\mathcal{L}^p(M,\Sigma,\mu)$. Also, if D is an interval, we may leave out the outer parentheses. That is, we may write $\mathcal{L}^p[a,b]$ for $\mathcal{L}^p([a,b])$, and so on.*

Aside from the capital $\mathcal{L}^p$ spaces, the lowercase l^p-spaces of Definition 15.14 below are sometimes used in analysis.

Definition 15.14 *For* $1 \leq p < \infty$, *define* $l^p := \left\{ \{x_j\}_{j=1}^{\infty} : x_j \in \mathbb{R}, \sum_{j=1}^{\infty} |x_j|^p < \infty \right\}$.

In this chapter, we will frequently translate results between "little l^p" and "big $\mathcal{L}^p$." This will not be a problem, because the computations are very similar. The only adjustment usually is that sums are interchanged with integrals and vice versa. As a first demonstration of the similarities between "little l^p" and "big $\mathcal{L}^p$" we leave the proof that l^p is a vector space to Exercise 15-2a. The two ways in which the analogy between l^p and $\mathcal{L}^p$ can be used will be presented in Sections 15.4 and 15.6, respectively. In Section 15.4, we will prove results for l^2 and then show how the proof is analogous for $\mathcal{L}^2$. In Section 15.6, we will prove results for $\mathcal{L}^p$ and show how results for l^p can be obtained as corollaries.

Exercises

15-1. Prove that if X is a vector space, then for all $x \in X$ we have $(-1)x = -x$.

15-2 Examples of vector spaces.

(a) Prove that for $1 \leq p < \infty$ the space l^p is a vector space.

(b) Let (M, Σ) be a measurable space Prove that the set $\mathcal{M}(M, \Sigma)$ of measurable real valued functions is a vector space

(c) Let $a < b$ and let $\mathcal{BV}[a, b]$ denote the set of all functions of **bounded variation** on $[a, b]$ (see Exercise 8-12). Prove that $\mathcal{BV}[a, b]$ with pointwise addition and scalar multiplication is a vector space.

15-3 **Containments of $\mathcal{L}^p$-spaces.**

(a) Prove that $l^1 \subset l^2 \subset l^\infty$ Then give concrete examples to show the containment is proper. *Hint* Harmonic series.

(b) Prove that if $1 \leq p < q$, then $l^p \subset l^q$ Give an example to show the containment is proper

(c) Let (M, Σ, μ) be a measure space with $\mu(M) < \infty$ Prove that if $1 \leq p < q$, then $\mathcal{L}^p(M, \Sigma, \mu) \supset \mathcal{L}^q(M, \Sigma, \mu)$. Give an example that shows that the containment is proper *Hint.* Exercise 9-32.

(d) Give an example of a measure space (M, Σ, μ) such that for all $1 \leq p < q$ we have that $\mathcal{L}^p(M, \Sigma, \mu) \not\subseteq \mathcal{L}^q(M, \Sigma, \mu)$ and $\mathcal{L}^p(M, \Sigma, \mu) \not\supseteq \mathcal{L}^q(M, \Sigma, \mu)$. *Hint.* By part 15-3b the measure space cannot be discrete and by part 15-3c it cannot be finite.

15.2 Representation of Elements: Bases and Dimension

It is often useful to represent vectors in terms of certain standard vectors. The most familiar example is the standard coordinate system in d-dimensional space.

Definition 15.15 *Let X be a vector space. A subset $S \subseteq X \setminus \{0\}$ is called* **linearly independent** *iff for all finite subsets $\{x_1, \ldots, x_n\} \subseteq S$ and all sets of scalars $\{\alpha_1, \ldots, \alpha_n\} \subseteq \mathbb{R}$ the equation $\sum_{i=1}^{n} \alpha_i x_i = 0$ implies $\alpha_1 = \alpha_2 = \cdots = \alpha_n = 0$.*

A sum $\sum_{i=1}^{n} \alpha_i x_i$ with $\alpha_i \in \mathbb{R}$ and $x_i \in X$ is also called a **linear combination** of $x_1, \ldots, x_n$.

Definition 15.16 *A linearly independent set $B \subseteq X$ such that for every $x \in X$ there are a finite subset $\{b_1, \ldots, b_n\} \subseteq B$ and a set of scalars $\{\alpha_1, \ldots, \alpha_n\} \subseteq \mathbb{R}$ so that $x = \sum_{i=1}^{n} \alpha_i b_i$ is called a* **base** *of a vector space.*

Obviously, linear independence and bases are finitary notions. They are usually investigated in linear algebra. For our purposes, they are important to establish the difference between finite dimensional and infinite dimensional spaces.

Example 15.17 *In $\mathbb{R}^d$, let e_i denote the vector such that the ith component is 1 and all other components are zero. Then $\{e_1, \ldots, e_d\}$ is a base of $\mathbb{R}^d$.*

To prove linear independence, for each $i = 1, \ldots, d$ let $e_i^{(j)}$ denote the jth component of e_i. Then for any $\alpha_1, \ldots, \alpha_d$ the vector equation $\sum_{i=1}^{d} \alpha_i e_i = 0$ leads with $j = 1, \ldots, d$ to the scalar equations $\sum_{i=1}^{d} \alpha_i e_i^{(j)} = 0$, which for each j simply state that $\alpha_j = 0$ as was to be proved.

Regarding the representation of elements, for each $x = (x_1, \ldots, x_d) \in \mathbb{R}^d$ we have that $x = (x_1, \ldots, x_d) = \sum_{i=1}^{d} x_i e_i$. □

The nice thing about bases is that the representation of elements is unique.

Proposition 15.18 *Let X be a vector space and let $B \subseteq X$ be a base. Then for each $x \in X$ the $b_1, \ldots, b_n \in B$ and $\alpha_1, \ldots, \alpha_n \in \mathbb{R}$ in Definition 15.16 are unique, except that any vector of $B \setminus \{b_1, \ldots, b_n\}$ can be added to the b_i with a coefficient zero and that any b_i with $\alpha_i = 0$ can be omitted.*

Proof. Suppose for a contradiction that there is an $x \in X$ with two different representations. Let $\{b_1, \ldots, b_n\} \subseteq B$ be the set of base vectors that occur in either representation. Then there are two distinct n-tuples $(\alpha_1, \ldots, \alpha_n) \neq (\beta_1, \ldots, \beta_n)$ in $\mathbb{R}^n$ so that $\sum_{i=1}^{n} \alpha_i b_i = x = \sum_{i=1}^{n} \beta_i b_i$. But then $\sum_{i=1}^{n} (\alpha_i - \beta_i) b_i = 0$, and hence by linear independence of B, for all $i \in \{1, \ldots, n\}$ we conclude $\alpha_i - \beta_i = 0$, that is, $\alpha_i = \beta_i$, a contradiction. ■

Every vector space is either finite dimensional, that is, it has a finite base, or not. Finite dimensional spaces are investigated linear algebra and in analysis. Infinite dimensional spaces are mostly investigated in (functional) analysis.

Proposition 15.19 *Let X be a vector space with a finite base F. Then every linearly independent subset L of X has at most as many elements as F. Moreover, all bases of X have as many elements as F.*

Proof. Let $F = \{f_1, \ldots, f_n\}$ be a finite base of X. Suppose for a contradiction that there is a linearly independent set $L \subseteq X$ that has more elements than F. Without loss of generality we can assume that L is such that $|L \cap F|$ is maximal. That is, if $\tilde{L}$ is another linearly independent subset of X with more elements than F, then $\left|\tilde{L} \cap F\right| \leq |L \cap F|$. Now let $b \in L \setminus F$ and consider the linearly independent sets $L \setminus \{b\}$ and F. By Exercise 15-4a, there is a subset $H \subseteq F \setminus L$ so that $C := \big(L \setminus \{b\}\big) \cup H$ is a base of X. By Exercise 15-4b, $L \setminus \{b\}$ is not a base of X, so $H \neq \emptyset$. Hence, C has at least as many elements as L and in particular it has more elements than F. Because $b \notin F$, we obtain $\left|\big[(L \setminus \{b\}) \cup H\big] \cap F\right| = \left|(L \cap F) \cup H\right| > |L \cap F|$, a contradiction.

Thus no linearly independent subset of X has more elements than F. In particular, if B is another base of X, then $|B| \leq |F|$. Therefore B is finite and with the same argument as above, if $L \subseteq X$ is linearly independent, then $|L| \leq |B|$. Because F is linearly independent, we obtain $|F| \leq |B|$, and hence $|F| = |B|$. ■

The fact that if one base is finite then all bases have the same size allows us to assign *one* number as the dimension.

Definition 15.20 *A vector space X is called* **finite dimensional** *iff it has a finite base. In this case the* **dimension** *of X is the number of vectors in a finite base. If X has no finite base it is called* **infinite dimensional**.

Proposition 15.21 *A vector space $X \neq \{0\}$ is infinite dimensional iff it contains an infinite linearly independent set.*

Proof. For "$\Rightarrow$," note that no finite linearly independent subset of X is a base. Construct a sequence of finite linearly independent subsets F_n of X as follows. Let $F_1 := \{f_1\}$ with $f_1 \in X \setminus \{0\}$ arbitrary. Once $F_n = \{f_1, \ldots, f_n\}$ is chosen, note that because F_n is not a base of X, there is an $f_{n+1} \in X$ so that there are no $\alpha_1, \ldots, \alpha_n \in \mathbb{R}$ so that $\sum_{j=1}^{n} \alpha_j f_j = f_{n+1}$. This means that $F_{n+1} := F_n \cup \{f_{n+1}\}$ is linearly independent (Exercise 15-5). But then the infinite set $\bigcup_{n=1}^{\infty} F_n$ is linearly independent in X.

For "$\Leftarrow$" let $A \subseteq X$ be an infinite linearly independent subset. Then by Proposition 15.19 X cannot have a finite base. ■

Our first example of an infinite dimensional space is easy to come by.

Example 15.22 *In l^1 let e_i be the sequence with j^{th} entry $e_i^{(j)} = \begin{cases} 0; & \text{for } j \neq i, \\ 1; & \text{for } j = i. \end{cases}$ Then $\{e_i : i \in \mathbb{N}\}$ is linearly independent, and hence l^1 is infinite dimensional.*

The reader will verify this in Exercise 15-7. □

Example 15.22 shows a shortcoming of the definition of a base that is ultimately addressed analytically in Banach and Hilbert Spaces. The set of "unit vectors" in Example 15.22 has everything we want in a base, except that we would need to sum infinitely many vectors for the representations. To define "infinite sums" we need a notion of convergence, which will be introduced as soon as we investigate metric spaces. For now, we continue in our exploration of the spaces of abstract analysis and we conclude this section with a notation we will use in l^p spaces from here on.

Convention 15.23 *When working with sequences $\{a_n\}_{n=1}^{\infty}$ of objects in spaces whose elements are finite or infinite sequences themselves, we will move the sequence index for the elements into the exponent in parentheses. That is $a_n^{(j)}$ will denote the jth element in the sequence a_n, which is itself an element of the sequence $\{a_n\}_{n=1}^{\infty}$. The vector e_i with j^{th} entry $e_i^{(j)} = \begin{cases} 0; & \text{for } j \neq i, \\ 1; & \text{for } j = i, \end{cases}$ will be called the ith* **standard unit vector**. *For all $1 \leq p \leq \infty$, the standard unit vectors are contained in l^p.*

Exercises

15-4 The span of linearly independent sets Let X be a vector space and let $W \subseteq X$. Define the **span** of W to be $\operatorname{span}(W) := \left\{ v = \sum_{i=1}^{n} a_i w_i \quad a_i \in \mathbb{R}, w_i \in W \right\}$

(a) Let B and F be linearly independent sets and let F be finite Prove that if $\operatorname{span}(B \cup F) = X$, then there is a subset H of F so that $B \cup H$ is a base of X

Hint. Induction on the size of F In the induction step, let $f \in F$ and distinguish the cases $f \in \operatorname{span}(B)$ and $f \notin \operatorname{span}(B)$

(b) Let B be a linearly independent set and let $b \in B$ Prove that $\operatorname{span}(B) \neq \operatorname{span}\left(B \setminus \{b\}\right)$.

15-5 Let $F = \{f_1, \ldots, f_n\}$ be linearly independent in the vector space X Prove that if $f_{n+1} \in X$ is so that there are no $\alpha_1, \ldots, \alpha_n \in \mathbb{R}$ so that $\sum_{j=1}^{n} \alpha_j f_j = f_{n+1}$, then $F \cup \{f_{n+1}\}$ is linearly independent.

15-6. **Base Exchange Theorem** Prove that if X is a vector space and $\{v_1, \ldots, v_n\}$ and $\{w_1, \ldots, w_n\}$ are bases of X, then there is a $j \in \{1, \ldots, n\}$ so that $\{v_1, \ldots, v_{n-1}, w_j\}$ is a base of X

15-7. Prove the claim in Example 15 22.

15-8. Prove that for $1 < p \leq \infty$ the space l^p is infinite dimensional

15-9. Prove that for $1 \leq p < \infty$ the space $\mathcal{L}^p[0, 1]$ is infinite dimensional.

15-10 Prove that the space $C^0[0, 1]$ is infinite dimensional

15.3 Identification of Spaces: Isomorphism

Isomorphisms are structure preserving maps. Basically, isomorphic structures are for all intents and purposes "the same," as long as we only care about the operations and concepts that are preserved by the isomorphism.

Definition 15.24 *Let X, Y be vector spaces. Then the function $\Phi : X \to Y$ is called an* **isomorphism** *iff Φ is bijective and for all $u, v \in X$ and $\alpha \in \mathbb{R}$ we have that $\Phi(u + v) = \Phi(u) + \Phi(v)$ and $\Phi(\alpha u) = \alpha\Phi(u)$.*

Definition 15.24 and Exercise 15-11, which proves that the inverse of an isomorphism is again an isomorphism, show why isomorphic structures are considered to be "the same." Let Φ be an isomorphism from the vector space X to the vector space Y. Obviously, we have a one-for-one matching between the elements of X and Y. Moreover, as long as we only care about linear operations, it does not matter if we carry out the operations in X or in Y, because we can always map back and forth between the spaces and get the corresponding results.

The next result shows that from the point-of-view of linear algebra, all finite dimensional spaces are "the same." Theorem 16.76 will show that they are also "the same" from an analytical point-of-view.

Proposition 15.25 *Every d-dimensional vector space X is isomorphic to $\mathbb{R}^d$.*

Proof. Let $\{b_1, \ldots, b_d\}$ be a base of X. Then for each vector $x \in X$ there are unique numbers $x_1, \ldots, x_d \in \mathbb{R}$ so that $x = \sum_{i=1}^{d} x_i b_i$. Define the function $\Phi : X \to \mathbb{R}^d$ by $\Phi(x) = \Phi\left(\sum_{i=1}^{d} x_i b_i\right) := \sum_{i=1}^{d} x_i e_i$. Because the x_i are unique, Φ is well-defined. Moreover, for all $\alpha \in \mathbb{R}$ and all $x, y \in X$ with $x = \sum_{i=1}^{d} x_i b_i$ and $y = \sum_{i=1}^{d} y_i b_i$ we have

$$\begin{aligned}\Phi(x+y) &= \Phi\left(\sum_{i=1}^{d} x_i b_i + \sum_{i=1}^{d} y_i b_i\right) = \Phi\left(\sum_{i=1}^{d} (x_i + y_i) b_i\right) \\ &= \sum_{i=1}^{d} (x_i + y_i) e_i = \sum_{i=1}^{d} x_i e_i + \sum_{i=1}^{d} y_i e_i = \Phi(x) + \Phi(y),\end{aligned}$$

and

$$\begin{aligned}\Phi(\alpha x) &= \Phi\left(\alpha \sum_{i=1}^{d} x_i b_i\right) = \Phi\left(\sum_{i=1}^{d} (\alpha x_i) b_i\right) \\ &= \sum_{i=1}^{d} (\alpha x_i) e_i = \alpha \sum_{i=1}^{d} x_i e_i = \alpha \Phi(x),\end{aligned}$$

which means that Φ is has the linearity properties of an isomorphism. For surjectivity, note that if $y = \sum_{i=1}^{d} y_i e_i$ is in $\mathbb{R}^d$ then $\Phi\left(\sum_{i=1}^{d} y_i b_i\right) = \sum_{i=1}^{d} y_i e_i = y$. Finally, to prove that Φ is injective, let $x = \sum_{i=1}^{d} x_i b_i$ and $z = \sum_{i=1}^{d} z_i b_i$ be in X with $\Phi(x) = \Phi(z)$. Then $0 = \Phi(x) - \Phi(z) = \Phi(x - z) = \Phi\left(\sum_{i=1}^{d} (x_i - z_i) b_i\right) = \sum_{i=1}^{d} (x_i - z_i) e_i$, which implies $x_i = z_i$ for all i, and hence $x = z$. ■

Exercises

15-11. Let X and Y be vector spaces. Prove that if $\Phi : X \to Y$ is an isomorphism, then so is its inverse $\Phi^{-1} : Y \to X$.

15-12. Prove that for any $1 \leq p \leq \infty$ the space $\mathbb{R}^d$ is isomorphic to the subspace of l^p consisting of sequences $\{x_j\}_{j=1}^{\infty}$ so that for all $j > d$ we have $x_j = 0$.

15-13 Prove that $\Phi : X \to Y$ is an isomorphism iff Φ is bijective and for all $\alpha \in \mathbb{R}$ and $x, y \in X$ we have that $\Phi(\alpha x + y) = \alpha\Phi(x) + \Phi(y)$

15.4 Abstraction II: Inner Product Spaces

In three dimensional space there are two ways to multiply vectors. The vector (or cross) product, inspired by observations of magnetic forces on moving charges and also used in the definition of the torque, is native to three dimensional space. However, the scalar (or dot) product, which is inspired by the consideration of work done by a force and which can be used to measure angles, can be generalized to other vector spaces. The corresponding notion is called an inner product. Strictly speaking, Definition 15.26 below defines a *real* inner product space. For complex inner product spaces, consider Definition 15.81.

Definition 15.26 *An* **inner product space** *is a pair* $(X, \langle\cdot,\cdot\rangle)$ *of a vector space* X *and a function* $\langle\cdot,\cdot\rangle : X \times X \to \mathbb{R}$, *called the* **inner product** *or* **scalar product** *on* X, *with the following properties.*

1. *The inner product is* **positive definite**.
   That is, for all $x \in X$ *we have* $\langle x, x\rangle \in [0, \infty)$ *with* $\langle x, x\rangle = 0$ *iff* $x = 0$.
2. *The inner product is* **symmetric**.
   That is, for all $x, y \in X$ *we have* $\langle x, y\rangle = \langle y, x\rangle$.
3. *The inner product is* **linear** *in the first factor.*
   That is, for all scalars α *and all* $x, y \in X$ *we have* $\langle \alpha x, y\rangle = \alpha\langle x, y\rangle$ *and for all* $x, y, z \in X$ *we have* $\langle x + y, z\rangle = \langle x, z\rangle + \langle y, z\rangle$.

We will usually call X *itself an inner product space. When we do so, we implicitly assume that there is an inner product on* X, *which will usually be denoted* $\langle\cdot,\cdot\rangle$.

Exercise 15-14 shows that the inner product is also linear in the second factor. Rather than staying with d-dimensional space, in our first example we go directly to an infinite dimensional inner product space.

Example 15.27 *The set* $l^2 := \left\{ \{x_i\}_{i=1}^{\infty} : x_i \in \mathbb{R}, \sum_{i=1}^{\infty} x_i^2 < \infty \right\}$ *is a vector subspace of* $F(\mathbb{N}, \mathbb{R})$ *and with* $\left\langle \{x_i\}_{i=1}^{\infty}, \{y_i\}_{i=1}^{\infty} \right\rangle := \sum_{i=1}^{\infty} x_i y_i$ *it is an inner product space.*

To prove that l^2 is a subspace of $F(\mathbb{N}, \mathbb{R})$ first let $\{x_i\}_{i=1}^{\infty} \in l^2$ and $\alpha \in \mathbb{R}$. Then $\sum_{i=1}^{\infty}(\alpha x_i)^2 = \alpha^2 \sum_{i=1}^{\infty} x_i^2 < \infty$, and hence $\{\alpha x_i\}_{i=1}^{\infty} \in l^2$. To prove that l^2 is closed under addition, let $\{x_i\}_{i=1}^\infty, \{y_i\}_{i=1}^\infty \in l^2$ and note that (see Exercise 15-15) for all $x, y \in \mathbb{R}$ the inequality $2|xy| \leq x^2 + y^2$ holds. Then

$$\begin{aligned}\sum_{i=1}^\infty (x_i + y_i)^2 &\leq \sum_{i=1}^\infty x_i^2 + 2|x_i y_i| + y_i^2 \leq \sum_{i=1}^\infty x_i^2 + \sum_{i=1}^\infty \left(x_i^2 + y_i^2\right) + \sum_{i=1}^\infty y_i^2 \\ &\leq 2\sum_{i=1}^\infty x_i^2 + 2\sum_{i=1}^\infty y_i^2 < \infty,\end{aligned}$$

and hence $\{x_i\}_{i=1}^\infty + \{y_i\}_{i=1}^\infty = \{x_i + y_i\}_{i=1}^\infty \in l^2$. Moreover, the series in the proposed inner product satisfy $\sum_{i=1}^\infty |x_i y_i| \leq \frac{1}{2}\sum_{i=1}^\infty \left(x_i^2 + y_i^2\right)$. Therefore for all sequences $\{x_i\}_{i=1}^\infty, \{y_i\}_{i=1}^\infty \in l^2$ the series $\left\langle \{x_i\}_{i=1}^\infty, \{y_i\}_{i=1}^\infty \right\rangle = \sum_{i=1}^\infty x_i y_i$ converges absolutely, and hence $\langle \cdot, \cdot \rangle$ is defined on all of $l^2 \times l^2$.

The fact that $\langle \cdot, \cdot \rangle$ defines an inner product now follows easily from standard results about series. (Exercise 15-16.) □

Subspaces of inner product spaces clearly are inner product spaces themselves. Example 15.28 shows how d-dimensional space can be obtained as a subspace of l^2, which means d-dimensional space can be equipped with an inner product.

Example 15.28 *Let $d \in \mathbb{N}$. The set $\mathbb{R}^d := \left\{(x_1, \ldots, x_d) : x_i \in \mathbb{R}\right\}$ with termwise addition and scalar multiplication and $\left\langle (x_1, \ldots, x_d), (y_1, \ldots, y_d) \right\rangle := \sum_{i=1}^d x_i y_i$ is an inner product space.*

For $x = (x_1, \ldots, x_d) \in \mathbb{R}^d$, define $l(x)$ to be the sequence $\{z_n\}_{n=1}^\infty$ defined termwise by $z_n := \begin{cases} x_n; & \text{for } n \leq d, \\ 0; & \text{otherwise.} \end{cases}$ Then $l\left[\mathbb{R}^d\right]$ is a subspace of l^2, and hence it is an inner product space. Because $l : \mathbb{R}^d \to l^2$ is injective, preserves sums and scalar multiples and satisfies $\langle x, y \rangle = \left\langle l(x), l(y) \right\rangle$ for all $x, y \in \mathbb{R}^d$ we conclude that $\langle \cdot, \cdot \rangle$ on $\mathbb{R}^d$ must satisfy the properties of an inner product. Thus $\mathbb{R}^d$ is an inner product space. □

In fact, $\mathbb{R}^d$ is, as an inner product space, isomorphic to the subspace $l\left[\mathbb{R}^d\right] \subseteq l^2$, where the isomorphisms between inner product spaces are defined as follows.

Definition 15.29 *Let $\left(X, \langle \cdot, \cdot \rangle_X\right)$ and $\left(Y, \langle \cdot, \cdot \rangle_Y\right)$ be inner product spaces. Then the function $\Phi : X \to Y$ is called an* **isomorphism** *iff it is an isomorphism between the vector spaces X and Y and for all $u, v \in X$ we have $\langle u, v \rangle_X = \left\langle \Phi(u), \Phi(v) \right\rangle_Y$.*

Many important results in analysis come from the fact that $\mathcal{L}^2$ is almost an inner product space. There is one problem, which the reader will explain in Exercise 15-18. Yet Proposition 15.30 shows that most properties hold just fine. The problem with $\mathcal{L}^2$, and indeed with $\mathcal{L}^p$ in general, will be resolved in Section 15.8.

Proposition 15.30 *Let (M, Σ, μ) be a measure space. Consider the vector space $\mathcal{L}^2(M, \Sigma, \mu)$ of square integrable functions. Then $\langle f, g\rangle := \int_M f(x)g(x)\,d\mu$ is a function from $\mathcal{L}^2(M, \Sigma, \mu) \times \mathcal{L}^2(M, \Sigma, \mu)$ to $\mathbb{R}$ such that*

1. *For all $f \in \mathcal{L}^2(M, \Sigma, \mu)$, we have $\langle f, f\rangle \in [0, \infty)$. Moreover, $\langle 0, 0\rangle = 0$.*
2. *For all $f, g \in \mathcal{L}^2(M, \Sigma, \mu)$, we have $\langle f, g\rangle = \langle g, f\rangle$.*
3. *For all $\alpha \in \mathbb{R}$ and all $f, g \in \mathcal{L}^2(M, \Sigma, \mu)$, we have $\langle \alpha f, g\rangle = \alpha\langle f, g\rangle$, and for all $f, g, h \in \mathcal{L}^2(M, \Sigma, \mu)$ we have $\langle f + g, h\rangle = \langle f, h\rangle + \langle g, h\rangle$.*

Proof. The proof that $\langle f, g\rangle$ exists for all $f, g \in \mathcal{L}^2(M, \Sigma, \mu)$ is similar to the proof for l^2 and left to the reader as Exercise 15-17. For the properties of $\langle \cdot, \cdot\rangle$, let $f, g, h \in \mathcal{L}^2(M, \Sigma, \mu)$ and $\alpha \in \mathbb{R}$. Then

$$\begin{aligned}
\langle f, f\rangle &= \int_M f^2(x)\,d\mu \geq 0, \quad \langle 0, 0\rangle = \int_M 0\,d\mu = 0, \\
\langle f, g\rangle &= \int_M f(x)g(x)\,d\mu = \int_M g(x)f(x)\,d\mu = \langle g, f\rangle, \\
\langle \alpha f, g\rangle &= \int_M \alpha f(x)g(x)\,d\mu = \alpha\int_M f(x)g(x)\,d\mu = \alpha\langle f, g\rangle, \\
\langle f + g, h\rangle &= \int_M (f + g)(x)h(x)\,d\mu \\
&= \int_M f(x)h(x)\,d\mu + \int_M g(x)h(x)\,d\mu = \langle f, h\rangle + \langle g, h\rangle.
\end{aligned}$$

■

Note that on $\mathcal{L}^2[-\pi, \pi)$ we usually work with $\langle f, g\rangle := \frac{1}{\pi}\int_{[-\pi,\pi)} fg\,d\lambda$. The extra factor $\frac{1}{\pi}$ is motivated by Example 15.34 below. Inner product spaces will be investigated in more detail in Chapter 20, with groundwork laid in this chapter, Chapter 16 and Section 17.1.

Exercises

15-14 Prove that if X is an inner product space, then for all $x, y, z \in X$ and $\alpha \in \mathbb{R}$ we have $\langle x, y + z\rangle = \langle x, y\rangle + \langle x, z\rangle$ and $\langle x, \alpha y\rangle = \alpha\langle x, y\rangle$.

15-15. Prove that for all $x, y \in \mathbb{R}$ we have $2|xy| \leq x^2 + y^2$.

15-16. Finish Example 15.27 by proving that $\langle \cdot, \cdot\rangle$ is an inner product on l^2.

15-17. Prove that $\langle f, g\rangle$ as in Proposition 15.30 exists for all $f, g \in \mathcal{L}^2(M, \Sigma, \mu)$.

15-18 Explain why $\mathcal{L}^2(M, \Sigma, \mu)$ with $\langle f, g\rangle := \int_M f(x)g(x)\,d\mu$ is not necessarily an inner product space.

Hint. Part 3 of Theorem 14.35

15-19 Prove that $C^0[-\pi, \pi]$ with $\langle f, g\rangle := \frac{1}{\pi}\int_{[-\pi,\pi)} fg\,d\lambda$ is an inner product space.

15.5 Nicer Representations: Orthonormal Sets

An inner product allows the definition of right angles. This in turn leads to the desire to represent vectors in terms of "nice" systems in which any two vectors are at right angles.

Definition 15.31 *Let X be an inner product space. Then $x, y \in X$ are called* **orthogonal** *iff $\langle x, y\rangle = 0$.*

Definition 15.32 *Let X be an inner product space. A subset $S \subseteq X$ is called an* **orthonormal system** *iff for any two $a, b \in S$ we have $\langle a, b\rangle = \begin{cases} 0; & \text{for } a \neq b, \\ 1; & \text{for } a = b. \end{cases}$ A* **maximal orthonormal system** *is an orthonormal system S so that if $x \in X$ is such that $\langle x, s\rangle = 0$ for all $s \in S$, then $x = 0$.*

Example 15.33 *The standard unit vectors e_i of Convention 15.23 form a maximal orthonormal system in l^2.*

For all $i, j \in \mathbb{N}$ we have $\langle e_i, e_j\rangle = \sum_{k=1}^{\infty} e_i^{(k)} e_j^{(k)} = \begin{cases} 0; & \text{if } i \neq j, \\ 1; & \text{if } i = j. \end{cases}$ Hence, the standard unit vectors form an orthonormal system in l^2.

Now let $a = \left\{a^{(j)}\right\}_{j=1}^{\infty} \in l^2$. Then $\langle a, e_i\rangle = \sum_{j=1}^{\infty} a^{(j)} e_i^{(j)} = a^{(i)}$ for all $i \in \mathbb{N}$. Hence, if $a \in l^2$ is so that $\langle a, e_i\rangle = 0$ for all $i \in \mathbb{N}$, then $a = 0$. This means that $\{e_i : i \in \mathbb{N}\}$ is a maximal orthonormal system in l^2. □

Example 15.34 For all $m, n \in \mathbb{N}$, we conclude the following from Exercise 12-15a.

$$\int_{-\pi}^{\pi} \cos(mx)\cos(nx)\,dx = \begin{cases} 0; & \text{if } m \neq n, \\ \pi; & \text{if } m = n, \end{cases}$$

$$\int_{-\pi}^{\pi} \sin(mx)\sin(nx)\,dx = \begin{cases} 0; & \text{if } m \neq n, \\ \pi; & \text{if } m = n, \end{cases}$$

$$\int_{-\pi}^{\pi} \sin(mx)\cos(nx)\,dx = 0.$$

This means that $T := \left\{\sin(nx), \cos(mx) : n \geq 1, m \geq 1\right\} \cup \left\{\frac{1}{\sqrt{2}}\right\}$ is an orthonormal system in the space of continuous bounded functions on $[-\pi, \pi)$ with the inner product $\langle f, g\rangle := \frac{1}{\pi}\int_{-\pi}^{\pi} f(x)g(x)\,dx$. Even more is true, as this set is a maximal orthonormal system. The proof is quite sophisticated however. We will present it in Theorem 20.12. Note that the extra factor in the inner product assures that for each function in T the inner product with itself is 1. Also note that the same properties hold in $\mathcal{L}^2[-\pi, \pi)$, except that $\mathcal{L}^2[-\pi, \pi)$ is not quite an inner product space. □

Orthonormal systems possess a key property for being a base.

Proposition 15.35 *Orthonormal systems are linearly independent.*

Proof. Let S be an orthonormal system in an inner product space X and let $c_1, \ldots, c_n \in S$ and $\alpha_1, \ldots, \alpha_n \in \mathbb{R}$ be such that $\sum_{i=1}^{n} \alpha_i c_i = 0$. Then for all indices $j = 1, \ldots, n$ we obtain

$$\alpha_j = \alpha_j \langle c_j, c_j \rangle = \sum_{i=1}^{n} \alpha_i \langle c_i, c_j \rangle = \left\langle \sum_{i=1}^{n} \alpha_i c_i, c_j \right\rangle = \langle 0, c_j \rangle = 0.$$

Therefore S is linearly independent. ■

This linear independence means that for finite dimensional spaces we should be able to find an **orthonormal base**, that is, a base that also is an orthonormal system.

Definition 15.36 *Let X be a vector space and let $W \subseteq X$. Define the* **span** *of W to be*

$$\operatorname{span}(W) := \left\{ v = \sum_{i=1}^{n} a_i w_i : a_i \in \mathbb{R},\ w_i \in W \right\}.$$

Theorem 15.37 *The* **Gram-Schmidt Orthonormalization Procedure**. *Every finite dimensional inner product space X has an orthonormal base.*

Proof. Let d be the dimension of X, let $\{b_1, \ldots, b_d\}$ be a base of X and for all vectors $x \in X$ use the notation $\|x\| := \sqrt{\langle x, x \rangle}$. Set $c_1 := \frac{b_1}{\|b_1\|}$. After $c_1, \ldots, c_{k-1}$ have been defined so that $\{c_1, \ldots, c_{k-1}\}$ is an orthonormal base for $\operatorname{span}(\{b_1, \ldots, b_{k-1}\})$, let $d_k := b_k - \sum_{i=1}^{k-1} \langle b_k, c_i \rangle c_i$. Because $\{b_1, \ldots, b_k\}$ is linearly independent, we infer $d_k \neq 0$. Define $c_k := \frac{d_k}{\|d_k\|}$. Then for all $j < k$ we obtain

$$\begin{aligned}
\langle c_k, c_j \rangle &= \left\langle \frac{d_k}{\|d_k\|}, c_j \right\rangle = \frac{1}{\|d_k\|} \left\langle b_k - \sum_{i=1}^{k-1} \langle b_k, c_i \rangle c_i, c_j \right\rangle \\
&= \frac{1}{\|d_k\|} \left[\langle b_k, c_j \rangle - \sum_{i=1}^{k-1} \langle b_k, c_i \rangle \langle c_i, c_j \rangle \right] \\
&= \frac{1}{\|d_k\|} \left[\langle b_k, c_j \rangle - \langle b_k, c_j \rangle \right] = 0.
\end{aligned}$$

Because $\|c_k\| = 1$, the set $\{c_1, \ldots, c_k\}$ is an orthonormal set and because all c_j are linear combinations of $b_1, \ldots, b_k$ it is contained in $\operatorname{span}(\{b_1, \ldots, b_k\})$. Because orthonormal sets are linearly independent, the set $\{c_1, \ldots, c_k\}$ is a base of the span of $b_1, \ldots, b_k$.

After $d - 1$ steps as above we obtain that $\{c_1, \ldots, c_d\}$ is a base of X. ■

The Gram-Schmidt Orthonormalization Procedure can be executed indefinitely if we start with a countably infinite linearly independent set (see Exercise 15-20). We will discuss in Section 20.1 how we need to adjust the notion of a base to obtain standard representations of elements of an infinite dimensional inner product space.

Exercises

15-20 Prove that every infinite dimensional inner product space contains an infinite orthonormal system.

15.6 Abstraction III: Normed Spaces

The definition of distances in vector spaces is a two step process. First, norms can be viewed as a generalization of absolute values. They still connect to the algebraic properties of the surrounding space. Second, metrics (discussed in Section 15.7) need not connect to these properties and can thus also be defined on subsets that are not vector spaces any more.

Definition 15.38 *A* **normed space** *is a pair* $(X, \|\cdot\|)$ *of a vector space* X *and a function* $\|\cdot\| : X \to [0, \infty)$, *called the* **norm** *on* X, *such that*

1. *For all* $x \in X$, *we have* $\|x\| = 0$ *iff* $x = 0$.
2. *For all scalars* α *and all* $x \in X$, *we have that* $\|\alpha x\| = |\alpha|\|x\|$.
3. *The* **triangular inequality** *holds.*
   That is, for all $x, y \in X$ *we have* $\|x + y\| \leq \|x\| + \|y\|$.

We will usually call X *itself a normed space. When we do so, we implicitly assume that there is a norm on* X, *which will usually be denoted* $\|\cdot\|$.

As we introduce more abstract settings, the spaces we are familiar with will be special cases of the more general spaces. For example, Proposition 15.40 below shows that every inner product space is a normed space, too. Because the function defined in Lemma 15.39 ultimately turns out to be a norm, we denote it *like* a norm right away. Note that until we establish that it *is* a norm, we are not using any of the properties of a norm.

Lemma 15.39 Cauchy-Schwarz inequality. *Let* X *be an inner product space and for all* $x \in X$ *let* $\|x\| := \sqrt{\langle x, x\rangle}$. *Then for all* $x, y \in X$ *we have* $|\langle x, y\rangle| \leq \|x\|\|y\|$.

Proof. The function $f(t) := \langle tx + y, tx + y\rangle = t^2\langle x, x\rangle + 2t\langle x, y\rangle + \langle y, y\rangle$ is a nonnegative quadratic function with absolute minimum at $t = -\frac{\langle x, y\rangle}{\langle x, x\rangle}$. But then

$$\begin{aligned} 0 &\leq f\left(-\frac{\langle x, y\rangle}{\langle x, x\rangle}\right) = \left(-\frac{\langle x, y\rangle}{\langle x, x\rangle}\right)^2 \langle x, x\rangle + 2\left(-\frac{\langle x, y\rangle}{\langle x, x\rangle}\right)\langle x, y\rangle + \langle y, y\rangle \\ &= \frac{\langle x, y\rangle^2}{\langle x, x\rangle} - 2\frac{\langle x, y\rangle^2}{\langle x, x\rangle} + \langle y, y\rangle = -\frac{\langle x, y\rangle^2}{\langle x, x\rangle} + \langle y, y\rangle, \end{aligned}$$

which implies $\langle x, y\rangle^2 \le \|x\|^2\|y\|^2$, and hence $|\langle x, y\rangle| \le \|x\|\|y\|$. ■

Now we can show that the norm notation is justified.

Theorem 15.40 *Let X be an inner product space. Then $\|x\| := \sqrt{\langle x, x\rangle}$ defines a norm on X. Therefore any inner product space is also a normed space.*

Proof. First, note that $\|x\| = 0$ iff $\sqrt{\langle x, x\rangle} = 0$ iff $\langle x, x\rangle = 0$ iff $x = 0$. Moreover, for all $\alpha \in \mathbb{R}$ and $x \in X$ we obtain $\|\alpha x\| = \sqrt{\langle \alpha x, \alpha x\rangle} = \sqrt{\alpha^2\langle x, x\rangle} = |\alpha|\|x\|$.

To prove the triangular inequality, let $x, y \in X$. Via the Cauchy-Schwarz inequality we infer the following.

$$\begin{aligned}\|x+y\|^2 &= \langle x+y, x+y\rangle = \langle x, x\rangle + 2\langle x, y\rangle + \langle y, y\rangle \\ &\le \|x\|^2 + 2\|x\|\|y\| + \|y\|^2 = \big(\|x\| + \|y\|\big)^2,\end{aligned}$$

which finishes the proof. ■

Definition 15.41 *On $\mathbb{R}^d$, the norm $\|x\|_2 := \sqrt{\langle x, x\rangle}$ that is induced by the inner product is called the* **Euclidean norm**.

Of course, a more general definition is only useful if it introduces new and interesting objects. The following examples are not inner product spaces.

Example 15.42 *The function $\left\|\{x_n\}_{n=1}^{\infty}\right\|_{\infty} := \sup\left\{|x_n| : n \in \mathbb{N}\right\}$ is a norm on l^{∞}.*

It is clear that $\left\|\{x_n\}_{n=1}^{\infty}\right\|_{\infty} \ge 0$ for all sequences $\{x_n\}_{n=1}^{\infty} \in l^{\infty}$. Moreover, we have $\left\|\{x_n\}_{n=1}^{\infty}\right\|_{\infty} = 0$ iff $\sup\left\{|x_n| : n \in \mathbb{N}\right\} = 0$, which is the case iff all x_n are zero, which is the case iff $\{x_n\}_{n=1}^{\infty} = \{0\}_{n=1}^{\infty} \in l^{\infty}$.

For $\alpha \in \mathbb{R}$ and $\{x_n\}_{n=1}^{\infty} \in l^{\infty}$, note that

$$\left\|\{\alpha x_n\}_{n=1}^{\infty}\right\|_{\infty} = \sup\left\{|\alpha||x_n| : n \in \mathbb{N}\right\} = |\alpha| \left\|\{x_n\}_{n=1}^{\infty}\right\|_{\infty}.$$

Finally, for $\{x_n\}_{n=1}^{\infty}, \{y_n\}_{n=1}^{\infty} \in l^{\infty}$ note that

$$\begin{aligned}\left\|\{x_n\}_{n=1}^{\infty} + \{y_n\}_{n=1}^{\infty}\right\|_{\infty} &= \left\|\{x_n + y_n\}_{n=1}^{\infty}\right\|_{\infty} = \sup\left\{|x_n + y_n| : n \in \mathbb{N}\right\} \\ &\le \sup\left\{|x_n| + |y_n| : n \in \mathbb{N}\right\} \\ &\le \sup\left\{|x_n| : n \in \mathbb{N}\right\} + \sup\left\{|y_n| : n \in \mathbb{N}\right\} \\ &= \left\|\{x_n\}_{n=1}^{\infty}\right\|_{\infty} + \left\|\{y_n\}_{n=1}^{\infty}\right\|_{\infty},\end{aligned}$$

where the last inequality is true because for each natural number $m \in \mathbb{N}$ we have that $|x_m| \le \sup\left\{|x_n| : n \in \mathbb{N}\right\}$ and $|y_m| \le \sup\left\{|y_n| : n \in \mathbb{N}\right\}$. □

Example 15.43 *Let D be a set. The function $\|f\|_{\infty} := \sup\left\{|f(x)| : x \in D\right\}$ defines a norm on $\mathcal{B}(D, \mathbb{R})$, the space of bounded functions on $[a, b]$. The norm $\|\cdot\|_{\infty}$ is also called the* **uniform norm**.

This is proved in Exercise 15-21. □

Example 15.43 gives access to two more commonly used spaces.

Example 15.44 The function $\left\|(x_1, \ldots, x_d)\right\|_\infty := \max\left\{|x_j| : j = 1, \ldots, d\right\}$ defines a norm on $\mathbb{R}^d = \mathcal{B}(\{1, \ldots, d\}, \mathbb{R})$. □

Example 15.45 Because linear subspaces of normed spaces are normed spaces, too, the space $\left(C^0[a, b], \|\cdot\|_\infty\right)$ is a normed space. □

We can now turn our attention to the $\mathcal{L}^p$ spaces once more.

Definition 15.46 *Let (M, Σ, μ) be a measure space and let $p \geq 1$. Then for all $f, g \in \mathcal{L}^p(M, \Sigma, \mu)$ we define $\|f\|_p := \left(\int_M |f|^p \, d\mu\right)^{\frac{1}{p}}$.*

Although $\|\cdot\|_p$ is not quite a norm, we have good reason to use norm notation. Part 1 of Theorem 15.50 will identify the only part of the definition of a norm that is not satisfied by $\|\cdot\|_p$ and we will resolve this problem in Section 15.8. The biggest immediate challenge is to prove the triangular inequality for $\|\cdot\|_p$, which is done in Theorem 15.49. Hölder's inequality (Theorem 15.48) is a lemma leading up to Theorem 15.49, but it is also of independent interest. In $\mathcal{L}^p$, Hölder's inequality often serves as a substitute for the Cauchy-Schwarz inequality, which is only valid in inner product spaces.

Lemma 15.47 Young's inequality. *Let $1 < p, q < \infty$ be such that $\frac{1}{p} + \frac{1}{q} = 1$ and let $x, y \in [0, \infty)$. Then $xy \leq \frac{x^p}{p} + \frac{x^q}{q}$.*

Proof. The inequality is trivial for $x = 0$ or $y = 0$, so we can assume both numbers are positive. We first perform some substitutions that simplify the inequality. With $u = x^p$ and $v = x^q$ the inequality is equivalent to the inequality $u^{\frac{1}{p}} v^{\frac{1}{q}} \leq \frac{u}{p} + \frac{v}{q}$ for all $u, v \in (0, \infty)$. With $t := \frac{u}{v}$ the inequality is equivalent to $t^{\frac{1}{p}} \leq \frac{t}{p} + \frac{1}{q}$, where we multiply or divide by v to go back and forth. But it is easy to verify with elementary calculus (see Exercise 15-22) that the function $f(t) := \frac{t}{p} + \frac{1}{q} - t^{\frac{1}{p}}$ has an absolute minimum value of 0 on $(0, \infty)$. ■

Theorem 15.48 Hölder's inequality *for integrals. Let (M, Σ, μ) be a measure space, let $1 < p, q < \infty$ satisfy the equality $\frac{1}{p} + \frac{1}{q} = 1$ and let $f \in \mathcal{L}^p(M, \Sigma, \mu)$ and $g \in \mathcal{L}^q(M, \Sigma, \mu)$. Then $fg \in \mathcal{L}^1(M, \Sigma, \mu)$ and $\|fg\|_1 \leq \|f\|_p \|g\|_q$.*

Proof. The computation below shows first of all that $\int_M fg \, d\mu = \|fg\|_1$ is finite, which establishes that $fg \in \mathcal{L}^1(M, \Sigma, \mu)$. Moreover, the claimed inequality can then be obtained by multiplying by $\|f\|_p\|g\|_q$.

$$\begin{aligned}\frac{\|fg\|_1}{\|f\|_p\|g\|_q} &= \int_M \frac{|f|}{\|f\|_p}\frac{|g|}{\|g\|_q}\,d\mu \\ &\leq \int_M \frac{1}{p}\frac{|f|^p}{\|f\|_p^p}+\frac{1}{q}\frac{|g|^q}{\|g\|_q^q}\,d\mu \\ &= \frac{\int_M |f|^p\,d\mu}{p\|f\|_p^p}+\frac{\int_M |g|^q\,d\mu}{q\|g\|_q^q} = \frac{1}{p}+\frac{1}{q} = 1.\end{aligned}$$

Use Young's inequality.

■

Theorem 15.49 Minkowski's inequality *for integrals. Let (M, Σ, μ) be a measure space and let $p \geq 1$. Then for all functions $f, g \in \mathcal{L}^p(M, \Sigma, \mu)$ we have the inequality $\|f+g\|_p \leq \|f\|_p + \|g\|_p$.*

Proof. For $p = 1$, the inequality is an easy consequence of the triangular inequality for absolute values. So for the remainder we can assume $1 < p < \infty$. Choose q such that $\frac{1}{p}+\frac{1}{q} = 1$. Then $p + q = pq$, and hence $(p-1)q = pq - q = p$. Thus, because we already know from Example 15.12 that $|f+g|^p$ is integrable, we conclude $|f+g|^{p-1} \in \mathcal{L}^q(M, \Sigma, \mu)$. Now via Hölder's inequality we obtain

$$\begin{aligned}\int_M |f+g|^p\,d\mu &\leq \int_M (|f|+|g|)|f+g|^{p-1}\,d\mu \\ &= \int_M |f||f+g|^{p-1}\,d\mu + \int_M |g||f+g|^{p-1}\,d\mu \\ &\leq \|f\|_p \left\||f+g|^{p-1}\right\|_q + \|g\|_p \left\||f+g|^{p-1}\right\|_q \\ &= (\|f\|_p + \|g\|_p)\left(\int_M |f+g|^p\,d\mu\right)^{\frac{1}{q}},\end{aligned}$$

which means $\left(\int_M |f+g|^p\,d\mu\right)^{1-\frac{1}{q}} \leq \|f\|_p + \|g\|_p$. Hence,

$$\|f+g\|_p = \left(\int_M |f+g|^p\,d\mu\right)^{\frac{1}{p}} = \left(\int_M |f+g|^p\,d\mu\right)^{1-\frac{1}{q}} \leq \|f\|_p + \|g\|_p,$$

which was to be proved. ■

We can now determine how close the $\|\cdot\|_p$ are to being norms. The only difference between the properties established in Theorem 15.50 and the properties of a norm is that $\|f\|_p = 0$ only implies that $f = 0$ almost everywhere, not everywhere. This minor nuisance will be remedied in Section 15.8.

Theorem 15.50 *Let (M, Σ, μ) be a measure space and let $p \geq 1$. Then the following hold.*

0. *For all* $f \in \mathcal{L}^p(M, \Sigma, \mu)$, *we have* $\|f\|_p \geq 0$.

1. *For all* $f \in \mathcal{L}^p(M, \Sigma, \mu)$, *we have* $\|f\|_p = 0$ *iff* $f = 0$ *a.e.*

2. *For all* $f \in \mathcal{L}^p(M, \Sigma, \mu)$ *and* $\alpha \in \mathbb{R}$, *we have* $\|\alpha f\|_p = |\alpha| \|f\|_p$.

3. *For all* $f, g \in \mathcal{L}^p(M, \Sigma, \mu)$, *we have* $\|f + g\|_p \leq \|f\|_p + \|g\|_p$.

Proof. Part 0 is trivial and part 3 is Minkowski's inequality. The remaining parts are left as Exercise 15-23. ■

The functions $\|\cdot\|_p$ can be defined on l^p also and on these spaces they do define a norm. This is the key advantage that l^p has over $\mathcal{L}^p$. It actually is a normed space.

Definition 15.51 *For* $1 \leq p < \infty$ *and* $\{x_n\}_{n=1}^{\infty} \in l^p$ *we define the p-norm of* $\{x_n\}_{n=1}^{\infty}$ *to be* $\left\| \{x_n\}_{n=1}^{\infty} \right\|_p := \left(\sum_{n=1}^{\infty} |x_n|^p \right)^{\frac{1}{p}}$.

Theorem 15.52 *For* $1 \leq p < \infty$ *the space* $\left(l^p, \|\cdot\|_p\right)$ *is a normed space.*

Proof. First note that with $\gamma_{\mathbb{N}}$ denoting counting measure on $\mathbb{N}$ the space l^p is the space $\mathcal{L}^p(\mathbb{N}, \mathcal{P}(\mathbb{N}), \gamma_{\mathbb{N}})$ and that the function $\|\cdot\|_p : l^p \to [0, \infty)$ is exactly the function $\|\cdot\|_p : \mathcal{L}^p(\mathbb{N}, \mathcal{P}(\mathbb{N}), \gamma_{\mathbb{N}}) \to [0, \infty)$ in Theorem 15.50. By Exercise 14-3a the only set of $\gamma_{\mathbb{N}}$-measure zero is the empty set. But this means that $\left\| \{x_n\}_{n=1}^{\infty} \right\|_p = 0$ iff $\{x_n\}_{n=1}^{\infty} = 0$ $\gamma_{\mathbb{N}}$-a.e., which is the case iff $x_n = 0$ for all $n \in \mathbb{N}$. Thus Theorem 15.50 proves that $\left(l^p, \|\cdot\|_p\right)$ is a normed space. ■

Note how the proof of Theorem 15.52 reveals that the l^p-spaces are special cases of the $\mathcal{L}^p$-spaces. Another consequence of Theorem 15.52 is that $\mathbb{R}^d$ equipped with the L^p norm $\left\| (x_1, \ldots, x_d) \right\|_p := \sqrt[p]{\sum_{i=1}^{d} |x_i|^p}$ is a normed space (Exercise 15-24). That is, *one* space such as $\mathbb{R}^d$ can be equipped with *several* norms. For d-dimensional space, this issue will be addressed in Section 16.6.

Normed spaces will be investigated in more detail in Chapter 17, with groundwork laid in this chapter and in Chapter 16.

Exercises

15-21. Finishing Example 15.43 Let $\mathcal{B}(D, \mathbb{R})$ be the set of bounded functions on the set D.

(a) Prove that with pointwise addition and scalar multiplication $\mathcal{B}(D, \mathbb{R})$ is a vector space.

(b) Prove that the function $\|f\|_\infty := \sup\left\{ \left| f(x) \right| : x \in D \right\}$ defines a norm on $\mathcal{B}(D, \mathbb{R})$.

15-22. Finish the proof of Young's inequality (Lemma 15.47) by proving that $f(t) := \frac{t}{p} + \frac{1}{q} - t^{\frac{1}{p}}$ has an absolute minimum value of 0 on $(0, \infty)$.

15-23. Finish the proof of Theorem 15.50. That is,

(a) Prove part 1 of Theorem 15.50, and

(b) Prove part 2 of Theorem 15.50.

15-24. For $p \geq 1$ define $\|\ \|_p : \mathbb{R}^d \to [0, \infty)$ by $\|(x_1, \ldots, x_d)\|_p = \sqrt[p]{\sum_{i=1}^{d} |x_i|^p}$ and prove that $\|\cdot\|_p$ is a norm on $\mathbb{R}^d$.

15-25. Let $a < b$ and let $\mathcal{BV}_0[a, b]$ denote the vector space of all functions of bounded variation on $[a, b]$ so that $f(a) = 0$. Prove that

$$\|f\|_{BV} = \sup\left\{\sum_{i=1}^{n} |f(a_i) - f(a_{i-1})| : a = a_0 < a_1 < \cdots < a_n = b\right\}$$

defines a norm on $\mathcal{BV}_0[a, b]$.

15-26. "Concrete" versions of the inequalities in this chapter.

(a) Let $x_1, \ldots, x_d, y_1, \ldots, y_d \in \mathbb{R}$ and let $p, q \in (1, \infty)$ with $\frac{1}{p} + \frac{1}{q} = 1$. Prove the inequalities

$$\sum_{j=1}^{d} |x_j y_j| \leq \left(\sum_{j=1}^{d} x_j^2\right)^{\frac{1}{2}} \left(\sum_{j=1}^{d} y_j^2\right)^{\frac{1}{2}} \text{ and } \sum_{j=1}^{d} |x_j y_j| \leq \left(\sum_{j=1}^{d} |x_j|^p\right)^{\frac{1}{p}} \left(\sum_{j=1}^{d} |y_j|^q\right)^{\frac{1}{q}}.$$

(b) Let $p \in [1, \infty)$. Prove that $\|f\|_p = \left(\int_a^b |f(x)|^p \, dx\right)^{\frac{1}{p}}$ defines a norm on $C^0[a, b]$.

(c) State the Cauchy-Schwarz, Hölder and Minkowski inequalities *in integral notation* for the norms from part 15-26b.

15-27. **Young's inequality** revisited. Let $1 < p, q < \infty$ be such that $\frac{1}{p} + \frac{1}{q} = 1$, let $\varepsilon > 0$ and let $x, y \in [0, \infty)$. Prove that $xy \leq \varepsilon \frac{x^p}{p} + \varepsilon^{-\frac{q}{p}} \frac{y^q}{q}$.

Hint. Absorb the ε into x and y.

15-28. Give a direct proof of Theorem 15.52. The first two parts of the definition of a norm are straightforward. For the triangular inequality, proceed as follows.

(a) First prove **Hölder's inequality** for series. Let $1 < p, q < \infty$ be such that $\frac{1}{p} + \frac{1}{q} = 1$ and let $\{x_n\}_{n=1}^{\infty} \in l^p$ and $\{y_n\}_{n=1}^{\infty} \in l^q$. Prove that then $\{x_n y_n\}_{n=1}^{\infty} \in l^1$ and we have the inequality $\left\|\{x_n y_n\}_{n=1}^{\infty}\right\|_1 \leq \left\|\{x_n\}_{n=1}^{\infty}\right\|_p \left\|\{y_n\}_{n=1}^{\infty}\right\|_q$.

(b) Now prove **Minkowski's inequality** for series. Prove that for all $\{x_n\}_{n=1}^{\infty}, \{y_n\}_{n=1}^{\infty} \in l^p$ we have $\left\|\{x_n\}_{n=1}^{\infty} + \{y_n\}_{n=1}^{\infty}\right\|_p \leq \left\|\{x_n\}_{n=1}^{\infty}\right\|_p + \left\|\{y_n\}_{n=1}^{\infty}\right\|_p$.

Hint. Use the proofs of the corresponding inequalities for integrals as guidance.

15-29. Let X be a normed space. Prove that $\left\|\sum_{i=1}^{n} x_i\right\| \leq \sum_{i=1}^{n} \|x_i\|$ for any n elements $x_1, \ldots, x_n \in X$.

15-30. When is a normed space an inner product space?

(a) **Parallelogram law**. Let X be an inner product space. Prove that for all $x, y \in X$ we have $\|x + y\|^2 + \|x - y\|^2 = 2\|x\|^2 + 2\|y\|^2$.

(b) **Polarization identity**. Let X be an inner product space. Prove that for all $x, y \in X$ we have $\|x + y\|^2 - \|x - y\|^2 = 4\langle x, y\rangle$.

(c) Let X be a normed space in which the parallelogram law holds Prove that the equation $\langle x, y\rangle := \frac{1}{4}\left(\|x+y\|^2 - \|x-y\|^2\right)$ defines an inner product on X with $\|x\| = \sqrt{\langle x, x\rangle}$ for all $x \in X$.

(d) State an equivalent formulation for the statement "X is a normed space and the norm of X is induced by an inner product"

(e) Give an example of functions $f, g \in C^0[a, b]$ that do not satisfy the parallelogram law for $\|\cdot\|_\infty$, thus proving that $\left(C^0[a, b], \|\cdot\|_\infty\right)$ is not an inner product space

15-31. Let (M, Σ, μ) be a measure space with $\mu(M) < \infty$. Prove that if $1 \le p < q < \infty$, then $\mathcal{L}^p(M, \Sigma, \mu) \supseteq \mathcal{L}^q(M, \Sigma, \mu)$ and for all $f \in \mathcal{L}^p(M, \Sigma, \mu)$ we have $\|f\|_p \le \|f\|_q \mu(M)^{\frac{1}{p}-\frac{1}{q}}$.
Hint. Hölder's inequality

15-32 **Jensen's inequality**

(a) Prove for all $x = (x_1, \ldots, x_d) \in \mathbb{R}^d$ and all $1 \le p < q < \infty$ that $\|x\|_p \ge \|x\|_q \ge \|x\|_\infty$,

that is, $$\left(\sum_{i=1}^{d} |x_i|^p\right)^{\frac{1}{p}} \ge \left(\sum_{i=1}^{d} |x_i|^q\right)^{\frac{1}{q}} \ge \max\left\{|x_i| : i = 1, \ldots, d\right\}.$$

Hint. The second inequality is simple. For the first inequality consider $\dfrac{x}{\|x\|_\infty}$.

(b) Prove that for all $x \in l^p$ and all $1 \le p < q < \infty$ we have $\|x\|_p \ge \|x\|_q \ge \|x\|_\infty$.

(c) Prove that Jensen's inequality can fail for continuous functions. That is, find $p < q$, $a < b$ and a continuous function $g : [a, b] \to \mathbb{R}$ so that $\|g\|_p < \|g\|_q$ and $\|g\|_p < \|g\|_\infty$.
Hint. A straight line rising from 0 to 1 on a sufficiently short interval will do.

15-33 The norm $\|\cdot\|_\infty$ **as limit of the** $\|\cdot\|_p$**-norms**

(a) Prove that for all $x \in \mathbb{R}^d$ we have $\lim_{p\to\infty} \|x\|_p = \|x\|_\infty$.

(b) Prove that for all $x \in l^1$ we have $\lim_{p\to\infty} \|x\|_p = \|x\|_\infty$

(c) Let $f : [a, b] \to \mathbb{R}$ be continuous Prove that for every $\varepsilon > 0$ there is a $\delta > 0$ so that for all

$p \ge 1$ we have $$\left(\|f\|_\infty - \varepsilon\right)\delta^{\frac{1}{p}} \le \left(\int_a^b |f(x)|^p \, dx\right)^{\frac{1}{p}} \le \|f\|_\infty (b-a)^{\frac{1}{p}}.$$

(d) Prove that for all $f \in C^0[a, b]$ we have $\lim_{p\to\infty} \|f\|_p = \|f\|_\infty$

15.7 Abstraction IV: Metric Spaces

Norms are used to measure distances in a vector space. Natural phenomena are often modeled in bounded subsets of d-dimensional space, which means sums and constant multiples do not necessarily stay in the subset. When there is no linear structure, distances are measured with metrics. The properties of a metric are inspired by the real life properties of distances. Distances are nonnegative, distinct objects have a nonzero distance from each other, the distance is independent of whether we go from point A to point B or vice versa, and detours through a third point cannot provide a shortcut.

Definition 15.53 *A* **metric space** *is a pair (X, d) of a set X (without additional properties; in particular, X need not be a vector space) and a function $d : X \times X \to [0, \infty)$, called the* **metric** *on X, with the following properties.*

1. *For all $x, y \in X$, we have that $d(x, y) = 0$ iff $x = y$.*
2. *For all $x, y \in X$, we have $d(x, y) = d(y, x)$.*
3. *For all $x, y, z \in X$, we have $d(x, z) \leq d(x, y) + d(y, z)$.*

We will usually call X itself a metric space. When we do so, we implicitly assume that there is a metric on X, which will usually be denoted d.

Normed spaces are metric spaces, too. So once more we have generalized a known concept.

Proposition 15.54 *Let X be a normed space. Then $d(x, y) := \|x - y\|$ defines a metric on X.*

Proof. Clearly, $d(x, y) = \|x - y\| \geq 0$ for all $x, y \in X$ and $d(x, y) = 0$ iff $\|x - y\| = 0$ iff $x - y = 0$ iff $x = y$. Also $d(x, y) = \|x - y\| = \|y - x\| = d(y, x)$. Finally, $d(x, z) = \|x - z\| \leq \|x - y\| + \|y - z\| = d(x, y) + d(y, z)$. ■

With metric spaces we are no longer tied to the linear structure of a vector space. In fact, *any* subset of a metric space is again a metric space.

Proposition 15.55 *Let (X, d) be a metric space and let $S \subseteq X$ be any subset of X. Let $d_s := d|_{S \times S}$ be the restriction of the metric d to the subset S. Then (S, d_S) is a metric space.*

Proof. Clearly, any property of d that holds for all elements of X will also hold for all elements of S. ■

Proposition 15.55 provides a wide range of examples of metric spaces. For example, we can now consider intervals on the real line and subsets of $\mathbb{R}^d$ as metric spaces.

Definition 15.56 *Let X be a metric space. Then we will automatically consider any subset $S \subseteq X$ to be a metric space, also called a* **metric subspace**, *carrying the metric d_S of Proposition 15.55. Said metric will usually also be denoted d. It is sometimes called the* **induced metric** *or the* **relative metric**.

Example 15.57 Not every metric space is a metric subspace of a normed space. On $\mathbb{R}^2$, for $x \neq y$ let $d(x, y) := \sqrt{x_1^2 + x_2^2} + \sqrt{y_1^2 + y_2^2}$, which is the sum of the lengths of the straight line segments from x to 0 and from 0 to y, and for $x = y$ let $d(x, y) := 0$. Then d is a metric on $\mathbb{R}^2$ that is not induced by a norm. This metric models distances in a situation in which all travel must go through a central hub. □

For more examples of metric spaces that are not subspaces of normed spaces, consider Exercise 15-34. We conclude this short section by proving the reverse triangular inequality for metric spaces. We could have proved it in normed spaces first, but with this approach we obtain it for normed spaces as a corollary.

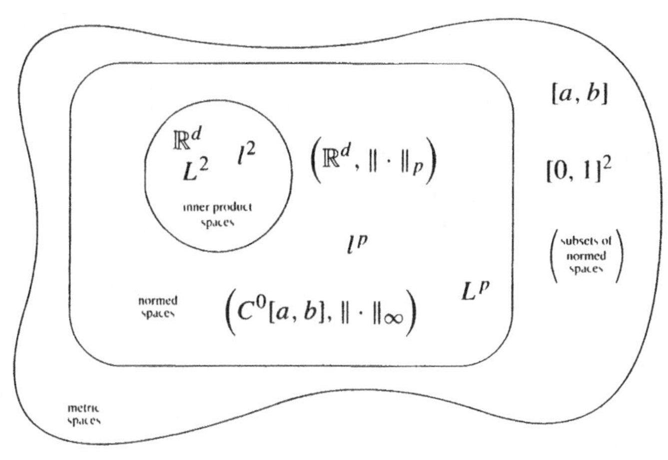

Figure 32: A hierarchy of structures for analysis. The L^p spaces will be introduced in the next section.

Theorem 15.58 *The* **reverse triangular inequality** *for metric spaces. Let X be a metric space. Then for all $x, y, z \in X$ we have $|d(x,y) - d(y,z)| \leq d(x,z)$.*

Proof. Let $x, y, z \in X$ and without loss of generality assume $d(y,z) \leq d(x,y)$. Then the inequality $d(x,y) \leq d(x,z) + d(z,y)$ implies the reverse triangular inequality $|d(x,y) - d(y,z)| = d(x,y) - d(y,z) \leq d(x,z)$. ■

Corollary 15.59 *The* **reverse triangular inequality** *for normed spaces. Let X be a normed space. Then for all $x, y \in X$ we have $\big|\|x\| - \|y\|\big| \leq \|x - y\|$.* ■

The only spaces more abstract than metric spaces that occur frequently in mathematics are topological spaces. In analysis, spaces usually carry a metric. Therefore we will not present topological spaces in this text. Metric spaces will be investigated in more detail in Chapter 16. Figure 32 shows the hierarchy of structures that arise in analysis.

To avoid notational confusion when working in normed spaces, we will use norm notation for the metrics on subsets of normed spaces. This is sensible, because all metrics that we will consider on subsets of normed spaces are induced by a norm.

Exercises

15-34. Examples of metric spaces.

(a) Let X be a set and for $x, y \in X$ let $d(x,y) := \begin{cases} 0; & \text{if } x = y, \\ 1; & \text{if } x \neq y \end{cases}$ Prove that d is a metric on X.

Note. d is called the **discrete metric** on X.

(b) Prove that $d(x,y) := \dfrac{|x-y|}{1+|x-y|}$ defines a metric on $\mathbb{R}$.

Hint. For the triangular inequality, expand with $\dfrac{1}{|x-y|}$.

(c) Consider the surface of a sphere in $\mathbb{R}^d$ with the usual distance function. Let the distance between two points be the length of the shortest path (on the sphere) between these two points. Explain why this distance function defines a metric between the points on the sphere and why this metric is not the metric induced by the usual distance function.

15-35 Explain why the metric induced on $\mathbb{R}^2$ by $\|\cdot\|_1$ is called the **taxicab metric**.

15-36 Let $1 \leq p \leq \infty$. In $\mathbb{R}^d$ equipped with the metric induced by $\|\cdot\|_p$, compute the distance from the origin to the point $(1, \ldots, 1)$. In which metric does the cube $[0, 1]^d$ have the longest diagonal? In which metric is the diagonal shortest?

15-37. The following, purportedly true, story illustrates the importance of having examples for an abstract notion. *Warning. The following notion is absolutely useless. We simply debunk it here. The story itself may well be a "mathematical urban legend."* A mathematician once spent a lot of time proving abstract properties about so-called "anti-metric" spaces. An "anti-metric" $d : X \times X \to [0, \infty)$ is a function so that for all $x \in X$ we have $d(x, x) = 0$ iff $x = 0$, for all $x, y \in X$ we have that $d(x, y) = d(y, x)$ and for all $x, y, z \in X$ we have that $d(x, z) \geq d(x, y) + d(y, z)$. So, all that has changed from metric spaces is that the triangular inequality is reversed. Prove that an anti-metric space can have at most one point.

The mathematician could have simplified all his proofs by using that these spaces have at most one point. Spaces with at most one point have lots of properties, *but* they are not very interesting. So if you try to "be wise, generalize," make sure that your generalization/modification still has models (examples) that are useful.

15.8 L^p Spaces

Part 1 of Theorem 15.50 shows that $\mathcal{L}^p$ spaces fail to be metric spaces because it is possible for distinct objects to have distance zero from each other. When all other properties of a metric are given, we speak of a semimetric space.

Definition 15.60 *A* **semimetric space** *is a pair (X^s, d^s) of a set X^s and a function $d^s : X^s \times X^s \to [0, \infty)$ with the following properties.*

1. *For all $x \in X^s$, we have $d(x, x) = 0$.*
2. *For all $x, y \in X^s$, we have $d^s(x, y) = d^s(y, x)$.*
3. *For all $x, y, z \in X^s$, we have $d^s(x, z) \leq d^s(x, y) + d^s(y, z)$.*

It would be cumbersome to develop a theory of semimetric spaces parallel to that of metric spaces. It is also more appropriate to work with metric spaces, because practical observation tells us that two distinct objects cannot occupy the same space at the same time. To overcome the minor deficiency of a semimetric, objects that have distance zero from each other are combined to become single points. This process produces a metric space in natural fashion.

Theorem 15.61 *Let (X^s, d^s) be a semimetric space. Then $\sim \subseteq X^s \times X^s$ defined by $x \sim y$ iff $d^s(x, y) = 0$ is an equivalence relation (see Definition C.5 in Appendix C.2). If we denote the equivalence classes of $\sim$ with $[x]$, then the set $X := \{[x] : x \in X^s\}$ equipped with $d([x], [y]) := d^s(x, y)$ is a metric space.*

Proof. It is clear that the relation $\sim$ is reflexive, symmetric and transitive. The function d is defined for equivalence classes, but it is defined in terms of representatives of each class. Therefore we must prove that d is well defined. That is, we must show that the value of d does not depend on which representatives are used. Let $[x], [y] \in X$, let $x_1, x_2 \in [x]$ and let $y_1, y_2 \in [y]$. Then

$$d^s(x_1, y_1) \leq d^s(x_1, x_2) + d^s(x_2, y_2) + d^s(y_2, y_1) = d^s(x_2, y_2)$$

and we prove the reversed inequality similarly. Hence, $d^s(x_1, y_1) = d^s(x_2, y_2)$ and the definition of d is independent of the representatives chosen from each equivalence class.

Now $d\big([x], [y]\big) = 0$ implies $d^s(x, y) = 0$, that is, $x \sim y$ and thus $[x] = [y]$ (Exercise 15-38a). Conversely, if $[x] = [y]$, then $d\big([x], [y]\big) = d^s(x, x) = 0$. Hence, $d\big([x], [y]\big) = 0$ is equivalent to $[x] = [y]$ and the first condition for being a metric is satisfied. Antisymmetry and the triangular inequality for d are easily verified (Exercises 15-38b and 15-38c), so d is a metric. ■

Theorem 15.61 allows us to define metric spaces of "p-integrable functions." Formally these spaces consist of classes of p-integrable functions, but the distinction blurs at times. Metric considerations are taken care of in $L^p(M, \Sigma, \mu)$, while integral equations etc. are proved in $\mathcal{L}^p(M, \Sigma, \mu)$.

Definition 15.62 *Let (M, Σ, μ) be a measure space. For $1 \leq p < \infty$, we denote by $L^p(M, \Sigma, \mu)$ the metric space obtained from $\mathcal{L}^p(M, \Sigma, \mu)$ via Theorem 15.61.*

Exercises 15-39 and 15-40 show that the L^p spaces actually are normed spaces and that the L^2 spaces are inner product spaces.

We conclude this section by defining a space similar to l^∞ on measure spaces.

Definition 15.63 *Let (M, Σ, μ) be a measure space. We define*

$$\mathcal{L}^\infty(M, \Sigma, \mu) := \Big\{ f \in F(M, \mathbb{R}) : \big(\exists B \in \mathbb{R} : |f(x)| \leq B \ \mu\text{-a.e.}\big) \Big\}.$$

For $f \in \mathcal{L}^\infty(M, \Sigma, \mu)$, we define $\|f\|_\infty := \inf\big\{ B \in \mathbb{R} : |f(x)| \leq B \ \mu\text{-a.e.} \big\}$.

Note that sometimes the μ-a.e. in the definition of $\mathcal{L}^\infty$ is replaced with "μ-locally a.e." Exercise 15-45 shows that for σ-finite measure spaces null sets and locally null sets are the same. We consider σ-finite measure spaces in this text, so the author chose the simpler definition.

Proposition 15.64 *Let (M, Σ, μ) be a measure space. Then $\mathcal{L}^\infty(M, \Sigma, \mu)$ equipped with $d_\infty(f, g) := \|f - g\|_\infty$ is a semimetric space.*

Proof. The reader will prove a little more in Exercise 15-41. ■

Definition 15.65 *Let (M, Σ, μ) be a measure space. We denote by $L^\infty(M, \Sigma, \mu)$ the metric space obtained from $\mathcal{L}^\infty(M, \Sigma, \mu)$ via Theorem 15.61.*

Note that by Exercises 15-39 and 15-41 the space $L^\infty(M, \Sigma, \mu)$ is actually a normed space.

Notation 15.66 *If $M = D$ is a Lebesgue measurable subset of $\mathbb{R}^d$, Σ is the σ-algebra of Lebesgue measurable subsets of D and μ is Lebesgue measure restricted to Σ we will also write $L^p(D)$ for $L^p(M, \Sigma, \mu)$ and if D is an interval we also write $L^p[a, b]$ for $L^p([a, b])$, and so on.*

Notation 15.67 *Certain spaces are usually assumed to carry a certain norm or metric. The space $C^0[a, b]$ is usually assumed to be normed by $\|\cdot\|_\infty$. The spaces L^p are usually assumed to be normed by $\|\cdot\|_p$. The space L^∞ is usually assumed to be normed by $\|\cdot\|_\infty$. Unless otherwise stated, we will assume that each of the above spaces carries the mentioned norm and that any subset of these spaces carries the metric induced by the mentioned norm. For finite dimensional spaces, Theorem 16.76 will show that although many norms are available, for most purposes any one of them can be used.*

Exercises

15-38 Fill in the remaining details in the proof of Theorem 15.61.

(a) Prove that if $\sim$ is an equivalence relation with equivalence classes denoted by $[\cdot]$ and $x \sim y$, then $[x] = [y]$.

(b) Prove that $d : X \times X \to [0, \infty)$ as in Theorem 15.61 is satisfies $d\left([x], [y]\right) = d\left([y], [x]\right)$ for all $[x], [y] \in X$

(c) Prove that $d : X \times X \to [0, \infty)$ as in Theorem 15.61 satisfies the triangular inequality.

15-39. A **seminormed space** is a pair $\left(X^s, \|\cdot\|_s\right)$ of a vector space X^s and a function $\|\ \|^s : X^s \to [0, \infty)$, called the **seminorm** such that the following hold.

- $\|0\|^s = 0$.
- For all real numbers α and all $x \in X^s$ we have $\|\alpha x\|^s = |\alpha|\|x\|^s$.
- For all $x, y \in X^s$ we have $\|x + y\|^s \leq \|x\|^s + \|y\|^s$

(a) Prove that $\sim\subseteq X^s \times X^s$ defined by $x \sim y$ iff $\|x - y\|^s = 0$ is an equivalence relation.

(b) Prove that if $[x]$ denotes the equivalence class of x under $\sim$, then $X := \left\{[x] \cdot x \in X^s\right\}$ equipped with $\left\|\,[x]\,\right\| := \|x\|^s$ is a normed space.

Be careful. You must also prove that X is a vector space.

15-40 A **semi-inner product space** is a pair $\left(X^s, \langle\cdot,\cdot\rangle^s\right)$ consisting of a vector space X^s and a function $\langle\cdot,\cdot\rangle^s : X^s \times X^s \to \mathbb{R}$, called the **semi-inner product** such that the following hold.

- $\langle 0, 0\rangle^s = 0$.
- For all $x \in X^s$, we have $\langle x, x\rangle^s \geq 0$
- For all $x, y \in X^s$, we have $\langle x, y\rangle^s = \langle y, x\rangle^s$
- For all real numbers α and all $x, y \in X^s$, we have $\langle \alpha x, y\rangle^s = \alpha\langle x, y\rangle^s$
- For all $x, y, z \in X^s$, we have $\langle x + y, z\rangle^s = \langle x, z\rangle^s + \langle y, z\rangle^s$.

(a) Prove that with $\|x\|^s := \sqrt{\langle x, x\rangle^s}$ the pair $\left(X^s, \|\ \|^s\right)$ is a seminormed space.

(b) Prove that $\sim\subseteq X^s \times X^s$ defined by $x \sim y$ iff $\|x - y\|^s = 0$ is an equivalence relation.

(c) Prove that if $[x]$ denotes the equivalence class of x under $\sim$, then the set $X := \left\{[x] : x \in X^s\right\}$ equipped with $\left\langle [x], [y] \right\rangle := \langle x, y\rangle^s$ is an inner product space.

15-41 Prove that $\left(\mathcal{L}^\infty(M, \Sigma, \mu), \|\ \|_\infty\right)$ in Proposition 15 64 is a seminormed space

15-42 Let (M, Σ, μ) be a finite measure space. For any two measurable functions $f, g : M \to \mathbb{R}$, define $d(f, g) := \int_M \frac{|f-g|}{1+|f-g|}\, d\mu$ Prove that d is a semimetric on the space of measurable functions.

15-43 **Hölder's inequality** for $p = 1$, $q = \infty$. Let (M, Σ, μ) be a measure space Prove that for all functions $f \in \mathcal{L}^1(M, \Sigma, \mu)$ and $g \in \mathcal{L}^\infty(M, \Sigma, \mu)$ we have $fg \in \mathcal{L}^1(M, \Sigma, \mu)$ and the inequality $\|fg\|_1 \leq \|f\|_1 \|g\|_\infty$ holds.
Hint. Use that g is bounded a.e. by $\|g\|_\infty$.

15-44 Let (M, Σ, μ) be a measure space with $\mu(M) < \infty$.

(a) Prove that $\mathcal{L}^\infty(M, \Sigma, \mu) \subseteq \bigcap_{p\in[1,\infty)} \mathcal{L}^p(M, \Sigma, \mu)$.

(b) Give an example that shows that the containment is proper.

(c) Prove that for all $f \in \mathcal{L}^\infty(M, \Sigma, \mu)$ we have $\lim_{p\to\infty} \|f\|_p = \|f\|_\infty$

15-45 Let (M, Σ, μ) be a measure space. A set $L \subseteq \Sigma$ is called **locally μ-null** iff for all $S \in \Sigma$ with $\mu(S) < \infty$ we have $\mu(S \cap L) = 0$.

(a) Prove that every null set is locally μ-null.

(b) Prove that if (M, Σ, μ) is σ-finite, then every locally μ-null set is a null set.

(c) Consider the function $\mu(A) := \begin{cases} \lambda(A), & \text{if } 0 \notin A, \\ \infty; & \text{if } 0 \in A, \end{cases}$ defined on the Lebesgue measurable subsets of $\mathbb{R}$

i Prove that μ is a measure.

ii. Prove that $\mathbb{Q}$ is locally μ-null, but not μ-null.

15.9 Another Number Field: Complex Numbers

Complex numbers are often used in analysis, typically when "square roots of negative numbers" are needed. Exercise 15-52 shows that there is a price to be paid, because we lose the order relation. That in itself is not a problem. This section shows that the field axioms and an absolute value function remain available. Moreover, Theorem 15.75 will show that the convergence of Cauchy sequences, which is fundamental to analysis, is also preserved. Consequently, in abstract analysis $\mathbb{R}$ and $\mathbb{C}$ are often used interchangeably.

Definition 15.68 *The* **complex numbers** $\mathbb{C}$ *are the set* $\mathbb{R} \times \mathbb{R}$ *equipped with addition and multiplication defined as follows. For all complex numbers* $(a, b), (c, d) \in \mathbb{C}$, *we set* $(a, b) + (c, d) := (a + c, b + d)$ *and* $(a, b) \cdot (c, d) := (ac - bd, ad + bc)$. *We define* $i := (0, 1)$ *and* $1 := (1, 0)$ *and write complex numbers also in the form* $(a, b) = a \cdot 1 + b \cdot i = a + ib$. *For* $z = a + ib \in \mathbb{C}$, *the number* a *is also called the* **real part** *of* z, *denoted* $\Re(z)$, *and the number* b *is also called the* **imaginary part** *of* z, *denoted* $\Im(z)$.

The algebraic properties of $\mathbb{C}$ are summarized in Theorems 15.69 and 15.70.

Theorem 15.69 *The complex numbers* $\mathbb{C}$ *are a field.*

Proof. The field axioms are easily verified with $0 = 0 + 0i$ being neutral for addition, $1 = 1 + 0i$ being neutral for multiplication, $-(a+bi) = (-a)+(-b)i$ being the additive inverse and $(a+ib)^{-1} = \frac{a}{a^2+b^2} - \frac{b}{a^2+b^2}i$ being the multiplicative inverse. (Exercise 15-46.) ∎

The special element i serves as the closest we can get to a "square root of (-1)."

Theorem 15.70 $i^2 = -1$.

Proof. $i^2 = (0 + 1i) \cdot (0 + 1i) = (0 \cdot 0 - 1 \cdot 1) + (0 \cdot 1 + 1 \cdot 0)i = -1$. ∎

Aside from the above algebraic properties, we need to know how to measure distances in the complex numbers. This is done via the absolute value function.

Definition 15.71 *For $z = a + ib \in \mathbb{C}$, the* **absolute value** *of z is $|z| := \sqrt{a^2 + b^2}$.*

Theorem 15.72 *Properties of the absolute value.*

0. *For all $z \in \mathbb{C}$, we have $|z| \geq 0$.*
1. *For all $z \in \mathbb{C}$, we have $|z| = 0$ iff $z = 0$.*
2. *For all $z_1, z_2 \in \mathbb{C}$, we have $|z_1 z_2| = |z_1||z_2|$.*
3. *The* **triangular inequality** *holds.*
   That is, for all $z_1, z_2 \in \mathbb{C}$ we have $|z_1 + z_2| \leq |z_1| + |z_2|$.

Proof. Exercise 15-47. ∎

If we switch the sign of the imaginary part of a complex number we obtain the complex conjugate.

Definition 15.73 *For $z = a + ib \in \mathbb{C}$, the* **complex conjugate** *of z is $\overline{z} := a - ib$.*

Absolute value and complex conjugate are related via a simple equation. This equation can be used to express multiplicative inverses.

Proposition 15.74 *For all $z \in \mathbb{C}$, the equalities $z + \overline{z} = 2\Re(z)$ and $|z|^2 = z\overline{z}$ hold. Moreover, for all $z \in \mathbb{C} \setminus \{0\}$ the multiplicative inverse is $\frac{1}{z} = \frac{\overline{z}}{|z|^2}$.*

Proof. Exercise 15-48. ∎

The definitions of **complex valued sequences**, **convergence** in $\mathbb{C}$, and **Cauchy sequences** in $\mathbb{C}$ are similar to the corresponding definitions in $\mathbb{R}$ (Exercise 15-49).

Theorem 15.75 *Every complex valued Cauchy sequence converges in $\mathbb{C}$.*

Proof. Let $\{z_n\}_{n=1}^{\infty}$ be a Cauchy sequence of complex numbers. For each $n \in \mathbb{N}$, let $a_n := \Re(z_n)$ and $b_n := \Im(z_n)$ so that $z_n = a_n + ib_n$. Then for all $\varepsilon > 0$ there is an $N \in \mathbb{N}$ so that for all $m, n \geq N$ we have $|z_n - z_m| < \varepsilon$. Thus for all $m, n \geq N$ we obtain

$$|a_n - a_m| \leq \sqrt{(a_n - a_m)^2 + (b_n - b_m)^2} = |z_n - z_m| < \varepsilon,$$

so $\{a_n\}_{n=1}^{\infty}$ is a Cauchy sequence in $\mathbb{R}$. Similarly, $\{b_n\}_{n=1}^{\infty}$ is shown to be a Cauchy sequence in $\mathbb{R}$. Let $a := \lim_{n\to\infty} a_n$ and $b := \lim_{n\to\infty} b_n$, where the limits are taken in $\mathbb{R}$. Let $z := a + ib$ and let $\varepsilon > 0$. Then there is an $N \in \mathbb{N}$ so that for all $n \geq N$ we have $|a - a_n| < \frac{\varepsilon}{\sqrt{2}}$ and $|b - b_n| < \frac{\varepsilon}{\sqrt{2}}$. Therefore, for all $n \geq N$ we obtain

$$|z - z_n| = \sqrt{(a - a_n)^2 + (b - b_n)^2} < \sqrt{\frac{\varepsilon^2}{2} + \frac{\varepsilon^2}{2}} = \varepsilon,$$

which means that $z = \lim_{n\to\infty} z_n$ in $\mathbb{C}$. ■

Throughout this text we will rely on the properties that $\mathbb{C}$ has in common with $\mathbb{R}$. It is a field in which Cauchy sequences converge and its absolute value function has similar properties as the absolute value function for $\mathbb{R}$. The fact that $\mathbb{C}$ contains an element i such that $i^2 = -1$ is what sometimes requires the explicit use of $\mathbb{C}$ instead of $\mathbb{R}$. As long as we do not use this fact, we could use $\mathbb{C}$ and $\mathbb{R}$ interchangeably. In particular, unless otherwise indicated, all theorems on normed real vector spaces also hold for normed complex vector spaces. Moreover, because no linear structure is used in metric spaces, all results for metric spaces hold in real spaces as well as in complex spaces. In the remainder of this section, we highlight the main adjustments that need to be made to examples and definitions when using $\mathbb{C}$ instead of $\mathbb{R}$ as the underlying field.

All results for **series**, except for the alternating series test (obviously), and all results for **power series** can be translated verbatim to results for series and power series of complex numbers. The definitions of the exponential and trigonometric functions retain their familiar form.

Definition 15.76 *For all $z \in \mathbb{C}$ we define*

1. *The complex* **exponential function** $e^z := \sum_{k=0}^{\infty} \frac{z^k}{k!}$.

2. *The complex* **sine function** $\sin(z) := \sum_{k=0}^{\infty} \frac{(-1)^k z^{2k+1}}{(2k+1)!}$.

3. *The complex* **cosine function** $\cos(z) := \sum_{k=0}^{\infty} \frac{(-1)^k z^{2k}}{(2k)!}$.

These functions have the same properties as were proved for their real counterparts in Sections 12.1 and 12.2. Moreover, they are related via the Euler identities.

Theorem 15.77 Euler identities. *For all $z \in \mathbb{C}$ we have $e^{iz} = \cos(z) + i\sin(z)$. Equivalently, for all $z \in \mathbb{C}$ we have $\sin(z) = \frac{e^{iz} - e^{-iz}}{2i}$ and $\cos(z) = \frac{e^{iz} + e^{-iz}}{2}$.*

Proof. Exercise 15-50. ■

To see how the all-important function spaces can be translated into complex vector spaces, we need to define integrability for complex valued functions. A **complex vector space** is of course defined similar to a real vector space, the only difference being that the scalars are taken from the field $\mathbb{C}$ of complex numbers rather than the field $\mathbb{R}$ of real numbers.

Definition 15.78 *Let (M, Σ, μ) be a measure space. Then a function $f : X \to \mathbb{C}$ will be called* **measurable** *iff its real part defined by $\Re(f)(x) := \Re\big(f(x)\big)$ and its imaginary part defined by $\Im(f)(x) := \Im\big(f(x)\big)$ are both measurable. It will be called* **integrable** *iff both real and imaginary parts are integrable and the* **integral** *will be $\int_M f\,d\mu := \int_M \Re(f)\,d\mu + i\int_M \Im(f)\,d\mu$.*

One of the main reasons why little changes for the integration of complex valued functions is that a complex valued function is integrable iff its absolute value is integrable.

Proposition 15.79 *Let (M, Σ, μ) be a measure space. Then a complex valued function $f : M \to \mathbb{C}$ is integrable iff the absolute value $|f| : M \to \mathbb{R}$ (of course taken pointwise) is integrable.*

Proof. For "$\Leftarrow$," note that $\big|\Re(f)\big| \leq |f|$ and $\big|\Im(f)\big| \leq |f|$, and apply part 2 of Theorem 14.35.

For "$\Rightarrow$," note that $|f| = \sqrt{\big(\Re(f)\big)^2 + \big(\Im(f)\big)^2} \leq \sqrt{2}\max\big\{\big|\Re(f)\big|, \big|\Im(f)\big|\big\}$, where the maximum is taken pointwise, and apply part 1 of Theorem 14.35. ■

Definition 15.80 *If D is a set, then $F(D, \mathbb{C})$ is the space of all functions from D to $\mathbb{C}$. Spaces l^p, $\mathcal{L}^p(M, \Sigma, \mu)$ and $L^p(M, \Sigma, \mu)$ are defined similar to their counterparts for real valued functions and they have similar properties.*

The similar properties mentioned above include being a normed space or an inner product space. A **complex normed space** is defined similar to a real normed space, with the only difference being that we use a complex vector space instead of a real vector space. In particular, *the norm on a complex normed space still maps into* $[0, \infty) \subseteq \mathbb{R}$. The most significant adjustment when going from real vector spaces to complex vector spaces is in the definition of an inner product space.

Definition 15.81 *A* **complex inner product space** *is a pair $\big(X, \langle\cdot,\cdot\rangle\big)$ of a complex vector space X and a function $\langle\cdot,\cdot\rangle : X \times X \to \mathbb{C}$, called the* **inner product** *on X, with the following properties.*

1. *The inner product is* **positive definite**.
   That is, for all $x \in X$ we have $\langle x, x\rangle \in [0, \infty)$ with $\langle x, x\rangle = 0$ iff $x = 0$.

2. *For all $x, y \in X$, we have $\langle x, y\rangle = \overline{\langle y, x\rangle}$.*

3. *The inner product is* **linear** *in the first factor.*
   That is, for all scalars α and all $x, y \in X$ we have $\langle \alpha x, y\rangle = \alpha\langle x, y\rangle$ and for all $x, y, z \in X$ we have $\langle x+y, z\rangle = \langle x, z\rangle + \langle y, z\rangle$.

We will usually call X itself an inner product space. When we do so, we implicitly assume that there is an inner product on X, which will usually be denoted $\langle\cdot,\cdot\rangle$.

The adjustment in part 2 of Definition 15.81 guarantees that the Cauchy-Schwarz inequality still holds and that complex inner product spaces are also complex normed spaces (see Exercise 15-51). To make $L^2(M, \Sigma, \mu)$ into a complex inner product space, we use the product $\langle f, g\rangle := \int_M f\,\overline{g}\,d\mu$ on $\mathcal{L}^2(M, \Sigma, \mu)$ and then identify functions whose distance from each other is zero as was done in Theorem 15.61.

Throughout the rest of this text, unless we explicitly demand the space to be a real or complex space, it is usually safe to pretend the underlying field is $\mathbb{R}$. But since we will only use field axioms, absolute values and convergence of Cauchy sequences in our proofs, the results will hold for complex vector spaces as well as for real vector spaces.

Exercises

15-46 Prove that $\mathbb{C}$ is a field. That is, prove each of the following.

(a) Prove that for all $x, y, z \in \mathbb{C}$ we have $(x+y)+z = x+(y+z)$.

(b) Prove that for all $x, y \in \mathbb{C}$ we have $x+y = y+x$.

(c) Prove that with $0 = 0+0i$ for all $x \in \mathbb{C}$ we have $x+0 = x$.

(d) Prove that for every element $x = a+ib \in \mathbb{C}$ the element $(-x) = (-a)+i(-b)$ is so that $x+(-x) = 0$

(e) Prove that for all $x, y, z \in \mathbb{C}$ we have $(x \cdot y)\cdot z = x \cdot (y \cdot z)$.

(f) Prove that for all $x, y \in \mathbb{C}$ we have $x \cdot y = y \cdot x$

(g) Prove that the element $1 := 1+i0$ is so that for all $x \in \mathbb{R}$ we have $1 \cdot x = x$

(h) Prove that for every element $x = a+ib \in \mathbb{C}$ the element $x^{-1} = \frac{a}{a^2+b^2} - i\frac{b}{a^2+b^2}$ is so that $x \cdot x^{-1} = 1$

(i) Prove that for all $\alpha, x, y \in \mathbb{C}$ we have $\alpha \cdot (x+y) = \alpha \cdot x + \alpha \cdot y$.

15-47. Prove Theorem 15.72. That is, prove each of the following.

(a) Prove that for all $z \in \mathbb{C}$ we have $|z| \geq 0$.

(b) Prove that for all $z \in \mathbb{C}$ we have $|z| = 0$ iff $z = 0$.

(c) Prove that for all $z_1, z_2 \in \mathbb{C}$ we have $|z_1 z_2| = |z_1||z_2|$.

(d) Prove that for all $z_1, z_2 \in \mathbb{C}$ we have $|z_1 + z_2| \leq |z_1| + |z_2|$

Hint For the triangular inequality, use the Cauchy-Schwarz inequality for $\mathbb{R}^2$

15-48 Prove Proposition 15.74.

15-49. Define each of the following for complex numbers.

(a) Sequences of complex numbers

(b) Convergence of sequences of complex numbers.

(c) Cauchy sequences of complex numbers

15-50 Prove the Euler identities

(a) First, use the power series to prove that $e^{iz} = \cos(z) + i\sin(z)$ for all $z \in \mathbb{C}$

(b) Use part 15-50a and the appropriate version of Exercise 12-13c to prove that for all $z \in \mathbb{C}$ we have $\sin(z) = \dfrac{e^{iz} - e^{-iz}}{2i}$ and $\cos(z) = \dfrac{e^{iz} + e^{-iz}}{2}$

(c) Prove that the two identities in part 15-50b imply the identity in part 15-50a

15-51. Let X be a complex inner product space with inner product $\langle \cdot , \cdot \rangle$

(a) Prove that for all $x, y, z \in X$ and all $\alpha \in \mathbb{C}$ we have $\langle x, y+z\rangle = \langle x, y\rangle + \langle x, z\rangle$ and $\langle x, \alpha y\rangle = \overline{\alpha}\langle x, y\rangle$

(b) Prove that $\left| \Re\left(\langle x, y\rangle \right) \right| \leq \sqrt{\langle x, x\rangle\langle y, y\rangle}$ holds for all $x, y \in X$

Hint. Use the proof of Lemma 15 39 as guidance and realize that
$f(t) = \langle tx + y, tx + y\rangle = t^2\langle x, x\rangle + 2t\Re\left(\langle x, y\rangle \right) + \langle y, y\rangle$.

(c) Prove that every complex inner product space is a complex normed space

Hint. Adjust the proof of Theorem 15 40 appropriately

(d) Prove the **Cauchy-Schwarz inequality** That is, prove that $\left| \langle x, y\rangle \right| \leq \sqrt{\langle x, x\rangle\langle y, y\rangle}$ holds for all $x, y \in X$

Hint $\left| \langle x, y\rangle \right| = \Re\left(\dfrac{\overline{\langle x, y\rangle}}{|\langle x, y\rangle|}\langle x, y\rangle \right)$.

15-52. Prove that $\mathbb{C}$ cannot be ordered.

Hint. Suppose there was a subset $\mathbb{C}^+$ of $\mathbb{C}$ with properties as in Axiom 1.6 for the real numbers Prove that it must contain both 1 and -1, which is not possible Use that it must contain i or $-i$

15-53. Representation of complex numbers and n^{th} roots

(a) Prove that for every complex number $z = x + iy$ there are real number $r \geq 0$ and $\theta \in [0, 2\pi]$ so that $z = re^{i\theta}$.

Hint. Exercise 15-50a

(b) Prove that for every complex number $z \neq 0$ and every $n \in \mathbb{N}$ there are n distinct complex numbers $w_1, \ldots, w_n$ with $w_i^n = z$

Chapter 16

The Topology of Metric Spaces

In Chapter 15, we started with d-dimensional space as our underlying motivation and visualization. By systematically stripping away specific properties, we obtained the successively more general structures of inner product spaces, normed spaces and metric spaces. We will now journey back from metric spaces in this chapter, to normed spaces in Chapter 17, and to inner product spaces in Chapter 20. Because we now move from more general to more specific structures, the results we prove here and in Chapter 17 will also apply to structures discussed in later chapters.

16.1 Convergence of Sequences

A metric provides a distance function on a set, which is enough to discuss convergence, the central notion of analysis. We start by defining convergence of sequences and by investigating examples and properties of convergent sequences. Note that convergence is defined exactly as for sequences of real numbers in Definition 2.2. A **sequence** in X is of course just a function $f : \mathbb{N} \to X$, denoted as before by $\{x_n\}_{n=1}^{\infty}$.

Definition 16.1 *Let $\{x_n\}_{n=1}^{\infty}$ be a sequence in the metric space X. Then $L \in X$ is called* **limit** *of $\{x_n\}_{n=1}^{\infty}$ iff for all $\varepsilon > 0$ there is an $N \in \mathbb{N}$ so that for all $n \geq N$ we have $d(x_n, L) < \varepsilon$. A sequence that has a limit will be called* **convergent**, *a sequence that does not have a limit will be called* **divergent**.

As in the real numbers, limits are unique and the proof is similar.

In this chapter, the reader will frequently be asked to translate proofs from the real numbers to a more abstract setting. In this fashion, the reader reviews the proof technique and becomes more familiar with the abstract setting. For this first result, we present the proof to demonstrate the similarities.

Proposition 16.2 *Let $\{x_n\}_{n=1}^{\infty}$ be a sequence in the metric space X. If both L and M are limits of $\{x_n\}_{n=1}^{\infty}$, then $L = M$.*

Proof. We mimic the proof of Proposition 2.4. Let $\{x_n\}_{n=1}^{\infty}$ be a sequence in X and let L and M be limits of $\{x_n\}_{n=1}^{\infty}$. We need to prove that $L = M$.

Let $\varepsilon > 0$ be arbitrary but fixed. Then there is an $N_1 \in \mathbb{N}$ such that for all $n \geq N_1$ we have $d(x_n, L) < \frac{\varepsilon}{2}$. There also is an $N_2 \in \mathbb{N}$ such that for all $n \geq N_2$ we have $d(x_n, M) < \frac{\varepsilon}{2}$. Let $N := \max\{N_1, N_2\}$. Then for all $n \geq N$ we infer $d(x_n, L) < \frac{\varepsilon}{2}$ and $d(x_n, M) < \frac{\varepsilon}{2}$. Hence, with $n = N$ we obtain

$$d(L, M) \leq d(L, x_N) + d(x_N, M) < \frac{\varepsilon}{2} + \frac{\varepsilon}{2} = \varepsilon.$$

Because $\varepsilon > 0$ was arbitrary, by Theorem 1.37 we conclude that $d(L, M) = 0$, and hence $L = M$. ■

The similarity to the proof of Proposition 2.4 is obvious. It seems the main change is that we work with $d(x_n, L)$ instead of $|a_n - L|$. It would be naive to hope that all proofs translate this easily from the real line to metric spaces. But, especially for results early in this chapter, the proof of the corresponding result for the real line will provide more than adequate guidance.

Definition 16.3 *Because the limit is unique, we speak of* the *limit of a sequence. The notation* $\lim_{n\to\infty} x_n = L$ *will indicate that the limit of* $\{x_n\}_{n=1}^{\infty}$ *exists and is equal to* L.

Although we are proving results for metric spaces in this chapter, if a metric $d(\cdot, \cdot)$ is induced by a norm $\|\cdot\|$ we will write $\|x - y\|$ instead of $d(x, y)$. We can do this because every normed space is a metric space, which means that all results proved here are also valid for normed spaces (and inner product spaces, too).

As a first example, we consider the convergence of sequences in d-dimensional space. It turns out that sequences in $\mathbb{R}^d$ converge iff they converge componentwise. In Theorem 16.4 we prove this for the metric induced by the uniform norm $\|\cdot\|_{\infty}$. Section 16.6 and specifically Theorem 16.78 will show that the same is true for all norms on $\mathbb{R}^d$. Until Section 16.6 we will consider only the uniform norm on $\mathbb{R}^d$. For the notation for the component sequences, recall Convention 15.23.

Theorem 16.4 *Consider* $\mathbb{R}^d$ *with the metric induced by the uniform norm* $\|\cdot\|_{\infty}$. *Then a sequence* $\{x_n\}_{n=1}^{\infty}$ *converges to* L *in* $\left(\mathbb{R}^d, \|\cdot\|_{\infty}\right)$ *iff for all* $k = 1, \ldots, d$ *the* k^{th} *component sequence* $\left\{x_n^{(k)}\right\}_{n=1}^{\infty}$ *converges in* $\mathbb{R}$ *to the* k^{th} *component* $L^{(k)}$ *of* L.

Proof. We have $\lim_{n\to\infty} x_n = L$ iff $\lim_{n\to\infty} \|x_n - L\|_{\infty} = 0$. For all $k \in \{1, \ldots, d\}$, we infer $\left|x_n^{(k)} - L^{(k)}\right| \leq \|x_n - L\|_{\infty}$, so convergence of the vectors implies convergence of the components in $\mathbb{R}$. (*Mentally fill in the argument.*)

Conversely, assume that for all $k \in \{1, \ldots, d\}$ we have $\lim_{n\to\infty} \left|x_n^{(k)} - L^{(k)}\right| = 0$. Let $\varepsilon > 0$. For each $k \in \{1, \ldots, d\}$ there is an N_k such that for all $n \geq N_k$ we have

$\left|x_n^{(k)} - L^{(k)}\right| < \varepsilon$. Let $N := \max\{N_k : k = 1, \ldots, d\}$. Then for all $n \geq N$ we obtain $\left|x_n^{(k)} - L^{(k)}\right| < \varepsilon$ for *all* $k \in \{1, \ldots, d\}$, and hence $\|x_n - L\|_\infty < \varepsilon$. ■

In other spaces, componentwise convergence is not equivalent to convergence.

Theorem 16.5 *Let $a < b$ be real numbers and consider the space $C^0[a,b]$ with the metric induced by the uniform norm $\|\cdot\|_\infty$. A sequence $\{f_n\}_{n=1}^\infty$ converges to f in $\left(C^0[a,b], \|\cdot\|_\infty\right)$ iff it converges uniformly to f on $[a,b]$.*

Proof. For "$\Leftarrow$," assume that $\{f_n\}_{n=1}^\infty$ converges uniformly to f and let $\varepsilon > 0$. Then there is an $N \in \mathbb{N}$ so that for all $n \geq N$ and all points $x \in [a,b]$ we have that $\left|f_n(x) - f(x)\right| < \varepsilon$. But this means $\|f_n - f\|_\infty \leq \varepsilon$ for all $n \geq N$, and hence $\{f_n\}_{n=1}^\infty$ converges to f in $\left(C^0[a,b], \|\cdot\|_\infty\right)$.

For "$\Rightarrow$," assume that $\{f_n\}_{n=1}^\infty$ converges to f in $\left(C^0[a,b], \|\cdot\|_\infty\right)$ and let $\varepsilon > 0$. Then there is an $N \in \mathbb{N}$ so that for all $n \geq N$ we have $\|f_n - f\|_\infty < \varepsilon$. But then for all $n \geq N$ and all $x \in [a,b]$ the inequality $\left|f_n(x) - f(x)\right| < \varepsilon$ holds, which means that $\{f_n\}_{n=1}^\infty$ converges uniformly to f. ■

Remark 16.6 *Theorem 16.5 is why the $\|\cdot\|_\infty$ norm is also called the* **uniform norm**.

Exercise 16-1a shows that pointwise convergence does not imply convergence in the uniform norm. Exercises 16-1b, 16-1c and 16-1d investigate the relationship between convergence and componentwise convergence in l^∞. (Note that in spaces like l^∞ sequences are in fact sequences of sequences, so we must be careful which index is which.)

In many spaces, there is no equivalent formulation of convergence in terms of components. For example, we typically consider the elements of L^1 to be functions. Formally, we would need to pick a representative of the equivalence class, which is a small step. Now, if a sequence $\{f_n\}_{n=1}^\infty$ converges to a function f in $L^1(M, \Sigma, \mu)$, then $\{f_n\}_{n=1}^\infty$ converges to f in mean. By Proposition 14.47, $\{f_n\}_{n=1}^\infty$ converges to f in measure and by Proposition 14.49 a subsequence, but not necessarily the whole sequence, converges to f pointwise a.e. Conversely, the Dominated Convergence Theorem shows that pointwise a.e. convergence only implies convergence in mean if there is an integrable function that dominates all functions in the sequence.

The above indicates that convergence in a specific space can be complicated. Keeping track of all the details about the elements in a space can become quite confusing. This is where results about abstract metric spaces help. A proof for abstract metric spaces cannot use any specific properties of an example. This means by working with abstract properties we do not need to keep track of the potentially complicated detailed properties of specific spaces.

We conclude this section with some general properties of convergent sequences. This will show how concrete results from the single variable setting translate to the abstract setting. For example, every convergent sequence of real numbers is bounded. In metric spaces we do not have a fixed point of reference such as $0 \in \mathbb{R}$. This is not a problem as the following definition shows.

Definition 16.7 *A metric space X is called* **bounded** *iff there is a point $c \in X$ and an $M \in \mathbb{R}$ so that for all $x \in X$ we have $d(x, c) \leq M$.*

We will frequently work with subsets of metric spaces. Because it would be inefficient to define properties for metric spaces and also for subsets of metric spaces we recall that any subset of a metric space is a metric space in its own right.

Definition 16.8 *Any property of metric spaces can also be viewed as a* **property of the subsets** *of metric spaces. Formally, we say that a subset B of a metric space X has property $\mathcal{P}$ iff B as a metric space with the induced metric $d|_{B\times B}$ has property $\mathcal{P}$.*

Proposition 16.9 *Let $\{x_n\}_{n=1}^{\infty}$ be a convergent sequence in the metric space X. Then $\{x_n : n \in \mathbb{N}\}$ is bounded.*

Proof. Adapt the proof of Proposition 2.34. (See Exercise 16-2.) ■

Subsequences are defined just as for sequences of real numbers.

Definition 16.10 *(Compare with Definition 2.39.) Let X be a metric space, let $\{x_n\}_{n=1}^{\infty}$ be a sequence of elements of X and let $\{n_k\}_{k=1}^{\infty}$ be a sequence of natural numbers so that $n_k < n_{k+1}$ for all $k \in \mathbb{N}$. Then $\left\{x_{n_k}\right\}_{k=1}^{\infty}$ is called a* **subsequence** *of $\{x_n\}_{n=1}^{\infty}$.*

As for real numbers, subsequences of convergent sequences have the same limit.

Proposition 16.11 *Let X be a metric space and let $\{x_n\}_{n=1}^{\infty}$ be a convergent sequence with limit L. Then every subsequence $\left\{x_{n_k}\right\}_{k=1}^{\infty}$ also converges to L.*

Proof. Mimic the proof of Proposition 2.40. (See Exercise 16-3.) ■

Exercises

16-1 Componentwise convergence is not equivalent to convergence

(a) Prove that the sequence $f_n(x) := \begin{cases} 0, & \text{for } 0 \leq x \leq 1 - \frac{1}{n}, \\ 2n\left[x - \left(1 - \frac{1}{n}\right)\right]; & \text{for } 1 - \frac{1}{n} \leq x \leq 1 - \frac{1}{2n}, \\ 2n(1-x); & \text{for } 1 - \frac{1}{2n} \leq x \leq 1, \end{cases}$

converges pointwise to $f(x) = 0$, but it does not converge to f in $\left(C^0[0,1], \|\cdot\|_\infty\right)$. In particular you need to prove that the f_n are continuous

(b) Prove that convergence in l^∞ implies componentwise convergence

Hint. Mimic the appropriate part of the proof of Theorem 16 4

(c) Construct a sequence $\{x_n\}_{n=1}^{\infty}$ in l^∞ so that for all $i \in \mathbb{N}$ the component sequence $\left\{x_n^{(i)}\right\}_{n=1}^{\infty}$ converges, but $\{x_n\}_{n=1}^{\infty}$ does not converge in l^∞.

(d) Explain why the proof of Theorem 16.4 cannot be generalized to apply to l^∞ That is, determine which part of the proof fails for l^∞

16-2 Prove Proposition 16.9.

16-3. Prove Proposition 16.11

16-4. Let X be a metric space and let $\{x_n\}_{n=1}^{\infty}$ and $\{y_n\}_{n=1}^{\infty}$ be sequences with $\lim_{n\to\infty} d(x_n, y_n) = 0$. Prove that if $\{x_n\}_{n=1}^{\infty}$ converges then so does $\{y_n\}_{n=1}^{\infty}$ and $\lim_{n\to\infty} x_n = \lim_{n\to\infty} y_n$.

16-5 Let X be a subset of $\mathbb{R}^d$ with the metric induced by the Euclidean norm $\|\cdot\|_2$ on $\mathbb{R}^d$. Prove that a sequence $\{x_n\}_{n=1}^{\infty}$ converges to L in X iff for all $k = 1, \ldots, d$ the kth component sequence $\left\{x_n^{(k)}\right\}_{n=1}^{\infty}$ converges in $\mathbb{R}$ to the kth component $L^{(k)}$ of L.

Hint. For the "$\Leftarrow$" direction, the difference in each component is best chosen to be $\frac{\varepsilon}{\sqrt{d}}$.

16-6. Let (M, Σ, μ) be a finite measure space and let $\mathcal{M}(M, \sigma, \mu)$ be the metric space obtained as in Theorem 15.61 from the measurable functions with the semimetric $d(f, g) = \int_M \frac{|f-g|}{1+|f-g|}\,d\mu$ from Exercise 15-42. Prove that the sequence $\left\{[f_n]\right\}_{n=1}^{\infty}$ converges to $[g] \in \mathcal{M}(M, \sigma, \mu)$ iff $\{f_n\}_{n=1}^{\infty}$ **converges in measure** to g

16-7 Let X be a metric space. Prove that the following are equivalent.

(a) X is bounded,

(b) For all $c \in X$ there is an $M_c \in \mathbb{R}$ so that for all $x \in X$ we have $d(x, c) \leq M_c$,

(c) There is an $M \in \mathbb{R}$ so that for all $x, y \in X$ we have $d(x, y) \leq M$.

16-8. Let X be a metric space, let $x \in X$ and let $\{x_n\}_{n=1}^{\infty}$ be a sequence in X such that every subsequence has a subsequence that converges to x. Prove that $\{x_n\}_{n=1}^{\infty}$ converges to x

16-9. Because every normed space is also a metric space, we have also defined convergence in normed spaces. So let X be a normed space, let $\{x_n\}_{n=1}^{\infty}$ and $\{y_n\}_{n=1}^{\infty}$ be sequences in X and let $\{c_n\}_{n=1}^{\infty}$ be a sequence in $\mathbb{R}$

(a) Prove that if $\lim_{n\to\infty} x_n = x$ and $\lim_{n\to\infty} y_n = y$, then $\lim_{n\to\infty} x_n + y_n = x + y$

(b) Prove that if $\lim_{n\to\infty} x_n = x$ and $\lim_{n\to\infty} y_n = y$, then $\lim_{n\to\infty} x_n - y_n = x - y$

(c) Prove that if $\lim_{n\to\infty} x_n = x$ and $\lim_{n\to\infty} c_n = c$, then $\lim_{n\to\infty} c_n x_n = cx$.

Hint. Mimic the proofs of the appropriate parts of Theorem 2.14.

16-10. Because every inner product space is also a metric space, we have also defined convergence in inner product spaces So let X be an inner product space and let $\{x_n\}_{n=1}^{\infty}$ and $\{y_n\}_{n=1}^{\infty}$ be sequences in X Prove that if $\lim_{n\to\infty} x_n = x$ and $\lim_{n\to\infty} y_n = y$, then $\lim_{n\to\infty} \langle x_n, y_n\rangle = \langle x, y\rangle$.

16-11. Let X be a metric space and let $\{a_n\}_{n=1}^{\infty}$ and $\{b_n\}_{n=1}^{\infty}$ be convergent sequences in X with limits a and b, respectively Prove that $\lim_{n\to\infty} d(a_n, b_n) = d(a, b)$

16.2 Completeness

By Theorem 2.27, every convergent sequence of real numbers is a Cauchy sequence. The definition of Cauchy sequences easily translates to metric spaces and the implication "convergence implies Cauchy" still holds.

Definition 16.12 *Let $\{x_n\}_{n=1}^{\infty}$ be a sequence in the metric space X. Then $\{x_n\}_{n=1}^{\infty}$ is called a* **Cauchy sequence** *iff for all $\varepsilon > 0$ there is an $N \in \mathbb{N}$ so that for all $m, n \geq N$ we have $d(x_n, x_m) < \varepsilon$.*

Theorem 16.13 *If the sequence $\{x_n\}_{n=1}^{\infty}$ in the metric space X converges, then it is a Cauchy sequence.*

Proof. Mimic part of the proof of Theorem 2.27. (See Exercise 16-12.) ■

Conversely, the fact that Cauchy sequences converge is one of the fundamental properties of the real numbers. Recall that by Exercise 2-25 this property is equivalent to the Completeness Axiom (Axiom 1.19). Hence, without this property analysis on the real line would be all but impossible. Convergence of Cauchy sequences is equally important in metric spaces. For example, in numerical considerations, the limit usually is not known. Therefore, to prove convergence of a numerical scheme it is common to prove that the scheme produces a Cauchy sequence and then use convergence of Cauchy sequences to conclude that the scheme will produce a limit (for an example, see the proof of Theorem 13.14). On the theoretical side, convergence of Cauchy sequences is crucial for the proofs of such fundamental results as the differentiability of the inversion operator (Theorem 17.32), Banach's Fixed Point Theorem (Theorem 17.64), the Implicit Function Theorem (Theorem 17.65), Riesz' Representation Theorem (Theorem 20.26), Picard and Lindelöf's Existence and Uniqueness Theorem (Theorem 22.6), and the Lax-Milgram Lemma (Lemma 23.4). Hence, we will always give special attention to spaces in which Cauchy sequences converge. Note that in a normed space we automatically assume the metric is induced by the norm.

Definition 16.14 *A metric space X is called* **complete** *iff all Cauchy sequences in X converge. A complete normed space is called a* **Banach space** *and a complete inner product space is called a* **Hilbert space**.

The real numbers are the simplest example of a complete metric space. Unsurprisingly, the spaces that are most commonly used in analysis are complete.

Theorem 16.15 $\left(\mathbb{R}^d, \|\cdot\|_\infty\right)$ *is complete, that is, it is a Banach space.*

Proof. We need to prove that every Cauchy sequence in $\mathbb{R}^d$ converges. *A typical completeness proof starts with a Cauchy sequence, constructs an element that should be the limit and then proves that the element is indeed the limit.*

Let $\{x_n\}_{n=1}^\infty$ be a Cauchy sequence in $\mathbb{R}^d$ and let $\varepsilon > 0$. There is an $N \in \mathbb{N}$ so that for all $m, n \geq N$ we have $\|x_m - x_n\|_\infty < \varepsilon$. But then for all $j = 1, \ldots, d$ and all $m, n \geq N$ we obtain $\left|x_m^{(j)} - x_n^{(j)}\right| \leq \|x_m - x_n\|_\infty < \varepsilon$, which means that each component sequence $\left\{x_n^{(j)}\right\}_{n=1}^\infty$ is a Cauchy sequence. Because $\mathbb{R}$ is complete, each component sequence $\left\{x_n^{(j)}\right\}_{n=1}^\infty$ has a limit $L^{(j)}$. Let $L = \left(L^{(1)}, \ldots, L^{(d)}\right)$. By Theorem 16.4 $\{x_n\}_{n=1}^\infty$ converges to L. Because $\{x_n\}_{n=1}^\infty$ was arbitrary, every Cauchy sequence in $\mathbb{R}^d$ converges and the space is complete. ■

Completeness of $\left(C^0[a,b], \|\cdot\|_\infty\right)$ will be proved in Exercise 16-13.

To prove that the L^p spaces are Banach spaces, it is helpful to first analyze convergence of series in normed spaces. Series, convergence of series and absolute convergence are defined just like for real numbers. Lemma 16.18 characterizes Banach spaces in terms of convergent series.

Definition 16.16 *(Compare with Definition 6.1.) Let X be a normed space and let $\{a_j\}_{j=1}^{\infty}$ be a sequence in X. The* **partial sums** *of the* **series** $\sum_{j=1}^{\infty} a_j$ *are defined to be $s_n := \sum_{j=1}^{n} a_j$. The series is said to* **converge** *iff the sequence of partial sums converges and it is said to* **diverge** *otherwise. For a convergent series, the limit is denoted $\sum_{j=1}^{\infty} a_j$.*

Definition 16.17 *(Compare with Definition 6.12.) Let X be a normed space. Then a series $\sum_{j=1}^{\infty} a_j$ in X* **converges absolutely** *iff the series $\sum_{j=1}^{\infty} \|a_j\|$ converges in $\mathbb{R}$.*

In normed spaces, absolutely convergent series do not automatically converge. In fact, they converge if and only if the space is a Banach space.

Lemma 16.18 *A normed space X is a Banach space iff every absolutely convergent series in X also converges in X.*

Proof. For "$\Rightarrow$," mimic the proof of Proposition 6.13. (Exercise 16-14.)

For "$\Leftarrow$," let X be a normed space in which every absolutely convergent series converges and let $\{a_n\}_{n=1}^{\infty}$ be a Cauchy sequence. Set $a_0 := 0$, $N_0 := 0$ and inductively for each $\varepsilon_k := \frac{1}{2^k}$ find an $N_k > N_{k-1}$ so that for all $n, m \geq N_k$ the inequality $\|a_n - a_m\| \leq \frac{1}{2^k}$ holds. For all $k \in \mathbb{N}$, let $d_k := a_{N_k} - a_{N_{k-1}}$. By construction of the N_k, we obtain $\sum_{j=1}^{\infty} \|d_j\| \leq \|a_{N_1}\| + \sum_{j=1}^{\infty} \frac{1}{2^j} = \|a_{N_1}\| + 1$. Therefore $\sum_{j=1}^{\infty} d_j$ converges absolutely, and hence, by hypothesis, it converges to a limit $L \in X$. But for all $k \in \mathbb{N}$ we have $\sum_{j=1}^{k} d_j = a_{N_k}$, so $\{a_{N_k}\}_{k=1}^{\infty}$ is a convergent subsequence of $\{a_n\}_{n=1}^{\infty}$.

To prove that $L := \lim_{k\to\infty} a_{N_k}$ is the limit of $\{a_n\}_{n=1}^{\infty}$ let $\varepsilon > 0$. Then there is a $K \in \mathbb{N}$ so that for all $k \geq K$ we have $\|a_{N_k} - L\| < \frac{\varepsilon}{2}$ and there is an $M \in \mathbb{N}$ so that for all $n, m \geq M$ we have $\|a_n - a_m\| < \frac{\varepsilon}{2}$. But then for all $n \geq M$ we can find a $k \geq K$ so that $N_k \geq M$. Then for all $n \geq M$ we obtain the inequality $\|a_n - L\| \leq \|a_n - a_{N_k}\| + \|a_{N_k} - L\| < \frac{\varepsilon}{2} + \frac{\varepsilon}{2} = \varepsilon$. Hence, $\{a_n\}_{n=1}^{\infty}$ converges to L and X is a Banach space. ■

Now we are ready to tackle L^p spaces.

Theorem 16.19 *Let (M, Σ, μ) be a measure space and let $1 \leq p \leq \infty$. Then $L^p(M, \Sigma, \mu)$ is complete, that is, it is a Banach space.*

Proof. By Lemma 16.18 we need to prove that every absolutely convergent series converges in L^p. So let $\sum_{j=1}^{\infty}[f_j]$ be a series in $L^p(M,\Sigma,\mu)$ with $\sum_{j=1}^{\infty}\left\|[f_j]\right\|_p < \infty$. For each $j \in \mathbb{N}$, the function f_j will be a *fixed* representative of $[f_j]$.

First, consider the case $p=\infty$. For all natural numbers $j \in \mathbb{N}$, we define the set $N_j := \{x \in M : f_j(x) > \|f_j\|_\infty\}$. Then $N := \bigcup_{j=1}^{\infty} N_j$ is a μ-null set and $\sum_{j=1}^{\infty} f_j$ converges on $M \setminus N$. Define $f(x) := \begin{cases} \sum_{j=1}^{\infty} f_j(x); & \text{for } x \in M \setminus N, \\ 0; & \text{for } x \in N. \end{cases}$ Because $\bigcup_{j=1}^{\infty} N_j$ is μ-null we infer

$$\lim_{n\to\infty}\left\|[f]-\sum_{j=1}^{n}[f_j]\right\|_\infty \le \lim_{n\to\infty}\sum_{j=n+1}^{\infty}\|f_j\|_\infty = \lim_{n\to\infty}\sum_{j=n+1}^{\infty}\left\|[f_j]\right\|_\infty = 0,$$

and hence by Lemma 16.18 $L^\infty(M,\Sigma,\mu)$ is complete.

This leaves the case $p \in [1,\infty)$. For all $x \in M$ set $g(x) := \left(\sum_{j=1}^{\infty}|f_j(x)|\right)^p$. By Minkowski's Inequality, for each $n \in \mathbb{N}$ we obtain

$$\left(\int_M \left(\sum_{j=1}^{n}|f_j|\right)^p d\mu\right)^{\frac{1}{p}} = \left\|\sum_{j=1}^{n}|f_j|\right\|_p \le \sum_{j=1}^{n}\|f_j\|_p \le \sum_{j=1}^{\infty}\|f_j\|_p .$$

Thus by the Monotone Convergence Theorem

$$\int_M g\,d\mu = \lim_{n\to\infty}\int_M \left(\sum_{j=1}^{n}|f_j|\right)^p d\mu \le \left(\sum_{j=1}^{\infty}\|f_j\|_p\right)^p < \infty,$$

and hence g is integrable. In particular, g is finite μ-a.e., which means that for almost every $x \in M$ the series $\sum_{j=1}^{\infty} f_j(x)$ converges absolutely. Define the function f by $f(x) := \begin{cases} \sum_{j=1}^{\infty} f_j(x); & \text{if } g(x) < \infty, \\ 0; & \text{otherwise.} \end{cases}$ Then the function f is measurable and it satisfies $|f|^p \le g$, so $f \in \mathcal{L}^p(M,\Sigma,\mu)$. Because $\lim_{n\to\infty}\left|f(x)-\sum_{j=1}^{n} f_j(x)\right| = 0$ μ-a.e. and $\left|f(x)-\sum_{j=1}^{n} f_j(x)\right|^p \le \left(\sum_{j=n+1}^{\infty}|f_j(x)|\right)^p \le g(x)$ μ-a.e., by the Dominated Convergence Theorem we conclude $\lim_{n\to\infty}\left\|f-\sum_{j=1}^{n} f_j\right\|_p = 0$. Therefore the series $\sum_{j=1}^{\infty}[f_j]$

converges to $[f]$. By Lemma 16.18, this means that $L^p(M, \Sigma, \mu)$ is complete. ■

The most natural example of a metric space that is not complete are the rational numbers $\mathbb{Q}$. To see why $\mathbb{Q}$ is not complete, recall that by Proposition 1.49 $\sqrt{2}$ is not a rational number. Use Theorem 1.36 to construct a sequence $\{q_n\}_{n=1}^{\infty}$ of rational numbers that (in $\mathbb{R}$) converges to $\sqrt{2}$ (for a concrete example, consider Exercise 2-17). Then, because $\{q_n\}_{n=1}^{\infty}$ converges in $\mathbb{R}$, $\{q_n\}_{n=1}^{\infty}$ is a Cauchy sequence. But $\{q_n\}_{n=1}^{\infty}$ does not have a limit in $\mathbb{Q}$, because if $L \in \mathbb{Q} \subseteq \mathbb{R}$ was a limit of $\{q_n\}_{n=1}^{\infty}$, then by the uniqueness of limits in $\mathbb{R}$ we would obtain $L = \sqrt{2} \notin \mathbb{Q}$, a contradiction.

The above argument shows the typical way in which metric spaces fail to be complete. Incomplete spaces contain sequences that appear to converge to an element that "just is not there." Theorem 16.89 will show that this is indeed the only way in which a space can fail to be complete.

Further examples of spaces that are not complete are given in Exercise 16-15. In particular, Exercise 16-15d shows that if we use the method that produced L^1 with the Riemann integral rather than the Lebesgue integral, then we do not obtain a complete space. This is the reason why for theoretical considerations the Lebesgue integral is preferred over the Riemann integral.

Exercises

16-12. Prove Theorem 16.13.

16-13 Prove that $C^0[a, b]$ with the $\|\cdot\|_\infty$ norm is a Banach space.
Hint. First, prove that if $\{f_n\}_{n=1}^{\infty}$ is a Cauchy sequence, then every sequence $\{f_n(x)\}_{n=1}^{\infty}$ is a Cauchy sequence Then prove that the pointwise limit f actually is a uniform limit Conclude that f is continuous (Corollary 11 8) Then prove that $\|f_n - f\|_\infty$ goes to zero.

16-14. Prove the "$\Rightarrow$" part of Lemma 16 18.

16-15. Metric spaces that are not complete.

(a) Prove that the interval $(0, 1)$ is not a complete metric space.

(b) Prove that $C^1(-1, 1)$ with the uniform norm $\|\cdot\|_\infty$ is not a Banach space.
Hint. Example 11.10.

(c) Prove that $C^0[0, 1]$ with the norm $\|f\|_1 := \int_0^1 |f(x)|\,dx$ is not a Banach space.

(d) Let $(R^1[0, 1], \|\cdot\|_1)$ be the space of Riemann integrable functions on $[0, 1]$ with functions f, g with $\int_0^1 |f(x) - g(x)|\,dx = 0$ identified into one object like in $L^1[0, 1]$ and with the norm $\|f\|_1 := \int_0^1 |f(x)|\,dx$. Prove that this space is not a Banach space.

Hint Let C^Q be a **Cantor set** with positive measure (see Exercise 8-3e) and let C_n^Q be the sets used to construct C^Q as in Definition 7.22. Prove that $\left\{\left[\mathbf{1}_{C_n^Q}\right]\right\}_{n=1}^{\infty}$ is a Cauchy sequence in $L^1[0, 1]$ that converges to $\left[\mathbf{1}_{C^Q}\right]$ (which is in $L^1[0, 1]$ by Exercise 9-29) in the L^1 norm. Use the fact that $R^1[0, 1]$ is a subspace of $L^1[0, 1]$ to conclude that $\left\{\left[\mathbf{1}_{C_n^Q}\right]\right\}_{n=1}^{\infty}$ is a Cauchy sequence in $R^1[0, 1]$ Then use that $\mathbf{1}_{C^Q}$ is not Riemann integrable (Exercise 8-15) to prove that $\left\{\left[\mathbf{1}_{C_n^Q}\right]\right\}_{n=1}^{\infty}$ cannot have a limit in $R^1[0, 1]$. In the last step some care is needed because formally the objects we are working with are equivalence classes.

(e) Let $1 < p < \infty$. Let $(R^p[0,1], \|\ \|_p)$ be the Riemann integrable functions on $[0,1]$ with f, g so that $\int_0^1 |f(x) - g(x)|^p\, dx = 0$ identified into one object like in $L^p[0,1]$ and with norm $\|f\|_p := \left(\int_0^1 |f(x)|^p\, dx\right)^{\frac{1}{p}}$ Prove that this space is not a Banach space.

(f) Prove that l^1 with the norm $\left\|\{x_n\}_{n=1}^{\infty}\right\|_{\infty} := \sup\left\{ |x_n| : n \in \mathbb{N} \right\}$ is not a Banach space.

16-16. Let X be a metric space and let $\{a_n\}_{n=1}^{\infty}$ be a Cauchy sequence. Prove that if $\{a_n\}_{n=1}^{\infty}$ has a convergent subsequence, then it converges.

16-17 Let $a < b$ Prove that on the space $B := \left\{ f \in C^1(a,b) : \|f\|_\infty < \infty, \left\|f'\right\|_\infty < \infty \right\}$ the function $\|f\|_1 := \|f\|_\infty + \left\|f'\right\|_\infty$ defines a norm and that $(B, \|\cdot\|_1)$ is a Banach space.

16-18. Give two proofs that for $1 \leq p \leq \infty$ the space l^p is a Banach space

(a) Prove it by applying Theorem 16.19

(b) Give a direct proof.

Hint Mimic the proof of Theorem 16 19.

16-19. Unconditional convergence of **series** in Banach spaces.

(a) Define what it means when a series **converges unconditionally** in a normed space.

(b) Prove that in Banach spaces absolute convergence implies unconditional convergence.

Hint. Mimic the "$\Rightarrow$" part of the proof of Theorem 6 18.

(c) In Banach spaces, unconditional convergence does not imply absolute convergence. Prove that the series $\sum_{k=1}^{\infty} \frac{1}{k} e_k$ converges unconditionally, but not absolutely, in l^∞.

16.3 Continuous Functions

Once convergence has been defined, it is possible to talk about continuity. After all, continuous functions are exactly those functions that preserve convergent sequences. We will define continuity directly, that is, not via limits of functions. The technicalities that would make a definition via limits of functions tedious are discussed at the end of this section,

Definition 16.20 **ε-δ formulation of continuity**. *Let X, Y be metric spaces and let $f : X \to Y$. Then f is called* **continuous** *at $x \in X$ iff for all $\varepsilon > 0$ there is a $\delta > 0$ such that for all $z \in X$ with $d_X(z, x) < \delta$ we have $d_Y\big(f(z), f(x)\big) < \varepsilon$. A function that is continuous at every $x \in X$ is simply called* **continuous** *on X.*

Notation 16.21 When the distance between two objects is computed, the space in which the objects reside determines which metric needs to be used. Therefore, unlike in Definition 16.20, we typically do not index metrics and norms with the space on which they act. Metrics and norms will only be indexed when we are working with several metrics or norms on the same space (for examples, see Section 16.6) or when a particular norm is needed for a proof (Section 18.4 is a good example). □

The first examples are (of course) all continuous functions $f : I \to \mathbb{R}$, where I is an interval. Next note that the natural projections on $\mathbb{R}^d$ are continuous.

Example 16.22 *For $k \in \{1, \ldots, d\}$, define the* **natural projection** $\pi_k : \mathbb{R}^d \to \mathbb{R}$ *on the* k^{th} *coordinate by* $\pi(x_1, \ldots, x_d) := x_k$. *Then* π_k *is continuous on* $\mathbb{R}^d$ *when* $\mathbb{R}^d$ *is equipped with the* $\|\cdot\|_\infty$ *norm.*

Let $x = (x_1, \ldots, x_d) \in \mathbb{R}^d$ and let $\varepsilon > 0$. Then, with $\delta := \varepsilon$, for all elements $z \in \mathbb{R}^d$ with $\|z - x\|_\infty < \varepsilon$ we obtain $\left|\pi_k(z) - \pi_k(x)\right| = |z_k - x_k| \leq \|z - x\|_\infty < \varepsilon$, and hence π_k is continuous at x. ∎

To obtain more examples, we need to establish some properties of continuous functions. To make some of these results easier to prove, we first connect continuity to sequences.

Theorem 16.23 Sequence formulation of continuity. *Let X, Y be metric spaces and let $x \in X$. Then $f : X \to Y$ is continuous at $x \in X$ iff for all sequences $\{z_n\}_{n=1}^\infty$ in X with $\lim_{n\to\infty} z_n = x$ we have $\lim_{n\to\infty} f(z_n) = f(x)$.*

Proof. Adapt the proof of Theorem 3.25 (Exercise 16-20). ∎

Continuity is compatible with algebraic operations, if they are available, and it is preserved by compositions.

Theorem 16.24 *Let X be a metric space, let Y be a normed space and let the functions $f, g : X \to Y$ be continuous. Then $f + g$ and $f - g$ are continuous.*

Proof. Adjust the appropriate parts of the proof of Theorem 3.27. This can be done via Theorem 16.23 and Exercises 16-9a and 16-9b or with a direct ε-δ argument that mimics the appropriate part of the proof of Theorem 2.14. (Exercise 16-21.) ∎

Theorem 16.25 *Let X be a metric space, let Y be a normed space, and let the functions $f : X \to Y$ and $\alpha : X \to \mathbb{R}$ be continuous. Then $\alpha f : X \to Y$ is continuous.*

Proof. Adjust the appropriate part of the proof of Theorem 3.27. This can be done via Theorem 16.23 and Exercise 16-9c or with a direct ε-δ argument that mimics the appropriate part of the proof of Theorem 2.14. (Exercise 16-22.) ∎

Theorem 16.26 *Let X be a metric space, let Y be an inner product space, and let $f, g : X \to Y$ be continuous. Then $\langle f, g\rangle : X \to \mathbb{R}$ is continuous.*

Proof. Adjust the appropriate part of the proof of Theorem 3.27. This can be done via Theorem 16.23 and Exercise 16-10 or with a direct ε-δ argument that mimics the appropriate part of the proof of Theorem 2.14. (Exercise 16-23.) ∎

Theorem 16.27 *Let X, Y, Z be metric spaces and let $g : X \to Y$ and $f : Y \to Z$ be continuous functions. Then $f \circ g : X \to Z$ is continuous.*

Proof. Mimic the proof of Theorem 3.30. (See Exercise 16-24.) ∎

The above shows that functions on $\mathbb{R}^d$, like $f(x, y) := xy + \cos\left(x^2 - y^2\right)$ (see Exercise 16-25), which combine the components using known continuous operations and functions are continuous. Our definition of continuity also gives access to more complicated situations. For example, it is well known that integration is a "continuous operation." The following example makes this observation more precise.

Example 16.28 *Let (M, Σ, μ) be a measure space, let $1 \leq p \leq \infty$, let q be such that $\frac{1}{p} + \frac{1}{q} = 1$ and let $[g] \in L^q(M, \Sigma, \mu)$. Then $I_{[g]}([f]) := \int_M fg \, d\mu$ defines a continuous function $I_{[g]} : L^p(M, \Sigma, \mu) \to \mathbb{R}$.*

Because the elements of $L^p(M, \Sigma, \mu)$ and $L^q(M, \Sigma, \mu)$ are equivalence classes and $I_{[g]}$ is defined in terms of representatives of the equivalence classes, we must first prove that the functions $I_{[g]}$ are well defined. For functions $f \in \mathcal{L}^p(M, \Sigma, \mu)$ and $g \in \mathcal{L}^q(M, \Sigma, \mu)$ set $I_g(f) := \int_M fg \, d\mu$. By Hölder's inequality (Theorem 15.48 and Exercise 15-43), the integral in $I_g(f)$ exists for all functions $f \in \mathcal{L}^p(M, \sigma, \mu)$ and $g \in \mathcal{L}^q(M, \sigma, \mu)$ and the inequality $|I_g(f)| \leq \|f\|_p \|g\|_q$ holds. In particular this means if $f_1, f_2 \in [f]$ and $g_1, g_2 \in [g]$, then

$$\begin{aligned} \left|I_{g_1}(f_1) - I_{g_2}(f_2)\right| &\leq \left|I_{g_1}(f_1) - I_{g_1}(f_2)\right| + \left|I_{g_1}(f_2) - I_{g_2}(f_2)\right| \\ &= \left|I_{g_1}(f_1 - f_2)\right| + \left|I_{g_1 - g_2}(f_2)\right| \\ &\leq \|f_1 - f_2\|_p \|g_1\|_q + \|f_2\|_p \|g_1 - g_2\|_q = 0. \end{aligned}$$

That is, the value of $I_g(f)$ does not change if different representatives of the equivalence class of $[f] \in L^p(M, \sigma, \mu)$ and $[g] \in L^q(M, \sigma, \mu)$ are chosen. Now if the functions $f \in \mathcal{L}^p(M, \Sigma, \mu)$ and $g_1, g_2 \in \mathcal{L}^q(M, \Sigma, \mu)$ satisfy $[g_1] = [g_2]$, then $\left|I_{g_1}(f) - I_{g_2}(f)\right| = 0$. Therefore, for $f \in \mathcal{L}^p(M, \Sigma, \mu)$ and $[g] \in L^q(M, \Sigma, \mu)$ setting $J_{[g]}(f) := I_g(f)$ defines a unique function on $\mathcal{L}^p(M, \Sigma, \mu)$. Moreover, if the functions $f_1, f_2 \in \mathcal{L}^p(M, \Sigma, \mu)$ satisfy $[f_1] = [f_2]$ and $[g] \in L^q(M, \Sigma, \mu)$, then $\left|J_{[g]}(f_1) - J_{[g]}(f_2)\right| = \left|I_g(f_1) - I_g(f_2)\right| = 0$. Therefore $I_{[g]}([f]) := J_{[g]}(f)$ defines a well-defined function on $L^p(M, \Sigma, \mu)$.

Now let $[g] \in L^q(M, \Sigma, \mu)$. To see that $I_{[g]}$ is continuous, first note that by Hölder's inequality we infer $\left|I_{[g]}([f])\right| \leq \left\|[f]\right\|_p \left\|[g]\right\|_q$ for all $[f] \in L^p(M, \sigma, \mu)$ and $[g] \in L^q(M, \sigma, \mu)$. Let $[f_1] \in L^p(M, \sigma, \mu)$ and let $\varepsilon > 0$. Then for all elements $[f_2] \in L^p(M, \sigma, \mu)$ with $\left\|[f_1] - [f_2]\right\|_p < \delta := \frac{\varepsilon}{\left\|[g]\right\|_q + 1}$ we obtain

$$\begin{aligned} \left|I_{[g]}([f_1]) - I_{[g]}([f_2])\right| &= \left|I_{[g]}([f_1 - f_2])\right| \leq \left\|[f_1 - f_2]\right\|_p \left\|[g]\right\|_q \\ &\leq \frac{\varepsilon}{\left\|[g]\right\|_q + 1} \left\|[g]\right\|_q < \varepsilon. \end{aligned}$$

Hence, $I_{[g]}$ is continuous. □

Note that each $I_{[g]}$ is actually a pretty complicated function. It maps equivalence classes of functions to numbers. For such complicated constructions, it is important to establish some type of mental hierarchy so that the functions that are elements of the space are distinguished from the functions that act on the space itself. The best way to achieve this is to practice with the complicated settings. (Consider, for example, Exercise 16-26, where functions map sequences to numbers.) When working with such functions, the following notations can also help establish a mental hierarchy.

Definition 16.29 *A continuous function into the real numbers is also referred to as a* **functional**. *A continuous function between metric spaces is also called an* **operator**.

Note that in Example 16.28 we have proved a bit more than just continuity. We have established the inequality $\left|I_{[g]}([f_1]) - I_{[g]}([f_2])\right| \leq \left\|[g]\right\|_q \left\|[f_1 - f_2]\right\|_p$, for all $[f_1], [f_2] \in L^p(M, \Sigma, \mu)$, where $\left\|[g]\right\|_q$ actually is a constant for the function $I_{[g]}$. Functions that satisfy such a condition are called Lipschitz continuous.

Definition 16.30 *Let X, Y be metric spaces. A function $f : X \to Y$ is called* **Lipschitz continuous** *iff there is an $L > 0$, called the* **Lipschitz constant** *of f, so that for all $x, y \in X$ we have $d\big(f(x), f(y)\big) \leq Ld(x, y)$.*

It is easy to see that Lipschitz continuous functions are continuous.

Proposition 16.31 *Let X, Y be metric spaces and let $f : X \to Y$ be a Lipschitz continuous function. Then f is continuous.*

Proof. Exercise 16-30a. ■

16.3.1 The Trouble with Limits of Functions

The familiar Definition 3.23 from calculus was not used to define continuity in metric spaces because a technical problem allows us to only define limits at accumulation points of the space.

Definition 16.32 *Let X be a metric space. Then $x \in X$ is called an* **accumulation point** *of X iff for all $\delta > 0$ there is a $z \in X \setminus \{x\}$ such that $d(z, x) < \delta$.*

Definition 16.33 **ε-δ formulation of the limit of a function.** *Let X, Y be metric spaces, let $f : X \setminus \{x\} \to Y$ and let x be an accumulation point of X. Then $L \in Y$ is called the* **limit** *of f at $x \in X$ iff for all $\varepsilon > 0$ there is a $\delta > 0$ such that for all $z \in X \setminus \{x\}$ with $d(z, x) < \delta$ we have $d\big(f(z), L\big) < \varepsilon$.*

We require that the point x is an accumulation point, because otherwise the limit of a function would not be unique. Indeed, if x is not an accumulation point (that is, if x is an **isolated point**), then for some $\delta > 0$ there is no $z \in X \setminus \{x\}$ with $d(z, x) < \delta$. Because there are no such points, it is vacuously true that for all such points z and any $\varepsilon > 0$ and $L \in Y$ we have $d\big(f(z), L\big) < \varepsilon$, *no matter what L is.* So at an isolated point x any element of Y would be a limit of f, which is a rather pathological situation. To avoid this pathology, we demand that limits are only taken at accumulation points.

Note that for functions $f : I \setminus \{x\} \to \mathbb{R}$, where I is an open interval and $x \in I$, the definition of the limit via Definition 16.33 is consistent with Definition 3.1, because x clearly is an accumulation point of $I \setminus \{x\}$ and because we can apply Theorem 3.7. Exercises 16-28 and 16-29 show that Definition 16.33 removes the formal problems we had identified for Definition 3.1.

Regarding continuity, it is reassuring to note that there is no need to worry about isolated points. The definition of continuity states that as the inputs get close to x, the outputs must get close to $f(x)$. If the inputs cannot get close to x, then the outputs vacuously get close to $f(x)$. This is still a silly notion, but there is no logical problem and usually no one worries about continuity at an isolated point. The definition implies that if a function is defined at an isolated point, then it is continuous there. So be it.

Aside from the problem with isolated points, limits work very much like for functions on intervals (a, b).

Theorem 16.34 Sequence formulation of the limit of a function. *Let X, Y be metric spaces, let x be an accumulation point of X and let $L \in Y$. Then $\lim_{z\to x} f(z) = L$ iff for all sequences $\{z_n\}_{n=1}^{\infty}$ in $X \setminus \{x\}$ with $\lim_{n\to\infty} z_n = x$ we have $\lim_{n\to\infty} f(z_n) = L$.*

Proof. Mimic the proof of Theorem 3.7. (Exercise 16-27.) ■

Exercises

16-20. Prove Theorem 16.23.

16-21. Prove Theorem 16.24.

16-22. Prove Theorem 16.25.

16-23. Prove Theorem 16.26.

16-24 Prove Theorem 16.27.

16-25 Prove that $f : \mathbb{R}^2 \to \mathbb{R}$ defined by $f(x, y) := xy + \cos\left(x^2 - y^2\right)$ is continuous.
Hint. Break up the way the function is computed into elementary steps that are continuous.

16-26. Let $1 \leq p \leq \infty$. For $k \in \mathbb{N}$, define the projection on the kth component $\pi_k : l^p \to \mathbb{R}$ by $\pi_k\left(\left\{a^{(j)}\right\}_{j=1}^{\infty}\right) := a^{(k)}$. Prove that for all p and k the function $\pi_k : l^p \to \mathbb{R}$ is continuous

16-27. Prove Theorem 16.34.

16-28. Exercise 3-8 revisited

(a) Prove that $\lim_{z\to 0} \sqrt{z} = 0$. In particular, this means that with Definition 16.33 there are no formal problems with the limit of the square root function at 0

(b) Explain why, despite the result in Exercise 16-28a, Definition 16.33 does *not* supersede the definition of one-sided limits.
Hint. Step functions

16-29. Prove that the function from Exercise 3-32 does not have a limit at $x = 0$

16-30. Let X, Y be metric spaces and let $f : X \to Y$ be a function.

(a) Prove that if f is Lipschitz continuous, then f is continuous.

(b) Define uniform continuity for functions $g : X \to Y$.

(c) Prove that if f is Lipschitz continuous, then f is uniformly continuous.

16-31. Let X be a metric space and let $z \in X$. Prove that $d(\cdot, z) : X \to \mathbb{R}$ is continuous.

16-32 Let X be a metric space and let $f, g : X \to \mathbb{R}$ be continuous functions. Prove that if $g(x) \neq 0$ for all $x \in X$, then $\frac{f}{g}$ is continuous on X.
Hint. Use the proof of Theorem 2.14 as guidance

16-33. Paths and continuity.

(a) Let C be a subset of a normed space X, let $x \in C$ be such that there is an $\varepsilon > 0$ so that $\left\{z \in X : \|z - x\| < \varepsilon\right\} \subseteq C$ and let Y be a metric space Prove that $f : C \to Y$ is continuous at x iff for all continuous functions $p : [0, 1] \to X$ so that $p(0) = x$, the composition $f \circ p : [0, 1] \to Y$ is continuous at 0

(b) Prove that $f:\mathbb{R}^2\to\mathbb{R}$ defined by $f(x,y):=\begin{cases}\frac{xy}{x^2+y^2}, & \text{for } (x,y)\neq(0,0),\\ 0, & \text{for } x=y=0,\end{cases}$

is not continuous at the origin

16-34 A **polynomial** $p:\mathbb{R}^d\to\mathbb{R}$ is a finite sum of finite products of the variables. That is, there are an $n\in\mathbb{N}$, $c_1,\ldots,c_n\in\mathbb{R}$ and $\left(\alpha_1^i,\ldots,\alpha_d^i\right)\in\left(\mathbb{N}\cup\{0\}\right)^d$ so that for all $(x_1,\ldots,x_d)\in\mathbb{R}^d$ we have $p(x_1,\ldots,x_d)=\sum_{i=1}^{n}c_i x_1^{\alpha_1^i}\cdots x_d^{\alpha_d^i}$. Prove that every polynomial $p:\mathbb{R}^d\to\mathbb{R}$ is continuous

Hint. Induction. First, for $n=1$ on the sum of the αs to take care of the products, then on n.

16-35. Let X and Y be metric spaces. A sequence of functions $\{f_n\}_{n=1}^{\infty}$ from X to Y is called **uniformly convergent** to the function $f:X\to Y$ iff for all $\varepsilon>0$ there is an $N\in\mathbb{N}$ so that for all $n\geq N$ and all $x\in X$ we have $d\left(f_n(x),f(x)\right)<\varepsilon$

Let $\{f_n\}_{n=1}^{\infty}$ be a sequence of functions $f_n:X\to Y$ that converges uniformly to $f:X\to Y$ and let all f_n be continuous at $c\in X$. Prove that f is continuous at c.

Hint Mimic the proof of Theorem 11 7

16-36. For natural numbers $j\geq 2$, let $g_{\left[\frac{1}{j+1},\frac{1}{j-1}\right]}(x):=\begin{cases} j(j+1)\left(x-\frac{1}{j+1}\right); & \text{for } \frac{1}{j+1}\leq x\leq\frac{1}{j},\\ j(j-1)\left(\frac{1}{j-1}-x\right); & \text{for } \frac{1}{j}\leq x\leq\frac{1}{j-1},\\ 0, & \text{otherwise.}\end{cases}$

Prove that the function $f:\mathbb{R}\to l^{\infty}$ defined by $f:=\sum_{j=2}^{\infty}g_{\left[\frac{1}{j+1},\frac{1}{j-1}\right]}e_j$ is continuous on $\mathbb{R}\setminus\{0\}$, bounded, and for any sequence $x_n\to 0$ with $x_n>0$ the limit $\lim_{n\to\infty}f(x_n)$ does not exist.

Note. This function shows that in infinite dimensional spaces functions can have discontinuities that are neither removable, nor jumps, nor infinite, nor by oscillation

16.4 Open and Closed Sets

Results, such as the Intermediate Value Theorem (Theorem 3.34) or the fact that continuous functions assume absolute maxima and minima (Theorem 3.44), refer specifically to the ordering of the real numbers. This ordering is not available in metric spaces. To obtain more general versions of these theorems, we must first establish properties that will take the place of the ordering of the real numbers. Connectedness (see Definition 16.90), which is needed to prove a version of the Intermediate Value Theorem, is defined purely in topological terms. Compactness, which allows us to prove a version of Theorem 3.44 for metric spaces, can be defined in terms of sequences, but it is more commonly defined in topological terms (see Theorem 16.72). Thus, before we continue, we introduce the requisite topological ideas. In a nutshell, topology revolves around open and closed sets. Figure 34 on page 304 summarizes the main properties of these sets.

With a metric envisioned as measuring distances, a ball must be the set of points within a certain distance of a center point.

Definition 16.35 *Let X be a metric space, let $x\in X$ and let $\varepsilon>0$. Then the* **open ball** *of* **radius** *ε about x is defined to be $B_\varepsilon(x):=\left\{p\in X:d(p,x)<\varepsilon\right\}$.*

An open set contains for each of its points a small open ball around that point. So, just like in an open interval, at every point in an open set we have a certain amount of room to go in any direction (also see Figure 34(a)).

Figure 33: Visualization of the proof of Proposition 16.37.

Definition 16.36 *Let X be a metric space. A subset $O \subseteq X$ is called* **open** *iff for all $x \in O$ there is an $\varepsilon_x > 0$ such that $B_{\varepsilon_x}(x) \subseteq O$.*

Open intervals (a, b) on the real line are open sets as in Definition 16.36, because for each point $x \in (a, b)$ with $\varepsilon := \min\{b - x, x - a\}$ we obtain the containment $B_\varepsilon(x) = (x - \varepsilon, x + \varepsilon) \subseteq (a, b)$. Thus the nomenclature is consistent. In general, open balls are the prototypical examples of open sets. Although it is convenient to envision these balls as round entities, we should bear in mind that they need not be round at all. Balls with respect to the uniform norm $\|\cdot\|_\infty$ on $\mathbb{R}^d$ are cubes (also see Figure 37 on page 317).

Proposition 16.37 *Let X be a metric space, let $z \in X$ and let $\varepsilon > 0$. Then the open ball $B_\varepsilon(z)$ is open.*

Proof. *To prove that a set is open, we prove that it contains a small ball around each of its points.*

Let $x \in B_\varepsilon(z)$ be arbitrary. Set $\varepsilon_x := \varepsilon - d(x, z) > 0$. Then for all $y \in B_{\varepsilon_x}(x)$ we have that $d(y, z) \leq d(y, x) + d(x, z) < \varepsilon - d(x, z) + d(x, z) = \varepsilon$. Hence, $B_{\varepsilon_x}(x) \subseteq B_\varepsilon(z)$ (see Figure 33) and since x was arbitrary, $B_\varepsilon(z)$ is open. ∎

We can summarize the most important properties of open sets as follows.

Theorem 16.38 *Let X be a metric space. Then the following results hold:*

1. *Both $\emptyset$ and X are open subsets of X.*

2. *If $O_1, \ldots, O_n$ are open subsets of X, then $\bigcap_{k=1}^{n} O_k$ is open.*

3. *If $\mathcal{O}$ is a family of open subsets of X then $\bigcup \mathcal{O}$ is open.*

Proof. Part 1 is trivial.

For part 2, let $O_1, \ldots, O_n$ be open subsets of X, and let $x \in \bigcap_{k=1}^{n} O_k$ be arbitrary. Then for each integer $k \in \{1, \ldots, n\}$ there is an $\varepsilon_k > 0$ such that $B_{\varepsilon_k}(x) \subseteq O_k$. Let $\varepsilon := \min\{\varepsilon_k : k = 1, \ldots, n\}$. Then for all integers $k \in \{1, \ldots, n\}$ the containments $B_\varepsilon(x) \subseteq B_{\varepsilon_k}(x) \subseteq O_k$ hold. Hence, $B_\varepsilon(x) \subseteq \bigcap_{k=1}^{n} O_k$. Since $x \in \bigcap_{k=1}^{n} O_k$ was arbitrary, the intersection $\bigcap_{k=1}^{n} O_k$ is open.

For part 3, let $\mathcal{O}$ a family of open subsets of X and let $x \in \bigcup \mathcal{O}$ be arbitrary. Then there is an $O \in \mathcal{O}$ such that $x \in O$ and there is an $\varepsilon > 0$ so that $B_\varepsilon(x) \subseteq O$. But then $B_\varepsilon(x) \subseteq O \subseteq \bigcup \mathcal{O}$. Since $x \in \bigcup \mathcal{O}$ was arbitrary, $\bigcup \mathcal{O}$ is open. ■

Example 16.39 *Arbitrary intersections of open sets need not be open.*

For any real numbers $a < b$, we have that $[a, b] = \bigcap_{n=1}^{\infty} \left(a - \frac{1}{n}, b + \frac{1}{n} \right)$, but $[a, b]$ is not an open subset of $\mathbb{R}$. □

Once open sets are defined we can briefly summarize the fundamental concepts of topology. A **topology** is a family $\mathcal{T}$ of subsets of a set X such that the three conditions in Theorem 16.38 hold. The pair $(X, \mathcal{T})$ is called a **topological space** and the sets $O \in \mathcal{T}$ are called **open** sets. Any idea that can be stated purely in terms of open sets is actually a topological idea. In this text, we will focus on metric spaces and only touch upon the topological notions and vocabulary as necessary. As an example consider how continuous functions would be defined in terms of open sets.

Theorem 16.40 Topological formulation of continuity. *Let X and Y be metric spaces and let $f : X \to Y$ be a function. Then f is continuous on X iff for all open subsets $V \subseteq Y$ the inverse image $f^{-1}[V] \subseteq X$ is open.*

Proof. For "$\Rightarrow$," let f be continuous and let $V \subseteq Y$ be open. Let $x \in f^{-1}[V]$. Then there is an $\varepsilon > 0$ so that $B_\varepsilon\big(f(x)\big) \subseteq V$. Because f is continuous, there is a $\delta > 0$ such that for all $z \in X$ with $d(z, x) < \delta$ we have $d\big(f(z), f(x)\big) < \varepsilon$. Therefore, for all $z \in B_\delta(x)$ we obtain $f(z) \in B_\varepsilon\big(f(x)\big) \subseteq V$, and hence $B_\delta(x) \subseteq f^{-1}[V]$. Because $x \in f^{-1}[V]$ was arbitrary, $f^{-1}[V]$ is open.

For "$\Leftarrow$," let the function f be such that for all open subsets $V \subseteq Y$ the inverse image $f^{-1}[V] \subseteq X$ is open, let $x \in X$ and let $\varepsilon > 0$. Then $f^{-1}\big[B_\varepsilon\big(f(x)\big)\big]$ is open. Hence, there is a $\delta > 0$ with $B_\delta(x) \subseteq f^{-1}\big[B_\varepsilon\big(f(x)\big)\big]$. But this means for any $z \in X$ with $d(z, x) < \delta$ the inequality $d\big(f(z), f(x)\big) < \varepsilon$ holds, and hence f is continuous at x. Because $x \in X$ was arbitrary, f is continuous. ■

We will often be concerned with subsets of metric spaces that contain an open set around a given point x. To abbreviate the wording and because open sets contain a little "padding" around each of their points (see Figure 34(a)), we call such sets neighborhoods.

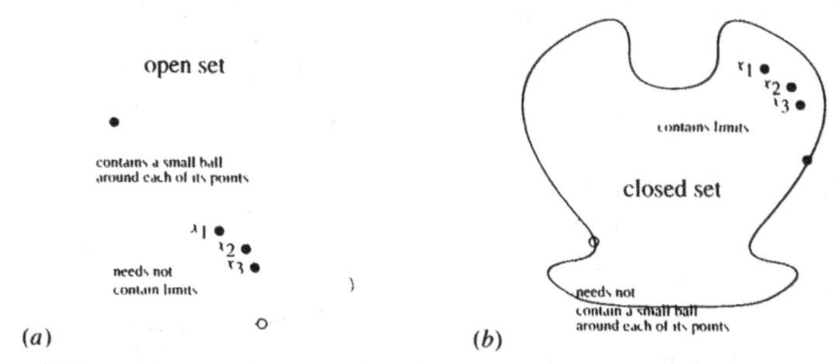

Figure 34: Open sets contain a certain amount of "padding" around each of their points (a), while closed sets contain all their limit points (b).

Definition 16.41 *Let X be a metric space and let $x \in X$. Then $N \subseteq X$ is called a* **neighborhood** *of x iff $x \in N$ and there is an open set $O \subseteq X$ so that $x \in O \subseteq N$. If $S \subseteq X$ and N is a neighborhood of each $s \in S$, then N is also called a* **neighborhood** *of S.*

The complements of open sets are called closed sets.

Definition 16.42 *Let X be a metric space. A subset $C \subseteq X$ is called* **closed** *iff its complement $X \setminus C$ is open.*

Closed intervals on the real line are also closed sets as in Definition 16.42. Once again, the nomenclature is consistent.

Example 16.43 Although it is tempting to hope that open and closed are mutually exclusive and exhaustive properties, sets can actually be open, closed, both, or neither.

1. The interval $[0, 1]$ is a closed subset of the real line.

2. In any metric space, the sets $\emptyset$ and X are both open and closed.

3. The interval $[0, 1)$ as a subset of the real line is neither open nor closed. □

Although properties of closed sets can in theory be obtained through complementation, closed sets are interesting in their own right, because closed sets "keep their limits" (also see Figure 34(b)).

Definition 16.44 *Let X be a metric space and let $A \subseteq X$. Then $x \in X$ is called a* **limit point** *of A iff there is a sequence $\{a_n\}_{n=1}^{\infty}$ with $a_n \in A$ for all $n \in \mathbb{N}$ so that $x = \lim\limits_{n\to\infty} a_n$.*

Limit points are different from accumulation points in that the sequence that converges to a limit point x can assume x as a value (and even be constant with value x), which is forbidden for accumulation points.

Theorem 16.45 *Let X be a metric space. Then $C \subseteq X$ is closed iff for all convergent sequences $\{z_n\}_{n=1}^{\infty}$ with $z_n \in C$ for all $n \in \mathbb{N}$ we have $\lim_{n\to\infty} z_n \in C$.*

Proof. For "$\Rightarrow$," let $C \subseteq X$ be closed, let $\{z_n\}_{n=1}^{\infty}$ be a convergent sequence with $z_n \in C$ for all $n \in \mathbb{N}$ and let $x := \lim_{n\to\infty} z_n$. For a contradiction, suppose that $x \notin C$. Because $X \setminus C$ is open, there is an $\varepsilon > 0$ so that $B_\varepsilon(x) \subseteq X \setminus C$. But then there is an $N \in \mathbb{N}$ so that for all $n \geq N$ the point z_n is in $B_\varepsilon(x) \subseteq X \setminus C$ and in C, a contradiction.

For "$\Leftarrow$," suppose for a contradiction that $X \setminus C$ is not open. Then there is an $x \in X \setminus C$ so that for every $n \in \mathbb{N}$ there is a point $z_n \in B_{\frac{1}{n}}(x) \cap C$. But then $\{z_n\}_{n=1}^{\infty}$ is a sequence of points in C that converges to a point $x \notin C$, contradiction. ■

Remark 16.46 Definition 16.36 and Theorem 16.45 provide descriptive characterizations of open and closed sets, respectively. Note (also see Figure 34) that sequences in open sets need not take their limit in the open set (consider the sequence $\left\{\frac{1}{n}\right\}_{n=1}^{\infty}$ in the interval $(0, 1)$) and that closed sets need not contain a small ball around each of their points (consider the point 0 in $[0, 1]$). □

Because completeness is important, we should note that closed subsets of complete spaces will again be complete.

Corollary 16.47 *Let X be a complete metric space and let $C \subseteq X$ be closed. Then C with the induced metric is a complete metric space.*

Proof. Exercise 16-37. ■

16.4.1 The Interior and The Closure

For each subset A of a metric space, there are a largest open set contained in A and a smallest closed set that contains A. These sets are called the interior and the closure, respectively.

Definition 16.48 *Let X be a metric space and let $A \subseteq X$. The* **interior** *A° of A is the set of all points $x \in A$ so that a small ball around the point is also in A, that is, $A^\circ := \{x \in A : (\exists \varepsilon > 0 : B_\varepsilon(x) \subseteq A)\}$. The points in A° are also called the* **interior points** *of A.*

Proposition 16.49 *Let X be a metric space and let $A \subseteq X$. Then A° is an open set that contains all open subsets of A. Moreover, $A^\circ = \bigcup\{O \subseteq A : O \text{ is open }\}$.*

Proof. To see that A° is open, let $x \in A^\circ$. Then $x \in A$ and there is an $\varepsilon > 0$ so that $B_\varepsilon(x) \subseteq A$. By Proposition 16.37, for every $z \in B_\varepsilon(x)$ there is an $\varepsilon_z > 0$ so that $B_{\varepsilon_z}(z) \subseteq B_\varepsilon(x) \subseteq A$. But then $B_\varepsilon(x) \subseteq A^\circ$, which proves that A° is open.

Now let $U \subseteq A$ be an open subset of A. Then for all $x \in U$ there is an $\varepsilon > 0$ so that $B_\varepsilon(x) \subseteq U \subseteq A$, which means $x \in A^\circ$. Hence, $U \subseteq A^\circ$ for all open subsets of A.

Finally to prove the equation, note that the set on the right is an open subset of A, which means (by what we just proved) that the set on the right is contained in A°. Conversely, A° is an open set contained in A, so A° is contained in the union on the right, thus establishing the equation. ■

Example 16.50

1. The interior of an open ball $B_r(x)$ obviously is the ball itself, $\left(B_r(x)\right)^\circ = B_r(x)$.

2. The interior of the rational numbers as a subset of the real numbers is $\mathbb{Q}^\circ = \emptyset$, because every ball around a rational number contains an irrational number. □

Because every singleton set $\{x\}$ is contained in all open balls of radius $r > 0$ about x, it is easy to see that there is no smallest open set that contains a given set.

Definition 16.51 *Let X be a metric space and let $A \subseteq X$. The* **closure** *A^- (or $\overline{A}$) of A is defined to be the set of all limit points of A, that is,*

$$\overline{A} := A^- := \left\{x \in X : \left(\exists \{a_n\}_{n=1}^\infty : \lim_{n\to\infty} a_n = x \text{ and } \forall n \in \mathbb{N} : a_n \in A\right)\right\}.$$

Proposition 16.52 *Let X be a metric space and let $A \subseteq X$. Then A^- is closed and it is contained in all closed supersets of A. Moreover, $A^- = \bigcap\{C \supseteq A : C \text{ is closed }\}$.*

Proof. To prove that A^- is closed, let $x \in X$ and let $\{x_n\}_{n=1}^\infty$ be a sequence with $x_n \in A^-$ for all $n \in \mathbb{N}$ and $\lim_{n\to\infty} x_n = x$. Then for each $n \in \mathbb{N}$ there is an $a_n \in A$ with $d(x_n, a_n) < \frac{1}{n}$. But then $\lim_{n\to\infty} a_n = \lim_{n\to\infty} x_n = x$ and $x \in A^-$. Hence, A^- is closed.

Now let $C \supseteq A$ be a closed superset of A and let $x \in A^-$. Then there is a sequence $\{a_n\}_{n=1}^\infty$ so that $a_n \in A$ for all $n \in \mathbb{N}$ and $\lim_{n\to\infty} a_n = x$. But then, because C is closed and contains A we conclude $x = \lim_{n\to\infty} a_n \in C$, and hence $A^- \subseteq C$.

Finally, to prove the equation, note that the set on the right is a closed superset of A (use Exercise 16-41c), so A^- must be contained in the set on the right. Conversely, because A^- is a closed superset of A it must contain the intersection on the right, which establishes the equality. ■

Example 16.53

1. The closure of an open ball is $\overline{B_r(x)} = \left\{p \in X : d(p, x) \leq r\right\}$ (Exercise 16-48a).

2. The closure of the rational numbers in the real numbers is $\overline{\mathbb{Q}} = \mathbb{R}$, because every real number is the limit of a sequence of rational numbers. □

Because every open interval is the union of all its closed subintervals, there is no largest closed set that is contained in a given set. The set that is geometrically between the interior of a set and the interior of the set's complement is called the boundary.

Definition 16.54 *Let X be a metric space and let $A \subseteq X$. The* **boundary** *δA of A is defined to be $\delta A := A^- \setminus A^\circ$.*

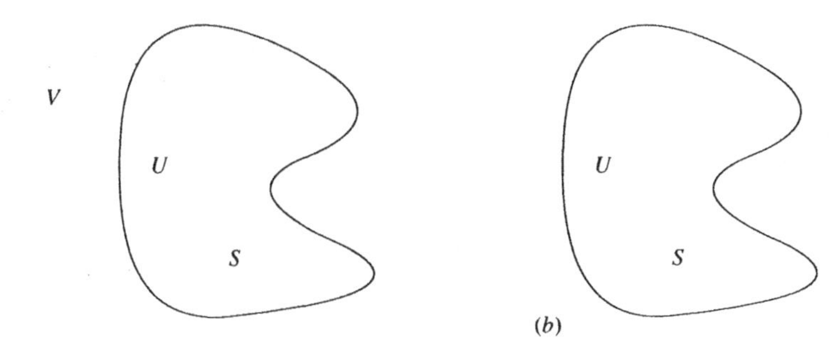

Figure 35: Visualization of Proposition 16.56. Relatively open sets U are intersections of the subset with open sets V in the space (a). In particular, as subsets of the space itself, they need not be open (b).

Example 16.55

1. The boundary of an open ball is $\delta B_r(x) = \{p \in X : d(p,x) = r\}$ (Exercise 16-48b).

2. The boundary of the rational numbers in the real numbers is $\delta\mathbb{Q} = \mathbb{R}$. □

Further properties of the interior, the closure, and the boundary of a set will be investigated in Exercises 16-43–16-47.

16.4.2 Relatively Open Sets

When subspaces S of a metric space X are investigated, subsets of S can be open in the subspace S, as well as in the space X itself. However, it is important to realize that a set $U \subset S$ can be open in S and still it may not be open in X. For example, the interval $[0, 1)$ is an *open* subset of the metric space $[0, 2]$. Indeed, for every $x \in [0, 1)$ there is a small $\varepsilon > 0$ so that the ball *in [0,2]* around x of radius ε is contained in $[0, 1)$. Proposition 16.56 describes the relation between open sets in the space X and open sets in a subset S. Open sets in a subset are also called **relatively open**.

Proposition 16.56 *Let X be a metric space and let $S \subseteq X$ be a subset. Then $U \subseteq S$ is open with respect to the induced metric on S iff there is a set $V \subseteq X$ that is open with respect to the metric on X and such that $U = V \cap S$. (Also see Figure 35.)*

Proof. To prove "⇒," for each element $u \in U$ find $\varepsilon_u > 0$ so that the containment $B^S_{\varepsilon_u}(u) := \{x \in S : d(x,u) < \varepsilon_u\} \subseteq U$ holds. Let

$$V := \bigcup_{u \in U} B^X_{\varepsilon_u}(u) = \bigcup_{u \in U} \{x \in X : d(x,u) < \varepsilon_u\}.$$

Then V is open and we claim that $U = S \cap V$. Clearly, $U \subseteq S \cap V$. Now let $v \in S \cap V$. Then there is a $u \in U$ so that $v \in B^X_{\varepsilon_u}(u)$. Because $v \in S$ we infer $v \in B^S_{\varepsilon_u}(u) \subseteq U$, and hence $S \cap V \subseteq U$.

The part "$\Leftarrow$" is left as Exercise 16-38. ■

A similar result is proved for closed sets in Exercise 16-49.

Exercises

16-37 Prove Corollary 16.47.

16-38 Finish the proof of Proposition 16.56 by proving the "$\Leftarrow$" part.

16-39 Let X be a metric space. Prove that for all $x \in X$ the set $X \setminus \{x\}$ is open.

16-40. Let X be a metric space. Prove that $U \subseteq X$ is open iff U is a union of open balls $B_{\varepsilon_x}(x)$.

16-41. Closed sets. Let X be a metric space.

(a) Prove that both $\emptyset$ and X are closed subsets of X.

(b) Prove that if $C_1, \ldots, C_n$ are closed subsets of X, then $\bigcup_{k=1}^{n} C_k$ is closed.

(c) Prove that if $\mathcal{C}$ is a family of closed subsets of X then $\bigcap \mathcal{C}$ is closed.

(d) Give an example of an infinite union of closed sets that is not closed.

16-42. Let $C \subseteq \mathbb{R}^d$ be so that for all $n \in \mathbb{N}$ the set $C \cap \prod_{i=1}^{d} [-n, n]$ is closed. Prove that C is closed.

16-43 The interior of a set. Let X be a metric space.

(a) Prove that $A^{\circ\circ} = A^\circ$ for all $A \subseteq X$.

(b) Prove that if $A_1, \ldots, A_n \subseteq X$, then $\left(\bigcap_{j=1}^{n} A_j\right)^\circ = \bigcap_{j=1}^{n} A_j^\circ$.

(c) Prove that if $A_j \subseteq X$ for all $j \in \mathbb{N}$, then $\left(\bigcap_{j=1}^{\infty} A_j\right)^\circ \subseteq \bigcap_{j=1}^{\infty} A_j^\circ$ and give an example that shows that the containment can be proper.

(d) Prove that $U \subseteq X$ is open iff $U^\circ = U$.

16-44 The closure of a set. Let X be a metric space.

(a) Prove that $A^{--} = A^-$ for all $A \subseteq X$.

(b) Prove that if $A_1, \ldots, A_n \subseteq X$, then $\overline{\bigcup_{j=1}^{n} A_j} = \bigcup_{j=1}^{n} \overline{A_j}$.

(c) Prove that if $A_j \subseteq X$ for all $j \in \mathbb{N}$, then $\overline{\bigcup_{j=1}^{\infty} A_j} \supseteq \bigcup_{j=1}^{\infty} \overline{A_j}$ and give an example that shows that the containment can be proper.

(d) Prove that $C \subseteq X$ is closed iff $C^- = C$.

16-45. The boundary of a set. Let X be a metric space and let $A \subseteq X$.

(a) Prove that the boundary δA of A is closed.

(b) Find a set B in a metric space X so that B, δB and $\delta(\delta B)$ are three distinct sets.

(c) Prove that if A is closed, then $\overline{(X \setminus A) \cup A^\circ} = X$.

(d) Prove that for any set A we have $\delta(\delta A) = \delta\left(\delta(\delta A)\right)$.

16-46. Closure, interior and the boundary. Let X be a metric space and let $A \subseteq X$

(a) Prove that $X \setminus \left(A^{\circ}\right) = (X \setminus A)^{-}$.

(b) Prove that $X \setminus \left(A^{-}\right) = (X \setminus A)^{\circ}$

(c) Prove that $\delta A = A^{-} \cap (X \setminus A)^{-}$.

(d) Prove that $A^{-\circ-} \subseteq A^{-}$ and show that the containment can be proper

(e) Prove that $A^{\circ-\circ} \supseteq A^{\circ}$ and show that the containment can be proper.

16-47 Let X be a normed space and let $Y \subseteq X$ be a normed subspace. Prove that Y^{-} also is a normed subspace of X

16-48 Let X be a metric space, let $x \in X$ and let $r > 0$.

(a) Prove that $\overline{B_r(x)} = \left\{ p \in X : d(p,x) \leq r \right\}$.

(b) Prove that $\delta B_r(x) = \left\{ p \in X : d(p,x) = r \right\}$.

16-49. Let X be a metric space and let $S \subseteq X$ be a subset. Prove that $C \subseteq S$ is closed with respect to the induced metric on S iff there is a set $D \subseteq X$ that is closed with respect to the metric on X and such that $C = D \cap S$.

Hint Use Theorem 16.45. For "$\Rightarrow$," let D be the set of all limits of sequences in C

16-50 Let X, Y be metric spaces and let $f : X \to Y$. Prove that f is continuous at $x \in X$ iff for all open subsets $V \subseteq Y$ that contain $f(x)$ the inverse image $f^{-1}[V] \subseteq X$ contains an open ball around x.

16-51. Let x be a point in a metric space X. Can a neighborhood of x be closed? Explain.

16-52. Let X be a normed space and let $\Omega \subseteq X$ be an open subset. Prove that every $x \in \Omega$ is an accumulation point of Ω.

16-53. Let X, Y be metric spaces and let $f : X \to Y$ be a function. Define the **oscillation** of f over the open set $U \subseteq X$ as $\omega_f(U) := \sup \left\{ d\left(f(y), f(z)\right) : y, z \in U \right\}$ Define the **oscillation** of f at $x \in X$ as $\omega_f(x) := \inf \left\{ \omega_f(U) : x \in U, U \text{ open} \right\}$.

(a) Prove that f is continuous at x iff $\omega_f(x) = 0$.

(b) Prove that for all $p \geq 0$, the set $\left\{ x \in X : \omega_f(x) \geq p \right\}$ is closed.

(c) Prove that for all $p \geq 0$, the set $\left\{ x \in X : \omega_f(x) < p \right\}$ is open.

(d) Prove that $\omega_f(x) = \lim_{n\to\infty} \omega_f\left(B_{\frac{1}{n}}(x)\right)$.

16-54 Prove that Cantor sets are closed.

16.5 Compactness

The Bolzano-Weierstrass Theorem plays a key role in the proofs of several important results for functions of a single variable. For example, it is used to prove that a continuous function $f : [a,b] \to \mathbb{R}$ always assumes an absolute maximum value (see Theorem 3.44), as well as to show that continuous functions $f : [a,b] \to \mathbb{R}$ are uniformly continuous (see Lemma 5.19). It therefore is natural to investigate spaces that satisfy the conclusion of the Bolzano-Weierstrass Theorem.

Definition 16.57 Bolzano-Weierstrass formulation of compactness. *A metric space X is called* **compact** *iff every sequence $\{x_n\}_{n=1}^{\infty}$ of elements in X has a convergent subsequence.*

Compactness is usually formulated in topological terms. We will investigate this formulation, which is reminiscent of the Heine-Borel Theorem (see Theorem 8.4), in Theorem 16.72. In metric spaces, the Bolzano-Weierstrass formulation of compactness is equivalent to the topological formulation, but we need to be careful. In general topological spaces, the Bolzano-Weierstrass formulation of compactness is called **sequential compactness**. It is a consequence of (topological) compactness, but it is not equivalent to it.

Closed and bounded subsets of finite dimensional spaces are the prototypical examples of compact sets (also see Theorem 16.80).

Example 16.58 *Let $r > 0$ and let $\overline{C_r(0)} := \left\{x \in \mathbb{R}^d : \|x\|_\infty \leq r\right\}$. Then $\overline{C_r(0)}$ is compact when equipped with the metric induced by the $\|\cdot\|_\infty$-norm.*

A typical compactness proof with the Bolzano-Weierstrass formulation will take a sequence in the space and produce a convergent subsequence.

Let $\{x_n\}_{n=1}^\infty$ be a sequence in $\overline{C_r(0)}$. Then each component sequence $\left\{x_n^{(i)}\right\}_{n=1}^\infty$ is bounded. In particular, $\left\{x_n^{(1)}\right\}_{n=1}^\infty$ has a convergent subsequence $\left\{x_{n_k^1}^{(1)}\right\}_{k=1}^\infty$. Now let $1 \leq j < d$ and assume $\left\{n_m^j\right\}_{m=1}^\infty$ is a strictly increasing sequence of integers such that for $1 \leq i \leq j$ the subsequences $\left\{x_{n_m^j}^{(i)}\right\}_{m=1}^\infty$ converge. Then $\left\{x_{n_m^j}^{(j+1)}\right\}_{m=1}^\infty$ has a convergent subsequence. That is, there is a strictly increasing sequence of integers $\left\{n_l^{j+1}\right\}_{l=1}^\infty$ such that for all $1 \leq i \leq j+1$ the subsequences $\left\{x_{n_l^{j+1}}^{(i)}\right\}_{l=1}^\infty$ converge. Inductively we conclude that there is a subsequence $\left\{x_{n_k^d}\right\}_{k=1}^\infty$ such that for $1 \leq i \leq d$ the subsequences $\left\{x_{n_k^d}^{(i)}\right\}_{k=1}^\infty$ converge. Call each limit $x^{(i)}$ and let $x := \sum_{i=1}^d x^{(i)} e_i$. Then by Theorem 16.4 $\left\{x_{n_k^d}\right\}_{k=1}^\infty$ is a convergent subsequence of $\{x_n\}_{n=1}^\infty$ with limit x. Moreover, $x \in \overline{C_r(0)}$, because $\overline{C_r(0)}$ is closed. Since $\{x_n\}_{n=1}^\infty$ was arbitrary this means that $\overline{C_r(0)}$ is compact. □

Compact subspaces of $\mathbb{R}^d$ will be investigated in detail in Section 16.6. As Example 16.58 indicates, compact spaces are usually subsets of larger metric spaces.

Definition 16.59 *Let X be a metric space. A subset $C \subseteq X$ is called* **compact** *iff C with the induced metric is a compact metric space. Equivalently, C is compact iff every sequence in C has a convergent subsequence* whose limit is in C.

Compact subsets of metric spaces are closed and bounded

Proposition 16.60 *Let X be a metric space and let C be a compact subset of X. Then C is closed and bounded.*

Proof. Let $C \subseteq X$ be compact. To prove that C is closed, suppose for a contradiction that C is not closed. Then there is a sequence $\{x_n\}_{n=1}^{\infty}$ of elements of C that converges in X to a limit $x \notin C$. But then by Proposition 16.11 all subsequences of $\{x_n\}_{n=1}^{\infty}$ converge (in X) to x, and hence no subsequence of $\{x_n\}_{n=1}^{\infty}$ has a limit in C, a contradiction.

To prove that C is bounded, suppose for a contradiction that C is not bounded. Let $x_1 \in C$. Once $x_1, \ldots, x_n \in C$ have been chosen, we can find an $x_{n+1} \in C$ so that $d(x_{n+1}, x_n) \geq (n+1) + \max\left\{d(x_k, x_n) : k = 1, \ldots, n-1\right\}$. But then the inequality $d(x_{n+1}, x_k) \geq d(x_{n+1}, x_n) - d(x_n, x_k) \geq n+1$ holds for all $k = 1, \ldots, n$, and hence all subsequences of the inductively constructed sequence $\{x_n\}_{n=1}^{\infty}$ are unbounded. Now by Proposition 16.9 no subsequence of $\{x_n\}_{n=1}^{\infty}$ has a limit in C (or even in X), a contradiction. ■

Exercise 16-55 shows that the converse of Proposition 16.60 does not hold in general. Next, we note that the closed subsets of a compact space inherit the compactness.

Proposition 16.61 *Let X be a compact metric space and let $C \subseteq X$ be closed. Then C is compact.*

Proof. Let $\{a_n\}_{n=1}^{\infty}$ be a sequence of elements of C. Because $C \subseteq X$ there is a subsequence $\left\{a_{n_k}\right\}_{k=1}^{\infty}$ that converges in X to a limit L. But then by Theorem 16.45 $L \in C$, and hence C is compact. ■

The general version of Theorem 3.44 is now the following.

Theorem 16.62 *Let X be a compact metric space, let Y be a metric space and let $f : X \to Y$ be continuous and surjective. Then Y is compact.*

Proof. Let $\{y_n\}_{n=1}^{\infty}$ be a sequence in Y. For each $n \in \mathbb{N}$ let $x_n \in X$ be such that $f(x_n) = y_n$. Because X is compact, there is a subsequence $\left\{x_{n_k}\right\}_{k=1}^{\infty}$ of $\{x_n\}_{n=1}^{\infty}$ and an $x \in X$ such that $\lim_{k\to\infty} x_{n_k} = x$. But then $y = f(x) \in Y$, $\left\{y_{n_k}\right\}_{k=1}^{\infty}$ is a subsequence of $\{y_n\}_{n=1}^{\infty}$ and $\lim_{k\to\infty} y_{n_k} = \lim_{k\to\infty} f(x_{n_k}) = f(x) = y$. Therefore Y is compact. ■

Corollary 16.63 *(Compare with Theorem 3.44.) Let X be a compact metric space and let $f : X \to \mathbb{R}$ be continuous. Then f assumes its* **absolute maximum** *on X. That is, there is an $x \in X$ so that $f(x) \geq f(z)$ for all $z \in X$.*

Proof. The function f is surjective onto $f[X] \subseteq \mathbb{R}$. By Theorem 16.62 this means $f[X]$ is compact and by Proposition 16.60 this means that $f[X]$ is closed and bounded. Then $M := \sup\left(f[X]\right)$ is an element of $f[X]$. Therefore there is an $x \in X$ so that $f(x) = M$, and M is greater than or equal to all other values f assumes. ■

It is also easy to see that the inverses of continuous functions on compact metric spaces are continuous.

Theorem 16.64 *(Compare with Theorem 3.38.) Let X be a compact metric space, let Y be a metric space and let the function $f : X \to Y$ be continuous and injective. Then the inverse function $f^{-1} : f[X] \to X$ is continuous, too.*

Proof (sketch). Let $\{y_n\}_{n=1}^{\infty}$ be a sequence in Y that converges to $y \in f[X]$. Prove that every subsequence of $\left\{f^{-1}(y_n)\right\}_{n=1}^{\infty}$ has a subsequence that converges to $f^{-1}(y)$ and use Exercise 16-8.

The full proof is left to the reader as Exercise 16-56. ■

An important metric property of compact spaces is that they are complete.

Theorem 16.65 *Let X be a compact metric space. Then X is complete.*

Proof. Exercise 16-57. ■

Another important consequence of compactness is that continuous functions are uniformly continuous on compact metric spaces.

Definition 16.66 *Let X, Y be metric spaces. Then the function $f : X \to Y$ is called* **uniformly continuous** *iff for every $\varepsilon > 0$ there is a $\delta > 0$ such that for all $u, v \in X$ with $d(u, v) < \delta$ we have that $d\big(f(u), f(v)\big) < \varepsilon$.*

Lemma 16.67 *Let X be a compact metric space, let Y be a metric space and let the function $f : X \to Y$ be continuous. Then f is uniformly continuous.*

Proof. Mimic the proof of Lemma 5.19. (Exercise 16-58.) ■

Compactness typically is not formulated in terms of the Bolzano-Weierstrass Theorem, but in terms similar to the Heine-Borel Theorem (see Theorem 8.4). For metric spaces, both formulations are equivalent. Because the Heine-Borel formulation is exclusively in terms of open sets it is the topological (and thus more general) description of compactness. Recall that the Heine-Borel Theorem said that each open cover of a closed and bounded interval has a finite subcover.

Definition 16.68 *A* **cover** *of a metric space X is a family $\mathcal{C}$ of sets such that $X \subseteq \bigcup \mathcal{C}$. An* **open cover** *$\mathcal{C}$ is a cover such that all sets in $\mathcal{C}$ are open.*

For a subset S of a metric space X, it is usually more natural to cover S with sets that are open in X, even if this means that the sets are not contained in S. Hence, if $S \subseteq X$, we will also call a family $\mathcal{C}$ an **open cover** *of S iff all sets in $\mathcal{C}$ are open (in X) and $S \subseteq \bigcup \mathcal{C}$.*

For a visualization of open covers, consider Figure 36.

Example 16.69 *Open covers.*

1. $\left\{B_n(0) : n \in \mathbb{N}\right\}$ *is an open cover of* $\mathbb{R}^d$.
2. $\left\{B_{\frac{1-\|x\|}{2}}(0) : x \in B_1(0)\right\}$ *is an open cover of* $B_1(0) \subseteq \mathbb{R}^d$.
3. *For any $a, b \in (0, 1)$, the set* $\left\{\left(\frac{1}{n}, 1 - \frac{1}{n}\right) : n \in \mathbb{N}\right\} \cup \left\{[0, a), (b, 1]\right\}$ *is an open cover of the interval* $[0, 1]$.

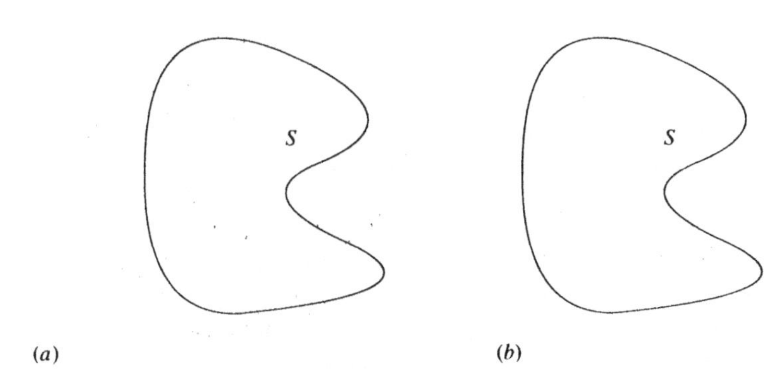

Figure 36: Because compact sets are usually subsets of other spaces, open covers are typically visualized with sets that are open in the surrounding space (a) rather than with relatively open sets (b).

Moreover, the set $\left\{\left(\frac{1}{n}, 1-\frac{1}{n}\right) : n \in \mathbb{N}\right\} \cup \left\{(-1, a), (b, 2)\right\}$ *is an open cover of the interval* $[0, 1]$ *if we consider* $[0, 1]$ *as a subspace of* $\mathbb{R}$.

We have encountered the proof technique of finding finite subcovers of open covers in Proposition 8.5, Exercise 8-3c and Lemma 8.11. Definition 14.17 of the d-dimensional outer Lebesgue measure also relies on open covers and the remarks after the proof of Proposition 14.60 suggest that we need a version of the Heine-Borel Theorem to prove that d-dimensional Lebesgue measure is a product of lower dimensional Lebesgue measures. Indeed, in this text, the main use of finite subcovers of open covers is in connecting topology with measure theory. Theorem 16.72 below shows that compactness provides an abstract version of the Heine-Borel Theorem.

Lemma 16.70 *Let* X *be a compact metric space. Then for every* $\varepsilon > 0$ *there are* $x_1, \ldots, x_n \in X$ *so that* $X \subseteq \bigcup_{j=1}^{n} B_\varepsilon(x_j)$.

Proof. Let $\varepsilon > 0$. Let $x_1 \in X$ be arbitrary. If $X \subseteq B_\varepsilon(x_1)$ stop. Otherwise continue as follows. If $x_1, \ldots, x_{n-1} \in X$ are so that $X \not\subseteq \bigcup_{j=1}^{n-1} B_\varepsilon(x_j)$, choose $x_n \notin \bigcup_{j=1}^{n-1} B_\varepsilon(x_j)$. If $X \subseteq \bigcup_{j=1}^{n} B_\varepsilon(x_j)$, stop, otherwise continue. This process cannot continue indefinitely, because if it did, $\{x_n\}_{n=1}^{\infty}$ would be a sequence such that for any distinct $m, n \in \mathbb{N}$ we have $d(x_m, x_n) \geq \varepsilon$. This would mean that $\{x_n\}_{n=1}^{\infty}$ has no convergent subsequence, a contradiction to the compactness of X. The $x_1, \ldots, x_n$ for which the construction stops are as desired. ∎

Definition 16.71 *Let X be a metric space and let $\mathcal{O}$ be an open cover. A finite subset $\{O_1, \ldots, O_n\} \subseteq \mathcal{O}$ so that $X \subseteq \bigcup_{j=1}^{n} O_j$ is also called a* **finite subcover**.

Theorem 16.72 Heine-Borel formulation of compactness. *A metric space X is* **compact** *iff every open cover $\mathcal{O}$ of X has a finite subcover.*

Proof. For "$\Leftarrow$," we will prove the contrapositive. So assume that X is not compact. Let $\{x_n\}_{n=1}^{\infty}$ be a sequence that does not have a convergent subsequence. Then (Exercise 16-59) for every $x \in X$ there is an $\varepsilon_x > 0$ so that $\left\{n \in \mathbb{N} : x_n \in B_{\varepsilon_x}(x)\right\}$ is finite. But then $C = \left\{B_{\varepsilon_x}(x) : x \in X\right\}$ is an open cover of X that cannot have a finite subcover.

For "$\Rightarrow$," let X be compact and let $\mathcal{O}$ be an open cover. We first prove that there is an $\varepsilon > 0$ so that for every $x \in X$ there is an $O \in \mathcal{O}$ so that $B_\varepsilon(x) \subseteq O$. For a contradiction, suppose that this is not the case. Then for each $n \in \mathbb{N}$ there is an $x_n \in X$ such that $B_{\frac{1}{n}}(x_n)$ is not contained in any set $O \in \mathcal{O}$. Because X is compact, $\{x_n\}_{n=1}^{\infty}$ has a convergent subsequence $\left\{x_{n_k}\right\}_{k=1}^{\infty}$. Let $x := \lim_{k\to\infty} x_{n_k}$. Then there is an $O \in \mathcal{O}$ so that $x \in O$. Moreover, there is an $\varepsilon > 0$ so that $B_\varepsilon(x) \subseteq O$. Now let $k \in \mathbb{N}$ be such that $\frac{1}{n_k} < \frac{\varepsilon}{2}$ and such that $d\left(x_{n_k}, x\right) < \frac{1}{n_k}$. Then for all $y \in B_{\frac{1}{n_k}}(x_{n_k})$ we have that $d(y, x) \leq d(y, x_{n_k}) + d(x_{n_k}, x) < \frac{1}{n_k} + \frac{1}{n_k} < \frac{\varepsilon}{2} + \frac{\varepsilon}{2} = \varepsilon$. Consequently, the containments $B_{\frac{1}{n_k}}(x_{n_k}) \subseteq B_\varepsilon(x) \subseteq O$ provide the desired contradiction.

Now let $\varepsilon > 0$ be so that for every $x \in X$ there is an $O \in \mathcal{O}$ so that $B_\varepsilon(x) \subseteq O$. By Lemma 16.70, there are finitely many $x_1, \ldots, x_n \in X$ so that $X \subseteq \bigcup_{j=1}^{n} B_\varepsilon(x_j)$. For each $j = 1, \ldots, n$, let $O_j \in \mathcal{O}$ be such that $B_\varepsilon(x_j) \subseteq O_j$. Then $X \subseteq \bigcup_{j=1}^{n} B_\varepsilon(x_j) \subseteq \bigcup_{j=1}^{n} O_j$ and $\{O_j\}_{j=1}^{n}$ is the desired finite subcover of $\mathcal{O}$. ■

Typically, when we invoke compactness we obtain a finite subcover of a cover with sets that are open in a surrounding space. It is not necessary to explicitly prove that a subset S of a metric space is compact iff every open cover $\mathcal{O}$ (with sets that are open in the surrounding space) has a finite subcover. The translation is easily made by finding a finite subcover $\{O_1 \cap S, \ldots, O_n \cap S\}$ of the corresponding cover $\widetilde{\mathcal{O}} := \{O \cap S : O \in \mathcal{O}\}$ with relatively open sets and then going back to the sets $\{O_1, \ldots, O_n\}$ that are open in X (also see Figure 36).

Exercises

16-55. Prove that $B_1 := \left\{x \in l^\infty : \|x\|_\infty \leq 1\right\}$ is a closed and bounded subset of l^∞ that is not compact

16-56. Prove Theorem 16.64

16-57. Prove Theorem 16.65
Hint Exercise 16-16

16-58 Prove Lemma 16.67

16-59. Prove that if X is a metric space, $\{x_n\}_{n=1}^{\infty}$ is a sequence and $x \in X$ is such that for all $\varepsilon > 0$ the set $\{ n \in \mathbb{N} : x_n \in B_\varepsilon(x) \}$ is infinite, then $\{x_n\}_{n=1}^{\infty}$ has a convergent subsequence.

16-60. Let X be a metric space. Prove that if $C \subseteq X$ is not closed, then there exists an open cover without a finite subcover.

16-61 Let X be a metric space. Prove that if $C \subseteq X$ is not bounded, then there exists an open cover without a finite subcover

16-62. Let X be a metric space, let $K \subseteq X$ be compact, and let $O \subseteq X$ be open such that $K \subseteq O$. Prove that there is an $\varepsilon > 0$ such that for all $x \in K$ we have $B_\varepsilon(x) \subseteq O$.

16-63. Let X be a metric space such that all closed and bounded subsets are compact and let $\{a_n\}_{n=1}^{\infty}$ be a sequence. Prove that $\{a_n\}_{n=1}^{\infty}$ diverges if and only if there is a subsequence $\{a_{n_k}\}_{k=1}^{\infty}$ that is unbounded, *or* there are two subsequences $\{a_{n_k}\}_{k=1}^{\infty}$ and $\{a_{l_m}\}_{m=1}^{\infty}$ such that $\lim_{m\to\infty} a_{l_m}$ and $\lim_{k\to\infty} a_{n_k}$ both exist, but are not equal.

16-64 More on Lemma 16.70

(a) Prove that the conclusion of Lemma 16.70 is not equivalent to compactness by showing that the open interval $(0, 1)$ satisfies the conclusion of Lemma 16.70.

(b) Prove that a metric space X is compact iff

i. X is complete and

ii. For every $\varepsilon > 0$, there are $x_1, \ldots, x_n \in X$ so that $X \subseteq \bigcup_{j=1}^{n} B_\varepsilon(x_j)$

Hint. We only need to prove "$\Leftarrow$." For this direction, let $\{y_j\}_{j=1}^{\infty}$ be a sequence in X. Construct a Cauchy sequence of points $\{z_k\}_{k=1}^{\infty}$ in X so that $\{ j \in \mathbb{N} \cdot y_j \in B_{\frac{1}{k}}(z_k) \}$ is infinite You will need to take a subsequence of a subsequence when constructing z_{k+1} after obtaining $z_1, \ldots, z_k$.

16-65. **Continuous functions need not be bounded on noncompact closed and bounded sets.** For each $n \in \mathbb{N}$ define the function $f : l^{\infty} \to \mathbb{R}$ to be $f(x) = n\left(1 - \frac{1}{100}\left(x^{(n)} - 1\right)^2\right)$ on the ball $B_{\frac{1}{10}}(e_n)$ around the nth unit vector and zero on $l^{\infty} \setminus \bigcup_{n=1}^{\infty} B_{\frac{1}{10}}(e_n)$ Prove that f is continuous on l^{∞} and unbounded on $\overline{B_1(0)} \subseteq l^{\infty}$.

16-66 The tangent function can be inverted on the interval $\left(-\frac{\pi}{2}, \frac{\pi}{2}\right)$. Its inverse is the arctangent function $\arctan : \mathbb{R} \to \left(-\frac{\pi}{2}, \frac{\pi}{2}\right)$. Extend $\arctan(\cdot)$ to the interval $[-\infty, \infty]$ by defining $\arctan(\infty) := \frac{\pi}{2}$ and $\arctan(-\infty) := -\frac{\pi}{2}$. For $x, y \in [-\infty, \infty]$ define $d_c(x, y) := \left|\arctan(x) - \arctan(y)\right|$

(a) Prove that $\left([-\infty, \infty], d_c\right)$ is a compact metric space

(b) Prove that if $\{x_n\}_{n=1}^{\infty}$ is a sequence of real numbers and $x \in \mathbb{R}$, then $\lim_{n\to\infty} d_c(x_n, x) = 0$ iff $\lim_{n\to\infty} |x_n - x| = 0$.

(c) Prove that if $\{x_n\}_{n=1}^{\infty}$ is a sequence of real numbers, then we have $\lim_{n\to\infty} d_c(x_n, \infty) = 0$ iff $\lim_{n\to\infty} x_n = \infty$ in the sense of Definition 2.42

16-67. Prove that if $f : [c, d] \times [a, b] \to \mathbb{R}$ is continuous, then the function $g : [a, b] \to \mathbb{R}$ defined by $g(t) := \int_c^d f(x, t)\, dx$ is continuous.

Hint Use the uniform continuity of f

16-68 Lemmas for the Stone-Weierstrass Theorem.

(a) **Dini's Theorem**. Let $\{f_n\}_{n=1}^{\infty}$ be a nondecreasing sequence of continuous real-valued functions on $[a, b]$ that converges pointwise to the continuous function $f : [a, b] \to \mathbb{R}$. Prove that $\{f_n\}_{n=1}^{\infty}$ converges uniformly to f.
Hint. Cover the interval with open intervals $(x - \delta_x, x + \delta_x)$ on which f is close to a function f_{n_x}. Then take a finite subcover.

(b) Prove that the sequence $\{P_n\}_{n=0}^{\infty}$ of polynomials defined recursively for all $x \in [0, 1]$ by $P_0(x) := 0$ and $P_{n+1}(x) := P_n(x) + \frac{1}{2}\left(x - P_n^2(x)\right)$ converges uniformly on $[0, 1]$ to the function $f(x) = \sqrt{x}$.
Hint. Same approach as in Exercise 2-42.

(c) Prove that for any $M > 0$ there is a sequence of polynomials that converges uniformly to $f(x) = \sqrt{x}$ on $[0, M]$.

16.6 The Normed Topology of $\mathbb{R}^d$

So far, on d-dimensional space we have only worked with the uniform norm $\|\cdot\|_\infty$. This norm is easy to work with, but it does not measure the usual Euclidean distance. Therefore we will now investigate how norms in d-dimensional space relate to each other. It turns out that all norms on finite dimensional spaces are equivalent and that they induce the same notion of convergence. This means there is no loss of generality in working with the uniform norm on a finite dimensional space.

Lemma 16.73 *Let $\|\cdot\|$ be a norm on $\mathbb{R}^d$. Then for all $x = (x_1, \ldots, x_d) \in \mathbb{R}^d$ the inequality $\|x\| \leq \|x\|_\infty \sum_{i=1}^{d} \|e_i\|$ holds, where e_i denotes the ith unit vector in $\mathbb{R}^d$.*

Proof. Let $x = (x_1, \ldots, x_d) \in \mathbb{R}^d$. Then

$$\|x\| = \left\|\sum_{i=1}^{d} x_i e_i\right\| \leq \sum_{i=1}^{d} |x_i|\,\|e_i\| \leq \sum_{i=1}^{d} \|x\|_\infty \|e_i\| = \|x\|_\infty \sum_{i=1}^{d} \|e_i\|. \qquad \blacksquare$$

Theorem 16.74 *If both $\|\cdot\|_1$ and $\|\cdot\|_2$ are norms on $\mathbb{R}^d$, then there are real numbers $c, C > 0$ such that for all $x \in \mathbb{R}^d$ the inequalities $c\|x\|_1 \leq \|x\|_2 \leq C\|x\|_1$ hold.*

Proof. Let $\|\cdot\|$ be an arbitrary norm on $\mathbb{R}^d$. By Lemma 16.73, for all points $x, y \in \mathbb{R}^d$ we infer $\big|\|x\| - \|y\|\big| \leq \|x - y\| \leq \|x - y\|_\infty \sum_{i=1}^{d} \|e_i\|$. Thus $\|\cdot\|$ is continuous with respect to $\|\cdot\|_\infty$. By Example 16.58, Proposition 16.61, and Corollary 16.63 we conclude that the norm $\|\cdot\|$ assumes an absolute minimum and an absolute maximum on the compact set $B := \left\{y \in \mathbb{R}^d : \|y\|_\infty = 1\right\}$. Moreover, the absolute minimum cannot be zero, because $\|x\| = 0$ implies $x = 0$ and $0 \notin B$.

The result will be proved if we can show that for any norms $\|\cdot\|_1$ and $\|\cdot\|_2$ on $\mathbb{R}^d$ there is a $C > 0$ such that for all $x \in \mathbb{R}^d \setminus \{0\}$ we have that $\|x\|_2 \leq C\|x\|_1$. Let $M := \max\left\{\|y\|_2 : y \in B\right\}$ and $m := \min\left\{\|y\|_1 : y \in B\right\}$, and let $x \in \mathbb{R}^d \setminus \{0\}$

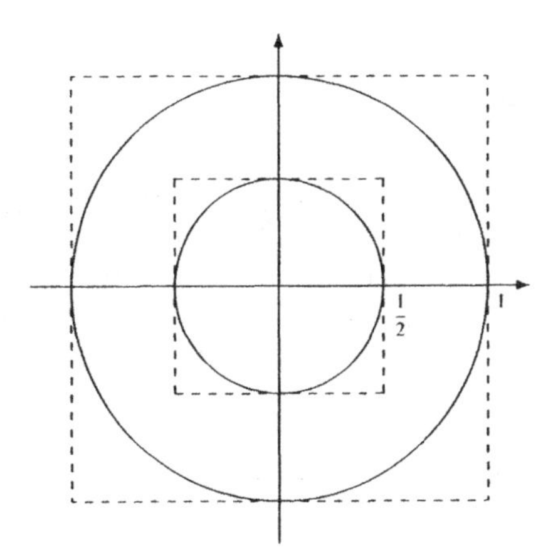

Figure 37: Geometrically, Theorem 16.76 says that on a finite dimensional space inside any ball with respect to one norm we can find a ball with respect to any other norm and with the same center. Moreover, this smaller ball contains a ball with respect to the first norm with the same center. The figure shows this nesting for balls with respect to the three most common norms on $\mathbb{R}^2$, the uniform norm $\|\cdot\|_\infty$ (dashed), the Euclidean norm $\|\cdot\|_2$ (solid), and the taxicab norm $\|\cdot\|_1$ (dotted).

be arbitrary. Then $\|x\|_2 = \frac{\|x\|_2}{\|x\|_\infty}\frac{\|x\|_\infty}{\|x\|_1}\|x\|_1 = \left\|\frac{x}{\|x\|_\infty}\right\|_2 \frac{1}{\left\|\frac{x}{\|x\|_\infty}\right\|_1}\|x\|_1 \leq \frac{M}{m}\|x\|_1$, which establishes the result. ■

Norms that satisfy the conclusion of Theorem 16.74 are also called equivalent. Theorem 16.76 below shows that any two norms on a finite dimensional vector space are equivalent. Figure 37 provides a visualization of equivalence for norms.

Definition 16.75 *Let X be a vector space and let $\|\cdot\|_1$ and $\|\cdot\|_2$ be norms on X. Then $\|\cdot\|_1$ and $\|\cdot\|_2$ are called* **equivalent** *iff there are real numbers $c, C > 0$ such that for all $x \in X$ we have $c\|x\|_1 \leq \|x\|_2 \leq C\|x\|_1$.*

Theorem 16.76 *Let X be a finite dimensional vector space. Then all norms on X are equivalent.*

Proof. Let $\|\cdot\|_1$ and $\|\cdot\|_2$ be two norms on X. Let $\{b_1, \ldots, b_d\}$ be a base of X and let $\Phi : X \to \mathbb{R}^d$ be the isomorphism that maps each b_i to e_i. For $k = 1, 2$, define $\left\|\sum_{i=1}^d x^{(i)}e_i\right\|_{k,\mathbb{R}^d} := \left\|\sum_{i=1}^d x^{(i)}b_i\right\|_k$. Both $\|\cdot\|_{1,\mathbb{R}^d}$ and $\|\cdot\|_{2,\mathbb{R}^d}$ are norms on $\mathbb{R}^d$. Hence, there are $c, C > 0$ so that for all $x \in \mathbb{R}^d$ we have $c\|x\|_{1,\mathbb{R}^d} \leq \|x\|_{2,\mathbb{R}^d} \leq C\|x\|_{1,\mathbb{R}^d}$.

But then for all $x := \sum_{i=1}^{d} x^{(i)} b_i \in X$ we infer

$$\begin{aligned}
c\|x\|_1 &= c\left\|\sum_{i=1}^{d} x^{(i)} b_i\right\|_1 = c\left\|\sum_{i=1}^{d} x^{(i)} e_i\right\|_{1,\mathbb{R}^d} \leq \left\|\sum_{i=1}^{d} x^{(i)} e_i\right\|_{2,\mathbb{R}^d} \\
&= \left\|\sum_{i=1}^{d} x^{(i)} b_i\right\|_2 = \|x\|_2 = \left\|\sum_{i=1}^{d} x^{(i)} b_i\right\|_2 = \left\|\sum_{i=1}^{d} x^{(i)} e_i\right\|_{2,\mathbb{R}^d} \\
&\leq C\left\|\sum_{i=1}^{d} x^{(i)} e_i\right\|_{1,\mathbb{R}^d} = C\left\|\sum_{i=1}^{d} x^{(i)} b_i\right\|_1 = C\|x\|_1.
\end{aligned}$$

■

Equivalent norms induce the same notion of convergence.

Proposition 16.77 *Let X be a vector space, let $\|\cdot\|_1$ and $\|\cdot\|_2$ be two equivalent norms on X, let $\{x_n\}_{n=1}^{\infty}$ be a sequence in X and let $x \in X$. Then $\lim_{n\to\infty} x_n = x$ in $(X, \|\cdot\|_1)$ iff $\lim_{n\to\infty} x_n = x$ in $(X, \|\cdot\|_2)$.*

Proof. Exercise 16-69. ■

Because equivalent norms induce the same notion of convergence, sequences in d-dimensional space converge iff their component sequences converge.

Theorem 16.78 *Let X be a finite dimensional normed space and let $\{b_1, \ldots, b_d\}$ be a base of X. For each element x in X, let $x^{(1)}, \ldots, x^{(d)}$ be the components such that $x = \sum_{i=1}^{d} x^{(i)} b_i$. Then a sequence $\{x_n\}_{n=1}^{\infty}$ converges to L in X iff all component sequences $\left\{x_n^{(i)}\right\}_{n=1}^{\infty}$ converge to $L^{(i)}$.*

Proof. Use Theorems 16.4, 16.76, and Proposition 16.77. (Exercise 16-70.) ■

In particular, we obtain that all finite dimensional normed spaces are complete.

Theorem 16.79 *Let X be a finite dimensional normed space. Then X is complete.*

Proof. Exercise 16-71. ■

Moreover, in finite dimensional spaces compactness is equivalent to being closed and bounded.

Theorem 16.80 *A subset C of a finite dimensional normed space X is compact iff it is closed and bounded.*

Proof. The direction "$\Rightarrow$" follows from Proposition 16.60.

For "$\Leftarrow$," let $\{b_1, \ldots, b_d\}$ be a base of X and for each $x \in X$ let $x^{(1)}, \ldots, x^{(d)}$ be the components so that $x = \sum_{i=1}^{d} x^{(i)} b_i$. Let $C \subseteq X$ be closed and bounded and let $\{x_n\}_{n=1}^{\infty}$ be a sequence in C. Because the component sequence $\left\{x_n^{(1)}\right\}_{n=1}^{\infty}$ is bounded in $\mathbb{R}$ there is a convergent subsequence $\left\{x_{n_j^1}^{(1)}\right\}_{j=1}^{\infty}$ with limit $x^{(1)}$. Now suppose n_j^i has been chosen so that for $m = 1, \ldots, i$ the sequence $\left\{x_{n_j^i}^{(m)}\right\}_{j=1}^{\infty}$ has a limit $x^{(m)}$. Then we can choose a subsequence $\left\{n_j^{i+1}\right\}_{j=1}^{\infty}$ of $\left\{n_j^i\right\}_{j=1}^{\infty}$ so that $\left\{x_{n_j^{i+1}}^{(i+1)}\right\}_{j=1}^{\infty}$ has a limit $x^{(i+1)}$. But then for $m = 1, \ldots, i+1$ the sequence $\left\{x_{n_j^{i+1}}^{(m)}\right\}_{j=1}^{\infty}$ has a limit $x^{(m)}$. Continue this selection process up to $i = d$. By Theorem 16.78 the subsequence $\left\{x_{n_j^d}\right\}_{j=1}^{\infty}$ converges to $x := \sum_{i=1}^{d} x^{(i)} b_i$, and because C is closed, $x \in C$. ■

Theorem 16.80 and the Heine-Borel formulation of compactness allow us to prove that if $m, n, d \in \mathbb{N}$ with $m + n = d$, then for all sets $S \in \Sigma_{\lambda_m} \times \Sigma_{\lambda_n}$ the d-dimensional Lebesgue measure $\lambda_d(S)$ is equal to the product measure $\lambda_m \times \lambda_n(S)$ of m-dimensional and n-dimensional Lebesgue measure. In particular, this completes the investigation started in Proposition 14.60.

Theorem 16.81 *Let $d \in \mathbb{N}$ and for $i = 1, \ldots, d$ let J_i be an interval of finite length. Then $\lambda_d\left(\prod_{i=1}^{d} J_i\right) = \prod_{i=1}^{d} |J_i|$. Moreover, if the numbers $m, n \in \mathbb{N}$ satisfy $m + n = d$, then $\lambda_d|_{\Sigma_{\lambda_m} \times \Sigma_{\lambda_n}} = \lambda_m \times \lambda_n$, that is, the restriction of the Lebesgue measure on $\mathbb{R}^d$ to $\Sigma_{\lambda_m} \times \Sigma_{\lambda_n}$ is equal to the product of the Lebesgue measures on $\mathbb{R}^m$ and $\mathbb{R}^n$.*

Proof. The inequality $\lambda_d\left(\prod_{i=1}^{d} J_i\right) \leq \prod_{i=1}^{d} |J_i|$ follows directly from the definition of outer Lebesgue measure. To prove the reversed inequality, we proceed as follows.

For $i = 1, \ldots, d$, let a_i be the left endpoint of J_i and let b_i be the right endpoint. Let $K := \prod_{i=1}^{d} [a_i, b_i]$. It is easy to prove (see Exercise 16-72) the equality $\lambda_d\left(\prod_{i=1}^{d} J_i\right) = \lambda_d(K)$. Now let $\varepsilon > 0$ and let $\{D_j\}_{j=1}^{\infty}$ be a sequence of dyadic open boxes so that $K \subseteq \bigcup_{j=1}^{\infty} D_j$ and $\lambda_d(K) + \varepsilon > \sum_{j=1}^{\infty} |D_j|$. (By Exercise 14-17, such a sequence exists.) Because K is compact, we can assume without loss of generality that there is an $N \in \mathbb{N}$ so that $K \subseteq \bigcup_{j=1}^{N} D_j$. For each $j \in \{1, \ldots, N\}$, let $D_j = \prod_{i=1}^{d} \left(a_i^j, b_i^j\right)$. Let M be the largest integer so that 2^M is the denominator of any of the completely simplified dyadic rational numbers a_i^j and b_i^j. Let C_M be the set of all cubes of the form $\prod_{i=1}^{d} \left(\frac{c_i}{2^M}, \frac{c_i + 1}{2^M}\right)$, where the c_i are integers. Then for each D_j the equality $|D_j| = \sum_{E \in C_M, E \cap D_j \neq \emptyset} |E|$ holds (see Exercise 16-73). For $i = 1, \ldots, d$, let l_i be the largest integer so that $\frac{l_i}{2^M} < a_i$ and let r_i be the smallest integer so that $\frac{r_i}{2^M} > b_i$. Then for every $E \in C_M$ that is contained in $Q := \prod_{i=1}^{d} \left(\frac{l_i}{2^M}, \frac{r_i}{2^M}\right)$ there is a $j \in 1, \ldots, N$ so that $E \subseteq D_j$. Therefore

$$\begin{aligned}
\lambda_d\left(\prod_{i=1}^{d} J_i\right) + \varepsilon &= \lambda_d(K) + \varepsilon > \sum_{j=1}^{\infty} |D_j| \geq \sum_{j=1}^{N} |D_j| \\
&\geq \sum_{E \in C_M, \exists j \in \{1,\ldots,N\}: E \cap D_j \neq \emptyset} |E| \geq \sum_{E \in C_M, E \cap Q \neq \emptyset} |E| \\
\boxed{\text{Use Exercise 16-73.}} \quad &= |Q| = \prod_{i=1}^{d} \left(\frac{r_i}{2^M} - \frac{l_i}{2^M}\right) \geq \prod_{i=1}^{d} (b_i - a_i) = \prod_{i=1}^{d} |J_i|.
\end{aligned}$$

Because ε was arbitrary we conclude $\lambda_d\left(\prod_{i=1}^{d} J_i\right) \geq \prod_{i=1}^{d} |J_i|$, and hence the two sides are equal.

In particular, this means that if A is an m-dimensional open box and B is an n-dimensional open box, then $\lambda_d(A \times B) = \lambda_m(A)\lambda_n(B)$. By Proposition 14.60, this equation also holds when one of A or B is a null set and the other is an arbitrary Lebesgue measurable set. But then, because the (m- and n-dimensional) Lebesgue measurable sets are the σ-algebra generated by the open boxes and the null sets, the above and Theorem 14.57 prove that $\lambda_d|_{\Sigma_{\lambda_m} \times \Sigma_{\lambda_n}} = \lambda_m \times \lambda_n$. ■

Exercises

16-69. Prove Proposition 16.77.

16-70. Prove Theorem 16.78.

16-71. Prove Theorem 16.79.

16-72. Let $d \in \mathbb{N}$ and for $i = 1, \ldots, d$ let J_i be a, not necessarily closed, interval of finite length with left endpoint a_i and right endpoint b_i. Prove that $\lambda_d\left(\prod_{i=1}^{d} J_i\right) = \lambda_d\left(\prod_{i=1}^{d} [a_i, b_i]\right)$.

16-73. Let $M \in \mathbb{N}$ and let C_M be the set of all cubes of the form $\prod_{i=1}^{d}\left(\frac{c_i}{2^M}, \frac{c_i+1}{2^M}\right)$, where the c_i are integers. Prove that for each dyadic open box of the form $D = \prod_{i=1}^{d}\left(\frac{l_i}{2^M}, \frac{r_i}{2^M}\right)$ the equality $|D| = \sum_{E \in C_M, E \cap D \neq \emptyset} |E|$ holds.

16-74 The **Fundamental Theorem of Algebra**. Let $P : \mathbb{C} \to \mathbb{C}$ defined by $P(z) := \sum_{k=0}^{n} a_k z^k$ be a nonconstant complex polynomial.

(a) Prove that P is continuous.

(b) Prove that $|P| : \mathbb{C} \to [0, \infty)$ assumes an absolute minimum in $\mathbb{C}$.
Hint. Recall that $\mathbb{C}$ is (as a metric space) isomorphic to $\mathbb{R}^2$. Prove that for any $M > 0$ there is an $r > 0$ so that for all $|z| > r$ the inequality $\left|P(z)\right| > M$ holds.

(c) Now prove the **Fundamental Theorem of Algebra**. That is, prove that there must be a $z \in \mathbb{C}$ so that $P(z) = 0$
Hint. Suppose there is no such z, let the absolute minimum of $|P|$ be assumed at z_0 and consider $Q(z) := \frac{1}{P(z_0)} P(z + z_0)$. Then $Q(z) = 1 + b_m z^m + \sum_{j=m+1}^{n} b_j z^j$ for some $m \in \mathbb{N}$. Apply the triangular inequality and find a z with $\left|Q(z)\right| < 1$.

(d) Prove that there are, not necessarily distinct, $z_1, \ldots, z_n \in \mathbb{C}$ so that $P(z) = a_n \prod_{j=1}^{n}(z - z_j)$ for all $z \in \mathbb{C}$.

16-75 **Partial Fraction Decompositions**.

(a) Let P be a polynomial with real coefficients. Prove that if $z \in \mathbb{C}$ is so that $P(z) = 0$, then $P(\bar{z}) = 0$

(b) Use the Fundamental Theorem of Algebra to prove that each polynomial with real coefficients can be written as a product of the leading coefficient, linear factors $(x - c)$ and irreducible quadratic factors $\left((x-a)^2 + b^2\right)$, where all constants a, b, and c are real.

(c) Prove that every rational function with real coefficients can be written as the sum of a polynomial and a linear combination of horizontally shifted rational functions as in Exercises 12-11, 12-17d, and 12-18.
Hint. Induction on the degree of the denominator.

(d) Explain why (at least in principle) it is possible to find a symbolic antiderivative for every rational function with real coefficients.

16-76 Prove that on l^2 the norms $\|\cdot\|_2$ and $\|\cdot\|_\infty$ are not equivalent.

16-77 Prove that in a finite dimensional normed space a **series** converges unconditionally iff it converges absolutely.

16-78 Let X be an infinite dimensional inner product space. Prove that $\left\{x \in X : \|x\| \leq 1\right\}$ is closed and bounded, but not compact.
Hint. Apply the Gram-Schmidt Orthonormalization Procedure to a sequence $\{b_n\}_{n=1}^{\infty}$ of linearly independent vectors in X to obtain a countable orthonormal system in $\left\{x \in X : \|x\| \leq 1\right\}$.

16-79. Proceed as follows to prove that a normed space X is finite dimensional iff compactness is equivalent to being closed and bounded.

(a) Briefly explain why we only need to prove "$\Leftarrow$"

For "$\Leftarrow$" we prove the contrapositive. So for the remainder, let X be an infinite dimensional normed space and let $\{b_n\}_{n=1}^{\infty}$ be a sequence in X so that any finite subset is linearly independent. Even though we do not necessarily have an inner product in X, we can adapt the idea from Exercise 16-78.

(b) Let $A \subseteq X$ be a nonempty subset. For all $x \in X$, define the **distance** from x to A as $\operatorname{dist}(x, A) := \inf\left\{ d(x, a) : a \in A \right\}$. (We will investigate this function in Section 16.9.) For $v_1, \ldots, v_n \in X$ and $r \geq 0$ let $B_r^{\operatorname{span}(v_1,\ldots,v_n)} := \left\{ x \in \operatorname{span}(v_1, \ldots, v_n) : \|x\| \leq r \right\}$. Prove that for any element $w \notin \operatorname{span}(v_1, \ldots, v_n)$ there is an $a(w) \in B_1^{\operatorname{span}(v_1,\ldots,v_n)}$ so that $\left\| w - a(w) \right\| = \operatorname{dist}\left(w, B_1^{\operatorname{span}(v_1,\ldots,v_n)} \right) > 0$.

(c) Prove that $\left\| a(w) \right\| < 2\|w\|$.

(d) Prove that if $\|w\| < \frac{1}{4}$, then $\operatorname{dist}\left(w - a(w), B_1^{\operatorname{span}(v_1,\ldots,v_n)} \right) = \left\| w - a(w) \right\|$.

(e) Prove that if $\|w\| < \frac{1}{4}$, then $\operatorname{dist}\left(\frac{w - a(w)}{\|w - a(w)\|}, B_1^{\operatorname{span}(v_1,\ldots,v_n)} \right) = 1$.

(f) Construct a sequence $\{v_n\}_{n=1}^{\infty}$ in X so that $\|v_n\| = 1$ for all $n \in \mathbb{N}$ and so that for all distinct $i, j \in \mathbb{N}$ we have $\|v_i - v_j\| \geq 1$.

(g) Finish the proof of the contrapositive of "$\Leftarrow$."

16-80. **Fubini's Theorem** revisited. Let λ_2 be Lebesgue measure on $\mathbb{R}^2$ and let λ_x and λ_y denote Lebesgue measure on the x- and y-axes, respectively.

(a) Prove that if the function $f : \mathbb{R}^2 \to [-\infty, \infty]$ is Lebesgue integrable and $\lambda_x \times \lambda_y$-measurable, then for almost all elements $x \in \mathbb{R}$ the function $f_x(y) = f(x, y)$ is Lebesgue integrable, for almost all elements $y \in \mathbb{R}$ the function $f^y(x) := f(x, y)$ is Lebesgue integrable, the function $x \mapsto \begin{cases} \int_{\mathbb{R}} f_x \, d\lambda_y; & \text{if } f_x \text{ is Lebesgue integrable,} \\ 0; & \text{otherwise,} \end{cases}$ is Lebesgue integrable and the function $y \mapsto \begin{cases} \int_{\mathbb{R}} f^y \, d\lambda_x, & \text{if } f^y \text{ is Lebesgue integrable,} \\ 0, & \text{otherwise} \end{cases}$ is Lebesgue integrable. Moreover,

$$\int_{\mathbb{R}^2} f(x, y) \, d\lambda_2 = \int_{\mathbb{R}} \left(\int_{\mathbb{R}} f(x, y) \, d\lambda_y \right) d\lambda_x = \int_{\mathbb{R}} \left(\int_{\mathbb{R}} f(x, y) \, d\lambda_x \right) d\lambda_y$$

(b) Compute the following integrals:

i. $$\int_{\mathbb{R}^2} xe^x \mathbf{1}_{[0,1]}(x) y \mathbf{1}_{[0,2]}(y) \, d\lambda_2$$

ii. $$\int_{\{(x,y)\in\mathbb{R}^2 : 0\leq x\leq 2, x^2\leq y\leq 4\}} \left(1 + y^2\right)^{10} x \, d\lambda_2$$

(c) State and prove a result similar to the result in part 16-80a for the Lebesgue integral on $\mathbb{R}^3$, representing it as an iteration of three single variable Lebesgue integrals.

(d) Compute the following integrals.

i. $$\int_{\{(x,y,z)\in\mathbb{R}^3 : 0\leq x\leq 1, 0\leq y\leq 3, 0\leq z\leq x\}} y \, d\lambda_3$$

ii. $$\int_{\{(x,y,z)\in\mathbb{R}^3 : -1\leq z\leq 1, 0\leq y\leq z, 0\leq x\leq y\}} e^{z^3} \, d\lambda_3$$

16.7 Dense Subspaces

Recall that the integral of a nonnegative measurable function is defined as a supremum of integrals of simple functions. This means that for every integrable function there should be simple functions arbitrarily "close" to it. The concept of a dense subset expresses this idea in precise terms.

Definition 16.82 *Let X be a metric space. A set $S \subseteq X$ is called* **dense** *in X iff for every $\varepsilon > 0$ and every $x \in X$ there is an $s \in S$ so that $d(x, s) < \varepsilon$.*

So a subset S of a metric space X is dense iff every neighborhood of every point of X contains a point in S. In terms of approximating elements, we can say the following.

Proposition 16.83 *Let X be a metric space. Then $S \subseteq X$ is dense in X iff for all $x \in X$ there is a sequence $\{s_n\}_{n=1}^{\infty}$ of elements in S so that $\lim_{n\to\infty} s_n = x$.*

Proof. Use Standard Proof Technique 3.8. (Exercise 16-81.) ■

The simplest example of a dense subset are the rational numbers as a subset of the real numbers.

Theorem 16.84 *$\mathbb{Q}$ is dense in $\mathbb{R}$.*

Proof. Use Theorem 1.36. (Exercise 16-82.) ■

Once we take care of the usual problem of equality almost everywhere, the above mentioned simple functions can be considered "dense in L^p."

Theorem 16.85 *Let (M, Σ, μ) be a measure space and let $1 \leq p < \infty$. Then the set $S := \Big\{[s] : s \in F(M, \mathbb{R}) \text{ is simple } \Big\}$ is dense in $L^p(M, \Sigma, \mu)$.*

Proof. First, consider a nonnegative function $g \in \mathcal{L}^p(M, \Sigma, \mu)$. From the proof of Theorem 14.29 (see proof of Theorem 9.19), we infer that there is a sequence $\{s_n\}_{n=1}^{\infty}$ of nonnegative simple functions that converges pointwise to g with $0 \leq s_n \leq g$ for all $n \in \mathbb{N}$. Hence, the sequence $\Big\{|s_n - g|^p\Big\}_{n=1}^{\infty}$ converges pointwise to zero and it is bounded by $g^p \in \mathcal{L}^p(M, \Sigma, \mu)$. Thus by the Dominated Convergence Theorem we obtain $\lim_{n\to\infty} \int_M |s_n - g|^p \, d\mu = 0$, that is, $\lim_{n\to\infty} \Big\|[s_n] - [g]\Big\|_p = 0$.

Now let $f \in \mathcal{L}^p(M, \Sigma, \mu)$. Let $\{s_n\}_{n=1}^{\infty}$ be a sequence of simple functions so that $\lim_{n\to\infty} \left\|s_n - f^+\right\|_p = 0$ and let $\{t_n\}_{n=1}^{\infty}$ be a sequence of simple functions so that $\lim_{n\to\infty} \left\|t_n - f^-\right\|_p = 0$. Then $\{s_n - t_n\}_{n=1}^{\infty}$ is a sequence of simple functions and we conclude $0 \leq \lim_{n\to\infty} \Big\|[s_n - t_n] - [f]\Big\|_p \leq \lim_{n\to\infty} \left\|s_n - f^+\right\|_p + \left\|t_n - f^-\right\|_p = 0$.

By Proposition 16.83, S is dense in $L^p(M, \Sigma, \mu)$. ■

Although simple functions can be defined for arbitrary measure spaces, when additional structure is available it would be desirable to have dense subsets of functions with properties related to that structure. The next result shows that the continuous functions are "dense in $L^p[a, b]$." For some L^p spaces, we will find an even nicer dense subspace in Theorem 18.12.

Theorem 16.86 *Let $a < b$, let $C[a, b] := \Big\{[f] : f \in C^0[a, b]\Big\}$ and let $1 \leq p < \infty$. Then $C[a, b]$ is dense in $L^p[a, b]$.*

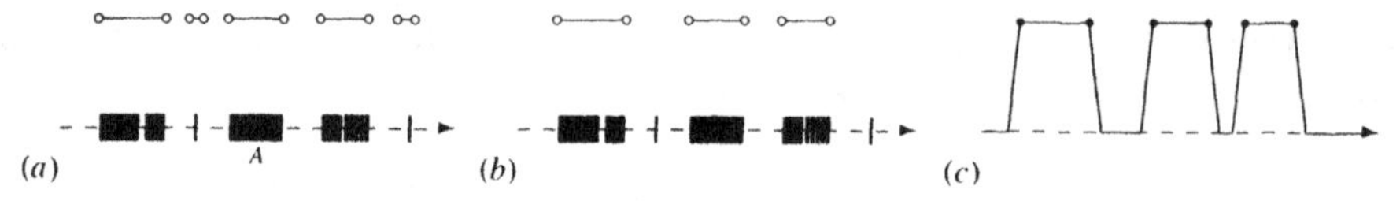

Figure 38: Illustration of the approximation of indicator functions in the proof of Theorem 16.86. First (a) cover the set A with open intervals so that the measure of the union of the intervals is close to $\lambda(A)$. Then (b) discard all but finitely many intervals but do not decrease the measure of the union too much. Then (c) approximate the indicator function of each interval with a continuous function so that the integrals remain close.

Proof. By Theorems 5.20 and 9.26, $C[a,b]$ is a subset of $L^p[a,b]$. We are done if we can show that for any $\varepsilon > 0$ and $f \in \mathcal{L}^p(\mathbb{R})$ there is a continuous $g \in \mathcal{L}^p(\mathbb{R})$ so that $\|f-g\|_p < \varepsilon$. The result for $L^p[a,b]$ will follow because $\mathcal{L}^p[a,b]$ is embedded in $\mathcal{L}^p(\mathbb{R})$ by setting each function equal to zero outside $[a,b]$ and because the restriction of a continuous function on $\mathbb{R}$ to $[a,b]$ is continuous on $[a,b]$.

First, let (l,r) be an open interval on the real line and let $\varepsilon > 0$. Define

$$h_{(l,r),\varepsilon}(x) := \begin{cases} 0; & \text{for } x \notin \left(l-\left(\frac{\varepsilon}{2}\right)^p, r+\left(\frac{\varepsilon}{2}\right)^p\right), \\ \frac{2^p}{\varepsilon^p}\left(x-\left(l-\left(\frac{\varepsilon}{2}\right)^p\right)\right); & \text{for } x \in \left[l-\left(\frac{\varepsilon}{2}\right)^p, l\right], \\ \frac{2^p}{\varepsilon^p}\left(\left(r+\left(\frac{\varepsilon}{2}\right)^p\right)-x\right); & \text{for } x \in \left[r, r+\left(\frac{\varepsilon}{2}\right)^p\right], \\ 1; & \text{for } x \in (l,r). \end{cases}$$

Then each $h_{(l,r),\varepsilon}$ is continuous on $\mathbb{R}$ and the following inequalities hold.

$$\begin{aligned} &\left(\int_{\mathbb{R}} \left|h_{(l,r),\varepsilon} - \mathbf{1}_{(l,r)}\right|^p \, d\mu\right)^{\frac{1}{p}} \\ &\leq \left(\int_{l-\left(\frac{\varepsilon}{2}\right)^p}^{l} \left|\frac{2^p}{\varepsilon^p}\left(x-\left(l-\left(\frac{\varepsilon}{2}\right)^p\right)\right)\right|^p dx\right)^{\frac{1}{p}} \\ &\quad + \left(\int_{r}^{r+\left(\frac{\varepsilon}{2}\right)^p} \left|\frac{2^p}{\varepsilon^p}\left(\left(r+\left(\frac{\varepsilon}{2}\right)^p\right)-x\right)\right|^p dx\right)^{\frac{1}{p}} \\ &< \left(\int_{l-\left(\frac{\varepsilon}{2}\right)^p}^{l} 1\,dx\right)^{\frac{1}{p}} + \left(\int_{r}^{r+\left(\frac{\varepsilon}{2}\right)^p} 1\,dx\right)^{\frac{1}{p}} = \left(\left(\frac{\varepsilon}{2}\right)^p\right)^{\frac{1}{p}} + \left(\left(\frac{\varepsilon}{2}\right)^p\right)^{\frac{1}{p}} = \varepsilon. \end{aligned}$$

We now prove that for every measurable set A and every $\varepsilon > 0$ there is a continuous function $g_{A,\varepsilon}$ so that $\|\mathbf{1}_A - g_{A,\varepsilon}\|_p < \varepsilon$. For the idea, consider Figure 38. Let $\{I_j\}_{j=1}^{\infty}$ be a sequence of open intervals so that $A \subseteq \bigcup_{j=1}^{\infty} I_j$ and $\sum_{j=1}^{\infty} |I_j| < \lambda(A) + \left(\frac{\varepsilon}{3}\right)^p$. Let $n \in \mathbb{N}$ be such that $\left(\sum_{j=n+1}^{\infty} |I_j|\right)^{\frac{1}{p}} < \frac{\varepsilon}{3}$. Then the function $g_{A,\varepsilon} := \sum_{j=1}^{n} h_{I_j,\frac{1}{n}\frac{\varepsilon}{3}}$ is continuous on $\mathbb{R}$ and

$$\left\| \sum_{j=1}^{n} h_{I_j,\frac{1}{n}\frac{\varepsilon}{3}} - \mathbf{1}_A \right\|_p$$

$$\leq \left\| \sum_{j=1}^{n} h_{I_j,\frac{1}{n}\frac{\varepsilon}{3}} - \sum_{j=1}^{n} \mathbf{1}_{I_j} \right\|_p + \left\| \sum_{j=1}^{n} \mathbf{1}_{I_j} - \sum_{j=1}^{\infty} \mathbf{1}_{I_j} \right\|_p + \left\| \sum_{j=1}^{\infty} \mathbf{1}_{I_j} - \mathbf{1}_A \right\|_p$$

$$< \sum_{j=1}^{n} \left\| h_{I_j,\frac{1}{n}\frac{\varepsilon}{3}} - \mathbf{1}_{I_j} \right\|_p + \frac{\varepsilon}{3} + \frac{\varepsilon}{3} < \sum_{j=1}^{n} \frac{1}{n}\frac{\varepsilon}{3} + 2\frac{\varepsilon}{3} = \varepsilon.$$

Now we prove that for every simple function s and every $\varepsilon > 0$ there is a continuous function $g_{s,\varepsilon}$ so that $\|g_{s,\varepsilon} - s\|_p < \varepsilon$. Let $s = \sum_{j=1}^{n} a_j \mathbf{1}_{A_j}$ be a simple function on $\mathbb{R}$ and let $\varepsilon > 0$. Then $g_{s,\varepsilon} := \sum_{j=1}^{n} a_j g_{A_j,\frac{\varepsilon}{n(|a_j|+1)}}$ is continuous on $\mathbb{R}$ and

$$\left\| \sum_{j=1}^{n} a_j g_{A_j,\frac{\varepsilon}{n(|a_j|+1)}} - \sum_{j=1}^{n} a_j \mathbf{1}_{A_j} \right\|_p \leq \sum_{j=1}^{n} |a_j| \left\| g_{A_j,\frac{\varepsilon}{n(|a_j|+1)}} - \mathbf{1}_{A_j} \right\|_p$$

$$\leq \sum_{j=1}^{n} |a_j| \frac{\varepsilon}{n(|a_j|+1)} < \varepsilon.$$

Now finally let $f \in \mathcal{L}^p(\mathbb{R})$ and let $\varepsilon > 0$. By Theorem 16.85 there is a simple function s so that $\|f - s\|_p < \frac{\varepsilon}{2}$. Moreover, $g_{s,\frac{\varepsilon}{2}}$ is continuous, $\left\| s - g_{s,\frac{\varepsilon}{2}} \right\|_p < \frac{\varepsilon}{2}$ and

$$\left\| f - g_{s,\frac{\varepsilon}{2}} \right\|_p \leq \|f - s\|_p + \left\| s - g_{s,\frac{\varepsilon}{2}} \right\|_p < \frac{\varepsilon}{2} + \frac{\varepsilon}{2} = \varepsilon.$$

Hence, for every $[f] \in L^p(\mathbb{R})$ and every $\varepsilon > 0$ there is a $g \in C(\mathbb{R})$ so that $\big\| [f] - [g] \big\|_p < \varepsilon$, which proves that $C[a,b]$ is dense in $L^p[a,b]$. ■

It is worth noting that S as well as $C[a,b]$ are actually linear subspaces of the normed spaces $L^p(M, \Sigma, \mu)$ and $L^p[a,b]$, respectively. In finite dimensional spaces, proper linear subspaces cannot be dense in the whole space. In infinite dimensional spaces there can be many dense linear subspaces comprised of "nice" elements.

For integrable functions, it is standard practice to prove results for a dense subset of functions with nice properties and then use a limit argument to get the result for all functions. The proof of Theorem 18.37 is a prime example of this approach. Theorem 16.87 gives a first impression how an equality on a dense subset translates to an equality on the whole space.

Theorem 16.87 *Let X, Y be metric spaces, let $D \subseteq X$ be dense and let the functions $f, g : X \to Y$ be continuous with $f|_D = g|_D$. Then $f = g$.*

Proof. Let $x \in X \setminus D$ and let $\{d_n\}_{n=1}^{\infty}$ be a sequence in D so that $\lim_{n\to\infty} d_n = x$. Then $f(x) = f\left(\lim_{n\to\infty} d_n\right) = \lim_{n\to\infty} f(d_n) = \lim_{n\to\infty} g(d_n) = g\left(\lim_{n\to\infty} d_n\right) = g(x)$. Because $f(x) = g(x)$ for all $x \in D$ this proves $f = g$. ■

Because completeness is such a useful analytical property, we conclude this section by proving that every metric space can be viewed as a dense subspace of a complete metric space.

> **The proof of Theorem 16.89 is quite technical. Most applications rely on the fact *that* any metric space can be isometrically and densely embedded in a complete space rather than on the details of the construction. Hence, the reader could postpone reading the details of the proof of Theorem 16.89.**

Definition 16.88 *Let X, Y be metric spaces. A function $f : X \to Y$ is called an* **isometry** *iff for all $x, x' \in X$ we have $d(f(x), f(x')) = d(x, x')$. If there is an isometry $f : X \to Y$, we will also say that X can be* **isometrically embedded** *into Y.*

Theorem 16.89 *Every metric space X can be isometrically embedded as a dense subspace into a complete metric space $C(X)$.*

Proof. For this proof, let d_X denote the metric on X. Define

$$C(X) := \left\{\left\{x^{(i)}\right\}_{i=1}^{\infty} : \left\{x^{(i)}\right\}_{i=1}^{\infty} \text{ is a Cauchy sequence in } X\right\}.$$

Let $\left\{x^{(i)}\right\}_{i=1}^{\infty}, \left\{y^{(i)}\right\}_{i=1}^{\infty} \in C(X)$. We will first show $\left\{d_X\left(x^{(i)}, y^{(i)}\right)\right\}_{i=1}^{\infty}$ is a Cauchy sequence. To do this let $i, j \in \mathbb{N}$. Assume without loss of generality that $d_X\left(x^{(i)}, y^{(i)}\right) \geq d_X\left(x^{(j)}, y^{(j)}\right)$. Then

$$d_X\left(x^{(i)}, y^{(i)}\right) \leq d_X\left(x^{(i)}, x^{(j)}\right) + d_X\left(x^{(j)}, y^{(j)}\right) + d_X\left(y^{(j)}, y^{(i)}\right),$$

and hence for all $i, j \in \mathbb{N}$ we obtain

$$\begin{aligned}\left|d_X\left(x^{(i)}, y^{(i)}\right) - d_X\left(x^{(j)}, y^{(j)}\right)\right| &= d_X\left(x^{(i)}, y^{(i)}\right) - d_X\left(x^{(j)}, y^{(j)}\right) \\ &\leq d_X\left(x^{(i)}, x^{(j)}\right) + d_X\left(y^{(j)}, y^{(i)}\right).\end{aligned}$$

Let $\varepsilon > 0$. Then there are $N_x, N_y \in \mathbb{N}$ so that for all $i, j \geq N_x$ we have that $d_X\left(x^{(i)}, x^{(j)}\right) < \frac{\varepsilon}{2}$ and for all $i, j \geq N_y$ we have that $d_X\left(y^{(i)}, y^{(j)}\right) < \frac{\varepsilon}{2}$. Define $N := \max\{N_x, N_y\}$. Then for all $i, j \geq N$ we infer

$$\left|d_X\left(x^{(i)}, y^{(i)}\right) - d_X\left(x^{(j)}, y^{(j)}\right)\right| \leq d_X\left(x^{(i)}, x^{(j)}\right) + d_X\left(y^{(j)}, y^{(i)}\right) < \frac{\varepsilon}{2} + \frac{\varepsilon}{2} = \varepsilon.$$

Thus for any $\left\{x^{(i)}\right\}_{i=1}^{\infty}, \left\{y^{(i)}\right\}_{i=1}^{\infty} \in C(X)$ the sequence $\left\{d_X\left(x^{(i)}, y^{(i)}\right)\right\}_{i=1}^{\infty}$ is a Cauchy sequence, and hence it has a limit. For $\left\{x^{(i)}\right\}_{i=1}^{\infty}, \left\{y^{(i)}\right\}_{i=1}^{\infty} \in C(X)$, define $d^s\left(\left\{x^{(i)}\right\}_{i=1}^{\infty}, \left\{y^{(i)}\right\}_{i=1}^{\infty}\right) := \lim_{i\to\infty} d_X\left(x^{(i)}, y^{(i)}\right)$.

We claim that d^s is a semimetric on the set $\mathcal{C}(X)$. Clearly, for all $x, y \in \mathcal{C}(X)$ we have $d^s(x, y) \geq 0$, $d^s(x, x) = 0$ and $d^s(x, y) = d^s(y, x)$. Now consider three elements $\left\{x^{(i)}\right\}_{i=1}^{\infty}, \left\{y^{(i)}\right\}_{i=1}^{\infty}, \left\{z^{(i)}\right\}_{i=1}^{\infty} \in \mathcal{C}(X)$. Then

$$\begin{aligned} d^s\left(\left\{x^{(i)}\right\}_{i=1}^{\infty}, \left\{z^{(i)}\right\}_{i=1}^{\infty}\right) &= \lim_{i\to\infty} d_X\left(x^{(i)}, z^{(i)}\right) \leq \lim_{i\to\infty} d_X\left(x^{(i)}, y^{(i)}\right) + d_X\left(y^{(i)}, z^{(i)}\right) \\ &= d^s\left(\left\{x^{(i)}\right\}_{i=1}^{\infty}, \left\{y^{(i)}\right\}_{i=1}^{\infty}\right) + d^s\left(\left\{y^{(i)}\right\}_{i=1}^{\infty}, \left\{z^{(i)}\right\}_{i=1}^{\infty}\right). \end{aligned}$$

Let $\sim$ be the equivalence relation on $\mathcal{C}(X)$ as in Theorem 15.61. Let $C(X)$ be the set and let d be the metric on $C(X)$ obtained from $(\mathcal{C}(X), d^s)$ via Theorem 15.61. As in Theorem 15.61 we will denote the elements of $C(X)$ by $[x]$, where $x \in \mathcal{C}(X)$.

We claim for every $\left[\left\{x^{(i)}\right\}_{i=1}^{\infty}\right] \in C(X)$ and every $n \in \mathbb{N}$ there is an equivalent $\left\{y^{(i)}\right\}_{i=1}^{\infty} \sim \left\{x^{(i)}\right\}_{i=1}^{\infty}$ so that for all $i, j \in \mathbb{N}$ we have $d_X\left(y^{(i)}, y^{(j)}\right) < \frac{1}{n}$. Let $n \in \mathbb{N}$. There is an $m \in \mathbb{N}$ so that for $i, j \geq m$ we have that $d_X\left(x^{(i)}, x^{(j)}\right) < \frac{1}{n}$. The sequence $\left\{y^{(i)}\right\}_{i=1}^{\infty} := \left\{x^{(m+i)}\right\}_{i=1}^{\infty}$ is a Cauchy sequence and clearly for all $i, j \in \mathbb{N}$ we have $d_X\left(y^{(i)}, y^{(j)}\right) < \frac{1}{n}$. Moreover, because $\left\{x^{(i)}\right\}_{i=1}^{\infty}$ is a Cauchy sequence, we infer $\lim_{i\to\infty} d_X\left(y^{(i)}, x^{(i)}\right) = \lim_{i\to\infty} d_X\left(x^{(i+m)}, x^{(i)}\right) = 0$, and hence $\left\{y^{(i)}\right\}_{i=1}^{\infty} \sim \left\{x^{(i)}\right\}_{i=1}^{\infty}$.

To prove that $C(X)$ is complete, let $\{[x_n]\}_{n=1}^{\infty}$ be a Cauchy sequence in $C(X)$. By the above, for each $n \in \mathbb{N}$ we can assume without loss of generality that the sequence $x_n = \left\{x_n^{(i)}\right\}_{i=1}^{\infty}$ is such that for all $i, j \in \mathbb{N}$ we have $d_X\left(x_n^{(i)}, x_n^{(j)}\right) < \frac{1}{n}$.

Define $x := \left\{x_i^{(1)}\right\}_{i=1}^{\infty}$. We claim that x is a Cauchy sequence. Let $\varepsilon > 0$. Then there is an $N \in \mathbb{N}$ so that $\frac{1}{N} < \frac{\varepsilon}{3}$ and for all $m, n \geq N$ we have $d([x_m], [x_n]) < \frac{\varepsilon}{3}$. Let $n, m \geq N$ be fixed. Because $\lim_{i\to\infty} d_X\left(x_m^{(i)}, x_n^{(i)}\right) = d([x_m], [x_n]) < \frac{\varepsilon}{3}$, there is a $k \in \mathbb{N}$ so that $d_X\left(x_m^{(k)}, x_n^{(k)}\right) < \frac{\varepsilon}{3}$. Hence, for all $m, n \geq N$ we obtain

$$\begin{aligned} d_X\left(x_m^{(1)}, x_n^{(1)}\right) &\leq d_X\left(x_m^{(1)}, x_m^{(k)}\right) + d_X\left(x_m^{(k)}, x_n^{(k)}\right) + d_X\left(x_n^{(k)}, x_n^{(1)}\right) \\ &< \frac{1}{m} + \frac{\varepsilon}{3} + \frac{1}{n} < \varepsilon. \end{aligned}$$

Because $m, n \geq N$ were arbitrary, x is a Cauchy sequence.

We now claim that $\lim_{n\to\infty} [x_n] = [x]$. Let $\varepsilon > 0$. With $N \in \mathbb{N}$ so that $\frac{1}{N} < \frac{\varepsilon}{3}$ we obtain that for all $n, i \geq N$ there is a $k \geq N$ with $d_X\left(x_n^{(k)}, x_i^{(k)}\right) < \frac{\varepsilon}{3}$, and hence

$$\begin{aligned} d_X\left(x_n^{(i)}, x_i^{(1)}\right) &\leq d_X\left(x_n^{(i)}, x_n^{(k)}\right) + d_X\left(x_n^{(k)}, x_i^{(k)}\right) + d_X\left(x_i^{(k)}, x_i^{(1)}\right) \\ &< \frac{1}{n} + \frac{\varepsilon}{3} + \frac{1}{i} < \varepsilon. \end{aligned}$$

Therefore $d([x_n],[x]) = \lim_{i\to\infty} d_X\left(x_n^{(i)}, x_i^{(1)}\right) < \varepsilon$ and we conclude, as was claimed, that $\lim_{n\to\infty} d([x_n],[x]) = 0$.

Clearly, the map $f : X \to C(X)$ defined by $f(x) := \left[\{x\}_{i=1}^{\infty}\right]$ is well-defined and for all $x, y \in X$ we have $d_X(x, y) = d(f(x), f(y))$. This means that X can be isometrically embedded into $C(X)$.

To prove $f[X]$ is dense in $C(X)$, let $[x] \in C(X)$. Fix a representative $\left\{x^{(i)}\right\}_{i=1}^{\infty}$ of $[x]$. Then the sequence $\left\{f\left(x^{(i)}\right)\right\}_{i=1}^{\infty}$ converges to $[x]$ in $C(X)$ (Exercise 16-83). ■

A complete superspace Y of a metric space X so that X is dense in Y is also called the **completion** of the space. Any two completions are isometrically isomorphic (see Exercise 16-84d), so we can really speak of *the* completion. Thus Theorem 16.84 says that the real numbers are the completion of the rational numbers. Indeed, one way to construct the real numbers from the rational numbers in set theory is to apply the construction in the proof of Theorem 16.89 to $\mathbb{Q}$ (see Exercise 16-93).

When considering function spaces, Theorem 16.86 says that $L^p[a,b]$ is the completion of $C[a,b]$ with respect to the $\|\cdot\|_p$ norm. Because continuous functions are Riemann integrable, $L^p[a,b]$ is also the completion of the set of (equivalence classes of) Riemann integrable functions with respect to the $\|\cdot\|_p$ norm. It is possible to introduce the L^p spaces as these completions, but by explicitly defining the integrals the elements of the completion are better understood.

Finally, note that Exercises 16-94 and 16-95 show that the completions of normed and inner product spaces are Banach and Hilbert spaces, respectively.

Exercises

16-81. Prove Proposition 16.83.

16-82 Prove Theorem 16.84

16-83 Finish the proof of Theorem 16 89 as follows Let $[x] \in C(X)$. Fix a representative $\left\{x^{(i)}\right\}_{i=1}^{\infty}$ of $[x]$ Prove that the sequence $\left\{f\left(x^{(i)}\right)\right\}_{i=1}^{\infty}$ converges to $[x]$ in $C(X)$

16-84 Extensions of continuous functions.

(a) Let X be a metric space, Y a complete metric space, $D \subseteq X$ dense and let $f : D \to Y$ be uniformly continuous Prove that f can be extended to a unique continuous function $F\ X \to Y$ so that $F|_D = f$

(b) Prove that there are continuous functions $f : (0,1] \to \mathbb{R}$ that do *not* have a continuous extension $F : [0,1] \to \mathbb{R}$ as in Exercise 16-84a

(c) Give an example that shows that the space Y in Exercise 16-84a must be complete

(d) Prove that if X, Y are metric spaces, $D_X \subseteq X$ is dense in X, $D_Y \subseteq Y$ is dense in Y and $f : D_X \to D_Y$ is an isometry, then there is an isometry between X and Y.

16-85 Examples of dense subspaces.

(a) Let (M, Σ, μ) be a measure space. Prove that the space of "simple functions" $S := \{[s] : s \in F(M, \mathbb{R}) \text{ is simple }\}$ is dense in $L^{\infty}(M, \Sigma, \mu)$.

(b) Prove that for $1 \le p < \infty$, $C^1 := \{[f] : f \in C^1(a,b)\}$ is dense in $L^p(a,b)$.
Hint You can defer most of the proof to corresponding parts of the proof of Theorem 16 86. Only the start needs to be modified.

16-86. Prove that $C[a, b]$ is *not* dense in $L^\infty[a, b]$

16-87 Dense subspaces of $\left(C^0(X, \mathbb{R}), \|\ \|_\infty\right)$ Let X be a compact metric space and let $C^0(X, \mathbb{R})$ be the space of continuous functions from X to $\mathbb{R}$ with the uniform norm $\|\cdot\|_\infty$. A vector subspace L of $C^0(X, \mathbb{R})$ is called a **sublattice** of $C^0(X, \mathbb{R})$ iff for all $f \in L$ we have that $\max\{f, g\} \in L$ and $\min\{f, g\} \in L$. A subset S of $C^0(X, \mathbb{R})$ is called **point-separating** iff for all $x, y \in X$ there is a $g \in S$ with $g(x) \neq g(y)$. A vector subspace A of $C^0(X, \mathbb{R})$ is called a **subalgebra** of $C^0(X, \mathbb{R})$ iff for all $f, g \in A$ we have that $fg \in A$.

(a) Prove that if L is a point-separating sublattice of $C^0(X, \mathbb{R})$ that contains the constant functions, then for all $\varepsilon > 0$, all functions $f \in C^0(X, \mathbb{R})$ and all $x \in X$ there is a $g \in L$ so that $g(x) = f(x)$ and for all $y \in X$ we have $g(y) \leq f(y) + \varepsilon$.

Hint. Set $h_x(z) := f(x)$ and for $y \in X \setminus \{x\}$ find $g \in L$ with $g(y) \neq g(x)$ and set $h_y(z) := f(x) + \left(f(y) - f(x)\right)\dfrac{g(z) - g(x)}{g(y) - g(x)}$. Cover X with open sets I_y on which the inequality $h_y(z) \leq f(z) + \varepsilon$ holds for all $z \in I_y$. Then use a finite open subcover and minima.

(b) Prove that if L is a point-separating sublattice of $C^0(X, \mathbb{R})$ that contains the constant functions, then L is dense in $C^0(X, \mathbb{R})$.

Hint Let $f \in C^0(X, \mathbb{R})$ For each $x \in X$ find an open subset $I_x \ni x$ and a function $g_x \in L$ so that $g_x(y) < f(y) + \varepsilon$ for all $y \in X$ and so that $\left|g_x(z) - f(z)\right| < \varepsilon$ for all $z \in I_x$. Then use a finite open subcover and maxima.

(c) Prove that if L is a subspace of $C^0(X, \mathbb{R})$ so that for all $f \in L$ we have that $|f| \in L$, then L is a sublattice of $C^0(X, \mathbb{R})$

(d) Let A be a subalgebra of $C^0(X, \mathbb{R})$ and let $P : \mathbb{R} \to \mathbb{R}$ be a polynomial. Prove that for all $f \in A$ the function $P \circ f$ is also in A.

(e) Let A be a subalgebra of the space $C^0(X, \mathbb{R})$. Prove that the subspace $\overline{A}$ of $C^0(X, \mathbb{R})$ is a sublattice.

Hint. First prove that $\overline{A}$ is a subalgebra. Then prove that for each $f \in \overline{A}$ the function $|f| = \sqrt{f^2}$ is also in $\overline{A}$ To prove that the square root can be taken, use Exercise 16-68c.

(f) **Stone-Weierstrass Theorem** Let X be compact. Prove that if A is a point-separating subalgebra of $C^0(X, \mathbb{R})$ that contains the constant functions, then A is dense in $C^0(X, \mathbb{R})$

(g) Prove that the subspace of all polynomials on $[a, b]$ is dense in $C^0[a, b]$.

(h) Let $\varepsilon > 0$. Prove that the subspace of all trigonometric polynomials on $[-\pi, \pi - \varepsilon]$ is dense in $C^0[-\pi, \pi - \varepsilon]$.

Hint. Exercise 12-15b

(i) Why are the trigonometric polynomials *not* dense in $C^0[-\pi, \pi]$?

(j) Let $C := [-\pi, \pi)$ with the metric $d(x, y) := \left\| \left(\cos(x), \sin(x)\right) - \left(\cos(y), \sin(y)\right) \right\|_2$.

i. Prove that C is isometrically isomorphic to $\left\{z \in \mathbb{R}^2 : \|z\|_2 = 1\right\}$.

ii. Prove that C is compact.

iii. Prove that the trigonometric polynomials are dense in $\left(C^0(C, \mathbb{R}), \|\cdot\|_\infty\right)$.

16-88. Density of polynomials does not imply that all Taylor series converge

(a) Let $X \subseteq Y \subseteq Z$ be metric spaces so that X is dense in Z. Prove that X is dense in Y.

(b) Prove that the set of polynomials is dense in $\left(C^\infty[a, b], \|\cdot\|_\infty\right)$ Derivatives on the boundary are understood to be one-sided. You may use Exercise 16-87g.

(c) Explain why part 16-88b does *not* imply that every function in the space $\left(C^\infty[-1, 1], \|\cdot\|_\infty\right)$ is the limit of its Taylor series about zero.

(Lemma 18 8 will ultimately provide an example of a function whose Taylor series does not converge to the function)

16-89. Use Egoroff's Theorem (see Exercise 14-37c) and Theorem 16.86 to prove that for every function $f \in L^p[a,b]$ and every $\varepsilon > 0$ there is a $B \subseteq [a,b]$ with $\lambda(B) > (b-a) - \varepsilon$ and a continuous function $g : B \to \mathbb{R}$ so that $g|_B = f$.

16-90. A metric space with a countable dense subset is called **separable**.

(a) Prove that every open subset in a separable metric space is a *countable* union of open balls.

(b) Prove that a compact metric space is separable.

Hint. Use Lemma 16.70 for $\varepsilon_k := \frac{1}{k}$ and all $k \in \mathbb{N}$.

(c) Let $O \subseteq \mathbb{R}^d$ be an open set. Prove that O is separable.

Hint. $\mathbb{Q}^d$ is countable.

(d) Give an example of a complete, separable metric space that is not compact.

(e) Prove that if X, Y are metric spaces, X is separable and $f : X \to Y$ is continuous and surjective, then Y is separable.

16-91. Let X be a metric space and let $D \subseteq X$. Prove that D is dense in X iff $\overline{D} = X$.

16-92. Let X be a metric space. List all closed dense subspaces of X.

16-93. Constructing the real numbers from the rational numbers. Prove that the metric space obtained when applying the construction in the proof of Theorem 16.89 to the rational numbers is an ordered field.

Hint. Define the operations $+$ and $\cdot$ for sequences termwise and prove that they induce well-defined operations on the equivalence classes. Then prove that these operations satisfy the field axioms. Define $\mathbb{R}^+$ as the set of equivalence classes of sequences for which there is a positive rational number r so that the terms are eventually greater than r. Then prove that Axiom 1.6 is satisfied.

The above, together with Theorem 16.89, shows that the space is a complete, ordered field. The Completeness Axiom as stated in this text can be proved using convergence of Cauchy sequences as in Exercise 2-25.

16-94. Prove that the completion of a normed space is a Banach space.

16-95. Prove that the completion of an inner product space is a Hilbert space.

16.8 Connectedness

Intuitively, a metric space should be connected iff it is possible to get from any place to any other place by going along an unbroken path. Unfortunately, this would define connectedness in terms of itself, because the only interpretation of "unbroken" is "connected." Using continuous paths to connect points also leads to some problems, which will be explained in Example 16.96. Therefore, we define connectedness by what it means to be disconnected. Sensibly, being disconnected should mean that the space can be split into two nonempty pieces that are separate from each other. Two disjoint open sets can be considered to be separate entities, because each element in each set has nonzero distance to the respective other set.

Definition 16.90 *A metric space X is called* **disconnected** *iff there are two disjoint nonempty open sets $U, V \subseteq X$ such that $U \cup V = X$. A metric space X is called* **connected** *iff no such disjoint nonempty open sets exist.*

Note that if X is disconnected, then the sets U and V are closed as well as open (Exercise 16-96). Such sets are sometimes called **clopen**.

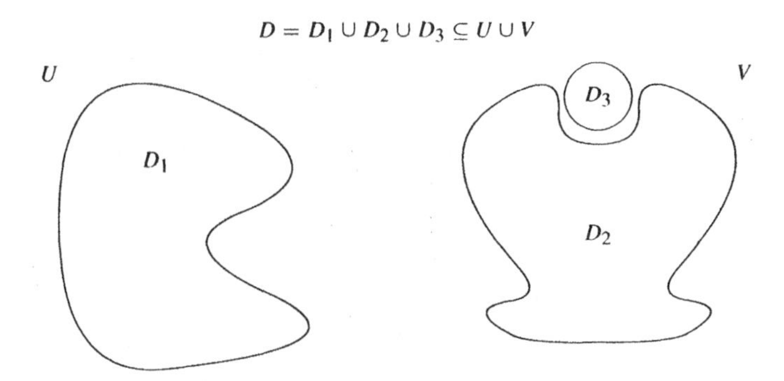

Figure 39: A disconnected subset D of a metric space can be separated by two open sets U and V into at least two nonempty components (D_1 and $D_2 \cup D_3$ in this figure).

As with any topological notion, subsets of metric spaces are disconnected iff they are disconnected as metric spaces. Figure 39 gives the idea and Exercise 16-97 investigates the properties of sets that disconnect a subset of a metric space. Most importantly, subspaces can only be disconnected by open subsets of the surrounding space.

The most natural example of a connected set is an interval. It turns out that intervals are the only connected subsets of the real numbers.

Theorem 16.91 *A subset of the real line is connected iff it is an interval.*

Proof. For "$\Rightarrow$," let S be a connected subset of $\mathbb{R}$ and suppose for a contradiction that it is not an interval. Then there are $l, r \in S$ with $l < r$ such that there is an $m \in \mathbb{R} \setminus S$ with $l < m < r$. But then $U := S \cap (-\infty, m)$ and $V := S \cap (m, \infty)$ are both open in S, nonempty and disjoint. This implies that S is disconnected, contradiction.

For "$\Leftarrow$," let I be an interval. Suppose for a contradiction that I is not connected and let $U, V \subseteq I$ be disjoint nonempty open subsets of I so that $U \cup V = I$. Let $c \in I$. Then $c \in U$ or $c \in V$. Without loss of generality assume that $c \in U$. Then there is an element $v \in V$ so that $v > c$ or $v < c$. Without loss of generality assume $v > c$. Consider the set $W := \{x \in V : x > c\}$, which is nonempty and bounded below by c. Let $b := \inf W$. Then because $c \notin V$ and U is open we infer $c < b$.

We claim that $b \notin W$. Indeed, otherwise $b \in V$ and then there is an $\varepsilon > 0$ so that $I \cap (b - \varepsilon, b + \varepsilon) \subseteq V$. Because $c < b$ and $c \notin V$ we obtain $c \leq b - \varepsilon$. Because I is an interval, we infer $(b - \varepsilon, b] \subseteq I$, and then $(b - \varepsilon, b] \subseteq W$, contradicting the choice of b.

Thus $b \notin W$. Because $v > b > c$ and I is an interval we obtain $b \in I$, and hence $b \in U$. But then there is an $\varepsilon > 0$ so that $I \cap (b - \varepsilon, b + \varepsilon) \subseteq U$. In particular, this means that $\inf(W) \geq b + \varepsilon$, contradicting the choice of b. ∎

Continuous functions satisfy an abstract version of the Intermediate Value Theorem, with connected spaces taking the place of the intervals on the real line.

Theorem 16.92 *Let X, Y be metric spaces and let $f : X \to Y$ be continuous. If X is connected, then the image $f[X]$ is a connected subspace of Y.*

Proof. Suppose for a contradiction that $f[X]$ is not connected. Then there are disjoint nonempty sets $U, V \subseteq f[X]$ that are open in $f[X]$ and satisfy $U \cup V = f[X]$. But then $f^{-1}[U]$ and $f^{-1}[V]$ are disjoint nonempty open subsets of X so that the equality $f^{-1}[U] \cup f^{-1}[V] = X$ holds, contradicting the assumption that X is connected. Thus $f[X]$ must be connected. ■

The Intermediate Value Theorem is more readily recognized in the following result.

Corollary 16.93 Intermediate Value Theorem. *Let X be a connected metric space and let $f : X \to \mathbb{R}$ be continuous. Then for all $a, b \in X$ with $f(a) \neq f(b)$ and all t between $f(a)$ and $f(b)$ there is a $c \in X$ with $f(c) = t$.*

Proof. Exercise 16-98. ■

By using images of intervals in Theorem 16.92 we obtain the idea of connectedness that was mentioned in the introduction.

Definition 16.94 *A metric space X is called* **pathwise connected** *iff for any two points $x, y \in X$ there is a continuous function $f : [0, 1] \to X$ such that $f(0) = x$ and $f(1) = y$.*

Some examples of pathwise connected sets are given in Exercise 16-99. The next result shows that these are also examples of connected sets.

Theorem 16.95 *Let X be a pathwise connected metric space. Then X is connected.*

Proof. Suppose for a contradiction that X is not connected. Then there are two nonempty disjoint open subsets U and V of X so that $U \cup V = X$. Let $u \in U$ and $v \in V$ and let $f : [0, 1] \to X$ be continuous with $f(0) = u$ and $f(1) = v$. Then $f^{-1}[U]$ and $f^{-1}[V]$ are disjoint nonempty open subsets of $[0, 1]$, which is impossible because $[0, 1]$ is connected. ■

Exercise 16-100 will show that connectedness and pathwise connectedness are equivalent for open subsets of normed spaces. However, in general pathwise connectedness is strictly stronger than connectedness.

Example 16.96 *The space $X := \big\{(x, 0) : x \leq 0\big\} \cup \left\{\left(x, \sin\left(\frac{1}{x}\right)\right) : x > 0\right\}$ is connected, but it is not pathwise connected.*

The space X is not pathwise connected, because there is no continuous function $f : [0, 1] \to X$ such that $f(0) = \left(\frac{1}{2\pi}, 0\right)$ and $f(1) = (0, 0)$. On the other hand, both $\big\{(x, 0) : x \leq 0\big\}$ and $\left\{\left(x, \sin\left(\frac{1}{x}\right)\right) : x > 0\right\}$ are pathwise connected, and hence connected. Thus, if X was disconnected, there would be open sets $U, V \subseteq \mathbb{R}^2$ so that $\big\{(x, 0) : x \leq 0\big\} \subseteq U$ and $\left\{\left(x, \sin\left(\frac{1}{x}\right)\right) : x > 0\right\} \subseteq V$. But $(0, 0) \in U$ implies that U intersects $\left\{\left(x, \sin\left(\frac{1}{x}\right)\right) : x > 0\right\}$, so this is not possible. □

Exercises

16-96. Prove that a metric space X is disconnected iff there are disjoint nonempty closed sets $C, D \subseteq X$ so that $C \cup D = X$.

16-97. Disconnected subspaces. Let X be a metric space and let $S \subseteq X$.

(a) Prove that S is disconnected iff there are disjoint nonempty subsets $U, V \subseteq X$ *that are open in* X such that $S \subseteq U \cup V$.

(b) Give an example that shows that a subspace S can be disconnected and there are *no* disjoint nonempty subsets $C, D \subseteq X$ *that are closed in* X such that $S \subseteq C \cup D$.

16-98. Prove Corollary 16.93.

16-99. Let X be a normed space.

(a) Prove that for any two points $a, b \in X$ the function $f(t) := a + t(b - a)$ is continuous.

(b) Prove that X is pathwise connected.

(c) Prove that for all $x \in X$ and all $\varepsilon > 0$ the ball $B_\varepsilon(x)$ is pathwise connected.

16-100. Let X be a normed space and let $\Omega \subseteq X$ be open. Prove that Ω is connected iff it is pathwise connected.

16-101. Let X, Y be metric spaces and let $f : X \to Y$ be continuous. Prove that if X is pathwise connected, then $f[X]$ is a pathwise connected subspace of Y.

16-102. Unit "circles."

(a) Prove that $\left\{ (x, y) \in \mathbb{R}^2 : x^2 + y^2 = 1 \right\}$ is connected.
Hint. $c(t) = \cos(t)e_1 + \sin(t)e_2$.

(b) Let $1 \leq p < \infty$. Prove that $\left\{ (x, y) \in \mathbb{R}^2 : \|(x, y)\|_p = 1 \right\}$ is connected.

(c) Prove that $\left\{ (x, y) \in \mathbb{R}^2 : \|(x, y)\|_\infty = 1 \right\}$ is connected.

16-103. Connected components. Let X be a metric space. A subset $C \subseteq X$ is called a **component** of X iff C is connected and there is no proper superset $D \supset C$ that is connected.

(a) Prove that if $A, B \subseteq X$ are connected and $A \cap B \neq \emptyset$, then $A \cup B$ is connected.

(b) Prove that if C_1, C_2 are components of X then either $C_1 = C_2$ or $C_1 \cap C_2 = \emptyset$.

(c) Prove that every $x \in X$ is contained in a component of X.

(d) Prove that every open subset of $\mathbb{R}$ is a countable union of pairwise disjoint open intervals.

16.9 Locally Compact Spaces

Readers interested in differentiation could move on to Chapter 17 and pick the results in this section up when they are needed before Section 19.5.

The goal of this section is to construct (families of) continuous functions that are equal to one on a specified set and equal to zero on another set. To construct such functions, we introduce the distance function.

Definition 16.97 *Let X be a metric space and let $A \subseteq X$ be nonempty. For all $x \in X$, we define* $\text{dist}(x, A) := \inf\left\{d(x, a) : a \in A\right\}$ *and call it the* **distance** *from x to A.*

Lemma 16.98 *Let X be a metric space and let $A \subseteq X$ be nonempty. Then the function* $\text{dist}(\cdot, A)$ *is Lipschitz continuous.*

Proof. Let $x, y \in X$, let $\varepsilon > 0$ and let $a \in A$ be so that $d(y,a) \leq \text{dist}(y, A) + \varepsilon$. Then $\text{dist}(x, A) \leq d(x,a) \leq d(x, y) + d(y,a) \leq d(x, y) + \text{dist}(y, A) + \varepsilon$, and hence $\text{dist}(x, A) - \text{dist}(y, A) \leq d(x, y) + \varepsilon$. Because $\varepsilon > 0$ was arbitrary this means that $\text{dist}(x, A) - \text{dist}(y, A) \leq d(x, y)$. We can prove $\text{dist}(y, A) - \text{dist}(x, A) \leq d(x, y)$ in similar fashion. Hence, $\left|\text{dist}(x, A) - \text{dist}(y, A)\right| \leq d(x, y)$ and $\text{dist}(\cdot, A)$ is Lipschitz continuous with Lipschitz constant 1 ■

Lemma 16.99 below says that for any closed set C that is contained in an open set U the distance function allows us to slip another open set V between C and U so that the closure of V is also between C and U. For an illustration of Lemma 16.99, see Figure 40 on page 336.

Lemma 16.99 *Let X be a metric space, let $C \subseteq X$ be closed and let $U \subseteq X$ be open so that $C \subseteq U$. Then there is a continuous function $f : X \to [0, 1]$ so that $f|_C = 1$ and $f|_{X\setminus U} = 0$. Moreover, there is an open set V so that $C \subseteq V \subseteq \overline{V} \subseteq U$.*

Proof. For each $x \in X$, let $f(x) := \dfrac{\text{dist}(x, X \setminus U)}{\text{dist}(x, X \setminus U) + \text{dist}(x, C)}$. Because C and $X \setminus U$ are disjoint closed sets, the denominator is greater than zero for all $x \in X$ (see Exercise 16-104b). Thus f is continuous on X. Moreover, (see Exercise 16-105) $f|_C = 1$ and $f|_{X\setminus U} = 0$.

To prove the claim about the sets, let $V := f^{-1}\left[\left(\frac{1}{2}, 1\right]\right]$. Because $\left(\frac{1}{2}, 1\right]$ is open in $[0, 1]$ and f is continuous, V is an open set in X and because $f|_C = 1$ it contains C. Moreover, because f is continuous, for all $x \in \overline{V}$ we infer $f(x) \geq \frac{1}{2}$, which means that $\overline{V} \subseteq U$. ■

For compact sets C, we would like to separate C from its neighborhood U with an open set V whose closure is compact. While this is not possible in general (see Exercise 16-108), it is possible in spaces with sufficiently many compact subsets. Local compactness guarantees that locally there are enough compact subsets by demanding that every point has a compact neighborhood.

Definition 16.100 *A metric space X is called* **locally compact** *iff every $x \in X$ has a compact neighborhood.*

Proposition 16.101 *A metric space X is locally compact iff for every $x \in X$ there is an $\varepsilon > 0$ so that $\overline{B_\varepsilon(x)}$ is compact.*

Proof. For "⇒," let X be locally compact and let $x \in X$. Then x has a compact neighborhood N. Let $\varepsilon > 0$ be so that $\overline{B_\varepsilon(x)} \subseteq N$. Then by Proposition 16.61 the set $\overline{B_\varepsilon(x)}$ is compact.

Conversely, let X be so that for every $x \in X$ there is an $\varepsilon > 0$ so that $\overline{B_\varepsilon(x)}$ is compact. Then for every $x \in X$ the neighborhood $N := \overline{B_\varepsilon(x)}$ is compact. ■

It is easy to infer from Proposition 16.101 that all open subsets of $\mathbb{R}^d$ and all closed subsets of $\mathbb{R}^d$ are locally compact. In particular, we obtain that surfaces like the unit sphere $\left\{x \in \mathbb{R}^d : \|x\|_2 = 1\right\}$ are locally compact. More generally, we can say the following.

Definition 16.102 *Let X, Y be metric spaces. Then $f : X \to Y$ is called a* **homeomorphism** *iff f is continuous and bijective and its inverse is continuous, too.*

Example 16.103 Any metric space for which each point has a neighborhood that is homeomorphic to an open set in $\mathbb{R}^d$ is locally compact. If the dimension d does not depend on the point, we call the space a manifold. Manifolds are discussed in detail in Chapter 19. Surfaces such as the unit sphere are manifolds. For much of the following, solids and surfaces in $\mathbb{R}^d$ are a good visualization and motivation. □

In locally compact spaces, between any compact set C and any open neighborhood U of C we can slip a compact neighborhood of C (also see Figure 40(a)).

Lemma 16.104 *Let X be a locally compact metric space, let $C \subseteq X$ be compact and let $U \subseteq X$ be open with $C \subseteq U$. Then C has a neighborhood V so that $\overline{V}$ is compact and contained in U.*

Proof. For each $c \in C$, there is an $\varepsilon_c > 0$ so that $\overline{B_{\varepsilon_c}(c)}$ is compact and contained in U. Because $C \subseteq \bigcup_{c \in C} B_{\varepsilon_c}(c)$ and C is compact, there are $c_1, \ldots, c_n \in C$ so that $C \subseteq \bigcup_{j=1}^{n} B_{\varepsilon_{c_j}}(c_j)$. But then $V := \bigcup_{j=1}^{n} B_{\varepsilon_{c_j}}(c_j)$ is a neighborhood of C so that by Exercises 16-44b and 16-110 the closure $\overline{V} = \overline{\bigcup_{j=1}^{n} B_{\varepsilon_{c_j}}(c_j)} = \bigcup_{j=1}^{n} \overline{B_{\varepsilon_{c_j}}(c_j)}$ is compact. Moreover, because all $\overline{B_{\varepsilon_c}(c)}$ are contained in U, the union is contained in U, too. ■

Standard Proof Technique 16.105 The argument in the proof of Lemma 16.104 is a typical application of the Heine-Borel formulation of compactness. We start with an open cover and because the notion we want to preserve is only preserved by finite unions, we use a **finite subcover**. □

Local compactness only applies near individual points. As it turns out, for connected spaces we can turn this local idea into a property that allows us to use compactness in a more global fashion.

Definition 16.106 *A metric space X is called* **σ-compact** *iff X is the union of countably many compact sets.*

Clearly, closed subsets of $\mathbb{R}^d$ are σ-compact, because their intersections with the closed balls $\overline{B_n(0)}$ are compact. More specifically we can say the following.

Theorem 16.107 *Let X be a connected, locally compact metric space. Then X is σ-compact.*

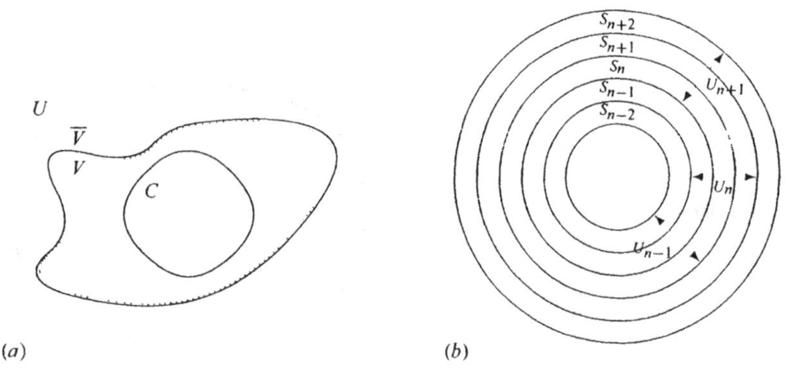

Figure 40: Part (a) illustrates Lemmas 16.99 and 16.104, which say that between any closed (compact) set C and any open set U surrounding C there is an open set V so that V and $\overline{V}$ are "between" C and U. Part (b) shows a σ-compact space and the idea for the proof of Theorem 16.112. The concentric circles depict a compact exhaustion (the sets K_n in the proof of Theorem 16.112). The shells S_n and their neighborhoods U_n are set up so that only finitely may of the U_n can intersect.

Proof. For each $x \in X$, let $r_x := \sup\left\{r > 0 : \overline{B_r(x)} \text{ is compact}\right\}$. Because X is locally compact, each r_x is greater than zero. If $x \in X$ is so that r_x is infinity, then for all $n \in \mathbb{N}$ the set $\overline{B_n(x)}$ is compact, and hence $X = \bigcup_{n=1}^{\infty} \overline{B_n(x)}$ is σ-compact. This leaves the case in which each r_x is finite.

In this case, we first prove that the function $x \mapsto r_x$ is continuous. Let $x, z \in X$ be so that $d(x, z) < r_x$. Then for all $r \in \big(d(x, z), r_x\big)$ we infer $\overline{B_{r-d(x,z)}(z)} \subseteq \overline{B_r(x)}$ and the ball on the right is compact. Hence, $r_z \geq r_x - d(x, z)$, that is, $r_x - r_z \leq d(x, z)$. Moreover, for $d(x, z) < \frac{r_x}{2}$ we have $r_z > \frac{r_x}{2}$ and, reversing the roles of x and z, we can also prove $r_z - r_x \leq d(x, z)$. Hence $|r_x - r_z| \leq d(x, z)$ for all z with $d(x, z) < \frac{r_x}{2}$ and $x \mapsto r_x$ is continuous at each $x \in X$.

Now for each compact subset C of X we define $N(C) := \bigcup_{c \in C} \overline{B_{\frac{r_c}{2}}(c)}$. We claim that the set $N(C)$ is compact. Let $\{x_n\}_{n=1}^{\infty}$ be a sequence in $N(C)$. For each x_n there is a $c_n \in C$ so that $x_n \in \overline{B_{\frac{r_{c_n}}{2}}(c_n)}$. Because C is compact, $\{c_n\}_{n=1}^{\infty}$ has a convergent subsequence with limit $c \in C$. Without loss of generality we can assume that $\{c_n\}_{n=1}^{\infty}$ itself converges to c. Then there is an $N \in \mathbb{N}$ so that for all $n \geq N$ the inequalities $d(c_n, c) < \frac{r_c}{4}$ and $\left|r_{c_n} - r_c\right| < \frac{r_c}{4}$ hold. But then for all $n \geq N$ we have $\frac{r_{c_n}}{2} < \frac{5r_c}{8}$, and hence $d(x_n, c) \leq d(x_n, c_n) + d(c_n, c) < \frac{r_{c_n}}{2} + \frac{r_c}{4} < \frac{7r_c}{8}$, so $x_n \in \overline{B_{\frac{7r_c}{8}}(c)}$. Because $\overline{B_{\frac{7r_c}{8}}(c)}$ is compact, $\{x_n\}_{n=1}^{\infty}$ has a convergent subsequence. This proves that $N(C)$ is compact.

Now let $x_0 \in X$ be arbitrary and let $C_1 := \{x_0\}$. Recursively define $C_{n+1} := N(C_n)$

for $n \in \mathbb{N}$. Let $H := \bigcup_{n=1}^{\infty} C_n$. Then because each C_n is compact, H is σ-compact. Moreover, because each C_{n+1} is a neighborhood of every element of C_n, H is open. To see that H is closed, let $x \in \overline{H}$. Then there is an $h \in H$ so that $d(h, x) < \frac{r_x}{4}$ and $|r_h - r_x| < \frac{r_x}{4}$. But then $\frac{r_h}{2} > \frac{3r_x}{8} > \frac{r_x}{4}$, which means that if $n \in \mathbb{N}$ is so that $h \in C_n$, then $x \in N(C_n)$. Hence, $H = \overline{H}$ is closed and so $X \setminus H$ is open. Because X is connected, this means that $X = H$ and since H is σ-compact the result is proved. ■

We conclude this section by proving that locally compact spaces have a partition of unity (see Definition 16.110 below) with certain properties.

Definition 16.108 *The cover $\mathcal{O}$ of the metric space X is called* **locally finite** *iff each $p \in X$ has a neighborhood that intersects only finitely many elements of $\mathcal{O}$.*

Definition 16.109 *Let X be a metric space and let $f : X \to \mathbb{R}$. Then the* **support** *of f is defined to be* $\text{supp}(f) := \overline{\{x \in X : f(x) \neq 0\}}$.

Definition 16.110 *A family $\{\varphi_j\}_{j \in J}$ of continuous functions on a metric space X is called a* **partition of unity** *iff*

1. *The collection $\left\{\{x \in X : \varphi_j(x) \neq 0\}\right\}_{j \in J}$ is a locally finite cover of X.*

2. *For all $x \in X$ we have that $\sum_{j \in J} \varphi_j(x) = 1$. (By part 1 for each $x \in X$ this sum has only finitely many nonzero terms.)*

If $\mathcal{O}$ is an open cover of X and for each $j \in J$ the containment $\text{supp}(\varphi_j) \subset U$ *holds for some $U \in \mathcal{O}$, then the partition of unity is called* **subordinate** *to $\mathcal{O}$.*

The importance of partitions of unity will become clear in Section 19.5. Until then, consider the following. Many surfaces in $\mathbb{R}^d$ cannot be parametrized with just one function that has an open domain. (Open domains are needed, because differentiable functions typically have open domains, see Chapter 17.) For example, a parametrization of the unit sphere, say, with spherical coordinates, will always either hit a few points twice or it will miss at least a "seam." This is because a parametrization must be a homeomorphism and the unit sphere is compact, which means it cannot be homeomorphic to an open subset of $\mathbb{R}^d$. However, roughly speaking, a function is integrated over the sphere by integrating its composition with the right parametrization. Thus, it is problematic to double count points or miss points. Either case would distort the integral. This problem does not arise for functions that are zero except on some small open set. For such functions, we could simply use a parametrization for which the missed seam does not intersect the support of the function. A partition of unity $\{\varphi_j\}_{j \in J}$ allows us to represent arbitrary functions f as sums $\sum_{j \in J} \varphi_j f$ of functions $\varphi_j f$ whose supports are contained in "small" open sets. We can then integrate these functions separately and the overall sum will be the integral of f. There is still a tremendous amount of detail left to be considered (think about independence of the parametrization), and this is why we will later need partitions of unity subordinate to open covers that have further nice properties.

Because not every cover is locally finite, we need the notion of a refinement.

Definition 16.111 *Let $\mathcal{O}$ be a cover of the metric space X. A cover $\widetilde{\mathcal{O}}$ is called a* **refinement** *of $\mathcal{O}$ iff for all $U \in \widetilde{\mathcal{O}}$ there is a $V \in \mathcal{O}$ so that $U \subseteq V$.*

Theorem 16.112 *Let $\mathcal{O}$ be an open cover of the locally compact, σ-compact metric space X. Then $\mathcal{O}$ has a countable locally finite open refinement. Moreover, the closures of the sets in the refinement are compact.*

Proof. There is nothing to prove if X is compact. If X is not compact, let $\{C_j\}_{j=1}^{\infty}$ be a sequence of compact sets with $X = \bigcup_{j=1}^{\infty} C_j$. Let $K_1 := C_1$ and once $K_1, \ldots, K_n$ are defined, let K_{n+1} be a compact neighborhood of $K_n \cup C_n$ so that $K_n \cup C_n \subseteq K_{n+1}^{\circ}$. Then $X = \bigcup_{n=1}^{\infty} K_n$ and for all $n \in \mathbb{N}$ we have the containment $K_n \subseteq K_{n+1}^{\circ}$.

Let $K_{-1} := K_0 := \emptyset$. For all $n \in \mathbb{N}$ let $S_n := K_n \setminus K_{n-1}^{\circ}$ and $U_n := K_{n+1}^{\circ} \setminus K_{n-2}$. Then $X = \bigcup_{j=1}^{\infty} S_j$ and for all $n \in \mathbb{N}$ the set S_n is compact, U_n is open and $S_n \subseteq U_n$. For $|n - m| > 1$, we have $S_n \cap S_m = \emptyset$ and for all $|n - m| > 2$ we have $U_n \cap U_m = \emptyset$ (also see the right part of Figure 40). For each $x \in S_n$, let N_x be an open neighborhood of X that is contained in U_n and in some $O \in \mathcal{O}$. Then $S_n \subseteq \bigcup_{x \in S_n} N_x$ and because S_n is compact, there are $x_1^{(n)}, \ldots, x_{k_n}^{(n)} \in S_n$ so that $S_n \subseteq \bigcup_{j=1}^{k_n} N_{x_j^{(n)}}$.

We define $\widetilde{\mathcal{O}} := \left\{ N_{x_j^{(n)}} : n \in \mathbb{N}, j = 1, \ldots, k_n \right\}$. Clearly, $\widetilde{\mathcal{O}}$ is countable. Because $X = \bigcup_{j=1}^{\infty} S_j$ and $S_n \subseteq \bigcup_{j=1}^{k_n} N_{x_j^{(n)}}$ for all $n \in \mathbb{N}$ we conclude that $\widetilde{\mathcal{O}}$ is a cover of X. All $N_{x_j^{(n)}}$ are open and contained in an $O \in \mathcal{O}$, so $\widetilde{\mathcal{O}}$ is an open refinement of $\mathcal{O}$. Finally, to see that $\widetilde{\mathcal{O}}$ is locally finite, let $x \in X$. Then there is a $k \in \mathbb{N}$ so that $x \notin U_m$ unless $m \in \{k, k+1, k+2\}$. Any $N_{x_j}^{(n)}$ that intersects $U_k \cup U_{k+1} \cup U_{k+2}$ must be contained in one of $U_{k-2}, \ldots, U_{k+4}$. These sets contain finitely many $N_{x_j}^{(n)}$ each. Therefore x can be in at most finitely many $N_{x_j}^{(n)}$, and hence $\widetilde{\mathcal{O}}$ is locally finite. Finally, because each N_x is contained in a $U_n \subseteq K_{n+1}$, the closure of each N_x is compact. ■

As it turns out, any locally finite open cover of a σ-compact space is countable.

Proposition 16.113 *Let $\mathcal{O}$ be a locally finite open cover of the σ-compact metric space X. Then $\mathcal{O}$ is countable.*

Proof. Exercise 16-109. ■

To construct a partition of unity subordinate to a locally finite open cover we need to find a way to define functions that are supported inside the open sets so that the sets where the functions are not zero cover the whole metric space. To do that, we need to be able to shrink every set in the open cover a little bit while making sure that we still cover the whole space.

Theorem 16.114 Shrinking Lemma. *Let $\mathcal{O}$ be a locally finite open cover of the locally compact, σ-compact metric space X. Then for each $U \in \mathcal{O}$ there is an open set V_U so that $\overline{V_U} \subset U$ and so that $\widetilde{\mathcal{O}} := \{V_U : U \in \mathcal{O}\}$ is a locally finite open cover of X.*

Proof. By Proposition 16.113, $\mathcal{O}$ is at most countable, so let $\{U_n : n \in \mathbb{N}\} := \mathcal{O}$. (For finite covers $\mathcal{O}$, the construction below terminates in finitely many steps with the desired cover $\widetilde{\mathcal{O}}$.) The set $C_1 := U_1 \setminus \bigcup_{n=2}^{\infty} U_n = X \setminus \bigcup_{n=2}^{\infty} U_n$ is closed and contained in U_1. By Lemma 16.99, there is an open set V_1 so that $C_1 \subseteq V_1 \subseteq \overline{V_1} \subseteq U_1$. Then $\mathcal{O}_1 := \{V_1\} \cup \{U_n : n > 1\}$ is an open cover of X and $\overline{V_1} \subseteq U_1$.

Once an open cover $\mathcal{O}_k = \{V_1, \ldots, V_k\} \cup \{U_n : n > k\}$ has been constructed so that $\overline{V_j} \subseteq U_j$ holds for $j = 1, \ldots, k$, let $C_{k+1} := U_{k+1} \setminus \left(\bigcup_{j=1}^{k} V_j \cup \bigcup_{n=k+2}^{\infty} U_n \right)$. Then C_{k+1} is closed and contained in U_{k+1}. By Lemma 16.99, there is an open set V_{k+1} so that $C_k \subseteq V_{k+1} \subseteq \overline{V_{k+1}} \subseteq U_{k+1}$. Let $\mathcal{O}_{k+1} := \{V_1, \ldots, V_k, V_{k+1}\} \cup \{U_n : n > k+1\}$. Then $\mathcal{O}_{k+1}$ is an open cover of X and $\overline{V_j} \subseteq U_j$ for all $j = 1, \ldots, k+1$.

Now consider $\widetilde{\mathcal{O}} := \{V_j : j \in \mathbb{N}\}$. Because $\mathcal{O}$ is locally finite, for each $x \in X$ there is an $N \in \mathbb{N}$ so that $x \notin U_n$ for all $n \geq N$. But then, because $\mathcal{O}_N$ was an open cover of X, there must be a $j < N$ so that $x \in V_j$. Hence, $\widetilde{\mathcal{O}}$ is a cover of X. By construction, for all $j \in \mathbb{N}$ the set V_j is open and satisfies $\overline{V_j} \subseteq U_j$. Because $\mathcal{O}$ is locally finite, $\widetilde{\mathcal{O}}$ must also be locally finite and thus the result is proved. ■

Theorem 16.115 *Let $\mathcal{O}$ be an open cover of the locally compact, σ-compact metric space X. Then there is a partition of unity subordinate to $\mathcal{O}$.*

Proof. Let $\mathcal{U}$ be a locally finite open refinement of $\mathcal{O}$ as guaranteed by Theorem 16.112. Let $\widetilde{\mathcal{U}}$ be a locally finite open cover so that for every $U \in \mathcal{U}$ there is a $V_U \in \widetilde{\mathcal{U}}$ so that $\overline{V_U} \subseteq U$, as guaranteed by the Shrinking Lemma. For each $U \in \mathcal{U}$, let W_U be an open set so that $\overline{V_U} \subseteq W_U \subseteq \overline{W_U} \subseteq U$ as provided by Lemma 16.99 and let $\psi_U : X \to [0, 1]$ be a continuous function so that $\psi_U|_{V_U} = 1$ and $\psi_U|_{X \setminus W_U} = 0$ as provided by Lemma 16.99. Because $\mathcal{U}$ is locally finite, each $x \in X$ has a neighborhood V so that for all $v \in V$ the equality $\psi_U(v) = 0$ holds for all but finitely many $U \in \mathcal{U}$. Because $\widetilde{\mathcal{U}}$ is a cover of X, for at least one $U \in \mathcal{U}$ we have $\psi_U(x) \neq 0$. Hence, for all $x \in X$ the sum $\psi(x) := \sum_{U \in \mathcal{U}} \psi_U(x)$ is a positive real number. Moreover, for each

$x \in X$ on a neighborhood V of x the function ψ is the sum of finitely many continuous functions. Hence, ψ is continuous on this neighborhood of x and so ψ is continuous at x. Because x was arbitrary, ψ is continuous on X.

For each $U \in \mathcal{U}$, we have $\left\{x \in X : \frac{\psi_u(x)}{\psi(x)} \neq 0\right\} = \{x \in X : \psi_U(x) \neq 0\}$. For all $U \in \mathcal{U}$ define $\varphi_U := \frac{\psi_u}{\psi}$. Then $\{\{x \in X : \varphi_U(x) \neq 0\}\}_{U \in \mathcal{U}}$ is locally finite. Moreover, for all $x \in X$ we have $\sum_{U \in \mathcal{U}} \varphi_U(x) = \sum_{U \in \mathcal{U}} \frac{\psi_U(x)}{\psi(x)} = \frac{\psi(x)}{\psi(x)} = 1$. Hence, $\{\varphi_U\}_{U \in \mathcal{U}}$ is a partition of unity. For each U, we have $\text{supp}(\varphi_U) \subseteq U \subseteq O$ for some $O \in \mathcal{O}$, so the partition of unity $\{\varphi_U\}_{U \in \mathcal{U}}$ is subordinate to $\mathcal{O}$. ■

Exercises

16-104. Let X be a metric space and let $A \subseteq X$ be a nonempty subset

(a) Prove that $\text{dist}(x, A) = 0$ iff there is a sequence $\{a_n\}_{n=1}^{\infty}$ in A with $\lim_{n \to \infty} a_n = x$.

(b) Let $C \subseteq X$ be closed and nonempty Prove that $\text{dist}(x, C) = 0$ iff $x \in C$.

16-105 Let a, b be distinct nonnegative numbers that are not both zero Prove that we have $\frac{a}{a+b} \in [0, 1]$, that $\frac{a}{a+b} = 0$ iff $a = 0$, and that $\frac{a}{a+b} = 1$ iff $b = 0$.

16-106. Let X be a metric space and let $C \subseteq X$ be a nonempty compact subset Prove that for all $x \in X$ there is a $c_x \in C$ so that $d(x, c_x) = \text{dist}(x, C)$

16-107 Let X be a metric space. For any two nonempty subsets $A, B \subseteq X$, define the **distance** from A to B as $\text{dist}(A, B) = \inf \{ \text{dist}(a, B) . a \in A \}$.

(a) Prove that for all nonempty subsets $A, B \subseteq X$ we have $\text{dist}(A, B) = \text{dist}(B, A)$.

(b) Give an example of two closed, disjoint, nonempty sets A, B such that $\text{dist}(A, B) = 0$.

(c) Prove that the function in Lemma 16.99 need not be uniformly continuous, and hence in particular it need not be Lipschitz continuous.

(d) Prove that if B and C are not empty and C is compact, then there is a $c \in C$ so that $\text{dist}(C, B) = \text{dist}(c, B)$

16-108. Give an example of a metric space in which Lemma 16.104 fails That is, give an example of a metric space in which no neighborhood of a compact set is compact.

16-109. Prove Proposition 16.113

16-110. Let X be a metric space and let $C_1, \ldots, C_n$ be compact subsets of X. Prove that $\bigcup_{j=1}^{n} C_j$ is compact.

16-111 Give an example of a bijective continuous linear function whose inverse is not continuous

Chapter 17

Differentiation in Normed Spaces

To discuss differentiation, we need algebraic operations and a metric. Normed spaces have the algebraic structure of a vector space and a metric induced by the norm (see Proposition 15.54). The vector space structure is typically discussed in linear algebra. To keep the text self-contained, we introduce the requisite concepts and ideas in this chapter. Moreover, we freely use metric concepts discussed in Chapter 16. The presentation will be coordinate-free and valid for arbitrary (including infinite) dimensions. Although derivatives are mainly used in finite dimensional spaces and computed through partial derivatives along coordinate axes, this abstraction does not make the proofs more complicated. Instead, the omission of coordinates allows us to focus on the conceptual core of differentiation. The relevant results for finite dimensional spaces are given as consequences of the general theory.

To understand differentiation in multidimensional spaces, we must first adjust our expectation what a derivative should be. The derivative cannot be a number or a slope, because it is not clear in which direction this slope would go. Instead, differentiation is defined similar to Theorem 4.5, which says that the derivative at x determines the unique straight line through $\big(x, f(x)\big)$ for which the difference (at z) between the function and the line goes to zero faster than $|z - x|$. Geometrically, (hyper)planes will take the place of lines. Linear functions are the analytical tool used to define (hyper)planes. We start our investigation with linear functions in Sections 17.1 and 17.2. Derivatives and partial derivatives are introduced in Sections 17.3 and 17.5. In between, Section 17.4 introduces the Mean Value Theorem, which is crucial for using derivatives to estimate differences. Section 17.6 introduces tensors, which are needed to represent higher derivatives in Section 17.7. We conclude in Section 17.8 with the Implicit Function Theorem, which provides important examples of manifolds.

17.1 Continuous Linear Functions

The definition of a linear function is entirely algebraic. It simply states that the function is compatible with the vector space operations.

Definition 17.1 *Let X, Y be vector spaces. A function $L : X \to Y$ is called* **linear** *iff for all $x_1, x_2 \in X$ we have $L[x_1 + x_2] = L[x_1] + L[x_2]$ and for all $\alpha \in \mathbb{R}$ and $x \in X$ we have $L[\alpha x] = \alpha L[x]$. A linear function is also sometimes called a* **linear operator**.

Notation 17.2 Derivatives will be functions that map points to linear functions. To distinguish the various evaluations, throughout the text we will **enclose the argument of a linear function in square brackets** rather than round parentheses. To avoid confusion with the square brackets which indicate that the elements of L^p spaces are equivalence classes, we will henceforth **omit the brackets around the elements of L^p spaces**, as is customary in analysis. □

Differentiation and integration both lead to examples of linear functions.

Example 17.3

1. Let (M, Σ, μ) be a measure space, let $1 \leq p, q \leq \infty$ with $\frac{1}{p} + \frac{1}{q} = 1$ and let $g \in L^q(M, \Sigma, \mu)$. By Example 16.28, $I_g[f] := \int_M fg \, d\mu$ defines a continuous function $I_g : L^p(M, \sigma, \mu) \to \mathbb{R}$ and it is easy to see via Theorem 14.39 that I_g is linear.

2. Let $1 \leq p < \infty$ and let X be the subspace of functions $f \in C^1(a, b) \cap L^p(a, b)$ so that $f' \in L^p(a, b)$, too. When it is not mentioned explicitly that elements of $L^p(a, b)$ are equivalence classes, it is customary to use intersections like $C^1(a, b) \cap L^p(a, b)$ rather than the more complicated notation from Theorem 16.86. We claim that the function $D : X \to L^p(a, b)$ defined by $D[f] := f'$ is linear, but not continuous.

   If two functions in X are equal almost everywhere, then they must be equal (see Exercise 8-5). That is, if $f, g \in X \subseteq L^p(a, b)$ and $f = g$ a.e., then $f = g$, which implies that D is well-defined. (Just because we do not mention that the elements of L^p are equivalence classes does not mean we do not need to pay attention to this fact when we define a function.) Linearity of D follows easily from the corresponding linearity properties of the derivative (see Theorem 4.6). To see that D is not continuous, we will only consider the case $a = 0$ and $b = 1$ here. The general case is left to the reader in Exercise 17-1. Because

$$\lim_{n\to\infty} \|x^n\|_p = \lim_{n\to\infty} \left(\int_0^1 (x^n)^p \, d\lambda \right)^{\frac{1}{p}} = \lim_{n\to\infty} \left(\frac{1}{np+1} \right)^{\frac{1}{p}} = 0, \text{ while}$$

$$\lim_{n\to\infty} \left\| D\left[x^n\right] \right\|_p = \lim_{n\to\infty} \left\| nx^{n-1} \right\|_p = \lim_{n\to\infty} \left(\int_0^1 \left(nx^{n-1}\right)^p \, d\lambda \right)^{\frac{1}{p}}$$

$$= \lim_{n\to\infty}\left(\frac{n^p}{np-p+1}\right)^{\frac{1}{p}} = \begin{cases}\infty; & \text{if } p>1,\\ 1; & \text{if } p=1,\end{cases}$$

the function D is not continuous at 0. □

It is a good rule of thumb that integration usually defines continuous linear functions and differentiation usually defines discontinuous linear functions. We have to be careful, though. Exercise 17-10 shows that integration need not always define a continuous linear function and Exercise 17-13 shows that differentiation can be continuous.

The multiplication of a matrix with column vectors is another fundamental example of a linear function. We will investigate the connection between the "abstract" concept of continuous linear functions and the more "concrete" idea of matrix multiplication in Section 17.2. Before then, we need to investigate continuity for linear functions.

In Example 17.3, we have only proved that D is not continuous at the origin. Loosely speaking, for linear functions the behavior at the origin is duplicated at every point. In particular, Exercise 17-2 shows that D is discontinuous at every point of $L^p(a,b)$. As a positive result, Theorem 17.4 below shows that for linear functions continuity at the origin is equivalent to continuity everywhere. Note how the proof uses linearity to reduce every situation to a configuration near the origin.

Theorem 17.4 *Let X, Y be normed spaces and let $L : X \to Y$ be a linear function. The following are equivalent:*

1. *L is continuous on X.*
2. *L is continuous at 0.*
3. *L is bounded on $\overline{B_1(0)} \subset X$.*
4. *There is a $c \in \mathbb{R}$ so that for all $x \in X$ we have $\|L[x]\| \leq c\|x\|$.*

Proof. The implication "1⇒2" is trivial. For "2⇒3," let $\varepsilon > 0$. Because L is continuous at 0, there is a $\delta > 0$ so that for all $x \in X$ with $\|x\| = \|x-0\| \leq \delta$ we have $\|L[x]\| < \varepsilon$. Therefore for all points $x \in X$ with $\|x\| = \|x-0\| \leq 1$ we obtain $\|L[x]\| = \left\|L\left[\frac{1}{\delta}\delta x\right]\right\| = \frac{1}{\delta}\|L[\delta x]\| \leq \frac{1}{\delta}\varepsilon$. Hence, L is bounded on $\overline{B_1(0)}$.

For "3⇒4," let $c \in \mathbb{R}$ be such that for all $x \in \overline{B_1(0)}$ we have $\|L[x]\| \leq c$. Then for all $x \in X$ we infer $\|L[x]\| = \left\|L\left[\frac{x}{\|x\|}\right]\right\| \|x\| \leq c\|x\|$.

For "4⇒1," let $c > 0$ be such that for all $x \in X$ we have $\|L[x]\| \leq c\|x\|$. Let $\varepsilon > 0$ and set $\delta := \frac{\varepsilon}{c}$. Then for all $x, y \in X$ with $\|x-y\| < \delta$ we obtain

$$\|L[x]-L[y]\| = \|L[x-y]\| \leq c\|x-y\| < c\frac{\varepsilon}{c} = \varepsilon.$$ ■

In particular, Theorem 17.4 shows that every continuous linear function is Lipschitz continuous.

Definition 17.5 *Because of part 3 of Theorem 17.4, continuous linear functions are also called* **bounded** *linear functions.*

It is now easy to show that any linear function on a finite dimensional normed space is continuous.

Corollary 17.6 *Let X be a finite dimensional normed space, let Y be a normed space and let $L : X \to Y$ be a linear function. Then L is continuous.*

Proof. Let $\{v_1, \ldots, v_d\}$ be a base of X. Then for all $x \in X$ there are unique coefficients $c_1, \ldots, c_d$ so that $x = \sum_{i=1}^{d} c_i v_i$. Denote the norm on X by $\|\cdot\|$ and note that $\|x\|_\infty := \max\left\{|c_i| : i = 1, \ldots, d\right\}$ is another norm on X. By Theorem 16.76, $\|\cdot\|_\infty$ and the original norm $\|\cdot\|$ of X are equivalent. Let $r > 0$ be such that for all $x \in X$ we have $\|x\|_\infty \le r\|x\|$ and let $c := r\sum_{i=1}^{d} \left\|L[v_i]\right\|$. Then for all $x = \sum_{i=1}^{d} c_i v_i \in X$ we obtain

$$\left\|L[x]\right\| = \left\|L\left[\sum_{i=1}^{d} c_i v_i\right]\right\| \le \sum_{i=1}^{d} |c_i| \left\|L[v_i]\right\| \le \left(\sum_{i=1}^{d} \left\|L[v_i]\right\|\right) \|x\|_\infty \le c\|x\|,$$

which means that L is continuous. ■

Linear functions can be added pointwise and they can be multiplied with real numbers. Hence, the linear functions from one normed space to another form a vector space. Part 4 of Theorem 17.4 enables us to define a norm on this space.

Definition 17.7 *Let X, Y be normed spaces. We define $\mathcal{L}(X, Y)$ to be the set of all continuous linear functions from X to Y.*

Theorem 17.8 *Let X, Y be normed spaces. Then, with pointwise addition and scalar multiplication, $\mathcal{L}(X, Y)$ is a vector space. Moreover, the function*

$$\|L\| := \min\left\{c \ge 0 : \left(\forall x \in X : \|L[x]\| \le c\|x\|\right)\right\}$$

is a norm on $\mathcal{L}(X, Y)$ such that for all $x \in X$ we have $\left\|L[x]\right\| \le \|L\|\|x\|$.

Proof. The proof that $\mathcal{L}(X, Y)$ is a vector space is left to Exercise 17-3a. To prove that $\mathcal{L}(X, Y)$ is a normed space, we start by defining for all $L \in \mathcal{L}(X, Y)$ the quantity $\|L\| := \inf\left\{c \ge 0 : \left(\forall x \in X : \|L[x]\| \le c\|x\|\right)\right\}$. This is necessary, because we do not know *a priori* that the infimum is assumed, as is implicitly claimed in the definition of the norm on $\mathcal{L}(X, Y)$ in the statement of the theorem.

To prove that the $\|L\|$ defined above is a norm, first note that if $L : X \to Y$ is linear and bounded, then $\|L\| = 0$ iff $\inf\left\{c \ge 0 : \left(\forall x \in X : \|L[x]\| \le c\|x\|\right)\right\} = 0$ iff for all

$x \in X$ we have $L[x] = 0$ iff $L = 0$. Moreover, if $L \in \mathcal{L}(X, Y)$ and $\alpha \in \mathbb{R}$, then

$$\begin{aligned} \|\alpha L\| &= \inf\{c \geq 0 : (\forall x \in X : \|\alpha L[x]\| \leq c\|x\|)\} \\ &= \inf\left\{c \geq 0 : \left(\forall x \in X : \|L[x]\| \leq \frac{c}{|\alpha|}\|x\|\right)\right\} \\ &= |\alpha| \inf\{d \geq 0 : (\forall x \in X : \|L[x]\| \leq d\|x\|)\} \\ &= |\alpha|\|L\|. \end{aligned}$$

For the triangular inequality, let $L, M \in \mathcal{L}(X, Y)$. Then

$$\begin{aligned} \|L + M\| &= \inf\{c \geq 0 : (\forall x \in X : \|L[x] + M[x]\| \leq c\|x\|)\} \\ &\leq \inf\{c \geq 0 : (\forall x \in X : \|L[x]\| + \|M[x]\| \leq c\|x\|)\} \\ &\leq \inf\{a \geq 0 : (\forall x \in X : \|L[x]\| \leq a\|x\|)\} \\ &\quad + \inf\{b \geq 0 : (\forall x \in X : \|M[x]\| \leq b\|x\|)\} \\ &= \|L\| + \|M\|. \end{aligned}$$

By Exercise 17-3b, we have $\|L[x]\| \leq \|L\|\|x\|$ for all $x \in X$. In particular, the infimum that is used to define the norm is actually a minimum, as claimed. ■

Unless otherwise indicated, throughout this text the norm on any space of continuous linear functions will be assumed to be the norm from Theorem 17.8.

Definition 17.9 *The norm from Theorem 17.8 is also called the* **operator norm** *of the continuous linear function L.*

Another way to represent the norm of a continuous linear function is the following.

Proposition 17.10 *Let X, Y be normed spaces and let $L : X \to Y$ be a continuous linear function. Then $\|L\| = \sup\left\{\|L[x]\| : x \in \overline{B_1(0)}\right\}$.*

Proof. Exercise 17-4. ■

The completeness of the spaces $\mathcal{L}(X, Y)$ solely depends on the completeness of the image space Y.

Theorem 17.11 *Let X be a normed space and let Y be a Banach space. Then $\mathcal{L}(X, Y)$ is a Banach space.*

Proof. Let $\{L_n\}_{n=1}^{\infty}$ be a Cauchy sequence in $\mathcal{L}(X, Y)$. We first claim that for all $x \in X$ the sequence $\{L_n[x]\}_{n=1}^{\infty}$ is a Cauchy sequence. To prove this claim, let $x \in X$ and let $\varepsilon > 0$. Find an $N \in \mathbb{N}$ so that for all $m, n \geq N$ we have the inequality $\|L_m - L_n\| \leq \frac{\varepsilon}{\|x\| + 1}$. Then for all $m, n \geq N$ we infer

$$\|L_n[x] - L_m[x]\| = \|(L_m - L_n)[x]\| \leq \|L_m - L_n\|\|x\| \leq \frac{\varepsilon}{\|x\| + 1}\|x\| < \varepsilon$$

and $\left\{L_n[x]\right\}_{n=1}^{\infty}$ is a Cauchy sequence.

Because Y is complete, for all $x \in X$ we can define $L[x] := \lim_{n\to\infty} L_n[x]$. It is easy to prove that L is linear (see Exercise 17-5). Because the reverse triangular inequality $\left|\|L_m\| - \|L_n\|\right| \leq \|L_m - L_n\|$ holds for all $m, n \in \mathbb{N}$ we obtain that $\left\{\|L_n\|\right\}_{n=1}^{\infty}$ is a Cauchy sequence, and hence $\lim_{n\to\infty} \|L_n\|$ exists. Now for all $x \in X$ the inequality $\left\|L[x]\right\| = \lim_{n\to\infty} \left\|L_n[x]\right\| \leq \left(\lim_{n\to\infty} \|L_n\|\right) \|x\|$ holds, and hence $L \in \mathcal{L}(X, Y)$. To prove that L is the limit of $\{L_n\}_{n=1}^{\infty}$ in $\mathcal{L}(X, Y)$ let $\varepsilon > 0$. Let $N \in \mathbb{N}$ be such that for all $m, n \geq N$ we have $\|L_m - L_n\| < \varepsilon$. Then for all $n \geq N$ we obtain

$$\begin{aligned}
\|L - L_n\| &= \sup\left\{\left\|(L - L_n)[x]\right\| : x \in \overline{B_1(0)}\right\} \\
&= \sup\left\{\left\|\lim_{m\to\infty} L_m[x] - L_n[x]\right\| : x \in \overline{B_1(0)}\right\} \\
&\leq \sup\left\{\limsup_{m\to\infty} \|L_m - L_n\| \, \|x\| : x \in \overline{B_1(0)}\right\} \leq \varepsilon.
\end{aligned}$$

Thus for all $n \geq N$ we have $\|L - L_n\| < \varepsilon$ and $\{L_n\}_{n=1}^{\infty}$ converges to L. ∎

For our investigation of derivatives, it is important to realize that there is a simple bijective correspondence between the elements of a normed space Y and the linear functions from $\mathbb{R}$ to Y.

Theorem 17.12 *Let Y be a normed space. For $f \in Y$ define $L_f : \mathbb{R} \to Y$ by $L_f[x] := xf$. Then the function $I[f] := L_f$ defines an isometric isomorphism from Y to $\mathcal{L}(\mathbb{R}, Y)$.*

Proof. Clearly, for all $f \in Y$ the function L_f is linear and $\|L_f\| = \|f\|$. Linearity of I is trivial. Because $\|L_f\| = \|f\|$, I is an isometry, and hence it is injective. To prove that I is surjective, let $L \in \mathcal{L}(\mathbb{R}, Y)$. Set $f := L[1]$. Then for all $x \in \mathbb{R}$ we have $L[x] = L[x1] = xL[1] = xf$, which means that $L = L_f$ and I is surjective. ∎

We conclude this section by noting that for continuous linear functions we could always assume that the domain is a Banach space. Recall that by Theorem 16.89 and Exercise 16-94 every normed space can be densely embedded into a Banach space. Theorem 17.13 shows that any linear function $L : X \to Y$ can be extended to a unique continuous linear function from the completion of X to the completion of Y. This means that as long as we work with continuous linear functions, which we do exclusively in this chapter, it would be no loss of generality to assume (as is often done) that domain and range are Banach spaces.

Theorem 17.13 *Let X be a normed space, let Y be a Banach space, let $D \subseteq X$ be a dense linear subspace of X and let $L : D \to Y$ be a continuous linear function. Then there is a unique continuous linear function $M : X \to Y$ such that $M|_D = L$.*

Proof. Exercise 17-6. ∎

Exercises

17-1. Let $1 \le p < \infty$, let $a < b$ and let $X := \{ f \in C^1(a,b) \cap L^p(a,b) : f' \in L^p(a,b) \}$. Prove that $D : X \to L^p(a,b)$ defined by $D[f] := [f']$ is not continuous.

17-2. Let X, Y be normed spaces and let $L : X \to Y$ be linear. Prove that if L is discontinuous at the origin, then it is discontinuous at every $x \in X$.

17-3. Finish the proof of Theorem 17.8. Let X, Y be normed spaces.

(a) Prove that $\mathcal{L}(X,Y)$ with pointwise addition and scalar multiplication is a vector space.

(b) Let $L : X \to Y$ be a continuous linear function. Prove that $\| L[x] \| \le \|L\| \|x\|$ for all $x \in X$.

17-4. Prove Proposition 17.10.

17-5. Finish the proof of Theorem 17.11 by proving that the function $L[x] := \lim_{n\to\infty} L_n[x]$ defined in the proof is linear.

17-6. Prove Theorem 17.13.

17-7. Let X, Y be vector spaces and let $\mathcal{B}$ be a base of X.

(a) Let $L, M : X \to Y$ be linear. Prove that if $L(b) = M(b)$ for all $b \in \mathcal{B}$, then $L = M$.

(b) Prove that if $f : \mathcal{B} \to Y$ is a function defined on the base $\mathcal{B}$, then there is a unique linear function $L_f : X \to Y$ so that $L_f|_{\mathcal{B}} = f$.

17-8. Let $1 \le p \le \infty$, let q be so that $\frac{1}{p} + \frac{1}{q} = 1$ and let $\{a_j\}_{j=1}^{\infty} \in l^q$. Prove that the function $S_{\{a_j\}_{j=1}^{\infty}}\left(\{x_j\}_{j=1}^{\infty}\right) := \sum_{j=1}^{\infty} a_j x_j$ is a continuous linear operator from l^p to $\mathbb{R}$.

17-9. Let X, Y be normed spaces and let $L : X \to Y$ be linear. Prove that if L is continuous at some $x \in X$, then L is continuous on X.

17-10. Integration need not always define a continuous linear function. Let X be the space of all continuous functions $f : \mathbb{R} \to \mathbb{R}$ so that $\{ x : f(x) \neq 0 \}$ is bounded.

(a) Prove that X is a normed subspace of $\left(C(\mathbb{R}, \mathbb{R}), \| \cdot \|_\infty \right)$.

(b) Prove that $L : X \to \mathbb{R}$ defined by $L[f] := \int_{\mathbb{R}} x f(x)\, dx$ is a linear function on X.

(c) Prove that L is not bounded on $\overline{B_1(0)}$.

(d) Let $f \in X$. Find a sequence $\{f_n\}_{n=1}^{\infty}$ of elements of X that converges to f and such that $\{ L[f_n] \}_{n=1}^{\infty}$ does not converge to $L[f]$.

(e) Prove that X is not a Banach space.

(f) Prove that X is *not* dense in $\left(C(\mathbb{R}, \mathbb{R}), \| \cdot \|_\infty \right)$.

17-11. Let X be the space of polynomials of order at most 3 on the interval $(0,1)$. Prove that $D[p] := p'$ is a continuous linear function from X to X.

17-12. Let X, Y and Z be normed spaces and let $K : X \to Y$ and $L : Y \to Z$ be continuous linear functions.

(a) Prove that $\|L \circ K\| \le \|L\| \|K\|$.

(b) Prove that the inequality in part 17-12a can be strict.

17-13. Prove that if $C^1(a,b)$ is equipped with the norm $\|f\| := \|f\|_\infty + \|f'\|_\infty$ (see Exercise 16-17), then $f \mapsto f'$ is a continuous mapping from $C^1(a,b)$ to $C^0(a,b)$.

Readers interested in inner product spaces could read Chapter 20 before continuing.

17.2 Matrix Representation of Linear Functions

The coordinate-free introduction in Section 17.1 provides a concise description of linear functions. Without specifics about a coordinate system, an abstract notion can usually be investigated more easily, because there are fewer details to keep track of. On the other hand, coordinates bridge the gap between abstract notions and concrete applications. Therefore, the connection between a concept and its coordinatized version should be investigated very carefully. This section shows that coordinatization of linear functions is done by carefully reinterpreting some natural coefficients.

A **coordinate system** in a vector space ultimately is nothing but a base, because any vector can be expressed as a unique linear combination of the base vectors (see Proposition 15.18). The coefficients in the base representation of the vector can be viewed as the **coordinates**. In this section, we investigate how a linear function between finite dimensional spaces maps the coordinates of its input vectors to the coordinates of its output vectors. Because all finite dimensional spaces are isomorphic to some $\mathbb{R}^d$ (see Proposition 15.25) we will work with spaces $\mathbb{R}^m$, $\mathbb{R}^n$ etc. throughout this section.

The choice of a base in domain and range leads to the connection between linear functions and matrices.

Proposition 17.14 *Let $m, n \in \mathbb{N}$, let $L : \mathbb{R}^n \to \mathbb{R}^m$ be linear, let $\{v_1, \ldots, v_n\}$ be a base of $\mathbb{R}^n$ and let $\{w_1, \ldots, w_m\}$ be a base of $\mathbb{R}^m$. For all $j = 1, \ldots, n$, let a_{ij} with $i = 1, \ldots, m$ be such that $L[v_j] = \sum_{i=1}^{m} a_{ij} w_i$. Because $\{v_1, \ldots, v_n\}$ is a base of $\mathbb{R}^n$, for all $x \in \mathbb{R}^n$, there are unique coefficients $c_1, \ldots, c_n$ so that $x = \sum_{j=1}^{n} c_j v_j$. The image of x under L is $L\left[\sum_{j=1}^{n} c_j v_j\right] = \sum_{i=1}^{m} \sum_{j=1}^{n} a_{ij} c_j w_i$.*

Proof. Exercise 17-14. ∎

Proposition 17.14 shows that, once we fix bases in $\mathbb{R}^n$ and $\mathbb{R}^m$, for each linear function $L : \mathbb{R}^n \to \mathbb{R}^m$ there is a unique rectangular array of real numbers a_{ij}, with indices $i = 1, \ldots, m$, $j = 1, \ldots, n$ that can be used to represent L. Such rectangular arrays of numbers are called matrices. The set of matrices with the natural addition and scalar multiplication is a vector space.

Definition 17.15 *Let $m, n \in \mathbb{N}$. A real $m \times n$-**matrix** with m rows and n columns is a function $A : \{1, \ldots, m\} \times \{1, \ldots, n\} \to \mathbb{R}$, denoted $A = (a_{ij})_{\substack{i=1,\ldots,m \\ j=1,\ldots,n}}$. The index i is called the **row index** and the index j is called the **column index**. We define $M(m \times n, \mathbb{R})$ to be the set of all real $m \times n$-matrices.*

Proposition 17.16 *Let $m, n \in \mathbb{N}$, let $(a_{ij})_{\substack{i=1,\ldots,m \\ j=1,\ldots,n}}$, $(b_{ij})_{\substack{i=1,\ldots,m \\ j=1,\ldots,n}}$ be real*

$m \times n$ matrices and let $c \in \mathbb{R}$. With addition defined by

$$\left(a_{ij}\right)_{\substack{i=1,\ldots,m\\ j=1,\ldots,n}} + \left(b_{ij}\right)_{\substack{i=1,\ldots,m\\ j=1,\ldots,n}} := \left(a_{ij}+b_{ij}\right)_{\substack{i=1,\ldots,m\\ j=1,\ldots,n}}$$

and with scalar multiplication defined by

$$c\left(a_{ij}\right)_{\substack{i=1,\ldots,m\\ j=1,\ldots,n}} := \left(ca_{ij}\right)_{\substack{i=1,\ldots,m\\ j=1,\ldots,n}}$$

the set of matrices $M(m \times n, \mathbb{R})$ is a vector space.

Proof. Exercise 17-15. ■

The coefficients from Proposition 17.14 immediately lead to an isomorphism between $\mathcal{L}\left(\mathbb{R}^n, \mathbb{R}^m\right)$ and $M(m \times n, \mathbb{R})$, where both are considered as vector spaces. Note that the specific isomorphism will depend on which bases we choose in $\mathbb{R}^n$ and $\mathbb{R}^m$.

Theorem 17.17 *Let $m, n \in \mathbb{N}$, and let $\{v_1, \ldots, v_n\}$ and $\{w_1, \ldots, w_m\}$ be bases of $\mathbb{R}^n$ and $\mathbb{R}^m$, respectively. For each $L \in \mathcal{L}\left(\mathbb{R}^n, \mathbb{R}^m\right)$, let $A(L) = \left(a_{ij}\right)_{\substack{i=1,\ldots,m\\ j=1,\ldots,n}}$ be the matrix with coefficients a_{ij} provided by Proposition 17.14. Then the function $A : \mathcal{L}\left(\mathbb{R}^n, \mathbb{R}^m\right) \to M(m \times n, \mathbb{R})$ thus defined is a vector space isomorphism.*

Proof. Exercise 17-16. ■

Similar to Theorem 17.12, for $m = n = 1$ Theorem 17.17 shows that linear functions from $\mathbb{R}$ to $\mathbb{R}$ are in bijective correspondence with numbers (considered as "1×1 matrices").

Composition of functions can be used to define a multiplication on $\mathcal{L}\left(\mathbb{R}^n, \mathbb{R}^m\right)$. Exercise 17-17 shows that this multiplication is compatible with addition and scalar multiplication. For $M(m \times n, \mathbb{R})$, we can define multiplication of matrices as follows.

Definition 17.18 Matrix multiplication. *Let $m, n, p \in \mathbb{N}$. For the real matrices $A = \left(a_{jk}\right)_{\substack{j=1,\ldots,m\\ k=1,\ldots,n}} \in M(m \times n, \mathbb{R})$ and $B = \left(b_{ij}\right)_{\substack{i=1,\ldots,p\\ j=1,\ldots,m}} \in M(p \times m, \mathbb{R})$, we*

define the **product** $BA := \left(\sum_{j=1}^{m} b_{ij}a_{jk}\right)_{\substack{i=1,\ldots,p\\ k=1,\ldots,n}} \in M(p \times n, \mathbb{R}).$

Theorems 17.20 and 17.21 below show the connection between matrix multiplication and the evaluation and composition of linear functions. For the remainder of this chapter, the base used in any space $\mathbb{R}^d$ is the standard base $\{e_1, \ldots, e_d\}$.

Proposition 17.19 *Let $m \in \mathbb{N}$. The function $V_m\left(v_{ij}\right)_{\substack{i=1,\ldots,m\\ j=1}} := \sum_{i=1}^{m} v_{i1}e_i$ is an isomorphism from $M(m \times 1, \mathbb{R})$ to $\mathbb{R}^m$.*

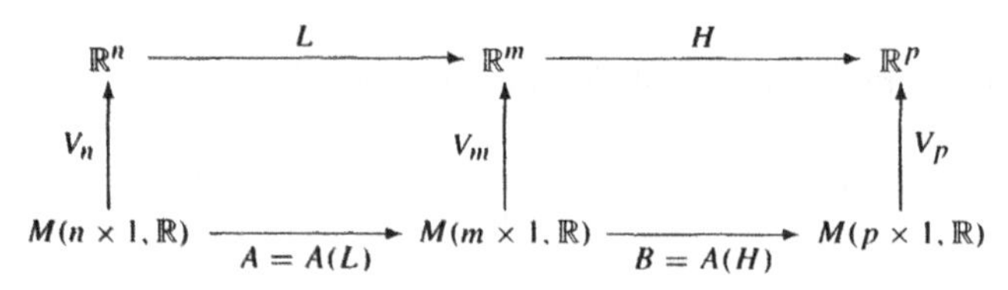

Figure 41: The connection between linear functions and their matrix representations. Note that instead of representing linear functions between spaces $\mathbb{R}^d$ with the standard basis, we could have also represented linear functions between arbitrary finite dimensional vector spaces using any base in each space.

Proof. Exercise 17-18. ■

The isomorphisms V_m are the key to representing evaluations and compositions of linear functions as matrix multiplications and vice versa. It is customary to drop the second index for elements of $M(m \times 1, \mathbb{R})$ and we will do so in the following. The representation of elements of $\mathbb{R}^m$ as in Proposition 17.19 is also called the representation with **column vectors**. The corresponding representation with $1 \times m$ matrices is called the representation with **row vectors**.

Theorem 17.20 *Matrix multiplication and evaluation of linear functions. Let $m, n \in \mathbb{N}$, let $L : \mathbb{R}^n \to \mathbb{R}^m$ be a linear function and let $A := A(L) \in M(m \times n, \mathbb{R})$ be the matrix obtained from Theorem 17.17, using the standard bases in $\mathbb{R}^n$ and $\mathbb{R}^m$. Then for all $v \in \mathbb{R}^n$ we have $L[v] = V_m\left[AV_n^{-1}[v]\right]$, where A and $V_n^{-1}[v]$ are multiplied as matrices.*

Proof. Exercise 17-19. ■

Exercise 17-20 shows that matrix multiplication is associative, which means we can write a product of three or more matrices without parentheses indicating which pairs of matrices are multiplied first.

Theorem 17.21 *Matrix multiplication and composition of linear functions. Let $m, n, p \in \mathbb{N}$, let $L : \mathbb{R}^n \to \mathbb{R}^m$ and $H : \mathbb{R}^m \to \mathbb{R}^p$ be linear functions and let $A := A(L) \in M(m \times n, \mathbb{R})$ and $B := A(H) \in M(p \times m, \mathbb{R})$ be the matrices obtained from Theorem 17.17, using the standard bases in $\mathbb{R}^n$, $\mathbb{R}^m$, and $\mathbb{R}^p$. Then for all $v \in \mathbb{R}^n$ we have $H \circ L[v] = V_p\left[BAV_n^{-1}[x]\right]$, where B, A and $V_n^{-1}[x]$ are multiplied as matrices.*

Proof. Exercise 17-21. ■

The connection between linear functions and the matrices that represent them is also expressed in Figure 41. Diagrams as in this figure are often called **commutative diagrams**, because the order in which the arrows are followed can be interchanged and the arrows representing isomorphisms can be reversed.

To further familiarize ourselves with the correspondence between matrix multiplication and composition of linear functions, let us cast the well-known Gauss-Jordan algorithm from linear algebra in the language of linear functions. We will focus the result on bijective linear functions because we need this reinterpretation for Lemma 18.35 in the proof of the Multidimensional Substitution Formula.

Definition 17.22 *Elementary row operation functions. Let $d \in \mathbb{N}$ and let every vector $x \in \mathbb{R}^d$ be represented as $x = \sum_{i=1}^{d} x_i e_i =: (x_1, \ldots, x_d)$.*

1. *$D : \mathbb{R}^d \to \mathbb{R}^d$ is called a* **diagonal operator** *iff there are $c_1, \ldots, c_d \in \mathbb{R}$ so that $D(x_1, \ldots, x_d) = (c_1 x_1, \ldots, c_d x_d)$.*
2. *$A : \mathbb{R}^d \to \mathbb{R}^d$ is called a* **row addition operator** *iff there are a number $a \in \mathbb{R}$ and distinct indices $i, j \in \{1, \ldots, d\}$ so that for all $x = (x_1, \ldots, x_d) \in \mathbb{R}^d$ we have $A(x_1, \ldots, x_d) = (x_1, \ldots, x_{i-1}, x_i + ax_j, x_{i+1}, \ldots, x_d)$.*
3. *$T : \mathbb{R}^d \to \mathbb{R}^d$ is called a* **row transposition operator** *iff there are indices $i, j \in \{1, \ldots, d\}$ with $i < j$ such that for all $x = (x_1, \ldots, x_d) \in \mathbb{R}^d$ we have $T(x_1, \ldots, x_d) = (x_1, \ldots, x_{i-1}, x_j, x_{i+1}, \ldots, x_{j-1}, x_i, x_{j+1}, \ldots, x_d)$.*

Theorem 17.23 Gauss-Jordan Algorithm. *Let $d \in \mathbb{N}$. Every bijective linear function $L : \mathbb{R}^d \to \mathbb{R}^d$ is a composition of one diagonal operator so that all $c_i \neq 0$ with row addition and row transposition operators.*

Proof. We will provide an outline here and leave the full proof to Exercise 17-22. The proof is an induction on the dimension d. The base step $d = 1$ is trivial. For the induction step, let $A^0 := A(L) \in M(d \times d, \mathbb{R})$ be the matrix obtained from Theorem 17.17 using the standard base in $\mathbb{R}^d$. Because L is bijective, there is a coefficient a_{i1}^0 that is not equal to zero (explain). This means the transposition of rows 1 and i produces a matrix A^1 with $a_{11}^1 \neq 0$. Execute $d - 1$ row additions to produce a matrix A^2 with $a_{11}^2 \neq 0$ and $a_{i1}^2 = 0$ for $i = 2, \ldots, d$. Now consider the matrix B obtained from A^2 by erasing the first row and the first column. This matrix is the image $B = A\left(L'\right)$ of a bijective linear function $L' : \mathbb{R}^{d-1} \to \mathbb{R}^{d-1}$ (explain). Therefore, by induction hypothesis (explain) there is a sequence of row additions and transpositions that turns B into a diagonal matrix C. The corresponding row additions and transpositions turn A^2 into a matrix A^3 whose only nonzero entries are in the first row and on the diagonal (explain). Moreover, all of the entries on the diagonal are not zero (explain). Perform the appropriate row additions to obtain a matrix A^4 whose only nonzero entries are on the diagonal. (explain).

For the above constructed sequence of row transpositions and additions, let the operators $M_1, \ldots, M_n$ be the corresponding row transposition and row addition operators in the order in which the operations were performed. Then $N := M_n \circ \cdots \circ M_1 \circ L$ is a linear function so that $A(N) = A^4$ (explain; a commutative diagram may help). That is, N is a diagonal operator and $L = M_1^{-1} \circ \cdots \circ M_n^{-1} \circ N$. Row transposition operators are their own inverses and the inverse of a row addition operator is another row addition operator (explain by stating the inverse). Thus we have proved the theorem. ■

Note how the whole proof of Theorem 17.23 depends on the fluent translation between matrix operations and their interpretations as compositions of linear operators.

Exercise 17-24 shows another application of this translation by assuring that the solution x of a uniquely solvable system of equations $Ax = b$ depends continuously on the coefficients of A and b.

Exercises

17-14. Prove Proposition 17 14

17-15 Prove Proposition 17.16

17-16. Prove Theorem 17.17

17-17. Let V, X, Y, Z be vector spaces, let $L\ Y \to Z$, $M, N\ X \to Y$ and $K : V \to X$ be linear and let $c \in \mathbb{R}$. Prove each of the following.

(a) $L \circ (M + N) = L \circ M + L \circ N$

(b) $(M + N) \circ K = M \circ K + N \circ K$

(c) $(cL) \circ M = c(L \circ M) = L \circ (cM)$

17-18. Prove Proposition 17.19

17-19. Prove Theorem 17.20.

17-20. Prove that matrix multiplication is associative. That is, prove that if $A \in M(m \times n, \mathbb{R})$, $B \in M(p \times m, \mathbb{R})$, and $C \in M(q \times p, \mathbb{R})$, then $C(BA) = (CB)A$

17-21. Prove Theorem 17.21

17-22. Prove Theorem 17.23

17-23. Interpret the $m \times n$ matrix A with entries a_{ij} as a linear function from $\left(\mathbb{R}^n, \|\cdot\|_2\right)$ to $\left(\mathbb{R}^m, \|\ \|_2\right)$.

Prove that the operator norm of A satisfies $\|A\| \leq \left(\sum_{i=1}^{m} \sum_{j=1}^{n} |a_{ij}|^2\right)^{\frac{1}{2}}$.

17-24 The continuous dependence of solutions x of uniquely solvable systems of linear equations $Ax = b$ on the entries of the coefficient matrix A and the right side b.

(a) Let X, Y be normed spaces and let $\mathcal{H}(X, Y)$ be the set of all invertible continuous linear functions from X to Y with continuous inverse ("linear homeomorphisms"). Prove that the function $J \cdot \mathcal{H}(X, Y) \to \mathcal{H}(Y, X)$, which maps each $A \in \mathcal{H}(X, Y)$ to $A^{-1} \in \mathcal{H}(Y, X)$ is continuous on $\mathcal{H}(X, Y)$

Hint. Prove that for all $A, B \in \mathcal{H}(X, Y)$ that are close enough together, the inequality $\left\|A^{-1} - B^{-1}\right\| \leq \dfrac{\left\|A^{-1}\right\|^2}{1 - \left\|A^{-1}\right\| \|B - A\|} \|B - A\|$ holds and prove that this implies that J is continuous at A.

(b) Prove that if A is an invertible $d \times d$ matrix and $b \in \mathbb{R}^d$, then $x = J(A)b$ is the unique solution of the system of equations $Ax = b$

(c) Prove that the map $S : \mathcal{H}\left(\mathbb{R}^d, \mathbb{R}^d\right) \times \mathbb{R}^d \to \mathbb{R}^d$ defined by $S(A, b) := J(A)b$ is continuous.

(d) Suppose an industrial process allows the measurement of the coefficient matrix A and of the right hand side b of a linear system of equations $Ax = b$. Moreover, suppose that the process requires the computation of the solution x Explain why (unavoidable) sufficiently small measurement errors are not likely to have a large effect on the computed solution x

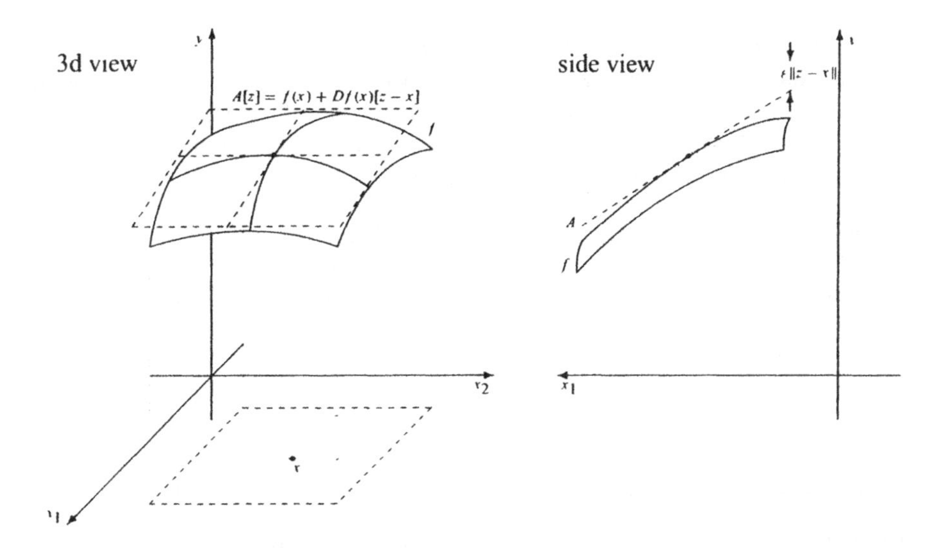

Figure 42: Differentiation is approximation with linear functions. The graph of a linear function from $\mathbb{R}^2$ to $\mathbb{R}$ is a plane. The figure shows that the difference between the graph of a differentiable function and the graph of an appropriately shifted linear function fits into arbitrarily small "cones." In the side view, the plane is seen "edge-on" and we only indicate the sides of the cone with dotted lines.

17.3 Differentiability

Although it may be geometrically intuitive, defining the derivative of a multivariable function in terms of partial derivatives can become a notational nightmare. Just consider the indices and notations in the matrix representation of a linear function in Section 17.2 and imagine them as part of a more complex definition. To circumvent this level of detail, which might obscure the forest for the trees, we introduce derivatives with a coordinate-free definition. Aside from simpler notation, we avoid the pathology presented in Exercises 17-60 and 17-61 and we gain conceptual insights into a theory that is not bound to finite dimensional spaces. Indeed, all results in this section hold for infinite dimensional spaces, too. Interestingly enough, restriction to finite dimensional spaces would not simplify the proofs. This is similar to Sections 14.2–14.4, where proofs originally designed for functions of one real variable translated verbatim to the setting of measure spaces. In this chapter, we state the proofs in the abstract setting and provide results for d-dimensional space as corollaries. By staying with this level of generality, we will ultimately produce some rather elegant proofs of important results. For examples, consider the proof of Leibniz' Rule in Exercise 17-58, as well as the ends of Sections 17.7 and 17.8. In each case, important results for the familiar d-dimensional setting are obtained as corollaries of coordinate-free results on differentiation.

The key to differentiation in higher dimensional spaces lies in Figure 9 and its analytical formulation in Theorem 4.5. A function should be differentiable iff it can be approximated very closely by a shifted linear function. As in Figure 9, the idea in Definition 17.24 below is that near the point where the derivative is taken, for any multidimensional analogue of a cone, both the function and the approximating linear entity should ultimately be in the same "cone" (see Figure 42). Similar to the single variable setting, the natural domains for differentiable functions are open sets.

Definition 17.24 *Let X, Y be normed spaces, let $\Omega \subseteq X$ be open, let $f : \Omega \to Y$ be a function and let $x \in \Omega$. Then f is called* **differentiable** *at x iff there is a continuous linear function $L : X \to Y$ so that for all $\varepsilon > 0$ there is a $\delta > 0$ such that for all $z \in X$ with $\|z - x\| < \delta$ we have*

$$\left\| f(z) - f(x) - L[z - x] \right\| \leq \varepsilon \|z - x\|.$$

In this case, we set $Df(x) := L$ and call it the **derivative** *of f at x.*

By Exercise 17-25b, the derivative is unique, so we are can speak of *the* derivative.

Notation 17.25 The argument in round parentheses behind a derivative Df will denote the point at which the derivative is taken and the argument in square brackets will denote the place where the derivative (remember that it is a linear function) is evaluated. That is, for $f : \Omega \to Y$, $Df(x)[a]$ will denote the derivative of f taken at $x \in \Omega$ and evaluated at $a \in X$. □

The function $A[z] := f(x) + Df(x)[z - x]$ is also often called the **linear approximation** of f at x. The name should be clear. The function A is "linear" (affine linear to be precise, but that distinction is not always made) and it approximates the function f (see Figure 42). Exercise 17-25 investigates the quality of the approximation and it also gives a geometric interpretation for functions into the real numbers.

Theorem 17.12 expresses the connection between linear functions $L : \mathbb{R} \to Y$ and vectors and Theorem 17.17 expresses (among other things) the connection between linear functions $L : \mathbb{R} \to \mathbb{R}$ and numbers. These connections are the reason why derivatives of functions defined on intervals (a, b) are usually considered to be numbers or (tangent) vectors. The formal justification is the following.

Proposition 17.26 *Let Y be a normed space and let $a < b$ be real numbers. The function $f : (a, b) \to Y$ is differentiable at x in the sense of Definition 17.24 iff the limit $f'(x) := \lim_{h \to 0} \frac{f(x+h) - f(x)}{h}$ exists. In this case, we call $f'(x)$ the* **velocity vector**[1] *and we have $Df(x)[a] = f'(x)a$ for all $a \in \mathbb{R}$.*

Proof. Mimic the proof of Theorem 4.5. (Exercise 17-26.) ■

Continuous linear functions are the simplest example of differentiable functions that do not follow the pattern of Proposition 17.26

Proposition 17.27 *Let X, Y be normed spaces and let $L : X \to Y$ be continuous and linear. Then L is differentiable at every $x \in X$ with $DL(x)[a] = L[a]$.*

[1] The name comes from the fact that if $f : (a, b) \to \mathbb{R}^3$ gives the position of a particle at time t, then $f'(t)$ is the velocity of the particle.

Proof. Exercise 17-27. ■

Note that the derivative of a continuous linear function actually is a constant function (whose value at every point is $L[\cdot]$)! This is similar to the derivative of $f(x) = cx$ being $f'(x) = c$. More examples of differentiable functions will be encountered throughout this chapter. For the rest of this section, we will work with derivatives in their full generality. First note that differentiability still implies continuity.

Theorem 17.28 *Let X, Y be normed spaces, let $\Omega \subseteq X$ be an open set, and let the function $f : \Omega \to Y$ be differentiable at $x \in \Omega$. Then f is continuous at x.*

Proof. Exercise 17-28 ■

We conclude this section with differentiation rules.

Theorem 17.29 *Let X, Y be normed spaces, let $\Omega \subseteq X$ be open, let the functions $f, g : \Omega \to Y$ be differentiable at $x \in \Omega$ and let $\alpha \in \mathbb{R}$. Then the sum $f + g$ is differentiable at x with $D(f+g)(x) = Df(x) + Dg(x)$ and the scalar multiple αf is differentiable at x with $D(\alpha f)(x) = \alpha Df(x)$.*

Proof. When the derivative is given, differentiability is proved directly with the definition. Consider αf. For given $\varepsilon > 0$, find $\delta > 0$ so that for all $z \in \Omega$ with $\|z - x\| < \delta$ we have $\big\|f(z) - f(x) - Df(x)[z - x]\big\| \leq \frac{\varepsilon}{|\alpha| + 1}\|z - x\|$. Then for all $z \in \Omega$ with $\|z - x\| < \delta$ we infer

$$\begin{aligned}\big\|(\alpha f)(z) - (\alpha f)(x) - \alpha Df(x)[z - x]\big\| &= \big\|\alpha\big(f(z) - f(x) - Df(x)[z - x]\big)\big\| \\ &\leq |\alpha|\frac{\varepsilon}{|\alpha| + 1}\|z - x\| \leq \varepsilon\|z - x\|,\end{aligned}$$

which means that f is differentiable at x and the derivative is $\alpha Df(x)$.

The claim about the sum is proved similarly (Exercise 17-29). ■

The Chain Rule retains its familiar form from calculus, except that the multiplication is replaced by composition. This is natural, because the composition of two linear functions from $\mathbb{R}$ to $\mathbb{R}$ corresponds to the multiplication of the numbers that represent the functions (see Theorem 17.21).

Theorem 17.30 Chain Rule. *Let X, Y, Z be normed spaces, let $\Omega_1 \subseteq X$ and $\Omega_2 \subseteq Y$ be open, let $g : \Omega_1 \to \Omega_2$ be differentiable at x and let $f : \Omega_2 \to Z$ be differentiable at $g(x)$. Then $f \circ g$ is differentiable at x with derivative $D(f \circ g)(x) = Df(g(x)) \circ Dg(x)$.*

Proof. Let $\varepsilon > 0$. Find $\delta_1 > 0$ so that for all $y \in Y$ with $\|y - g(x)\| < \delta_1$ we have $\big\|f(y) - f(g(x)) - Df(g(x))[y - g(x)]\big\| \leq \frac{\varepsilon}{2(\|Dg(x)\| + 1)}\|y - g(x)\|$. Then find $\delta > 0$ so that for all points $z \in X$ with $\|z - x\| < \delta$ we have the inequalities $\|g(z) - g(x)\| < \delta_1$, $\big\|g(z) - g(x) - Dg(x)[z - x]\big\| \leq \frac{\varepsilon}{2(\|Df(g(x))\| + 1)}\|z - x\|$ and $\|g(z) - g(x)\| \leq \|Dg(x)\|\|z - x\| + \|z - x\|$, where the last inequality stems from applying the reverse triangular inequality to the condition for differentiability of g at x with $\varepsilon = 1$. Then for all $z \in X$ with $\|z - x\| < \delta$ we conclude the following:

$$\begin{aligned}
&\left\|(f \circ g)(z) - (f \circ g)(x) - \left(Df(g(x)) \circ Dg(x)\right)[z - x]\right\| \\
&\leq \left\|f(g(z)) - f(g(x)) - Df(g(x))[g(z) - g(x)]\right\| \\
&\quad + \left\|Df(g(x))\left[g(z) - g(x) - Dg(x)[z - x]\right]\right\| \\
&\leq \frac{\varepsilon}{2\left(\|Dg(x)\|+1\right)}\|g(z)-g(x)\|+\|Df(g(x))\|\frac{\varepsilon}{2\left(\|Df(g(x))\|+1\right)}\|z-x\| \\
&\leq \frac{\varepsilon}{2\left(\|Dg(x)\| + 1\right)}\left(\|Dg(x)\|\|z - x\| + \|z - x\|\right) + \frac{\varepsilon}{2}\|z - x\| \\
&= \varepsilon\|z - x\|,
\end{aligned}$$

which proves the Chain Rule. ■

Similar to the Chain Rule, the rule for the differentiation of inverse functions retains its overall form. The reciprocal in Theorem 4.21 is replaced with the inversion of the derivative. This is once again natural, because if two linear functions from $\mathbb{R}$ to $\mathbb{R}$ are inverses of each other, then the numbers that represent the functions are reciprocals of each other. Unlike in Theorem 4.21 we must demand that the image of the domain of f is open and that the inverse is continuous. This is because the argument at the beginning of the proof of Theorem 4.21 is not easily translated to the setting of normed spaces. Corollary 17.66 will show that the translation is possible, but it requires the Implicit Function Theorem.

Theorem 17.31 *Let X, Y be normed spaces, let $\Omega \subseteq X$, $\widetilde{\Omega} \subseteq Y$ be open and let $f : \Omega \to \widetilde{\Omega}$ be a continuous bijective function with continuous inverse. If f is differentiable at x_0 and $Df(x_0)$ is continuous, bijective and linear with continuous inverse, then f^{-1} is differentiable at $y_0 := f(x_0)$ and $D\left(f^{-1}\right)(y_0) = \left(Df\left(f^{-1}(y_0)\right)\right)^{-1}$.*

Proof. First note that there is a $\delta_1 > 0$ so that for all $x \in X$ with $\|x - x_0\| < \delta_1$ we have $\left\|f(x) - f(x_0) - Df(x_0)[x - x_0]\right\| \leq \frac{1}{2\left\|(Df(x_0))^{-1}\right\|}\|x - x_0\|$. *(Why is the denominator not zero?)* Hence, for all $x \in X$ with $\|x - x_0\| < \delta_1$ the following holds:

$$\begin{aligned}
&\left\|f(x) - f(x_0)\right\| \\
&= \left\|Df(x_0)[x - x_0] + f(x) - f(x_0) - Df(x_0)[x - x_0]\right\| \\
&\geq \left\|Df(x_0)[x - x_0]\right\| - \left\|f(x) - f(x_0) - Df(x_0)[x - x_0]\right\| \\
&= \frac{\left\|(Df(x_0))^{-1}\right\|}{\left\|(Df(x_0))^{-1}\right\|}\left\|Df(x_0)[x - x_0]\right\| - \left\|f(x) - f(x_0) - Df(x_0)[x - x_0]\right\| \\
&\geq \frac{1}{\left\|(Df(x_0))^{-1}\right\|}\left\|(Df(x_0))^{-1}\left[Df(x_0)[x-x_0]\right]\right\| - \frac{1}{2\left\|(Df(x_0))^{-1}\right\|}\|x-x_0\| \\
&= \frac{1}{2\left\|(Df(x_0))^{-1}\right\|}\|x - x_0\|.
\end{aligned}$$

In particular, for all elements $x \in X$ with $\|x - x_0\| < \delta_1$ we have the inequality $\|x - x_0\| \leq 2\left\|(Df(x_0))^{-1}\right\| \left\|f(x) - f(x_0)\right\|$. Now let $\varepsilon > 0$. Find $\delta_2 \in (0, \delta_1)$ so that for all $x \in X$ with $\|x - x_0\| < \delta_2$ we have

$$\left\|f(x) - f(x_0) - Df(x_0)[x - x_0]\right\| \leq \frac{\varepsilon}{2\left\|(Df(x_0))^{-1}\right\|^2}\|x - x_0\|.$$

Because f^{-1} is continuous, there is a $\delta > 0$ so that for all $y \in Y$ with $\|y - y_0\| < \delta$ we have $\left\|f^{-1}(y) - f^{-1}(y_0)\right\| < \delta_2$. Then for all $y \in Y$ with $\|y - y_0\| < \delta$ the following holds. For simpler notation, we use $x := f^{-1}(y)$.

$$\begin{aligned}
&\left\|f^{-1}(y) - f^{-1}(y_0) - (Df(x_0))^{-1}[y - y_0]\right\| \\
\leq\ & \left\|(Df(x_0))^{-1}\right\| \left\|Df(x_0)\left[f^{-1}(y) - f^{-1}(y_0) - (Df(x_0))^{-1}[y - y_0]\right]\right\| \\
=\ & \left\|(Df(x_0))^{-1}\right\| \left\|y - y_0 - Df(x_0)\left[f^{-1}(y) - f^{-1}(y_0)\right]\right\| \\
=\ & \left\|(Df(x_0))^{-1}\right\| \left\|f(x) - f(x_0) - Df(x_0)[x - x_0]\right\| \\
\leq\ & \left\|(Df(x_0))^{-1}\right\| \frac{\varepsilon}{2\left\|(Df(x_0))^{-1}\right\|^2}\|x - x_0\| \\
\leq\ & \frac{\varepsilon}{2\left\|(Df(x_0))^{-1}\right\|^2} 2\left\|(Df(x_0))^{-1}\right\| \|y - y_0\| \\
=\ & \varepsilon\|y - y_0\|.
\end{aligned}$$

■

The derivative of the inverse of a function must not be confused with the derivative of the function that maps an invertible linear function to its inverse, which is considered in the following. (Recall that in Exercise 17-24a we have already proved that inversion is a continuous operation. Theorem 17.32 provides another proof of this fact.) Note that for the theorem to make sense, we first must also prove that the linear homeomorphisms (invertible continuous linear functions with continuous inverse) form an open subset of the space $\mathcal{L}(X, Y)$.

Theorem 17.32 *Let X be a Banach space, let Y be a normed space and let $\mathcal{H}(X, Y)$ be the set of linear homeomorphisms from X to Y. Then $\mathcal{H}(X, Y)$ is an open subset of $\mathcal{L}(X, Y)$ and $J(A) := A^{-1}$ is a differentiable function from $\mathcal{H}(X, Y)$ to $\mathcal{H}(Y, X)$ whose derivative at $A \in \mathcal{H}(X, Y)$ is $DJ(A)[F] = -A^{-1} \circ F \circ A^{-1}$.*

Proof. For $K \in \mathcal{L}(X, X)$ with $\|K\| < 1$, the series $\sum_{j=0}^{\infty}(-1)^j K^j$ converges absolutely. Because X is a Banach space, the series converges. Let $I : X \to X$ denote the identity. It can be verified directly that $(I + K)^{-1} = \sum_{j=0}^{\infty}(-1)^j K^j$. In particular, this means that if $\|K\| < 1$, then $I + K$ is invertible with continuous inverse and its norm is bounded by $\left\|(I+K)^{-1}\right\| \leq \frac{1}{1-\|K\|}$.

Now let $A \in \mathcal{H}(X,Y)$ and consider $F \in \mathcal{L}(X,Y)$ with $\|F\| < \frac{1}{\left\|A^{-1}\right\|}$. Then $A + F = A\left(I + A^{-1}F\right)$ and $\left\|A^{-1}F\right\| \leq \left\|A^{-1}\right\| \|F\| < 1$, which means that A and $I + A^{-1}F$ are invertible with continuous inverse. Hence, $A+F$ is invertible with continuous inverse. Thus for all $A \in \mathcal{H}(X,Y)$ we infer that $B_{\frac{1}{\|A^{-1}\|}}(A) \subseteq \mathcal{H}(X,Y)$ and $\mathcal{H}(X,Y)$ is open in $\mathcal{L}(X,Y)$.

To prove the differentiability of J at $A \in \mathcal{H}(X,Y)$ let $\varepsilon > 0$ and find a positive $\delta < \min\left\{\frac{\varepsilon}{2\left\|A^{-1}\right\|^3}, \frac{1}{\left\|A^{-1}\right\|}\right\}$ so that for all $F \in \mathcal{L}(X,Y)$ the inequality $\|F\| < \delta$ implies that $I + A^{-1}F$ is invertible with continuous inverse and $\sum_{j=0}^{\infty}\left\|A^{-1}F\right\|^j < 2$. Then for all functions $F \in \mathcal{L}(X,Y)$ with $\|F\| < \delta$ we conclude that $A+F$ is invertible with continuous inverse and

$$\begin{aligned}
\left\|(A+F)^{-1} - A^{-1} + A^{-1}FA^{-1}\right\| &= \left\|\left[A\left(I + A^{-1}F\right)\right]^{-1} - \left(I - A^{-1}F\right)A^{-1}\right\| \\
&\leq \left\|\left(I + A^{-1}F\right)^{-1} - \left(I - A^{-1}F\right)\right\| \left\|A^{-1}\right\| \\
&= \left\|\sum_{j=2}^{\infty}(-1)^j\left(A^{-1}F\right)^j\right\| \left\|A^{-1}\right\| \\
&\leq \left\|A^{-1}F\right\|^2 \left\|\sum_{j=0}^{\infty}(-1)^j\left(A^{-1}F\right)^j\right\| \left\|A^{-1}\right\| \\
&\leq \|F\|^2\left(\sum_{j=0}^{\infty}\left\|A^{-1}F\right\|^j\right)\left\|A^{-1}\right\|^3 \\
&\leq \frac{\varepsilon}{2\left\|A^{-1}\right\|^3} 2\left\|A^{-1}\right\|^3 \|F\| = \varepsilon\|F\|,
\end{aligned}$$

which proves the result. ■

Exercises

17-25. Let X, Y be normed spaces, let $\Omega \subseteq X$ be open and let $f, g : \Omega \to Y$ be functions. We will say that g is **tangential** to f at $x \in \Omega$ iff $\lim_{z \to x} \frac{\|f(z) - g(z)\|}{\|z - x\|} = 0$. (Note that we can work with limits, because by Exercise 16-52 every point of Ω is an accumulation point.)

(a) Prove that the function $f : \Omega \to Y$ is differentiable at $x \in \Omega$ with derivative L iff the function $g(z) := f(x) + L[z - x]$ is tangential to f at x.

(b) Prove that if two continuous linear functions $L_1, L_2 : X \to Y$ are so that both shifted functions $f(x) + L_i[\cdot - x]$ are tangential to f at x, then $L_1 = L_2$

(c) Let $f : \mathbb{R}^2 \to \mathbb{R}$ be differentiable at the point $(x_0, y_0) \in \mathbb{R}^2$. Explain why the function $P(x, y) = f(x_0, y_0) + Df(x_0, y_0)[(x - x_0, y - y_0)]$ is also called the **tangent plane** of f at (x_0, y_0).

Hint Represent $Df(x_0, y_0)$ as a matrix.

(d) Let X be a normed space, let $\Omega \subseteq X$ be open and let $f : \Omega \to \mathbb{R}$ be differentiable at $x \in \Omega$ Define the **tangent hyperplane** of $f : \Omega \to \mathbb{R}$ at x

17-26 Prove Proposition 17.26.

17-27. Prove Proposition 17.27

17-28 Prove Theorem 17.28

Hint. Add and subtract $Df(x)[z - x]$ inside $\| f(z) - f(x) \|$.

17-29 Finish the proof of Theorem 17 29. That is, let X, Y be normed spaces, let $\Omega \subseteq X$ be open, let $f, g : \Omega \to Y$ be differentiable at $x \in \Omega$ and prove that $f + g$ is differentiable at x with derivative $D(f + g)(x) = Df(x) + Dg(x)$.

17-30 Explain why we must use "less than or equal" in Definition 17.24 while we can use the strict inequality in Theorem 4.5.

17-31 Let $\Omega \subseteq \mathbb{R}^n$ be open and let $x \in \Omega$ Explain why a function $f : \Omega \to \mathbb{R}^m$ is differentiable at $x \in \Omega$ iff there is a matrix $A \in M(m \times n, \mathbb{R})$ so that for every $\varepsilon > 0$ there is a $\delta > 0$ so that for all $z \in \Omega$ with $\|z - x\| < \delta$ we have $\| f(z) - f(x) - A(z - x) \| < \varepsilon\|z - x\|$, where $A(z - x)$ is the matrix product of A with the representation of $z - x$ as a column vector with respect to the standard base.

17-32 Let (M, Σ, μ) be a measure space Define $I : L^2(M, \Sigma, \mu) \to \mathbb{R}$ by $I(f) := \int_M f^2 \, d\mu$. Prove that I is differentiable with $DI(f)[u] = \int_M 2fu \, d\mu$.

17-33 *Use Theorem 17 32* to prove that $\frac{d}{dx}\left(\frac{1}{x}\right) = -\frac{1}{x^2}$

17-34 Compare the proofs of Theorems 4 10 and 17 30 Decide which is simpler and explain your decision.

17-35 Let X, Y be normed spaces, let $\Omega \subseteq X$ be open, let $f : \Omega \to \mathbb{R}$ be differentiable at $a \in \Omega$, let $y \in Y$ and let $(fy)(x) := f(x)y$ for all $x \in \Omega$ Prove that $D(fy)(a)[\cdot] = Df(a)[\cdot]y$.

17-36. Let X, Y be normed spaces, let $\Omega \subseteq X$ be open and let $f : \Omega \to Y$ be continuous at the point $x \in \Omega$. Prove that if there is a linear function $L : X \to Y$ (*we do not assume L is continuous*) so that for all $\varepsilon > 0$ there is a $\delta > 0$ so that for all $z \in \Omega$ with $\|z - x\| < \delta$ we have $\| f(z) - f(x) - L[z - x] \| \le \varepsilon\|z - x\|$, then f is differentiable at x.

17-37 Let $(X, \|\cdot\|_X)$, $(Y, \|\cdot\|_Y)$ be normed spaces, let $\Omega \subseteq X$ be open, let $f : \Omega \to Y$, let $x \in \Omega$, let $\|\cdot\|'_X$ be a norm on X that is equivalent to $\|\cdot\|_X$ and let $\|\cdot\|'_Y$ be a norm on Y that is equivalent to $\|\cdot\|_Y$ Prove that f considered as a function from a subset of $(X, \|\cdot\|_X)$ to $(Y, \|\cdot\|_Y)$ is differentiable at x iff f considered as a function from a subset of $(X, \|\cdot\|'_X)$ to $(Y, \|\cdot\|'_Y)$ is differentiable at x

17-38. Let X be a normed space, let $\Omega \subseteq X$ be open and let $f : \Omega \to \mathbb{R}$ be differentiable at $x \in \Omega$ Prove that if there is a $\delta > 0$ so that for all $z \in \Omega$ with $\|z - x\| < \delta$ we have that $f(z) \le f(x)$ (that is, there is a **local maximum** at x), then $Df(x) = 0$.

Hint. Suppose the contrary and use an $a \in X$ so that $Df(x)[a] > 0$.

17-39 Let X, Y be normed spaces, let $\Omega \subseteq X$ be open and for all $n \in \mathbb{N}$ let $f_n : \Omega \to Y$ be differentiable on Ω with continuous derivative $Df_n : \Omega \to \mathcal{L}(X, Y)$. Prove that if $\{f_n\}_{n=1}^{\infty}$ converges pointwise to a function $f : \Omega \to Y$ and $\{ Df_n \}_{n=1}^{\infty}$ converges uniformly to a function $\overline{f} : \Omega \to \mathcal{L}(X, Y)$, then f is differentiable and $Df = \overline{f}$

Hint First define pointwise and uniform convergence

17.4 The Mean Value Theorem

The Mean Value Theorem (Theorem 4.18) cannot be translated directly to higher dimensional settings. Exercise 17-40 shows that for functions $f : [a, b] \to X$, where X is a normed space, the derivative (velocity vector) need not be parallel to $f(b) - f(a)$ at any $t \in (a, b)$. However, the Mean Value Theorem is mainly used to bound the difference $f(b) - f(a)$ by the product of $b - a$ with the supremum of the derivative. Such a result can be proved in a more general setting and we will call the general result "Mean Value Theorem," too. A natural idea for the proof is to first use the Fundamental Theorem of Calculus for an appropriately defined integral and to then use the triangular inequality (Exercises 17-41 and 17-42). This approach ultimately requires the continuity of the derivative or a technical integrability condition. Conditions of this kind can be avoided by working with compactness instead.

Theorem 17.33 Mean Value Theorem. *Let X, Y be normed spaces, let $\Omega \subseteq X$ be open, let $f : \Omega \to X$ be differentiable and let $a, b \in \Omega$ be distinct points so that for all $t \in [0, 1]$ we have $a + t(b - a) \in \Omega$. Then*

$$\left\| f(b) - f(a) \right\| \leq \sup\left\{ \left\| Df(a + t(b - a)) \right\| : t \in [0, 1] \right\} \|b - a\|.$$

Proof. Consider $g(t) := f(a+t(b-a))$, defined on an interval $(-\nu, 1+\nu)$ for some $\nu > 0$. By the Chain Rule, we obtain $g'(t) = \frac{d}{dt} g(t) = Df(a + t(b - a))[b - a]$. Let $\varepsilon > 0$. For every $t \in [0, 1]$, there is a $\delta_t > 0$ so that for all $z \in X$ with $\|z\| < \delta_t \|b-a\|$ we have

$$\left\| f(a+t(b-a)+z) - f(a+t(b-a)) - Df(a+t(b-a))[z] \right\| \leq \frac{\varepsilon}{\|b-a\|} \|z\|.$$

Therefore, with $c := \sup\left\{ \left\| Df(a + t(b - a)) \right\| : t \in [0, 1] \right\}$, for all $t, x \in [0, 1]$ with $|x - t| < \delta_t$ we infer

$$\begin{aligned}
&\left\| g(x) - g(t) \right\| \\
&\leq \left\| g(x) - g(t) - g'(t)(x - t) \right\| + \left\| g'(t)(x - t) \right\| \\
&= \left\| f(a+x(b-a)) - f(a+t(b-a)) - Df(a+t(b-a))[(x-t)(b-a)] \right\| \\
&\quad + \left\| Df(a+t(b-a))[(x-t)(b-a)] \right\| \\
&\leq \frac{\varepsilon}{\|b - a\|} \|(x - t)(b - a)\| + \left\| Df(a+t(b-a)) \right\| \|b-a\| |x-t| \\
&\leq \left(c\|b-a\| + \varepsilon \right) |x - t|.
\end{aligned}$$

Now $\left\{ (t - \delta_t, t + \delta_t) : t \in [0, 1] \right\}$ is an open cover of the compact set $[0, 1]$, which means there is a finite subcover $\left\{ \left(t_i - \delta_{t_i}, t_i + \delta_{t_i} \right) : i = 1, \ldots, n \right\}$. Without loss of generality, we can assume that $t_1 < t_2 < \cdots < t_n$, $t_1 - \delta_{t_1} < 0$, $1 < t_n + \delta_{t_n}$ and $t_{i+1} - \delta_{t_{i+1}} < t_i + \delta_{t_i}$ for $i = 1, \ldots, n-1$. Set $x_0 := 0$, $x_n := 1$ and for $i = 1, \ldots, n-1$ let $x_i \in \left(t_{i+1} - \delta_{t_{i+1}}, t_i + \delta_{t_i} \right) \cap (t_i, t_{i+1})$. Then, using a telescoping sum,

$$\left\| f(b) - f(a) \right\|$$

$$\leq \sum_{i=1}^{n} \|g(x_i) - g(x_{i-1})\| \leq \sum_{i=1}^{n} \|g(x_i) - g(t_i)\| + \|g(t_i) - g(x_{i-1})\|$$

$$\leq \sum_{i=1}^{n} \big(c\|b-a\| + \varepsilon\big)(x_i - t_i) + \big(c\|b-a\| + \varepsilon\big)(t_i - x_{i-1}) = c\|b-a\| + \varepsilon.$$

Because $\varepsilon > 0$ was arbitrary, the theorem is proved. ■

The requirement that there is a straight line connection between the points a and b feels limiting, but it cannot be dropped. In particular, mere connectedness of the domain is not enough. (See Exercise 17-62.)

Exercises

17-40. There is **no direct translation of the Mean Value Theorem** to vector valued functions. Let the function $f : [0, \pi] \to \mathbb{R}^3$ be defined by $f(t) := \big(\cos(t), \sin(t), t^2(\pi - t)\big)$. Prove that there is no $c \in (0, \pi)$ so that $f(\pi) - f(0) = Df(c)[\pi - 0]$.

17-41. The **Riemann integral for Banach space valued functions** Let X be a Banach space and let $f : [a, b] \to X$ be a function.

(a) Define what it means for f to be Riemann integrable.

(b) Prove that if f is continuous, then f is Riemann integrable.

Hint. Prove that for a sufficiently fine partition P any two Riemann sums $R(f, P, T_1)$ and $R(f, P, T_2)$ are close to each other. Then prove that for any refinement Q the Riemann sums for P and for Q are close to each other. Prove that for P_n being the equidistant partition with n intervals, the Riemann sums converge. Then prove that all Riemann sums converge.

(c) Prove the **Antiderivative Form of the Fundamental Theorem of Calculus**. That is, prove that if $f : [a, b] \to X$ is differentiable and f' is integrable, then $\int_a^b f'(x)\,dx = f(b) - f(a)$.

Hint. Cover $[a, b]$ with intervals $(x - \delta_x, x + \delta_x)$ so that for all $z \in [a, b]$ with $|z - x| < \delta_x$ we have $\|f(z) - f(x) - f'(x)(z - x)\| \leq \frac{\varepsilon}{b - a}|z - x|$ Take a finite subcover and construct a partition

(d) **Triangular inequality**. Prove that $\left\|\int_a^b f(x)\,dx\right\| \leq \int_a^b \|f(x)\|\,dx$ if both integrals exist.

(e) Prove the **Derivative Form of the Fundamental Theorem of Calculus**. That is, prove that if $f : [a, b] \to X$ is continuous, then $\frac{d}{dt}\int_a^t f(x)\,dx = f(t)$ for all $t \in [a, b]$.

17-42. Use Exercises 17-41c and 17-41d to prove that if X, Y are normed spaces, $\Omega \subseteq X$ is an open subset, $f : \Omega \to X$ is differentiable, Df is continuous, and $a, b \in \Omega$ are so that for all $t \in [0, 1]$ we have $a + t(b - a) \in \Omega$, then $\|f(b) - f(a)\| \leq \sup\big\{\|Df(a + t(b - a))\| : t \in [0, 1]\big\}\|b - a\|$. Then explain why we cannot avoid the continuity hypothesis in this approach.

17-43 Use Exercises 17-41c and 17-41d to prove that if X, Y are normed spaces, $\Omega \subseteq X$ is an open subset, $f : \Omega \to X$ is differentiable, Df is continuous, and $a, b \in \Omega$ and $\gamma > 0$ are so that for all $t \in [0, 1]$ we have $a + t(b - a) \in \Omega$ and $\|Df(a + t(b - a)) - Df(a)\| \leq \gamma t\|b - a\|$, then the inequality $\|f(b) - f(a) - Df(a)[b - a]\| \leq \frac{\gamma}{2}\|b - a\|^2$ holds.

Hint. Mimic the proof of Lemma 13 13 and use Exercise 17-41c.

17-44. **Newton's method** in several variables. Let X be a Banach space, let $\Omega \subseteq X$ be open and let $f : \Omega \to X$ be a continuously differentiable function so that $Df(x) \in \mathcal{H}(X, X)$ for all $x \in \Omega$. Prove that if there are $x_0 \in \Omega$ and $\alpha, \beta, \gamma > 0$ so that $\left\|Df(x_0)^{-1}\left[f(x_0)\right]\right\| \leq \alpha$, so that for all $x \in \Omega$

we have $\left\| Df(x)^{-1} \right\| \leq \beta$, so that for all $x, z \in \Omega$ we have $\| Df(z) - Df(x) \| \leq \gamma \|z - x\|$, so that $h := \frac{\alpha\beta\gamma}{2} < 1$ and so that with $r := \frac{\alpha}{1-h}$ we have $\overline{B_r(x_0)} \subseteq \Omega$, then

(a) Each recursively defined point $x_{n+1} := x_n - Df(x_n)^{-1} \left[f(x_n) \right]$ is in $B_r(x_0)$.

(b) The sequence $\{x_n\}_{n=0}^{\infty}$ converges to a point $u \in \overline{B_r(x_0)}$ with $f(u) = 0$

(c) For all $n \geq 0$ we have $\|u - x_n\| \leq \alpha \frac{h^{2^n - 1}}{1 - h^{2^n}}$.

Hint Mimic the proof of Theorem 13.14

17.5 How Partial Derivatives Fit In

Partial derivatives are usually defined in the direction of a coordinate axis. Rather than tying our presentation to $\mathbb{R}^d$ and its coordinate axes, we consider a product $\prod_{i=1}^{d} X_i$ of normed spaces.

Proposition 17.34 *For $i = 1, \ldots, d$, let $(X_i, \|\cdot\|_i)$ be a normed space and let $\|\cdot\|_{\mathbb{R}^d}$ be a norm on $\mathbb{R}^d$. Then $\prod_{i=1}^{d} X_i$ with componentwise addition and scalar multiplication is a vector space and $\left\|(x_1, \ldots, x_d)\right\| := \left\|\left(\|x_1\|_1, \ldots, \|x_d\|_d\right)\right\|_{\mathbb{R}^d}$ defines a norm on it.*

Proof. Exercise 17-45. ■

Definition 17.35 *For $i = 1, \ldots, d$, let $(X_i, \|\cdot\|_i)$ be a normed space and let $\|\cdot\|_{\mathbb{R}^d}$ be a norm on $\mathbb{R}^d$. The norm $\left\|(x_1, \ldots, x_d)\right\| := \left\|\left(\|x_1\|_1, \ldots, \|x_d\|_d\right)\right\|_{\mathbb{R}^d}$ is called a* **product norm**. *From now on, we will assume that any product $\prod_{i=1}^{d} X_i$ of normed spaces is equipped with a product norm and we will call it a* **product space**. *Because all norms on $\mathbb{R}^d$ are equivalent, all product norms are equivalent. Hence, unless otherwise stated, we will use* $\max\left\{\|x_1\|_1, \ldots, \|x_d\|_d\right\}$ *as the product norm.*

Exercise 17-46 shows that all product norms are equivalent, so we are free to interchange specific norms in $\mathbb{R}^d$ in the definition of the product norm we use. The definition of product norms implies that the product of Banach spaces is again a Banach space and that the natural projections are continuous.

Proposition 17.36 *For $i = 1, \ldots, d$, let $(X_i, \|\cdot\|_i)$ be a normed space. The product $\prod_{i=1}^{d} X_i$ is a Banach space iff all factor spaces $(X_i, \|\cdot\|_i)$ are Banach spaces.*

Proof. Exercise 17-47. ■

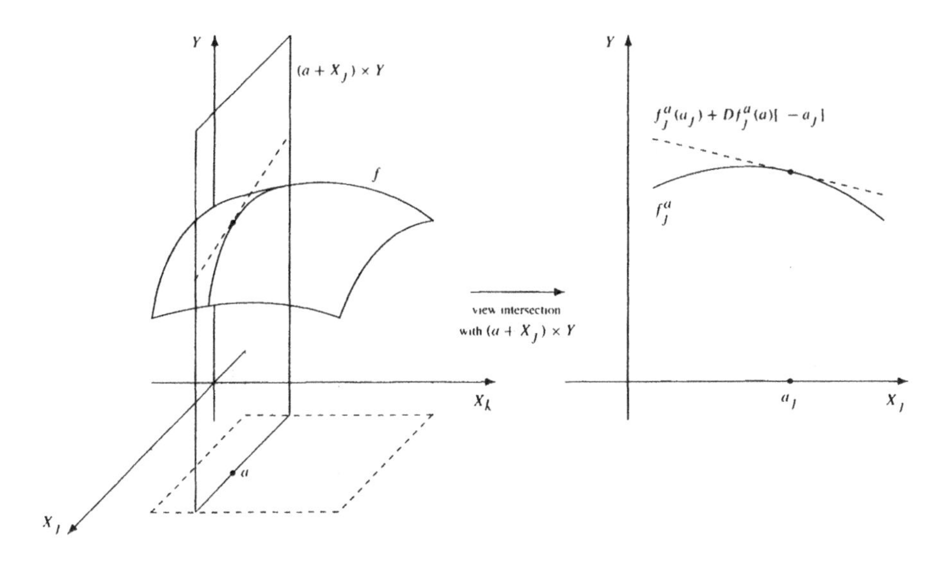

Figure 43: For partial derivatives, we restrict our attention to an appropriately translated subspace and take the derivative in that subspace.

Proposition 17.37 *For $i = 1, \ldots, d$ let $(X_i, \|\cdot\|_i)$ be a normed space and let $\prod_{i=1}^{d} X_i$ be the product space. Then the* **natural projections** *$\pi_j : \prod_{i=1}^{d} X_i \to X_i$ defined by $\pi_j(x_1, \ldots, x_d) := x_j$ are continuous.*

Proof. Exercise 17-48. ■

Exercise 17-49 now shows that not every norm on a product of vector spaces is a product norm. Proposition 17.38 assures that as we consider partial derivatives with respect to a factor space, the domains of the requisite functions are open.

Proposition 17.38 *Let $X = \prod_{i=1}^{d} X_i$ be a product space, let $\Omega \subseteq X$ be open, and let $a = (a_1, \ldots, a_d) \in \Omega$. Then for all $j \in \{1, \ldots, d\}$ the set*

$$\Omega_j^a := \left\{x_j \in X_j : (a_1, \ldots, a_{j-1}, x_j, a_{j+1}, \ldots, a_d) \in \Omega\right\}$$

is open in $(X_j, \|\cdot\|_j)$.

Proof. Exercise 17-50. ■

Definition 17.39 *Let $X = \prod_{i=1}^{d} X_i$ be a product space, let Y be a normed space, let $\Omega \subseteq X$ be open, let $a = (a_1, \ldots, a_d) \in \Omega$, and let $j \in \{1, \ldots, d\}$. For $f : \Omega \to Y$,* *define $f_j^a : \Omega_j^a \to Y$ by $f_j^a(x_j) := f(a_1, \ldots, a_{j-1}, x_j, a_{j+1}, \ldots, a_d)$. If f_j^a is differentiable at a_j, then the derivative of f_j^a at a_j is denoted $D_j f(a)$ and it is called the* **partial derivative** *of f with respect to X_j at a or the* **partial derivative** *of f with respect to the jth variable at a. For a visualization, consider Figure 43.*

It is easy to see that if f is differentiable at $a \in \prod_{i=1}^{d} X_i$, then the restriction of its derivative to X_j is the partial derivative with respect to X_j.

Proposition 17.40 *Let $X = \prod_{i=1}^{d} X_i$ be a product space, let Y be a normed space, let $\Omega \subseteq X$ be open and let $f : \Omega \to Y$. If f is differentiable at $a \in \Omega$, then for all integers $j \in \{1, \ldots, d\}$ the partial derivative at a with respect to X_j exists and is equal to $D_j f(a) = Df(a)|_{\{0\}\times\cdots\times\{0\}\times X_j\times\{0\}\times\cdots\times\{0\}}$. Moreover, for all $u \in X$ we have*

$$Df(a)[u_1, \ldots, u_d] = \sum_{j=1}^{d} D_j f(a)[u_j].$$

Proof. Exercise 17-51. ■

Unfortunately, the existence of partial derivatives is not sufficient for differentiability (in fact, not even for continuity) as Exercises 17-60 and 17-61 show. The pathology exhibited in these exercises is why we build the theory around derivatives as in Definition 17.24 instead of partial derivatives. Nonetheless, it is possible to construct the derivative from partial derivatives, as long as the partial derivatives are continuous.

Theorem 17.41 *Let $X = \prod_{i=1}^{d} X_i$ be a product space, let Y be a normed space, let $\Omega \subseteq X$ be open and let $f : \Omega \to Y$. If for every $j \in \{1, \ldots, d\}$ the function f is differentiable with respect to the jth variable at every $x \in \Omega$ and the function $x \mapsto D_j f(x)$ is continuous at $a \in \Omega$, then f is differentiable at a and the derivative of f at a is*

$$Df(a)[u_1, \ldots, u_d] = \sum_{j=1}^{d} D_j f(a)[u_j].$$

Proof. Recall that we said we would use $\|x\| := \max\left\{\|x_i\|_i : i = 1, \ldots, d\right\}$ as the product norm. Let $\varepsilon > 0$. Find $\delta > 0$ so that for all $j = 1, \ldots, d$ and all $x \in B_\delta(a)$ we have $\left\|D_j(x) - D_j(a)\right\| < \frac{\varepsilon}{d}$. Then for all $(z_1, \ldots, z_d) \in B_\delta(a_1, \ldots, a_d)$ we obtain the following.

$$\left\| f(z_1, \ldots, z_d) - f(a_1, \ldots, a_d) - \sum_{j=1}^{d} D_j f(a_1, \ldots, a_d)[z_j - a_j] \right\|$$

Turn the difference into a telescoping sum in which the differences are between terms that differ in only one component.

$$\leq \sum_{j=1}^{d} \Big\| f(z_1,\ldots,z_j,a_{j+1},\ldots a_d) - f(z_1,\ldots,z_{j-1},a_j,\ldots,a_d)$$
$$-D_j f(a_1,\ldots,a_d)[z_j - a_j] \Big\|$$

Apply the Mean Value Theorem (Theorem 17.33) to $f(z_1,\ldots,z_{j-1},\cdot,a_{j+1},\ldots,a_d) - D_j f(a_1,\ldots,a_d)[\cdot - a_j]$.

$$\leq \sum_{j=1}^{d} \|z_j - a_j\| \sup \Big\{ \big\| D_j f(z_1,\ldots,z_{j-1},a_j + t(z_j - a_j),a_{j+1},\ldots,a_d)$$
$$-D_j f(a_1,\ldots,a_d) \big\| : t \in [0,1] \Big\}$$
$$\leq \varepsilon \|z - a\|.$$

The above proves that $Df(a)[u_1,\ldots,u_d] = \sum_{j=1}^{d} D_j f(a)[u_j]$. ■

With the general results established, we can now turn to the familiar partial derivatives on $\mathbb{R}^n$. Because $\mathbb{R}^n = \prod_{i=1}^{n} \mathbb{R}$ is the prototypical product space, all results proved so far apply to $\mathbb{R}^n$. Therefore we can concentrate on translating the abstract notions to the more concrete setting of n-dimensional space. Partial derivatives in $\mathbb{R}^n$ are typically defined in the direction of a coordinate vector.

Definition 17.42 *Let the set $\Omega \subseteq \mathbb{R}^n$ be open, let $f : \Omega \to \mathbb{R}^m$ be a function and let $a \in \Omega$. Then the* **partial derivative** *of f with respect to x_j at a is defined to be $\frac{\partial f}{\partial x_j}(a) := \lim_{h \to 0} \frac{f(a + he_j) - f(a)}{h}$ if this limit exists.*

The connection between these partial derivatives and the derivative is an exercise in reinterpreting abstract concepts as matrices.

Theorem 17.43 *Let $\Omega \subseteq \mathbb{R}^n$ be an open set, let $f : \Omega \to \mathbb{R}^m$ be differentiable at $a \in \Omega$, for $i = 1,\ldots,m$ let $f_i := \pi_i \circ f$ and let $u \in \mathbb{R}^n$ be so that $u = \sum_{j=1}^{n} u_j e_j$. Then $Df(a)[u] = \sum_{i=1}^{m} e_i \sum_{j=1}^{n} \frac{\partial f_i}{\partial x_j}(a) u_j$. That is, the matrix $\left(\frac{\partial f_i}{\partial x_j}(a)\right)_{\substack{i=1,\ldots,m \\ j=1,\ldots,n}}$ represents $Df(a)$ with respect to the standard bases in $\mathbb{R}^n$ and $\mathbb{R}^m$. This matrix is also called the* **Jacobian matrix** *of f at a.*

Conversely, if $f : \Omega \to \mathbb{R}^m$ is so that for all $i = 1,\ldots,m$ and $j = 1,\ldots,n$ and all $x \in \Omega$ the partial derivative $\frac{\partial f_i}{\partial x_j}(x)$ exists and if all these partial derivatives are continuous at a, then f is differentiable at a.

Proof. For $j = 1, \ldots, n$, let $\mathbb{R}_j := \mathbb{R}$ so that $\mathbb{R}^n = \prod_{j=1}^{n} \mathbb{R}_j$. (In this fashion, we can distinguish the partial derivatives in the coordinate directions.) For $g : \Omega \to \mathbb{R}$ differentiable at a, we infer by Proposition 17.40 that $Dg(a)[u] = \sum_{j=1}^{n} D_{\mathbb{R}_j} g(a)[u_j]$. By Proposition 17.26 for all $j = 1, \ldots, n$ we obtain $D_{\mathbb{R}_j} g(a)[u_j] = \frac{\partial g}{\partial x_j}(a) u_j$, and hence $Dg(a)[u] = \sum_{j=1}^{n} \frac{\partial g}{\partial x_j}(a) u_j$. Now for $f : \Omega \to \mathbb{R}^m$ by Exercise 17-35 we conclude $Df(a)[u] = D\left(\sum_{i=1}^{m} f_i e_i\right)(a)[u] = \sum_{i=1}^{m} e_i Df_i(a)[u] = \sum_{i=1}^{m} e_i \sum_{j=1}^{n} \frac{\partial f_i}{\partial x_j}(a) u_j$.
The last statement follows from Theorem 17.41. ∎

For $m = 1$, the Jacobian matrix of a differentiable function $f : \mathbb{R}^n \to \mathbb{R}$ is given special attention because of its physical meaning.

Definition 17.44 *Let $\Omega \subseteq \mathbb{R}^n$ be open and let $f : \Omega \to \mathbb{R}$ be so that all partial derivatives of f exist at $a \in \Omega$. Then we define the* **gradient** *of f at a to be* $\operatorname{grad}(f)(a) := \nabla f(a) := \begin{pmatrix} \frac{\partial f}{\partial x_1}(a) \\ \vdots \\ \frac{\partial f}{\partial x_n}(a) \end{pmatrix}$. *With $\nabla := \begin{pmatrix} \frac{\partial}{\partial x_1} \\ \vdots \\ \frac{\partial}{\partial x_n} \end{pmatrix}$ this looks like a formal multiplication of f with the "vector" ∇. The (purely formal) "vector" ∇ is also called the* **nabla operator**.

The physical meaning of the gradient is easily explained with what we have derived so far. Let $f : \Omega \to \mathbb{R}$ be differentiable at a. Then by Theorem 17.43 for all $u \in \mathbb{R}^d$ we obtain $Df(a)[u] = \langle \nabla f(a), u \rangle$. Moreover, the derivative describes the local behavior of f near a. In particular, for any vector u with $\|u\| = 1$ we infer $\lim_{t \to 0} \frac{f(a + tu) - f(a)}{t} = \langle \nabla f(a), u \rangle$ (Exercise 17-52). That is, the inner product $\langle \nabla f(a), u \rangle$ gives the derivative of f in the direction u. This inner product is largest when $u = \frac{\nabla f(a)}{\|\nabla f(a)\|}$, so the gradient vector gives the **direction of steepest ascent** with its norm being the slope in that direction. Similarly, $-\nabla f(a)$ gives the **direction of steepest descent** with the negative of its norm being the slope in that direction. Physical systems usually strive for equilibrium. Whenever there is an imbalance in a physical quantity described by f in a homogeneous medium, there will be a flow in the direction of $-\nabla f$ that tries to restore equilibrium. For more on the physical ideas, consider Section 21.2.

Exercises

17-45 Prove Proposition 17.34 as follows

(a) For $i = 1, \ldots, d$, let X_i be a vector space. Prove that $\prod_{i=1}^{d} X_i$ with componentwise addition and scalar multiplication is a vector space.

(b) For $i = 1, \ldots, d$, let $\left(X_i, \|\cdot\|_i \right)$ be a normed space and let $\|\cdot\|_{\mathbb{R}^d}$ be a norm on $\mathbb{R}^d$. Prove that $\left\| (x_1, \ldots, x_d) \right\| := \left\| \left(\|x_1\|_1, \ldots, \|x_d\|_d \right) \right\|_{\mathbb{R}^d}$ defines a norm on $\prod_{i=1}^{d} X_i$

17-46. Prove that any two product norms are equivalent.
Note. This and Exercise 17-37 justify the free interchange of one product norm for another.

17-47. Prove Proposition 17.36.

17-48. Prove Proposition 17.37.

17-49. Not every norm on a product is a **product norm**.

(a) Let $X_1, \ldots, X_d$ be vector spaces and let $\|\cdot\|_X$ be a norm on $\prod_{j=1}^{d} X_j$.

i. Prove that for all $j = 1, \ldots, d$ the function $\|x_j\|_j' := \left\| (0, \ldots, 0, x_j, 0, \ldots, 0) \right\|_X$ is a norm on X_j.

ii. Prove that if for each $i \in \{1, \ldots, d\}$ we pick a fixed $x_i \in X_i \setminus \{0\}$, then the function $\left\| (a_1, \ldots, a_d) \right\| := \left\| (a_1 x_1, \ldots, a_d x_d) \right\|_X$ defines a norm on $\mathbb{R}^d$.

(b) Let $\left(X, \|\cdot\| \right)$ be a normed space, let $S \subseteq X$ be a dense subspace (like the simple functions in L^p, see Theorem 16.85) and let F be a subspace so that $S \cap F = \{0\}$ (with S being the set of simple functions in L^p, F could be the set of scalar multiples of a function $f \notin S$). Define $Y = S \times F$ and let $\left\| (s, f) \right\|_{np} := \|s + f\|$.

i. Prove that $\|\cdot\|_{np}$ is a norm on $S \times F$.

ii. Prove that the natural projection $\pi_S : S \times F \to S$ is not continuous.
Hint. Let $s_n \to f$ and consider $s_n - f$.

iii. Conclude that $\|\cdot\|_{np}$ is *not* a product norm on $S \times F$.

17-50. Prove Proposition 17.38.

17-51. Prove Proposition 17.40.

17-52. Let $\Omega \subseteq \mathbb{R}^d$ be open and let $f : \Omega \to \mathbb{R}$ be differentiable at $a \in \Omega$. Prove that for any vector u with $\|u\| = 1$ we have $\lim_{t \to 0} \frac{f(a + tu) - f(u)}{t} = \left\langle \nabla f(a), u \right\rangle$

17-53. Let $X = \prod_{i=1}^{d} X_i$ be a product space, let Y be a normed space, let $\Omega \subseteq X$ be open and let $f : \Omega \to Y$ be a function. Prove that if f is differentiable on Ω and Df is continuous, then all partial derivatives $D_j f$ exist for all $a \in \Omega$ and the functions $D_j f$ are continuous on Ω.

17-54. Let X be a vector space and let $E, F \subseteq X$ be vector subspaces of X. Then X is called the **direct sum** of E and F, denoted $E \oplus F$, iff $E \cap F = \{0\}$ and for all $x \in X$ there are $e \in E$ and $f \in F$ so that $x = e + f$.
Prove that if X is the direct sum of E and F, then $X = E \oplus F$ is isomorphic to $E \times F$.

17-55. Prove the **Multivariable Chain Rule**. That is, prove that if $f(x_1, \ldots, x_n) : \mathbb{R}^n \to \mathbb{R}$ is a differentiable function of n variables and the components are differentiable functions $x_j(t_1, \ldots, t_m)$ of m variables, then $\frac{\partial f}{\partial t_i} = \sum_{j=1}^{n} \frac{\partial f}{\partial x_j} \frac{\partial x_j}{\partial t_i} = \frac{\partial f}{\partial x_1} \frac{\partial x_1}{\partial t_i} + \frac{\partial f}{\partial x_2} \frac{\partial x_2}{\partial t_i} + \cdots + \frac{\partial f}{\partial x_n} \frac{\partial x_n}{\partial t_i}$.

17-56. **Coordinate transformations for differential operators.** Let (x, y) be rectangular coordinates on $\mathbb{R}^2$, let $r = \sqrt{x^2 + y^2}$ and let $\theta := \begin{cases} \arctan\left(\frac{y}{x}\right); & \text{if } x > 0, \\ \arctan\left(\frac{y}{x}\right) + \pi; & \text{if } x < 0, \end{cases}$ be polar coordinates on $\mathbb{R}^2$

(a) Prove that if $f : \mathbb{R}^2 \to \mathbb{R}$ is differentiable, then $\frac{\partial f}{\partial x} = \frac{\partial f}{\partial r}\cos(\theta) - \frac{\partial f}{\partial \theta}\frac{\sin(\theta)}{r}$.

Hint. Use the Chain Rule as stated in Exercise 17-55.

(b) Prove that if $f : \mathbb{R}^2 \to \mathbb{R}$ is differentiable and $\frac{\partial f}{\partial x}, \frac{\partial f}{\partial r}$ and $\frac{\partial f}{\partial \theta}$ are differentiable, then

$$\frac{\partial^2 f}{\partial x^2} = \frac{\partial^2 f}{\partial r^2}\cos^2(\theta) - 2\frac{\partial^2 f}{\partial r \partial \theta}\frac{\sin(\theta)\cos(\theta)}{r} + 2\frac{\partial f}{\partial \theta}\frac{\sin(\theta)\cos(\theta)}{r^2} + \frac{\partial f}{\partial r}\frac{\sin^2(\theta)}{r} + \frac{\partial^2 f}{\partial \theta^2}\frac{\sin^2(\theta)}{r^2}$$

The second partial derivatives simply denote partial derivatives of partial derivatives.

(c) Let f be a function whose partial derivatives $\frac{\partial f}{\partial x}, \frac{\partial f}{\partial y}, \frac{\partial f}{\partial r}$ and $\frac{\partial f}{\partial \theta}$ are differentiable. Derive an expression for $\frac{\partial^2 f}{\partial x^2} + \frac{\partial^2 f}{\partial y^2}$ in polar coordinates.

Hint. Derive a formula similar to that in part 17-56b for $\frac{\partial^2 f}{\partial y^2}$ and add the two.

17-57 Let W be a normed space, let $\prod_{i=1}^{d} X_i$ be a product space, let $\Omega \subseteq W$ be open and for $i = 1, \ldots, d$ let $f_i : \Omega \to X_i$ be differentiable at $x \in \Omega$. Prove that $f := (f_1, \ldots, f_d)$ is differentiable at x with $Df(x)[\cdot] = \big(Df_1(x)[\cdot], \ldots, Df_d(x)[\cdot]\big)$.

17-58. **Leibniz' Rule.** Let $a, b : (c, d) \to (l, u)$ be differentiable and let $g : (l, u) \times (c, d) \to \mathbb{R}$ be differentiable with respect to the second variable. Let $F \in L^1(l, u)$ be so that all $g(\cdot, t)$, $\frac{g(\cdot, t+h) - g(\cdot, t)}{h}$ and $\frac{\partial}{\partial t} g(\cdot, t)$ are bounded by $F(\cdot)$ and let all $g(\cdot, t)$ be continuous. Use the steps outlined below to prove that

$$\frac{d}{dt}\int_{a(t)}^{b(t)} g(x, t)\, dx = g\big(b(t), t\big)\frac{d}{dt}b(t) - g\big(a(t), t\big)\frac{d}{dt}a(t) + \int_{a(t)}^{b(t)} \frac{\partial}{\partial t} g(x, t)\, dx.$$

(a) Prove that $v : (c, d) \to (l, u) \times (l, u) \times L^1(l, u)$ defined by $v(t) := \begin{pmatrix} a(t) \\ b(t) \\ g(x, t) \end{pmatrix}$ is differentiable.

Hint. Use the result of Exercise 17-57. Use Proposition 17.26 and the Dominated Convergence Theorem for the differentiability of the third component.

(b) Prove that $s : (l, u) \times (l, u) \times C(l, u) \to \mathbb{R}$ defined by $s\begin{pmatrix} a \\ b \\ h \end{pmatrix} := \int_a^b h(x)\, dx$ is differentiable, where $C(l, u)$ is a normed subspace of $L^1(l, u)$.

Hint. Theorem 17.41. Use the linearity of the integral operator $h \mapsto \int_a^b h(x)\, dx$ on $L^1(l, u)$ for the partial derivative with respect to the third component.

(c) Prove Leibniz' Rule using the Chain Rule.

17-59. Let X be a normed space, let $D \subseteq X$ be a dense subspace, let Y be a Banach space, let $\Omega \subseteq D$ be open in D and let $f : \Omega \to Y$ be continuously differentiable with bounded uniformly continuous derivative. Prove that there is an open set $\widetilde{\Omega} \subseteq X$ and a unique continuously differentiable function $e : \widetilde{\Omega} \to Y$ so that $\widetilde{\Omega} \cap D = \Omega$ and so that $e|_{\Omega} = f$.

Hint. Use the Mean Value Theorem (Theorem 17.33).

17-60 Consider the function $f(x, y) = \begin{cases} \frac{xy}{x^2+y^2}; & \text{for } (x, y) \neq (0, 0), \\ 0 & \text{for } (x, y) = (0, 0). \end{cases}$

(a) Prove that $\frac{\partial f}{\partial x}(0, 0) = \frac{\partial f}{\partial y}(0, 0) = 0$.

(b) Prove that f is not continuous at $(0,0)$.

17-61. Consider the function $f(x,y)=\begin{cases}\frac{x^2y}{x^2+y^2}; & \text{for } (x,y)\neq(0,0),\\ 0; & \text{for } (x,y)=(0,0).\end{cases}$

(a) Prove that f is continuous at $(0,0)$.

(b) Let $X=\operatorname{span}(u)\subset\mathbb{R}^2$ be an arbitrary one dimensional linear subspace of $\mathbb{R}^2$ with $\|u\|=1$. Prove that $D_X f(0,0) := \lim_{t\to0}\frac{f(tu)-f(0,0)}{t}$ exists.

(c) Prove that f is not differentiable at $(0,0)$.
Hint. If f was differentiable at $(0,0)$, what would the derivatives in part 17-61b be equal to? Use Theorem 17.43.

17-62 An **unbounded function with bounded derivative and bounded, connected two-dimensional domain.** Let $A:=\left\{(\theta,\Delta r)\in\mathbb{R}^2:\theta>\frac{\pi}{4},\Delta r\in\left(0,\frac{1}{100}e^{-\theta}\right)\right\}$ and define $f:A\to\mathbb{R}^2$ by $f(\theta,\Delta r):=\left(\left(\frac{1}{\theta}+\Delta r\right)\cos(\theta),\left(\frac{1}{\theta}+\Delta r\right)\sin(\theta)\right)$.

(a) Prove that $S:=f[A]$ is open, connected and contained in $B_2(0,0)$.

(b) Sketch S and state the geometric meaning of θ and Δr.

(c) Prove that f is injective

(d) For $(x,y)\in S$, define $\theta(x,y)$ to be the first component of $f^{-1}(x,y)$ Prove that the function $(x,y)\mapsto\theta(x,y)$ is differentiable at every point of S by showing that on every open disk contained in S it differs from $\arctan\left(\frac{y}{x}\right)$ or $\arctan\left(-\frac{x}{y}\right)$ by a constant

(e) For $(x,y)\in S$, define $r(x,y):=\sqrt{x^2+y^2}$ and prove that r is differentiable on S.

(f) For $(x,y)\in S$ define $B(x,y):=\ln\big(\theta(x,y)\big)$. Prove that B is differentiable with bounded derivative
Hint Prove that the absolute values of the partial derivatives of θ are equal to one of the expressions $\frac{|x|}{x^2+y^2}=\frac{|\cos(\theta(x,y))|}{r(x,y)}$ or $\frac{|y|}{x^2+y^2}=\frac{|\sin(\theta(x,y))|}{r(x,y)}$ and use that the radius satisfies $r(x,y)\in\left(\frac{1}{\theta(x,y)},\frac{1}{\theta(x,y)}+\frac{1}{100}e^{-\theta(x,y)}\right)$.

(g) Prove that $\lim_{(x,y)\to(0,0)} B(x,y)=\infty$

17.6 Multilinear Functions (Tensors)

If $f:\Omega\to Y$ is differentiable at every $x\in\Omega$, then we can also try to differentiate the derivative $Df:\Omega\to\mathcal{L}(X,Y)$. If the thus computed second derivative exists, it would be a function that maps points $x\in\Omega$ to linear functions $D^2f(x)[\cdot]\in\mathcal{L}\big(X,\mathcal{L}(X,Y)\big)$ and these linear functions map points $u\in X$ to linear functions $D^2f(x)[u][\cdot]$. It turns out that such functions are linear in both square bracketed arguments (see Proposition 17.53 below). To simplify notation, higher derivatives are usually identified with functions that have several arguments and are linear in each one of them.

Definition 17.45 *Let* $X = \prod_{i=1}^{k} X_i$ *be a product space and let* Y *be a normed space. Then* $T : \prod_{i=1}^{k} X_i \to Y$ *is called* **multilinear** *or* k-**linear** *or a* k-**tensor** *iff for all indices* $j \in \{1, \ldots, k\}$, *all* $x, y \in X_j$ *and all* $\alpha, \beta \in \mathbb{R}$ *we have*

$$\begin{aligned} & T[x_1, \ldots, x_{j-1}, \alpha x + \beta y, x_{j+1}, \ldots, x_k] \\ = \;& \alpha T[x_1, \ldots, x_{j-1}, x, x_{j+1}, \ldots, x_k] + \beta T[x_1, \ldots, x_{j-1}, y, x_{j+1}, \ldots, x_k]. \end{aligned}$$

2-linear functions are also called **bilinear**.

As for linear functions, we enclose the argument of multilinear functions in square brackets instead of round parentheses. This is because as multilinear functions are identified with higher derivatives we will evaluate higher derivatives at a point x (enclosed in round parentheses) for an argument $[t_1, \ldots, t_m]$, which will be distinguished by being enclosed in square brackets.

Example 17.46 Examples of **bilinear** functions.

1. The function $m : \mathbb{R}^2 \to \mathbb{R}$ defined by $m[x, y] := xy$ is bilinear.
2. Let X be a *real* vector space and let $\langle \cdot, \cdot \rangle$ be an inner product on X. Then $\langle \cdot, \cdot \rangle$ is bilinear. An inner product on a complex vector space X is *not* bilinear, because for $x, y \in X$ and $\alpha \in \mathbb{C}$ we have $\langle x, \alpha y \rangle = \overline{\alpha} \langle x, y \rangle$. Complex inner products are also called **sesquilinear**.
3. The **cross product** $\begin{pmatrix} x_1 \\ y_1 \\ z_1 \end{pmatrix} \times \begin{pmatrix} x_2 \\ y_2 \\ z_2 \end{pmatrix} := \begin{pmatrix} y_1 z_2 - z_1 y_2 \\ -(x_1 z_2 - z_1 x_2) \\ x_1 y_2 - y_1 x_2 \end{pmatrix}$ is a bilinear function from $\mathbb{R}^3 \times \mathbb{R}^3$ to $\mathbb{R}^3$. □

Continuity of multilinear functions is characterized similar to continuity of linear functions.

Theorem 17.47 *Let* $X = \prod_{i=1}^{k} X_i$ *be a product space, let* Y *be a normed space and let* $T : \prod_{i=1}^{k} X_i \to Y$ *be k-linear. Then the following are equivalent:*

1. T *is continuous on* X.
2. T *is continuous at* $0 \in X$.
3. T *is bounded on* $\overline{B_1(0)} \subset X$.
4. *There is a* $c \in \mathbb{R}$ *so that for all elements* $(x_1, \ldots, x_k) \in X = \prod_{i=1}^{k} X_i$ *we have that* $\|T[x_1, \ldots, x_k]\| \leq c\|x_1\| \cdots \|x_k\|$.

Proof. Mimic the proof of Theorem 17.4 (Exercise 17-63). ∎

Corollary 17.48 *Let* $X = \prod_{i=1}^{k} X_i$ *be a finite dimensional product space, let* Y *be a normed space and let* $T : \prod_{i=1}^{k} X_i \to Y$ *be k-linear. Then* T *is continuous.*

Proof. Mimic the proof of Corollary 17.6 (Exercise 17-64). ∎

Similar to continuous linear functions, continuous multilinear functions form a normed space.

Definition 17.49 *Let* $X_1, \ldots, X_k, Y$ *be normed spaces. Define* $\mathcal{T}^k(X_1, \ldots, X_k, Y)$ *to be the set of all continuous k-linear functions from the product space* $\prod_{i=1}^{k} X_i$ *to* Y*. If* $X_1 = \cdots = X_k = X$*, we also write* $\mathcal{T}^k(X, Y)$ *instead of* $\mathcal{T}^k(X_1, \ldots, X_k, Y)$*.*

Theorem 17.50 *Let* $X_1, \ldots, X_k, Y$ *be normed spaces. Then, with pointwise addition and scalar multiplication,* $\mathcal{T}^k(X_1, \ldots, X_k, Y)$ *is a vector space and the function*

$$\|T\| := \min\left\{c \geq 0 : \left(\forall (x_1, \ldots, x_k) \in \prod_{i=1}^{k} X_i : \big\|T[x_1, \ldots, x_k]\big\| \leq c\|x_1\| \cdots \|x_k\|\right)\right\}$$

is a norm on $\mathcal{T}^k(X_1, \ldots, X_k, Y)$ *so that* $\big\|T(x_1, \ldots, x_k)\big\| \leq \|T\|\|x_1\| \cdots \|x_k\|$ *for all* $(x_1, \ldots, x_k) \in \prod_{i=1}^{k} X_i$*. Moreover, if* Y *is a Banach space, then so is* $\mathcal{T}^k(X_1, \ldots, X_k, Y)$*.*

Proof. Mimic the proofs of Theorems 17.8 and 17.11 (Exercise 17-65). ∎

Definition 17.51 *Similar to the operator norm of a continuous linear function, we will call the norm from Theorem 17.50 the* **tensor norm** *of the continuous k-tensor* T*.*

Continuing with similarities to continuous linear functions, multilinear functions are differentiable iff they are continuous.

Theorem 17.52 *Let* $X_1, \ldots, X_k$ *and* Y *be normed spaces and consider the function* $T \in \mathcal{T}^k(X_1, \ldots, X_k, Y)$*. Then* T *is differentiable and for each* $(x_1, \ldots, x_k) \in \prod_{i=1}^{k} X_i$ *the derivative is* $DT(x_1, \ldots, x_k)[u_1, \ldots, u_k] = \sum_{j=1}^{k} T[x_1, \ldots, x_{j-1}, u_j, x_{j+1}, \ldots, x_k]$*.*

Proof. The case $k = 1$ is Proposition 17.27. Hence, we will assume $k \geq 2$ throughout. We use a telescoping sum that is similar to the one in the proof of Theorem

17.41. Let $(x_1, \ldots, x_k) \in \prod_{i=1}^{k} X_i$. Then for all elements $(z_1, \ldots, z_k) \in \prod_{i=1}^{k} X_i$ so that $\left\|(z_1, \ldots, z_k)\right\| < 2\left\|(x_1, \ldots, x_k)\right\| + 1 =: M$ we obtain the following.

$$\begin{aligned}
&\left\| T[z_1, \ldots, z_k] - T[x_1, \ldots, x_k] - \sum_{j=1}^{k} T[x_1, \ldots, x_{j-1}, z_j - x_j, x_{j+1}, \ldots, x_k] \right\| \\
&= \left\| \sum_{j=1}^{k} T[z_1, \ldots, z_{j-1}, z_j - x_j, x_{j+1}, \ldots, x_k] \right. \\
&\qquad \left. - \sum_{j=1}^{k} T[x_1, \ldots, x_{j-1}, z_j - x_j, x_{j+1}, \ldots, x_k] \right\| \\
&= \left\| \sum_{j=2}^{k} \sum_{i=1}^{j-1} T[z_1, \ldots z_{i-1}, z_i - x_i, x_{i+1}, \ldots, x_{j-1}, z_j - x_j, x_{j+1}, \ldots, x_k] \right\| \\
&\leq \sum_{j=2}^{k} \sum_{i=1}^{j-1} \|T\| \|z_1\| \cdots \|z_{i-1}\| \|z_i - x_i\| \|x_{i+1}\| \cdots \|x_{j-1}\| \|z_j - x_j\| \|x_{j+1}\| \cdots \|x_k\| \\
&\leq \underbrace{\left(\frac{k}{2}(k-1)\|T\| M^{k-2} \right)}_{=C} \|z - x\|^2.
\end{aligned}$$

Now for any $\varepsilon > 0$ we can choose $\delta := \frac{\varepsilon}{C+1}$ to make the difference smaller than $\varepsilon\|z - x\|$. Hence, T is differentiable with the indicated derivative. ∎

In particular, Theorem 17.52 says that the derivative of a k-tensor is a sum of $(k-1)$-tensors. We conclude this section with the result that allows us to identify higher derivatives with continuous k-linear functions.

Proposition 17.53 *The spaces $\mathcal{L}\left(X, \mathcal{T}^k(X, Y)\right)$ and $\mathcal{T}^{k+1}(X, Y)$ are isomorphic via the map that sends the function $D : X \to \mathcal{T}^k(X, Y)$ in $\mathcal{L}\left(X, \mathcal{T}^k(X, Y)\right)$ to the multilinear function in $\mathcal{T}^{k+1}(X, Y)$ that maps $(u_0, u_1, \ldots, u_k)$ to $D[u_0][u_1, \ldots, u_k]$.*

Proof. Exercise 17-66. ∎

Starting with Exercise 17-69 below the exercises emphasize an important idea that was already used in Exercise 17-58. If we can write a complicated function as the appropriate composition of simpler functions, taking the derivative becomes a comparatively easy task of combining the Chain Rule and Exercise 17-57. This is one of the advantages of the coordinate free approach to differentiation.

Exercises

17-63. Prove Theorem 17.47.

17-64 Prove Corollary 17.48.

17-65 Prove Theorem 17.50.

17-66. Prove Proposition 17.53.

17-67 More examples of k-linear maps.

(a) Prove that $m : \mathbb{R}^k \to \mathbb{R}$ defined by $m[x_1, \ldots, x_k] := x_1 \cdots x_k$ is continuous and k-linear.

(b) Let (M, Σ, μ) be a measure space and let $1 \leq p, q \leq \infty$ with $\frac{1}{p} + \frac{1}{q} = 1$. Prove that $I : L^p(M, \Sigma, \mu) \times L^q(M, \Sigma, \mu) \to \mathbb{R}$ defined by $I[f, g] := \int_M fg \, d\mu$ is a continuous bilinear map.

(c) Let X, Y, Z be normed spaces and let $\circ : \mathcal{L}(Y, Z) \times \mathcal{L}(X, Y)$ be defined by $\circ[L, M] := L \circ M$. Prove that $\circ$ is a continuous bilinear map

17-68. Prove that for every 2-tensor $T : \mathbb{R}^d \times \mathbb{R}^d \to \mathbb{R}$ there is a $d \times d$-matrix A so that $T[v, w] = v^T Aw$, where v, w are column vectors with respect to the standard base and v^T is the transpose of v, that is, a row vector.

17-69. Prove each of the following *as a consequence of Theorem 17.52 and Exercise 17-57.*

(a) Let $f, g : (a, b) \to \mathbb{R}$ be differentiable Prove that $(f \cdot g)' = f'g + fg'$

(b) Let Y be an inner product space and let $f, g : (a, b) \to Y$ be differentiable. Prove that $\langle f, g\rangle' = \langle f', g\rangle + \langle f, g'\rangle$.

(c) Let $\Omega \subseteq \mathbb{R}^3$ be open, let $f, g : \Omega \times \Omega \to \mathbb{R}^3$ be differentiable and let $\times$ denote the cross product on $\mathbb{R}^3$. Prove that $(f \times g)' = f' \times g + f \times g'$.

(d) Let W, X, Y, Z be normed spaces, let $\Omega \subseteq W$ be open and let $L : \Omega \to \mathcal{L}(Y, Z)$ and $M : \Omega \to \mathcal{L}(X, Y)$ be differentiable. Prove that $D(L \circ M) = DL \circ M + L \circ DM$.

Hint. Exercise 17-67c.

(e) Explain why all the above product rules "look the same"

17-70 Derive a product rule for products of k functions $f_1, \ldots, f_k : (a, b) \to \mathbb{R}$.

17-71. Let $GL(n \times n, \mathbb{R})$ be the set of invertible $n \times n$ matrices and let $S : GL(n \times n, \mathbb{R}) \times \mathbb{R}^n \to \mathbb{R}^n$ be the function that maps the pair (A, b) of an invertible matrix A and a "right hand side vector" b to the solution x of the system of equations $Ax = b$

(a) Prove that S is differentiable.

Hint Theorem 17 32.

(b) Compute the derivative of S at an arbitrary (A, b).

17.7 Higher Derivatives

Now we are ready to investigate higher order derivatives. The underlying definition is the obvious one.

Definition 17.54 *Let X, Y be normed spaces and let $\Omega \subseteq X$ be open. The function $f : \Omega \to Y$ is called k* **times differentiable** *at x iff f is $(k-1)$ times differentiable on Ω and its $(k-1)^{\text{st}}$ derivative $D^{k-1}f$ is differentiable at x. The kth* **derivative** *of f at x is denoted $D^k f(x)$. The function f is called k times differentiable on Ω iff f is k times differentiable at every $x \in \Omega$. It is called k* **times continuously differentiable** *on Ω iff it is k times differentiable on Ω and $D^k f$ is continuous on Ω. Finally, the function f is called* **infinitely differentiable** *on Ω iff for all $k \in \mathbb{N}$ f is k times differentiable on Ω.*

Appropriate application of Proposition 17.53 allows us to identify k^{th} derivatives with k-tensors. This identification is common in analysis and we will use it throughout this text. That is, $D^k f(x)$ **will denote the k-tensor that is associated with the kth derivative of f at x**. The next result shows how higher derivatives of higher derivatives are related.

Proposition 17.55 *Let X, Y be normed spaces, let $m, n \in \mathbb{N}$, let $\Omega \subseteq X$ be open, let $f : \Omega \to Y$, and let $x \in \Omega$. Then f is $m+n$ times differentiable at x iff f is m times differentiable on Ω and $D^m f$ is n times differentiable at $x \in \Omega$. Moreover, $D^{m+n} f(x) = D^n \left(D^m f\right)(x)$ and for $(t_1, \ldots, t_{m+n}) \in X^{m+n}$ we have the identity $D^{m+n} f(x)[t_1, \ldots, t_{m+n}] = D^n \left(D^m f\right)(x)[t_1, \ldots, t_n][t_{n+1}, \ldots, t_{m+n}]$.*

Proof. Let $m \in \mathbb{N}$ be arbitrary. The proof is an induction on n, with the definition being the base case $n = 1$. For the representations of the tensors, note that for $(t_1, \ldots, t_{m+1}) \in X^{m+1}$ the derivative $D\left(D^m f\right)(x)[t_1][t_2, \ldots, t_{m+1}]$ is by Proposition 17.53 identified with the value $D^{m+1} f(x)[t_1, \ldots, t_{m+1}]$, where $D^{m+1} f(x)$ is the corresponding $(m+1)$-linear map.

For the induction step, note that f is $m+n+1$ times differentiable at x iff f is $m+1$ times differentiable on Ω and $D^{m+1} f$ is n times differentiable at $x \in \Omega$. This is the case iff f is m times differentiable on Ω, $D^m f$ is differentiable on Ω and $D^{m+1} f$ is n times differentiable at $x \in \Omega$, which by induction hypothesis is the case iff f is m times differentiable on Ω and $D^m f$ is $n+1$ times differentiable at $x \in \Omega$. For the representations of the tensors note that for all $(t_1, \ldots, t_{m+n+1}) \in X^{m+n+1}$ the following hold.

$$
\begin{aligned}
& D^{n+1}\left(D^m f\right)(x)[t_1, \ldots, t_{n+1}][t_{n+2}, \ldots, t_{m+n+1}] \\
&= D\left(D^n\left(D^m f\right)\right)(x)[t_1][t_2, \ldots, t_{n+1}][t_{n+2}, \ldots, t_{m+n+1}] \\
&= D\left(D^{n+m} f\right)(x)[t_1][t_2, \ldots, t_{m+n+1}] \\
&= D^{n+m+1} f(x)[t_1, \ldots, t_{m+n+1}].
\end{aligned}
$$

■

For k^{th} derivatives (or, more accurately, for the tensors associated with them) the order of the arguments does not matter. The key to this insight is to prove the result for second derivatives.

Theorem 17.56 Hermann Armandus Schwarz' Theorem. *Let X and Y be normed spaces, let $\Omega \subseteq X$ be open, and let $f : \Omega \to Y$ be twice differentiable at $x \in \Omega$. Then for all $(s, t) \in X^2$ we have $D^2 f(x)[s, t] = D^2 f(x)[t, s]$.*

Proof. *The main idea is that the sum $f(x+s+t) - f(x+t) - f(x+s) + f(x)$ should be close to $D^2 f(x)[s, t]$ and $D^2 f(x)[t, s]$. To understand where this expression comes from, recall that for single variable functions the difference quotient $\dfrac{f(x+s) - f(x)}{s}$ is close to $f'(x)$ for small enough s. Hence, the difference quotient*

$$
\frac{\frac{f(x+s+t)-f(x+t)}{s} - \frac{f(x+s)-f(x)}{s}}{t} = \frac{f(x+s+t) - f(x+t) - f(x+s) + f(x)}{st}
$$

should be close to $f''(x)$, and so $f(x+s+t)-f(x+t)-f(x+s)+f(x)$ should be close to the product $f''(x)st$, which for general f would be $D^2f(x)[s,t]$ or $D^2f(x)[t,s]$.

Let $\varepsilon > 0$ and let $\delta_{st} > 0$ be so that for all $s, t \in X$ with $\|s\| < \delta_{st}$ and $\|t\| < \delta_{st}$ we have $x+s+t \in \Omega$ and $\left\| Df(x+s+t) - Df(x) - D^2f(x)[s+t] \right\| \leq \frac{\varepsilon}{4}\|s+t\|$, where D^2f is interpreted as a linear map into $\mathcal{L}(X,Y)$. Then for all $s, t \in X$ with $\|s\| < \delta_{st}$ and $\|t\| < \delta_{st}$ we obtain

$$\begin{aligned}
&\left\| f(x+s+t) - f(x+t) - f(x+s) + f(x) - D^2f(x)[s,t] \right\| \\
&\quad\leq \left\| f(x+s+t) - f(x+t) - \big(Df(x+s)[t] - Df(x)[t]\big) - \big(f(x+s) - f(x)\big) \right\| \\
&\qquad + \left\| Df(x+s)[t] - Df(x)[t] - D^2f(x)[s,t] \right\|
\end{aligned}$$

Apply the Mean Value Theorem (Theorem 17.33) to the function $f(x+s+\cdot) - f(x+\cdot) - \big(Df(x+s)[\cdot] - Df(x)[\cdot]\big)$ with $a = 0, b = t$.

$$\begin{aligned}
&\quad\leq \sup\left\{ \left\| Df(x+s+\xi t) - Df(x+\xi t) - \big(Df(x+s) - Df(x)\big) \right\| : \xi \in [0,1] \right\} \|t\| \\
&\qquad + \left\| \Big(Df(x+s) - Df(x) - D^2f(x)[s] \Big)[t] \right\| \\
&\quad\leq \sup\left\{ \left\| Df(x+s+\xi t) - Df(x) - D^2f(x)[s+\xi t] \right\| \right. \\
&\qquad\qquad + \left\| Df(x+\xi t) - Df(x) - D^2f(x)[\xi t] \right\| \\
&\qquad\qquad \left. + \left\| Df(x+s) - Df(x) - D^2f(x)[s] \right\| : \xi \in [0,1] \right\} \|t\| \\
&\qquad + \left\| Df(x+s) - Df(x) - D^2f(x)[s] \right\| \|t\| \\
&\quad\leq \left(\frac{\varepsilon}{4}\big(\|s\| + \|t\|\big) + \frac{\varepsilon}{4}\|t\| + \frac{\varepsilon}{4}\|s\| \right) \|t\| + \frac{\varepsilon}{4}\|s\|\|t\| \\
&\quad= \frac{3\varepsilon}{4}\|s\|\|t\| + \frac{\varepsilon}{2}\|t\|^2 \leq \frac{\varepsilon}{2}\big(\|s\| + \|t\|\big)^2.
\end{aligned}$$

Similarly (Exercise 17-72), we can find a $\delta_{ts} > 0$ so that for all $s, t \in X$ with $\|s\| < \delta_{ts}$ and $\|t\| < \delta_{ts}$ we have

$$\left\| f(x+t+s) - f(x+s) - f(x+t) + f(x) - D^2f(x)[t,s] \right\| \leq \frac{\varepsilon}{2}\big(\|s\| + \|t\|\big)^2.$$

Let $\delta := \min\{\delta_{st}, \delta_{ts}\}$. For all $s, t \in X$ with $\|s\| < \delta$ and $\|t\| < \delta$, we infer

$$\begin{aligned}
&\left\| D^2f(x)[s,t] - D^2f(x)[t,s] \right\| \\
&\quad\leq \left\| f(x+s+t) - f(x+t) - f(x+s) + f(x) - D^2f(x)[s,t] \right\| \\
&\qquad + \left\| f(x+t+s) - f(x+s) - f(x+t) + f(x) - D^2f(x)[t,s] \right\| \\
&\quad\leq \frac{\varepsilon}{2}\big(\|s\| + \|t\|\big)^2 + \frac{\varepsilon}{2}\big(\|s\| + \|t\|\big)^2 = \varepsilon\big(\|s\| + \|t\|\big)^2.
\end{aligned}$$

But this implies the following for all $s, t \in X \setminus \{0\}$.

$$\begin{aligned}
&\left\| D^2 f(x)[s,t] - D^2 f(x)[t,s] \right\| \\
&= \frac{1}{\frac{\delta}{2\|s\|}\frac{\delta}{2\|t\|}} \left\| D^2 f(x)\left[\frac{\delta}{2\|s\|}s, \frac{\delta}{2\|t\|}t\right] - D^2 f(x)\left[\frac{\delta}{2\|t\|}t, \frac{\delta}{2\|s\|}s\right]\right\| \\
&\leq \frac{4}{\delta^2}\|s\|\|t\|\varepsilon\left(\frac{\delta}{2}+\frac{\delta}{2}\right)^2 = 4\varepsilon\|s\|\|t\|.
\end{aligned}$$

Therefore, for $A[s,t] := D^2 f(x)[s,t]$ and $B[s,t] := D^2 f(x)[t,s]$ the tensor norm of the difference is $\|A - B\| \leq 4\varepsilon$. Because $\varepsilon > 0$ was arbitrary this implies $\|A - B\| = 0$, and hence $D^2 f(x)[s,t] = D^2 f(x)[t,s]$ for all $s, t \in X$. ■

Definition 17.57 *A bijective function $\sigma : \{1, \ldots, k\} \to \{1, \ldots, k\}$ is also called a* **permutation**. *A k-tensor $T : \prod_{i=1}^{k} X_i \to Y$ is called* **symmetric** *iff for all k-tuples $(x_1, \ldots, x_k) \in \prod_{i=1}^{k} X_i$ and all permutations $\sigma : \{1, \ldots, k\} \to \{1, \ldots, k\}$ the equality $T[x_1, \ldots, x_k] = T[x_{\sigma(1)}, \ldots, x_{\sigma(k)}]$ holds.*

Corollary 17.58 *Let X, Y be normed spaces, let $\Omega \subseteq X$ be open, let $x \in \Omega$, and let $f : \Omega \to Y$ be k times differentiable at x. Then $D^k f(x)$ is symmetric.*

Proof. The proof is an induction on k with $k = 1$ being trivial and $k = 2$ proved in Theorem 17.56.

For the induction step, assume that $k > 2$ and that the result is proved for all $j < k$. Let $(t_1, \ldots, t_k) \in X^k$ and let $\sigma : \{1, \ldots, k\} \to \{1, \ldots, k\}$ be a permutation. If $\sigma(1) \neq 1$ let $\tau : \{2, \ldots, k\} \to \{2, \ldots, k\}$ be a permutation with $\tau(2) = \sigma(1)$. If $\sigma(1) = 1$ let $\tau(i) = i$ for all $i \in \{2, \ldots, k\}$ and skip the middle three lines in the computation below.

$$\begin{aligned}
D^k f(x)[t_1, \ldots, t_k] &= D\left(D^{k-1} f(x)\right)[t_1][t_2, \ldots, t_k] \\
&= D\left(D^{k-1} f(x)\right)[t_1][t_{\tau(2)}, \ldots, t_{\tau(k)}] \\
&= D^2\left(D^{k-2} f(x)\right)[t_1, t_{\tau(2)}][t_{\tau(3)}, \ldots, t_{\tau(k)}] \\
&= D^2\left(D^{k-2} f(x)\right)[t_{\tau(2)}, t_1][t_{\tau(3)}, \ldots, t_{\tau(k)}] \\
&= D\left(D^{k-1} f(x)\right)[t_{\sigma(1)}][t_1, t_{\tau(3)}, \ldots, t_{\tau(k)}] \\
&= D\left(D^{k-1} f(x)\right)[t_{\sigma(1)}][t_{\sigma(2)}, t_{\sigma(3)}, \ldots, t_{\sigma(k)}] \\
&= D^k f(x)[t_{\sigma(1)}, \ldots, t_{\sigma(k)}],
\end{aligned}$$

where in the second to last step we use $\left\{1, \tau(3), \ldots, \tau(k)\right\} = \left\{\sigma(2), \ldots, \sigma(k)\right\}$ and apply an appropriate permutation. ■

Clairaut's Theorem says that the order in which partial derivatives are taken is not important as long as the function is twice differentiable. It is usually stated for second partial derivatives of functions from $\mathbb{R}^d$ to $\mathbb{R}$. The corollary below is easily seen to imply Clairaut's Theorem (see Corollary 17.60). Note, however (Exercise 17-73), that mixed partial derivatives can exist and not be equal.

Corollary 17.59 *Let $\prod_{i=1}^{d} X_i$ be a product space, let Y be a normed space, let $\Omega \subseteq \prod_{i=1}^{d} X_i$ be open, let $x \in \Omega$, and let $f : \Omega \to Y$ be k times differentiable at x. Then for all indices $i_1, \ldots, i_k \in \{1, \ldots, d\}$ the partial derivative $D_{i_1} \cdots D_{i_k} f$ exists and for all permutations $\sigma : \{1, \ldots, k\} \to \{1, \ldots, k\}$ and all $(x_{i_1}, \ldots, x_{i_k})$ we have $D_{i_1} \cdots D_{i_k} f(x_{i_1}, \ldots, x_{i_k}) = D_{\sigma(i_1)} \cdots D_{\sigma(i_k)} f(x_{\sigma(i_1)}, \ldots, x_{\sigma(i_k)})$.*

Proof. For $j \in \{1, \ldots, d\}$, define $e[X_j] := \{(0, \ldots, 0, x_j, 0, \ldots, 0) : x_j \in X_j\}$, where the x_j is in the jth component of each vector. First we prove by induction on k that $D_{i_1} \cdots D_{i_k} f = D^k f|_{\prod_{j=1}^{k} e[X_{i_k}]}$. The case $k = 1$ follows from $D_i f = Df|_{e[X_i]}$ (see Proposition 17.40) wherever f is differentiable.

For the induction step, assume the result has been proved for all $j < k$. Then

$$\begin{aligned} D_{i_1} \cdots D_{i_k} f &= D_{i_1}\left(D^{k-1} f|_{\prod_{j=2}^{k} e[X_{i_k}]}\right) = D\left(D^{k-1} f|_{\prod_{j=2}^{k} e[X_{i_k}]}\right)\Big|_{e[X_{i_1}]} \\ &= D^k f|_{\prod_{j=1}^{k} e[X_{i_k}]}. \end{aligned}$$

Now Corollary 17.58 implies

$$\begin{aligned} D_{i_1} \cdots D_{i_k} f(x_{i_1}, \ldots, x_{i_k}) &= D^k f|_{\prod_{j=1}^{k} e[X_{i_k}]}(x_{i_1}, \ldots, x_{i_k}) \\ &= D^k f|_{\prod_{j=1}^{k} e[X_{\sigma(i_k)}]}(x_{\sigma(i_1)}, \ldots, x_{\sigma(i_k)}) \\ &= D_{\sigma(i_1)} \cdots D_{\sigma(i_k)} f(x_{\sigma(i_1)}, \ldots, x_{\sigma(i_k)}), \end{aligned}$$

which finishes the proof. ■

Corollary 17.60 Clairaut's Theorem. *Let $\Omega \subseteq \mathbb{R}^d$ be an open set and let the function $f : \Omega \to \mathbb{R}$ be twice differentiable at $x \in \Omega$. Then for all $i_1, i_2 \in \{1, \ldots, d\}$ we have $\frac{\partial^2 f}{\partial x_{i_1} \partial x_{i_2}}(x) = \frac{\partial^2 f}{\partial x_{i_2} \partial x_{i_1}}(x)$.*

If f is k times differentiable at x, then for all $i_1, \ldots, i_k \in \{1, \ldots, d\}$ and all permutations σ of $\{1, \ldots, k\}$ we have $\frac{\partial^k f}{\partial x_{i_1} \cdots \partial x_{i_k}}(x) = \frac{\partial^k f}{\partial x_{\sigma(i_1)} \cdots \partial x_{\sigma(i_k)}}(x)$.

Proof. Easy consequence of $\frac{\partial f}{\partial x_i}(x) \cdot t = D_{\mathbb{R}_i} f(x)[t]$ and Corollary 17.59. ■

Now that we have made the connection to partial derivatives, we can also give an explicit formula for the kth derivative in terms of partial derivatives that is similar to Theorem 17.43.

Corollary 17.61 *Let $\Omega \subseteq \mathbb{R}^d$ be open, let $x \in \Omega$ and let $f : \Omega \to \mathbb{R}$ be k times differentiable at x. Then*

$$D^k f(x)[h_1, \ldots, h_k] = \sum_{(i_1, \ldots, i_k) \in \{1, \ldots, d\}^k} \frac{\partial^k f}{\partial x_{i_1} \cdots \partial x_{i_k}}(x) \pi_{i_1}[h_1] \cdots \pi_{i_k}[h_k].$$

Moreover, for $h_1 = h_2 = \cdots = h_k = c = (c_1, \ldots, c_d)$ we have

$$D^k f(x)[c, \ldots, c] = \sum_{\alpha_1 + \cdots + \alpha_d = k} \frac{k!}{\alpha_1! \cdots \alpha_d!} \frac{\partial^k f}{\partial x_1^{\alpha_1} \cdots \partial x_d^{\alpha_d}}(x) c_1^{\alpha_1} \cdots c_d^{\alpha_d}.$$

Proof. Use that $h_j = \sum_{i_j = 1}^{d} \pi_{i_j}[h_j] e_{i_j}$ for all $j = 1, \ldots, k$. For the second part, note that because the kth partial derivatives do not depend on the order in which the partial derivatives are taken, we can sort the first sum by how often each kth partial derivative occurs. ■

We conclude this section with a proof that k-fold differentiability is preserved by compositions.

Proposition 17.62 Chain Rule. *Let X, Y, Z be normed spaces, let $\Omega_1 \subseteq X$, $\Omega_2 \subseteq Y$ be open, let $g : \Omega_1 \to \Omega_2$ be k times differentiable at $x \in \Omega_1$ and let $f : \Omega_2 \to Z$ be k times differentiable at $g(x) \in \Omega_2$. Then $f \circ g$ is k times differentiable at x.*

Proof. This proof is an induction on k, with the base step $k = 1$ being the Chain Rule (Theorem 17.30).

For the induction step $(k - 1) \to k$, let f and g be k times differentiable. Recall that by Theorem 17.30 we have $D(f \circ g)(x) = Df(g(x)) \circ Dg(x)$. By Proposition 17.55, $Dg(\cdot)$ is $k - 1$ times differentiable at x and by induction hypothesis $Df(g(\cdot))$ is $k - 1$ times differentiable at x. Therefore by an easy generalization of Exercise 17-57 the function $x \mapsto \big(Df(g(x)), Dg(x)\big)$ is $k - 1$ times differentiable. Moreover, the function $(L, M) \mapsto L \circ M$ from $\mathcal{L}(Y, Z) \times \mathcal{L}(X, Y)$ to $\mathcal{L}(X, Z)$ is continuous and bilinear, and hence $k - 1$ times differentiable by Exercise 17-74. But then by induction hypothesis, the composition $x \mapsto \big(Df(g(x)), Dg(x)\big) \mapsto Df(g(x)) \circ Dg(x)$ is $k - 1$ times differentiable, which completes the proof. ■

Exercises

17-72. Finish the proof of Theorem 17.56 by proving that for all $\varepsilon > 0$ we can find a $\delta_{ts} > 0$ so that for all $s, t \in X$ with $\|s\| < \delta_{ts}$ and $\|t\| < \delta_{ts}$ we have

$$\left\| f(x + t + s) - f(x + s) - f(x + t) + f(x) - D^2 f(x)[t, s] \right\| \leq \frac{\varepsilon}{2} \left(\|s\| + \|t\| \right)^2$$

17-73. Even when mixed partial derivatives exist, they need not be equal. Consider the function

$$f(x, y) = \begin{cases} \frac{xy\left(x^2 - y^2\right)}{x^2 + y^2}, & \text{for } (x, y) \neq (0, 0), \\ 0, & \text{for } (x, y) = (0, 0). \end{cases}$$

(a) Prove that both $\frac{\partial^2 f}{\partial x \partial y}(0,0)$ and $\frac{\partial^2 f}{\partial y \partial x}(0,0)$ exist and are not equal

(b) Prove that neither $\frac{\partial f}{\partial x}$ nor $\frac{\partial f}{\partial y}$ is differentiable at $(0,0)$.

(c) Prove that f is differentiable at every $(x,y) \in \mathbb{R}^2$.

(d) Prove that Df is not differentiable at $(0,0)$.

17-74. Prove that every continuous k-tensor T is infinitely differentiable with $D^j T = 0$ for $j > k$.

17-75 Let $\Omega \subseteq \mathbb{R}^d$ be open, let $f : \Omega \to \mathbb{R}$ be a twice differentiable function and let $x \in \Omega$

(a) Prove that $A := \left(D^2 f(x)[e_i, e_j]\right)_{\substack{i=1,\ldots,d \\ j=1,\ldots,d}}$ is such that $D^2 f(x)[v,w] = v^T A w$,

where v, w are column vectors with respect to the standard base and v^T is the transpose of v, that is, a row vector

(b) Prove that A is a symmetric matrix, that is, for all $i, j \in \{1, \ldots, d\}$ we have $a_{ij} = a_{ji}$

17-76. For each $f : \mathbb{R}^2 \to \mathbb{R}$, compute the second derivative Use the representation of Exercise 17-75.

(a) $f(x,y) = ye^x$ (b) $f(x,y) = x^3 + 3xy + y^2$

17-77 **Taylor's Formula** Let X, Y be Banach spaces, let $\Omega \subseteq X$ be open, let $f : \Omega \to Y$ be $k+1$ times continuously differentiable on Ω, and let $x \in \Omega$, $z \in X$ be so that for all $t \in [0,1]$ we have $x + tz \in \Omega$

(a) Prove $f(x+z) = \sum_{i=0}^{k} \frac{1}{i!} D^i f(x)[z, \ldots, z] + \int_0^1 \frac{(1-u)^k}{k!} D^{k+1} f(x+uz)\, du[z, \ldots, z]$.

Hint. Consider the function $t \mapsto f(x+tz)$. The Riemann integral for continuous Banach space valued functions is defined in Exercise 17-41. Use Theorem 13.3 as guidance.

(b) For $\Omega \subseteq \mathbb{R}^d$ and $Y = \mathbb{R}$ state Taylor's formula from part 17-77a in terms of partial derivatives. Then decide which of the two formulas is easier to use computationally and which formula is easier to read.

Hint Corollary 17.61.

(c) Let $\Theta(z) := \int_0^1 \frac{(1-u)^k}{k!} D^{k+1} f(x+uz)\, du[z, \ldots, z]$. Prove that $\lim_{\|z\| \to 0} \frac{\Theta(z)}{\|z\|^k} = 0$.

The function $T_k(z) := \sum_{i=0}^{k} \frac{1}{i!} D^i f(x)[z, \ldots, z]$ is called the kth **Taylor polynomial** of f at x.

17-78 For each function $f : \mathbb{R}^2 \to \mathbb{R}$ below, compute the second Taylor polynomial at $x = 0$. Use the representation of Exercise 17-75 for the second derivative.

(a) $f(x,y) = e^{xy^2}$ (b) $f(x,y) = x^2 + y^2$ (Is the result a surprise?)

17-79 Let X be a normed space. A bilinear function $T : X \times X \to \mathbb{R}$ is called **positive definite** iff for all $x \in X \setminus \{0\}$ we have $T[x,x] > 0$

(a) **Second Derivative Test**. Let $\Omega \subseteq X$ be open, let $f : X \to \mathbb{R}$ be twice continuously differentiable and let $u \in \Omega$ be so that $Df(u) = 0$

i. Prove that if $D^2 f(u)$ is positive definite, then there is an $\varepsilon > 0$ so that for all points $v \in X \setminus \{u\}$ with $\|u - v\| < \varepsilon$ we have $f(u) < f(v)$,

ii. Prove that if $-D^2 f(u)$ is positive definite, then there is an $\varepsilon > 0$ so that for all points $v \in X \setminus \{u\}$ with $\|u - v\| < \varepsilon$ we have $f(u) > f(v)$,

iii. Prove that if there are $x_1, x_2 \in X$ so that $D^2 f(u)[x_1, x_1] > 0$ and $D^2 f(u)[x_2, x_2] < 0$, then for every $\varepsilon > 0$ there are elements $v, w \in X \setminus \{u\}$ with $\|u - v\| < \varepsilon$, $\|u - w\| < \varepsilon$, $f(u) < f(v)$, and $f(u) > f(w)$.

Hint. Use Exercise 17-77

(b) Prove that if $X = \mathbb{R}^2$, then the second derivative $D^2 f(u)$ is positive definite iff $\frac{\partial^2 f}{\partial x^2}(u) > 0$ and $\frac{\partial^2 f}{\partial x^2}(u)\frac{\partial^2 f}{\partial y^2}(u) - \left(\frac{\partial^2 f}{\partial x \partial y}(u)\right)^2 > 0$.

Hint. Use Exercise 17-75.

(c) Prove that if $X = \mathbb{R}^2$ and $\frac{\partial^2 f}{\partial x^2}(u)\frac{\partial^2 f}{\partial y^2}(u) - \left(\frac{\partial^2 f}{\partial x \partial y}(u)\right)^2 < 0$, then there are $x_1, x_2 \in X$ so that $D^2 f(u)[x_1, x_1] > 0$ and $D^2 f(u)[x_2, x_2] < 0$.

(d) State and prove a result similar to part 17-79a for a k times continuously differentiable function $f : \Omega \to \mathbb{R}$ so that $Df(x) = 0, \ldots, D^{k-1} f(x) = 0$. (Distinguish even and odd k.)

17-80. A characterization of symmetry

(a) Prove that a continuous k-linear function $T : \prod_{i=1}^{k} X_i \to Y$ is symmetric iff for every $\varepsilon > 0$ there is a $\delta > 0$ so that for all k-tuples $(x_1, \ldots, x_k) \in \prod_{i=1}^{k} X_i$ that satisfy $\|(x_1, \ldots, x_k)\| < \delta$ and all permutations $\sigma : \{1, \ldots, k\} \to \{1, \ldots, k\}$ we have

$$\left\| T(x_1, \ldots, x_k) - T(x_{\sigma(1)}, \ldots, x_{\sigma(k)}) \right\| \leq \varepsilon \left(\sum_{i=1}^{k} \|x_i\| \right)^k .$$

(b) Consider the continuous bilinear map $T : \mathbb{R}^2 \times \mathbb{R}^2 \to \mathbb{R}$ defined by $T(x, y) := \pi_1(x)\pi_2(y)$ Prove that for every $\varepsilon > 0$ there is a $\delta > 0$ so that for all $(x, y) \in \mathbb{R}^2 \times \mathbb{R}^2$ with $\|(x, y)\| < \delta$ we have $\left\| T(x, y) - T(y, x) \right\| \leq \varepsilon \left(\|x\| + \|y\| \right)$, but T is not symmetric.

17-81. Let X be a Banach space, let Y be a normed space and let $\mathcal{H}(X, Y)$ be the set of linear homeomorphisms from X to Y Prove that the map $J(A) := A^{-1}$ is an infinitely differentiable function from $\mathcal{H}(X, Y)$ to $\mathcal{H}(Y, X)$.

17.8 The Implicit Function Theorem

To investigate the solution sets of equations $f(x, y) = 0$ in more detail, it is often helpful to represent y as a function $g(x)$. The Implicit Function Theorem says that under mild hypotheses this is possible and g has the same differentiability properties as f. The first step toward the Implicit Function Theorem is a result about fixed points.

Definition 17.63 *Let S be a set and let $f : S \to S$ be a function. Then $p \in S$ is called a* **fixed point** *of f iff $f(p) = p$.*

Fixed points are important throughout applied mathematics because many equations can be rewritten as fixed point equations. (Recall Newton's Method from Section 13.2.) Under certain conditions, fixed points must exist. That is, if a fixed point equation from a concrete application satisfies the right abstract condition, then it must have a solution. Banach's Fixed Point Theorem provides one such condition.

Theorem 17.64 Banach's Fixed Point Theorem *Let X be a complete metric space, let $0 \leq q < 1$ and let $f : X \to X$ be a function so that for all points $x, y \in X$ we have $d(f(x), f(y)) \leq q d(x, y)$ Then f has a unique fixed point $p \in X$.*

Proof. Let $x \in X$ and consider the sequence $\{f^n(x)\}_{n=1}^{\infty}$. An easy induction proves that for all $n \in \mathbb{N}$ we have $d\left(f^n(x), f^{n+1}(x)\right) \leq q^n d(x, f(x))$. Hence, for all $n, m \in \mathbb{N}$ we infer $d\left(f^n(x), f^m(x)\right) \leq \sum_{k=n}^{m-1} q^k d(x, f(x))$. Therefore, $\{f^n(x)\}_{n=1}^{\infty}$ is a Cauchy sequence. Let $p := \lim_{n\to\infty} f^n(x)$ and let $\varepsilon > 0$. Then there is an $n \in \mathbb{N}$ so that $d\left(p, f^n(x)\right) < \frac{\varepsilon}{3}$ and $d\left(f^n(x), f^{n+1}(x)\right) < \frac{\varepsilon}{3}$, which implies

$$\begin{aligned} d(p, f(p)) &\leq d\left(p, f^n(x)\right) + d\left(f^n(x), f^{n+1}(x)\right) + d\left(f^{n+1}(x), f(p)\right) \\ &\leq d\left(p, f^n(x)\right) + d\left(f^n(x), f^{n+1}(x)\right) + q d\left(f^n(x), p\right) \\ &< \frac{\varepsilon}{3} + \frac{\varepsilon}{3} + q\frac{\varepsilon}{3} < \varepsilon. \end{aligned}$$

Because $\varepsilon > 0$ was arbitrary we have proved that $p = f(p)$.

Regarding uniqueness, suppose that $\widetilde{p}$ is another fixed point of f. Then we have $d(p, \widetilde{p}) = d(f(p), f(\widetilde{p})) \leq q d(p, \widetilde{p})$, which implies $d(p, \widetilde{p}) = 0$, and hence we conclude $p = \widetilde{p}$ ∎

Now we are ready to state and prove the Implicit Function Theorem. Note that the function h in the proof is similar to applying Newton's Method in the Y-coordinate.

Theorem 17.65 Implicit Function Theorem. *Let X, Y and Z be Banach spaces, let $\Omega \subseteq X \times Y$ be an open set, let $(x_0, y_0) \in \Omega$, and let $f : \Omega \to Z$ be continuously differentiable so that $f(x_0, y_0) = 0$ and so that $D_Y f(x_0, y_0) : Y \to Z$ is a linear homeomorphism. Then there is an open neighborhood $N \subseteq X$ of x_0 such that there is a unique continuously differentiable function $g : N \to Y$ so that $g(x_0) = y_0$ and for all $x \in N$ we have $(x, g(x)) \in \Omega$ and $f(x, g(x)) = 0$. The derivative of g is $Dg(x) = -\left(D_Y f(x, g(x))\right)^{-1} \circ D_X f(x, g(x))$. Moreover, if f is k times continuously differentiable, then so is g.*

Proof. Let $L_0 := D_Y f(x_0, y_0)$, let $h(x, y) := y - L_0^{-1}\left[f(x, y)\right]$ and let $\delta > 0$ be so that for all $(x, y) \in X \times Y$ with $\left\|(x, y) - (x_0, y_0)\right\| \leq \delta$ we have that $(x, y) \in \Omega$, $D_Y f(x, y)$ is a linear homeomorphism and $\left\|D_Y f(x, y) - D_Y f(x_0, y_0)\right\| < \frac{1}{2\left\|L_0^{-1}\right\|}$, and so that $\left\|D_X f(x, y)\right\|$ is bounded on $B_\delta\left((x_0, y_0)\right)$. Then for all $(x, y_i) \in X \times Y$ with $\left\|(x, y_i) - (x_0, y_0)\right\| \leq \delta$ $(i = 1, 2)$ the following holds

$$\begin{aligned} &\left\|h(x, y_2) - h(x, y_1)\right\| \\ &= \left\|y_2 - L_0^{-1}[f(x, y_2)] - \left[y_1 - L_0^{-1}[f(x, y_1)]\right]\right\| \end{aligned}$$

$$= \left\| L_0^{-1}\left[D_Y f(x_0, y_0)[y_2 - y_1] - (f(x, y_2) - f(x, y_1))\right]\right\|$$

$$\leq \left\| L_0^{-1}\right\| \left\| f(x, y_2) - f(x, y_1) - D_Y f(x_0, y_0)[y_2 - y_1]\right\|$$

Apply the Mean Value Theorem (Theorem 17.33) to the function $f(x, \cdot) - D_Y f(x_0, y_0)[\cdot]$ with $a = y_1$, $b = y_2$. (The line segment from y_1 to y_2 is in $B_\delta\big((x_0, y_0)\big)$.)

$$\leq \left\| L_0^{-1}\right\| \sup\left\{\left\| D_Y f(x, y_1 + t(y_2 - y_1)) - D_Y f(x_0, y_0)\right\| : t \in [0, 1]\right\} \|y_2 - y_1\|$$

$$\leq \frac{1}{2}\|y_2 - y_1\|.$$

Let $N := \pi_X\left[B_\delta((x_0, y_0))\right] = B_\delta(x_0)$ (equality holds because so far we are using the norm $\max\left\{\|\cdot\|_X, \|\cdot\|_Y\right\}$ on $X \times Y$). By Theorem 17.64 for each point $x \in N$ the function $h(x, \cdot) : \overline{B_\delta(y_0)} \to \overline{B_\delta(y_0)}$ has a unique fixed point $g(x)$. Because we have $g(x) = h(x, g(x)) = g(x) - L_0^{-1}\left[f(x, g(x))\right]$ iff $f(x, g(x)) = 0$, a function $g : N \to Y$ with $f(x, g(x)) = 0$ exists on N and by Theorem 17.64 it is unique.

To prove that g is continuous, note that by the proof of Theorem 17.64 the functions g_n, where $g_0(x) := y_0$ and $g_n(x) := h(x, g_{n-1}(x))$ for $n \geq 1$, form a uniform Cauchy sequence. Thus g is the uniform limit of the functions g_n. Because the g_n are continuous, so is g (easy adaptation of the proof of Theorem 11.7, see Exercise 17-82).

To prove that g is differentiable, fix $x \in N$. To simplify the following argument, we will continue the proof with the product norm $\left\|(x, y)\right\| := \|x\| + \|y\|$ on $X \times Y$. Note that by continuity of g and Proposition 17.40 for every $x \in N$ and every $\varepsilon > 0$ with $\varepsilon < \dfrac{1}{2\left\|-\left(D_Y f(x, g(x))\right)^{-1}\right\|}$ there is a $\delta > 0$ so that for all $z \in N$ with $\|z - x\| < \delta$ we have

$$\begin{aligned}&\left\| f(z, g(z)) - f(x, g(x)) - D_X f(x, g(x))[z - x] - D_Y f(x, g(x))[g(z) - g(x)]\right\| \\ &\quad \leq \varepsilon\left[\ \|z - x\| + \|g(z) - g(x)\|\ \right].\end{aligned}$$

Because $f(z, g(z)) = f(x, g(x)) = 0$, this implies

$$\begin{aligned}&\left\| -D_X f(x, g(x))[z - x] - D_Y f(x, g(x))\left[g(z) - g(x)\right]\right\| \\ &\quad \leq \varepsilon\left[\ \|z - x\| + \|g(z) - g(x)\|\ \right],\end{aligned}$$

and hence

$$\begin{aligned}&\left\| g(z) - g(x) - \left[-\left(D_Y f(x, g(x))\right)^{-1} D_X f(x, g(x))[z - x]\right]\right\| \\ &\quad \leq \varepsilon\left\|-\left(D_Y f(x, g(x))\right)^{-1}\right\| \left[\ \|z - x\| + \|g(z) - g(x)\|\ \right]. \qquad (*)\end{aligned}$$

To prove that g is differentiable at x we need to prove that the right side of $(*)$ is bounded by a multiple of ε times $\|z - x\|$. To do this, let

$$a := 2\left\|-\left(D_Y f(x, g(x))\right)^{-1} D_X f(x, g(x))\right\| + 1.$$

Via $(*)$ and the reverse triangular inequality, for all z with $\|z-x\| < \delta$ we infer

$$\begin{aligned}
&\|g(z)-g(x)\| - \frac{a-1}{2}\|z-x\| \\
&\leq \left\| g(z)-g(x) - \left[-\left(D_Y f(x,g(x))\right)^{-1} D_X f(x,g(x))[z-x] \right] \right\| \\
&\leq \frac{1}{2}\left[\|z-x\| + \|g(z)-g(x)\| \right],
\end{aligned}$$

which implies that $\|g(z)-g(x)\| \leq a\|z-x\|$. Substituting this into the equation $(*)$ proves that g is differentiable with the derivative being as claimed.

To prove that g is k times continuously differentiable if f is, we proceed by induction. The base step was just proved. For the induction step $k \to (k+1)$, note that both $-\left(D_Y f(x,g(x))\right)^{-1}$ and $D_X f(x,g(x))$ are compositions of k times continuously differentiable functions, and hence k times continuously differentiable themselves (for the inversion, Exercise 17-81 is needed). But then, because the composition operator is continuous and bilinear from $\mathcal{L}(Z,Y) \times \mathcal{L}(X,Z)$ to $\mathcal{L}(X,Y)$ the derivative $Dg(x) = -\left(D_Y f(x,g(x))\right)^{-1} \circ D_X f(x,g(x))$ is k times continuously differentiable and therefore g is $k+1$ times continuously differentiable. ■

The Implicit Function Theorem allows us to show that in Theorem 17.31 the hypotheses that f is bijective, f^{-1} is continuous and the range is open can be dropped.

Corollary 17.66 *Let X, Y be Banach spaces, let $V \subseteq X$ be open, let $f : V \to Y$ be continuously differentiable and let $x_0 \in V$ be so that $Df(x_0) : X \to Y$ is a linear homeomorphism. Then there exists an open neighborhood U of x_0 so that $f|_U$ is a homeomorphism from U to an open neighborhood of $f(x_0)$. Moreover, if f is k times continuously differentiable in U, then the inverse f^{-1} is k times continuously differentiable in $f[U]$*

Proof. Define $F : V \times Y \to Y$ by $F(x,y) := f(x) - y$ and let $y_0 := f(x_0)$. By the Implicit Function Theorem, because $D_X f(x_0, y_0) = Df(x_0)$ there is a neighborhood N of y_0 and a k times continuously differentiable function g from N to X so that $F(g(y), y) = 0$ for all $y \in N$. Now g is the inverse of f and the result follows. ■

The Implicit Function Theorem also is fundamental for the theory of manifolds, because it allows us to express sets of the form $\{(x,y) : f(x,y) = 0\}$ as manifolds, or, as Corollaries 17.67 and 17.69 show, as embedded manifolds. We will revisit the results below when we prove Theorem 19.7.

Corollary 17.67 *Let X, Y, Z be Banach spaces, let $U \subseteq X$ and $V \subseteq Y$ be open, let $f : U \times V \to Z$ be k times continuously differentiable and let $(x_0, y_0) \in U \times V$ be so that $D_Y f(x_0, y_0)$ is a linear homeomorphism. Then there are open neighborhoods U_0 of x_0 and W_0 of $f(x_0, y_0)$ and a unique k times continuously differentiable function $g \cdot U_0 \times W_0 \to Y$ so that for all $(x, w) \in U_0 \times W_0$ we have $f(x, g(x, w)) = w$.*

Proof. Apply the Implicit Function Theorem to $H(x,y,w) := w - f(x,y)$ (Exercise 17-83). ■

Definition 17.68 *The* **rank** *of a matrix is the largest number of linearly independent column vectors in the matrix.*

Corollary 17.69 *Let $\Omega \subseteq \mathbb{R}^d$ be open and let $f : \Omega \to \mathbb{R}^{d-m}$ be a k times continuously differentiable function such that the matrix $\left(\frac{\partial f_i}{\partial x_j}\right)_{\substack{i=1,\ldots,d-m \\ j=1,\ldots,d}}$ has rank $d-m$. If $a \in \Omega$ is such that $f(a) = 0$, then there is an open set $G \subseteq \mathbb{R}^d$ and a k times continuously differentiable function $g : G \to \Omega$ with k times continuously differentiable inverse so that $g[G]$ is open, $a \in g[G] \subseteq \Omega$ and the equation $f \circ g(v_1, \ldots, v_d) = (v_{m+1}, \ldots, v_d)$ holds for all $(v_1, \ldots, v_d) \in G$.*

Proof. Let $j_1, \ldots, j_{d-m}$ be the indices of the columns so that the corresponding column vectors of $\left(\frac{\partial f_i}{\partial x_j}(a)\right)_{\substack{i=1,\ldots,d-m \\ j=1,\ldots,d}}$ are linearly independent. Represent $\mathbb{R}^d$ as $\operatorname{span}\{e_k : k \notin \{j_1, \ldots, j_{d-m}\}\} \times \operatorname{span}\{e_{j_i} : i = 1, \ldots, d-m\}$ (this permutes the components appropriately) and apply Corollary 17.67. The details are left to the reader as Exercise 17-84. ■

Exercises

17-82. Let X, Y be Banach spaces, let $\Omega \subseteq X$ and let $\{f_n\}_{n=1}^{\infty}$ be a sequence of continuous functions $f_n : \Omega \to Y$ so that for all $\varepsilon > 0$ there is an $N \in \mathbb{N}$ so that for all $m, n \geq N$ and all $x \in \Omega$ we have $\| f_n(x) - f_m(x) \| < \varepsilon$. Prove that then there is a continuous function $f : \Omega \to Y$ so that $\lim_{n\to\infty} f_n(x) = f(x)$ for all $x \in \Omega$

17-83. Prove Corollary 17.67.

17-84. Prove Corollary 17.69.

17-85. Explain why in the proof of the Implicit Function Theorem we must prove that g is continuous before we can prove that it is differentiable.

17-86. **Lagrange multipliers.**

(a) Let X, Y be finite dimensional normed spaces with $\dim(X) = n \geq m = \dim(Y)$, let $\Omega \subseteq X$ be open, let $f : \Omega \to \mathbb{R}$ be continuously differentiable, let $g : \Omega \to Y$ be continuously differentiable, and let $x \in \Omega$ be so that $g(x) = 0$, the rank of $Dg(x)$ is m, and there is an $\varepsilon > 0$ so that for all $z \in \Omega$ with $g(z) = 0$ and $\|z - x\| < \varepsilon$ we have that $f(x) \geq f(z)$. Prove that there is a continuous linear function $\varphi \in \mathcal{L}(Y, \mathbb{R})$ so that $Df(x) + \varphi \circ Dg(x) = 0$.
Hint. Let $X_1 := \{a \in X : Dg(x)[a] = 0\}$ and represent X as a product $X = X_1 \times X_2$ with $x = (x_1, x_2)$ Find a neighborhood $U \times V$ of x so that there is a continuously differentiable $\Phi : U \to V$ so that $g(x_1, \Phi(x_1)) = 0$ for all $x \in U$. Then consider $H(x) := f(x_1, \Phi(x_1))$.

(b) Prove that if $X = \mathbb{R}^n$ and $Y = \mathbb{R}^m$, then there are $\lambda_1, \ldots, \lambda_m \in \mathbb{R}$ so that we have

$$\operatorname{grad}(f)(x) + \sum_{i=1}^{m} \lambda_i \operatorname{grad}(\pi_i \circ g)(x) = 0$$

17-87. Let X be a complete metric space and let $f : X \to X$ be a function so that there is a sequence $\{b_n\}_{n=1}^{\infty}$ of nonnegative numbers such that $\sum_{n=1}^{\infty} b_n$ converges and for all $n \in \mathbb{N}$ and all $x, y \in X$ we have $d\left(f^n(x), f^n(y)\right) \leq b_n d(x, y)$. Prove that f has a unique fixed point p in X.

17-88. Let $\| \cdot \|_2$ be the Euclidean norm on $\mathbb{R}^3$. Find a map $f : \mathbb{R}^3 \to \mathbb{R}^3$ that has no fixed points and so that for all $x, y \in \mathbb{R}^3$ we have $\| f(x) - f(y) \|_2 = \|x - y\|_2$

Chapter 18

Measure, Topology, and Differentiation

Continuity, differentiation, and integration are the three main topics in analysis. Part I of the text has shown that the combination of these concepts can lead to powerful new insights, such as the Fundamental Theorem of Calculus or the Lebesgue criterion for Riemann integrability. For abstract spaces, we have seen in Chapter 14 how integration leads to measure theory, in Chapter 16 how the investigation of limits and continuity leads to topological concepts, and in Chapter 17 how differentiation is approximation with continuous linear functions. In this chapter, we investigate the connections between measure, topology, and differentiation in d-dimensional space. Section 18.1 characterizes Lebesgue measurable sets topologically by approximating them from the inside with closed sets and from the outside with open sets. Similarly, Section 18.2 shows that p-integrable functions can be approximated with infinitely differentiable functions. After a brief excursion into tensor algebra in Section 18.3 (placed there to keep the presentation self-contained), the chapter concludes with the proof of the Multidimensional Substitution Formula in Section 18.4.

18.1 Lebesgue Measurable Sets in $\mathbb{R}^d$

This section shows how Lebesgue measurable sets in $\mathbb{R}^d$ can be characterized almost exclusively with the topological ideas of openness and closedness. We start by proving that the most fundamental subsets of $\mathbb{R}^d$ are Lebesgue measurable.

Theorem 18.1 *Open and closed subsets of $\mathbb{R}^d$ are Lebesgue measurable.*

Proof. By Proposition 14.60, for any $x = (x_1, \ldots, x_d) \in \mathbb{R}^d$ and any $\varepsilon > 0$ the open cube $C_\varepsilon(x) := \prod_{i=1}^{d}(x_i - \varepsilon, x_i + \varepsilon)$ is Lebesgue measurable. Moreover, $C_\varepsilon(x)$ is the open ball of radius ε around x in the uniform norm $\|\cdot\|_\infty$.

Let $O \subseteq \mathbb{R}^d$ be an open set. Then $O_{\mathbb{Q}} := O \cap \mathbb{Q}^d$ is a countable dense subset of O. For all $x \in O_{\mathbb{Q}}$, let $\varepsilon_x := \sup\{\varepsilon < 1 : C_\varepsilon(x) \subseteq O\}$. Then clearly $\bigcup_{x \in O_{\mathbb{Q}}} C_{\varepsilon_x}(x) \subseteq O$. To prove the reversed inclusion, let $y \in O$. There are an $\varepsilon \in (0, 1)$ so that $C_\varepsilon(y) \subseteq O$ and an $x \in O_{\mathbb{Q}}$ so that $\|x - y\|_\infty < \frac{\varepsilon}{2}$. But then $y \in C_{\frac{\varepsilon}{2}}(x) \subseteq C_\varepsilon(y) \subseteq O$, which means $y \in \bigcup_{x \in O_{\mathbb{Q}}} C_{\varepsilon_x}(x)$. Because y was arbitrary this proves that $\bigcup_{x \in O_{\mathbb{Q}}} C_{\varepsilon_x}(x) \supseteq O$, and hence $\bigcup_{x \in O_{\mathbb{Q}}} C_{\varepsilon_x}(x) = O$. Because $O_{\mathbb{Q}}$ is countable and every $C_{\varepsilon_x}(x)$ is Lebesgue measurable we conclude that O is Lebesgue measurable because it is a countable union of Lebesgue measurable sets. Therefore, the open subsets of $\mathbb{R}^d$ are Lebesgue measurable.

Now let $C \subseteq \mathbb{R}^d$ be closed. Then $O = \mathbb{R}^d \setminus C$ is open and thus Lebesgue measurable. But then C is the complement of a Lebesgue measurable set, and hence it is Lebesgue measurable, too. ∎

Corollary 18.2 *Let $S \subseteq \mathbb{R}^d$ be Lebesgue measurable and let $f : S \to \mathbb{R}$ be continuous. Then f is Lebesgue measurable.*

Proof. Let $a \in \mathbb{R}$. Then the set (a, ∞) is open in $\mathbb{R}$. Because f is continuous, $f^{-1}\big[(a, \infty)\big]$ is relatively open. That is, there is an open set $O \subseteq \mathbb{R}^d$ so that $f^{-1}\big[(a, \infty)\big] = S \cap O$, which by Theorem 18.1 is Lebesgue measurable. Hence, f is Lebesgue measurable. ∎

Lebesgue measurable sets can now be characterized as the subsets of $\mathbb{R}^d$ that can be approximated by open sets from the outside and by closed sets from the inside so that the measure of the difference can be made arbitrarily small. We will first prove this result for sets of finite measure and then for sets of arbitrary measure.

Theorem 18.3 *A subset $S \subseteq \mathbb{R}^d$ with $\lambda(S) < \infty$ is Lebesgue measurable iff for every $\varepsilon > 0$ there are a compact set C and an open set O such that $C \subseteq S \subseteq O$ and $\lambda(O \setminus C) < \varepsilon$.*

Proof. For "⇒," let S be Lebesgue measurable with $\lambda(S) < \infty$ and let $\varepsilon > 0$. Then there is a family of open boxes $\{B_j\}_{j=1}^\infty$ with $S \subseteq \bigcup_{j=1}^\infty B_j$ and $\sum_{j=1}^\infty |B_j| < \lambda(S) + \frac{\varepsilon}{2}$.

Let $O := \bigcup_{j=1}^\infty B_j$. Then O is open, $S \subseteq O$, and $\lambda(O) \leq \sum_{j=1}^\infty |B_j| < \lambda(S) + \frac{\varepsilon}{2}$. For any $n \in \mathbb{N}$ let $K_n := \prod_{i=1}^d [-n, n]$. By Theorem 14.15, we obtain $\lambda(S) = \lim_{n\to\infty} \lambda(S \cap K_n)$. Hence, there is an $N \in \mathbb{N}$ so that $K := K_N$ satisfies $\lambda(S \cap K) \geq \lambda(S) - \frac{\varepsilon}{4}$. The set $K \setminus S$ is Lebesgue measurable and there is a family of open boxes $\{A_j\}_{j=1}^\infty$ so that

$K \setminus S \subseteq \bigcup_{j=1}^{\infty} A_j$ and $\sum_{j=1}^{\infty} |A_j| < \lambda(K \setminus S) + \frac{\varepsilon}{4}$. But then $C := K \setminus \bigcup_{j=1}^{\infty} A_j$ is closed and bounded, and hence by Theorem 16.80 it is compact. Moreover, the containment $C = K \setminus \bigcup_{j=1}^{\infty} A_j \subseteq K \setminus (K \setminus S) = S$ holds. Now

$$\begin{aligned}
&\lambda\left(S \cap \bigcup_{j=1}^{\infty} A_j\right) + \lambda\left(S' \cap \bigcup_{j=1}^{\infty} A_j\right) \\
&= \lambda\left(\bigcup_{j=1}^{\infty} A_j\right) \le \sum_{j=1}^{\infty} |A_j| < \lambda(K \setminus S) + \frac{\varepsilon}{4} \le \lambda\left(S' \cap \bigcup_{j=1}^{\infty} A_j\right) + \frac{\varepsilon}{4},
\end{aligned}$$

which means $\lambda\left(S \cap \bigcup_{j=1}^{\infty} A_j\right) < \frac{\varepsilon}{4}$. Hence,

$$\begin{aligned}
\lambda(C) &= \lambda\left(K \setminus \bigcup_{j=1}^{\infty} A_j\right) = \lambda\left((S \cap K) \setminus \bigcup_{j=1}^{\infty} A_j\right) + \lambda\left(\underbrace{(K \setminus S) \setminus \bigcup_{j=1}^{\infty} A_j}_{=\emptyset}\right) \\
&= \lambda\left((S \cap K) \setminus \bigcup_{j=1}^{\infty} A_j\right) = \lambda\left((S \cap K) \setminus \left(S \cap \bigcup_{j=1}^{\infty} A_j\right)\right)
\end{aligned}$$

> **For this step, note that $(S \cap K) \setminus \left(S \cap \bigcup_{j=1}^{\infty} A_j\right)$ and $S \cap \bigcup_{j=1}^{\infty} A_j$ are disjoint and their union contains $S \cap K$. This means that the sum of their measures is greater than or equal to $\lambda(S \cap K)$.**

$$\begin{aligned}
&\ge \lambda(S \cap K) - \lambda\left(S \cap \bigcup_{j=1}^{\infty} A_j\right) \\
&> \left(\lambda(S) - \frac{\varepsilon}{4}\right) - \frac{\varepsilon}{4} = \lambda(S) - \frac{\varepsilon}{2}.
\end{aligned}$$

Now $\lambda(O \setminus C) = \lambda(O) - \lambda(C) < \lambda(S) + \frac{\varepsilon}{2} - \left(\lambda(S) - \frac{\varepsilon}{2}\right) = \varepsilon$, which proves the direction "$\Rightarrow$."

Conversely, for "$\Leftarrow$" let $S \subseteq \mathbb{R}^d$ be such that $\lambda(S) < \infty$ and for every $\varepsilon > 0$ there are a compact set C and an open set O such that $C \subseteq S \subseteq O$ and $\lambda(O \setminus C) < \varepsilon$. Let $T \subseteq \mathbb{R}^d$ and let $\varepsilon > 0$. Then with O and C as described we infer

$$\begin{aligned}
\lambda(T) &= \lambda(O \cap T) + \lambda\left(O' \cap T\right) \\
&> \lambda(O \cap T) + \lambda\left(O' \cap T\right) + \lambda(O \setminus C) - \varepsilon \\
&\geq \lambda(O \cap T) + \lambda\left(O' \cap T\right) + \lambda((O \setminus C) \cap T) - \varepsilon \\
&\geq \lambda(O \cap T) + \lambda\left(O' \cap C' \cap T\right) + \lambda\left(\left(O \cap C'\right) \cap T\right) - \varepsilon
\end{aligned}$$

Apply Proposition 14.21 to $C' \cap T$ and O.

$$\begin{aligned}
&\geq \lambda(O \cap T) + \lambda\left(C' \cap T\right) - \varepsilon \\
&\geq \lambda(S \cap T) + \lambda\left(S' \cap T\right) - \varepsilon.
\end{aligned}$$

Because ε was arbitrary we obtain $\lambda(T) \geq \lambda(S \cap T) + \lambda\left(S' \cap T\right)$, and because T was arbitrary, this means that S is Lebesgue measurable. ■

Theorem 18.3 indicates that an "inner volume" or "inner Lebesgue measure" for a set S could be defined as the supremum of the (outer) Lebesgue measures of all compact sets contained in S (also see Exercise 18-3). A set would then be measurable iff the inner and outer volume are equal. Measurability in this sense turns out to be the same as Lebesgue measurability. However, this idea is quite complicated and it requires some topological structure, while Definition 14.19 of measurability can be used on arbitrary sets. This is why, even though it intuitively feels like a good idea, we do not work with "inner measures."

We can characterize Lebesgue measurable sets in $\mathbb{R}^d$ as follows.

Theorem 18.4 *Let $S \subseteq \mathbb{R}^d$. Then the following are equivalent.*

1. *S is Lebesgue measurable.*
2. *For every $\varepsilon > 0$, there are a closed set C and an open set O so that $C \subseteq S \subseteq O$ and $\lambda(O \setminus C) < \varepsilon$.*
3. *There is a sequence $\{O_n\}_{n=1}^{\infty}$ of open sets and a sequence $\{C_n\}_{n=1}^{\infty}$ of closed sets with $O_n \supseteq O_{n+1}$ and $C_n \subseteq C_{n+1}$ for all $n \in \mathbb{N}$ so that $\bigcup_{n=1}^{\infty} C_n \subseteq S \subseteq \bigcap_{n=1}^{\infty} O_n$ and so that $\lambda\left(\bigcap_{n=1}^{\infty} O_n \setminus \bigcup_{n=1}^{\infty} C_n\right) = 0$.*

Proof. To prove "1⇒2," for any natural number $n \in \mathbb{N}$ let $K_n := \prod_{i=1}^{d} [-n, n]$ and let $K_{-1} := K_0 := \emptyset$. For each $n \in \mathbb{N}$ the set $S \cap (K_n \setminus K_{n-1})$ is Lebesgue measurable with finite Lebesgue measure, so by Theorem 18.3 there are a closed set C_n and an open set O_n with $C_n \subseteq S \cap (K_n \setminus K_{n-1}) \subseteq O_n \subseteq K_{n+1}^{\circ} \setminus K_{n-1}$ and so that $\lambda(O_n \setminus C_n) < \dfrac{\varepsilon}{2 \cdot 2^n}$. Then the set $O := \bigcup_{n=1}^{\infty} O_n$ is open and by Exercise 16-42, the set $C := \bigcup_{n=1}^{\infty} C_n$ is closed.

Finally, with $O_0 := \emptyset$ we obtain the following.

$$\begin{aligned}
\lambda(O \setminus C) &= \sum_{n=1}^{\infty} \lambda\big((O \setminus C) \cap (K_n \setminus K_{n-1})\big) \\
&= \sum_{n=1}^{\infty} \lambda\Big(\big(O \cap (K_n \setminus K_{n-1})\big) \setminus \big(C \cap (K_n \setminus K_{n-1})\big)\Big) \\
&\le \sum_{n=1}^{\infty} \lambda(O_n \setminus C_n) + \lambda\big(O_{n-1} \cap (K_n \setminus K_{n-1})\big) \\
&< \sum_{n=1}^{\infty} \frac{\varepsilon}{2^n} = \varepsilon.
\end{aligned}$$

For "2⇒3," for each natural number $k \in \mathbb{N}$ let $\widetilde{O}_k$ be open and let $\widetilde{C}_k$ be closed with $\widetilde{C}_k \subseteq S \subseteq \widetilde{O}_k$ and $\lambda\big(\widetilde{O}_k \setminus \widetilde{C}_k\big) < \frac{1}{k}$. Then for each $n \in \mathbb{N}$, the set $O_n := \bigcap_{k=1}^{n} \widetilde{O}_k$ is open, the set $C_n := \bigcup_{k=1}^{n} \widetilde{C}_k$ is closed, $O_n \supseteq O_{n+1}$, $C_n \subseteq C_{n+1}$ and $C_n \subseteq S \subseteq O_n$. Therefore, $\bigcup_{n=1}^{\infty} C_n \subseteq S \subseteq \bigcap_{n=1}^{\infty} O_n$. Moreover, for all $k \in \mathbb{N}$ we infer

$$\lambda\left(\bigcap_{n=1}^{\infty} O_n \setminus \bigcup_{n=1}^{\infty} C_n\right) \le \lambda(O_k \setminus C_k) \le \lambda\big(\widetilde{O}_k \setminus \widetilde{C}_k\big) < \frac{1}{k},$$

and because $k \in \mathbb{N}$ is arbitrary we conclude $\lambda\left(\bigcap_{n=1}^{\infty} O_n \setminus \bigcup_{n=1}^{\infty} C_n\right) = 0$.

The part "3⇒1" follows from the fact that open sets, closed sets and null sets are Lebesgue measurable. ∎

Lebesgue measure is also often considered on Lebesgue measurable subsets of $\mathbb{R}^d$ (see Example 14.10). Let $A \subseteq \mathbb{R}^d$ be Lebesgue measurable. Because the intersection of Lebesgue measurable sets is again Lebesgue measurable, Theorem 18.4 also holds for Lebesgue measurable subsets of A. If we want all sets involved to be subsets of A, we need to replace the demand that the O_n are open with the demand that the O_n are open in A and we need to replace the demand that the C_n are closed with the demand that the C_n are closed in A. If A itself is open, this is not necessary.

Theorem 18.4 also shows that the σ-algebra generated by the open subsets of $\mathbb{R}^d$ is interesting in itself. It is investigated further in Exercise 18-7.

Exercises

18-1. Explain why the hypothesis $\lambda(S) < \infty$ in Theorem 18.3 cannot be dropped.

18-2. Prove that every open subset of $\mathbb{R}^d$ is a countable union of compact sets.

18-3 Let $S \subseteq \mathbb{R}^d$ be Lebesgue measurable.

(a) Prove that $\lambda(S) = \inf\left\{\lambda(O) : S \subseteq O, O \text{ is open }\right\}$

(b) Prove that $\lambda(S) = \sup\left\{\lambda(C) : C \subseteq S, C \text{ is compact }\right\}$.

(c) Prove that $\lambda(S) = \sup\left\{\lambda(C) : C \subseteq S, C \text{ is closed }\right\}$.

18-4. Prove that $S \subseteq \mathbb{R}^d$ with $\lambda(S) < \infty$ is Lebesgue measurable iff for every $\varepsilon > 0$ there is a compact set $C \subseteq S$ with $\lambda(C) > \lambda(S) - \varepsilon$

18-5 Lebesgue measurable functions. Let $\Omega \subseteq \mathbb{R}^d$ be open and let $f : \Omega \to \mathbb{R}$

(a) Prove that f is Lebesgue measurable iff for all open sets $O \subseteq \mathbb{R}$ the set $f^{-1}[O]$ is Lebesgue measurable.

(b) Prove that f is Lebesgue measurable iff for all closed sets $C \subseteq \mathbb{R}$ the set $f^{-1}[C]$ is Lebesgue measurable.

(c) Prove that f is Lebesgue measurable iff for all compact sets $K \subseteq \mathbb{R}$ the set $f^{-1}[K]$ is Lebesgue measurable

18-6 The **Derivative Form of the Fundamental Theorem of Calculus** for the Lebesgue integral states the following. If $h : [a, b] \to \mathbb{R}$ is Lebesgue integrable, then the function $H(x) := \int_a^x h(t)\, d\lambda(t)$ is differentiable a.e with derivative $\frac{d}{dx}\int_a^x h(t)\, d\lambda(t) = h(x)$ for almost all $x \in [a, b]$ In this exercise, we will prove the result with the steps given below. (Recall that by Exercise 10-7, H is differentiable a.e if h is nonnegative)

(a) Let $a_1 < b_1 < a_2 < b_2 < \cdots a_n < b_n$ and let $A := \bigcup_{j=1}^{n}(a_j, b_j)$. Prove the result for the indicator function $\mathbf{1}_A$

(b) Use Fubini's Theorem on limits of nondecreasing functions (see Exercise 11-20) and Exercise 16-103d to prove the result for indicator functions of open subsets of $[a, b]$.

(c) Prove the result for indicator functions of closed subsets of $[a, b]$

(d) Use Fubini's Theorem on limits of nondecreasing functions (see Exercise 11-20) and part 3 of Theorem 18.4 to prove the result for indicator functions of Lebesgue measurable subsets of $[a, b]$

(e) Prove the result for simple Lebesgue measurable functions on $[a, b]$.

(f) Use Fubini's Theorem on limits of nondecreasing functions (see Exercise 11-20) to prove the result for nonnegative Lebesgue integrable functions on $[a, b]$.

(g) Prove the result for all Lebesgue integrable functions on $[a, b]$.

18-7 Borel Sets. Let X be a metric space. The σ-algebra generated by the open sets in X is called the σ-algebra of **Borel sets** of X, denoted $\mathcal{B}(X)$. A function $f : X \to \mathbb{R}$ is called **Borel measurable** iff it is measurable as a function on the measurable space $\left(X, \mathcal{B}(X)\right)$.

(a) Prove that if X is a Lebesgue measurable subset of $\mathbb{R}^d$ considered as a metric subspace of $\mathbb{R}^d$, then every Borel set of X is also Lebesgue measurable.

(b) Let $X \subseteq \mathbb{R}^d$ be Lebesgue measurable. Prove that for every Lebesgue measurable set $S \subseteq X$ there are Borel sets $F, G \in \mathcal{B}(X)$ so that $F \subseteq S \subseteq G$ and $\lambda(G \setminus F) = 0$.

(c) Let $X \subseteq \mathbb{R}^d$ be Lebesgue measurable and let $f : X \to \mathbb{R}$ be Lebesgue measurable. Prove that there is a Borel measurable function $f_B : X \to \mathbb{R}$ so that $\lambda\left(\left\{x \in X : f(x) \neq f_B(x)\right\}\right) = 0$.

Note This result is the reason why the focus of measure theory can be restricted to Borel measurable functions when necessary

(d) Prove that if $m+n=d$ and $X \in \Sigma_{\lambda_m} \times \Sigma_{\lambda_n}$, then every Borel measurable function $f : X \to \mathbb{R}$ is also $\Sigma_{\lambda_m} \times \Sigma_{\lambda_n}$-measurable.

(e) A measure $\mu : \mathcal{B}\left(\mathbb{R}^d\right) \to [0,\infty]$ is called a **Borel measure**. Let the norm on $\mathbb{R}^d$ be the uniform norm. Prove that two Borel measures μ and ν are equal iff for all $x \in \mathbb{Q}^d$ and all $r \in (0,1) \cap \mathbb{Q}$ we have that $\mu\left(B_r(x)\right)$ is finite and $\mu\left(B_r(x)\right) = \nu\left(B_r(x)\right)$.

(f) Let the norm on $\mathbb{R}^d$ be the uniform norm. Prove that there are two distinct measures $\mu, \nu : \mathcal{B}\left(\mathbb{R}^d\right) \to [0,\infty]$ so that $\mu\left(B_r(x)\right) = \nu\left(B_r(x)\right)$ holds for all $x \in \mathbb{Q}^d$ and all $r \in (0,1) \cap \mathbb{Q}$.

18-8 The σ-algebra of subsets of $[a,b]$ generated by the open sets in $[a,b]$ (as a metric space in its own right) is also called the σ-algebra of **Borel sets** (of $[a,b]$). Let $g : [a,b] \to \mathbb{R}$ be nondecreasing and let λ_g be the Lebesgue-Stieltjes measure defined in Exercise 14-20.

(a) Prove that all Borel sets of $[a,b]$ are λ_g-measurable.

(b) Prove that λ_g is a finite measure on the Borel sets of $[a,b]$.

(c) Let $f : [a,b] \to \mathbb{R}$ be a function. Then f is called **Lebesgue-Stieltjes integrable** with respect to g iff f is λ_g-integrable.
Prove that if f is continuous, then f is Lebesgue-Stieltjes integrable with respect to g and $\int_a^b f\,d\lambda_g = \int_a^b f\,dg$, where the right side is the Riemann-Stieltjes integral of f, which exists by Exercise 5-19b.

18.2 C^∞ and Approximation of Integrable Functions

Theorem 16.86 makes a first connection between topological and measure theoretical notions by showing that, loosely speaking, the continuous functions are dense in $L^p[a,b]$ for $1 \leq p < \infty$. In this section, we show that even more is true. The set of infinitely often differentiable functions is dense in L^p, too. Because differentiability is usually defined on open sets, we work with open domains $\Omega \subseteq \mathbb{R}^d$ throughout this section. Recall that by Notation 15.66, with terminology as in Example 14.10, we have $\mathcal{L}^p(\Omega) = \mathcal{L}^p\left(\Omega, \Sigma_\lambda^\Omega, \lambda|_{\Sigma_\lambda^\Omega}\right)$ and $L^p(\Omega) = L^p\left(\Omega, \Sigma_\lambda^\Omega, \lambda|_{\Sigma_\lambda^\Omega}\right)$.

Definition 18.5 *Let $\Omega \subseteq \mathbb{R}^d$ be open and let $k \geq 0$. We define $C^k(\Omega)$ to be the set of all k times continuously differentiable functions $f : \Omega \to \mathbb{R}$. Moreover, we define*

$$C^\infty(\Omega) := \bigcap_{k=1}^{\infty} C^k(\Omega).$$

Note that infinite differentiability is compatible with the usual algebraic operations.

Theorem 18.6 *Let $\Omega \subseteq \mathbb{R}^d$ be an open subset of $\mathbb{R}^d$ and let $f, g \in C^\infty(\Omega)$. Then the functions $f+g$, $f-g$, fg are in $C^\infty(\Omega)$. Moreover, $\frac{f}{g} \in C^\infty\big(\Omega \setminus \{x : g(x)=0\}\big)$.*

Proof. We first show that for the first three operations, the result is a consequence of Proposition 17.62. If $f, g \in C^\infty(\Omega)$, then $x \mapsto \big(f(x), g(x)\big)$ is infinitely differentiable. Moreover, any bilinear function on $\mathbb{R}^2$ is continuous, and hence (use Theorem 17.52 and note that the derivative is the sum of two linear functions) infinitely differentiable. Therefore, by Proposition 17.62 for all $k \in \mathbb{N}$ we infer $f+g \in C^k(\Omega)$, because it is the composition of the k times differentiable functions $x \mapsto \big(f(x), g(x)\big)$ and $(y, z) \mapsto y + z$. Because k was arbitrary, $f + g \in C^\infty(\Omega)$. Differences and products are treated similarly.

For the quotient, note that the function $(y, z) \mapsto \left(y, \frac{1}{z}\right)$ is infinitely differentiable on $\mathbb{R} \times \mathbb{R} \setminus \{0\}$ and that the quotient is the composition of the functions $x \mapsto \big(f(x), g(x)\big)$, $(y, z) \mapsto \left(y, \frac{1}{z}\right)$ and $(u, v) \mapsto uv$. ■

We will prove more than that infinitely differentiable functions are dense in $L^p(\Omega)$. We will show that there is a dense subset consisting of infinitely differentiable functions that are zero outside a compact set.

Definition 18.7 *Let $\Omega \subseteq \mathbb{R}^d$ be open. A function $f \in C^\infty(\Omega)$ is called a* **test function** *iff* supp(f) *is compact. The set of all test functions on Ω is denoted $C_0^\infty(\Omega)$.*

The terminology "test function" comes from the role these functions play in the investigation of partial differential equations. One concrete example is the definition of weak derivatives (see Definition 23.11). With $C_0^\infty(\Omega)$ defined, our next step is to approximate the indicator function of an interval with a function in $C_0^\infty(\mathbb{R})$. The construction is visualized in Figure 44.

Lemma 18.8 *The C^∞ connector. The function $c(x) := \begin{cases} 0; & \text{for } x \leq 0, \\ e^{-\frac{1}{x}}; & \text{for } x > 0, \end{cases}$ is infinitely differentiable on $\mathbb{R}$.*

Proof. It is clear that $c \in C^\infty\big(\mathbb{R} \setminus \{0\}\big)$. It remains to be proved that c is infinitely differentiable at $x = 0$. For all $x < 0$, we have $c^{(n)}(x) = 0$. Therefore, if they exist, all derivatives of c at $x = 0$ must be zero. For all $n \in \mathbb{N} \cup \{0\}$, we have $\lim_{z \to 0^-} \frac{c^{(n)}(z) - 0}{z - 0} = 0$. If we can show that the right-sided limit also is 0, then we have proved that $c \in C^\infty(\mathbb{R})$ with $c^{(n)}(x) := \begin{cases} 0; & \text{for } x \in (-\infty, 0], \\ \frac{d^n}{dx^n} e^{-\frac{1}{x}}; & \text{for } x \in (0, \infty), \end{cases}$ for all $n \in \mathbb{N}$. *(Formally, this would require an induction on n, but it is not hard to mentally fill in.)*

An easy induction shows that for every $n \in \mathbb{N} \cup \{0\}$ there is a polynomial p_n so that $\frac{d^n}{dx^n} e^{-\frac{1}{x}} = p_n\left(\frac{1}{x}\right) e^{-\frac{1}{x}}$ for all $x > 0$ (see Exercise 18-9). For all $j \in \mathbb{N}$ by Exercise 12-27, we obtain $\lim_{x \to 0^+} \frac{e^{-\frac{1}{x}}}{x^j} = \lim_{x \to 0^+} \left(\frac{1}{x}\right)^j e^{-\frac{1}{x}} = \lim_{u \to \infty} u^j e^{-u} = 0$, which means that for every polynomial $p(x) = \sum_{j=0}^{k} a_j x^j$ we can conclude

$$\lim_{x \to 0^+} p\left(\frac{1}{x}\right) e^{-\frac{1}{x}} = \lim_{x \to 0^+} \sum_{j=0}^{k} a_j \left(\frac{1}{x}\right)^j e^{-\frac{1}{x}} = 0.$$

But then for all $n \in \mathbb{N} \cup \{0\}$ we obtain

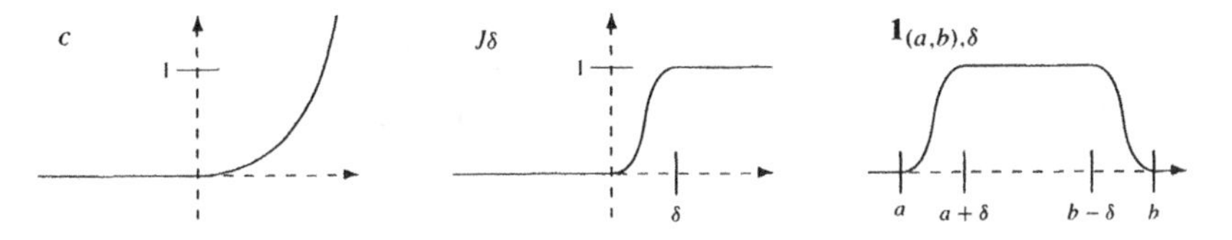

Figure 44: Constructing a C^∞ "indicator function" for intervals.

$$\lim_{z\to 0^+} \frac{\frac{d^n}{dz^n}e^{-\frac{1}{z}} - 0}{z-0} = \lim_{z\to 0^+} \frac{p_n\left(\frac{1}{z}\right)e^{-\frac{1}{z}} - 0}{z-0} = \lim_{z\to 0^+} \frac{1}{z} p_n\left(\frac{1}{z}\right)e^{-\frac{1}{z}} = 0. \quad \blacksquare$$

Note that the function in Lemma 18.8 also shows that not every C^∞ function can be represented as a power series. This is because $c^{(n)}(0) = 0$ for all $n \in \mathbb{N}$, but $c \neq 0$.

Lemma 18.9 *The C^∞ jump function. Let $\delta > 0$. Then $j_\delta(x) := \dfrac{c(x)}{c(\delta - x) + c(x)}$ is infinitely differentiable on $\mathbb{R}$. Moreover, it takes values in $[0,1]$, it is identical to zero on $(-\infty, 0]$, it is identical to 1 on $[\delta, \infty)$ and $j_\delta\big[(0,\delta)\big] \subseteq (0,1)$.*

Proof. Because $c \geq 0$ and for each $x \in \mathbb{R}$ at least one of $c(x)$ and $c(\delta - x)$ is not zero, Proposition 17.62 and Theorem 18.6 imply that $j_\delta \in C^\infty(\mathbb{R})$. Because $c \geq 0$ for all $x \in \mathbb{R}$ we obtain $c(x) \leq c(\delta - x) + c(x)$, which implies $0 \leq \dfrac{c(x)}{c(\delta - x) + c(x)} \leq 1$ for all $x \in \mathbb{R}$ and the inequalities are strict for $x \in (0, \delta)$. Finally, because $c(x) = 0$ for $x \leq 0$ we conclude $j_\delta(x) = 0$ for $x \leq 0$ and because $c(\delta - x) = 0$ for $x \geq \delta$ we conclude $j_\delta(x) = 1$ for $x \geq \delta$. ■

Lemma 18.10 *The C^∞ interval indicator. Let $a < b$ and $\delta > 0$ be real numbers. The function $\mathbf{1}_{(a,b),\delta}(x) := j_\delta(x-a)j_\delta(b-x)$ is infinitely differentiable on $\mathbb{R}$. Moreover, it takes values in $[0,1]$, it is identical to zero on $\mathbb{R} \setminus (a,b)$ and it is identical to 1 on $[a+\delta, b-\delta]$.*

Proof. Exercise 18-10. ■

C^∞ interval indicator functions are the key ingredient to constructing similar "indicator functions" for closed sets. These functions are then used to show that the compactly supported infinitely differentiable functions are dense in $L^p(\Omega)$.

Lemma 18.11 *Let $C \subseteq \mathbb{R}^d$ be closed and let $U \subseteq \mathbb{R}^d$ be open so that $C \subseteq U$. Then there is a function $\mathbf{1}_{C,U} \in C^\infty\left(\mathbb{R}^d\right)$ that takes values in $[0,1]$ and satisfies $\mathbf{1}_{C,U}\big|_C = 1$ and $\operatorname{supp}\left(\mathbf{1}_{C,U}\right) \subset U$.*

Proof. First consider the case that C is compact. Throughout, we will use the uniform norm $\|\cdot\|_\infty$. So, "balls" around points will actually be cubes. For each $x \in C$, let $\varepsilon_x > 0$ be so that $\overline{B_{\varepsilon_x}(x)} \subset U$. Then $\left\{B_{\frac{\varepsilon_x}{2}}(x)\right\}_{x\in C}$ is an open cover of C. Let $x_1, \ldots, x_n$ be so that $C \subseteq \bigcup_{j=1}^{n} B_{\frac{\varepsilon_{x_j}}{2}}(x_j)$. For each x_j, let $x_j^{(i)}$ denote the ith component of the representation of x_j with respect to the standard base. For $j = 1, \ldots, n$ let $h_j := \prod_{i=1}^{d} \mathbf{1}_{\left(x_j^{(i)} - \varepsilon_{x_j}, x_j^{(i)} + \varepsilon_{x_j}\right), \frac{\varepsilon_{x_j}}{2}}$. Then $\operatorname{supp}(h_j) \subseteq \overline{B_{\varepsilon_{x_j}}(x_j)} \subset U$ and $h_j \in C^\infty\left(\mathbb{R}^d\right)$ because it is a product of infinitely differentiable functions. Hence, for $h := \sum_{j=1}^{n} h_j$ we infer $\operatorname{supp}(h) \subset U$, $h \in C^\infty\left(\mathbb{R}^d\right)$ and for all $x \in C$ there is a $j \in \{1, \ldots, n\}$ so that $x \in B_{\frac{\varepsilon_{x_j}}{2}}(x_j)$, which means $h(x) \geq h_j(x) = 1$. With j_1 being a C^∞ jump function as in Lemma 18.9, define $\mathbf{1}_{C,U} := j_1 \circ h$. By Proposition 17.62, this function is in $C^\infty\left(\mathbb{R}^d\right)$. By the above, we have $\mathbf{1}_{C,U}\big|_C = 1$ and $\operatorname{supp}\left(\mathbf{1}_{C,U}\right) \subset U$, so the result is established for compact C.

Now consider the case that C is closed, but not necessarily compact. For each $n \in \mathbb{N}$, let $C_n := C \cap \left(\overline{B_n(0)} \setminus B_{n-1}(0)\right)$. Then each C_n is compact. For each $x \in C_n$, let $\varepsilon_x \in (0, 1)$ be so that $\overline{B_{\varepsilon_x}(x)} \subset U$ and let $U_n := \bigcup_{x \in C_n} B_{\varepsilon_x}(x)$. Then for each $n \in \mathbb{N}$, the set U_n is open and contains the compact set C_n. Let the functions $\mathbf{1}_{C_n, U_n}$ be as constructed above and let $g := \sum_{n=1}^{\infty} \mathbf{1}_{C_n, U_n}$. For each $x \in \mathbb{R}^d$, there is a neighborhood V of x so that at most four terms in this series are not equal to zero on V. Thus g is in $C^\infty\left(\mathbb{R}^d\right)$. Moreover, $g(x) \geq 1$ for all $x \in C$, and $\operatorname{supp}(g) \subset U$. Now $\mathbf{1}_{C,U} := j_1 \circ g$ is as desired. ∎

Note that Lemma 18.11 shows in particular that $C_0^\infty(\Omega)$ is not empty, which is not necessarily trivial. The next result shows that even more is true.

Theorem 18.12 *Let $1 \leq p < \infty$ and let $\Omega \subseteq \mathbb{R}^d$ be an open subset of $\mathbb{R}^d$. Then $C_0^\infty(\Omega)$, or more formally $T := \left\{[f] : f \in C_0^\infty(\Omega)\right\}$, is dense in $L^p(\Omega)$.*

Proof. The proof runs along the same lines as the proof of Theorem 16.86. The main change is that we need to approximate the indicator functions of open boxes contained in Ω (rather than those of intervals) with functions in $C_0^\infty(\Omega)$ (rather than with continuous functions). Let $B = \prod_{i=1}^{d} (l_i, r_i)$ be a box that is contained in Ω and let $\varepsilon > 0$. By Theorem 18.3, there is a compact set $C \subseteq B$ so that $\lambda(B \setminus C) < \varepsilon^p$. Let $\mathbf{1}_{C,B}$ be as in Lemma 18.11. Then $\operatorname{supp}\left(\mathbf{1}_{C,B}\right) \subset B \subseteq \Omega$, $\mathbf{1}_{C,B} \in C^\infty(\Omega)$ and

$$\left(\int_\Omega \left|\mathbf{1}_B - \mathbf{1}_{C,B}\right|^p \, d\lambda\right)^{\frac{1}{p}} \leq \left(\int_\Omega \mathbf{1}_{B \setminus C}^p \, d\lambda\right)^{\frac{1}{p}} \leq \left(\lambda(B \setminus C)\right)^{\frac{1}{p}} < \varepsilon.$$

We conclude that in any L^p-neighborhood of an indicator function of a box in Ω we can find a function in $C_0^\infty(\Omega)$.

Now that we can approximate indicator functions of boxes with $C_0^\infty(\Omega)$-functions, the proof proceeds just like the proof of Theorem 16.86. To approximate indicator

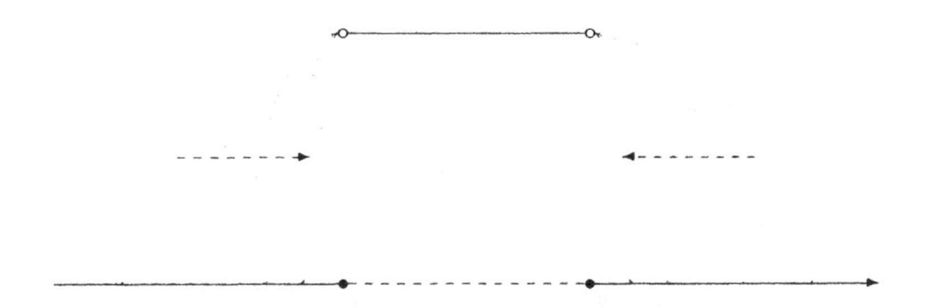

Figure 45: Illustration how the indicator function of an interval can be approximated in L^p with C^∞ functions. The area bounded by the difference shrinks to zero, which allows the approximation (in the L^p sense) of the discontinuities with infinitely smooth functions.

functions of sets, cover the set A with open boxes $B_i \subseteq \Omega$ such that the sum of the volumes of the boxes B_i is close to that of the set. Truncate the sum to obtain a finite number of boxes B_i so that the sum of the volumes of the finitely many boxes is close to the total sum of volumes. Approximate the indicator functions of these finitely many boxes with $C_0^\infty(\Omega)$ functions as indicated above to prove that indicator functions of sets can be approximated with $C_0^\infty(\Omega)$ functions. The details are to be given in Exercise 18-11a. Next, just like the proof of Theorem 16.86, approximate simple functions with functions in $C_0^\infty(\Omega)$ (Exercise 18-11b). Finally, just like the proof of Theorem 16.86, apply Theorem 16.85 (Exercise 18-11c). ■

Geometrically, Theorem 18.12 says that even highly discontinuous functions can be approximated arbitrarily well with infinitely smooth functions in L^p. Figure 45 shows such an approximation for the indicator function of an interval. This visualization illustrates the crucial part of the proof of Theorem 18.12 and it also shows that the result is not counterintuitive. The infinitely smooth approximations do not in any way "fix" the jumps at the discontinuities.

Exercises

18-9. Prove by induction that for every $n \in \mathbb{N}$ there is a polynomial p_n so that for all $x > 0$ we have $\frac{d^n}{dx^n} e^{-\frac{1}{x}} = p_n\left(\frac{1}{x}\right) e^{-\frac{1}{x}}$.

18-10. Prove Lemma 18.10.

18-11. Finish the proof of Theorem 18.12.

(a) Prove that for every measurable subset A of Ω and every $\varepsilon > 0$ there is a test function $g_{A,\varepsilon} \in C_0^\infty(\Omega)$ so that $\|\mathbf{1}_A - g_{A,\varepsilon}\|_p < \varepsilon$.

(b) Prove that for every simple function $s \in \mathcal{L}^p(\Omega)$ and every $\varepsilon > 0$ there is a test function $g_{s,\varepsilon} \in C_0^\infty(\Omega)$ so that $\|s - g_{s,\varepsilon}\|_p < \varepsilon$.

(c) Apply Theorem 16.85 to complete the proof.

18-12. Let $\Omega \subseteq \mathbb{R}^d$ be an open set. Prove that for every function $f \in L^1(\Omega)$ there is a sequence of functions in $C_0^\infty(\Omega)$ that converges a.e. to f.

18-13. Let $\Omega \subseteq \mathbb{R}^d$ be an open set. Prove that if $f \in L^1(\Omega)$ is so that $\int_\Omega f\varphi \, d\lambda = 0$ for all $\varphi \in C_0^\infty(\Omega)$, then $f = 0$ a.e.

18-14 A **half-open box** in $\mathbb{R}^d$ is a set H for which there are real numbers $a_i < b_i$, $i = 1, \ldots, d$ so that

$$H = \prod_{i=1}^{d} [a_i, b_i)$$

Let $O \subseteq \mathbb{R}^d$ be open and let $K \subseteq O$ be compact. Prove that for every $\varepsilon > 0$ there is a finite family $\{H_j\}_{j=1}^n$ of pairwise disjoint half-open cubes of side length less that ε so that $K \subseteq \bigcup_{j=1}^n H_j \subseteq O$.

Hint Work with the $\|\ \|_\infty$ norm, let $\delta := \frac{1}{2} \min\left\{\varepsilon, \operatorname{dist}\left(C, \mathbb{R}^d \setminus O\right)\right\}$ (need to prove $\delta > 0$) and use cubes of the form $\prod_{i=1}^{n} \left[l_i \delta, (l_i + 1)\delta\right)$.

18-15 Prove that the function $g : \mathbb{R}^d \to \mathbb{R}$ defined by $g(x) := \begin{cases} e^{\frac{1}{\sum_{j=1}^d x_j^2 - 1}}; & \text{for } \|x\|_2 < 1, \\ 0; & \text{for } \|x\|_2 \geq 1, \end{cases}$

is a C_0^∞ function.
Hint Chain Rule and Lemma 18 8

18-16 Let $\Omega \subseteq \mathbb{R}^d$ be open and let $f \in L^p(\Omega)$. With f extended to all of $\mathbb{R}^d$ by setting it equal to zero outside of Ω, prove that $\lim_{z \to 0} \int_\Omega \left| f(x) - f(x+z) \right|^p \, d\lambda(x) = 0$

Hints. First prove the result for C_0^∞ functions. Then prove the result in general by approximating f with C_0^∞ functions

18-17 Let $\Omega \subseteq \mathbb{R}^d$ be open, let $f \in L^p(\Omega)$, let $\varphi \in C_0^\infty(\Omega)$ and for all $x \in \mathbb{R}^d$ define the function f_φ by $f_\varphi(x) = \int_\Omega f(z)\varphi(x - z) \, d\lambda(z)$. (The integrals exist because $L^p(\operatorname{supp}(\varphi)) \subseteq L^1(\operatorname{supp}(\varphi))$.)

(a) Prove that f_φ is continuous on $\mathbb{R}^d$
Hint Dominated Convergence Theorem

(b) Prove that at every $x \in \mathbb{R}^d$ all first partial derivatives $\frac{\partial f_\varphi}{\partial x_i}$ of f_φ exist and are equal to $\frac{\partial f_\varphi}{\partial x_i} = \int_\Omega f(z) \frac{\partial \varphi}{\partial x_i}(x - z) \, d\lambda(z)$
Hint Dominated Convergence Theorem, Mean Value Theorem for functions of one variable and boundedness of the first partial derivatives of φ.

(c) Prove that all first partial derivatives $\frac{\partial f_\varphi}{\partial x_i}$ of f_φ are continuous at every $x \in \mathbb{R}^d$
Hint. Apply parts 18-17a and 18-17b

(d) Prove that $f_\varphi \in C^\infty\left(\mathbb{R}^d\right)$
Hint Apply part 18-17c and use induction.

Note The operation that produces f_φ from f and φ is also called the **convolution** of f and φ

18-18 Let $\Omega \subseteq \mathbb{R}^d$ be open and let $p \in [1, \infty)$

(a) Prove that $L^p(\Omega)$ has a countable dense subset consisting of simple functions
Hint Boxes with dyadic rational bounds

(b) Prove that $L^p(\Omega)$ has a countable dense subset consisting of C_0^∞ functions

18-19. Prove that for every Riemann integrable function $f : [a, b] \to \mathbb{R}$ on $[a, b]$ there is a sequence $\{g_n\}_{n=1}^\infty$ of infinitely differentiable functions on $[a, b]$ such that $g_n(a) = g_n(b) = 0$ for all $n \in \mathbb{N}$, $\lim_{n \to \infty} \int_a^b |f - g_n| \, dx = 0$ and for all $n \in \mathbb{N}$ we have $|g_n| \leq |f|$

Hint First find $a = x_0 < x_1 < \cdots < x_n = b$ and a step function $s(x) = \sum_{k=1}^n a_k \mathbf{1}_{[x_{k-1}, x_k)}$ so that $|s| \leq |f|$ and $\|s - f\|_1$ is small Then use C^∞ interval indicator functions

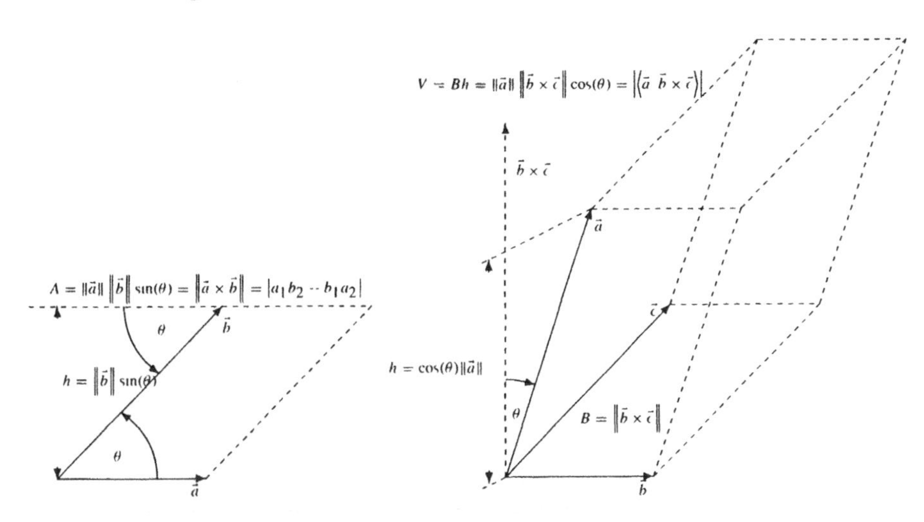

Figure 46: Geometric arguments that give the area of a parallelogram and the volume of a parallelepiped. (Formally, the cross product on the left is obtained by making $\vec{a}$ and $\vec{b}$ vectors in $\mathbb{R}^3$ whose third component is zero.)

18.3 Tensor Algebra and Determinants

To prove the Multidimensional Substitution Formula we need a volume function for n-dimensional parallelepipeds spanned by vectors $a_1, \ldots, a_n$. Moreover, for integration on manifolds we need functions that allow us to compute the lower dimensional volume of lower dimensional parallelepipeds in higher dimensional spaces. For example, we will be interested in the area (two dimensional volume) of a parallelogram (two dimensional parallelepiped) in three dimensional space.

We start by analyzing some formulas from geometry (also see Figure 46). It can be proved that the area of the parallelogram in $\mathbb{R}^2$ spanned by the vectors $a = \begin{pmatrix} a_1 \\ a_2 \end{pmatrix}$ and $b = \begin{pmatrix} b_1 \\ b_2 \end{pmatrix}$ is $A(a, b) = |a_1 b_2 - b_1 a_2|$. Similarly, the volume of the three dimensional parallelepiped spanned by the vectors a, b, c is $V(a, b, c) = \left| \langle a, b \times c \rangle \right|$. Note that if we drop the absolute values, the (oriented) area is bilinear and the (oriented) volume is 3-linear (see Definition 17.45) in the input vectors. Throughout, we will consider oriented volumes, that is, volumes that can be positive or negative. This is not much of a loss, because if we really want a nonnegative number, we simply take absolute values.

Because the two dimensional volume formula is incorporated in the three dimensional volume formula, the above suggests that we should try to construct a general tensor formalism that produces the volume function for parallelepipeds in arbitrary dimensions.

For the Multidimensional Substitution Formula, we only need a definition of the determinant and Corollary 18.27. Readers familiar with both (say, from linear algebra) may postpone this section until before Section 19.3.

We start by representing k-tensors on finite dimensional vector spaces. The representation in Theorem 18.16 below is already implicit in Corollary 17.61.

Definition 18.13 *Let V be a vector space. The set of real valued k-**tensors** on V is denoted by $\mathcal{T}^k(V)$. For a finite dimensional vector space V, the **dual space** is defined to be $V^* := \mathcal{T}^1(V) = \mathcal{L}(V, \mathbb{R})$.*

With pointwise addition and scalar multiplication $\mathcal{T}^k(V)$ is a vector space (see Exercise 18-20). The (higher dimensional) volume of a box is the product of the lower dimensional volumes of the projections: for a rectangle, area is the product of the lengths of the sides (length is one dimensional volume), and for a three dimensional box, the volume is the area of the base times the height. It is thus not surprising that tensors can be multiplied in the same simplistic fashion.

Definition 18.14 *Let V be a vector space and $S \in \mathcal{T}^k(V)$, $T \in \mathcal{T}^l(V)$. We define $S \otimes T[v_1, \ldots, v_k, v_{k+1}, \ldots, v_{k+l}] := S[v_1, \ldots, v_k]T[v_{k+1}, \ldots, v_{k+l}]$ and call it the **tensor product** of S and T. Clearly, $S \otimes T \in \mathcal{T}^{k+l}(V)$.*

The tensor product is not commutative, but it is associative (see Exercise 18-21). This means that while we need to be careful with the order of the factors, tensor products with more than two factors can be written without parentheses. Tensor products allow us to represent tensors in terms of a very natural base.

Theorem 18.15 *Let V be a finite dimensional vector space with base $\{v_1, \ldots, v_d\}$. Then the maps $\phi_i : V \to \mathbb{R}$ defined by $\phi_i\left[\sum_{j=1}^{d} a_j v_j\right] := a_i$ form a base of the dual space of V, called the **dual base** of $\{v_1, \ldots, v_d\}$.*

Proof. To see that $\{\phi_1, \ldots, \phi_d\}$ is linearly independent, let $\alpha_1, \ldots, \alpha_d \in \mathbb{R}$ be so that $\sum_{i=1}^{d} \alpha_i \phi_i = 0$. Then for all $j \in \{1, \ldots, d\}$ we infer $0 = \sum_{i=1}^{d} \alpha_i \phi_i [v_j] = \alpha_j$, and hence $\{\phi_1, \ldots, \phi_d\}$ is linearly independent.

To see that $\{\phi_1, \ldots, \phi_d\}$ is a base of V^*, let $\psi \in V^*$. Then for all $v = \sum_{i=1}^{d} a_i v_i$ in V we have $\psi[v] = \sum_{i=1}^{d} a_i \psi[v_i] = \sum_{i=1}^{d} \psi[v_i]\phi_i[v]$, which means that ψ is a linear combination of the ϕ_i. Hence, $\{\phi_1, \ldots, \phi_d\}$ is a base of V^*. ■

Theorem 18.16 *Let V be a finite dimensional vector space with base $\{v_1, \ldots, v_d\}$ and let $\{\phi_1, \ldots, \phi_d\}$ be the associated dual base as in Theorem 18.15. Then the set $\mathcal{B} := \left\{\phi_{i_1} \otimes \cdots \otimes \phi_{i_k} : i_1, \ldots, i_k \in \{1, \ldots, d\}\right\}$ is a base for $\mathcal{T}^k(V)$.*

Proof. The proof is similar to the proof of Theorem 18.15. (Exercise 18-22). ■

Both the area of a parallelogram as well as the volume of a three dimensional parallelepiped have multiple summands. Hence, the volume function for parallelepipeds cannot just be a simple tensor product. A key property of both geometric formulas is that if we switch any two vectors, the sign of the result changes. This observation leads us to alternating tensors.

Definition 18.17 *Let V be a vector space. A k-tensor $\omega \in \mathcal{T}^k(V)$ is called* **alternating** *iff for all $1 \le i < j \le k$ and all $v_1, \ldots, v_k \in V$ we have*

$$\begin{aligned}&\omega[v_1, \ldots, v_{i-1}, v_i, v_{i+1}, \ldots, v_{j-1}, v_j, v_{j+1}, \ldots, v_k] \\ &= -\omega[v_1, \ldots, v_{i-1}, v_j, v_{i+1}, \ldots, v_{j-1}, v_i, v_{j+1}, \ldots, v_k].\end{aligned}$$

The set of alternating k-tensors on V is denoted $\Lambda^k(V)$.

Exercise 18-23 shows that the set of alternating k-tensors forms a vector space. Because they have only one argument, linear functions $\phi : V \to \mathbb{R}$ are alternating 1-tensors. (The universal quantification in the definition of an "alternating 1-tensor" is over the empty set, so it is vacuously true.) For any tensor, we can define a corresponding alternating tensor. The idea is to sum all terms that can be obtained by permuting the entries in such a way that a transposition of two vectors will switch a positive summand with a negative summand.

Definition 18.18 *For $k \in \mathbb{N}$ we let S_k be the set of all permutations on $\{1, \ldots, k\}$. A* **transposition** *is a permutation τ that fixes all but two elements of the set. The* **sign** $\text{sgn}(\sigma)$ *of a permutation $\sigma \in S_k$ is 1 if σ is a composition of an even number of transpositions and it is -1 if σ is a composition of an odd number of transpositions. (Exercise 18-24 shows that the sign is well-defined.) If $T \in \mathcal{T}^k(V)$ we define*

$$\text{Alt}(T)[v_1, \ldots, v_k] := \frac{1}{k!} \sum_{\sigma \in S_k} \text{sgn}(\sigma) T[v_{\sigma(1)}, \ldots, v_{\sigma(k)}].$$

Theorem 18.19 shows that Alt(T) is an alternating tensor, thus explaining the notation. The proof also shows why we need the coefficient $\frac{1}{k!}$. Without it, Alt($\cdot$) applied to an alternating tensor would multiply the alternating tensor with a factor that is not equal to 1.

Theorem 18.19 *Let V be a vector space and let $k \in \mathbb{N}$. For all $T \in \mathcal{T}^k(V)$ we have* $\text{Alt}(T) \in \Lambda^k(V)$ *and for all $\omega \in \Lambda^k(V)$ we have* $\text{Alt}(\omega) = \omega$.

Proof. First note that a tensor $\omega \in \mathcal{T}^k(V)$ is alternating iff for all transpositions $\tau \in S_k$ and all $v_1, \ldots, v_k \in V$ we have $\omega[v_1, \ldots, v_k] = -\omega[v_{\tau(1)}, \ldots, v_{\tau(k)}]$. Moreover, note that for all transpositions $\tau \in S_k$ we have $\tau \circ \tau = id_{\{1,\ldots,k\}}$.

Now let $T \in \mathcal{T}^k(V)$ and let $v_1, \ldots, v_k \in V$. Then for all transpositions $\tau \in S_k$ we obtain the following.

$$\text{Alt}(T)[v_{\tau(1)}, \ldots, v_{\tau(k)}] = \frac{1}{k!} \sum_{\sigma \in S_k} \text{sgn}(\sigma) T[v_{\sigma(\tau(1))}, \ldots, v_{\sigma(\tau(k))}]$$

$$= \frac{1}{k!}\sum_{\sigma\in S_k} -\text{sgn}(\sigma\circ\tau)T[v_{\sigma\circ\tau(1)},\ldots,v_{\sigma\circ\tau(k)}]$$

Let $\sigma' = \sigma\circ\tau$ and note that the function $\sigma \mapsto \sigma\circ\tau$ bijective from S_k to S_k.

$$= -\frac{1}{k!}\sum_{\sigma'\in S_k} \text{sgn}\,(\sigma')\,T[v_{\sigma'(1)},\ldots,v_{\sigma'(k)}]$$

$$= -\text{Alt}(T)[v_1,\ldots,v_k].$$

Therefore Alt(T) is alternating.

For the second part, note that if $\omega \in \Lambda^k(V)$ and $\sigma \in S_k$, then for all $v_1,\ldots,v_k \in V$ we have $\omega[v_1,\ldots,v_k] = \text{sgn}(\sigma)\omega[v_{\sigma(1)},\ldots,v_{\sigma(k)}]$ (Exercise 18-25). Now

$$\text{Alt}(\omega)[v_1,\ldots,v_k] = \frac{1}{k!}\sum_{\sigma\in S_k}\text{sgn}(\sigma)\omega[v_{\sigma(1)},\ldots,v_{\sigma(k)}]$$

$$= \frac{1}{k!}\sum_{\sigma\in S_k}\omega[v_1,\ldots,v_k]$$

There are $k!$ permutations on a k-element set.

$$= \omega[v_1,\ldots,v_k].$$

■

It is also easy to see that Alt : $T^k(V) \to \Lambda^k(V)$ is linear (see Exercise 18-26). With Alt$(\cdot)$ we can now multiply alternating tensors in a way that produces a new alternating tensor.

Definition 18.20 *Let V be a vector space. For $\omega \in \Lambda^k(V)$ and $\eta \in \Lambda^l(V)$ define the* **wedge product** $\omega\wedge\eta$ *to be* $\omega\wedge\eta := \frac{(k+l)!}{k!l!}\text{Alt}(\omega\otimes\eta)$.

Example 18.21 The wedge product is the key to obtaining volume functions. Let $\{e_1,\ldots,e_d\}$ be the standard base in $\mathbb{R}^d$ and let $\{\pi_1,\ldots,\pi_d\}$ be the corresponding dual base.

1. For all $a,b \in \mathbb{R}^d$, we have $\pi_1\wedge\pi_2(a,b) = \pi_1(a)\pi_2(b) - \pi_1(b)\pi_2(a)$. That is, the wedge product of π_1 and π_2 maps any two vectors $a,b \in \mathbb{R}^d$ to the oriented area of the parallelogram spanned by the vectors made up of the first two components of a and b. More generally, the wedge product of π_i and π_j maps $a,b \in \mathbb{R}^d$ to the area of the parallelogram spanned by the vectors made up of the ith and j^{th} components of a and b.

2. For all $a,b,c \in \mathbb{R}^d$, we have

$$\big(\pi_1\wedge(\pi_2\wedge\pi_3)\big)(a,b,c) = \left\langle \begin{pmatrix}\pi_1(a)\\ \pi_2(a)\\ \pi_3(a)\end{pmatrix}, \begin{pmatrix}\pi_1(b)\\ \pi_2(b)\\ \pi_3(b)\end{pmatrix} \times \begin{pmatrix}\pi_1(c)\\ \pi_2(c)\\ \pi_3(c)\end{pmatrix}\right\rangle.$$

(The lengthy componentwise proof can be produced in Exercise 18-27.)

That is, the wedge product of π_1, π_2, and π_3 maps three vectors $a, b, c \in \mathbb{R}^d$ to the oriented volume of the parallelepiped spanned by the vectors made up of the first three components of a, b and c. Similarly, the wedge product of π_i, π_j and π_k maps any three vectors $a, b, c \in \mathbb{R}^d$ to the volume of the parallelepiped spanned by the vectors made up of the ith, j^{th} and k^{th} components of a, b and c.

The above indicates that the wedge product of k vectors of the dual base of the standard base $e_1, \ldots, e_d$ should map any k-tuple of vectors to the lower dimensional volume of the projection of the parallelepiped spanned by the vectors into the right lower dimensional subspace (see Exercise 18-28). Moreover, part 2 indicates that the wedge product of "projection volume functions" gives a higher dimensional volume function. Therefore the wedge product formalism should be the version of the familiar "base times height" formula for boxes that holds for parallelepipeds. □

With the wedge product accepted as the right formalism to compute volumes of parallelepipeds we need to investigate its properties. Bilinearity of the wedge product is established in Exercise 18-29. Moreover, the wedge product allows a base representation of alternating tensors on finite dimensional spaces similar to Theorem 18.16. For this base representation, we first need to establish associativity and the behavior when two factors are transposed. The first two parts of the following lemma are motivated by the proof of the wedge product's associativity.

Lemma 18.22 *Let V be a vector space.*

1. *If $S \in \mathcal{T}^k(V)$, $T \in \mathcal{T}^l(V)$ and $\text{Alt}(S) = 0$, then*

   $\text{Alt}(S \otimes T) = \text{Alt}(T \otimes S) = 0.$

2. *If $S \in \mathcal{T}^k(V)$, $T \in \mathcal{T}^l(V)$ and $U \in \mathcal{T}^m(V)$, then*

   $\text{Alt}\big(\text{Alt}(S \otimes T) \otimes U\big) = \text{Alt}(S \otimes T \otimes U) = \text{Alt}\big(S \otimes \text{Alt}(T \otimes U)\big).$

3. *If $\omega \in \Lambda^k(V)$, $\eta \in \Lambda^l(V)$ and $\theta \in \Lambda^m(V)$, then*

$$(\omega \wedge \eta) \wedge \theta = \frac{(k+l+m)!}{k!l!m!} \text{Alt}(\omega \otimes \eta \otimes \theta) = \omega \wedge (\eta \wedge \theta).$$

   In particular this means that the wedge product is **associative**.

4. *The wedge product is "***anti-commutative***." If $\omega \in \Lambda^k(V)$ and $\eta \in \Lambda^l(V)$, then $\omega \wedge \eta = (-1)^{kl} \eta \wedge \omega$.*

5. *If k is odd and $\phi \in \Lambda^k(V)$, then $\phi \wedge \phi = 0$.*

Proof. For part 1, we simply compute $\text{Alt}(S \otimes T)$ and represent the permutations $\sigma \in S_{k+l}$ as the composition of three permutations. The first sorts the elements of $\{1, \ldots, k+l\}$ into a set of size k and a set of size l, so that in each set the elements are listed in increasing order and the sets are listed as two consecutive blocks. The remaining two permutations then act on these sets.

$$\begin{aligned}
&\mathrm{Alt}(S\otimes T)[v_1,\ldots,v_{k+l}]\\
&= \frac{1}{(k+l)!}\sum_{\sigma\in S_{k+l}}\mathrm{sgn}(\sigma)S\otimes T[v_{\sigma(1)},\ldots,v_{\sigma(k+l)}]\\
&= \frac{1}{(k+l)!}\sum_{\sigma\in S_{k+l}}\mathrm{sgn}(\sigma)S[v_{\sigma(1)},\ldots,v_{\sigma(k)}]T[v_{\sigma(k+1)},\ldots,v_{\sigma(k+l)}]
\end{aligned}$$

For any U, W as indicated in the sum below, let $\delta_{U,W}$ be the permutation that maps $u_1\mapsto 1,\ldots, u_k\mapsto k$, $w_1\mapsto k+1$, $\ldots, w_l\mapsto k+l$. Then for any permutation $\sigma\in S_{k+l}$ there are unique sets U, W as indicated and permutations α' and β' that are constant on $\{k+1,\ldots,k+l\}$ and $\{1,\ldots,k\}$, respectively, so that $\sigma=\alpha'\circ\beta'\circ\delta_{U,W}$.

$$\begin{aligned}
&= \frac{1}{(k+l)!}\sum_{\substack{U\cup W=\{1,\ldots,k+l\}\\ U\cap W=\emptyset\\ U=\{u_1<\cdots<u_k\}\\ W=\{w_1<\cdots<w_l\}}}\sum_{\alpha\in S_k}\sum_{\beta\in S_l}\mathrm{sgn}(\alpha)\mathrm{sgn}(\beta)\mathrm{sgn}(\delta_{U,W})\times\\
&\qquad\qquad \times S\left[v_{u_{\alpha(1)}},\ldots,v_{u_{\alpha(k)}}\right]T\left[v_{w_{\beta(1)}},\ldots,v_{w_{\beta(l)}}\right]\\
&= \frac{1}{(k+l)!}\sum_{\substack{U\cup W=\{1,\ldots,k+l\}\\ U\cap W=\emptyset\\ U=\{u_1<\cdots<u_k\}\\ W=\{w_1<\cdots<w_l\}}}\mathrm{sgn}(\delta_{U,W})\sum_{\beta\in S_l}\mathrm{sgn}(\beta)T\left[v_{w_{\beta(1)}},\ldots,v_{w_{\beta(l)}}\right]k!\times\\
&\qquad\qquad \times\underbrace{\frac{1}{k!}\sum_{\alpha\in S_k}\mathrm{sgn}(\alpha)S\left[v_{u_{\alpha(1)}},\ldots,v_{u_{\alpha(k)}}\right]}_{=\mathrm{Alt}(S)\left[v_{u_1},\ldots,v_{u_k}\right]=0}\\
&= 0.
\end{aligned}$$

Part 2 is a straightforward application of the linearity of Alt(·), the multilinearity of tensor products and part 1.

$$\begin{aligned}
&\mathrm{Alt}\big(\mathrm{Alt}(S\otimes T)\otimes U\big)-\mathrm{Alt}(S\otimes T\otimes U)\\
&\quad= \mathrm{Alt}\big(\mathrm{Alt}(S\otimes T)\otimes U-(S\otimes T)\otimes U\big)\\
&\quad= \mathrm{Alt}\big([\mathrm{Alt}(S\otimes T)-(S\otimes T)]\otimes U\big)
\end{aligned}$$

Now use part 1 and that $\mathrm{Alt}\big[\mathrm{Alt}(S\otimes T)-(S\otimes T)\big]=0$.

$$= 0,$$

and the other equality is proved similarly. Part 3 now follows from part 2.

$$(\omega\wedge\eta)\wedge\theta = \left[\frac{(k+l)!}{k!l!}\mathrm{Alt}(\omega\otimes\eta)\right]\wedge\theta$$

$$
\begin{aligned}
&= \frac{(k+l+m)!}{(k+l)!m!}\text{Alt}\left(\left[\frac{(k+l)!}{k!l!}\text{Alt}(\omega\otimes\eta)\right]\otimes\theta\right)\\
&= \frac{(k+l+m)!}{k!l!m!}\text{Alt}\big(\text{Alt}(\omega\otimes\eta)\otimes\theta\big)\\
&= \frac{(k+l+m)!}{k!l!m!}\text{Alt}\,(\omega\otimes\eta\otimes\theta),
\end{aligned}
$$

and the other equality is proved similarly.

For part 4, represent $\omega \wedge \eta$ and $\eta \wedge \omega$ similar to the representation for part 1. Then note that $\delta_{U,W} \circ \delta_{W,U}^{-1}$, the permutation that transposes the "blocks" $\{1, \ldots, k\}$ and $\{k+1, \ldots, k+l\}$ can be represented as a composition of kl transpositions by going left to right in $\{k+1, \ldots, k+l\}$, transposing each element k times with its current predecessor. This establishes the claimed equality.

Part 5 immediately follows from part 4. ∎

With associativity of the wedge product established, we no longer need to include parentheses in the wedge multiplication of three or more alternating tensors.

Theorem 18.23 *Let V be a finite dimensional vector space with base $\{v_1, \ldots, v_d\}$, and let $\{\phi_1, \ldots, \phi_d\}$ be the associated dual base as in Theorem 18.15. Then the set $\mathcal{A} := \left\{\phi_{i_1} \wedge \cdots \wedge \phi_{i_k} : 1 \le i_1 < \cdots < i_k \le d\right\}$ is a base for $\Lambda^k(V)$.*

Proof. By Theorem 18.16, every $\phi \in \mathcal{T}^k(V)$ is a linear combination of tensor products $\phi_{i_1} \otimes \cdots \otimes \phi_{i_k}$. The function $\text{Alt} : \mathcal{T}^k(V) \to \Lambda^k(V)$ is surjective by Theorem 18.19 and by Exercise 18-26 it is linear. Therefore every $\omega \in \Lambda^k(V)$ is a linear combination of tensors $\text{Alt}(\phi_{i_1} \otimes \cdots \otimes \phi_{i_k})$. By associativity of the wedge product, $\text{Alt}(\phi_{i_1} \otimes \cdots \otimes \phi_{i_k})$ is a multiple of $\phi_{i_1} \wedge \cdots \wedge \phi_{i_k}$. If any two indices i_j are equal, then by part 5 of Lemma 18.22 $\phi_{i_j} \wedge \phi_{i_j} = 0$. Moreover, for any permutation $\sigma \in S_k$ by part 4 of Lemma 18.22 we infer that $\phi_{i_1} \wedge \cdots \wedge \phi_{i_k} \in \left\{\pm\, \phi_{\sigma(i_1)} \wedge \cdots \wedge \phi_{\sigma(i_k)}\right\}$. This means that every $\omega \in \Lambda^k(V)$ is a linear combination of tensors $\phi_{i_1} \wedge \cdots \wedge \phi_{i_k}$ with $i_1 < i_2 < \cdots < i_k$.

To see that these tensors are linearly independent, let the numbers $\alpha_{i_1\ldots i_k}$ be so that $\sum_{1\le i_1 < \cdots < i_k \le d} \alpha_{i_1\ldots i_k}\phi_{i_1} \wedge \cdots \wedge \phi_{i_k} = 0$. For fixed $1 \le j_1 < \cdots < j_k \le d$ note that

$$
\begin{aligned}
0 &= \sum_{1\le i_1<\cdots<i_k\le d} \alpha_{i_1\ldots i_k}\phi_{i_1} \wedge \cdots \wedge \phi_{i_k}[v_{j_1}, \ldots, v_{j_k}]\\
&= \alpha_{j_1\ldots j_k}\phi_{j_1} \wedge \cdots \wedge \phi_{j_k}[v_{j_1}, \ldots, v_{j_k}] = \alpha_{j_1\ldots j_k},
\end{aligned}
$$

which proves that the tensors $\phi_{i_1} \wedge \cdots \wedge \phi_{i_k}$ with $i_1 < i_2 < \cdots < i_k$ are linearly independent. ∎

In particular, Theorem 18.23 shows that $\Lambda^d(V)$ is one dimensional. Therefore all alternating d-tensors on d-dimensional space are multiples of *one* tensor. Geometrically this means that on d-dimensional space there is only one volume function for d-dimensional parallelepipeds, which is the only sensible possibility.

Definition 18.24 *The unique alternating d-tensor ω on $\mathbb{R}^d$ with $\omega(e_1,\ldots,e_d)=1$ is called the d-dimensional* **determinant**. *It is denoted* $\det(v_1,\ldots,v_d):=\omega(v_1,\ldots,v_d)$. *If $A=(a_{ij})_{\substack{i=1,\ldots,d\\ j=1,\ldots,d}}$ is a $d\times d$-matrix, we define the determinant of the matrix as*

$$\det(A):=\det\left((a_{i1})_{i=1,\ldots,d},\ldots,(a_{id})_{i=1,\ldots,d}\right):=\det\left(\sum_{i=1}^d a_{i1}e_i,\ldots,\sum_{i=1}^d a_{id}e_i\right).$$

The connection between determinants and volumes, which retroactively validates our geometric motivation, is made in Exercise 18-48. To be able to prove the results needed to get there we conclude this section with a few key properties of determinants.

Proposition 18.25 *Let V be a vector space with base $\{v_1,\ldots,v_d\}$ and let $\omega\in\Lambda^d(V)$. If $w_j=\sum_{i=1}^d a_{ij}v_i$, then* $\omega(w_1,\ldots,w_d)=\det\left((a_{ij})_{\substack{i=1,\ldots,d\\ j=1,\ldots,d}}\right)\omega(v_1,\ldots,v_d)$.

Proof. Because $\{v_1,\ldots,v_d\}$ is linearly independent, $\omega(v_1,\ldots,v_d)\neq 0$ (see Exercise 18-30) and so for $A=(a_{ij})_{\substack{i=1,\ldots,d\\ j=1,\ldots,d}}$ we can define

$$\eta(A):=\eta\left((a_{i1})_{i=1,\ldots,d},\ldots,(a_{id})_{i=1,\ldots,d}\right):=\frac{\omega\left(\sum_{i=1}^d a_{i1}v_i,\ldots,\sum_{i=1}^d a_{id}v_i\right)}{\omega(v_1,\ldots,v_d)}.$$

It is easy to see that η is d-linear in the columns of A, that the transposition of any two columns of A changes the sign of $\eta(A)$ and that $\eta(I)=1$ when I is the identity matrix. This means that if we represent each vector $w=\sum_{i=1}^d w_ie_i$ in $\mathbb{R}^d$ with the column vector $\begin{pmatrix} w_1\\ \vdots\\ w_d\end{pmatrix}$, then $\eta\in\Lambda^d\left(\mathbb{R}^d\right)$ and $\eta(e_1,\ldots,e_d)=1$. But this means that η is the determinant on $\mathbb{R}^d$, that is,

$$\eta\left((a_{i1})_{i=1,\ldots,d},\ldots,(a_{id})_{i=1,\ldots,d}\right)=\det\left(\sum_{i=1}^d a_{i1}e_i,\ldots,\sum_{i=1}^d a_{id}e_i\right)=\det(A).$$

The result now follows from applying the above to vectors $w_j=\sum_{i=1}^d a_{ij}v_i$. ■

Proposition 18.26 *Let V be a d-dimensional vector space and let $\omega\in\Lambda^d(V)$. If $\{w_1,\ldots,w_d\}$ is not linearly independent, then $\omega(w_1,\ldots,w_d)=0$.*

Proof. Without loss of generality assume that there are $a_1,\ldots,a_{d-1}\in\mathbb{R}$ so that

$w_d = \sum_{i=1}^{d-1} a_i w_i$. Then

$$\begin{aligned}\omega(w_1, \ldots, w_{d-1}, w_d) &= \omega\left(w_1, \ldots, w_{d-1}, \sum_{i=1}^{d-1} a_i w_i\right) \\ &= \sum_{i=1}^{d-1} a_i \omega(w_1, \ldots, w_{d-1}, w_i) = 0.\end{aligned}$$

■

Corollary 18.27 *Let A, B be real $d \times d$-matrices. Then* $\det(AB) = \det(A)\det(B)$.

Proof. If the columns of A are linearly independent, we prove the result similar to the proof of Proposition 18.25. The function $\eta(B) := \frac{\det(AB)}{\det(A)}$ is linear in each column of B, transposing two columns transposes two columns of AB and thus it changes the sign of η, and $\eta(I) = 1$, where I is the identity matrix. Hence, we infer $\eta(B) = \det(B)$ in this case. If the columns of A are not linearly independent, then both sides of the equation are zero. ■

When arguing in general, we usually work with linear functions, not matrices. Therefore, it is sensible to define the determinant of a linear function.

Definition 18.28 *Let $L : \mathbb{R}^d \to \mathbb{R}^d$ be a linear function and let A be its matrix representation with respect to the standard base. Then we define the* **determinant** *of L as* $\det(L) := \det(A)$.

Exercise 18-31 shows that the determinant of a linear function L does not depend on the specific base chosen, as long as we use the same base to represent arguments and images. Finally, because Exercise 18-48 shows that the determinant defines volumes in $\mathbb{R}^d$, we want to define volumes on more general spaces. As long as the space carries an inner product, this is sensible. With an inner product we can define orthonormal bases and cubes whose sides are defined by the vectors of an orthonormal base should have volume 1. To make a volume definition with an alternating d-tensor possible, we must assure that these tensors are constant on the orthonormal bases. The next result shows that this is almost the case.

Proposition 18.29 *Let V be a d-dimensional inner product space and let $\omega \in \Lambda^d(V)$. Then for any two orthonormal bases $\{v_1, \ldots, v_d\}$ and $\{w_1, \ldots, w_d\}$ of V the equation $|\omega(v_1, \ldots, v_d)| = |\omega(w_1, \ldots, w_d)|$ holds.*

Proof. For $i, j = 1, \ldots, d$ let $a_{ij} \in \mathbb{R}$ be so that $w_j = \sum_{i=1}^{d} a_{ij} v_i$. Then for all $j, k \in \{1, \ldots, d\}$ we have $\langle w_j, w_k \rangle = \left\langle \sum_{i=1}^{d} a_{ij} v_i, \sum_{l=1}^{d} a_{lk} v_l \right\rangle = \sum_{i=1}^{d} a_{ij} a_{ik}$. Hence, if A

is the matrix with entries a_{ij}, then the product AA^T of A and its transpose is the identity I. By Corollary 18.27 and Exercise 18-34 we infer that the determinant of A is 1 or -1 because $\left(\det(A)\right)^2 = \det(A)\det\left(A^T\right) = \det\left(AA^T\right) = \det(I) = 1$. Therefore by Proposition 18.25 we conclude

$$\left|\omega(w_1, \ldots, w_d)\right| = \left|\det(A)\omega(v_1, \ldots, v_d)\right| = \left|\omega(v_1, \ldots, v_d)\right|. \qquad \blacksquare$$

With d-forms evaluating to the same absolute value when applied to orthonormal bases, it makes sense to pick one of the two d-forms that assign values of ± 1 to orthonormal bases and designating it as the **volume element**. We will see in Section 19.5 how to choose one of these two to obtain a volume element for very general surfaces.

Exercises

18-20. Let V be a vector space. Prove that $\mathcal{T}^k(V)$ with pointwise addition and scalar multiplication is a vector space. That is, prove that

(a) If $\omega, \eta \in \mathcal{T}^k(V)$, then $\nu(v_1, \ldots, v_k) := \omega(v_1, \ldots, v_k) + \eta(v_1, \ldots, v_k)$ is a k-tensor on V.

(b) If $\omega \in \mathcal{T}^k(V)$ and $c \in \mathbb{R}$, then $\mu(v_1, \ldots, v_k) := c\omega(v_1, \ldots, v_k)$ is a k-tensor on V

18-21. Properties of the tensor product Let V be a vector space.

(a) Prove that for all tensors $S \in \mathcal{T}^k(V)$, $T \in \mathcal{T}^l(V)$ and $U \in \mathcal{T}^m(V)$ we have $(S \otimes T) \otimes U = S \otimes (T \otimes U)$.

(b) Prove that there are tensors S and T on $\mathbb{R}^2$ so that $S \otimes T \neq T \otimes S$

18-22. Prove Theorem 18.16.

18-23. Let V be a vector space and let $k \in \mathbb{N}$. Prove that $\Lambda^k(V)$ is a vector space.

18-24 Let $k \in \mathbb{N}$ Proceed as follows to prove that the sign of a permutation is well-defined.

(a) Prove that every $\sigma \in S_k$ can be represented as a composition of transpositions
Hint. Induction on k. In the induction step, transpose $\sigma(k)$ back to k.

(b) Prove that if σ can be represented as the composition of n transpositions and as the composition of $m > n$ transpositions, then $m - n$ is even
Hint. Again, induction on k

(c) Prove that the function sgn $S_k \to \{1, -1\}$ is well-defined

18-25 Let V be a vector space. Prove that if $\omega \in \Lambda^k(V)$ and $\sigma \in S_k$, then for all $v_1, \ldots, v_k \in V$ we have $\omega[v_1, \ldots, v_k] = \text{sgn}(\sigma)\omega[v_{\sigma(1)}, \ldots, v_{\sigma(k)}]$.
Hint. Induction on the number of transpositions that need to be composed to get σ

18-26 Prove that Alt : $\mathcal{T}^k(V) \to \Lambda^k(V)$ is linear, that is, for all tensors $S, T \in \mathcal{T}^k(V)$ and all scalars $\alpha \in \mathbb{R}$ we have $\text{Alt}(S + T) = \text{Alt}(S) + \text{Alt}(T)$ and $\text{Alt}(\alpha T) = \alpha\text{Alt}(T)$.

18-27 Prove the formula in part 2 of Example 18.21.

18-28 The wedge product $\pi_{i_1} \wedge \cdots \wedge \pi_{i_k}$ gives the k-dimensional volume of the projection of a parallelepiped into $\text{span}\{e_{i_1}, \ldots, e_{i_k}\}$.

(a) Let $a_1, \ldots, a_k \in \mathbb{R}^d$ be zero in all but the first k components. Prove that the wedge product $\pi_1 \wedge \cdots \wedge \pi_k(a_1, \ldots, a_k)$ gives the k-dimensional volume of the parallelepiped spanned by $a_1, \ldots, a_k$
Hint. Prove that $\pi_1 \wedge \cdots \wedge \pi_k\big|_{\{x \in \mathbb{R}^d : x_{k+1} = \cdots = x_d = 0\}^k}$ composed with the isomorphism between $\left(\mathbb{R}^k\right)^k$ and $\left\{x \in \mathbb{R}^d : x_{k+1} = \cdots = x_d = 0\right\}^k$ is equal to the k-dimensional determinant and use that the k-dimensional determinant is the volume function for parallelepipeds spanned by k vectors in k-dimensional space. (This fact will be established in Exercise 18-48 without using this exercise.)

(b) Let $a_1, \ldots, a_k \in \mathbb{R}^d$. Prove that wedge product $\pi_{i_1} \wedge \cdots \wedge \pi_{i_k}(a_1, \ldots, a_k)$ gives the k-dimensional volume of the projection of the parallelepiped spanned by $a_1, \ldots, a_k$ into $\text{span}\{e_{i_1}, \ldots, e_{i_k}\}$

18-29 Let V be a vector space. Prove that the wedge product is bilinear, that is, prove the following

(a) Prove that if $\omega_1, \omega_2 \in \Lambda^k(V)$ and $\eta \in \Lambda^l(V)$, then $(\omega_1 + \omega_2) \wedge \eta = \omega_1 \wedge \eta + \omega_2 \wedge \eta$.

(b) Prove that if $\omega \in \Lambda^k(V)$ and $\eta_1, \eta_2 \in \Lambda^l(V)$, then $\omega \wedge (\eta_1 + \eta_2) = \omega \wedge \eta_1 + \omega \wedge \eta_2$.

(c) Prove that if $\omega \in \Lambda^k(V)$, $\eta \in \Lambda^l(V)$ and $c \in \mathbb{R}$, then $(c\omega) \wedge \eta = c(\omega \wedge \eta) = \omega \wedge (c\eta)$

18-30 Let V be a d-dimensional vector space, let the set $\{v_1, \ldots, v_d\} \subseteq V$ be linearly independent and let $\omega \in \Lambda^d(V) \setminus \{0\}$. Prove that $\omega(v_1, \ldots, v_d) \neq 0$

Hint It suffices to prove the result for $\mathbb{R}^d$. Suppose for a contradiction that $\omega(v_1, \ldots, v_d) = 0$ and use linear combinations to derive that $\omega(e_1, \ldots, e_d) = 0$.

18-31 The determinant of a linear function is independent of the base with respect to which it is represented Let $L : \mathbb{R}^d \to \mathbb{R}^d$ be a linear function and let B be its matrix representation with respect to the base $\{v_1, \ldots, v_d\}$. Prove that $\det(B) = \det(L)$.

18-32 **Summation formula for the determinant.** Let $d \in \mathbb{N}$. Prove that the **determinant** of any real $d \times d$ matrix $A = (a_{ij})_{\substack{i=1,\ldots,d \\ j=1,\ldots,d}}$ equals $\det(A) = \sum_{\sigma \in S_d} \text{sgn}(\sigma) a_{1\sigma(1)} \cdots a_{d\sigma(d)}$.

Hint. Prove that the determinant is the image under Alt( ) of the tensor that multiplies $d!$ with the diagonal entries.

18-33 **Row Expansion of the Determinant.** Let $A \in M(d \times d, \mathbb{R})$. We claim that for any number $i \in \{1, \ldots, d\}$ we have $\det(A) = \sum_{j=1}^{d} (-1)^{i+j} a_{ij} \det(A_{ij})$, where $A_{ij} \in M(d-1 \times d-1, \mathbb{R})$ is obtained from A by erasing the ith row and the jth column

(a) Prove the claim using Exercise 18-32.

(b) Prove the claim by proving that the formula defines an alternating d-tensor that is equal to 1 for the identity matrix

18-34 The **transpose** of a matrix $A = (a_{ij})_{\substack{i=1,\ldots,d \\ j=1,\ldots,d}}$ is $A^T := (a_{ji})_{\substack{i=1,\ldots,d \\ j=1,\ldots,d}}$ Prove that for all real $d \times d$ matrices A we have $\det(A) = \det\left(A^T\right)$

Hint. Exercise 18-32

18-35. Determinants of certain linear functions

(a) Let $c_1, \ldots, c_d \in \mathbb{R}$ and let $D : \mathbb{R}^d \to \mathbb{R}^d$ be the **diagonal operator** $D(x_1, \ldots, x_d) := (c_1 x_1, \ldots, c_d x_d)$ Prove that $\det(D) = c_1 \cdots c_d$.

(b) Let $a \in \mathbb{R}$, let $i, j \in \{1, \ldots, d\}$ be distinct and let $A : \mathbb{R}^d \to \mathbb{R}^d$ be the **row addition operator** $A(x_1, \ldots, x_d) := (x_1, \ldots, x_{i-1}, x_i + a x_j, x_{i+1}, \ldots, x_d)$. Prove that $\det(A) = 1$.

(c) Let $i, j \in \{1, \ldots, d\}$ with $i < j$ and let $T : \mathbb{R}^d \to \mathbb{R}^d$ be the **row transposition operator** $T(x_1, \ldots, x_d) := (x_1, \ldots, x_{i-1}, x_j, x_{i+1}, \ldots, x_{j-1}, x_i, x_{j+1}, \ldots, x_d)$ Prove that $\det(T) = -1$

18.4 Multidimensional Substitution

The d-dimensional substitution formula (see Theorem 18.37) is a surprisingly deep result. Its proof draws on topology, differential calculus and measure theory as well as on linear algebra. However, the underlying idea is fairly simple. Consider the image of a small cube under a differentiable function g. If the cube is sufficiently small, then on the cube the function g is approximately equal to its derivative L. The image of the cube under L is a parallelepiped. This means that the g-images of our most elementary volume elements (cubes) are approximately parallelepipeds.

It can be shown geometrically that the absolute value of the determinant of a 3×3 matrix is the volume of the parallelepiped spanned by the column vectors. So, in three dimensions, the volume of the image of the cube spanned by the standard unit vectors under a linear function L is the absolute value of the determinant of L. Because translations do not affect the volume (see Lemma 18.31) and linear factors in the columns can be factored out of the determinant, we expect that the absolute value of the determinant of a linear function L is exactly the factor by which the volume of a cube is multiplied to obtain the volume of its image under L. Lemma 18.35 shows that this is the case for arbitrary sets in arbitrary finite dimensions.

Lemma 18.36 then shows that for continuously differentiable functions the absolute value of the determinant of the derivative is the local volume distortion factor and Theorem 18.37 pulls everything together to establish the Multidimensional Substitution Formula. Throughout this section we work with the uniform norm $\|\cdot\|_\infty$ on $\mathbb{R}^d$ because the "balls" in this norm are cubes.

Lemma 18.30 *Let $\Omega_1, \Omega_2 \subseteq \mathbb{R}^d$ be open subsets of $\mathbb{R}^d$ and let $g : \Omega_1 \to \Omega_2$ be a continuously differentiable function. If $S \subseteq \Omega_1$ is Lebesgue measurable, then $g[S]$ is Lebesgue measurable, too.*

Proof. We first prove that g maps null sets to null sets. Let $N \subset \Omega_1$ be a null set. For $n \in \mathbb{N}$, let $K_n := \left\{ x \in \Omega_1 : \operatorname{dist}\left(x, \mathbb{R}^d \setminus \Omega_1\right) \geq \frac{1}{n} \text{ and } \|x\| \leq n \right\}$. Then all K_n are closed and bounded and $\Omega_1 = \bigcup_{n=1}^{\infty} K_n$. It is enough to prove that each $g[N \cap K_n]$ is a null set. Because K_{2n} is compact and Dg is continuous, we can find an upper bound $L > 0$ for $\|Dg\|$ on K_{2n}, with all distances measured with the $\|\cdot\|_\infty$-norm on $\mathbb{R}^d$. Let $\varepsilon > 0$ and let $\left\{\widetilde{B}_j\right\}_{j=1}^{\infty}$ be a family of open boxes such that $N \cap K_n \subseteq \bigcup_{j=1}^{\infty} \widetilde{B}_j$ and $\sum_{j=1}^{\infty} \left|\widetilde{B}_j\right| < \frac{\varepsilon}{2L^d}$. From these boxes, construct a family of open cubes $\left\{B_j\right\}_{j=1}^{\infty}$ so that $N \cap K_n \subseteq \bigcup_{j=1}^{\infty} B_j$, $\sum_{j=1}^{\infty} |B_j| < \frac{\varepsilon}{L^d}$ and the diameter of each B_j is at most $\frac{1}{2n}$ (see Exercise 18-36a). Because we can discard any cubes that do not intersect K_n we can assume that all B_j intersect K_n. Then $\bigcup_{j=1}^{\infty} B_j \subseteq K_{2n}$. By Theorem 17.33, this means that g is Lipschitz continuous on each B_j and the Lipschitz constant is L when distances are measured with the $\|\cdot\|_\infty$-norm on $\mathbb{R}^d$. Hence, each $g[B_j]$ is contained in a cube

C_j with $\lambda(C_j) \leq L^d \lambda(B_j)$ (see Exercise 18-36b). Therefore, $g[N \cap K_n] \subseteq \bigcup_{j=1}^{\infty} C_j$ and $\lambda\big(g[N \cap K_n]\big) \leq \sum_{j=1}^{\infty} \lambda(C_j) \leq \sum_{j=1}^{\infty} L^d \lambda(B_j) < L^d \frac{\varepsilon}{L^d} = \varepsilon$. Because ε was arbitrary, we obtain $\lambda\big(g[N \cap K_n]\big) = 0$ and we have proved that g maps null sets to null sets.

Now let $S \subseteq \Omega_1$ be Lebesgue measurable. Then by Theorem 18.4 there is a sequence $\{F_i\}_{i=1}^{\infty}$ of closed sets and a sequence $\{G_i\}_{i=1}^{\infty}$ of open sets so that for all $i \in \mathbb{N}$ we have $F_i \subseteq F_{i+1}$ and $G_i \supseteq G_{i+1}$ and so that with $F := \bigcup_{i=1}^{\infty} F_i$ and $G := \bigcap_{i=1}^{\infty} G_i$ we have $F \subseteq S \subseteq G \subseteq \Omega_1$ and $\lambda(G \setminus F) = 0$. Clearly, $g[F] \subseteq g[S] \subseteq g[G] \subseteq \Omega_2$ and by the above $\lambda\big(g[G] \setminus g[F]\big) \leq \lambda\big(g[G \setminus F]\big) = 0$.

For $i \in \mathbb{N}$, set $\widetilde{F}_i := F_i \cap K_i$. Then each $\widetilde{F}_i$ is closed and bounded, and hence compact and $\bigcup_{i=1}^{\infty} \widetilde{F}_i = \bigcup_{i=1}^{\infty} (F_i \cap K_i) = \bigcup_{i=1}^{\infty} F_i \cap \bigcup_{i=1}^{\infty} K_i = \bigcup_{i=1}^{\infty} F_i \cap \Omega_1 = \bigcup_{i=1}^{\infty} F_i$. By Theorem 16.62, all $g\left[\widetilde{F}_i\right]$ are compact, and hence closed, so $g[F]$ is the union of a sequence of closed sets $g\left[\widetilde{F}_i\right]$ with $g\left[\widetilde{F}_i\right] \subseteq g\left[\widetilde{F}_{i+1}\right]$ for all $i \in \mathbb{N}$. This means that $g[F]$ is Lebesgue measurable. Because $\lambda\big(g[S] \setminus g[F]\big) \leq \lambda\big(g[G] \setminus g[F]\big) = 0$ and null sets are Lebesgue measurable, this means that $g[S] \setminus g[F]$ is Lebesgue measurable, and hence $g[S]$ is Lebesgue measurable. ■

Exercise 18-54 shows that continuity alone is not enough to assure that the images of null sets are null sets. Unsurprisingly, translations do not affect the Lebesgue measure of a set, as the next result shows.

Lemma 18.31 *The effect of a translation on Lebesgue measure. For $S \subseteq \mathbb{R}^d$ and $x \in \mathbb{R}^d$, define $x + S := \left\{x + s \in \mathbb{R}^d : s \in S\right\}$. If S is Lebesgue measurable, then $x + S$ is Lebesgue measurable and $\lambda(x + S) = \lambda(S)$.*

Proof. Exercise 18-37. ■

The key to establishing Lemma 18.35 is to prove it for sufficiently basic linear functions from which arbitrary linear functions can be built. The first step are linear functions that stretch or shrink objects parallel to the coordinate axes.

Lemma 18.32 *The effect of a diagonal operator on Lebesgue measure. Let the set $S \subseteq \mathbb{R}^d$ be Lebesgue measurable, let $c_1, \ldots, c_d \in \mathbb{R} \setminus \{0\}$ and let the function $D : \mathbb{R}^d \to \mathbb{R}^d$ be defined by $D(x_1, \ldots, x_d) := (c_1 x_1, \ldots, c_d x_d)$. Then $D[S]$ is Lebesgue measurable with $\lambda\big(D[S]\big) = |c_1 \ldots c_d| \lambda(S) = \big|\det(D)\big| \lambda(S)$.*

Proof. The Lebesgue measurability of $D[S]$ follows from Lemma 18.30. Now let $\varepsilon > 0$. To compute $\lambda\big(D[S]\big)$, let $\{B_j\}_{j=1}^{\infty}$ be a family of open boxes so that $S \subseteq \bigcup_{j=1}^{\infty} B_j$

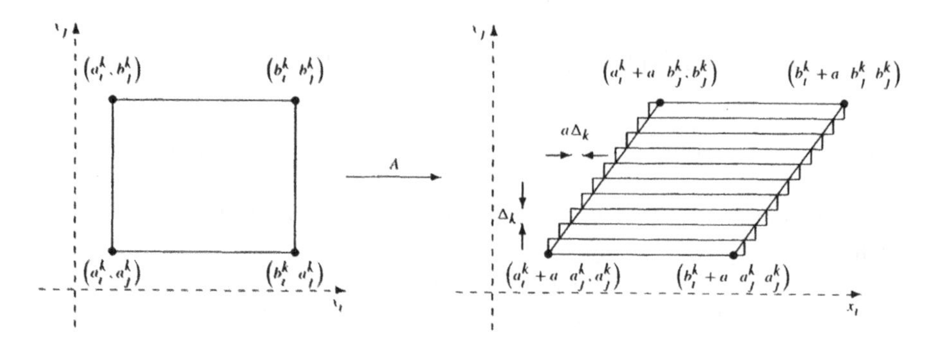

Figure 47: Visualization of the proof of Lemma 18.33. Maps as in Lemma 18.33 only affect two coordinates, x_i and x_j. In these coordinates, they enforce a uniform "shear" on the space. This shear turns rectangles into parallelograms without affecting the area. These parallelograms can be covered with rectangles whose sides are parallel to the coordinate axes.

and so that $\sum_{j=1}^{\infty} |B_j| < \lambda(S) + \frac{\varepsilon}{|c_1 \cdots c_n| + 1}$. Then $\{D[B_j]\}_{j=1}^{\infty}$ is a family of open boxes so that $D[S] \subseteq \bigcup_{j=1}^{\infty} D[B_j]$ and for each $j \in \mathbb{N}$ we have $|D[B_j]| = |c_1 \cdots c_n||B_j|$ (Exercise 18-38). This means that

$$\begin{aligned}\lambda(D[S]) &\leq \sum_{j=1}^{\infty} |D[B_j]| \leq \sum_{j=1}^{\infty} |c_1 \cdots c_n||B_j| \\ &\leq |c_1 \cdots c_n| \left(\lambda(S) + \frac{\varepsilon}{|c_1 \cdots c_n| + 1}\right) \leq |c_1 \cdots c_n|\lambda(S) + \varepsilon.\end{aligned}$$

Because $\varepsilon > 0$ was arbitrary we infer $\lambda(D[S]) \leq |c_1 \cdots c_d|\lambda(S)$. A similar result holds for $D^{-1}(x_1, \ldots, x_d) = \left(\frac{1}{c_1}x_1, \ldots, \frac{1}{c_d}x_d\right)$. For all Lebesgue measurable sets A the set $D^{-1}[A]$ is Lebesgue measurable and $\lambda\left(D^{-1}[A]\right) \leq \frac{1}{|c_1 \cdots c_d|}\lambda(A)$. So $\lambda(S) = \lambda\left(D^{-1}[D[S]]\right) \leq \frac{1}{|c_1 \cdots c_d|}\lambda(D[S])$, that is, $\lambda(D[S]) \geq |c_1 \cdots c_d|\lambda(S)$, which establishes $\lambda(D[S]) = |c_1 \cdots c_d|\lambda(S)$. Finally, it was shown in Exercise 18-35a that $\det(D) = c_1 \cdots c_d$. ■

The next step are functions that shear objects in one coordinate direction. Figure 47 shows the geometric idea for the proof of Lemma 18.33.

Lemma 18.33 *The effect of row addition on Lebesgue measure. Let $S \subseteq \mathbb{R}^d$ be a Lebesgue measurable set, let $a \in \mathbb{R}$, let $i, j \in \{1, \ldots, d\}$ be distinct and define $A : \mathbb{R}^d \to \mathbb{R}^d$ by $A(x_1, \ldots, x_d) := (x_1, \ldots, x_{i-1}, x_i + ax_j, x_{i+1}, \ldots, x_d)$. Then $A[S]$ is Lebesgue measurable with $\lambda(A[S]) = \lambda(S) = |\det(A)|\lambda(S)$.*

Proof. By Lemma 18.30 $A[S]$ is Lebesgue measurable. Without loss of generality assume $i < j$ and $a > 0$. Let $\varepsilon > 0$. To compute $\lambda(A[S])$, let $\{B_k\}_{k=1}^{\infty}$ be a family of open boxes so that $S \subseteq \bigcup_{k=1}^{\infty} B_k$ and $\sum_{k=1}^{\infty} |B_k| < \lambda(S) + \frac{\varepsilon}{2}$. For each k, let $\prod_{l=1}^{d}\left(a_l^k, b_l^k\right) := B_k$, let $n_k \in \mathbb{N}$ be so that $a\frac{\left(b_j^k - a_j^k\right)^2}{n_k}\prod_{l\neq i,j}\left(b_l^k - a_l^k\right) < \frac{\varepsilon}{2^{k+1}}$ and let $\Delta_k := \frac{b_j^k - a_j^k}{n_k}$. Then for every $x = (x_1, \ldots, x_d) \in B_k$ there is an $m \in \{1, \ldots, n_k\}$ so that $x_j \in \left[a_j^k + (m-1)\Delta_k, a_j^k + m\Delta_k\right]$. This means that the ith component of $A(x)$ satisfies

$$\begin{aligned} x_i + ax_j &\in \left[x_i + a\left(a_j^k + (m-1)\Delta_k\right), x_i + a\left(a_j^k + m\Delta_k\right)\right] \\ &\subseteq \left[a_i^k + a\left(a_j^k + (m-1)\Delta_k\right), b_i^k + a\left(a_j^k + m\Delta_k\right)\right]. \end{aligned}$$

Hence,

$$\begin{aligned} A[B_k] &\subseteq \bigcup_{m=1}^{n_k}\prod_{l=1}^{i-1}\left[a_l^k, b_l^k\right] \times \left[a_i^k + aa_j^k + (m-1)a\Delta_k, b_i^k + aa_j^k + ma\Delta_k\right] \\ &\quad \times \prod_{l=i+1}^{j-1}\left[a_l^k, b_l^k\right] \times \left[a_j^k + (m-1)\Delta_k, a_j^k + m\Delta_k\right] \times \prod_{l=j+1}^{d}\left[a_l^k, b_l^k\right] \end{aligned}$$

(see Figure 47), and therefore

$$\begin{aligned} \lambda(A[B_k]) &\leq \sum_{m=1}^{n_k}\left(b_i^k - a_i^k + a\Delta_k\right)\cdot \Delta_k \cdot \prod_{l\neq i,j}\left(b_l^k - a_l^k\right) \\ &= \sum_{m=1}^{n_k}\left(b_i^k - a_i^k\right)\Delta_k\prod_{l\neq i,j}\left(b_l^k - a_l^k\right) + a\Delta_k^2\sum_{m=1}^{n_k}\prod_{l\neq i,j}\left(b_l^k - a_l^k\right) \\ &= \lambda(B_k) + n_k a\frac{\left(b_j^k - a_j^k\right)^2}{n_k^2}\prod_{l\neq i,j}\left(b_l^k - a_l^k\right) \\ &< \lambda(B_k) + \frac{\varepsilon}{2^{k+1}}. \end{aligned}$$

We obtain

$$\lambda(A[S]) \leq \sum_{k=1}^{\infty}\lambda(A[B_k]) < \sum_{k=1}^{\infty}\left(\lambda(B_k) + \frac{\varepsilon}{2^{k+1}}\right) < \lambda(S) + \frac{\varepsilon}{2} + \frac{\varepsilon}{2} = \lambda(S) + \varepsilon.$$

Because $\varepsilon > 0$ was arbitrary we infer $\lambda(A[S]) \leq \lambda(S)$. We can obtain the same inequality for $A^{-1}(x_1, \ldots, x_d) = (x_1, \ldots, x_{i-1}, x_i - ax_j, x_{i+1}, \ldots, x_d)$. But this means $\lambda(S) = \lambda\left(A^{-1}[A[S]]\right) \leq \lambda\big(A[S]\big)$, and hence $\lambda(S) = \lambda\big(A[S]\big)$. Finally, it was shown in Exercise 18-35b that $\det(A) = 1$. ■

Lemma 18.34 *The effect of row transposition on Lebesgue measure. Let $S \subseteq \mathbb{R}^d$ be Lebesgue measurable, let $i, j \in \{1, \ldots, d\}$ with $i < j$ and let $T : \mathbb{R}^d \to \mathbb{R}^d$ be $T(x_1, \ldots, x_d) := (x_1, \ldots, x_{i-1}, x_j, x_{i+1}, \ldots, x_{j-1}, x_i, x_{j+1}, \ldots, x_d)$. Then $T[S]$ is Lebesgue measurable with $\lambda\big(T[S]\big) = \lambda(S) = \big|\det(T)\big|\lambda(S)$.*

Proof. Exercise 18-39. ■

Lemma 18.35 *The effect of a bijective linear operator on Lebesgue measure. Let the function $L : \mathbb{R}^d \to \mathbb{R}^d$ be linear and bijective and let $S \subseteq \mathbb{R}^d$ be Lebesgue measurable. Then $L[S]$ is Lebesgue measurable and $\lambda\big(L[S]\big) = \big|\det(L)\big|\lambda(S)$.*

Proof. The Gauss-Jordan algorithm from Theorem 17.23 shows that there is a diagonal operator D with nonzero diagonal entries and a sequence of row transposition and row addition operators $A_1, \ldots, A_n$ so that $L = A_n A_{n-1} \cdots A_1 D$. Because the determinant of a composition is the product of the determinants of the factors (see Corollary 18.27) we infer that $\big|\det(L)\big| = \left(\prod_{i=1}^{n} \big|\det(A_i)\big|\right) \big|\det(D)\big|$, and hence

$$\begin{aligned} \lambda\big(L[S]\big) &= \lambda\big(A_n A_{n-1} \cdots A_1 D[S]\big) = \big|\det(A_n)\big|\lambda\big(A_{n-1} \cdots A_1 D[S]\big) \\ &= \cdots = \left(\prod_{i=1}^{n} \big|\det(A_i)\big|\right) \big|\det(D)\big|\lambda(S) = \big|\det(L)\big|\lambda(S). \end{aligned}$$

■

Lemma 18.35 confirms our initial geometric idea for arbitrary dimensions. If a parallelepiped is spanned by vectors $v_1, \ldots, v_d$, then it is the image of the unit cube under the linear map whose matrix representation is the matrix A with columns $v_1, \ldots, v_d$. Lemma 18.35 shows that the volume of this parallelepiped is the determinant of A. (Details are left to Exercise 18-48.)

For the remaining results, we will work with **half-open cubes**, which are cubes of the form $\prod_{i=1}^{d} [a_i, b_i)$ with all $b_i - a_i$ being equal. The **radius** of a half-open cube is half its side length.

Lemma 18.36 *The effect of a diffeomorphism on Lebesgue measure. Let Ω_1 and Ω_2 be open subsets of $\mathbb{R}^d$, let $K \subseteq \Omega_1$ be compact and let $g : \Omega_1 \to \Omega_2$ be a continuously differentiable bijective function with $\det\big(Dg(x)\big) \neq 0$ for all $x \in \Omega_1$. Then for every $\varepsilon > 0$ there is a $\delta > 0$ so that for every half-open cube $B \subseteq K$ with center point x and with radius less than δ we have $\lambda\big(g[B]\big) - \big|\det(Dg(x))\big|\lambda(B) < \varepsilon\lambda(B)$.*

Proof. Because K is compact, the derivative Dg is uniformly continuous on K and the norm $\left\|Dg(x)^{-1}\right\|$ is uniformly bounded on K by an $M > 0$. Moreover, we can find a $\nu > 0$ such that $(1+\nu)^d - 1 < \dfrac{\varepsilon}{\max\left\{\left|\det Dg(x)\right| : x \in K\right\}}$. There is a $\delta > 0$ so that for all $x, y \in K$ with $\|y - x\|_\infty < \delta$ we have $\left\|Dg(y) - Dg(x)\right\| < \dfrac{\nu}{M}$. Now let B be a half-open cube of radius less than δ, contained in K and centered at x. Then for all $z \in B$ the following holds.

$$\left\|g(z) - \big(g(x) + Dg(x)[z - x]\big)\right\|_\infty$$

Apply the Mean Value Theorem (Theorem 17.33) to the function $g(\cdot) - Dg(x)[\cdot - x]$ with $a = x$ and $b = z$.

$$\begin{aligned}
&\le \sup\left\{\left\|Dg(x + t(z - x)) - Dg(x)\right\| : t \in [0, 1]\right\} \|z - x\|_\infty \\
&\le \frac{\nu}{M}\|z - x\|_\infty,
\end{aligned}$$

which implies

$$\begin{aligned}
&\left\|Dg(x)^{-1}\big[g(z) - g(x)\big]\right\|_\infty - \|z - x\|_\infty \\
&\le \left\|Dg(x)^{-1}\big[g(z) - g(x)\big] - [z - x]\right\|_\infty \\
&= \left\|Dg(x)^{-1}\big[g(z) - g(x)\big] - Dg(x)^{-1}\big[Dg(x)[z - x]\big]\right\|_\infty \\
&\le \left\|Dg(x)^{-1}\right\| \left\|g(z) - g(x) - Dg(x)[z - x]\right\|_\infty \\
&\le \left\|Dg(x)^{-1}\right\| \frac{\nu}{M}\|z - x\|_\infty \\
&= \nu\|z - x\|_\infty.
\end{aligned}$$

Now for $A \subseteq \mathbb{R}^d$, $y \in \mathbb{R}^d$ and $c > 0$ define the sets $A - y := (-y) + A$ and $cA := \{ca : a \in A\}$. The above shows that for any half-open cube $B \subseteq K$ with center x and radius less than δ we have $\left\|Dg(x)^{-1}\big[g(z) - g(x)\big]\right\|_\infty \le (1 + \nu)\|z - x\|_\infty$ for all $z \in B$, and hence $Dg(x)^{-1}\big[g[B] - g(x)\big] \subseteq (1 + \nu)(B - x)$, which means that $g[B] - g(x) \subseteq Dg(x)\big[(1 + \nu)(B - x)\big]$. But then

$$\begin{aligned}
&\lambda\big(g[B]\big) - \left|\det(Dg(x))\right|\lambda(B) \\
&= \lambda\big(g[B] - g(x)\big) - \left|\det(Dg(x))\right|\lambda(B) \\
&\le \lambda\left(Dg(x)\big[(1 + \nu)(B - x)\big]\right) - \left|\det(Dg(x))\right|\lambda(B) \\
&= \left|\det(Dg(x))\right|\lambda\big((1 + \nu)(B - x)\big) - \left|\det(Dg(x))\right|\lambda(B) \\
&= \left|\det(Dg(x))\right|(1 + \nu)^d\lambda(B) - \left|\det(Dg(x))\right|\lambda(B) \\
&= \Big((1 + \nu)^d - 1\Big)\left|\det(Dg(x))\right|\lambda(B) \\
&< \varepsilon\lambda(B),
\end{aligned}$$

which establishes the result. ■

Now we are ready to prove the Multidimensional Substitution Formula. Because we work with individual functions, we state the result for functions f in $\mathcal{L}^1$. It should be clear that a similar result holds for L^1.

Theorem 18.37 Multidimensional Substitution Formula. *Let $\Omega_1, \Omega_2 \subseteq \mathbb{R}^d$ be open subsets of $\mathbb{R}^d$ and let $g : \Omega_1 \to \Omega_2$ be a continuously differentiable bijective function so that for all $x \in \Omega_1$ we have $\det\big(Dg(x)\big) \neq 0$. Then for all $f \in \mathcal{L}^1(\Omega_2)$ we have $f \circ g \in \mathcal{L}^1(\Omega_1)$ and $\int_{\Omega_2} f\, d\lambda = \int_{\Omega_1} (f \circ g)\big|\det(Dg)\big|\, d\lambda$.*

Proof. By Corollary 17.66, g^{-1} is also continuously differentiable. The measurability of $f \circ g$ now follows from Lemma 18.30 applied to g^{-1}. We will first prove the identity for all nonnegative $f \in C_0^\infty(\Omega_2)$. Let $f \in C_0^\infty(\Omega_2)$. Then $g^{-1}\big[\mathrm{supp}(f)\big]$ is compact. For every $x \in g^{-1}\big[\mathrm{supp}(f)\big]$ find $\varepsilon_x > 0$ so that $\overline{B_{\varepsilon_x}(x)} \subseteq \Omega_1$. Then $\left\{B_{\frac{\varepsilon_x}{3}}(x) : x \in g^{-1}\big[\mathrm{supp}(f)\big]\right\}$ is an open cover of the set $g^{-1}[\mathrm{supp}(f)]$, and hence there are $x_1, \ldots, x_k$ so that $g^{-1}\big[\mathrm{supp}(f)\big] \subseteq \bigcup_{i=1}^{k} B_{\frac{\varepsilon_{x_i}}{3}}(x_i)$. Then $K := \overline{\bigcup_{i=1}^{k} B_{\varepsilon_{x_i}}(x_i)}$ is a compact subset of Ω_1. Moreover, if we define $\delta_{\mathrm{supp}} := \min\left\{\frac{\varepsilon_{x_i}}{3} : i = 1, \ldots, n\right\}$, then every half-open cube of radius less than δ_{supp} that intersects $g^{-1}\big[\mathrm{supp}(f)\big]$ is contained in K. (Recall that we are working with $\|\cdot\|_\infty$.)

Let $\varepsilon > 0$. The function f is uniformly continuous on the compact set $g[K]$. Therefore there is a $\delta_f > 0$ so that for all $x, y \in g[K]$ with $\|y - x\| < \delta_f$ we have $\big|f(y) - f(x)\big| < \frac{\varepsilon}{3\lambda\big(g[K]\big)}$. The functions $(f \circ g)\big|\det(Dg)\big|$ and g are uniformly continuous on K. Therefore there is a $\delta_{f\circ g} > 0$ so that for all points $x, y \in K$ with $\|y - x\| < \delta_{f\circ g}$ we have $\Big|f(g(x))\big|\det(Dg(x))\big| - f(g(y))\big|\det(Dg(y))\big|\Big| < \frac{\varepsilon}{3\lambda(K)}$ and $\big\|g(y) - g(x)\big\| < \delta_f$. Let $M := \max\big\{|f(y)| : y \in \Omega_2\big\} + 1$. By Lemma 18.36, there is a $\delta \in \big(0, \min\{\delta_{f\circ g}, \delta_{\mathrm{supp}}\}\big)$ so that for every half-open cube $B \subseteq K$ centered at x with radius less than δ we have $\lambda\big(g[B]\big) - \big|\det(Dg(x))\big|\lambda(B) < \frac{\varepsilon}{3M\lambda(K)}\lambda(B)$.

Let $B_1, \ldots, B_n$ be all half-open cubes of the form $\prod_{i=1}^{d}\big[l_i\delta, (l_i + 1)\delta\big)$ that intersect $g^{-1}\big[\mathrm{supp}(f)\big]$. Then $g^{-1}\big[\mathrm{supp}(f)\big] \subseteq \bigcup_{i=1}^{n} B_i$ and $\mathrm{supp}(f) \subseteq \bigcup_{i=1}^{n} g[B_i]$. Because $\delta < \delta_{\mathrm{supp}}$, all B_i are contained in K. Let x_i be the center point of B_i. Then for all points $y \in B_i$ we have $\Big|f(g(y))\big|\det(Dg(y))\big| - f(g(x_i))\big|\det(Dg(x_i))\big|\Big| < \frac{\varepsilon}{3\lambda(K)}$, $\big\|g(y) - g(x_i)\big\| < \delta_f$ and for all $z \in g[B_i]$ we have $\big|f(z) - f(g(x_i))\big| < \frac{\varepsilon}{3\lambda\big(g[K]\big)}$.

Therefore

$$\begin{aligned}
&\int_{\Omega_2} f\, d\lambda - \int_{\Omega_1} (f \circ g)\big|\det(Dg)\big|\, d\lambda \\
&= \int_{\Omega_2} f\, d\lambda - \sum_{i=1}^{n} f(g(x_i))\lambda\big(g[B_i]\big)
\end{aligned}$$

$$\begin{aligned}
&+\sum_{i=1}^{n} f(g(x_i))\lambda(g[B_i]) - \sum_{i=1}^{n} f(g(x_i))\left|\det(Dg(x_i))\right|\lambda(B_i) \\
&+\sum_{i=1}^{n} f(g(x_i))\left|\det(Dg(x_i))\right|\lambda(B_i) - \int_{\Omega_1} (f\circ g)\left|\det(Dg)\right| \, d\lambda \\
\leq\ & \int_{\Omega_2} \left| f - \sum_{i=1}^{n} f(g(x_i))\mathbf{1}_{g[B_i]} \right| d\lambda \\
&+\sum_{i=1}^{n} f(g(x_i)) \Big(\lambda(g[B_i]) - \left|\det(Dg(x_i))\right|\lambda(B_i)\Big) \\
&+\int_{\Omega_1} \left| \sum_{i=1}^{n} f(g(x_i))\left|\det(Dg(x_i))\right|\mathbf{1}_{B_i} - (f\circ g)\left|\det(Dg)\right| \right| d\lambda
\end{aligned}$$

For the middle sum, first use that $f(g(x_i)) \geq 0$ and the estimate for the difference, then use that $f(g(x_i)) \leq M$.

$$\begin{aligned}
<\ & \int_{\Omega_2} \frac{\varepsilon}{3\lambda(g[K])}\mathbf{1}_{g[K]}\, d\lambda + \sum_{i=1}^{n} M\frac{\varepsilon}{3M\lambda(K)}\lambda(B_i) + \int_{\Omega_1} \frac{\varepsilon}{3\lambda(K)}\mathbf{1}_K \, d\lambda \\
=\ & \varepsilon .
\end{aligned}$$

Because ε was arbitrary the inequality $\int_{\Omega_2} f \, d\lambda \leq \int_{\Omega_1} (f\circ g)\left|\det(Dg)\right| \, d\lambda$ holds for all nonnegative $f \in C_0^\infty(\Omega_2)$. The same argument with the roles of domain and range reversed proves the reversed inequality as follows. By Theorem 17.31, we have that $Dg^{-1} = \left((Dg)\circ g^{-1}\right)^{-1}$, so that

$$\left(\det(Dg)\circ g^{-1}\right)\det\left(Dg^{-1}\right) = \det\left[\left((Dg)\circ g^{-1}\right)\left((Dg)\circ g^{-1}\right)^{-1}\right] = 1,$$

and hence

$$\begin{aligned}
&\int_{\Omega_1} (f\circ g)\left|\det(Dg)\right| \, d\lambda \\
\leq\ & \int_{\Omega_2} \left[(f\circ g)\circ g^{-1}\right]\left|\det(Dg)\circ g^{-1}\right|\left|\det\left(Dg^{-1}\right)\right| \, d\lambda = \int_{\Omega_2} f \, d\lambda ,
\end{aligned}$$

which establishes the result for all nonnegative $f \in C_0^\infty(\Omega_2)$.

Now let f be a nonnegative simple function on Ω_2 whose support is contained in the interior of a compact set $K \subseteq \Omega_2$ and for which there is a $\delta > 0$ so that $\widetilde{K} := \overline{\bigcup_{x\in K} B_\delta(x)}$ is compact and contained in Ω_2. By Theorem 18.12, there is a sequence $\{f_n\}_{n=1}^\infty$ in $C_0^\infty(\Omega_2)$ so that $\lim_{n\to\infty} \int_{\Omega_2} |f - f_n| \, d\lambda = 0$. By Lemma 18.11, we can assume without loss of generality that $\text{supp}(f_n) \subseteq \widetilde{K}$ for all natural numbers

$n \in \mathbb{N}$. Let $U := \max\{f(z) : z \in \Omega_2\}$. Because we can subtract the function $h_n(x) := \big(f_n(x) - U\big) j_1\big(f_n(x) - U\big)$ (see Lemma 18.9 for the definition of j_1) from f_n without affecting the convergence, we can assume without loss of generality that the f_n are uniformly bounded by a multiple of $\mathbf{1}_{\widetilde{K}}$. Clearly, $\mathbf{1}_{\widetilde{K}}$ is integrable. By Proposition 14.49 we can assume without loss of generality that $\{f_n\}_{n=1}^{\infty}$ also converges pointwise a.e. to f. Because g^{-1} maps null sets to null sets (see proof of Lemma 18.30) $\big\{(f_n \circ g)\big|\det(Dg)\big|\big\}_{n=1}^{\infty}$ converges pointwise a.e. to $(f \circ g)\big|\det(Dg)\big|$. Moreover, all these functions are bounded by a multiple of $(\mathbf{1}_{\widetilde{K}} \circ g)\big|\det(Dg)\big|$, which is integrable. Thus the result follows for f via the Dominated Convergence Theorem.

For nonnegative functions $f \in \mathcal{L}^1(\Omega_2)$, the result now follows via the Monotone Convergence Theorem (Exercise 18-40a). Finally, for arbitrary functions $f \in \mathcal{L}^1(\Omega_2)$ the result follows by considering f^+ and f^- separately (Exercise 18-40b). ■

We conclude this section by assigning the function g and the volume distortion factor $\det(Dg)$ their commonly used names.

Definition 18.38 *Let $\Omega_1, \Omega_2 \subseteq \mathbb{R}^d$ be open subsets of $\mathbb{R}^d$ and let $g : \Omega_1 \to \Omega_2$ be a continuously differentiable bijective function so that for all $x \in \Omega_1$ we have* $\det\big(Dg(x)\big) \neq 0$. *Then g is also called a* **coordinate transformation** *and the determinant* $\det\big(Dg(x)\big)$ *is also called the* **Jacobian** *of g.*

Exercises

18-36 Finishing the proof of Lemma 18.30

(a) Let $B = \prod_{i=1}^{d}(a_i, b_i)$ be an open box and let $\varepsilon, \delta > 0$ Prove that there is a finite set of open cubes $B_1, \ldots, B_n$ so that $B \subseteq \bigcup_{j=1}^{n} B_j$, $\lambda\left(\bigcup_{j=1}^{n} B_j\right) < \lambda(B) + \varepsilon$ and the diameter of each B_j (with respect to $\|\ \|_\infty$) is at most δ

Hint Define $\nu := \min\left\{\frac{\delta}{2}, \frac{\varepsilon}{4\sum_{j=1}^{d}\prod_{i=1}^{j-1}(b_i - a_i)\prod_{i=j+1}^{d}(b_i - a_i)}\right\}$ and let the sets $B_1, \ldots, B_n$ be all the cubes of the form $\prod_{i=1}^{d}\big(l_i\nu, (l_i + 1)\nu\big)$ that intersect B. Show that the measure of the union satisfies $\lambda\left(\bigcup_{j=1}^{n} B_j\right) < \lambda(B) + \frac{\varepsilon}{2}$ Then slightly enlarge each cube to get the result

(b) Let $B = \prod_{i=1}^{d}(a_i, b_i)$ be an open cube, let $f \cdot B \to \mathbb{R}^d$ be a Lipschitz continuous function and let L be its Lipschitz constant when all distances are measured with respect to $\|\cdot\|_\infty$. Prove that with $(c_1, \ldots, c_d) := f\left(\frac{b_1 + a_1}{2}, \ldots, \frac{b_d + a_d}{2}\right)$ and $r := \frac{b_i - a_i}{2}$ (remember that all $b_i - a_i$ are equal), we have $f[B] \subseteq \prod_{i=1}^{d}(c_i - Lr, c_i + Lr)$.

(c) Explain why it is *not* possible to prove that if $B = \prod_{i=1}^{d} (a_i, b_i)$ is an open box and $f : B \to \mathbb{R}^d$ is Lipschitz continuous with Lipschitz constant L when all distances are measured *with respect to the Euclidean norm* $\|\cdot\|_2$, then with $(c_1, \ldots, c_d) := f\left(\frac{b_1 + a_1}{2}, \ldots, \frac{b_d + a_d}{2}\right)$ and $r_i := \frac{b_i - a_i}{2}$, we have $f[B] \subseteq \prod_{i=1}^{d} (c_i - Lr_i, c_i + Lr_i)$.

Hint. Rotations.

18-37. Prove Lemma 18.31.

Hint. Use the proof of Lemma 18.32 as guidance.

18-38. Let D be as in Lemma 18.32 and let $B = \prod_{i=1}^{d} [a_i, b_i]$ be a d-dimensional box. Prove that $D[B]$ is a box with $\big| D[B] \big| = |c_1 \cdots c_n||B|$.

18-39 Prove Lemma 18.34

Hint. Use the proof of Lemma 18.32 as guidance.

18-40 Finishing the proof of the Multivariable Substitution Theorem.

(a) Prove Theorem 18.37 for nonnegative $f \in \mathcal{L}^1(\Omega_2)$, assuming that it holds for nonnegative simple functions $f \in \mathcal{L}^1(\Omega_2)$ whose support is contained in the interior of a compact set $K \subseteq \Omega_2$ for which there is a $\delta > 0$ so that $\widetilde{K} := \overline{\bigcup_{x \in K} B_\delta(x)}$ is compact and contained in Ω_2.

(b) Prove Theorem 18.37 for arbitrary $f \in \mathcal{L}^1(\Omega_2)$, assuming that it holds for nonnegative functions $f \in \mathcal{L}^1(\Omega_2)$

18-41 Let $\Omega_1 \subseteq \mathbb{R}^d$ be an open subset of $\mathbb{R}^d$. Prove that if $f : \Omega_1 \to \mathbb{R}^d$ is Lipschitz continuous, then there is a constant $L > 0$ so that for all $S \subseteq \Omega_1$ we have $\lambda\big(f[S]\big) \leq L\lambda(S)$.

18-42 Compute the Jacobians of the following coordinate transformations g

(a) **Polar Coordinates**. The coordinate transformation $s_2 : (0, \infty) \times (0, 2\pi) \to \mathbb{R}^2$ is defined by $s_2(r, \theta) = \big(r\cos(\theta), r\sin(\theta)\big)$.

(b) **Cylindrical Coordinates** The coordinate transformation $c : (0, \infty) \times (0, 2\pi) \times \mathbb{R} \to \mathbb{R}^3$ is defined by $c(r, \theta, z) = \big(r\cos(\theta), r\sin(\theta), z\big)$.

(c) **Spherical Coordinates**. The coordinate transformation $s_3 : (0, \infty) \times (0, 2\pi) \times (0, \pi) \to \mathbb{R}^3$ is defined by $s_3(\rho, \theta, \varphi) = \big(\rho\cos(\theta)\sin(\varphi), \rho\sin(\theta)\sin(\varphi), \rho\cos(\varphi)\big)$

18-43 Compute the following integrals where dx, dy, dz denote the Lebesgue integrals along the x-, y- and z-axes, respectively. Use the coordinate transforms from Exercise 18-42 as appropriate

(a) $$\int_{-1}^{0} \int_{0}^{\sqrt{1-y^2}} \left(x^2 + y^2\right)^{\frac{3}{2}} \, dx \, dy$$

(b) $$\int_{-1}^{1} \int_{-\sqrt{1-x^2}}^{0} \int_{-\sqrt{1-x^2-y^2}}^{\sqrt{1-x^2-y^2}} y\sqrt{x^2 + y^2 + z^2} \, dz \, dy \, dx$$

(c) $$\int_{0}^{4} \int_{-\sqrt{16-x^2}}^{\sqrt{16-x^2}} \int_{-1}^{1} x \, dy \, dz \, dx$$

(d) $$\int_{\{(x,y)\in\mathbb{R}^2 : 1 \leq x^2+y^2 \leq 4\}} -x^2 \, d\lambda_2$$

18-44. The integral of $N_{\mu,\sigma}(x) = \frac{1}{\sqrt{2\pi}\sigma} e^{-\frac{(x-\mu)^2}{2\sigma^2}}$.

(a) Compute the integral of $f(x, y) = e^{-\frac{x^2+y^2}{2}}$ over a disk of radius a centered at the origin. Then take the limit of the integrals as $a \to \infty$

(b) Compute $\int_{-\infty}^{\infty} e^{-\frac{x^2}{2}} d\lambda(x)$ by proving $\left(\int_{-\infty}^{\infty} e^{-\frac{x^2}{2}} d\lambda(x)\right)^2 = \int_{\mathbb{R}\times\mathbb{R}} e^{-\frac{x^2+y^2}{2}} d\lambda(x, y)$ and then computing the latter integral
Hint. Proposition 14.65 and part 18-44a.

(c) Prove that $\int_{-\infty}^{\infty} \frac{1}{\sqrt{2\pi}\,\sigma} e^{-\frac{(x-\mu)^2}{2\sigma^2}} d\lambda(x) = 1$

18-45. The **volume of d-dimensional balls**. We first need to define **d-dimensional Spherical Coordinates.** Two and three dimensional spherical coordinates s_2 are defined in Exercises 18-42a and 18-42c. For $d \geq 3$ with $s_{d-1} : (0, \infty) \times (0, 2\pi) \times (0, \pi)^{d-3} \to \mathbb{R}^{d-1}$ denoting $d - 1$ dimensional spherical coordinates, let

$$s_d(\rho, \theta, \varphi_1, \ldots, \varphi_{d-2}) := \Big(\pi_1\big(s_{d-1}(\rho, \theta, \varphi_1, \ldots, \varphi_{d-3})\big)\sin(\varphi_{d-2}), \ldots, \pi_{d-1}\big(s_{d-1}(\rho, \theta, \varphi_1, \ldots, \varphi_{d-3})\big)\sin(\varphi_{d-2}), \rho\cos(\varphi_{d-2})\Big)$$

(a) Let J_d be the Jacobian of s_d. Prove that $|J_d| = \rho\left(\sin(\varphi_{d-2})\right)^{d-2} |J_{d-1}|$ for $d \geq 3$.
Hint. The last row of the matrix (which contains the derivatives of the last coordinate) has $\cos(\varphi_{d-2})$ as its first entry and $-\rho\sin(\varphi_{d-2})$ as its last entry. Expand the determinant with respect to the last row (see Exercise 18-33).

(b) Prove that the volume of the d-dimensional Euclidean ball of radius r about the origin is $V\left(B_r^d(0)\right) = 2\pi \frac{r^d}{d} \prod_{i=1}^{d-2} \int_0^{\pi} \left(\sin(\varphi_i)\right)^i d\varphi_i$

(c) Prove the area formula for circles in $\mathbb{R}^2$

(d) Prove the volume formula for balls in $\mathbb{R}^3$

(e) Prove that $\lim_{d\to\infty} V\left(B_r^d(0)\right) = 0$
Hint. It is not necessary to compute the integrals

18-46 Let (M, Σ, μ) be a σ-finite measure space, let $f : M \to [0, \infty]$ be Σ-measurable and let $(\mathbb{R}, \Sigma_\lambda, \lambda)$ be the real numbers with Lebesgue measure.

(a) Prove that $\int_M f^p \, d\mu = \int_0^{\infty} pt^{p-1}\mu\left(\{x \in M : f(x) > t\}\right) dt$ for all $1 \leq p < \infty$.
(Measurability is not an issue because of Exercise 14-51a.)

(b) Prove that if $g : (0, \infty) \to (0, \infty)$ is differentiable and $g'(x) > 0$ for all x, then $\int_M g \circ f \, d\mu = \int_0^{\infty} g'(t)\mu\left(\{x \in M : f(x) > t\}\right) dt.$

18-47 Let $f, g \in \mathcal{L}^1\left(\mathbb{R}^d\right)$.

(a) Prove that $C(x, t) := f(x - t)g(t)$ is Lebesgue measurable.

(b) Prove that for almost all $x \in \mathbb{R}^d$ the function $c_x(t) := f(x - t)g(t)$ is Lebesgue integrable and that the function $f * g(x) := \begin{cases} \int_{\mathbb{R}^d} f(x - t)g(t)\, dt, & \text{if } c_x \text{ is Lebesgue integrable,} \\ 0; & \text{otherwise,} \end{cases}$
is Lebesgue integrable and $\|f * g\|_1 \leq \|f\|_1 \|g\|_1$.
The function $f * g$ is also called the **convolution** of f and g

(c) Prove that if $h \in \mathcal{L}^1(\mathbb{R}^d)$, then $k(x) := h(-x)$ also is Lebesgue integrable

(d) Prove that $f * g = g * f$.

(e) Prove that if g is bounded, then $f * g$ is continuous.

Hint. First prove the result for continuous f, then use that the continuous functions are dense in $L^1(\mathbb{R}^d)$.

18-48 Prove that the determinant $\det(v_1, \ldots, v_d)$ is the oriented n-dimensional volume of the parallelepiped spanned by the column vectors $v_1, \ldots, v_d$.

Hint Map a box to the parallelepiped and use the Multivariable Substitution Formula.

18-49. Prove that the hypothesis that L is bijective can be dropped from Lemma 18.35.

Hint. Prove that if the linear function L is not bijective, then the Lebesgue measure of every image of a Lebesgue measurable set is 0.

18-50. Multidimensional Substitution with weaker hypotheses. Let $\Omega_1, \Omega_2 \subseteq \mathbb{R}^d$ be open subsets of $\mathbb{R}^d$ and let $g : \Omega_1 \to \Omega_2$ be a continuously differentiable function so that for almost all $x \in \Omega_1$ we have $\det\left(Dg(x)\right) \neq 0$ and so that $\left\{x \in \Omega_1 : \left(\exists z \in \Omega_1 : f(x) = f(z)\right)\right\}$ is a null set Then for all functions $f \in \mathcal{L}^1(\Omega_2)$ we have $f \circ g \in \mathcal{L}^1(\Omega_1)$ and $\int_{\Omega_2} f \, d\lambda = \int_{\Omega_1} (f \circ g) \left|\det(Dg)\right| d\lambda$.

Hint. The set $\left\{x \in \Omega_1 : \det(Dg(x)) = 0\right\}$ is closed in Ω_1. Prove that if K is compact, then the set $\left\{x \in K : \left(\exists z \in K . f(x) = f(z)\right)\right\}$ is closed, too. Apply Theorem 18.37 to the appropriate subset of Ω_1 to first prove the result for compactly supported functions.

18-51 On the injectivity hypothesis of Theorem 18.37. Consider the function $g : \mathbb{R}^2 \to \mathbb{R}^2$ defined by $g(x, y) = \left(x^2 - y^2, 2xy\right)$ (this function interprets (x, y) as a complex number and squares it).

(a) Prove that for all $(x, y) \in \mathbb{R}^2$ the function g is differentiable at (x, y) and that $\det\left(Dg(x, y)\right) = 4x^2 + 4y^2$.

(b) Prove that for all $(x, y) \in \mathbb{R}^2$ we have $\| g(x, y) \|_2 = x^2 + y^2$.

(c) Prove that for all $(a, b) \neq (0, 0)$ we have $g\left(\sqrt{\frac{a+\sqrt{a^2+b^2}}{2}}, \frac{b}{\sqrt{2\left(a+\sqrt{a^2+b^2}\right)}}\right) = (a, b)$

(d) Prove that $g\left[B_1(0, 0)\right] = B_1(0, 0)$.

(e) Prove that for the function $f(x, y) = 1$ we have $\int_{B_1(0,0)} f \, d\lambda = \pi$, but the transformed integral is $\int_{B_1(0,0)} f \circ g \left|\det(Dg)\right| d\lambda = 2\pi$. Then explain why this result does not contradict Theorem 18.37.

Hint. You may use the result of Exercise 18-42.

18-52. The effect of a diffeomorphism on Lebesgue measure. Let Ω_1 and Ω_2 be open subsets of $\mathbb{R}^d$, let $K \subseteq \Omega_1$ be compact and let $g : \Omega_1 \to \Omega_2$ be a continuously differentiable bijective function with $\det\left(Dg(x)\right) \neq 0$ for all $x \in \Omega_1$. Then for every $\varepsilon > 0$ there is a $\delta > 0$ so that for every box $B \subseteq K$ with center point x and with diameter less than δ we have $\left|\lambda\left(g[B]\right) - \left|\det(Dg(x))\right| \lambda(B)\right| < \varepsilon\lambda(B)$.

Hint. The difference between this result and Lemma 18.36 are the absolute value signs and that we can work with boxes rather than cubes. Use Theorem 18.37 with $f = 1$.

18-53 **Line integrals and surface integrals are independent of the parametrization.**

(a) Let $m \leq d$, let $\Omega_1, \Omega_2 \subseteq \mathbb{R}^m$ be open and let $r_1 : \Omega_1 \to \mathbb{R}^d$ and $r_2 : \Omega_2 \to \mathbb{R}^d$ be injective and continuously differentiable so that $r_1[\Omega_1] = r_2[\Omega_2]$ and so that all derivatives $Dr_i(x)$ are injective. Prove that $r_2^{-1} \circ r_1$ is a differentiable function from Ω_1 to Ω_2

Hint. Let $x \in \Omega_1$. Find $v_{m+1}, \ldots, v_d$ so that $\left\{ \frac{\partial}{\partial x_1} r_1(x), \ldots, \frac{\partial}{\partial x_m} r_1(x), v_{m+1}, \ldots, v_d \right\}$ is a base of $\mathbb{R}^d$ and on $\Omega_1 \times R^{d-m}$ set $R_1(z_1, \ldots, z_d) := r_1(z_1, \ldots, z_m) + \sum_{j=m+1}^{d} z_j v_j$. Use Corollary 17.66 to prove that R_1 is differentiable with differentiable inverse on a neighborhood of $(x_1, \ldots, x_m, 0, \ldots, 0)$. Define a similar function R_2 for r_2.

(b) Let $\delta > 0$ and let $r_1 : (a_1 - \delta, b_1 + \delta) \to \mathbb{R}^d$ and $r_2 : (a_2 - \delta, b_2 + \delta) \to \mathbb{R}^d$ be continuously differentiable functions with $r_1\left[[a_1, b_1]\right] = r_2\left[[a_2, b_2]\right]$, $r_1(a_1) = r_2(a_2)$, $r_1(b_1) = r_2(b_2)$ and so that all derivatives of the r_i are not zero. Let $\Omega \subseteq \mathbb{R}^d$ be an open set that contains $r_1\left[[a_1, b_1]\right]$ and let $F : \Omega \to \mathbb{R}^d$ be continuously differentiable. Prove that
$$\int_{a_1}^{b_1} \left\langle F\left(r_1(t)\right), \frac{d}{dt} r_1(t) \right\rangle dt = \int_{a_2}^{b_2} \left\langle F\left(r_2(s)\right), \frac{d}{ds} r_2(s) \right\rangle ds$$
Hint. $r_1 = r_2 \circ r_2^{-1} \circ r_1$.

(c) Let Ω_1 and Ω_2 be open subsets of $\mathbb{R}^2$ and let $r_1 : \Omega_1 \to \mathbb{R}^3$ and $r_2 : \Omega_2 \to \mathbb{R}^3$ be continuously differentiable with $r_1[\Omega_1] = r_2[\Omega_2]$ so that for all $(x, y) \in \Omega_1$ we have $\frac{\partial}{\partial x} r_1(x, y) \times \frac{\partial}{\partial y} r_1(x, y) \neq 0$, for all $(u, v) \in \Omega_2$ we have $\frac{\partial}{\partial u} r_2(u, v) \times \frac{\partial}{\partial v} r_2(u, v) \neq 0$, and so that there are points $(x, y) \in \Omega_1$ and $(u, v) \in \Omega_2$ so that $r_1(x, y) = r_2(u, v)$ and $\frac{\partial}{\partial x} r_1(x, y) \times \frac{\partial}{\partial y} r_1(x, y) = \lambda \frac{\partial}{\partial u} r_2(u, v) \times \frac{\partial}{\partial v} r_2(u, v)$ for some $\lambda > 0$. Let $\Omega \subseteq \mathbb{R}^3$ be an open set that contains $r_1[\Omega_1]$ and let $F : \Omega \to \mathbb{R}^3$ be continuously differentiable. Prove that
$$\iint_{\Omega_1} \left\langle F\left(r_1(x, y)\right), \frac{\partial}{\partial x} r_1(x, y) \times \frac{\partial}{\partial y} r_1(x, y) \right\rangle dy\, dx$$
$$= \iint_{\Omega_2} \left\langle F\left(r_2(u, v)\right), \frac{\partial}{\partial u} r_2(u, v) \times \frac{\partial}{\partial v} r_2(u, v) \right\rangle dv\, du$$

Hint. $r_1 = r_2 \circ r_2^{-1} \circ r_1$. Then work out componentwise that the effect on the cross product is a multiplication with the determinant of the appropriate 2×2 matrix.

18-54. **Continuous images of null sets can have nonzero measure**. For this exercise, let the function $\psi : [0, 1] \to [0, 1]$ be Lebesgue's singular function from Exercise 11-22.

(a) Prove that if $g : [a, b] \to \mathbb{R}$ is a nondecreasing function, $h(x) = g(x) + x$ and A is a set with $\lambda\left(g[A]\right) > 0$, then $\lambda\left(h[A]\right) > 0$.

(b) Prove that $F(x) = \psi(x) + x$ (where ψ is Lebesgue's singular function from Exercise 11-22) defines a continuous bijective function from $[0, 1]$ to $[0, 2]$ so that F^{-1} also is continuous and $\lambda\left(F\left[C^{\mathcal{Q}}\right]\right) > 0$

(c) Prove that for all $d \in \mathbb{N}$ there is a null set N in $\mathbb{R}^d$ and a continuous bijective function G on an open subset of $\mathbb{R}^d$ that contains N so that $\lambda\left(G[N]\right) > 0$.

18-55. Prove that if $f : [a, b] \to \mathbb{R}$ is **absolutely continuous** and nondecreasing and $N \subseteq [a, b]$ is a null set, then $f[N]$ also is a null set

18-56. Uniform limits of continuous functions with additional properties need not have these additional properties

(a) Use Exercises 18-54 and 18-55 to prove that if $\{f_n\}_{n=1}^{\infty}$ is a uniformly convergent sequence of **absolutely continuous** functions on $[a, b]$, then the limit need not be absolutely continuous.

(b) Prove that if $\{f_n\}_{n=1}^{\infty}$ is a uniformly convergent sequence of Lipschitz continuous functions on $[a, b]$, then the limit need not be Lipschitz continuous

Chapter 19

Introduction to Differential Geometry

Differential geometry is by its nature a bit technical. Readers willing to accept Theorems 19.69 and 19.70 for "sufficiently nice" objects, could postpone a detailed reading of this chapter.

In applications, it is often necessary to describe surfaces, like the d-dimensional spheres $\left\{p \in \mathbb{R}^d : \|p\|_2 = r\right\}$ of radius r, or, more generally, lower dimensional subsets S of a higher dimensional space. To describe such subsets or surfaces, we use parametrizations, that is, bijective, continuously differentiable functions $g : \Omega \to S$ with injective derivatives $Dg(x)$, where Ω is an open subset of $\mathbb{R}^m$ for an appropriate m. Unfortunately, for many surfaces such an overall parametrization cannot be defined. For example, for the spheres of radius 1 it can be shown (Exercise 19-12) that a parametrization g as mentioned must have a continuous inverse function $g^{-1} : S \to \Omega$. But S is compact and Ω is not compact, which is impossible by Theorem 16.62. On the other hand, the Implicit Function Theorem guarantees that for many surfaces that are solution sets of equations $f(x, y) = 0$, where both x and y can be in Banach spaces, we can *locally* find parametrizations. This is the idea behind the definition of a manifold.

This chapter introduces manifolds in Section 19.1 and their tangent spaces and differentiable functions in Section 19.2. Sections 19.3, 19.4, and 19.5 build the integration theory on manifolds and the chapter culminates in Section 19.6 with Stokes' Theorem.

19.1 Manifolds

The fundamental idea behind a manifold is that it "locally looks like m-dimensional space." This reflects our intuition that differentiable surfaces, despite being curved, locally look like two-dimensional space.

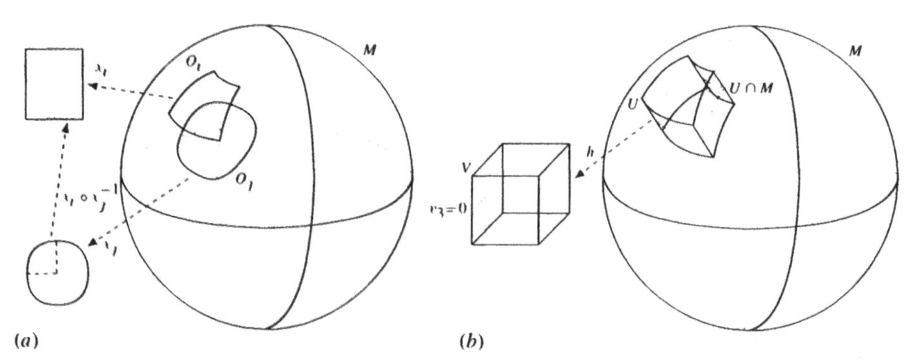

Figure 48: Manifolds (a) are spaces that locally look like $\mathbb{R}^m$. The atlas of a C^∞-manifold is a collection of homeomorphisms $\{x_i\}_{i\in I}$ into the appropriate $\mathbb{R}^m$ so that the compositions $x_i \circ x_j^{-1}$ are diffeomorphisms wherever they are defined. Embedded manifolds (b) are subspaces of $\mathbb{R}^d$ so that each point of the manifold has a neighborhood in $\mathbb{R}^d$ that can be transformed so that the image of the intersection of the manifold with the neighborhood lies in the hyperplane in which the $(m+1)^{\text{st}}$ through d^{th} coordinates are zero.

Definition 19.1 *Let $m \in \mathbb{N}$. A metric space M is called an m-dimensional (topological)* **manifold** *iff for each $p \in M$ there is an open set $O \subseteq M$ so that $p \in O$ and O is homeomorphic to $\mathbb{R}^m$. (See Figure 48.)*

Equivalently, in the definition of a manifold we could demand that each O is homeomorphic to an open subset of $\mathbb{R}^m$ (see Exercise 19-1). For further investigations, we also need differentiability properties. Because M is just a metric space, these properties are introduced through differentiability properties of compositions of the homeomorphisms from the definition.

Definition 19.2 *Let $U, V \subseteq \mathbb{R}^d$ be open. A bijective, infinitely differentiable function $h : U \to V$ with infinitely differentiable inverse is called a* **diffeomorphism**.

Definition 19.3 *Let $m \in \mathbb{N}$ and let M be an m-dimensional manifold. A family $\{x_i\}_{i\in I}$ is called an* **atlas** *for M iff each x_i is a homeomorphism from an open subset O_i of M to an open subset of $\mathbb{R}^m$, for each $p \in M$ there is an $i \in I$ so that $p \in O_i$ and for all $i, j \in I$ the composition $x_i \circ x_j^{-1} : x_j[O_i \cap O_j] \to x_i[O_i \cap O_j]$ is a diffeomorphism. The functions x_i are also called* **charts** *or* **coordinate systems**. *(See Figure 48.)*

The inverse x^{-1} of a coordinate system $x : U \to \mathbb{R}^m$ can be interpreted as a **parametrization** of the subset U of M.

Definition 19.4 *A pair $\left(M, \{x_i\}_{i\in I}\right)$ of an m-dimensional manifold M and an atlas $\{x_i\}_{i\in I}$ is also called an m-dimensional* C^∞**-manifold**. *As for spaces, we typically refer to C^∞-manifolds through the set M, implicitly assuming that an atlas is given.*

Of course, every C^∞-manifold is a manifold. A C^k**-diffeomorphism** is a bijective, k times continuously differentiable function with k times continuously differentiable inverse. By using C^k-diffeomorphisms instead of diffeomorphisms, it is possible to define **C^k-manifolds**. Working with C^k-manifolds would require us to keep track of detailed differentiability conditions. Hence, throughout this chapter we will work with C^∞-manifolds and we will simply refer to them as **manifolds**. Results similar to the ones derived in this chapter also hold for C^k-manifolds.

Trivially $\left(\mathbb{R}^d, \left\{id_{\mathbb{R}^d}\right\}\right)$ is a manifold. Whenever we work with $\mathbb{R}^d$ as a manifold we will assume that the atlas is $\left\{id_{\mathbb{R}^d}\right\}$. Our first nontrivial examples of manifolds are subsets of d-dimensional space. These "embedded manifolds" will be of particular interest throughout. They arise frequently in applications and we will use them as examples and for motivation of abstract definitions.

Definition 19.5 *Let $m, d \in \mathbb{N}$ be so that $m \leq d$. A set $M \subseteq \mathbb{R}^d$ is called an m-dimensional* **embedded manifold** *iff for every $p \in M$ there is an open neighborhood $U \subseteq \mathbb{R}^d$ of p, an open set $V \subseteq \mathbb{R}^d$ and a diffeomorphism $h : U \to V$ for which $h[U \cap M] = \{v \in V : v_{m+1} = \cdots = v_d = 0\}$. (See Figure 48.)*

Proposition 19.6 *Every m-dimensional embedded manifold M is an m-dimensional manifold.*

Proof. For each $p \in M$, let U_p be an open subset of $\mathbb{R}^d$ that contains p and for which there is a diffeomorphism $h_p : U_p \to V_p$ with $V_p \subseteq \mathbb{R}^d$ open and so that $h_p[U_p \cap M] = \{v \in V_p : v_{m+1} = \cdots = v_d = 0\}$. Let $\pi_{\mathbb{R}^m} : \mathbb{R}^d \to \mathbb{R}^m$ be the projection onto the first m coordinates. For all $p \in M$ let $x_p := \pi_{\mathbb{R}^m} \circ h_p\big|_{U_p \cap M}$. Clearly, each x_p is a homeomorphism from the set $O_p := U_p \cap M$ to the subset $\pi_{\mathbb{R}^m}\left[\{v \in V_p : v_{m+1} = \cdots = v_d = 0\}\right]$ of $\mathbb{R}^m$. For all $p, q \in M$, the composition $h_q \circ h_p^{-1} : h_p[U_p \cap U_q] \to h_q[U_p \cap U_q]$ is a diffeomorphism. Moreover, the sets $x_p[O_p \cap O_q] = \pi_{\mathbb{R}^m} \circ h_p[U_p \cap U_q \cap M]$ and $x_q[O_p \cap O_q] = \pi_{\mathbb{R}^m} \circ h_q[U_p \cap U_q \cap M]$ are projections of the intersections of open subsets with the subspace $\mathbb{R}^m \times \{0\}^{d-m}$ of $\mathbb{R}^d$, which means they are open subsets of $\mathbb{R}^m$. With $e_{\mathbb{R}^m} : \mathbb{R}^m \to \mathbb{R}^d$ being the natural embedding that maps $\mathbb{R}^m$ to $\mathbb{R}^m \times \{0\}^{d-m}$ the composition $x_q \circ x_p^{-1} = \pi_{\mathbb{R}^m} \circ h_q \circ h_p^{-1} \circ e_{\mathbb{R}^m}\Big|_{\pi_{\mathbb{R}^m}\left[h_p[U_p \cap U_q \cap M]\right]}$ is a diffeomorphism. Therefore $\{x_p\}_{p \in M}$ is an atlas, and hence M is an m-dimensional manifold. ■

Whenever we work with an embedded manifold, we will assume that its atlas was generated as in the proof of Proposition 19.6. Embedded manifolds arise naturally when solving equations.

Theorem 19.7 *Let $\Omega \subseteq \mathbb{R}^d$ be open and let $f : \Omega \to \mathbb{R}^{d-m}$ be an infinitely differentiable function so that for all $p \in \Omega$ with $f(p) = 0$ the matrix $Df(p)$ has rank $d - m$. Then $f^{-1}(0)$ is an m-dimensional embedded manifold in $\mathbb{R}^d$.*

Proof. Let $p \in \Omega$ be so that $f(p) = 0$. Apply Corollary 17.69 to obtain an open set $G \subseteq \mathbb{R}^d$ and a diffeomorphism $g : G \to g[G]$ so that $p \in g[G] \subseteq \Omega$ and $f \circ g(v_1, \ldots, v_d) = (v_{m+1}, \ldots, v_d)$ for all $(v_1, \ldots, v_d) \in G$. Then for all $(v_1, \ldots, v_d) \in g^{-1}\left[f^{-1}(0) \cap g[G]\right]$ we obtain $f \circ g(v_1, \ldots, v_d) = 0$, which means

$v_{m+1} = \cdots = v_d = 0$. With $U := g[G]$, $V = G$ and $h = g^{-1}$ we see that, because p was arbitrary, $f^{-1}(0)$ is an m-dimensional embedded manifold. ■

Proposition 19.6 and Theorem 19.7 provide a multitude of concrete examples.

Example 19.8 Examples of (embedded) manifolds.

1. Trivially, every open subset Ω of $\mathbb{R}^d$ is a d-dimensional manifold with atlas $\{i_\Omega\}$, where $i_\Omega(p) = p$ for all $p \in \Omega$. This observation is trivial, but it will help with integration over embedded manifolds.

2. Let $r > 0$. Because $f : \mathbb{R}^d \to \mathbb{R}$ defined by $f(x_1, \cdots, x_d) := r^2 - \sum_{j=1}^{d} x_j^2$ is infinitely differentiable, every sphere $\left\{p \in \mathbb{R}^d : f(p) = 0\right\}$ centered around the origin is a $(d-1)$-dimensional manifold.

3. A function $a : \mathbb{R}^d \to \mathbb{R}$ is called **affine linear** iff there is a nonzero linear function $L : \mathbb{R}^d \to \mathbb{R}$ and an $r \in \mathbb{R}$ so that $a(x) = r + L[x]$ for all $x \in \mathbb{R}^d$. Because affine linear functions are infinitely differentiable, the intersection of any hyperplane $\left\{p \in \mathbb{R}^d : a(p) = 0\right\}$ with an open set is a $(d-1)$-dimensional manifold.

4. More generally, the level surfaces of $f(x) = k$ of any infinitely differentiable function $f : \mathbb{R}^d \to \mathbb{R}$ with nonzero derivatives $Df(x)$ are $(d-1)$-dimensional manifolds. Under mild hypotheses on how the surfaces intersect, for $n < d$ the intersection of n level surfaces $f_1(x) = k_1, \ldots, f_n(x) = k_n$ is a $(d-n)$-dimensional manifold, because we can use $F(x) := \big(f_1(x) - k_1, \ldots, f_n(x) - k_n\big)$ and apply Proposition 19.6 and Theorem 19.7. □

Exercise 19-2 provides further examples. We can also obtain examples of manifolds by considering subspaces.

Definition 19.9 *Let M be a manifold and let $U \subseteq M$ be an open subset of M. Then U is called an open* **submanifold** *of M.*

Exercise 19-3 shows that open submanifolds are indeed manifolds themselves.

Some interesting sets cannot be described as manifolds. For example, the closed ball $\overline{B_1(0)} \subset \mathbb{R}^d$ (with respect to the Euclidean norm) is not a manifold, because none of the boundary points has a neighborhood that is isomorphic to $\mathbb{R}^d$. The observation that these points have neighborhoods that are isomorphic to a half-space leads to the idea of a manifold with boundary.

Definition 19.10 *Let $m \in \mathbb{N}$. A metric space M is called an m-dimensional (topological)* **manifold with boundary** *iff for each point $p \in M$ there is an open neighborhood $O \subseteq M$ of p that is homeomorphic to Euclidean space $\mathbb{R}^m$ or to the upper half space $\mathbb{H}^m := \left\{(x_1, \ldots, x_m) \in \mathbb{R}^m : x_m \geq 0\right\}$. Points $p \in M$ that do not have neighborhoods isomorphic to $\mathbb{R}^m$ are also called* **boundary points** *and the set of all these points is called the* **boundary** *∂M of M.*

To define atlases for manifolds with boundary, we define the following.

Definition 19.11 *Let $\Omega \subseteq \mathbb{R}^d$ be an open subset of $\mathbb{R}^d$ and let $B \subseteq \mathbb{R}^d$ be so that $\Omega \subseteq B \subseteq \overline{\Omega}$. Then $C^k(B)$ denotes the set of restrictions to B of functions that are k times differentiable on a neighborhood of B. Similarly, $C^\infty(B)$ denotes the set of restrictions to B of functions that are infinitely differentiable on a neighborhood of B.*

Because the ranges of the coordinate systems of manifolds with boundary are sets as in Definition 19.11, **atlases** for manifolds with boundary, C^k-manifolds with boundary and C^∞-manifolds with boundary are defined similar to atlases for manifolds, C^k-manifolds, and C^∞-manifolds, respectively. We will also refer to C^∞-manifolds with boundary simply as **manifolds with boundary**. Exercise 19-4 shows that the boundary of a manifold with boundary is a manifold.

As for manifolds, subsets of $\mathbb{R}^d$ are of particular interest.

Definition 19.12 *Let $m, d \in \mathbb{N}$ be so that $m \leq d$. A set $M \subseteq \mathbb{R}^d$ is called an m-dimensional* **embedded manifold with boundary** *iff for every $p \in M$ there is an open set $U \subseteq \mathbb{R}^d$ containing p and an open set $V \subseteq \mathbb{R}^d$ such that either*

1. *There is a diffeomorphism $h : U \to V$ so that*
   $h(U \cap M) = \{v \in V : v_{m+1} = \cdots = v_d = 0\}$, *or*
2. *There is a diffeomorphism $h : U \to V$ so that the mth component of $h(p)$ is zero and $h(U \cap M) = \{v \in V : v_m \geq 0, v_{m+1} = \cdots = v_d = 0\}$.*

The reader will prove in Exercise 19-5 that every embedded manifold with boundary is a manifold with boundary.

Corners cannot be described with differentiable functions. Therefore, some interesting sets, such as the cube $[0, 1]^d$, do not have a satisfactory description as embedded manifolds with boundary (see Exercise 19-13b). To include these sets, we define manifolds with corners similar to manifolds with boundary.

Definition 19.13 *Let $m \in \mathbb{N}$. A metric space M is called an m-dimensional (topological)* **manifold with corners** *iff each $p \in M$ has an open neighborhood $O \subseteq M$ that is homeomorphic to a subspace $C_k := \left\{(x_1, \ldots, x_m) \in \mathbb{R}^m : x_k \geq 0, \ldots, x_m \geq 0\right\}$ with $k \in \{1, \ldots, m\}$ or to $\mathbb{R}^m$. Points $p \in M$ that do not have neighborhoods isomorphic to $\mathbb{R}^m$ are also called* **boundary points** *and the set of all these points is called the* **boundary** *∂M of M.*

Atlases for manifolds with corners, C^k-manifolds with corners and C^∞-manifolds with corners are defined similar to atlases for manifolds with boundary, C^k-manifolds with boundary, and C^∞-manifolds with boundary, respectively. We will also refer to C^∞-manifolds with corners simply as **manifolds with corners**. For a C^∞-manifold with corners, we will say that a point $p \in M$ is contained **in a corner** iff there is a homeomorphism x from a neighborhood of p to a space C_k with $k < m$ and $x(p) = 0$. Embedded manifolds with corners are touched upon in Exercise 19-6. We should also note that, formally, every manifold with boundary is also a manifold with corners. We will typically not unify the two concepts, even when proving results valid for manifolds with corners, because "manifold with corners" should explicitly indicate the presence of corners, while a manifold with boundary is assumed to be smooth

Exercises

19-1 Let M be a metric space, let $p \in M$, let $O \subseteq M$ be an open neighborhood of p, let $V \subseteq \mathbb{R}^m$ be open and let $x : O \to V$ be a homeomorphism Prove that there is an open neighborhood U of p and a homeomorphism y $U \to \mathbb{R}^m$

Hint. Consider the restriction of x to the inverse image of a small ball around $x(p)$ and then map that ball diffeomorphically to $\mathbb{R}^m$

19-2 More examples of manifolds

(a) Let a, b $c \in (0, \infty)$ Prove that the ellipsoid $\frac{x^2}{a^2} + \frac{y^2}{b^2} + \frac{z^2}{c^2} = 1$ in $\mathbb{R}^3$ is a 2-dimensional manifold. Also construct an atlas.

(b) Let $\Omega \subseteq \mathbb{R}^2$ and let $f : \Omega \to \mathbb{R}$ be infinitely differentiable with $Df(x) \neq 0$ for all $x \in \Omega$. Prove that every level curve $f(x, y) = k$ is a one-dimensional manifold

(c) Prove that $SL(n, \mathbb{R}) := \{ A \in M(n \times n, \mathbb{R})$ $\det(A) = 1 \}$ is an $n^2 - 1$ dimensional manifold

19-3 Prove that if M is a manifold and U is an open submanifold, then U is also a manifold.

19-4. Prove that if M is a manifold with boundary, then ∂M is a manifold.

19-5. Let $M \subseteq \mathbb{R}^d$ be an embedded manifold with boundary

(a) Prove that M is a manifold with boundary

(b) Prove that the two conditions in the definition of an embedded manifold with boundary are mutually exclusive. That is, every point of M satisfies either 1 or 2, but not both

Hint Assume a point satisfies both and use Corollary 17 66

(c) Prove that the points of M that satisfy condition 2 are the boundary points of M.

(d) Prove that if M is the closure of an open subset of $\mathbb{R}^d$, then $\partial M = \delta M$, that is, the boundary of M as a manifold with boundary equals its topological boundary

(e) Prove that ∂M is an embedded manifold

19-6 Define **embedded manifolds with corners** and prove results similar to those in Exercise 19-5.

19-7. Prove that if M is a manifold with corners, then ∂M is a union of manifolds with corners

19-8 Prove that every connected manifold is pathwise connected

19-9 Prove that every manifold is locally compact and that every connected manifold is σ-compact

19-10 Let $\Omega_1, \Omega_2 \subseteq \mathbb{R}^m$ be open sets Prove that if g . $\Omega_1 \to \Omega_2$ is a C^1-diffeomorphism, then $\det\left(Dg(x) \right) \neq 0$ for all $x \in \Omega_1$

19-11 Prove that a set $M \subseteq \mathbb{R}^d$ is an m-dimensional embedded manifold in $\mathbb{R}^d$ iff for each $p \in M$ there are an open neighborhood $G \subseteq \mathbb{R}^d$, an open set $N \subseteq \mathbb{R}^m$ and an injective infinitely differentiable function f . $N \to G$ so that $f[N] = M \cap G$, $Df(z)$ has rank m for all $z \in N$ and f^{-1} . $f[N] \to N$ is continuous

Hint For "$\Leftarrow$," let $x \in N$ be so that $f(x) = p$ and let the vectors v_{m+1} . $v_d \in \mathbb{R}^d$ be so that the set $\left\{ \frac{\partial}{\partial x_1} f(x), \quad . \frac{\partial}{\partial x_m} f(x) \; v_{m+1}, \quad . v_d \right\}$ is a base of $\mathbb{R}^d$ Then apply Corollary 17 66 to the function $F(z_1 \quad . z_d) = f(z_1, \quad z_m) + \sum_{j=m+1}^{d} z_j v_j$

19-12 Let $\Omega \subseteq \mathbb{R}^2$ be open. Prove that no function $f : \Omega \to \left\{ p \in \mathbb{R}^2 : \|p\|_2 = 1 \right\}$ can be bijective, continuously differentiable and so that all $Dg(x)$ are injective.
Hint Use an idea similar to that for Exercise 19-11 to prove that f^{-1} is continuous.

19-13 Corners cannot be described with embedded manifolds. Let $d \geq 2$

(a) Prove that the graph $\left\{ (t, |t|, 0, \ldots, 0) : t \in (-1, 1) \right\}$ of the absolute value function is not an embedded manifold in $\mathbb{R}^d$.
Hint. Use Exercise 19-11

(b) Prove that the cube $[0, 1]^d$ is not an embedded manifold with boundary in $\mathbb{R}^d$.
Hint Use Exercise 19-5e and arguments similar to part 19-13a.

19-14 Let M be a manifold with corners.

(a) Prove that if $p \in M$ is in a corner isomorphic to the origin in C_k, then there are continuous functions $c_k, \ldots, c_m : [0, 1] \to M$ so that each $c_i \left[[0, 1] \right]$ is contained in a corner and the intersection of any two distinct sets $c_i \left[[0, 1] \right]$ is $\{p\}$.

(b) Prove that the set $C_M = \{p \in M : p \text{ is in a corner}\}$ is closed.

(c) Prove that if M is compact, then there are finitely many continuous functions $c_1, \ldots, c_n$ so that $C_M = \bigcup_{i=1}^{n} c_i \left[[0, 1] \right]$.

19-15. Prove that if $\mathcal{A}$ is an atlas of the manifold M, then the set $\bigcup \left\{ \widetilde{\mathcal{A}} : \mathcal{A} \subseteq \widetilde{\mathcal{A}} \text{ and } \widetilde{\mathcal{A}} \text{ is an atlas of } M \right\}$ is also an atlas of M.
Note. This atlas is called the **maximal atlas** of M

19.2 Tangent Spaces and Differentiable Functions

The definition of differentiable functions on manifolds (see Definition 19.14 below) is motivated by the fact that compositions of differentiable functions are again differentiable. However, rather than using differentiability of the factors to obtain differentiability of the composition, differentiability of the composition is used to define differentiability of the middle factor. Exercise 19-16a shows that this idea is consistent with the idea of differentiability according to Definition 17.24.

Definition 19.14 *Let M, N be manifolds and let $f : M \to N$ be a function. Then f is called* **differentiable** *iff for all x in the atlas of M and all y in the atlas of N the composition $y \circ f \circ x^{-1}$ is differentiable.*

Note that Definition 19.14 does not depend on the atlases used for M and N, as long as there is a containment relation between the atlases for each manifold. This is because if x_1 and x_2 are coordinate systems of M with overlapping domains, then the composition $x_1 \circ x_2^{-1}$ is differentiable and similar for coordinate systems y_1, y_2 for N. Therefore differentiability of $y_1 \circ f \circ x_1^{-1}$ implies differentiability of $y_2 \circ f \circ x_2^{-1}$ on a subset of its domain and the domain of $y_2 \circ f \circ x_2^{-1}$ can be pieced together from overlaps with domains of similar compositions using functions $\widetilde{x}_1$ and $\widetilde{y}_1$ from smaller atlases of M and N, respectively. The details are left to Exercise 19-16b. Proposition 19.15 and Exercise 19-17 show that differentiable functions on manifolds behave as we would expect differentiable functions to behave when it comes to domain restrictions and compositions.

Proposition 19.15 *Let M, N be manifolds, let $U \subseteq M$ be an open submanifold and let $f : M \to N$ be a differentiable function. Then $f|_U : U \to N$ also is differentiable*

Proof. Exercise 19-18. ■

Before we show that there are indeed many examples of differentiable functions, we introduce higher orders of differentiability. *Throughout, unless otherwise stated, the dimension of a manifold M will be m and the dimension of a manifold N will be n.*

Definition 19.16 *Let M, N be manifolds, let $k \in \mathbb{N}$ and let $U \subseteq M$ be open. Then $f : M \to N$ is a C^k function on U iff for all coordinate systems $x : V \to \mathbb{R}^m$ and $y : W \to \mathbb{R}^n$ the composition $y \circ f \circ x^{-1}\Big|_{x[U \cap V \cap f^{-1}[W]]}$ is a C^k function. A function that is C^k for all $k \in \mathbb{N}$ is called C^∞. A bijective C^k function whose inverse also is C^k is called a C^k-**diffeomorphism***

Proposition 19.17 *Let M be a manifold. A function $f : M \to \mathbb{R}$ is C^∞ iff for each $p \in M$ there is a neighborhood U_p of p so that f is C^∞ on U_p*

Proof. Exercise 19-20. ■

The next result is a translation of Lemma 18.11 to manifolds.

Theorem 19.18 *Let M be a manifold, let $C \subseteq M$ be compact and let $U \subseteq M$ be open so that $C \subseteq U$. Then there is a C^∞ function $f : M \to [0, 1]$ so that $f|_C = 1$ and* $\text{supp}(f) \subseteq U$

Proof. Let $p \in C$ and let $x : V \to \mathbb{R}^m$ be a coordinate system around p. Then there is an $\varepsilon_p > 0$ so that $\overline{B_{\varepsilon_p}(p)} \subseteq U \cap V$. Because x^{-1} is continuous, the image $U_{\mathbb{R}^m} := x\left[B_{\frac{2\varepsilon_p}{3}}(p)\right]$ is open and $C_{\mathbb{R}^m} := x\left[\overline{B_{\frac{\varepsilon_p}{3}}(p)} \cap C\right]$ is closed. With $\mathbf{1}_{C_{\mathbb{R}^m}, U_{\mathbb{R}^m}}$ as in Lemma 18.11 let $f_p(q) := \mathbf{1}_{C_{\mathbb{R}^m}, U_{\mathbb{R}^m}} \circ x(q)$ for all $q \in U \cap V$ and let $f_p(q) := 0$ for all other $q \in M$. It is easy to see that $\text{supp}(f_p) \subseteq \overline{B_{\frac{2\varepsilon_p}{3}}(p)} \subseteq U$.

To prove that f_p is C^∞, let $q \in M$. If f_p is identical to zero in a neighborhood of q, then clearly f_p is C^∞ in that neighborhood. If f_p is not identical to zero in any neighborhood of q, then $q \in \text{supp}(f_p) \subseteq \overline{B_{\frac{2\varepsilon_p}{3}}(p)}$. Let y be any coordinate system around q and let $\varepsilon \in \left(0, \frac{\varepsilon_p}{3}\right)$ be so that $B_\varepsilon(q)$ is contained in the domain of y. Because $\varepsilon < \frac{\varepsilon_p}{3}$ and $q \in \text{supp}(f_p)$, $B_\varepsilon(q)$ is also contained in the domain of x. Thus $f_p \circ y^{-1}\Big|_{y[B_\varepsilon(q)]} = \left(f_p \circ x^{-1}\right) \circ \left(x \circ y^{-1}\right)\Big|_{y[B_\varepsilon(q)]}$ is C^∞. Because $q \in M$ was arbitrary, f_p is C^∞.

Because $C \subseteq \bigcup_{p \in C} B_{\frac{\varepsilon_p}{3}}(p)$ and C is compact, there are points $p_1, \ldots, p_n \in C$ so that $C \subseteq \bigcup_{i=1}^{n} B_{\frac{\varepsilon_{p_i}}{3}}(p_i)$. Now $g := \sum_{i=1}^{n} f_{p_i}$ is C^∞, $g|_C \geq 1$ and $\text{supp}(g) \subseteq U$. Hence, $f := j_1 \circ g$ (see Lemma 18.9 for the definition of j_1) is as desired. ■

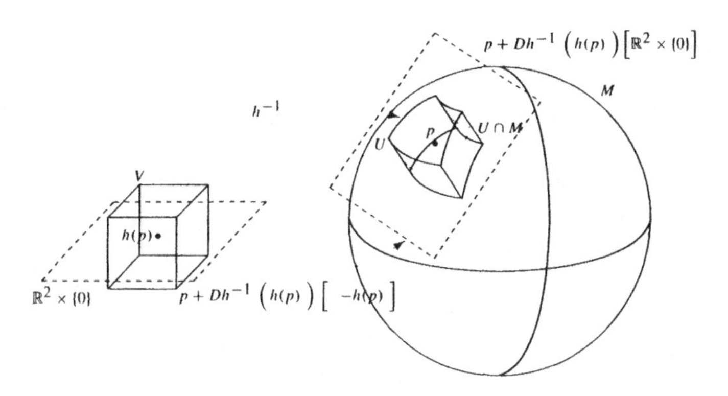

Figure 49: If M is an embedded manifold and h is as in Definition 19.5, then the derivative of h^{-1} maps the horizontal space into which $h[U \cap M]$ is embedded to a space that is tangential to M.

Although Definition 19.14 defines differentiability on a manifold, it is not satisfactory by itself. After all, differentiable functions have a derivative and Definition 19.14 does not tell us what the derivative is or what it should be. The problem is that a manifold does not have any linear structure. Without linear structure there is no way to shift a linear function $L[\cdot - x]$ so that $f(x) + L[\cdot - x]$ is locally a good approximation of f, as we did in Definition 17.24 (also see Exercise 17-25). In fact, we cannot even *define* linear functions, which means on a manifold we must start from scratch.

For embedded manifolds, there is a notion of differentiability in the surrounding space. Hence, we will use embedded manifolds as guidance. Throughout, we will make sure that our newly defined notions are consistent with what we should expect for embedded manifolds.

So consider an embedded manifold $M \subseteq \mathbb{R}^d$ and let $h : U \to V$ be as in Definition 19.5. Then for all $p \in U \cap M$ the function h^{-1} is differentiable at $h(p)$ and near p the affine function $h^{-1}\big(h(p)\big) + Dh^{-1}\big(h(p)\big)\big[\cdot - h(p)\big]$ is tangential to h^{-1} (see Figure 49). With $\mathbb{R}^m \times \{0\}^{d-m} := \left\{v \in \mathbb{R}^d : v_{m+1} = \cdots = v_d = 0\right\}$ there is an open set V in $\mathbb{R}^d$ so that h maps $U \cap M$ to the set $V \cap \left(\mathbb{R}^m \times \{0\}^{d-m}\right)$. It follows directly from the definition of differentiability that for every $\varepsilon > 0$ there is a $\delta > 0$ so that for all $z \in V \cap \left(\mathbb{R}^m \times \{0\}^{d-m}\right)$ with $\big\|z - h(p)\big\| < \delta$ we have $\left\|h^{-1}(z) - h^{-1}\big(h(p)\big) - Dh^{-1}\big(h(p)\big)\big[z - h(p)\big]\right\| \leq \varepsilon\big\|z - h(p)\big\|$. Therefore, the function $p + Dh^{-1}\big(h(p)\big)\big[\cdot - h(p)\big]\Big|_{\mathbb{R}^m \times \{0\}^{d-m}}$ is tangential to $h^{-1}\Big|_{V \cap (\mathbb{R}^m \times \{0\}^{d-m})}$. Geometrically, this means that the set of points $p + Dh^{-1}\big(h(p)\big)\left[\mathbb{R}^m \times \{0\}^{d-m}\right]$ is the tangent space of M at p (see Figure 49). It can be shown via the Chain Rule that this tangent space does not depend on the choice of h (see Exercise 19-19). Moreover, if M and N are embedded manifolds and f is a differentiable function on a neighborhood of M so that $f[M] \subseteq N$, then the derivative of f at p maps the above defined tangent space of M at p to the similarly defined tangent space of N at $f(p)$.

This means we can reinterpret derivatives as functions that map the right tangent spaces to each other. But for an abstract manifold there is no surrounding space in which to define tangent spaces on which the derivative of a function $f : M \to N$ could operate. Thus we first need to create tangent spaces. There are numerous ways in which tangent spaces can be defined in differential geometry and they are all equivalent in a certain sense. We choose a simple definition here and reinterpret it when this becomes conducive to the investigation in Section 19.4.

If x is a coordinate system for the manifold M, then for every element p in the domain of x the space $\mathbb{R}^m$ is the tangent space (of $\mathbb{R}^m$ itself) at $x(p)$. This is because the only space that could be tangential to an open set is the surrounding space itself. Proposition 19.19 shows that if x and y are coordinate systems of M and p is in the domains of x and y, then the tangent vectors in the tangent spaces at points $x(p)$ and $y(p)$ can be identified in such a way that they are useable as tangent vectors for M at p. Proposition 19.20 shows that this idea is consistent with the idea of a tangent space for embedded manifolds as described above.

The remaining results in this section will involve a lot of work with equivalence relations. Because every point of a manifold can be in the domain of several coordinate systems, at each step we must assure that our definitions do not depend on the specific coordinate system we use. This level of detail would be hard to work with on a regular basis. It would be similar to recall that formally every real number is an equivalence class of Cauchy sequences of rational numbers (see remarks after the proof of Theorem 16.89) whenever we work with real numbers. Clearly, this would be overkill. Therefore, we will establish that tangent spaces and the functions that take the place of derivatives behave as they should and we will subsequently use these properties, with the details of the definitions only to be used when this is unavoidable.

Proposition 19.19 *Let M be an m-dimensional manifold and let $p \in M$. For all $v, w \in \mathbb{R}^m$ and for all coordinate systems x, y so that p is in the domains of x and y, define the relation $\sim_p$ by $(x, v) \sim_p (y, w)$ iff $w = D\left(y \circ x^{-1}\right)(x(p))[v]$. Then the relation $\sim_p$ is an equivalence relation. Moreover, if we denote the equivalence classes by $[x, v]_p$ and the set of equivalence classes by M_p, then the binary operations $[x, u]_p + [x, v]_p := [x, u + v]_p$ and $\alpha[x, v]_p := [x, \alpha v]_p$, where $\alpha \in \mathbb{R}$, are well-defined and with these operations M_p is a vector space that is isomorphic to $\mathbb{R}^m$.*

Proof. Left to the reader as Exercise 19-21. The proof that $\sim_p$ is well-defined relies on the formula for the derivative of the inverse function for symmetry and on the Chain Rule for transitivity. ■

Proposition 19.20 *Let $M \subseteq \mathbb{R}^d$ be an m-dimensional embedded manifold, let $p \in M$ and let $x : U \to \mathbb{R}^m$ be a coordinate system around p as constructed in Proposition 19.6. Then x^{-1} is differentiable in the sense of Definition 17.24 and the "embedded tangent space" $M_p^{emb} := \{p\} \times Dx^{-1}(x(p))\left[\mathbb{R}^m\right]$ is isomorphic to M_p via the isomorphism $F\big((p, v)\big) := \left[x, \left(Dx^{-1}(x(p))\right)^{-1}[v]\right]_p$.*

Proof. Exercise 19-22. ■

Proposition 19.19 introduces an object M_p that could serve as a tangent space for M at p and Proposition 19.20 shows that for embedded manifolds there is a natural isomorphism between M_p and the space that we would expect to be the tangent space. Thus we call M_p the tangent space of M at p.

Definition 19.21 *The space M_p of Proposition 19.19 will be called the* **tangent space** *of M at p. The set $TM := \bigcup_{p\in M} M_p$ is called the* **tangent bundle** *of M.*

Simplistically speaking, at every $p \in M$ Definition 19.21 merely tacks a tangent space on to the manifold. However, even if this was all, this approach is consistent with an important physical motivation. Vectors in physics have magnitude and direction, just like vectors in mathematics. But vectors in physics also have a point of action. For example, consider a car that moves in a straight line at constant speed. The magnitude and direction of the force that the car exerts on the particles in front of it stay the same, independent of whether these particles are air or whether they constitute another car. Obviously, the effect is different in either situation. If the force acts on a set of air molecules, the car travels regularly. If it acts on another car, the car crashes. To give vectors in mathematics a point of action, we need to incorporate the point of action into the definition of the vector. This is done in the definition of TM.

For $\left(\mathbb{R}^d, \left\{id_{\mathbb{R}^d}\right\}\right)$, note that $\mathbb{R}^d_p$ is just $\left\{[id_{\mathbb{R}^d}, v]_p : v \in \mathbb{R}^d\right\}$, which is consistent with the idea that our vectors now have a point of action p. Similarly, for an open set $\Omega \subseteq \mathbb{R}^d$ considered as a manifold (Ω, i_Ω) we obtain $\Omega_p = \left\{[i_\Omega, v]_p : v \in \mathbb{R}^d\right\}$ for the tangent space. Although these realizations are trivial, they will be very useful when we consider integration over embedded manifolds.

Now that tangent spaces are defined, we can tackle the definition of a "derivative" on M. Consider embedded manifolds M and N and a differentiable function f defined on a neighborhood of M so that $f[M] \subseteq N$. Then the Chain Rule implies $Df(p)\left[Dh^{-1}(h(p))\left[\mathbb{R}^m \times \{0\}^{d-m}\right]\right] = D\left(f \circ h^{-1}\right)(h(p))\left[\mathbb{R}^m \times \{0\}^{d-m}\right]$, which is contained in the tangent space of N at $f(p)$, and it is equal to the tangent space if we assume that f is a diffeomorphism. Therefore derivatives on manifolds should map the tangent space at a point into the tangent space at the image point (see Figure 50). Proposition 19.22 shows that it is possible to define such a mapping on the tangent vectors and Proposition 19.24 shows that this map is consistent with what we expect it to do for embedded manifolds.

Proposition 19.22 *Let M, N be manifolds, let the function $f : M \to N$ be differentiable, let $p \in M$, let x be a coordinate system around p and let y be a coordinate system around $f(p)$. Then $f_{*p}\left([x, v]_p\right) := \left[y, D\left(y \circ f \circ x^{-1}\right)(x(p))[v]\right]_{f(p)}$ defines a linear function $f_{*p} : M_p \to N_p$.*

Proof. We first need to prove that f_{*p} is well-defined. Let $(x_1, v) \sim_p (x_2, w)$ and let y_1 and y_2 be coordinate systems about $f(p)$. Then

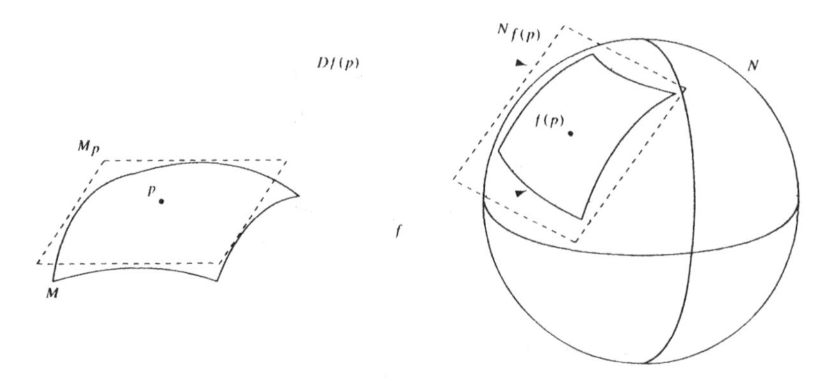

Figure 50: The tangent space M_p of a manifold M at a point p (see Definition 19.21) can be viewed as a tangential plane attached at p. For embedded manifolds, M_p can be considered to be the image of $\mathbb{R}^m$ under the derivative of the right parametrization (see Proposition 19.20). For a differentiable function $f : M \to N$ the map f_* (see Proposition 19.22) plays the role of the derivative. If M and N are both embedded manifolds and f is a differentiable function from a neighborhood of M to a neighborhood of N, then f_* combines the function and the derivative (see Proposition 19.24).

$$
\begin{aligned}
& D\left(y_1 \circ f \circ x_1^{-1}\right)(x_1(p))[v] \\
&= D\left(y_1 \circ y_2^{-1} \circ y_2 \circ f \circ x_2^{-1} \circ x_2 \circ x_1^{-1}\right)(x_1(p))[v] \\
&= D\left(y_1 \circ y_2^{-1}\right)(y_2(f(p))) \circ D\left(y_2 \circ f \circ x_2^{-1}\right)(x_2(p)) \circ D\left(x_2 \circ x_1^{-1}\right)(x_1(p))[v] \\
&= D\left(y_1 \circ y_2^{-1}\right)(y_2(f(p)))\left[D\left(y_2 \circ f \circ x_2^{-1}\right)(x_2(p))[w]\right],
\end{aligned}
$$

so $\left(y_1, D\left(y_1 \circ f \circ x_1^{-1}\right)(x_1(p))[v]\right) \sim_{f(p)} \left(y_2, D\left(y_2 \circ f \circ x_2^{-1}\right)(x_2(p))[w]\right)$, which proves that f_{*p} is well-defined. Linearity is now trivial. ∎

Definition 19.23 *We denote the function from TM to TN whose restriction to each M_p is f_{*p} by f_*. Moreover, unless necessary, we will not explicitly mention p and thus denote f_{*p} by f_*, too.*

Proposition 19.24 *Let M, N be embedded manifolds, let $p \in M$ and let f be a differentiable function from a neighborhood of M to a neighborhood of N. Then for all $(p, v) \in M_p^{\text{emb}}$ the function $f_*^{emb}((p, v)) := \big(f(p), Df(p)[v]\big)$ satisfies the equation $f_* = F_N \circ f_*^{\text{emb}} \circ F_M^{-1}$, where F_M and F_N are the isomorphisms from Proposition 19.20 for M and N, respectively.*

Proof. Let x be the coordinate system around p that is used to construct F_M and let y be the coordinate system around $f(p)$ that is used to construct F_N. Then

$$
\begin{aligned}
&F_N \circ f_*^{\text{emb}} \circ F_M^{-1}\left([x, v]_p\right) \\
&= F_N \circ f_*^{\text{emb}}\left(p, Dx^{-1}(x(p))[v]\right) \\
&= F_N\left(f(p), Df\left(x^{-1}(x(p))\right) \circ Dx^{-1}(x(p))[v]\right) \\
&= F_N\left(f(p), D\left(f \circ x^{-1}\right)(x(p))[v]\right) \\
&= \left[y, \left(Dy^{-1}(y(f(p)))\right)^{-1} \circ D\left(f \circ x^{-1}\right)(x(p))[v]\right]_{f(p)} \\
&= \left[y, Dy(f(p)) \circ D\left(f \circ x^{-1}\right)(x(p))[v]\right]_{f(p)} \\
&= \left[y, D\left(y \circ f \circ x^{-1}\right)(x(p))[v]\right]_{f(p)} \\
&= f_*\left([x, v]_p\right).
\end{aligned}
$$

■

As noted in the beginning, the details of chasing these compositions through the equivalence classes are too cumbersome to do on a regular basis. To avoid this level of detail, we accept that TM really defines the tangent spaces and that f_* is the "derivative" of a function $f : M \to N$, and we subsequently use the fundamental properties of these entities rather than their definitions.

For manifolds with boundary and manifolds with corners, we can also define tangent spaces. Because differentiable functions on sets that are not closed are restrictions of differentiable functions on larger sets, the derivatives $D\left(y \circ x^{-1}\right)$ are also defined for boundary points. This means we can say the following.

Definition 19.25 *If M is a manifold with boundary or a manifold with corners, the tangent bundle TM is defined in the same way as for manifolds.*

For points on the boundary of the manifold, there are two kinds of tangent vectors.

Definition 19.26 *Let M be a manifold with boundary and let $p \in \partial M$. A tangent vector $[x, v]_p$ will be called* **outward pointing** *iff $v_m < 0$ and it will be called* **inward pointing** *iff $v_m > 0$. For p being in the boundary of a manifold with corners the vector $[x, v]_p$ will be called* **outward pointing** *iff $x(p) + v \notin \overline{C_k}$. The vector $[x, v]_p$ will be called* **inward pointing** *iff $x(p) + v \in \left(C_k\right)^{\circ}$.*

With tangent spaces defined, it is now easy to define (tangential) vector fields.

Definition 19.27 *Let M be a manifold. A function $F : M \to TM$ so that $F(p) \in M_p$ for all $p \in M$ is called a* **vector field** *on M.*

Finally, note that even though we assumed throughout that our manifolds were C^{∞} manifolds, everything in this section can be defined and proved for C^1 manifolds.

Exercises

19-16 Consistency of Definition 19.14 with the original definition of differentiability and with itself.

(a) Let $\Omega_1, \Omega_1' \subseteq \mathbb{R}^m$ and $\Omega_2, \Omega_2' \subseteq \mathbb{R}^n$ be open sets, let $f : \Omega_1 \to \Omega_2$ be a function and let $\varphi : \Omega_1 \to \Omega_1'$ and $\psi : \Omega_2 \to \Omega_2'$ be differentiable bijective functions with differentiable inverse. Prove that f is differentiable iff $\psi \circ f \circ \varphi^{-1}$ is differentiable.

(b) Let M be a manifold with atlas $\mathcal{A}_M$, let N be a manifold with atlas $\mathcal{A}_N$ and let $f : M \to N$ be so that for all $x_A \in \mathcal{A}_M$ and all $y_A \in \mathcal{A}_N$ the composition $y_A \circ f \circ x_A^{-1}$ is differentiable. Let $\widetilde{\mathcal{A}}_M \supseteq \mathcal{A}_M$ and $\widetilde{\mathcal{A}}_N \supseteq \mathcal{A}_N$ be atlases of M and N, let $x : U \to \mathbb{R}^m$ be in $\widetilde{\mathcal{A}}_M$ and let $y : V \to \mathbb{R}^n$ be in $\widetilde{\mathcal{A}}_N$ so that the composition $y \circ f \circ x^{-1}$ has nonempty domain. Prove that $y \circ f \circ x^{-1}$ is differentiable.

19-17. Let M, N, O be manifolds and let $f : M \to N$ and $g : N \to O$ be differentiable. Prove that $g \circ f$ is differentiable and that $(g \circ f)_* = g_* \circ f_*$.

19-18. Prove Proposition 19.15.

19-19. Prove that if M is an embedded manifold and $h : U \to V$ and $k : \widetilde{U} \to \widetilde{V}$ are as in Definition 19.5, then $Dh^{-1}\left(h(p)\right)\left[\mathbb{R}^m \times \{0\}^{d-m}\right] = Dk^{-1}\left(k(p)\right)\left[\mathbb{R}^m \times \{0\}^{d-m}\right]$.

19-20. Prove Proposition 19.17.

19-21. Prove Proposition 19.19. That is, prove that the relation $\sim_p$ is an equivalence relation and that the vector addition and scalar multiplication are well-defined.

19-22. Prove Proposition 19.20.

19-23. Let M be a manifold. Prove that if $id : M \to M$ is the identity, then $id_* : TM \to TM$ also is the identity.

19-24. Let M, N be manifolds and let $U \subseteq M$ be an open submanifold of M.

(a) Prove that the tangent bundle $TU = \bigcup_{p \in U} U_p$ of U is equal to $\bigcup_{p \in U} M_p$.

(b) Prove that $(f|_U)_* = f_*|_U$ for all differentiable functions $f : M \to N$.

19-25. Let M be an m-dimensional manifold. Prove that TM is a $2m$-dimensional manifold.

19-26. Prove that if $M \subseteq \mathbb{R}^d$ is an embedded C^∞ manifold and Ω is an open neighborhood of M, then for every $f \in C^k(\Omega)$ the restriction $f|_M$ is C^k on M.

19-27. Let M be a connected manifold, let $C \subseteq M$ be closed (but not necessarily compact) and let $U \subseteq M$ be open so that $C \subseteq U$. Prove that there is a C^∞ function $f : M \to [0, 1]$ so that $f|_C = 1$ and $\text{supp}(f) \subseteq U$.

19.3 Differential Forms, Integrals Over the Unit Cube

As we start our investigation of integration on manifolds, we first note the following three shortcomings of the tools we currently have available.

First, although vector fields as in Definition 19.27 are important, they have a mortal flaw for applications in fluid mechanics and field theory. Any vector field that is defined as a map from the manifold into the tangent bundle is necessarily tangential to the manifold. In fluid mechanics and field theory, manifolds are test surfaces and vector fields usually go *through* these test surfaces or at least they have a component that causes some "transfer" through the surface. Obviously such vector fields cannot be modeled with the tangent bundle.

Second, a typical application of the interplay between fields and surfaces is the computation how much "matter" a vector field transfers through a test surface. To compute this quantity, we need to integrate the field over the manifold. In $\mathbb{R}^d$, integrals usually are computed with Fubini's Theorem and we do not give much thought to the fact that the coordinate directions are provided by the standard base. On a manifold there is no standard base and each neighborhood of a point has infinitely many parametrizations via coordinate systems. Thus we cannot just pick one coordinate system and use the standard base in its image space.

Third, lower dimensional objects, like two dimensional surfaces in three dimensional space, typically have measure zero in the higher dimensional space. Therefore, we need a notion of integration that gives nonzero integrals, even if our manifold resides in a higher dimensional space.

The above only reemphasizes that the requisite definitions will require a lot of attention to detail. Differential forms will allow us to model vector fields that are not necessarily parallel to the manifold. The integral of such a form over a manifold will be pieced together from simpler integrals. The simplest such integral is the integral of a differential form over a cube, which is presented in this section. In Section 19.4, cubes in $\mathbb{R}^m$ are lifted to the manifold as k-cubes and the differential forms will be k-forms on the manifold. Finally, Section 19.5 defines the integral of a form over the whole manifold. Some definitions will be repeated in this presentation, but the insight gained by first working out the details in a simple setting will be well worth it. (Compare with the double coverage of Lebesgue integration in Chapters 9 and 14.) For starters, recall that $\Lambda^k(V)$ denotes the space of alternating k-tensors on V (Definition 18.17).

Definition 19.28 *Let Ω be a subset of $\mathbb{R}^d$. A function $\omega : \Omega \to \bigcup_{p\in\Omega} \Lambda^k\left(\mathbb{R}^d_p\right)$ so that $\omega(p) \in \Lambda^k\left(\mathbb{R}^d_p\right)$ for all $p \in \Omega$ is called a k-**form** on Ω or simply a **differential form**. We formally set $\Lambda^0\left(\mathbb{R}^d_p\right) := \mathbb{R}$, so that a 0-**form** simply is a function.*

Because it is natural to identify each $[id_{\mathbb{R}^d}, v]_p \in \mathbb{R}^d_p$ with v, we will denote tangent vectors to $\mathbb{R}^d$ by single letters in this section. Because alternating k-tensors are associated with k-dimensional volumes (see Example 18.21) we introduce the following notation for the dual base.

Definition 19.29 *Let $\{e_1, \ldots, e_d\}$ be the standard base for $\mathbb{R}^d_p$. We will denote the dual base $\{\pi_1, \ldots, \pi_d\}$ as $\{dx_1, \ldots, dx_d\}$ where $dx_i(e_j) = \pi_i(e_j)$.*

The notation is similar to that for integrals because forms are connected to integrals of vector fields as follows. A differentiable function $r : [a,b] \to \mathbb{R}^3$ with $r'(t) \neq 0$ for all $t \in [a,b]$ can be interpreted as describing the position $r(t)$ of a **traveling particle** at time t. If Ω contains $r\big[[a,b]\big]$ and the vector field $F : \Omega \to T\Omega$ is so that $F(x,y,z)$ describes a force that is acting on a particle at the point (x,y,z), then the work that is done as the particle travels from $r(a)$ to $r(b)$ is the **line integral** $W = \int_a^b \left\langle F\big(r(t)\big), \frac{d}{dt}r(t)\right\rangle dt$, where we assume that each $F\big(r(t)\big)$ was projected back from $\Omega_{r(t)}$ to $\mathbb{R}^3$ via $[i_\Omega, v]_{r(t)} \mapsto v$. By Exercise 18-53b, the value of this integral depends only on the geometric shape of $r\big[[a,b]\big]$ and the direction of travel, but not on the speed at which the particle travels from $r(a)$ to $r(b)$ along this path. With $F(t) = \left[i_\Omega, \begin{pmatrix} P(t) \\ Q(t) \\ R(t) \end{pmatrix} \right]_{r(t)}$ we can write the integral as

$$W = \int_a^b P\big(r(t)\big) r_1'(t)\, dt + \int_a^b Q\big(r(t)\big) r_2'(t)\, dt + \int_a^b R\big(r(t)\big) r_3'(t)\, dt.$$

Each $r_i'(t)$ can be obtained from $r'(t)$ by applying the form dx_i. The integral is thus also abbreviated as $W = \int_a^b P\, dx_1 + \int_a^b Q\, dx_2 + \int_a^b R\, dx_3$, where dx_i stands for $dx_i\big(r'(t)\big)\, dt$. The form $\omega = P\, dx_1 + Q\, dx_2 + R\, dx_3$ can be interpreted as the differential amount of work that is done as the particle moves a differential step $\begin{pmatrix} dx_1 \\ dx_2 \\ dx_3 \end{pmatrix}$ from its current position $r(t)$ to $r(t) + r'(t)\, dt = r(t) + \begin{pmatrix} dx_1 \\ dx_2 \\ dx_3 \end{pmatrix}$. The integration formalism that we will define in Section 19.5 will indeed locally reduce the integral of the form ω to the above line integral.

Similarly, an injective differentiable function $r : [a, b] \times [c, d] \to \mathbb{R}^3$ so that $\left(\frac{\partial r}{\partial x} \times \frac{\partial r}{\partial y} \right)(x, y) \neq 0$ for all $(x, y) \in [a, b] \times [c, d]$ can be interpreted as a **surface** in $\mathbb{R}^3$. If we interpret F as a flow field, then (with the right hypotheses) the throughput of F through the surface defined by r is the **surface integral**

$$\begin{aligned} T &= \int_a^b \int_c^d \left\langle F\big(r(x, y)\big), \frac{\partial r}{\partial x}(x, y) \times \frac{\partial r}{\partial y}(x, y) \right\rangle dy\, dx \\ &= \int_a^b \int_c^d P\big(r(x, y)\big) \left(\frac{\partial r_2}{\partial x}\frac{\partial r_3}{\partial y} - \frac{\partial r_2}{\partial y}\frac{\partial r_3}{\partial x} \right) dy\, dx \\ &\quad + \int_a^b \int_c^d Q\big(r(x, y)\big) \left(\frac{\partial r_3}{\partial x}\frac{\partial r_1}{\partial y} - \frac{\partial r_3}{\partial y}\frac{\partial r_1}{\partial x} \right) dy\, dx \\ &\quad + \int_a^b \int_c^d R\big(r(x, y)\big) \left(\frac{\partial r_1}{\partial x}\frac{\partial r_2}{\partial y} - \frac{\partial r_1}{\partial y}\frac{\partial r_2}{\partial x} \right) dy\, dx. \end{aligned}$$

Exercise 18-53c shows that this integral also only depends on the geometric shape of the surface as long as the parametrizations respect the orientation of the surface. (Details on the rather subtle notion of orientation will be addressed later.) With the wedge product from Definition 18.20 the factors behind the components of F can be obtained by applying the forms $dx_2 \wedge dx_3$, $dx_3 \wedge dx_1$ and $dx_1 \wedge dx_2$ to the vectors $\frac{\partial r}{\partial x}$ and $\frac{\partial r}{\partial y}$. Thus $\omega = P\, dx_2 \wedge dx_3 + Q\, dx_3 \wedge dx_1 + R\, dx_1 \wedge dx_2$ can be interpreted as the differential throughput of the vector field F through a parallelogram spanned by the differential vectors $\frac{\partial r}{\partial x} dx$ and $\frac{\partial r}{\partial y} dy$ attached at $r(x, y)$ and located on the surface defined by r (also see Figure 51 on page 442).

Finally, for a scalar function $f : [a,b] \times [c,d] \times [l,h] \to \mathbb{R}$ the integral is $\int_l^h \int_c^d \int_a^b f(x,y,z)\,dx\,dy\,dz$. This time there is no extra factor involved. The form $f(p)\,dx \wedge dy \wedge dz$ can be interpreted as the differential contribution of $f(p)$ to the overall integral. To stay consistent with the above, we could say that the box is parametrized by the identity. The scaling factor, which is 1, is obtained by applying $dx \wedge dy \wedge dz$ to the partial derivatives of the identity with respect to x, y, and z.

The above indicates that differential forms should allow us to describe line integrals of vector fields, surface integrals of vector fields over surfaces with rectangular parameter domain, and integrals of scalar functions over boxes with *one* formalism. Moreover, in this formalism the forms carry *almost all* the information needed for the integral. The components of the field as well as the directions in which we integrate are part of the form. Only the parameter domain is not given and it can be supplied by a coordinate system. Therefore, this approach should be the right idea and, after taking care of the details, we will indeed see that forms on manifolds can be used to model vector fields on surfaces.

Definition 19.30 *Let $\Omega \subseteq \mathbb{R}^d$ be open and let $\omega : \Omega \to \bigcup_{p\in\Omega} \Lambda^k(\Omega_p)$ be a k-form on Ω. By Theorem 18.23, at every $p \in \Omega$ there is a base representation of the form $\omega(p) = \sum_{1\leq i_1<\cdots<i_k\leq d} \omega_{i_1,\ldots,i_k}(p)\,dx_{i_1} \wedge \cdots \wedge dx_{i_k}$. We will call ω* **differentiable** *at p iff each of the $\omega_{i_1,\ldots,i_k}$ is differentiable at p. If ω is differentiable at each $p \in \Omega$ we will call ω* **differentiable** *on Ω.*

Recall that by Theorem 17.43 for differentiable functions $f : \mathbb{R}^d \to \mathbb{R}$ the derivative Df is $Df = \frac{\partial f}{\partial x_1}\,dx_1 + \cdots + \frac{\partial f}{\partial x_d}\,dx_d$ and it is only a small abuse of notation to set $df := Df$. We define the differential of a k-form similar to the $k+1^{\text{st}}$ derivative of regular functions (see Corollary 17.61), except that we work with wedge products instead of tensor products. In this fashion, we will retain the connection to differential contributions to integrals.

Definition 19.31 *Let $\Omega \subseteq \mathbb{R}^d$ be open and let $f : \Omega \to \mathbb{R}$ be differentiable. We define $df := \frac{\partial f}{\partial x_1}\,dx_1 + \cdots + \frac{\partial f}{\partial x_d}\,dx_d$. If $\omega : \Omega \to \bigcup_{p\in\Omega} \Lambda^k(\Omega_p)$ is differentiable and*

$$\omega(p) = \sum_{1\leq i_1<\cdots<i_k\leq d} \omega_{i_1,\ldots,i_k}(p)\,dx_{i_1} \wedge \cdots \wedge dx_{i_k},$$

then we define the $(k+1)$-form $d\omega$ by

$$\begin{aligned} d\omega(p) &:= \sum_{1\leq i_1<\cdots<i_k\leq d} d\omega_{i_1,\ldots,i_k}(p)\,dx_{i_1} \wedge \cdots \wedge dx_{i_k} \\ &= \sum_{1\leq i_1<\cdots<i_k\leq d} \sum_{\alpha=1}^{d} \frac{\partial \omega_{i_1,\ldots,i_k}(p)}{\partial x_\alpha}\,dx_\alpha \wedge dx_{i_1} \wedge \cdots \wedge dx_{i_k} \end{aligned}$$

and call it the **differential** *of ω.*

Definition 19.31 allows us to make some connections that may be familiar from multivariable calculus or physics.

Example 19.32

1. *If* $\omega = P\,dx_2 \wedge dx_3 + Q\,dx_3 \wedge dx_1 + R\,dx_1 \wedge dx_2$, *then*
$$d\omega = \left(\frac{\partial P}{\partial x_1} + \frac{\partial Q}{\partial x_2} + \frac{\partial R}{\partial x_3}\right) dx_1 \wedge dx_2 \wedge dx_3.$$
2. *If* $\omega = P\,dx_1 + Q\,dx_2 + R\,dx_3$, *then*
$$\begin{aligned} d\omega &= \left(\frac{\partial R}{\partial x_2} - \frac{\partial Q}{\partial x_3}\right) dx_2 \wedge dx_3 + \left(\frac{\partial P}{\partial x_3} - \frac{\partial R}{\partial x_1}\right) dx_3 \wedge dx_1 \\ &\quad + \left(\frac{\partial Q}{\partial x_1} - \frac{\partial P}{\partial x_2}\right) dx_1 \wedge dx_2. \end{aligned}$$

Both equalities are a direct application of the definition of the differential and of $dx_i \wedge dx_i = 0$ (Exercise 19-28). □

The connection to vector fields defined as functions from a subset of $\mathbb{R}^3$ to $\mathbb{R}^3$ is now quite simple.

Definition 19.33 *Let* $\Omega \subseteq \mathbb{R}^d$ *be open and let* $F(p) = \left[i_\Omega, \begin{pmatrix} F_1 \\ \vdots \\ F_d \end{pmatrix} \right]_p$ *be a vector field on* Ω. *We define the* **divergence** *of* F *as* $\text{div}(F) := \sum_{j=1}^{d} \frac{\partial F_j}{\partial x_j}$. *The divergence is also denoted by* ∇F, *because the sum is the formal scalar product of the column vector in* F *with the* **nabla operator**.

Definition 19.34 *Let* $\Omega \subseteq \mathbb{R}^3$ *be open and let* $F = \left[i_\Omega, \begin{pmatrix} P \\ Q \\ R \end{pmatrix} \right]$ *be a vector field on* Ω. *We define*

$$\text{curl}(F) := \left(\frac{\partial R}{\partial x_2} - \frac{\partial Q}{\partial x_3}\right) [i_\Omega, e_1]_p + \left(\frac{\partial P}{\partial x_3} - \frac{\partial R}{\partial x_1}\right) [i_\Omega, e_2]_p + \left(\frac{\partial Q}{\partial x_1} - \frac{\partial P}{\partial x_2}\right) [i_\Omega, e_3]_p$$

and call it the **curl** *of* F. *The curl is also denoted by* $\nabla \times F$, *because the components can be obtained as the formal cross product of the column vector in* F *with the* **nabla operator**.

Example 19.32 shows that the differential of the appropriate form encodes the divergence and the curl of a vector field together with the appropriate volume or area element that is needed to integrate the divergence or the curl. Also recall that the **gradient** of a differentiable function $f : \Omega \to \mathbb{R}$ is $\text{grad}(f) = \sum_{j=1}^{d} \frac{\partial f}{\partial x_j} e_j$. Hence, the differential of a 0-form (a function) encodes the gradient together with the differential length element that is needed for a line integral. We now turn to the properties of differential forms.

Theorem 19.35 *Let $\Omega \subseteq \mathbb{R}^d$ be open, let ω be a differentiable k-form and let η be a differentiable l-form on Ω. Then the following hold.*

1. *If $l = k$, then $d(\omega + \eta) = d\omega + d\eta$.*
2. $d(\omega \wedge \eta) = (d\omega) \wedge \eta + (-1)^k \omega \wedge d\eta$.
3. *If ω is twice differentiable, then $d(d\omega) = 0$.*

Proof. Part 1 is trivial. Moreover, part 1 shows that we only need to consider forms $\omega = f\, dx_{i_1} \wedge \cdots \wedge dx_{i_k}$ and $\eta = g\, dx_{j_1} \wedge \cdots \wedge dx_{j_l}$ for parts 2 and 3. For such forms, we obtain the following for part 2.

$$\begin{aligned} d(\omega \wedge \eta) &= d\left[\left(f\, dx_{i_1} \wedge \cdots \wedge dx_{i_k}\right) \wedge \left(g\, dx_{j_1} \wedge \cdots \wedge dx_{j_l}\right)\right] \\ &= d\left[(fg)\, dx_{i_1} \wedge \cdots \wedge dx_{i_k} \wedge dx_{j_1} \wedge \cdots \wedge dx_{j_l}\right] \\ &= d(fg) \wedge dx_{i_1} \wedge \cdots \wedge dx_{i_k} \wedge dx_{j_1} \wedge \cdots \wedge dx_{j_l} \quad \boxed{\text{Use Exercise 19-29.}} \\ &= \big((df)g + f(dg)\big) \wedge dx_{i_1} \wedge \cdots \wedge dx_{i_k} \wedge dx_{j_1} \wedge \cdots \wedge dx_{j_l} \\ &= (df) \wedge dx_{i_1} \wedge \cdots \wedge dx_{i_k} \wedge g\, dx_{j_1} \wedge \cdots \wedge dx_{j_l} \\ &\quad +(-1)^k f\, dx_{i_1} \wedge \cdots \wedge dx_{i_k} \wedge dg \wedge dx_{j_1} \wedge \cdots \wedge dx_{j_l} \\ &= (d\omega) \wedge \eta + (-1)^k \omega \wedge d\eta. \end{aligned}$$

Finally, part 3 is proved as follows with D_α denoting the αth partial derivative.

$$\begin{aligned} d\big(d\omega(p)\big) &= d\left(\sum_{\alpha=1}^{d} D_\alpha f(p)\, dx_\alpha \wedge dx_{i_1} \wedge \cdots \wedge dx_{i_k}\right) \\ &= \sum_{\alpha=1}^{d} d\big(D_\alpha f(p)\big)\, dx_\alpha \wedge dx_{i_1} \wedge \cdots \wedge dx_{i_k} \\ &= \sum_{\alpha=1}^{d} \sum_{\beta=1}^{d} D^2_{\alpha,\beta} f(p)\, dx_\beta \wedge dx_\alpha \wedge dx_{i_1} \wedge \cdots \wedge dx_{i_k} \\ &= \sum_{\substack{\alpha<\beta \\ \alpha,\beta \notin \{i_1,\ldots,i_k\}}} D^2_{\alpha,\beta} f(p)\, dx_\beta \wedge dx_\alpha \wedge dx_{i_1} \wedge \cdots \wedge dx_{i_k} \\ &\qquad\qquad + D^2_{\beta,\alpha} f(p)\, dx_\alpha \wedge dx_\beta \wedge dx_{i_1} \wedge \cdots \wedge dx_{i_k} \end{aligned}$$

> **By Theorem 17.56, $D^2_{\alpha,\beta} f(p) = D^2_{\beta,\alpha} f(p)$ and by part 4 of Lemma 18.22 the wedge products satisfy $dx_\beta \wedge dx_\alpha \wedge dx_{i_1} \wedge \cdots \wedge dx_{i_k} = -dx_\alpha \wedge dx_\beta \wedge dx_{i_1} \wedge \cdots \wedge dx_{i_k}$, so the sum must be zero.**

$$= 0.$$

The above completes the proof. ■

Exercise 19-31 shows how Theorem 19.35 includes certain equations from vector analysis as special cases. We are now ready to define the integral of a k-form over the cube $[0, 1]^k$.

Definition 19.36 *If ω is a k-form on $[0, 1]^k$, then (by Theorem 18.23) there is a unique function f so that $\omega = f\, dx_1 \wedge \ldots \wedge dx_k$ and we define* $\displaystyle\int_{[0,1]^k} \omega := \int_{[0,1]^k} f\, d\lambda$.

In the following, we will often work with k-tuples from which a certain element, say, the lth, has been removed. To simplify notation, we define the following.

Definition 19.37 *If $a_1 \ldots a_k$ is a sequence of k mathematical symbols denoting elements of a set and we want to investigate the sequence from which the lth entry has been removed, we write $a_1 \ldots \widehat{a_l} \ldots a_k$ instead of $a_1 \ldots a_{l-1}a_{l+1} \ldots a_k$.*

Stokes' Theorem now says that the integral of the differential of a $(k-1)$-form (which is a k-form) over the cube $[0, 1]^k$ can be expressed as a sum of integrals of the $(k-1)$-form over the faces of the cube. We first state the result in its raw form and subsequently reduce the rather complicated right side to something more manageable.

Theorem 19.38 Stokes' Theorem *for* $[0, 1]^k$. *Let $k \in \mathbb{N}$ and let ω be a continuously differentiable $(k-1)$ form that is defined on a neighborhood of $[0, 1]^k$. Then*

$$\begin{aligned}\int_{[0,1]^k} d\omega &= \sum_{\alpha=1}^{k}(-1)^{\alpha} \int_{[0,1]^{k-1}} \omega_{1,\ldots,\hat{\alpha},\ldots,k}(p_1, \ldots, p_{\alpha-1}, 0, p_{\alpha+1}, \ldots, p_k) \\ &\qquad - \omega_{1,\ldots,\hat{\alpha},\ldots,k}(p_1, \ldots, p_{\alpha-1}, 1, p_{\alpha+1}, \ldots, p_k)\, d\lambda_{k-1},\end{aligned}$$

where for $k = 1$ the integral on the right (in this case there is only one "integral") simply evaluates the functions at the given point.

Proof. With $\omega(p) = \displaystyle\sum_{1 \le i_1 < \cdots < i_{k-1} \le k} \omega_{i_1,\ldots,i_k}(p)\, dx_{i_1} \wedge \cdots \wedge dx_{i_{k-1}}$ for each subset $\{i_1 < \cdots < i_{k-1}\} \subset \{1, \ldots, k\}$ there is a unique index $\beta \in \{1, \ldots, k\}$ so that the equality $\{i_1 < \cdots < i_{k-1}\} = \{1, \ldots, k\} \setminus \{\beta\}$ holds. Hence,

$$\begin{aligned}d\omega(p) &= \sum_{1 \le i_1 < \cdots < i_{k-1} \le k} \sum_{\alpha=1}^{k} D_\alpha \omega_{i_1,\ldots,i_{k-1}}(p)\, dx_\alpha \wedge dx_{i_1} \wedge \cdots \wedge dx_{i_{k-1}} \\ &= \sum_{\beta=1}^{k} \sum_{\alpha=1}^{k} D_\alpha \omega_{1,\ldots,\hat{\beta},\ldots,k}(p)\, dx_\alpha \wedge dx_1 \wedge \cdots \wedge \widehat{dx_\beta} \wedge \cdots \wedge dx_k \\ &= \sum_{\alpha=1}^{k} (-1)^{\alpha-1} D_\alpha \omega_{1,\ldots,\hat{\alpha},\ldots,k}(p)\, dx_1 \wedge \cdots \wedge dx_k.\end{aligned}$$

Via Fubini's Theorem and the Fundamental Theorem of Calculus we obtain

$$\begin{aligned}
\int_{[0,1]^k} d\omega &= \sum_{\alpha=1}^{k}(-1)^{\alpha-1}\int_{[0,1]^k} D_\alpha\omega_{1,\ldots,\hat{\alpha},\ldots,k}(p)\,d\lambda_k \\
&= \sum_{\alpha=1}^{k}(-1)^{\alpha-1}\int_{[0,1]^{k-1}} \omega_{1,\ldots,\hat{\alpha},\ldots,k}(p_1,\ldots,p_{\alpha-1},1,p_{\alpha+1},\ldots,p_k) \\
&\qquad -\omega_{1,\ldots,\hat{\alpha},\ldots,k}(p_1,\ldots,p_{\alpha-1},0,p_{\alpha+1},\ldots,p_k)\,d\lambda_{k-1} \\
&= \sum_{\alpha=1}^{k}(-1)^{\alpha}\int_{[0,1]^{k-1}} \omega_{1,\ldots,\hat{\alpha},\ldots,k}(p_1,\ldots,p_{\alpha-1},0,p_{\alpha+1},\ldots,p_k) \\
&\qquad -\omega_{1,\ldots,\hat{\alpha},\ldots,k}(p_1,\ldots,p_{\alpha-1},1,p_{\alpha+1},\ldots,p_k)\,d\lambda_{k-1}.
\end{aligned}$$

■

Geometrically, the sum of integrals on the right side of the equation in Theorem 19.38 runs over all the faces of the cube $[0,1]^k$. We thus would like to abbreviate it as the integral of the form ω over the surface of the cube. To incorporate the negative signs we have picked up, we attach the negative signs to the pieces of the surface. This is similar to how we incorporated the integration variables into forms. Moreover, in anticipation of the next section, we define our cubes and boundary pieces as functions.

Definition 19.39 *We define I^k as the identity function from $[0,1]^k$ to $\mathbb{R}^k$. The $(i,0)$-face of I^k is $I^k_{(i,0)}(x) := (x_1,\ldots,x_{i-1},0,x_{i+1},\ldots,x_k)$, and the $(i,1)$-face of I^k is $I^k_{(i,1)}(x) := (x_1,\ldots,x_{i-1},1,x_{i+1},\ldots,x_k)$, where $I^k_{(i,0)}$ and $I^k_{(i,1)}$ are functions from $[0,1]^{k-1}$ to $\mathbb{R}^k$. The (formal) sum $\partial I^k = \sum_{i=1}^{k}(-1)^i\left(I^k_{(i,0)} - I^k_{(i,1)}\right)$ is called the* **boundary** *of I^k. For a $k-1$ form $\omega(p) = \sum_{\alpha=1}^{k}\omega_{1,\ldots,\hat{\alpha},\ldots,k}(p)\,dx_1\wedge\cdots\wedge\widehat{dx_\alpha}\wedge\cdots\wedge dx_k$ we define*

$$\begin{aligned}
\int_{\partial I^k}\omega &:= \int_{\sum_{i=1}^{k}(-1)^i\left(I^k_{(i,0)}-I^k_{(i,1)}\right)}\omega \\
&:= \sum_{\alpha=1}^{k}(-1)^{\alpha}\int_{[0,1]^{k-1}} \omega_{1,\ldots,\hat{\alpha},\ldots,k}(p_1,\ldots,p_{\alpha-1},0,p_{\alpha+1},\ldots,p_k) \\
&\qquad -\omega_{1,\ldots,\hat{\alpha},\ldots,k}(p_1,\ldots,p_{\alpha-1},1,p_{\alpha+1},\ldots,p_k)\,d\lambda_{k-1}.
\end{aligned}$$

Clearly, the integral of ω over ∂I^k is defined so that the inconvenient sum from Theorem 19.38 is pushed into the formal sum that defines the domain of integration. More importantly, this formalism has a solid physical intuition behind it. Let us interpret ω as a vector field, say a flow of a fluid, that is decomposed into components so

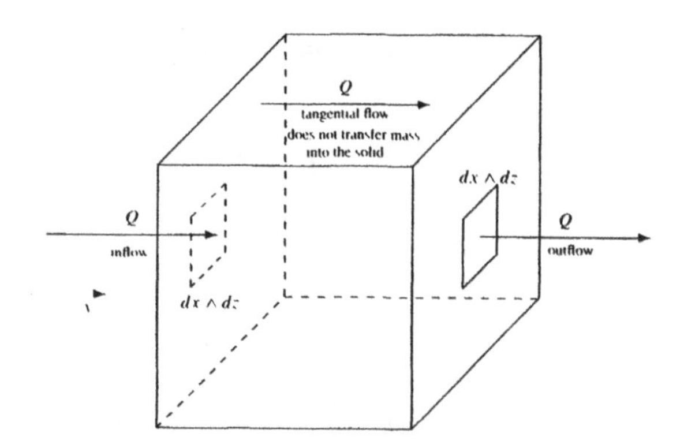

Figure 51: Visualization of the 2-form $Q\,dx \wedge dz$ as a vector field parallel to the y-axis. If $Q\,dx \wedge dz$ is interpreted as a flow field, then there is no transfer of material through the bottom, the top, the front or the back face of the cube. Moreover, if we assume that Q is a constant, then the transfer through the left face is $Q \cdot \langle\text{area of face}\rangle$ into the cube and the transfer through the right face is $Q \cdot \langle\text{area of face}\rangle$ out of the cube

that the $dx_1 \wedge \cdots \wedge \widehat{dx_\alpha} \wedge \cdots \wedge dx_k$-component is perpendicular to any plane of the form $x_\alpha = c$, where $c \in \mathbb{R}$ is a constant (also see Figure 51). Components that are parallel to the surface do not contribute to the surface integral, because flow parallel to a membrane does not transport material through it. On the other hand, components that are perpendicular to the surface contribute with their magnitude, because a flow of velocity $\vec{v}$ that is perpendicular to a small surface S transports per unit time a material volume of $\|\vec{v}\|A(S)$ through the surface, where $A(S)$ is the area of the surface S. Therefore, to compute the net flux of ω across the surface of $[0,1]^k$, for each $\alpha \in \{1,\ldots,k\}$ we only need the integrals of $\omega_{1,\ldots,\hat{\alpha},\ldots,k}$ over the parts of the surface of the cube for which $x_\alpha = 0$ or $x_\alpha = 1$. The opposite signs of $\omega_{1,\ldots,\hat{\alpha},\ldots,k}$ for $x_\alpha = 0$ and $x_\alpha = 1$ are needed to assure that the integral counts net transfer of material into or out of the cube, because a flow direction that enters the cube at $x_\alpha = 0$ is exiting the cube at $x_\alpha = 1$.

Corollary 19.40 *With Definition 19.39 the equation in* **Stokes' Theorem** *for* $[0,1]^k$ *becomes the much more readable equation* $\int_{[0,1]^k} d\omega = \int_{\partial I^k} \omega$. ■

Exercises

19-28. Prove the claims in Example 19.32.

19-29. Prove that if f and g are 0-forms, then $d(fg) = (df)g + f(dg)$.

19-30. Let $\Omega \subseteq \mathbb{R}^d$ and let ω be a k-form on Ω. Prove that for all $i_1 < \cdots < i_k$ the function $\omega_{i_1,\ldots,i_k}$ from Definition 19.30 is $\omega\left[e_{i_1},\ldots,e_{i_k}\right]$.

19-31. Let $\Omega \subseteq \mathbb{R}^3$ be open. *Use Theorem 19.35* to prove the following:

(a) If $F : \Omega \to \mathbb{R}^3$ is a twice differentiable vector field, then $\operatorname{div}\left(\operatorname{curl}(F)\right) = 0$.

(b) If $f : \Omega \to \mathbb{R}$ is a twice differentiable function, then $\operatorname{curl}\left(\operatorname{grad}(f)\right) = 0$.

(c) Why is is *not* possible to use Theorem 19 35 to prove the wrong equality grad $\left(\text{div}(F)\right) = 0$?

19-32. Prove that the function f in Definition 19.36 is $f(p) = \omega(p)[e_1, \ldots, e_k]$

19-33. **Divergence Theorem** for $[0,1]^3$.

(a) Prove that if F is a differentiable vector field defined on a neighborhood of the unit cube $[0,1]^3$, then

$$\begin{aligned}\int_{[0,1]^3} \text{div}(F)\, d\lambda &= \int_{[0,1]^2} \langle F(1,x_2,x_3), e_1 \rangle + \langle F(0,x_2,x_3), -e_1 \rangle d\lambda \\ &+ \int_{[0,1]^2} \langle F(x_1,1,x_3), e_2 \rangle + \langle F(x_1,0,x_3), -e_2 \rangle d\lambda \\ &+ \int_{[0,1]^2} \langle F(x_1,x_2,1), e_3 \rangle + \langle F(x_1,x_2,0), -e_3 \rangle d\lambda.\end{aligned}$$

Then explain why the integral on the right gives the net outflow from the unit cube.

(b) Integrate the vector field $F(x,y,z) = \begin{pmatrix} x^2 \\ y^2 \\ z^2 \end{pmatrix}$ over the surface of the unit cube $[0,1]^3$.

19.4 k-Forms and Integrals Over k-Chains

To define integration on manifolds, first note that a manifold locally looks like its tangent space. Hence, locally, volumes in the manifold are essentially volumes in the tangent space. Because volumes are computed via k-tensors we define a k-form on a manifold as a function into the spaces of k-tensors on the tangent spaces.

Definition 19.41 *A function* $\omega : M \to \bigcup_{p\in M} \Lambda^k\left(M_p\right)$ *such that* $\omega(p) \in \Lambda^k\left(M_p\right)$ *for each* $p \in M$ *is called a* ***k*-form** *on* M *or a* **differential form**.

Let $x : U \to \mathbb{R}^m$ be a coordinate system and let $p \in U$. To obtain a base in $\Lambda^k(M_p)$, let $\left\{[x,e_1]_p, \ldots, [x,e_m]_p\right\}$ be the "standard base" in M_p and denote its dual base in M_p^* by $\{dx_1, \ldots, dx_m\}$. Then the k-form ω on M can be written as $\omega(p) = \sum_{1\leq i_1<\cdots<i_k\leq m} \omega^x_{i_1,\ldots,i_k}(p)\, dx_{i_1} \wedge \cdots \wedge dx_{i_k}$, where the functions $\omega^x_{i_1,\ldots,i_k}$ are defined on U. Formally, the $\omega^x_{i_1,\ldots,i_k}$ depend on the choice of the coordinate system x. Similar to Exercise 19-16b it can be proved that if the $\omega^x_{i_1,\ldots,i_k}(p)$ in the representation with respect to the coordinate system x are differentiable at p, then for any other coordinate system y the $\omega^y_{i_1,\ldots,i_k}(p)$ in the representation with respect to y are also differentiable at p. Therefore, although the specific representation of the form depends on the coordinate system, the differentiability properties of the form are independent of the coordinate system chosen to represent it.

Definition 19.42 *A k-form* $\omega : M \to \bigcup_{p\in M} \Lambda^k(M_p)$ *will be called* **differentiable** *iff for every* $p \in M$ *there is a coordinate system* $x : U \to \mathbb{R}^m$ *so that* $p \in U$, *the form equals* $\omega = \sum_{1\leq i_1<\cdots<i_k\leq m} \omega^x_{i_1,\ldots,i_k}\, dx_{i_1} \wedge \cdots \wedge dx_{i_k}$ *on* U *and the* $\omega^x_{i_1,\ldots,i_k}$ *are differentiable at*

p. Similarly, we define C^j-forms for $j \in \mathbb{N}$ and C^∞-forms. Throughout, unless otherwise stated, k-forms will be assumed to be C^∞. We will also drop the superscript on the functions $\omega_{i_1 \ldots i_k}$ in the coordinate representation of a form.

Section 19.3 has shown that the differential of a form is useful in connecting integrals over solids with integrals over their boundary. The differential of a form on a manifold is defined similar to the differential of a form on $\mathbb{R}^m$. For the definition, we first need the differential of a function. To make the notation bearable, we introduce new notation for tangent vectors.

Let M be a manifold, let $p \in M$, let x, y be coordinate systems of M whose domains contain p, let $f : M \to \mathbb{R}$ be differentiable, and let $[x, v]_p, [y, w]_p \in M_p$ be so that $[x, v]_p = [y, w]_p$. Then

$$\begin{aligned} D\left(f \circ y^{-1}\right)(y(p))[w] &= D\left(f \circ y^{-1}\right)(y(p))\left[D\left(y \circ x^{-1}\right)(x(p))[v]\right] \\ &= D\left(f \circ x^{-1}\right)(x(p))[v]. \end{aligned}$$

Hence, for a manifold M, a point $p \in M$, a differentiable function $f : M \to \mathbb{R}$ and a tangent vector $[x, v]_p \in M_p$ the quantity $df(p)\big[[x, v]_p\big] := D\left(f \circ x^{-1}\right)(x(p))[v]$ does not depend on the representative (x, v). With $v = (v_1, \ldots, v_m)$ we obtain

$$\begin{aligned} df(p)\big[[x, v]_p\big] &= D\left(f \circ x^{-1}\right)(x(p))[v] = \sum_{j=1}^{m} v_j D\left(f \circ x^{-1}\right)(x(p))[e_j] \\ &= \sum_{j=1}^{m} v_j \frac{\partial f \circ x^{-1}}{\partial x_j}(x(p)). \end{aligned}$$

Because of the above we also denote the tangent vector $[x, e_j]_p$ as $\left.\frac{\partial}{\partial x_j}\right|_p$ and we also write $\frac{\partial f}{\partial x_j}(p) := \left.\frac{\partial}{\partial x_j}\right|_p f := \frac{\partial f \circ x^{-1}}{\partial x_j}(x(p))$. This notation expresses the fact that every tangent vector can be interpreted as a geometric object, but also as a differential operator. As was already hinted in the previous section, it is quite common that objects on manifolds have several interpretations. For practical purposes, it is sensible to remember that $\frac{\partial f}{\partial x_j}$ really is a partial derivative, namely, $\frac{\partial f \circ x^{-1}}{\partial x_j}(x(p))$, but that the composition with the parametrization x^{-1} is omitted in the notation.

Now we can define the differential.

Definition 19.43 *Let $f : M \to \mathbb{R}$ be differentiable and let $x : U \to \mathbb{R}^m$ be a coordinate system. Then for all $p \in U$ and $v = \sum_{j=1}^{m} v_j e_j$ we define*

$$df(p)\big[[x, v]_p\big] := \sum_{j=1}^{m} v_j D\left(f \circ x^{-1}\right)[e_j] = \sum_{j=1}^{m} dx_j[v] \left.\frac{\partial}{\partial x_j}\right|_p f = \sum_{j=1}^{m} \frac{\partial f}{\partial x_j}(p)\, dx_j[v].$$

If ω is a k-form on the manifold M and $x : U \to \mathbb{R}^m$ is a coordinate system on M, then for all $p \in U$ we define the $(k+1)$-form $d\omega$ by

$$\begin{aligned} d\omega(p) \quad &:= \sum_{1\leq i_1 < \cdots < i_k \leq m} d\omega_{i_1, \ldots, i_k}(p)\, dx_{i_1} \wedge \cdots \wedge dx_{i_k} \\ &= \sum_{1\leq i_1 < \cdots < i_k \leq m} \sum_{\alpha=1}^{m} \frac{\partial \omega_{i_1, \ldots, i_k}}{\partial x_\alpha}(p)\, dx_\alpha \wedge dx_{i_1} \wedge \cdots \wedge dx_{i_k} \end{aligned}$$

and call it the **differential** *of ω.*

To make sure everything is in order, we must prove that this definition is independent of the coordinate system used to define the differential. For 0-forms, this was already done before we defined the differential. Because the brute force computation for k-forms is quite tedious, we first establish some properties of d and then show that all operators with the same properties as d must be equal to d. This also establishes the properties of the differential that we will need later on.

Because the differential is defined locally, it has the same properties as the differential from Definition 19.31.

Theorem 19.44 *Let ω be a k-form and let η be an l-form on the manifold M. Then the following hold:*

1. *If $k = l$, then $d(\omega + \eta) = d\omega + d\eta$.*
2. $d(\omega \wedge \eta) = (d\omega) \wedge \eta + (-1)^k \omega \wedge d\eta$.
3. $d(d\omega) = 0$.

Proof. Mimic the proof of Theorem 19.35 (Exercise 19-34). ■

Theorem 19.45 *Let M be a manifold, let $x : U \to \mathbb{R}^m$ be a coordinate system and let d' be a map that takes k-forms on U to $(k+1)$-forms on U so that for all infinitely differentiable f on U the equation $d'f = \sum_{j=1}^{m} \frac{\partial f}{\partial x_j}(p)\, dx_j = df$ holds and so that the properties of Theorem 19.44 hold for d'. Then $d' = d$.*

Proof. By additivity of d and d' (see part 1, of Theorem 19.44) it is enough to show that $d\left(f\, dx_{i_1} \wedge \cdots \wedge x_{i_k}\right) = d'\left(f\, dx_{i_1} \wedge \cdots \wedge x_{i_k}\right)$ for all $i_1 < \ldots < i_k$ and f infinitely differentiable on U. With $f = x_i$ we have $d'x_i = \sum_{j=1}^{m} \frac{\partial x_i}{\partial x_j}(p)\, dx_j = dx_i$, but we should recall that the "partial derivatives" are not quite ordinary partial derivatives. Therefore, although it is simple, the reader should verify that the partials really vanish for $i \neq j$ and that the ith partial is 1 (Exercise 19-35).

We now prove by induction on k that $d'\left(dx_{\iota_1} \wedge \cdots \wedge dx_{\iota_k}\right) = 0$. For $k = 1$, this follows from $d'(dx_\iota) = d'\left(d'x_\iota\right) = 0$. For the step $k \to (k+1)$, we note

$$\begin{aligned} & d'\left(dx_{\iota_1} \wedge \cdots \wedge dx_{\iota_{k+1}}\right) \\ &= d'\left(d'x_{\iota_1} \wedge \left(d'x_{\iota_2} \wedge \cdots \wedge d'x_{\iota_{k+1}}\right)\right) \\ &= d'\left(d'x_{i_1}\right) \wedge \left(d'x_{i_2} \wedge \cdots \wedge d'x_{\iota_{k+1}}\right) - d'x_{\iota_1} \wedge d'\left(d'x_{\iota_2} \wedge \cdots \wedge d'x_{i_{k+1}}\right) \\ &= d'\left(d'x_{\iota_1}\right) \wedge \left(d'x_{\iota_2} \wedge \cdots \wedge d'x_{\iota_{k+1}}\right) - d'x_{\iota_1} \wedge d'\left(dx_{\iota_2} \wedge \cdots \wedge dx_{i_{k+1}}\right) \\ &= 0, \end{aligned}$$

where the last step is justified by $d'\left(d'x_i\right) = 0$ and by the induction hypothesis. Therefore, with f being a 0-form, we conclude the following for all k.

$$\begin{aligned} & d'\left(f\, dx_{\iota_1} \wedge \cdots \wedge dx_{\iota_k}\right) \\ &= d'f\, dx_{\iota_1} \wedge \cdots \wedge dx_{\iota_k} + (-1)^0 f \wedge d'\left(dx_{\iota_1} \wedge \cdots \wedge dx_{\iota_k}\right) \\ &= df\, dx_{\iota_1} \wedge \cdots \wedge x_{\iota_k} + 0 \\ &= d\left(f\, dx_{\iota_1} \wedge \cdots \wedge dx_{i_k}\right), \end{aligned}$$

which establishes the result. ■

Theorem 19.45 shows that Definition 19.43 does not depend on the coordinate system we choose. Indeed, on the domain of each coordinate system, any differential d defined as in Definition 19.43 with a specific coordinate system has the properties in Theorem 19.44. By Theorem 19.45, any two such operators must be equal on the intersection of the domains of the coordinate systems used in their definitions.

With forms and their differentials defined on manifolds, we can translate some results from the unit cube to manifolds. We first turn to the parametrizations of the domain of integration.

Definition 19.46 *Let M be a manifold, let $k \in \mathbb{N}$ and let x be a coordinate system whose range contains $[0,1]^k \times \{0\}^{m-k}$. A* **singular** *$k$*-**cube** *$c : [0,1]^k \to M$ in M is the composition of the restriction of the function x^{-1} to $[0,1]^k \times \{0\}^{m-k}$ with the natural bijection from $[0,1]^k$ to $[0,1]^k \times \{0\}^{m-k}$. For $k = 0$, we let a singular 0-cube be the function that maps 0 to $x^{-1}(0)$.*

A singular 0-cube is a point, a singular 1-cube is a curve and a singular 2-cube is a surface with square parameter domain. The **standard** k-**cube** is the identity function $I^k(x) = x$ on $[0,1]^k$ (as we already noted in Definition 19.39). Section 19.3 indicated the importance of chains, so we define them similar to Definition 19.39.

Definition 19.47 *Let M be a manifold and let S be the set of all singular k-cubes in M. Define the set C of all k-***chains** *in M to be the set of all functions $f : S \to \mathbb{Z}$ such that $f(c) = 0$ for all but finitely many c. The elements of C are usually written as $f = \sum_{c.f(c)\neq 0} a_c c$, where the a_c are integers and the sum has purely formal character.*

The definition of the boundary of a singular k-cube is motivated by Theorem 19.38 and the discussion thereafter.

Definition 19.48 *For a singular k-cube c in the manifold M, we define $c_{(i,0)} := c \circ I^k_{(i,0)}$, $c_{(i,1)} := c \circ I^k_{(i,1)}$, and we set the* **boundary** *of c equal to $\partial c = \sum_{i=1}^{k} (-1)^i \left(c_{(i,0)} - c_{(i,1)} \right)$. For a k-chain $\sum_{j=1}^{n} a_j c_j$, we define $\partial \left(\sum_{j=1}^{n} a_j c_j \right) = \sum_{j=1}^{n} a_j \partial \left(c_j \right)$.*

To integrate a form, we want to compose it with a singular k-chain and integrate the resulting form. That means we must first devise a way to compose forms with k-chains. For a linear map, it is easy to define a corresponding map from the k-tensors on the range to the k-tensors on the domain as follows.

Definition 19.49 *Let V, W be vector spaces. For $T \in \mathcal{T}^k(W)$, $f : V \to W$ linear and $v_1, \ldots, v_n \in V$ define $f^*T(v_1, \ldots, v_k) = T\big(f(v_1), \ldots, f(v_k)\big)$.*

For $f : M \to N$ differentiable, we know that f_* is a linear map from M_p to $N_{f(p)}$.

Definition 19.50 *Let M, N be manifolds, let $f : M \to N$ be differentiable and let ω be a k-form on N. Similar to Definition 19.49, for each $p \in M$ we define the k-form $f^*\omega$ on M by $\big(f^*\omega\big)(p)[v_1, \ldots, v_k] := \omega\big(f(p)\big)\big[f_{*p}[v_1], \ldots, f_{*p}[v_k]\big]$. In coordinates, this definition reads as follows.*

$$f^* \left(\sum_{1 \le i_1 < \cdots < i_k \le n} \omega_{i_1, \ldots, i_k}(\cdot)\, dy_{i_1} \wedge \cdots \wedge dy_{i_k} \right) (p)[v_1, \ldots, v_k]$$
$$:= \sum_{1 \le i_1 < \cdots < i_k \le n} \omega_{i_1, \ldots, i_k}\big(f(p)\big)\, dy_{i_1} \wedge \cdots \wedge dy_{i_k} \big[f_{*p}[v_1], \ldots, f_{*p}[v_k]\big].$$

For a 0-form g, we define $f^*g := g \circ f$.

Note the similarity between Definition 19.50 and the Multidimensional Substitution Formula. If the $\omega_{i_1, \ldots, i_k}$ are the functions and $dy_{i_1} \wedge \cdots \wedge dy_{i_k}$ are the volume elements, then the right side consists of the composition of the function with the diffeomorphism f multiplied with the transformed volume element. To make working with the coordinate representations easier we first state a simple lemma.

Lemma 19.51 *Let M, N be manifolds, let $x : U \to \mathbb{R}^m$ be a coordinate system, let $p \in U$, let $f : M \to N$ be differentiable and let $y : V \to \mathbb{R}^n$ be a coordinate system with $f(p) \in N$. Then $f_{*p}\left[\left. \frac{\partial}{\partial x_j} \right|_p \right] = \sum_{i=1}^{n} \frac{\partial (y_i \circ f)}{\partial x_j}(p) \left. \frac{\partial}{\partial y_i} \right|_{f(p)}$.*

Proof. Represent $D\left(y \circ f \circ x^{-1}\right)$ componentwise (Exercise 19-37). ■

Now we can show that the differential commutes with f^*, which is the key to Stokes' Theorem for k-chains.

Theorem 19.52 *Let M, N be manifolds and let ω be a k-form on N. Then the equality $f^*(d\omega) = d\left(f^*\omega\right)$ holds.*

Proof. It is enough to prove the result for forms $\omega = g\, dy_{i_1} \wedge \cdots \wedge dy_{i_k}$. For these forms, we proceed by induction on k. For $k = 0$, let x be a coordinate system around p and let y be a coordinate system around $f(p)$. It is enough to prove the claimed equality for the base vectors $\left.\frac{\partial}{\partial x_j}\right|_p$, $j = 1, \ldots, m$. Via Lemma 19.51 we obtain the following for all $j \in \{1, \ldots, m\}$.

$$\begin{aligned}
& f^*(dg)\left[\left.\frac{\partial}{\partial x_j}\right|_p\right] \\
&= f^*\left[\sum_{i=1}^{n} \frac{\partial g}{\partial y_i}(p)\, dy_i\right]\left[\left.\frac{\partial}{\partial x_j}\right|_p\right] = \sum_{i=1}^{n} \frac{\partial g}{\partial y_i}(f(p))\, dy_i\left[f_{*p}\left[\left.\frac{\partial}{\partial x_j}\right|_p\right]\right] \\
&= \sum_{i=1}^{n} \frac{\partial g}{\partial y_i}(f(p))\, dy_i\left[\sum_{l=1}^{n} \frac{\partial (y_l \circ f)}{\partial x_j}(p)\left.\frac{\partial}{\partial y_l}\right|_{f(p)}\right] \\
&= \sum_{i=1}^{n} \frac{\partial g}{\partial y_i}(f(p))\frac{\partial (y_i \circ f)}{\partial x_j}(p) = \frac{\partial (g \circ f)}{\partial x_j}(p) \quad \text{(By Chain Rule.)} \\
&= \sum_{l=1}^{n} \frac{\partial (g \circ f)}{\partial x_l}(p)\, dx_l\left[\left.\frac{\partial}{\partial x_j}\right|_p\right] = d(g \circ f)\left[\left.\frac{\partial}{\partial x_j}\right|_p\right] = d\left(f^*g\right)\left[\left.\frac{\partial}{\partial x_j}\right|_p\right].
\end{aligned}$$

For the induction step, we argue as follows.

$$\begin{aligned}
& d\left(f^*\left(g\, dy_{i_1} \wedge \cdots \wedge dy_{i_k} \wedge dy_{i_{k+1}}\right)\right) \\
&= d\left(f^*\left(g\, dy_{i_1} \wedge \cdots \wedge dy_{i_k}\right) \wedge f^*\left(dy_{i_{k+1}}\right)\right) \\
&= d\left(f^*\left(g\, dy_{i_1} \wedge \cdots \wedge dy_{i_k}\right)\right) \wedge f^*\left(dy_{i_{k+1}}\right) \\
&\quad +(-1)^k f^*\left(g\, dy_{i_1} \wedge \cdots \wedge dy_{i_k}\right) \wedge d\left(f^*\left(dy_{i_{k+1}}\right)\right) \\
&= f^*\left(d\left(g\, dy_{i_1} \wedge \cdots \wedge dy_{i_k}\right)\right) \wedge f^*\left(dy_{i_{k+1}}\right) \\
&\quad +(-1)^k f^*\left(g\, dy_{i_1} \wedge \cdots \wedge dy_{i_k}\right) \wedge d\left(d\left(f^* y_{i_{k+1}}\right)\right) \\
&= f^*\left(dg \wedge dy_{i_1} \wedge \cdots \wedge dy_{i_k}\right) \wedge f^*\left(dy_{i_{k+1}}\right) + 0 \\
&= f^*\left(dg \wedge dy_{i_1} \wedge \cdots \wedge dy_{i_k} \wedge dy_{i_{k+1}}\right),
\end{aligned}$$

which proves $d\left(f^*\omega\right) = f^*(d\omega)$ for $\omega = g\, dy_{i_1} \wedge \cdots \wedge dy_{i_k} \wedge dy_{i_{k+1}}$. This completes the induction step and the proof. ■

Now the following definition is sensible because open subsets of $\mathbb{R}^k$ are manifolds themselves. To assure everything is defined everywhere, we should assume that the function c in the singular k-cube is actually defined on a neighborhood of $[0, 1]^k$, but a look at Definition 19.46 reveals that this is not a problem. Conscientious readers may want to mentally substitute x^{-1} instead of c in the definition below.

Definition 19.53 *Let $k \in \mathbb{N}$, let c be a singular k-cube in M and let ω be a k-form on M. We define $\int_c \omega := \int_{[0,1]^k} c^*\omega$. For $k = 0$, we define $\int_c \omega := \omega(c(0))$.*

For a k-chain $c = \sum_{i=1}^{n} a_i c_i$ in M, we define $\int_c \omega := \sum_{i=1}^{n} a_i \int_{c_i} \omega$.

Theorem 19.54 Stokes' Theorem *for chains. If ω is a $(k-1)$-form on a manifold M and c is a k-chain in M, then $\int_c d\omega = \int_{\partial c} \omega$.*

Proof. Recall that by Theorem 19.38 the result holds for cubes I^k. (This is most easily seen with notation as in Corollary 19.40.) The idea for the proof therefore is to use Theorem 19.38 and we use it with the notation of Corollary 19.40.

$$\begin{aligned}
\int_c d\omega &= \int_{[0,1]^k} c^* d\omega = \int_{[0,1]^k} d\left(c^*\omega\right) \quad \boxed{\text{We used Theorem 19.52 in this step. Next we use Stokes' Theorem for } I^k.} \\
&= \int_{\partial([0,1]^k)} c^*\omega = \int_{\sum_{i=1}^k (-1)^i \left(I^k_{(i,0)} - I^k_{(i,1)}\right)} c^*\omega \\
&= \sum_{i=1}^k (-1)^i \left(\int_{I^k_{(i,0)}} c^*\omega - \int_{I^k_{(i,1)}} c^*\omega \right) \quad \boxed{\text{Next we use Definition 19.53 and Exercise 19-38.}} \\
&= \sum_{i=1}^k (-1)^i \left(\int_{[0,1]^{k-1}} \left(I^k_{(i,0)}\right)^* c^*\omega - \int_{[0,1]^{k-1}} \left(I^k_{(i,1)}\right)^* c^*\omega \right) \\
&= \sum_{i=1}^k (-1)^i \left(\int_{[0,1]^{k-1}} \left(c \circ I^k_{(i,0)}\right)^* \omega - \int_{[0,1]^{k-1}} \left(c \circ I^k_{(i,1)}\right)^* \omega \right) \\
&= \sum_{i=1}^k (-1)^i \left(\int_{c_{(i,0)}} \omega - \int_{c_{(i,1)}} \omega \right) \\
&= \int_{\sum_{i=1}^k (-1)^i \left(c_{(i,0)} - c_{(i,1)}\right)} \omega = \int_{\partial c} \omega.
\end{aligned}$$

■

Unfortunately, the integral of a form over a k-chain may depend on the coordinate system we chose to define the chain. Ultimately, we want an integral that only depends on the set $x^{-1}\left[[0,1]^k\right]$. Thus we need to investigate how integrals over k-cubes behave under a change of coordinates. The next result shows that a reparametrization f of the set $x^{-1}\left[[0,1]^k\right]$ should not affect the value of the integral, because the reparametrized form $f^*(\omega)$ will be multiplied by the factor that we recognize from the Multivariable Substitution Formula.

Theorem 19.55 *Let M and N be m-dimensional manifolds, let $f : M \to N$ be differentiable, let $x : U \to \mathbb{R}^m$ be a coordinate system of M let $y : V \to \mathbb{R}^m$ be a coordinate system of N with $f[U] \subseteq V$. Then for every differentiable function $g : N \to \mathbb{R}$ we have* $f^*(g\,dy_1 \wedge \cdots \wedge dy_m) = (g \circ f) \det\left(\frac{\partial(y_i \circ f)}{\partial x_j}\right) dx_1 \wedge \cdots \wedge dx_m$.

Proof. Let $p \in U$. To prove the claimed equation at p, we only need to consider the action of the functions on $\left[\left.\frac{\partial}{\partial x_1}\right|_p, \ldots, \left.\frac{\partial}{\partial x_m}\right|_p\right]$ (see Exercise 19-36). After rewriting f_* as $f_{*p}\left[\left.\frac{\partial}{\partial x_j}\right|_p\right] = \sum_{i=1}^{n} \frac{\partial(y_i \circ f)}{\partial x_j} \left.\frac{\partial}{\partial y_i}\right|_{f(p)}$ (see Lemma 19.51) we obtain the following.

$$\begin{aligned}
& f^*(g\,dy_1 \wedge \cdots \wedge dy_m)(p)\left[\left.\frac{\partial}{\partial x_1}\right|_p, \ldots, \left.\frac{\partial}{\partial x_m}\right|_p\right] \\
= \; & g(f(p))\,dy_1 \wedge \cdots \wedge dy_m \left[f_{*p}\left[\left.\frac{\partial}{\partial x_1}\right|_p\right], \ldots, f_{*p}\left[\left.\frac{\partial}{\partial x_m}\right|_p\right]\right] \\
= \; & g(f(p))\,dy_1 \wedge \cdots \wedge dy_m \left[\sum_{i=1}^{m} \frac{\partial(y_i \circ f)}{\partial x_1} \left.\frac{\partial}{\partial y_i}\right|_{f(p)}, \ldots, \sum_{i=1}^{m} \frac{\partial(y_i \circ f)}{\partial x_m} \left.\frac{\partial}{\partial y_i}\right|_p\right]
\end{aligned}$$

Now use Proposition 18.25.

$$\begin{aligned}
= \; & (g \circ f)(p) \det\left(\frac{\partial(y_i \circ f)}{\partial x_j}\right) dy_1 \wedge \cdots \wedge dy_m \left[\left.\frac{\partial}{\partial y_1}\right|_p, \ldots, \left.\frac{\partial}{\partial y_m}\right|_p\right] \\
= \; & (g \circ f)(p) \det\left(\frac{\partial(y_i \circ f)}{\partial x_j}\right) dx_1 \wedge \cdots \wedge dx_m \left[\left.\frac{\partial}{\partial x_1}\right|_p, \ldots, \left.\frac{\partial}{\partial x_m}\right|_p\right],
\end{aligned}$$

where the last step holds because both m-forms evaluate to 1 at the given vectors. ∎

Now we can prove that the integrals of certain m-forms over m-cubes are independent of the parametrization.

Theorem 19.56 *Let $c_1, c_2 : [0,1]^m \to M$ be two m-cubes in the m-dimensional manifold M so that* $\det\left(D\left(c_1^{-1} \circ c_2\right)\right) > 0$ *at every point in the domain of $c_1^{-1} \circ c_2$ and let ω be an m-form that vanishes outside $c_1\left[[0,1]^m\right] \cap c_2\left[[0,1]^m\right]$. Then $\int_{c_1} \omega = \int_{c_2} \omega$.*

Proof. Let x, y be coordinate systems on M so that the cubes are $c_1 = \left.x^{-1}\right|_{[0,1]^m}$ and $c_2 = \left.y^{-1}\right|_{[0,1]^m}$. Then with $\omega = h\,dx_1 \wedge \cdots \wedge dx_m$ and $\{e_1, \ldots, e_m\}$ the standard base for $\mathbb{R}^m$ we obtain

$$\begin{aligned}
\int_{c_1} \omega &= \int_{[0,1]^m} c_1^*\omega = \int_{[0,1]^m} h \circ c_1 \, d\lambda \quad \boxed{\text{Use the Multivariable Substitution Formula.}} \\
&= \int_{c_2^{-1} \circ c_1 \left[[0,1]^m\right]} h \circ c_1 \circ \left(c_1^{-1} \circ c_2\right) \det\left(D\left(c_1^{-1} \circ c_2\right)\right) \, d\lambda
\end{aligned}$$

Use Theorem 19.55 with $M = N = [0, 1]^m$ and $f = c_1^{-1} \circ c_2$. Also recall that ω vanishes outside $c_1\left[[0, 1]^m\right] \cap c_2\left[[0, 1]^m\right]$.

$$\begin{aligned}
&= \int_{[0,1]^m} \left(c_1^{-1} \circ c_2\right)^* \left(h \circ c_1 \, de_1 \wedge \cdots \wedge de_m\right) \\
&= \int_{[0,1]^m} \left(c_1^{-1} \circ c_2\right)^* c_1^*\omega = \int_{[0,1]^m} c_2^*\omega = \int_{c_2} \omega.
\end{aligned}$$

■

We conclude by noting that everything we have done in this section also works for manifolds with boundaries and for manifolds with corners. The only difference is that the k-cubes could actually touch the boundary in these settings. Hence, we will freely use the results established here for these entities, too.

Exercises

19-34. Prove Theorem 19.44.

19-35. Let M be a manifold, let $x : U \to \mathbb{R}^m$ be a coordinate system and let $x_i := \pi_i \circ x$ be the ith component of x. Prove that with the new definition of $\frac{\partial f}{\partial x_j}$ we have, as we would expect, that $\frac{\partial x_i}{\partial x_j} = 0$ for $i \neq j$ and $\frac{\partial x_i}{\partial x_i} = 1$.

19-36. Let $\omega : M \to \bigcup_{p \in M} \Lambda^k(M_p)$ be a k-form on M. Prove that each function $\omega^x_{i_1,\ldots,i_k}$ from Definition 19.42 satisfies $\omega^x_{i_1,\ldots,i_k}(p) = \omega(p)\left[[x, e_{i_1}]_p, \ldots, [x, e_{i_k}]_p\right]$.

19-37. Prove Lemma 19.51.

19-38. Let M, N, O be manifolds and let $g : M \to N$ and $f : N \to O$ be differentiable. Prove that $g^* f^* = (f \circ g)^*$.

19-39. The boundary of the boundary of a singular k-cube.

(a) Let $k \in \mathbb{N}$ be so that $k \geq 2$, let $1 \leq i \leq j \leq k - 1$ and let $\alpha, \beta \in \{0, 1\}$. Prove that $\left(I^k_{(i,\alpha)}\right)_{(j,\beta)} = \left(I^k_{(j+1,\beta)}\right)_{(i,\alpha)}$.

Hint. Straight substitution of a point $z \in [0, 1]^{k-2}$ into both sides.

(b) Prove that if c is a k-chain in the manifold M, then $\partial(\partial c) = 0$.

Hint. Represent the boundary operator as $\partial c = \sum_{i=1}^{k} \sum_{\alpha=0,1} (-1)^{i+\alpha} c_{(i,\alpha)}$ and use the translation of part 19-39a to singular k-cubes to prove that the summands of $\partial(\partial c)$ cancel in pairs.

19.5 Integration on Manifolds

Theorem 19.56 indicates that integration over m-cubes should make it possible to define integration "locally" on a manifold. To obtain a "global" notion of integration, we need to piece the local integrals together. With a partition of unity as in Section 16.9 we can reduce any globally defined function to a sum of "locally supported" functions. However, we need to assure that our transition from local to global integrals does not have any unintended consequences. Theorem 19.56 rules out the danger of local inconsistencies. Globally, one of the ideas behind the integral of a form (a vector field) over a manifold is to measure the throughput of a flow (modeled by the form/vector field) through a surface (modeled by the manifold). This idea of throughput is only sensible if the surface has a definite front and back, or, if it is closed, a definite inside and outside. Locally, it is always possible to define a front and a back, but an awkward global situation can cause problems. Consider the Möbius strip depicted in Figure 52. If we start as indicated in the figure, traveling on what we perceive as the "front" of the strip, then after one full traversal of the strip we find ourselves on what we perceive as the back of the strip, plus our directions of "up" and "down" have been reversed. Therefore it is not possible to globally define a positive direction of throughput through the Möbius strip. To avoid such problems, we must limit our attention to those manifolds for which we cannot lose our orientation like that. The first step is to recall that, because for d-dimensional vector spaces V the space $\Lambda^d(V)$ is one dimensional, we can represent the set of bases of V as a union of two disjoint subsets.

Definition 19.57 *Let V be a d-dimensional real vector space and let $\omega \in \Lambda^d(V)$. The two sets of bases* $\mathcal{P} := \{B : B = \{v_1, \ldots, v_d\}$ is a base of $V, \omega(v_1, \ldots, v_d) > 0\}$ *and* $\mathcal{N} := \{B : B = \{v_1, \ldots, v_d\}$ is a base of $V, \omega(v_1, \ldots, v_d) < 0\}$ *are called* **orientations** *of V and the orientation to which a base $\{v_1, \ldots, v_d\}$ belongs is denoted* $[v_1, \ldots, v_d]$.

For example, two orthonormal bases in $\mathbb{R}^2$ (with the order of the vectors assumed to be fixed) are in the same orientation iff each can be obtained from the other by rotating both base vectors by the same amount. This simple visualization shows the problem with the Möbius strip. If we choose an orthonormal base at our starting point (see Figure 52) and carry it with us in a way that tries to maintain the orientation (one such way is indicated), then, no matter what we do, after one full traversal of the strip, our base will be in the other orientation of $\mathbb{R}^d$. In Figure 52, this is easy to see because if we rotate the base labeled "problem" back onto our original base, we notice that the base vectors have changed roles. The originally horizontal vector is now vertical and vice versa. Such an interchange cannot be achieved with rotations alone. In terms of the form ω, the value of ω on the pair of vectors has changed sign, which means the bases are in opposite orientations.

For a general manifold, we cannot refer to surrounding space, but the tangent spaces allow us to model the idea described above. Simply speaking, we demand that in each domain of a chart x the orientation of the tangent spaces is given by the images of a fixed orientation of $\mathbb{R}^m$ under the parametrization x^{-1} of the manifold.

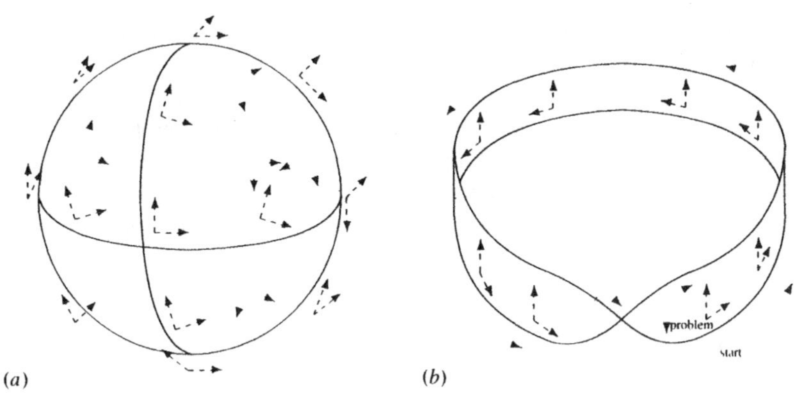

Figure 52: It is impossible to orient a Möbius strip (b). Because of the half-twist in the strip, any orientation that is carried around the strip in a continuous fashion will not arrive as the same orientation at the starting point. On the other hand, spheres are orientable as indicated by the consistent orientation in (a). (We need to be careful interpreting this figure. There is a difficult theorem in topology that says that on a sphere there is no continuous vector field without zeroes. Thus the bases indicated cannot be interpolated by vector fields. However, any two bases can be mapped to each other by translating tangentially along the sphere and then rotating and this process also works if we carry a base in any fashion around the sphere until we are back at the starting point.)

Definition 19.58 *A choice of an orientation μ_p for every tangent space M_p of the m-dimensional manifold M is called* **consistent** *iff for every chart $x : U \to \mathbb{R}^m$ and all $a, b \in U$ we have $\left[x_*^{-1}\left[[i_{x[U]}, e_1]_{x(a)}\right], \ldots, x_*^{-1}\left[[i_{x[U]}, e_m]_{x(a)}\right]\right] = \mu_a$ if and only if $\left[x_*^{-1}\left[[i_{x[U]}, e_1]_{x(b)}\right], \ldots, x_*^{-1}\left[[i_{x[U]}, e_m]_{x(b)}\right]\right] = \mu_b$. Manifolds with a consistent orientation are called* **orientable** *and the function μ is called an* **orientation**. *A manifold together with an orientation is called an* **oriented** *manifold. Charts $x : U \to \mathbb{R}^m$ such that for all $a \in U$ we have $\left[x_*^{-1}\left[[i_{x[U]}, e_1]_{x(a)}\right], \ldots, x_*^{-1}\left[[i_{x[U]}, e_m]_{x(a)}\right]\right] = \mu_a$ are called* **orientation preserving**.

Note that because k-cubes are defined in terms of coordinate systems, Definition 19.58 also defines orientation preserving k-cubes. For orientation preserving m-cubes c_1 and c_2, the determinant $\det\left(D\left(c_1^{-1} \circ c_2\right)\right)$ is always positive. Therefore Theorem 19.56 shows that the integral of an m-form over an orientation preserving m-cube only depends on the range $c\left[[0,1]^m\right]$ of the m-cube, not on the specific m-cube itself. That is, the integral over $c\left[[0,1]^m\right]$ does not depend on the parametrization, as long as we only use orientation preserving parametrizations (m-cubes).

Definition 19.59 *Let $c : [0,1]^m \to M$ be an orientation-preserving m-cube in the m-dimensional oriented manifold M and let ω be an m-form that vanishes outside $c\left[[0,1]^m\right]$. Then we define $\int_M \omega := \int_c \omega$.*

Orientations and forms are defined similarly for manifolds with boundary and for manifolds with corners.

Now that we have taken care of the orientability issue, the definition of the integral of a form over a manifold is straightforward. To avoid formal problems, we stay with connected manifolds. This is not a problem, because we will mostly be interested in manifolds that are the disjoint union of at most finitely many connected manifolds. To break up the integral, we need infinitely differentiable partitions of unity that are supported inside singular m-cubes. The following results show that such partitions exist.

Definition 19.60 *A partition of unity of the manifold M is called a C^∞* **partition of unity** *iff all functions in the partition are C^∞ functions.*

Theorem 19.61 *Let $\mathcal{O}$ be an open cover of the connected manifold M. Then there is a C^∞ partition of unity subordinate to $\mathcal{O}$.*

Proof. This proof is the same as for Theorem 16.115, except that we choose the ψ_U to be C^∞ instead of just continuous. By Theorem 19.18, this is possible. ■

Theorem 19.62 *Let M be an m-dimensional oriented manifold. Then there is an open cover $\mathcal{O}$ of M so that for each $U \in \mathcal{O}$ there is an orientation-preserving singular m-cube c_U with $U \subseteq c_U\left[[0,1]^m\right]$.*

Proof. *We provide a slightly more elaborate proof than necessary so that it is easy to see how to generalize the result to manifolds with boundary or corners.*

For each $p \in M$, let $x : V \to \mathbb{R}^m$ be an orientation-preserving coordinate system around p that maps p to the origin. Let $C_p \subseteq V$ be a compact cube in $\mathbb{R}^m$ so that $x(p)$ is in the relative interior (with respect to $x[V]$) of C_p. Then with $i_{C_p} : [0,1]^m \to C_p$ being the natural bijection between the cubes, the function $c_p := x^{-1} \circ i_{C_p}$ is an orientation preserving singular m-cube so that the M-interior of $c_p\left[[0,1]^p\right]$ contains p. Now let $W_p \subseteq C_p$ be a relatively open subset of $x[V]$ that contains p. Then $U_p := x^{-1}[W_p]$ is open, $p \in U_p \subseteq c_p\left[[0,1]^m\right]$ and $\mathcal{O} := \{U_p\}_{p\in M}$ covers M. ■

Versions of Theorems 19.61 and 19.62 can also be proved as follows for manifolds with boundary and manifolds with corners. For the C^∞ partition of unity, we just need to modify Theorem 19.18 appropriately for the boundary points. For the covers, the relatively open sets in the proof of Theorem 19.62 will not be open in $\mathbb{R}^m$, because for boundary points of manifolds with boundary one face of the cube C_p will be contained in the boundary of the space $\mathbb{H}^m$. For boundary points of manifolds with corners, several faces could be contained in the boundary of the range of the coordinate system.

With all machinery in place it is now easy to see (Exercise 19-41) that the definition below really defines only one number that does not depend on the parametrizations or the choice of partition of unity. The same definition also gives the integral over manifolds with boundary and manifolds with corners.

Definition 19.63 *Let M be an m-dimensional oriented manifold and let ω be a compactly supported m-form on M. Let Φ be a partition of unity subordinate to an open cover $\mathcal{O}$ of M so that for each $U \in \mathcal{O}$ there is an orientation-preserving singular m-cube c_U with $U \subseteq c_U\left[[0,1]^m\right]$. We define* $\int_M \omega := \sum_{\varphi\in\Phi} \int_M \varphi\omega.$

It is particularly noteworthy that because ω is compactly supported, the sum in Definition 19.63 actually is finite (see Exercise 19-40).

To visualize these ideas, consider integration over embedded manifolds in $\mathbb{R}^d$. For an embedded manifold M, we can consider every tangent space M_p to be a subspace of the tangent space $\mathbb{R}^d_p$ of $\mathbb{R}^d$ at p. The identification is done as follows. Let $p \in M$, let $U \subseteq \mathbb{R}^d$ be a neighborhood of p in $\mathbb{R}^d$, let $V \subseteq \mathbb{R}^d$ be open and let $h : U \to V$ be a diffeomorphism as in Definition 19.5. As in the proof of Proposition 19.6 let $x := \pi_{\mathbb{R}^m} \circ h|_{U\cap M}$ be the coordinate system around p obtained from h. The function $e : M_p \to \mathbb{R}^d_p$ defined by $e[x,v]_p := [h,v]_p$ is the desired isomorphism (see Exercise 19-42). The definition $\left\langle [id_{\mathbb{R}^d}, v]_p, [id_{\mathbb{R}^d}, w]_p\right\rangle_p := \langle v, w\rangle$, where $\langle\cdot,\cdot\rangle$ is the usual inner product in $\mathbb{R}^d$, produces a natural inner product on $\mathbb{R}^d_p$. (This inner product on $\mathbb{R}^d_p$ is well-defined, because the definition states that we will use the second component of *one* unique representative of the tangent vector.) Hence, via the above embedding, the tangent spaces M_p of embedded manifolds also carry an inner product. Recalling Proposition 18.29 we can define the following.

Definition 19.64 *Let M be an m-dimensional embedded oriented manifold in $\mathbb{R}^d$. Then for each $p \in M$ we define the* **volume element** *$\omega_V(p)$ to be the unique m-form so that for all orthonormal bases $\{v_1,\ldots,v_m\}$ in μ_p we have $\omega_V(p)[v_1,\ldots,v_m] = 1$.*

Example 19.65 Volume elements encode the integral over open subsets of $\mathbb{R}^d$, as well as the integrals over curves and surfaces.

1. Let $\Omega \subseteq \mathbb{R}^d$ be an open set considered as an oriented embedded manifold with $\mu_p = [e_1,\ldots,e_d]$ for all $p \in \Omega$. Then with $x : \Omega \to \mathbb{R}^d$ denoting the natural embedding, the volume element ω_V can be represented as the wedge product $\omega_V = dx_1 \wedge \cdots \wedge dx_d$. In particular, this means that for all $f \in C_0^\infty(\Omega)$ the integral of $f\omega_V$ from Definition 19.63 coincides with the Lebesgue integral of f over Ω. We will also denote this integral by $\int_\Omega f\, dV$.

2. Let M be a $d-1$ dimensional embedded oriented manifold and let x be an orientation-preserving coordinate system around p, constructed from a diffeomorphism $h : U \to V$ as in Definition 19.5. We define the **unit normal vector** to be the unique unit vector $n(p)$ in $\mathbb{R}^d_p$ so that $\left\{n(p), h_{*p}^{-1}(e_1),\ldots,h_{*p}^{-1}(e_{d-1})\right\}$ is a positively oriented base of $\mathbb{R}^d_p$. Because the hyperplane determined by $Dh^{-1}(h(p))[e_1],\ldots,Dh^{-1}(h(p))[e_{d-1}]$ (see Exercise 19-19) does not depend on the diffeomorphism h, the vector $n(p)$ does not depend on the coordinate system. Moreover, the functions $n_j : M \to \mathbb{R}$ that map each $p \in M$ to the jth coordinate of the unit normal vector are differentiable. To see this, in a neighborhood

of p, let $\widetilde{n}$ be the solution vector of the system of equations $\left\langle \widetilde{n}, Dh^{-1}[e_i] \right\rangle = 0$, $i = 1, \ldots, d-1$ and $\widetilde{n}_j = 1$ for some j so that $\frac{\partial h_j^{-1}}{\partial x_d} \neq 0$ in the h-image of the neighborhood of p. Then $\widetilde{n}$ is parallel or antiparallel to n. By Exercise 17-71 and the differentiability of the coefficients of the system, the components $\widetilde{n}_j$ are differentiable. The components of n are either the components of $\frac{\widetilde{n}}{\|\widetilde{n}\|}$ or of $-\frac{\widetilde{n}}{\|\widetilde{n}\|}$, so the n_j are differentiable, too.

The volume element of M is $\omega_V(p)[v_1, \ldots, v_{d-1}] = \det\left(n(p), v_1, \ldots, v_{d-1}\right)$, where in the determinant we use the coordinate representation of the vectors with respect to the base $\left\{[id, e_1]_p, \ldots, [id, e_d]_p\right\}$. Specifically for the case $d = 3$, if r is an orientation preserving singular 2-cube and $f : M \to \mathbb{R}$ is differentiable, then

$$\begin{aligned}
\int_r f\omega_V &= \int_{[0,1]^2} r^*(f\omega_V) = \int_{[0,1]^2} (f \circ r) \det\left(n(p), r_*[e_1], r_*[e_2]\right) d\lambda \\
&= \int_{[0,1]^2} (f \circ r) \left\| Dr[e_1] \times Dr[e_2] \right\| d\lambda \\
&= \int_{[0,1]^2} (f \circ r) \left\| \frac{\partial r}{\partial x_1} \times \frac{\partial r}{\partial x_2} \right\| d\lambda,
\end{aligned}$$

which is the parametric formula (from multivariable calculus) for the scalar surface integral of a function f over a parametric surface parametrized by r. For the integral of a vector field $F : \Omega \to T\mathbb{R}^d$ defined on a neighborhood Ω of M, we define the **surface integral** to be $\int_M \vec{F} \cdot d\vec{S} := \int_M \langle F, n \rangle \omega_V$. With the above, we obtain the following in case $d = 3$.

$$\begin{aligned}
\int_r \langle F, n \rangle \omega_V &= \int_{[0,1]^2} \left(\langle F, n \rangle \circ r\right) \left\| \frac{\partial r}{\partial x_1} \times \frac{\partial r}{\partial x_2} \right\| d\lambda \\
&= \int_{[0,1]^2} \left\langle (F \circ r), \frac{\frac{\partial r}{\partial x_1} \times \frac{\partial r}{\partial x_2}}{\left\| \frac{\partial r}{\partial x_1} \times \frac{\partial r}{\partial x_2} \right\|} \right\rangle \left\| \frac{\partial r}{\partial x_1} \times \frac{\partial r}{\partial x_2} \right\| d\lambda \\
&= \int_{[0,1]^2} \left\langle (F \circ r), \left(\frac{\partial r}{\partial x_1} \times \frac{\partial r}{\partial x_2} \right) \right\rangle d\lambda,
\end{aligned}$$

which is the parametric formula (from multivariable calculus) for the surface integral of a vector field F over a parametric surface parametrized by r (also see Exercise 18-53c).

3. Let M be a one dimensional embedded oriented manifold in $\mathbb{R}^d$, let $x : U \to \mathbb{R}$ be an orientation preserving coordinate system and let $a := x^{-1}$. Then (Exercise 19-43a) the volume element can be written as $\omega_V(t)[v] = \left\langle v, \frac{a'(t)}{\|a'(t)\|} \right\rangle$. The **line**

integral for scalar functions is defined in the obvious way. The **line integral** for vector fields is defined by $\int_M \vec{F} \cdot d\vec{r} := \int_M \left\langle F, \frac{a'}{\|a'\|} \right\rangle \omega_V$. As for the surface integral, these formulas reduce to formulas familiar from multivariable calculus when specific parametrizations are used (see Exercise 19-43).

The above examples show that the integral over manifolds encodes the integrals of scalar functions, as well as of vector fields over solids, hypersurfaces, and curves with one formalism. Moreover, this formalism shows that the integrals are independent of the parametrizations chosen for the objects in question, which is a big advantage for theoretical considerations. The computation of numerical values of these integrals still uses parametrizations and proceeds along the same lines as in calculus. □

Finally, note that the definitions presented here actually work in more general settings. The σ-algebra of **Borel sets** on M is the σ-algebra generated by the open subsets of M. The integral can actually be defined for m-forms $g\omega_V$ for which the function g is Borel measurable. Hence, if we use the measure ν that assigns to each Borel subset the integral of its indicator function, we can define the integral like we did on measure spaces. In particular, this means that we can talk about L^p **spaces** for which the domain is a manifold.

Regarding the domain of the integral we note that the manifold need not be C^∞. C^1 is sufficient and boundaries and corners are permissible.

Exercises

19-40. Prove that in any sum as in Definition 19 63 only finitely many summands are not zero
Hint. Use that the sets $\{ p : \varphi(p) \neq 0 \}$ form a locally finite open cover of M.

19-41. Let M be an m-dimensional oriented manifold, let ω be a compactly supported m-form on M, let Φ be a partition of unity subordinate to an open cover $\mathcal{O}$ of M so that for each $U \in \mathcal{O}$ there is an orientation-preserving singular m-cube c_U with $U \subseteq c_U\left[[0,1]^m\right]$ and let Ψ be a partition of unity subordinate to an open cover $\widetilde{\mathcal{O}}$ of M so that for each $V \in \widetilde{\mathcal{O}}$ there is an orientation-preserving singular m-cube c_V with $V \subseteq c_V\left[[0,1]^m\right]$. Prove that $\sum_{\varphi\in\Phi} \int_M \varphi\omega = \sum_{\psi\in\Psi} \int_M \psi\omega$.
Hints Use Theorem 19.56 to prove that the integrals of the $\phi\psi\omega$ do not depend on whether we choose the cube c_U or c_V. Then prove that $\int_{c_U} \varphi\omega = \sum_{\psi\in\Psi} \int_{c_V} \varphi\psi\omega$. To switch the order of summations use that by Exercise 19-40 only finitely many summands are not zero.

19-42. Let M be an m-dimensional embedded manifold, let $p \in M$, let $U \subseteq \mathbb{R}^d$ be a neighborhood of p in $\mathbb{R}^d$, let $V \subseteq \mathbb{R}^d$ be open and let $h : U \to V$ be a diffeomorphism as in Definition 19.5. As in the proof of Proposition 19.6 let $x := \pi_{\mathbb{R}^m} \circ h|_{U\cap M}$ be the coordinate system around p obtained from h. Prove that $e[x, v]_p := [h, v]_p$ defines an isomorphism from M_p to $\mathbb{R}^d_p$. You must also prove that e is well-defined.

19-43. Let M be a one dimensional embedded oriented manifold and let $r : [0,1] \to M$ be an orientation preserving singular 1-cube.

(a) Let $x \cdot U \to \mathbb{R}$ be an orientation preserving coordinate system and let $a := x^{-1}$. Prove that the volume element can be written as $\omega_V(t)[v] = \left\langle v, \frac{a'(t)}{\|a'(t)\|} \right\rangle$.

(b) Prove that $\int_r f\omega_V = \int_0^1 f\left(r(t)\right)\|r'(t)\|\, d\lambda(t)$ for all differentiable $f : M \to \mathbb{R}$.

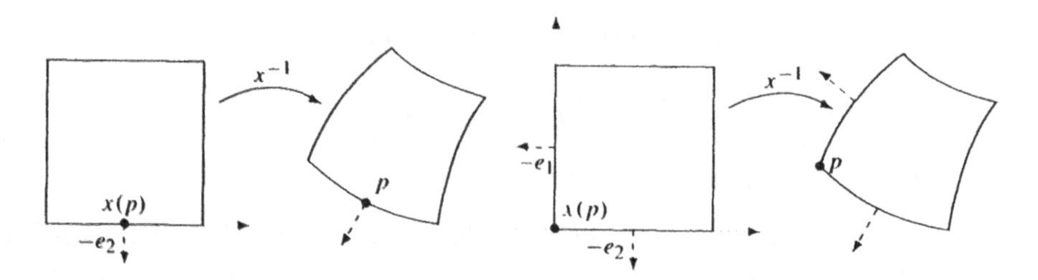

Figure 53: Visualization of the proof of Stokes' Theorem for embedded manifolds. The parametrization x^{-1} maps the outward normal direction of the cube in its domain to the outward normal direction of the k-cube that touches the boundary in the manifold.

(c) Prove that $\int_r \left\langle F, \frac{r'}{\|r'\|} \right\rangle \omega_V = \int_0^1 \langle F(r(t)), r'(t) \rangle \, d\lambda(t)$ for all vector fields $F : \Omega \to \mathbb{R}^3$ defined on a neighborhood of M

19.6 Stokes' Theorem

Because the tangent space $T\partial M$ of an oriented manifold with boundary or with corners is contained in TM, we can obtain an orientation for ∂M from the orientation of M.

Definition 19.66 *Let M be an oriented m-dimensional manifold with boundary or corners and let μ be its orientation. For every $p \in \partial M$ that is not contained in a corner, we let $[v_1, \ldots, v_{m-1}] \in (\partial\mu)_p$ iff $[w, v_1, \ldots, v_{m-1}] \in \mu_p$ for all outward pointing vectors w. The orientation $(\partial\mu)_p$ is called the* **induced orientation** *or the* **positive orientation** *on the boundary. For d-dimensional embedded manifolds in d-dimensional space, it is also called the* **outward orientation**, *because the associated normal vector literally points outward. We will always assume that the boundary carries the induced orientation.*

Formally, for a manifold with corners, the integral over the boundary is a sum of integrals over the pieces of the boundary that are manifolds with corners themselves. To ease notation, it will be understood that the integral over the boundary of a manifold with corners is such a sum.

Theorem 19.67 Stokes' Theorem. *Let M be an oriented m-dimensional manifold with boundary or corners, let ω be a compactly supported $(m-1)$-form on M and let ∂M carry the induced orientation. Then $\int_M d\omega = \int_{\partial M} \omega$.*

Proof. We first consider forms that are supported in the M-interior of an orientation-preserving m-cube. The result is trivial if ω is supported in the interior of an m-cube c so that $c\left[[0,1]^m\right] \subseteq M \setminus \partial M$ because then $\int_M d\omega = \int_c d\omega = \int_{\partial c} \omega = 0 = \int_{\partial M} \omega$.

For a form ω that is supported in the M-interior of an orientation preserving m-cube c in M that intersects the boundary but not the corners, note that $\partial M \cap c\left[[0,1]^m\right] \subseteq \partial c$

and that ω is zero on the parts of the boundary of c that do not intersect ∂M. Let $p \in \partial M \cap c\left[[0,1]^m\right]$ and let x be an orientation preserving coordinate system around p. (For the next statement, let a negative sign indicate the opposite orientation.) Then $x(p) \in \partial\mathbb{H}^m$ and $\mu_{x(p)} = [e_1, \ldots, e_m] = (-1)^m[-e_m, e_1, \ldots, e_{m-1}]$, where $-e_m$ is the outward unit normal vector. This means that the induced orientation on $\partial M \cap c\left[[0,1]^m\right]$ is $(-1)^m$ times the usual orientation of M_p. We can choose the orientation-preserving singular m-cube c so that $\partial M \cap c\left[[0,1]^m\right] = c_{(m,0)}\left[[0,1]^m\right]$. Note that by the above $c_{(m,0)} : [0,1]^{m-1} \to \partial M$ is orientation preserving for even m and orientation reversing for odd m. Hence, $\int_{c_{(m,0)}} \omega = (-1)^m \int_{\partial M} \omega$. But in ∂c the coefficient of $c_{(m,0)}$ is $(-1)^m$. (The left side of Figure 53 gives a visual idea of what we do here.) Hence, we obtain

$$\int_M d\omega = \int_c d\omega = \int_{\partial c} \omega = \int_{(-1)^m c_{(m,0)}} \omega = (-1)^m \int_{c_{(m,0)}} \omega = \int_{\partial M} \omega.$$

Finally, when c intersects the corners, a similar argument for the faces of c that are contained in the boundary gives the same result. (Exercise 19-44. The right side of Figure 53 gives a visual idea of what to do.)

When we integrate a general form, each form $\varphi\omega$ will be supported in the interior of an m-cube as indicated above. Thus we should be able to prove the general result by summation. To move the functions φ from the partition of unity into the differential d, we will use the equality $0 = 0 \wedge \omega = d(1) \wedge \omega = d\left(\sum_{\varphi\in\Phi} \varphi\right) \wedge \omega = \sum_{\varphi\in\Phi} (d\varphi) \wedge \omega$.

By part 2 of Theorem 19.44 with φ being a 0-form, we obtain $d(\varphi\omega) = (d\varphi)\wedge\omega + \varphi\, d\omega$. Hence, because the restrictions of the φ to the boundary are a partition of unity with the requisite properties for defining the integral on the boundary, we conclude

$$\begin{aligned} \int_M d\omega &= \sum_{\varphi\in\Phi} \int_M \varphi\, d\omega = \sum_{\varphi\in\Phi} \int_M (d\varphi) \wedge \omega + \varphi\, d\omega = \sum_{\varphi\in\Phi} \int_M d(\varphi\omega) \\ &= \sum_{\varphi\in\Phi} \int_{\partial M} \varphi\omega = \int_{\partial M} \omega. \end{aligned}$$

■

With the general result established, we can now present Stokes' Theorem in the forms that are familiar from calculus (also see Figure 54). To prove a Divergence Theorem in $\mathbb{R}^d$, we first need to consider the volume element of $(d-1)$-dimensional embedded manifolds.

Theorem 19.68 *Let M be an embedded oriented $(d-1)$-dimensional manifold in $\mathbb{R}^d$ and let n be its unit normal vector. Then the volume element can be represented as $\omega_V(p) = \sum_{j=1}^{d} (-1)^{1+j} n_j(p)\, dx_1 \wedge \cdots \wedge \widehat{dx_j} \wedge \cdots \wedge dx_d$. Moreover, the equality $n_j \omega_V = (-1)^{1+j} dx_1 \wedge \cdots \wedge \widehat{dx_j} \wedge \cdots \wedge dx_d$ holds.*

Proof. Let $v_1, \ldots, v_d \in M_p$. For the representation of ω_V, by Exercises 18-33 and 18-34 via expansion with respect to the first column, we obtain the following, where $\widetilde{\pi}_j$ denotes the projection that erases the jth component of each vector.

$$\begin{aligned}
&\omega_V(p)[v_1, \ldots, v_{d-1}] \\
&= \det\left(n(p), v_1, \ldots, v_{d-1}\right) = \sum_{j=1}^{d}(-1)^{1+j} n_j(p) \det\left(\widetilde{\pi}_j(v_1), \ldots, \widetilde{\pi}_j(v_{d-1})\right) \\
&= \sum_{j=1}^{d}(-1)^{1+j} n_j(p)\, dx_1 \wedge \cdots \wedge \widehat{dx_j} \wedge \cdots \wedge dx_d[v_1, \ldots, v_{d-1}].
\end{aligned}$$

For $n_j\omega_V$, we obtain with n being the unit normal vector,

$$\begin{aligned}
&n_j(p)\omega_V(p)[v_1, \ldots, v_{d-1}] \\
&= n_j(p) \det\left(n(p), v_1, \ldots, v_{d-1}\right) = \det\left((n_j(p)n(p) - e_j) + e_j, v_1, \ldots, v_{d-1}\right)
\end{aligned}$$

Now $\langle n_j(p)n(p) - e_j, n(p)\rangle = n_j(p) - \langle e_j, n(p)\rangle = 0$, and hence $n_j(p)n(p) - e_j \in \operatorname{span}(v_1, \ldots, v_{d-1})$.

$$\begin{aligned}
&= \det\left(e_j, v_1, \ldots, v_{d-1}\right) = (-1)^{1+j} \det\left(\widetilde{\pi}_j(v_1), \ldots, \widetilde{\pi}_j(v_{d-1})\right) \\
&= (-1)^{1+j} dx_1 \wedge \cdots \wedge \widehat{dx_j} \wedge \cdots \wedge dx_d.
\end{aligned}$$

■

Now we can prove the Divergence Theorem. For the remainder of this section, we adopt notation that is seen in physics and the sciences with vectors indicated by arrows on top of the letter and with **integrals over closed surfaces and curves** denoted with a circle in the integral sign(s).

Theorem 19.69 Gauss' Theorem, *also known as the* **Divergence Theorem.** *If the set $E \subseteq \mathbb{R}^d$ is an embedded oriented connected compact d-dimensional manifold with boundary or corners, if $S = \partial M$ with positive (outward) orientation and if $\vec{F}$ is a vector field with continuous partial derivatives on an open region that contains E, then $\oint_S \vec{F} \cdot d\vec{S} = \int_S \langle F, n\rangle\, \omega_V^S = \int_E \operatorname{div}\left(\vec{F}\right) d\omega_V^E = \int_E \operatorname{div}\left(\vec{F}\right) dV$ (also see Figure 54(a)).*

Proof. This is a consequence of Stokes' Theorem once we note the following:

$$\begin{aligned}
d\left(\langle F, n\rangle\, \omega_V^S\right) &= d\left(\sum_{j=1}^{d} F_j n_j\, \omega_V^S\right) \\
&= \sum_{j=1}^{d} d\left(F_j (-1)^{1+j} dx_1 \wedge \cdots \wedge \widehat{dx_j} \wedge \cdots \wedge dx_d\right) \\
&= \sum_{j=1}^{d} (-1)^{1+j} \sum_{i=1}^{d} \frac{\partial F_j}{\partial x_i}\, dx_i \wedge dx_1 \wedge \cdots \wedge \widehat{dx_j} \wedge \cdots \wedge dx_d
\end{aligned}$$

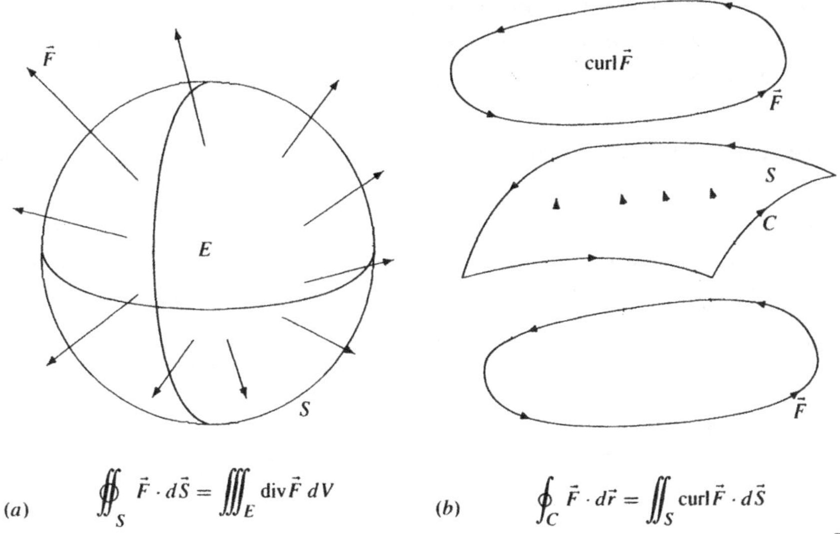

Figure 54: The Divergence Theorem (a) says that the integral of a vector field $\vec{F}$ over a closed surface S equals the integral of the field's divergence div$\vec{F}$ over the enclosed solid E. Stokes' Theorem (b) says that the line integral of a vector field $\vec{F}$ along a closed curve C equals the surface integral of curl$\vec{F}$ over any surface S bounded by the curve.

$$\begin{aligned} &= \sum_{j=1}^{d} (-1)^{1+j} \frac{\partial F_j}{\partial x_j} (-1)^{j-1} dx_1 \wedge \cdots \wedge dx_d \\ &= \left(\sum_{j=1}^{d} \frac{\partial F_j}{\partial x_j} \right) dx_1 \wedge \cdots \wedge dx_d = \operatorname{div}(F)\omega_V^E. \end{aligned}$$

■

The result that is typically called "Stokes' Theorem" is also a special case of Theorem 19.67. Note that integrals over two and three dimensional objects are often denoted with two and three integral signs, respectively.

Theorem 19.70 Stokes' Theorem for compact surfaces in $\mathbb{R}^3$. *If S is an embedded oriented connected compact two dimensional manifold with boundary or corners, if $C = \partial S$ with positive orientation and if $\vec{F}$ is a vector field with continuous partial derivatives in an open region of $\mathbb{R}^3$ that contains S, then (also see Figure 54(b))*

$$\oint_C \vec{F} \cdot d\vec{r} = \int_C \left\langle F, \frac{a'}{\|a\|} \right\rangle \omega_V^C = \int_S \langle \operatorname{curl}(F), n \rangle \, \omega_V^S = \iint_S \operatorname{curl}\left(\vec{F}\right) \cdot d\vec{S}.$$

Proof. Exercise 19-45. ■

We conclude with an important result for line integrals.

Theorem 19.71 Fundamental Theorem for Line Integrals. *Let the curve C be parametrized by the continuously differentiable function $\vec{r}(t)$, $a \leq t \leq b$. If f is differentiable and ∇f is continuous on C, then* $\int_C \nabla f \cdot d\vec{r} = f\big(\vec{r}(b)\big) - f\big(\vec{r}(a)\big)$.

Proof. This can be proved with manifolds or directly. (Exercise 19-46.) ■

Exercises

19-44. Let M be an m-dimensional manifold with corners, let $x : M \to C_k$ be a coordinate system and let $c = x^{-1}\Big|_{[0,1]^m} : [0,1]^m \to M$ be an order-preserving m-cube. Prove that for all $(m-1)$-forms ω that are supported in the M-interior of $c\left[[0,1]^m\right]$ the equality $\int_M d\omega = \int_{\partial M} \omega$ holds.

Hint. $\mu_{x(p)} = [e_1, \ldots, e_m] = (-1)^j\left[-e_j, e_1, \ldots, \widehat{e_j}, \ldots, e_m\right]$

19-45. Prove Theorem 19.70.

Hint. Prove that $\left\langle F, \frac{a'}{\|a\|}\right\rangle \omega_V^C = P\,dx_1 + Q\,dx_2 + R\,dx_3$. Then use Theorem 19.68 for the double integral.

19-46 Prove Theorem 19.71.

19-47 Some integrals

(a) Compute the integral of the vector field $\vec{F}(x, y, z) = \begin{pmatrix} x \\ x \\ z \end{pmatrix}$ over the unit circle $\left\{ (x, y, 0) \in \mathbb{R}^3 : x^2 + y^2 = 1 \right\}$

(b) Compute the integral of the vector field $\vec{F}(x, y, z) = \begin{pmatrix} x \\ y \\ z \end{pmatrix}$ over the line segment from $(0, 0, 1)$ to $(4, -3, 2)$

(c) Compute the integral of the vector field $\vec{F}(x, y, z) = \begin{pmatrix} yz^2 \\ x^2 + z^2 \\ z - y \end{pmatrix}$ over the surface of the cylinder $\left\{ (x, y, z) \in \mathbb{R}^3 . x^2 + y^2 \leq 1, -1 \leq z \leq 1 \right\}$.

19-48 A subset $\Omega \subseteq \mathbb{R}^m$ is called **convex** iff for all $x, y \in \Omega$ the line segment $\left\{ x + t(y - x) \quad t \in [0, 1] \right\}$ is contained in Ω.

(a) Prove that if $\Omega \subseteq \mathbb{R}^3$ is open and convex and $\vec{F} \quad \Omega \to \mathbb{R}^3$ satisfies $\operatorname{curl}\vec{F} = 0$ on Ω, then there is a differentiable function $\varphi : \Omega \to \mathbb{R}$ so that $\vec{F} = \nabla\varphi$

Hint. Let $x \in \Omega$ be fixed and define $\varphi(z)$ to be the line integral $\varphi(z) := \int_{[x,z]} \vec{F} \; d\vec{r}$, where the integral is over the line segment from x to z.

(b) A connected open subset $\Omega \subseteq \mathbb{R}^3$ is called **simply connected** iff every closed curve, that is parametrized by a continuous function $\vec{r} : [a, b] \to \mathbb{R}^3$ for which there is a $c \in [a, b]$ so that $\vec{r}|_{[a,c]}$ and $\vec{r}|_{[c,b]}$ are continuously differentiable, is the boundary of a compact embedded two dimensional manifold with corners that is contained in Ω.

Explain why the result from part 19-48a also holds for simply connected sets Ω

19-49 **Green's Theorem.** Let D be a two dimensional embedded connected compact oriented manifold with boundary or corners, let $C = \partial D$ be the boundary curve with positive orientation Let D be the region in the plane bounded by C Prove that if P and Q have continuous partial derivatives on an open set that contains D, then $\oint_C \begin{pmatrix} P \\ Q \end{pmatrix} d\vec{r} = \iint_D \left(\frac{\partial Q}{\partial x} - \frac{\partial P}{\partial y} \right) d\lambda$

Chapter 20

Hilbert Spaces

In addition to the topological structure of a metric space and the linear structure of a normed space, in an inner product space we can measure angles and in particular we can define orthogonality. This additional structure allows us to derive results that are not easily accessible otherwise. The properties of orthonormal bases investigated in Section 20.1 will allow us to establish the L^2-convergence of Fourier series in Section 20.2 and we conclude in Section 20.3 with Riesz' Representation Theorem for linear functionals on Hilbert spaces.

As noted in Section 15.9, the inner products of real and complex inner product spaces have slightly different properties. To avoid stating all results for real and for complex spaces, in Sections 20.1 and 20.3 we will assume that our inner product spaces are complex. The proofs will also work for real inner product spaces, because for a real number, the real part and the complex conjugate are equal to the number itself.

20.1 Orthonormal Bases

Because an inner product allows us to define orthogonality, we are interested in representing the elements of an inner product space as a sum of orthonormal vectors, similar to the base representation of vectors in d-dimensional space. This section presents the general results and Section 20.2 shows the consequences for the representation of functions with trigonometric polynomials.

We first need to make sure that in a representation with an orthonormal system there are not too many nonzero coefficients.

Proposition 20.1 Bessel's inequality. *Let S be an orthonormal system in the inner product space H and let $x \in H$. Then $\{s \in S : \langle x, s\rangle \neq 0\}$ is countable and $\sum_{s\in S} |\langle x, s\rangle|^2 \leq \|x\|^2$.*

Proof. First let $C \subseteq S$ be finite. Then

$$\begin{aligned}
0 &\leq \left\langle x - \sum_{c \in C} \langle x, c \rangle c, x - \sum_{\tilde{c} \in C} \langle x, \tilde{c} \rangle \tilde{c} \right\rangle \\
&= \left\langle x, x - \sum_{\tilde{c} \in C} \langle x, \tilde{c} \rangle \tilde{c} \right\rangle - \left\langle \sum_{c \in C} \langle x, c \rangle c, x - \sum_{\tilde{c} \in C} \langle x, \tilde{c} \rangle \tilde{c} \right\rangle \\
&= \langle x, x \rangle - \sum_{\tilde{c} \in C} \overline{\langle x, \tilde{c} \rangle} \langle x, \tilde{c} \rangle - \sum_{c \in C} \langle x, c \rangle \langle c, x \rangle + \left\langle \sum_{c \in C} \langle x, c \rangle c, \sum_{\tilde{c} \in C} \langle x, \tilde{c} \rangle \tilde{c} \right\rangle \\
&= \langle x, x \rangle - 2 \sum_{c \in C} \left| \langle x, c \rangle \right|^2 + \sum_{c \in C} \sum_{\tilde{c} \in C} \langle x, c \rangle \overline{\langle x, \tilde{c} \rangle} \langle c, \tilde{c} \rangle \\
&= \langle x, x \rangle - \sum_{c \in C} \left| \langle x, c \rangle \right|^2,
\end{aligned}$$

which means $\sum_{c \in C} \left| \langle x, c \rangle \right|^2 \leq \|x\|^2$. Now suppose for a contradiction that the set $\left\{ s \in S : \langle x, s \rangle \neq 0 \right\}$ is not countable. Then there are an $\varepsilon > 0$ and a set $B \subseteq S$ so that B is at least countably infinite and for all $b \in B$ the inequality $\left| \langle x, b \rangle \right|^2 > \varepsilon$ holds. Let $N \in \mathbb{N}$ be greater than $\left\lceil \frac{\|x\|^2}{\varepsilon} \right\rceil$ and let $B_N \subseteq B$ be an N-element subset of B. Then $\sum_{b \in B_N} \left| \langle x, b \rangle \right|^2 > N\varepsilon > \left\lceil \frac{\|x\|^2}{\varepsilon} \right\rceil \varepsilon \geq \|x\|^2$, a contradiction. Therefore the set $\left\{ s \in S : \langle x, s \rangle \neq 0 \right\}$ must be countable. The inequality follows from the inequality for finite subsets of S proved above. ■

Bessel's inequality shows that the sum $\sum_{s \in S} \langle x, s \rangle s$, which, under the right circumstances, should represent the element x, must converge in a Hilbert space (see Exercise 20-1). Thus we can define orthonormal bases.

Definition 20.2 *An orthonormal system S in an inner product space H is called an* **orthonormal base** *iff for all $x \in H$ the series $\sum_{s \in S} \langle x, s \rangle s$ converges to x. The numbers $\langle x, s \rangle$ are also called the* **Fourier coefficients** *of x with respect to S.*

The term "Fourier coefficients" is usually associated with the expansion of functions in terms of trigonometric functions. Section 20.2 will show that the results here generalize the original Fourier expansions.

Theorems 20.3 and 20.4 give several criteria for an orthonormal system to be an orthonormal base. Note that we will freely use the continuity of the inner product in both factors, which is guaranteed by the Cauchy-Schwarz inequality.

Theorem 20.3 Parseval's identity. *An orthonormal system S in an inner product space H is an orthonormal base iff for all $x \in H$ we have $\|x\|^2 = \sum_{s \in S} \left| \langle x, s \rangle \right|^2$.*

Proof. For "⇒," note that if S is an orthonormal base, then

$$\begin{aligned}\|x\|^2 &= \langle x, x\rangle = \left\langle \sum_{s\in S}\langle x, s\rangle s, \sum_{\widetilde{s}\in S}\langle x, \widetilde{s}\rangle \widetilde{s}\right\rangle = \sum_{s\in S}\sum_{\widetilde{s}\in S}\langle x, s\rangle\overline{\langle x, \widetilde{s}\rangle}\,\langle s, \widetilde{s}\rangle \\ &= \sum_{s\in S}\left|\langle x, s\rangle\right|^2.\end{aligned}$$

For "⇐," let $\|x\|^2 = \sum_{s\in S}\left|\langle x, s\rangle\right|^2$ for all $x \in H$. By Proposition 20.1 the set $S_x := \left\{s \in S : \langle x, s\rangle \neq 0\right\}$ is countable. Let $\{s_j\}_{j=1}^{\infty}$ be an enumeration of S_x, let $\varepsilon > 0$ and let $N \in \mathbb{N}$ be so that for all $n \geq N$ the inequality $\sum_{j=n+1}^{\infty}\left|\langle x, s_j\rangle\right|^2 < \varepsilon^2$ holds. Then for all $n \geq N$ we obtain the following:

$$\left\|x - \sum_{j=1}^{n}\langle x, s_j\rangle s_j\right\|^2 = \left\langle x - \sum_{j=1}^{n}\langle x, s_j\rangle s_j, x - \sum_{k=1}^{n}\langle x, s_k\rangle s_k\right\rangle$$

With a computation similar to the one at the beginning of the proof of Proposition 20.1 we obtain the expression below.

$$= \langle x, x\rangle - \sum_{j=1}^{n}\left|\langle x, s_j\rangle\right|^2 = \sum_{j=n+1}^{\infty}\left|\langle x, s_j\rangle\right|^2 < \varepsilon^2.$$

Hence, $\left\|x - \sum_{j=1}^{n}\langle x, s_j\rangle s_j\right\| < \varepsilon$ for all $n \geq N$ and the series converges to x. ■

Theorem 20.4 *Let S be an orthonormal system in a Hilbert space H. Then the following are equivalent:*

1. *S is an orthonormal base,*
2. *S is maximal,*
3. span(S) *is dense in H.*

Proof. "1⇒3" follows directly from the definition of orthonormal bases.

For "3⇒2," let span(S) be dense in H. For a contradiction, suppose S is not maximal. Then there is a $b \in H$ so that $\|b\| = 1$ and $\langle b, s\rangle = 0$ for all $s \in S$. But then $\langle b, c\rangle = 0$ for all $c \in$ span(S), and for all $c \in$ span(S) we conclude that $\|b - c\|^2 = \|b\|^2 - 2\Re\big(\langle b, c\rangle\big) + \|c\|^2 = \|b\|^2 + \|c\|^2 \geq 1$. This is a contradiction to span(S) being dense in H.

For "2⇒1," let S be a maximal orthonormal system in H. Suppose for a contradiction that there is an $x \in H$ so that $x \neq \sum_{s\in S}\langle x, s\rangle s$. Let $b := x - \sum_{s\in S}\langle x, s\rangle s \neq 0$. Then for all $t \in S$ we infer

$$\langle b, t\rangle = \left\langle x - \sum_{s\in S} \langle x, s\rangle s, t\right\rangle = \langle x, t\rangle - \sum_{s\in S} \langle x, s\rangle\langle s, t\rangle = \langle x, t\rangle - \langle x, t\rangle = 0,$$

which contradicts the maximality of S. ∎

A metric space is called **separable** iff it has a countable dense subset. Separable Hilbert spaces are important in quantum mechanics. The next result shows that, similar to Proposition 15.25 and Theorem 16.76 for finite dimensional spaces, all infinite dimensional separable Hilbert spaces are "the same."

Theorem 20.5 *Every infinite dimensional separable Hilbert space H is isomorphic to the space l^2.*

Proof. We will first prove that H has a countable orthonormal base. To do this let $C = \{c_n : n \in \mathbb{N}\} \subseteq H$ be a countable dense subset with $c_1 \neq 0$. Construct the subset $B \subseteq C$ recursively from C as follows. Let $c_1 \in B$. For all integers $n \geq 2$, let $c_n \in B$ iff $c_n \notin \text{span}(\{c_1, \ldots, c_{n-1}\})$. Then for all $n \in \mathbb{N}$ the set $\{c_k : k \leq n, c_k \in B\}$ is linearly independent and $\text{span}(\{c_k : k \leq n, c_k \in B\}) = \text{span}(\{c_1, \ldots, c_n\})$. Hence, B is linearly independent and $\text{span}(B) = \text{span}(C)$. The set B must be infinite, because otherwise, H has a finite dimensional dense subspace and is thus itself finite dimensional (Exercise 20-2). Now let $\{b_n : n \in \mathbb{N}\}$ be an enumeration of B and apply the Gram-Schmidt Orthonormalization Procedure indefinitely to obtain the orthonormal system $S = \{s_n : n \in \mathbb{N}\}$ with $\text{span}(S) = \text{span}(B) = \text{span}(C)$. Then S is an orthonormal system whose span is dense in H, which means S is an orthonormal base of H.

Because S is a countable orthonormal base, every element $x \in H$ has a unique representation $x = \sum_{j=1}^{\infty} \langle x, s_j\rangle s_j$. The map $I : H \to l^2$ with $I(x) := \sum_{j=1}^{\infty} \langle x, s_j\rangle e_j$ is the desired isomorphism (details left to Exercise 20-4a). ∎

Exercises

20-1 Prove that if H is a Hilbert space, $S \subseteq H$ is an orthonormal system and $x \in H$, then $\sum_{s\in S} \langle x, s\rangle s$ converges

20-2. Prove that if the Hilbert space H has a finite dimensional dense subspace F, then H is finite dimensional.

Hint Find an orthonormal base for F

20-3 A characterization of equality Let H be an inner product space, let $D \subseteq H$ be dense and let $u, f \in H$. Prove that $u = f$ iff for all $x \in D$ we have $\langle u, x\rangle = \langle f, x\rangle$

20-4 Finishing the proof of Theorem 20 5

(a) Prove that the function I defined at the end of the proof of Theorem 20 5 is well-defined, linear, bijective, and continuous and that its inverse is continuous, too.

(b) Explain why we need that H is a Hilbert space in Theorem 20 5

20-5 Let H be a Hilbert space and let S be an orthonormal base of H

(a) Prove that for all $x, y \in H$ we have $x = y$ iff $\langle x, s\rangle = \langle y, s\rangle$ for all $s \in S$

(b) Let Y be a Banach space. Prove that if $L, M : H \to Y$ are continuous linear functions and $L(s) = M(s)$ for all $s \in S$, then $L = M$.

20-6. Prove that every orthonormal system in a separable inner product space H is at most countable.
Hint. The distance between any two distinct elements in an orthonormal system is $\sqrt{2}$. Use that if C is dense, then for each element u in an uncountable orthonormal system, there must be an element $c_u \in C$ that is closer than $\frac{1}{2}$ to u.

20.2 Fourier Series

The representation of elements of inner product spaces is motivated by the corresponding representation in $\mathbb{R}^d$ as well as by the representation of functions via trigonometric polynomials. This representation is important, because it (and similar representations) arise naturally when solving partial differential equations (see Section 21.3). We will now use the tools we have introduced for inner product spaces to investigate the convergence of Fourier series of functions on $[-\pi, \pi)$. In this section, we work with the real normed spaces $L^p[-\pi, \pi) = L^p\big([-\pi, \pi), \mathbb{R}\big)$. The corresponding results for complex valued functions are proved in the exercises.

Definition 20.6 *Let $f \in L^1[-\pi, \pi)$. For $j \geq 0$ we define the* **Fourier cosine coefficients** *of f to be $a_j := \frac{1}{\pi}\int_{-\pi}^{\pi} f(t)\cos(jt)\,d\lambda(t)$ and for $j \geq 1$ we define the* **Fourier sine coefficients** *of f to be $b_j := \frac{1}{\pi}\int_{-\pi}^{\pi} f(t)\sin(jt)\,d\lambda(t)$. The coefficients a_j and b_j are also called the* **Fourier coefficients** *of f.*

If we consider $L^2[-\pi, \pi)$ with the inner product $\langle f, g\rangle := \frac{1}{\pi}\int_{-\pi}^{\pi} fg\,d\lambda$, then for all $f \in L^2[-\pi, \pi)$ we have $a_j = \big\langle f(t), \cos(jt)\big\rangle$ and $b_j = \big\langle f(t), \sin(jt)\big\rangle$. This explains why the coefficients obtained from an inner product are called Fourier coefficients in general. Throughout this section the space $L^2[-\pi, \pi)$ will be equipped with this inner product.

Definition 20.7 *Let $f \in L^1[-\pi, \pi)$ and let a_j $(j \geq 0)$ and b_j $(j \geq 1)$ be the Fourier coefficients of f. We call $F_n(x) := \frac{a_0}{2} + \sum_{j=1}^{n} a_j\cos(jx) + b_j\sin(jx)$ the nth* **Fourier polynomial** *of f. The function $F(x) := \frac{a_0}{2} + \sum_{j=1}^{\infty} a_j\cos(jx) + b_j\sin(jx)$ (defined at every point x at which this series converges) is called the* **Fourier series** *of f.*

It is often helpful to assume that functions on $[-\pi, \pi)$ are actually the restriction of 2π-periodic functions on $\mathbb{R}$. To obtain these functions, we define the periodic extension.

Definition 20.8 *The* **periodic extension** *of a function $f : [-\pi, \pi) \to \mathbb{R}$ is the function $f_p(x) := f\left(x - 2\pi\left\lfloor \frac{x+\pi}{2\pi} \right\rfloor\right)$. By abuse of notation, we will often not distinguish between f and its periodic extension.*

With periodic extensions in place we can represent Fourier polynomials as the integral of the product of the function f with another function D_n, called the Dirichlet kernel. The representation of a quantity as an integral of a product with a "kernel" is fairly common in analysis. (Recall the Peano kernel of Theorem 13.19.)

Theorem 20.9 *Let $f \in L^1[-\pi, \pi)$. The nth Fourier polynomial of f can be represented as the integral $F_n(x) = \frac{1}{\pi}\int_{-\pi}^{\pi} f(x-t)\frac{1}{2}\frac{\sin\left(\left(n+\frac{1}{2}\right)t\right)}{\sin\left(\frac{t}{2}\right)}\,d\lambda(t)$, where the function $D_n(t) := \frac{1}{2}\frac{\sin\left(\left(n+\frac{1}{2}\right)t\right)}{\sin\left(\frac{t}{2}\right)}$ is called the* **Dirichlet kernel**.

Proof. First note that the nth Fourier polynomial can be represented as follows.

$$\begin{aligned}
F_n(x) &= \frac{a_0}{2} + \sum_{j=1}^{n} a_j \cos(jx) + \sum_{j=1}^{n} b_j \sin(jx) \\
&= \frac{1}{2}\frac{1}{\pi}\int_{-\pi}^{\pi} f(t)\,d\lambda(t) + \sum_{j=1}^{n}\left(\frac{1}{\pi}\int_{-\pi}^{\pi} f(t)\cos(jt)\,d\lambda(t)\right)\cos(jx) \\
&\quad + \sum_{j=1}^{n}\left(\frac{1}{\pi}\int_{-\pi}^{\pi} f(t)\sin(jt)\,d\lambda(t)\right)\sin(jx) \\
&= \frac{1}{\pi}\int_{-\pi}^{\pi} f(t)\left(\frac{1}{2} + \sum_{j=1}^{n}\left[\cos(jt)\cos(jx) + \sin(jt)\sin(jx)\right]\right)d\lambda(t) \\
&= \frac{1}{\pi}\int_{-\pi}^{\pi} f(t)\left(\frac{1}{2} + \sum_{j=1}^{n}\cos\left(j(x-t)\right)\right)d\lambda(t).
\end{aligned}$$

Now, using the Euler identities, we obtain the following for all $z \in \mathbb{R}$ (actually for all $z \in \mathbb{C}$).

$$\begin{aligned}
\frac{1}{2} + \sum_{j=1}^{n}\cos(jz) &= \frac{1}{2} + \sum_{j=1}^{n}\frac{e^{ijz} + e^{-ijz}}{2} = \sum_{j=-n}^{n}\frac{e^{ijz}}{2} = \frac{1}{2}e^{-i(n+1)z}\sum_{k=1}^{2n+1} e^{ikz} \\
&= \frac{1}{2}e^{-i(n+1)z}e^{iz}\frac{e^{i(2n+1)z} - 1}{e^{iz} - 1} \\
&= \frac{1}{2}\frac{e^{i(n+1)z} - e^{-inz}}{e^{\frac{iz}{2}}\left(e^{\frac{iz}{2}} - e^{-\frac{iz}{2}}\right)} = \frac{1}{2}\frac{e^{i\left(n+\frac{1}{2}\right)z} - e^{-i\left(n+\frac{1}{2}\right)z}}{e^{\frac{iz}{2}} - e^{-\frac{iz}{2}}}
\end{aligned}$$

$$= \frac{1}{2}\frac{\sin\left(\left(n+\frac{1}{2}\right)z\right)}{\sin\left(\frac{z}{2}\right)}.$$

Note that $\lim_{z\to 0}\frac{1}{2}\frac{\sin\left(\left(n+\frac{1}{2}\right)z\right)}{\sin\left(\frac{z}{2}\right)} = n+\frac{1}{2}$, so $D_n \in L^\infty[-\pi,\pi)$. Moreover, D_n is 2π-periodic. Therefore,

$$\begin{aligned} F_n(x) &= \frac{1}{\pi}\int_{-\pi}^{\pi} f(t)D_n(x-t)\,d\lambda(t) = -\frac{1}{\pi}\int_{x+\pi}^{x-\pi} f_p(x-u)D_n(u)\,d\lambda(u) \\ &= \frac{1}{\pi}\int_{x-\pi}^{x+\pi} f_p(x-u)D_n(u)\,d\lambda(u) \end{aligned}$$

Without loss of generality assume $x \geq 0$ and use the 2π-periodicity of the integrand.

$$\begin{aligned} &= \frac{1}{\pi}\int_{x-\pi}^{\pi} f_p(x-u)D_n(u)\,d\lambda(u) + \frac{1}{\pi}\int_{\pi}^{x+\pi} f_p(x-u)D_n(u)\,d\lambda(u) \\ &= \frac{1}{\pi}\int_{x-\pi}^{\pi} f_p(x-u)D_n(u)\,d\lambda(u) + \frac{1}{\pi}\int_{-\pi}^{x-\pi} f_p(x-u)D_n(u)\,d\lambda(u) \\ &= \frac{1}{\pi}\int_{-\pi}^{\pi} f(x-u)D_n(u)\,d\lambda(u). \end{aligned}$$

■

The representation of Fourier polynomials with the Dirichlet kernel now allows us to prove that Fourier series converge pointwise for a large class of functions.

Definition 20.10 *A function $f:[-\pi,\pi)\to\mathbb{R}$ is called* **piecewise smooth** *iff there is a partition $P=\{-\pi=x_0<x_1<\cdots<x_n=\pi\}$ of $[-\pi,\pi]$ so that for all $j=1,\ldots,n$ the restriction $f|_{(x_{j-1},x_j)}$ of f to the interval (x_{j-1},x_j) is differentiable and its derivative is bounded.*

Note that in the definition of piecewise smooth functions, continuity of f at the points x_j is not demanded. The definition does imply however, that the periodic extension of a piecewise smooth function has left and right limits at every x_j (see Exercise 20-7). Therefore we can say the following.

Theorem 20.11 *If $f:[-\pi,\pi)\to\mathbb{R}$ is a piecewise smooth function, then at each point $x\in[-\pi,\pi)$ the Fourier series F of f converges to $\frac{1}{2}\left[\lim_{u\to x^-} f(u)+\lim_{u\to x^+} f(u)\right]$ (use f_p at $x=-\pi$). In particular, $F(x)=f(x)$ for all x at which f is continuous. Convergence is uniform to f_p on every closed subinterval of $[-\pi,\pi]$ so that f_p is continuous on a neighborhood of the interval (see Figure 55 for examples). In particular, if f_p is continuous, convergence is uniform to f on $[-\pi,\pi)$.*

Proof. For any constant function $g(x)=c$ all Fourier coefficients except a_0 are zero and $a_0=2c$. Hence, by Theorem 20.9 for all $c\in\mathbb{R}$ we infer, because g is constant

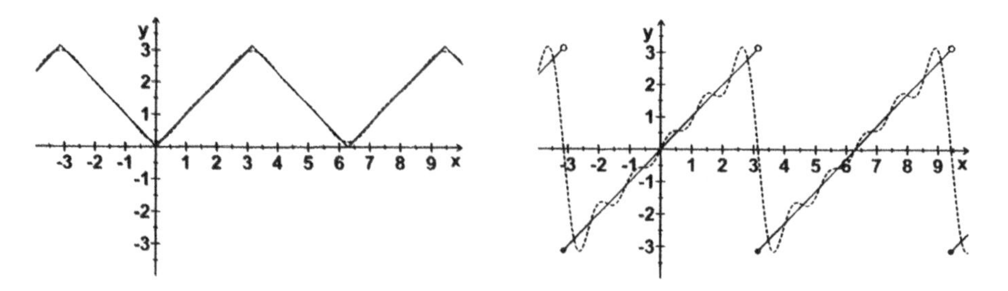

Figure 55: For a piecewise smooth function, Fourier series converge uniformly where the function is continuous (left) and they converge to the average of the left and right limits where the function is discontinuous (right).

and D_n is even,

$$c = \frac{1}{\pi}\int_{-\pi}^{\pi} g(x-t)\frac{1}{2}\frac{\sin\left(\left(n+\frac{1}{2}\right)t\right)}{\sin\left(\frac{t}{2}\right)}\,d\lambda(t) = \frac{1}{\pi}\int_{0}^{\pi}(2c)\frac{1}{2}\frac{\sin\left(\left(n+\frac{1}{2}\right)t\right)}{\sin\left(\frac{t}{2}\right)}\,d\lambda(t).$$

Now with $c := \frac{1}{2}\left[\lim_{u\to x^-} f(u) + \lim_{u\to x^+} f(u)\right]$ we obtain

$$\begin{aligned}
F_n(x) - c &= \frac{1}{\pi}\int_{-\pi}^{\pi} f(x-t)\frac{1}{2}\frac{\sin\left(\left(n+\frac{1}{2}\right)t\right)}{\sin\left(\frac{t}{2}\right)}\,d\lambda(t) - c \\
&= \frac{1}{\pi}\int_{0}^{\pi}\left(f(x-t)+f(x+t)-2c\right)\frac{1}{2}\frac{\sin\left(\left(n+\frac{1}{2}\right)t\right)}{\sin\left(\frac{t}{2}\right)}\,d\lambda(t) \\
&= \frac{1}{\pi}\int_{0}^{\pi}\frac{f(x-t)+f(x+t)-2c}{2\sin\left(\frac{t}{2}\right)}\sin\left(\left(n+\frac{1}{2}\right)t\right)\,d\lambda(t).
\end{aligned}$$

Let K be an upper bound for all existing values of f' and let $x \in [-\pi, \pi)$. There is a $\delta > 0$ so that for all $t \in (0, \delta)$ we have $x+t, x-t \notin \{x_0, \ldots, x_n\}$ and $\frac{t}{2\sin\left(\frac{t}{2}\right)} < 2$. By definition of piecewise smooth functions and the Mean Value Theorem, for all $t \in (0, \delta)$, independent of whether x is in $\{x_0, \ldots, x_n\}$ or not, the inequalities $\left|f(x+t) - \lim_{u\to x^+} f(u)\right| \leq Kt$ and $\left|f(x-t) - \lim_{u\to x^-} f(u)\right| \leq Kt$ hold. (Use f_p and appropriate additional hypotheses on δ if $x = -\pi$.) Hence, for all $t \in (0, \delta)$ we infer

$$\begin{aligned}
&\left|\frac{f(x-t)+f(x+t)-2c}{2\sin\left(\frac{t}{2}\right)}\right| \\
&= \left|\left[\frac{f(x-t)-\lim_{u\to x^-} f(u)}{t} + \frac{f(x+t)-\lim_{u\to x^+} f(u)}{t}\right]\frac{t}{2\sin\left(\frac{t}{2}\right)}\right| \\
&\leq (K+K)2 = 4K.
\end{aligned}$$

Therefore, the function $h(t) := \frac{f(x-t)+f(x+t)-2c}{2\sin\left(\frac{t}{2}\right)}\mathbf{1}_{[0,\pi)}$ is bounded and in particular it is in $L^2[-\pi,\pi)$. But then

$$
\begin{aligned}
&F_n(x) - c \\
&= \frac{1}{\pi}\int_0^{\pi} h(t)\sin\left(\left(n+\frac{1}{2}\right)t\right)\,d\lambda(t) \\
&= \frac{1}{\pi}\int_0^{\pi} h(t)\left[\sin(nt)\cos\left(\frac{1}{2}t\right)+\cos(nt)\sin\left(\frac{1}{2}t\right)\right]\,d\lambda(t) \\
&= \frac{1}{\pi}\int_0^{\pi} h(t)\cos\left(\frac{1}{2}t\right)\sin(nt)\,d\lambda(t)+\frac{1}{\pi}\int_0^{\pi} h(t)\sin\left(\frac{1}{2}t\right)\cos(nt)\,d\lambda(t)
\end{aligned}
$$

shows that the difference $F_n(x) - c$ is the sum of two n^{th} Fourier coefficients of the functions $h(t)\mathbf{1}_{[0,\pi)}\sin\left(\frac{1}{2}t\right)$ and $h(t)\mathbf{1}_{[0,\pi)}\cos\left(\frac{1}{2}t\right)$ in $L^2[-\pi,\pi)$. By Bessel's inequality, these Fourier coefficients converge to zero as $n \to \infty$, which means $F_n(x)$ converges to the claimed limit. Clearly, if f is continuous at x, then the limit is $f(x)$.

To prove the last part, let $[a,b] \subseteq [-\pi,\pi]$ be an interval so that f_p is continuous on a neighborhood (c,d) of the interval. We will keep using f instead of f_p and we assume that the x_j also include the translated points $x_j \pm 2\pi$. For $z \in [a,b]$, let $h_z(t) := \frac{f(z-t)+f(z+t)-2f(z)}{2\sin\left(\frac{t}{2}\right)}$. Let $\delta > 0$ be so that the distance between any two x_j is at least δ and so that $\frac{t}{2\sin\left(\frac{t}{2}\right)} < 2$ for $|t| < \delta$. Then, similar to the argument above, all h_z are bounded by $8K$ on the interval $(-\delta,\delta)$ (if an x_j is between z and $z+t$ or $z-t$, the argument needs to be split up at x_j). Hence, because f is bounded, there is a $B > 0$ so that all h_z with $z \in [a,b]$ are bounded by B. Moreover, for all $x \in [a,b]$ we have $\lim_{z\to x} h_z(t) = h_x(t)$ a.e. Therefore, by the Dominated Convergence Theorem we infer that $\lim_{z\to x}\|h_z - h_x\|_2 = 0$, where $\|\cdot\|_2$ is the norm induced by the inner product $\langle f,g\rangle := \frac{1}{\pi}\int_0^{\pi} f(t)g(t)\,d\lambda(t)$.

Now let $\varepsilon > 0$. For each $x \in [a,b]$, there is an $N_x \in \mathbb{N}$ so that for all $n \geq N_x$ we have $\left|F_n(x) - f(x)\right| < \frac{\varepsilon}{2}$. Let $\delta_x > 0$ be so that for all $z \in [a,b]$ with $|z-x| < \delta_x$ the inequality $\|h_z - h_x\|_2 < \frac{\varepsilon}{2}$ holds. Then for all $n \geq N_x$ and all $z \in [a,b]$ with $|z-x| < \delta_x$ we infer

$$
\begin{aligned}
\left|F_n(z) - f(z)\right| &\leq \left|(F_n(z)-f(z)) - (F_n(x)-f(x))\right| + \left|F_n(x)-f(x)\right| \\
&\leq \frac{1}{\pi}\int_0^{\pi}\left|h_z(t)-h_x(t)\right|\left|\sin\left(\left(n+\frac{1}{2}\right)t\right)\right|\,d\lambda(t)+\frac{\varepsilon}{2}
\end{aligned}
$$

Now use the Cauchy-Schwarz inequality.

$$
\leq \|h_z - h_x\|_2\left\|\sin\left(\left(n+\frac{1}{2}\right)t\right)\right\|_2+\frac{\varepsilon}{2}
$$

$$< \frac{\varepsilon}{2} + \frac{\varepsilon}{2} = \varepsilon.$$

Now $\{(x - \delta_x, x + \delta_x) : x \in [a, b]\}$ is an open cover of the compact interval $[a, b]$. Let $y_1, \ldots, y_k$ be so that $\{(y_j - \delta_{y_j}, y_j + \delta_{y_j}) : j = 1, \ldots, k\}$ is a finite subcover. Let $N := \max\{N_{y_j} : j = 1, \ldots, k\}$. Then for all $z \in [a, b]$ and all $n \geq N$ the inequality $F_n(z) - f(z) < \varepsilon$ holds, which means $\{F_n\}_{n=1}^{\infty}$ converges uniformly to f on $[a, b]$. ■

It now follows that Fourier series of L^2 functions converge in L^2.

Theorem 20.12 *The set* $T := \{\sin(nx), \cos(mx) : m, n \geq 0\} \cup \left\{\frac{1}{\sqrt{2}}\right\}$ *is a maximal orthonormal system in* $L^2[-\pi, \pi)$ *with the inner product* $\langle f, g\rangle := \frac{1}{\pi}\int_{-\pi}^{\pi} fg \, d\lambda$.

Therefore, for every $f \in L^2[-\pi, \pi)$ *the Fourier series of* f *converges to* f *in* L^2. *Moreover, it converges in measure and there is a sequence* $\{n_k\}_{k=1}^{\infty}$ *of indices so that* $\{F_{n_k}\}_{k=1}^{\infty}$ *converges pointwise a.e.*

Proof. We need to prove that the subspace span(T) is dense in $L^2[-\pi, \pi)$. First let $f \in L^2[-\pi, \pi)$ be continuously differentiable with $\lim_{x \to \pi^-} f(x) = f(-\pi)$. By Theorem 20.11, the Fourier series of this function converges uniformly to f. Therefore, for every $\varepsilon > 0$ there is a $p \in \text{span}(T)$ so that $\|p - f\|_\infty < \frac{\varepsilon}{\sqrt{2}}$, which means that

$$\|p - f\|_2 \leq \left(\frac{1}{\pi}\int_{-\pi}^{\pi} \|p - f\|_\infty^2 \, d\lambda\right)^{\frac{1}{2}} < \left(\frac{1}{\pi}\frac{\varepsilon^2}{2}2\pi\right)^{\frac{1}{2}} = \varepsilon.$$

By Theorem 18.12, the set $C_0^\infty(-\pi, \pi)$ is dense in $L^2(-\pi, \pi)$, which means that it is also dense in $L^2[-\pi, \pi)$. The above proves that for every $g \in C_0^\infty(-\pi, \pi)$ and every $\varepsilon > 0$ there is a $p \in \text{span}(T)$ with $\|p - g\|_2 < \varepsilon$. Because $C_0^\infty(-\pi, \pi)$ is dense in $L^2[-\pi, \pi)$, span(T) is dense in $L^2[-\pi, \pi)$.

By Theorem 20.4, this means that T is an orthonormal base in $L^2[-\pi, \pi)$. Hence, the Fourier series of any function $f \in L^2[-\pi, \pi)$ converges to f in L^2. By Proposition 14.47, the Fourier series converges in measure and by Proposition 14.49 there is a pointwise a.e. convergent subsequence. ■

Because the density proof works in arbitrary L^p-spaces we obtain the following.

Corollary 20.13 *The subspace* span(T) *is dense in* $L^p[-\pi, \pi)$ $(1 \leq p < \infty)$. ■

Note that the density of the trigonometric polynomials need not imply the convergence of Fourier series. We have encountered this situation with Taylor polynomials. By the Stone-Weierstrass Theorem (see Exercise 16-87f), the polynomials are dense in $C[-1, 1]$. Yet there are functions (see Lemma 18.8) for which the Taylor series do not converge to the function. Similarly, the Stone-Weierstrass Theorem (use Exercise 16-87j) can be used to prove that the trigonometric polynomials are dense in the continuous periodic functions, but there are examples of continuous periodic functions whose Fourier series do not converge in $L^\infty[-\pi, \pi)$. Moreover, there are functions in

$L^1[-\pi,\pi)$ whose Fourier series do not converge to the function in $L^1[-\pi,\pi)$. On the positive side, for $p\in(1,\infty)$ the Fourier series of functions in $L^p[-\pi,\pi)$ do converge to the function in $L^p[-\pi,\pi)$. The proofs of these results, which can be found in [30], are beyond the scope of this text. However, we can at least show that for L^1 functions the Fourier coefficients must converge to zero.

Corollary 20.14 Riemann-Lebesgue Theorem. *Let $f\in L^1[-\pi,\pi)$ and let a_j and b_j be its Fourier cosine and sine coefficients. Then $\lim_{j\to\infty} a_j = 0$ and $\lim_{j\to\infty} b_j = 0$.*

Proof. Let $f\in L^1[-\pi,\pi)$ and let $\varepsilon>0$. Then there a function $g\in C_0^\infty(-\pi,\pi)$ with $\|f-g\|_1 < \frac{\varepsilon}{2}$. Moreover, because $g\in L^2[-\pi,\pi)$ there is an $N\in\mathbb{N}$ so that for all $n\geq N$ we have $\left|\frac{1}{\pi}\int_{-\pi}^{\pi} g(t)\cos(jt)\,d\lambda(t)\right| < \frac{\varepsilon}{2}$. Thus for all $n\geq N$ we conclude via Hölder's inequality that

$$\begin{aligned}
|a_j| &= \left|\frac{1}{\pi}\int_{-\pi}^{\pi} f(t)\cos(jt)\,d\lambda(t)\right| \\
&\leq \frac{1}{\pi}\int_{-\pi}^{\pi} \big|f(t)-g(t)\big|\big|\cos(jt)\big|\,d\lambda(t) + \left|\frac{1}{\pi}\int_{-\pi}^{\pi} g(t)\cos(jt)\,d\lambda(t)\right| \\
&\leq \frac{1}{\pi}\|f-g\|_1\big\|\cos(j\cdot)\big\|_\infty + \left|\frac{1}{\pi}\int_{-\pi}^{\pi} g(t)\cos(jt)\,d\lambda(t)\right| \\
&< \frac{\varepsilon}{2}+\frac{\varepsilon}{2} = \varepsilon,
\end{aligned}$$

which means that the Fourier cosine coefficients of f converge to zero. The Fourier sine coefficients are handled similarly. ■

In applications, the integration and differentiation of Fourier series are important. Thus we conclude this section with two results that address these operations.

Theorem 20.15 *If $f\in L^2[-\pi,\pi)$ has Fourier coefficients a_n and b_n, then for all $x\in\mathbb{R}$ we have $\int_0^x f_p(t)\,d\lambda(t) = \frac{a_0}{2}x + \sum_{n=1}^{\infty}\frac{a_n}{n}\sin(nx) - \frac{b_n}{n}\big(\cos(nx)-1\big)$.*

Proof. Let $x\in\mathbb{R}$ and consider the linear function $\varphi_x : L^2[-\pi,\pi)\to\mathbb{R}$ defined by $\varphi_x(f) := \int_0^x f_p(t)\,d\lambda(t)$. This function is continuous, because by the Cauchy-Schwarz inequality the following inequality holds.

$$\begin{aligned}
|\varphi_x(f)| &\leq \left|\int_0^x f_p(t)\cdot\mathbf{1}\,d\lambda(t)\right| \leq \left|\int_0^x \big|f_p(t)\big|^2\,d\lambda(t)\right|^{\frac{1}{2}}\left|\int_0^x |\mathbf{1}|^2\,d\lambda(t)\right|^{\frac{1}{2}} \\
&\leq \sqrt{|x|}\left\lceil\frac{|x|}{2\pi}\right\rceil\|f\|_2.
\end{aligned}$$

The continuity of φ_x allows us to move L^2 limits out of the integral, which establishes the result as shown below.

$$\begin{aligned}\int_0^x f_p(t)\,d\lambda(t) &= \varphi_x\left(\frac{a_0}{2}+\sum_{n=1}^{\infty} a_n\cos(nt)+b_n\sin(nt)\right)\\ &= \varphi_x\left(\frac{a_0}{2}\right)+\sum_{n=1}^{\infty}\varphi_x\big(a_n\cos(nt)\big)+\varphi_x\big(b_n\sin(nt)\big)\\ &= \frac{a_0}{2}x+\sum_{n=1}^{\infty}\frac{a_n}{n}\sin(nx)-\frac{b_n}{n}\big(\cos(nx)-1\big).\end{aligned}$$

■

Theorem 20.16 *If f is a differentiable 2π-periodic function with continuous, piecewise smooth derivative f', then the Fourier series of f' converges uniformly and it can be obtained by termwise differentiation as $f'(x)=\sum_{n=1}^{\infty}-na_n\sin(nx)+nb_n\cos(nx)$.*

Proof. By Theorem 20.11, the Fourier series of f' converges uniformly to f'. Now $\frac{1}{\pi}\int_{-\pi}^{\pi} f'(x)\cos(nx)\,d\lambda(x)=\frac{1}{\pi}\left[f(x)\cos(nx)\Big|_{-\pi}^{\pi}+n\int_{-\pi}^{\pi} f(x)\sin(nx)\,d\lambda(x)\right]=nb_n.$ A similar computation yields the sine coefficients. ■

Theorem 20.15 is as robust as Theorem 20.16 is fragile. Exercise 20-8 shows that even for a reasonably simple function the termwise derivative of the Fourier series need not be the derivative of the function if the hypotheses of the theorem are not satisfied.

Exercises

20-7 Let $f:[-\pi,\pi)\to\mathbb{R}$ be a piecewise smooth function with $x_0,\ldots,x_n$ as in Definition 20.10. Prove that the periodic extension of f has left and right limits at every x_j.

20-8 The 2π-periodicity hypothesis in Theorem 20.16 is essential. Prove that that the termwise derivative of the Fourier series of $f(x)=x$ is not the derivative of f.

20-9. Prove that the Fourier series of functions in $L^2\big([-\pi,\pi),\mathbb{C}\big)$ converge in the L^2-norm.
Hint. Consider the real and imaginary parts of the functions separately.

20-10 Explain why the Riemann-Lebesgue Theorem is not a trivial consequence of Bessel's inequality. That is, explain why the "detour" through $C_0^\infty(-\pi,\pi)$ in the proof cannot be avoided.

20-11. Prove that the Fourier series of $f(x)=\sqrt{|x|}$ converges uniformly to f on $[-\pi,\pi)$.
Hint. Use the argument from the proof of Theorem 20.11 and the Riemann-Lebesgue Theorem to first prove convergence at 0.

20-12 Consider $L^2[-\pi,\pi)$ with the inner product $\langle f,g\rangle:=\frac{1}{\pi}\int_{-\pi}^{\pi} fg\,d\lambda$ and let $f\in L^2[-\pi,\pi)$ be a function with Fourier coefficients a_n and b_n.

(a) Prove that $\|f\|_2=\sqrt{\frac{a_0^2}{2}+\sum_{n=1}^{\infty}\left(a_n^2+b_n^2\right)}$. (**Parseval's identity**.)

(b) Prove that the series $\sum_{n=1}^{\infty}\frac{a_n}{n}$ and $\sum_{n=1}^{\infty}\frac{b_n}{n}$ both converge absolutely.
Hint. Cauchy-Schwarz inequality.

20-13. Some bounds for Fourier coefficients.

(a) Let $f : [-\pi, \pi] \to \mathbb{R}$ be continuous and twice continuously differentiable on $(-\pi, \pi)$ with $\int_{-\pi}^{\pi} |f''(x)|\, d\lambda < \infty$, $\lim_{x\to-\pi^+} |f'(x)| < \infty$, and $\lim_{x\to\pi^-} |f'(x)| < \infty$. Prove that for all $n \in \mathbb{N}$ we have $|a_n| \leq \frac{1}{n^2}\frac{1}{\pi}\left(\lim_{x\to\pi^-} |f'(x)| + \lim_{x\to-\pi^+} |f'(x)| + \int_{-\pi}^{\pi} |f''(x)|\, d\lambda\right)$.

(b) Let $f : [-\pi, \pi] \to \mathbb{R}$ be continuous and continuously differentiable on $(-\pi, \pi)$ so that $\int_{-\pi}^{\pi} |f'(x)|\, dx < \infty$. Prove that for all $n \in \mathbb{N}$ the Fourier sine coefficients satisfy the inequality $|b_n| \leq \frac{1}{n}\frac{1}{\pi}\left(|f(\pi)| + |f(-\pi)| + \int_{-\pi}^{\pi} |f'(x)|\, d\lambda\right)$.

20-14. Prove that if $f : [-\pi, \pi] \to \mathbb{R}$ is even, then its Fourier sine coefficients are zero and that if f is odd, its Fourier cosine coefficients are zero.

20-15. Convergence of Fourier series in other norms.

(a) Prove that if $f \in L^2[-\pi, \pi)$ and $p \in [1, 2)$, then the Fourier series of f also converges to f in $L^p[-\pi, \pi)$. *Hint.* Exercise 15-31.

(b) Explain why part 20-15a does not prove that Fourier series of all functions in $L^1[-\pi, \pi)$ converge in $L^1[-\pi, \pi)$.

20-16. For $f \in L^1\left([-\pi, \pi), \mathbb{C}\right)$ and $k \in \mathbb{Z}$ define $c_k := \frac{1}{2\pi}\int_{-\pi}^{\pi} f(t)e^{-ikt}\, d\lambda(t)$. The c_k are also called **Fourier coefficients** of f and $\sum_{k=-\infty}^{\infty} c_k e^{ikt} := \sum_{k=0}^{\infty} c_k e^{ikt} + \sum_{k=1}^{\infty} c_{-k}e^{-ikt}$ is also called the **Fourier series** of f. Prove that for $f \in L^2\left([-\pi, \pi), \mathbb{C}\right)$ the series $\sum_{k=-\infty}^{\infty} c_k e^{ikt}$ converges to f in L^2.

Hint. Prove that the series is equal to the Fourier series from Definition 20.6. Use the Euler identities.

20-17. Explain why the Stone-Weierstrass Theorem, and in particular Exercise 16-87(j)iii, does not prove that Fourier series of continuous functions converge uniformly.

20-18. Dense subspaces of $C^0\left([0, 2\pi], \mathbb{C}\right)$.

(a) Prove the complex version of the **Stone-Weierstrass Theorem**. That is, prove that if A is a point-separating subalgebra of $C^0\left([0, 2\pi], \mathbb{C}\right)$ that contains the constant functions and so that for each $f = u + iv \in A$ the conjugate $u - iv$ also is in A, then A is dense in $C^0\left([0, 2\pi], \mathbb{C}\right)$.
Hint. Prove that the regular Stone-Weierstrass Theorem can be applied to the sets of real and imaginary parts of functions in A. That is, prove that these sets satisfy the hypotheses of the Stone-Weierstrass Theorem.

(b) Prove that $A := \left\{\sum_{j=-n}^{n} a_j e^{ijx} : a_j \in \mathbb{C}, n \in \mathbb{N}\right\}$ is dense in $C^0\left([0, 2\pi], \mathbb{C}\right)$.

(c) Prove that the space from part 20-18b is dense in $L^p\left([0, 2\pi], \mathbb{C}\right)$, $1 \leq p < \infty$.

20.3 The Riesz Representation Theorem

The linear function that maps each element x of an inner product space to its zth Fourier coefficient $\langle x, z\rangle$ is easily seen to be continuous (Cauchy-Schwarz inequality). This section will show that on a Hilbert space H, *every* continuous linear function φ from H to the underlying field $\mathbb{R}$ or $\mathbb{C}$ is of this form. Throughout we will assume once more that H can be a real or a complex Hilbert space.

Lemma 20.17 *Let H be an inner product space and let $z \in H$. Then the function $\langle \cdot, z\rangle : H \to \mathbb{C}$ is a continuous linear function on H.*

Proof. Exercise 20-19. ■

The first step toward proving that every continuous linear function from a complex Hilbert space to $\mathbb{C}$ is of the form $\langle \cdot, z\rangle$ is to prove that in a closed linear subspace of H for every $x \in H$ there is a c in the subspace that is closest to x.

Definition 20.18 *Let X be a vector space. Then $K \subseteq X$ is called* **convex** *iff for all $x, y \in K$ the line segment $\{x + t(y - x) : t \in [0, 1]\}$ is contained in K.*

Theorem 20.19 Parallelogram law. *Let H be an inner product space. Then for all points $x, y \in H$ the equality $\|x + y\|^2 + \|x - y\|^2 = 2\|x\|^2 + 2\|y\|^2$ holds.*

Proof. For all $x, y \in H$, we compute

$$\begin{aligned} \|x + y\|^2 + \|x - y\|^2 &= \|x\|^2 + 2\Re\langle x, y\rangle + \|y\|^2 + \|x\|^2 - 2\Re\langle x, y\rangle + \|y\|^2 \\ &= 2\|x\|^2 + 2\|y\|^2. \end{aligned}$$

■

Theorem 20.20 *Let $K \neq \emptyset$ be a convex and complete subset of the inner product space H. Then for every point $x \in H$ there is an element $c \in K$ so that for all $a \in K$ we have $\|x - c\| \leq \|x - a\|$.*

Proof. Let $\mu := \inf\{\|x - a\| : a \in K\}$ and let $\{y_n\}_{n=1}^{\infty}$ be a sequence in K so that $\lim_{n\to\infty} \|x - y_n\| = \mu$. Then for all $m, n \in \mathbb{N}$ the point $\frac{y_m + y_n}{2}$ is in K. Let $\varepsilon > 0$. Then there is an $N \in \mathbb{N}$ so that for all $n \geq N$ we have $\|x - y_n\| < \mu + \frac{\min\{\varepsilon^2, 1\}}{16(\mu + 1)}$. Therefore, for all $m, n \geq N$ we conclude the following via the parallelogram law.

$$\begin{aligned} \|y_m - y_n\|^2 &= \left\|(x - y_m) - (x - y_n)\right\|^2 \\ &= 2\|x - y_m\|^2 + 2\|x - y_n\|^2 - \|x - y_m + x - y_n\|^2 \\ &= 2\|x - y_m\|^2 + 2\|x - y_n\|^2 - 4\left\|x - \frac{y_m + y_n}{2}\right\|^2 \\ &\leq 2\|x - y_m\|^2 + 2\|x - y_n\|^2 - 4\mu^2 \\ &< 2\left(\mu + \frac{\min\{\varepsilon^2, 1\}}{16(\mu + 1)}\right)^2 + 2\left(\mu + \frac{\min\{\varepsilon^2, 1\}}{16(\mu + 1)}\right)^2 - 4\mu^2 < \varepsilon^2. \end{aligned}$$

Hence, $\{y_n\}_{n=1}^{\infty}$ is a Cauchy sequence. Because the subset K is complete, the limit $c := \lim_{n\to\infty} y_n$ exists in K and $\mu = \|x - c\|$. ■

Definition 20.21 *If the $c \in K$ as in Theorem 20.20 is unique, it is also called the* **best approximation** *of x in K (also see Figure 56).*

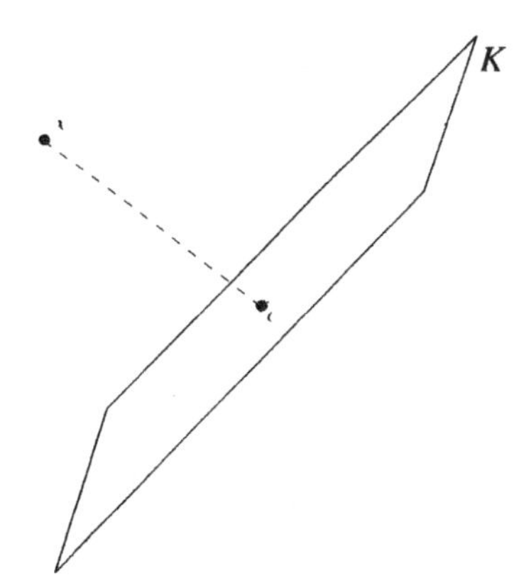

Figure 56: The best approximation of x in a linear subspace K is obtained via orthogonal projections (also see Corollary 20.23 and Exercise 20-23).

When K is a complete linear subspace, the best approximation is unique.

Theorem 20.22 *Let K be a complete linear subspace of the inner product space H and let $x \in H$. Then the c from Theorem 20.20 is unique. Moreover, x has a unique decomposition $x = k + o$, where $k \in K$ and o is orthogonal to K. The vector k in this decomposition is the unique c from Theorem 20.20.*

Proof. Let K be a complete linear subspace and let $c \in K$ be as in Theorem 20.20. We first prove that $x - c$ must be orthogonal to K. For a contradiction, suppose that $\langle x - c, y\rangle \neq 0$ for some $y \in K$. Then we can assume without loss of generality that $\|y\| = 1$ and, because $\left|e^{i\theta}\right| = 1$ for all real numbers θ, that $\Re\big(\langle x - c, y\rangle\big) \neq 0$. With $\delta := -\Re\big(\langle x - c, y\rangle\big)$ we obtain $c - \delta y \in K$ and

$$\begin{aligned}
\left\|x - (c - \delta y)\right\|^2 &= \|x - c + \delta y\|^2 \\
&= \|x - c\|^2 + 2\delta\Re\big(\langle x - c, y\rangle\big) + \delta^2\|y\|^2 \\
&= \|x - c\|^2 - 2\Re\big(\langle x - c, y\rangle\big)\Re\big(\langle x - c, y\rangle\big) + \Big(\Re\big(\langle x - c, y\rangle\big)\Big)^2 \\
&= \|x - c\|^2 - \Big(\Re\big(\langle x - c, y\rangle\big)\Big)^2 \\
&< \|x - c\|^2,
\end{aligned}$$

which is not possible. Thus $x - c$ must be orthogonal to all vectors in K.

To prove uniqueness of c, let $\widetilde{c} \in K$ be such that $\left\|x - \widetilde{c}\right\| = \|x - c\|$. Then $x - \widetilde{c}$ is orthogonal to all vectors in K, too, and hence $c - \widetilde{c} = \left(x - \widetilde{c}\right) - (x - c)$ is orthogonal to all vectors in K. But $c - \widetilde{c} \in K$, so $\left\|c - \widetilde{c}\right\|^2 = \left\langle c - \widetilde{c}, c - \widetilde{c}\right\rangle = 0$, and hence $c = \widetilde{c}$. Thus c is unique.

The decomposition of x now is $x = c + (x - c)$, that is, $o = x - c$. For uniqueness of the decomposition, note that for any other decomposition of x as $x = \widetilde{c} + \widetilde{o}$ with $\widetilde{c} \in K$ and $\widetilde{o}$ orthogonal to K, we obtain $0 = c - \widetilde{c} + o - \widetilde{o}$ with $c - \widetilde{c}$ being orthogonal to $o - \widetilde{o}$. But then $c - \widetilde{c} = o - \widetilde{o} = 0$ (Exercise 20-20) and we are done. ∎

If the subspace in Theorem 20.22 has an orthonormal base, then the best approximation has a very nice representation.

Corollary 20.23 *Let K be a complete linear subspace of the inner product space H and let S be an orthonormal base of K. Then for all $x \in H$ the sum $c := \sum_{s\in S} \langle x, s\rangle s$ is the best approximation of x in K.*

Proof. For all $y \in K$, we obtain

$$\begin{aligned}
\langle x - c, y\rangle &= \left\langle x - \sum_{s\in S} \langle x, s\rangle s, \sum_{\tilde{s}\in S} \langle y, \tilde{s}\rangle \tilde{s} \right\rangle \\
&= \sum_{\tilde{s}\in S} \overline{\langle y, \tilde{s}\rangle} \langle x, \tilde{s}\rangle - \sum_{s\in S}\sum_{\tilde{s}\in S} \langle x, s\rangle \overline{\langle y, \tilde{s}\rangle} \langle s, \tilde{s}\rangle \\
&= \sum_{\tilde{s}\in S} \overline{\langle y, \tilde{s}\rangle} \langle x, \tilde{s}\rangle - \sum_{s\in S} \langle x, s\rangle \overline{\langle y, s\rangle} = 0.
\end{aligned}$$

Therefore with $o := x - c$ the sum $x = c + o$ is a decomposition as in Theorem 20.22 with $c \in K$ and o orthogonal to K. Because this decomposition is unique, $c = \sum_{s\in S} \langle x, s\rangle s$ must be the best approximation of x in K. ■

It is possible to prove Corollary 20.23 without using Theorem 20.22 (see Exercise 20-21). Representations via orthonormal bases as in Corollary 20.23 are normally used when dealing with best approximations or with similar computations, as, for example, in the finite element method described in Chapter 23. Thus from a practical point-of-view it may have been worthwhile to prove Corollary 20.23 directly. However, if we had done so, we would not have had any guarantee that best approximations exist in every complete linear subspace unless we had proved that every such subspace has an orthonormal base. It is indeed true that every Hilbert space has an orthonormal base, but the proof would have required a detour into set theory and Zorn's Lemma. As this text is on analysis, the path given here seemed more appropriate. Another characterization of the best approximation is given in Exercise 20-22. Further properties of best approximations, or, more accurately, the maps that map points to their best approximations, are investigated in Exercise 20-23.

We are now ready to prove the representation of continuous linear maps from H into the underlying field.

Definition 20.24 *Let X be a real or complex normed space and let $\mathbb{F}$ be the real or complex numbers, respectively. Then $X^* := \mathcal{L}(X, \mathbb{F})$ is called the* **dual space** *of X. The elements of X^* are also called* **functionals**.

Proposition 20.25 *Let X be a real or complex normed space. Then X^* with pointwise addition and scalar multiplication is a (real or complex) vector space. Moreover,*

$$\|f\| := \min\left\{c \in \mathbb{R} : \left(\forall x \in X : \|f(x)\| \leq c\|x\|\right)\right\}$$

defines a norm on X^* that makes X^* a Banach space.

Proof. Easy corollary to Theorems 17.8 and 17.11 (or their versions for complex spaces). ■

Theorem 20.26 Riesz' Representation Theorem. *Let H be a real or complex Hilbert space and let $\mathbb{F}$ be the real or complex numbers, as appropriate. Then the function $S : H \to H^*,\, z \mapsto \langle \cdot, z\rangle$ is a surjective isometry and if H is a real Hilbert space, it is an isomorphism*

Proof. Clearly, each function $\langle \cdot, z\rangle$ is linear. Moreover, the H^* norm of $\langle \cdot, z\rangle$ is $\|z\|$, because for all $y \in H$ the Cauchy-Schwarz inequality guarantees $\left|\langle y, z\rangle\right| \leq \|y\|\|z\|$ and for $y = z$ equality holds. This means that $z \mapsto \langle \cdot, z\rangle$ is an isometry from H to H^*. This isometry is linear if H is real and it is "almost linear" if H is complex. To finish the proof we need to prove that the map is surjective.

Let $\varphi \in H^*$. Then $K := \varphi^{-1}\left[\{0\}\right]$ is a closed (and hence complete) linear subspace of H. If $K = H$, then φ is the map $\langle \cdot, 0\rangle$. If $K \neq H$ let $w \notin K$ and let $w = c + y$ be the unique orthogonal decomposition of w so that $c \in K$ and y is orthogonal to K. Because $w \notin K$ we infer $y \neq 0$ and because $y \notin K$ we obtain $\varphi(y) \neq 0$.

Let $y' \neq y$ be orthogonal to K. Then $\varphi\left(y' - \dfrac{\varphi(y')}{\varphi(y)} y\right) = \varphi\left(y'\right) - \dfrac{\varphi(y')}{\varphi(y)} \varphi\left(y\right) = 0$, which means $y' - \dfrac{\varphi(y')}{\varphi(y)} y \in K$, and hence, because $y' - \dfrac{\varphi(y')}{\varphi(y)} y$ is orthogonal to K, $y' - \dfrac{\varphi(y')}{\varphi(y)} y = 0$ Therefore every vector that is orthogonal to K is parallel to y. That is, for each $o \perp K$ there is an $\alpha \in \mathbb{F}$ so that $o = \alpha y$.

Let $z := \dfrac{\varphi(y)}{\|y\|^2} y$. Then for all $x \in H$ there are a unique $c \in K$ and a unique $\alpha \in \mathbb{F}$ so that

$$\begin{aligned} \varphi(x) &= \varphi(c + \alpha y) = \alpha\varphi(y) = \alpha\left\langle y, \frac{\varphi(y)}{\|y\|^2} y\right\rangle \\ &= \langle \alpha y, z\rangle = \langle c + \alpha y, z\rangle = \langle x, z\rangle, \end{aligned}$$

which means that φ is the function $\langle \cdot, z\rangle$. Because $\varphi \in H^*$ was arbitrary, the function $z \mapsto \langle \cdot, z\rangle$ is surjective, which concludes the proof. ■

Riesz' Representation Theorem allows a simple representation of the functionals on a Hilbert space as elements of the same Hilbert space. It would be nice to have a simple representation of the functionals on other Banach spaces also. Such representations do indeed exist. For example, if (M, Σ, μ) is a measure space and $1 < p, q < \infty$ are so that $\dfrac{1}{p} + \dfrac{1}{q} = 1$, then $I : L^q(M, \Sigma, \mu) \to \left(L^p(M, \Sigma, \mu)\right)^*$ defined by $I(g) := I_g$ with $I_g(f) = \displaystyle\int_M fg\, d\mu$ as in Example 16.28 (recall that we now omit brackets around elements of L^p) is an isometric isomorphism. If μ is σ-finite, the same holds for $\{p, q\} = \{1, \infty\}$. The proofs are beyond the scope of this text and can be found in [7]. The computation of the dual space of $C^0[a, b]$ in Exercise 20-24 gives an idea how complicated the arguments get.

The Lax-Milgram Lemma (Lemma 23.4) will show the use of Riesz' Representation Theorem in the approximation of solutions of partial differential equations.

Exercises

20-19. Prove Theorem 20 17.

20-20. Let H be an inner product space and let $u, v \in H$ be orthogonal to each other. Prove that $u + v = 0$ iff $u = v = 0$.

20-21 Prove Corollary 20.23 directly, that is, without referring to Theorem 20 22.

20-22. Another characterization of the best approximations in a complete subspace. Let H be an inner product space, let $f \in H$, let V be a complete subspace of H, let $u \in V$ and let $f_\perp$ be the best approximation of f in V. Prove that $\langle u, v\rangle = \langle f, v\rangle$ for all $v \in V$ iff $u = f_\perp$.

20-23. Let K be a complete linear subspace of the inner product space H and let $P_K : H \to K$ be the function so that $P_K(x)$ is the best approximation of x in K.

(a) Prove that P_K is a continuous linear function

(b) Prove that $P_K \circ P_K = P_K$.

Functions P_K as in this exercise are called **orthogonal projections**

20-24. The dual space of $C^0[a, b]$.

(a) Let $\varphi \in \left(C^0[a, b]\right)^*$ and let $f \in C^0[a, b]$ be so that $f(x) \geq 0$ for all $x \in [a, b]$. Define $\varphi_+(f) = \sup\left\{\varphi(g) : g \in C^0[a, b], 0 \leq g \leq f\right\}$.

i. Prove that $0 \leq \varphi_+(f) \leq \|\varphi\|\|f\|_\infty$

ii. Prove that for all $t \geq 0$ we have $\varphi_+(tf) = t\varphi_+(f)$

iii. Prove that for all $f_1, f_2 \in C^0[a, b]$ with $f_i(x) \geq 0$ for all $x \in [a, b]$ we have $\varphi_+(f_1 + f_2) = \varphi_+(f_1) + \varphi_+(f_2)$

(b) For all $f \in C^0[a, b]$ define $\varphi_+(f) := \varphi_+\left(f^+\right) - \varphi_+\left(f^-\right)$. Prove that $\varphi_+ \in \left(C^0[a, b]\right)^*$ and that for all $f \in C^0[a, b]$ with $f \geq 0$ (pointwise) we have $\varphi_+(f) \geq 0$.

Hint. For the additivity, use an argument similar to that at the end of the proof of Theorem 9.25 on page 163.

Note Functionals with the properties of φ_+ are called **positive**.

(c) Define $\varphi_- := \varphi_+ - \varphi$ and prove that $\varphi_- \in \left(C^0[a, b]\right)^*$, that φ_- is positive and that $\varphi = \varphi_+ - \varphi_-$.

(d) Prove that if $\varphi \in \left(C^0[a, b]\right)^*$ is positive and $f, g \in C^0[a, b]$ satisfy $f \leq g$ (pointwise), then $\varphi(f) \leq \varphi(g)$.

(e) Let $\varphi \in \left(C^0[a, b]\right)^*$ be a positive functional

i. For all $x \in [a, b]$ and $n \in \mathbb{N}$ define $\mathbf{1}_{[a,x],n}(t) := \begin{cases} 1; & \text{for } a \leq t \leq x, \\ 1 - n(t - x); & \text{for } x \leq t \leq x + \frac{1}{n}, \\ 0; & \text{for } t \geq x + \frac{1}{n}. \end{cases}$

Prove that $\lim_{n\to\infty} \varphi\left(\mathbf{1}_{[a,x],n}\right)$ exists for all $x \in [a, b]$

ii. Let $g_\varphi(a) := 0$ and for $x \in (a, b]$ define $g_\varphi(x) = \lim_{n\to\infty} \varphi\left(\mathbf{1}_{[a,x],n}\right)$. Prove that g is nondecreasing.

iii. Prove that g_φ is **right-continuous** on $(a, b]$, that is, for all $x \in (a, b]$ we have for the right limit that $\lim_{z\to x^+} g_\varphi(z) = g_\varphi(x)$

Hint For $\varepsilon > 0$ find n so that $\varphi\left(\mathbf{1}_{[a,x],n}\right) - g_\varphi(x) < \frac{\varepsilon}{2}$ Then use $\delta := \frac{\varepsilon}{2\left(\|\varphi\| + 1\right)n}$.

iv. Prove that for all $f \in C^0[a,b]$ we have $\varphi(f) = \int_a^b f \, d\lambda_{g_\varphi}$. (The integral exists by Exercise 18-8c. We could also use the Riemann-Stieltjes integral dg_φ.)
Hint. First consider $f \geq 0$. Find $a = x_0 < x_1 < \cdots < x_m = b$ and a step function $s_0 := a_0 \mathbf{1}_{\{a\}} + \sum_{i=1}^{m} a_i \mathbf{1}_{(x_{i-1}, x_i]}$ with distinct $a_i > 0$ so that $s_0 < f$ and $\|s_0 - f\|_\infty$ is small. Let $l_0 := \min\{a_0, \ldots, a_m\}$, $A_0 := [a,b]$, $s_1 := s_0 - l_0 \mathbf{1}_{A_0}$. Inductively, let $l_j := \min\left\{ s_j(x) : x \in [a,b], s_j(x) > 0 \right\}$, $A_j := \left\{ x \in [a,b] : s_j(x) \geq l_j \right\}$ and let $s_{j+1} := s_j - l_j \mathbf{1}_{A_j}$. Stop this process at the first $k \in \mathbb{N} \cup \{0\}$ so that $s_k = 0$. For all $j > 0$, the set A_j is a union of left-open, right-closed intervals $\left(c_N^j, d_N^j\right]$, $N = 1, \ldots, M_j$, one A_{j_a} will also have $\{a\}$ in the union, and l_j is small.
Set $s_c := l_0 \mathbf{1}_{[a,b]} + l_{j_a} \mathbf{1}_{\{a\},n} + \sum_{j=1}^{k} l_j \sum_{N=1}^{M_j} \left(\mathbf{1}_{\left[a, d_N^j\right],n} - \mathbf{1}_{\left[a, c_N^j\right],n} \right)$ for sufficiently large n. Then s_c is continuous, $s_c < f$ and s_c is close to f and to $s_0 = \sum_{j=0}^{k} l_j \mathbf{1}_{A_j}$ in the uniform norm. Therefore $\varphi(f)$ is close to $\varphi(s_c)$, which is close to $\int_a^b s_0 \, d\lambda_{g_\varphi}$, which in turn is close to $\int_a^b f \, d\lambda_{g_\varphi}$.

(f) Prove that for every $\varphi \in \left(C^0[a,b]\right)^*$ there are nondecreasing functions g_+ and g_- with $g_+(a) = g_-(a) = 0$ that are right-continuous on $(a,b]$ and so that for all $f \in C^0[a,b]$ we have $\varphi(f) = \int_a^b f \, d\lambda_{g_+} - \int_a^b f \, d\lambda_{g_-}$.
Hint. Part 20-24c.

(g) Let $\mathcal{BV}_0^r[a,b]$ be the space of functions of bounded variation (see Exercise 15-2c) on $[a,b]$ that are zero at a and right-continuous on $(a,b]$. Prove that $\mathcal{BV}_0^r[a,b]$ is isomorphic to $\left(C^0[a,b]\right)^*$.
Hint. Call the functions from Exercise 8-12a g_+ and g_-. Use them to construct a function T from $\mathcal{BV}_0^r[a,b]$ to $\left(C^0[a,b]\right)^*$ and prove that it is an isomorphism. For injectivity, note that if $g \in \mathcal{BV}_0^r[a,b]$, then $g_+, g_- \in \mathcal{BV}_0^r[a,b]$ and $T[g] = 0$ implies $T\left[g_+\right] = T\left[g_-\right]$, which can be shown to imply $g = 0$.

20-25 Let (M, Σ, μ) be a measure space, let $p \in [1, \infty]$ and let $\varphi \in \left(L^p(M, \sigma, \mu)\right)^*$. Prove that there are measures ν_+ and ν_- on Σ so that for all $f \in L^p(M, \Sigma, \mu)$ we have $\varphi(f) = \int_M f \, d\nu_+ - \int_M f \, d\nu_-$.
Hint. Decompose φ into a positive and a negative part similar to Exercises 20-24a, 20-24b and 20-24c. Then define $\nu_+(A) := \varphi_+(\mathbf{1}_A)$ for all A for which the right side is finite and use this to define a measure.

20-26 Representation of finite **Borel measures** on $[a,b]$.

(a) Let μ be a finite Borel measure on $[a,b]$ and let $\varphi_\mu(f) := \int_a^b f \, d\mu$. Prove that φ_μ is a continuous linear function from $C^0[a,b]$ to $\mathbb{R}$ and prove that φ_μ is positive.

(b) Prove that there is a unique nondecreasing function $g : [a,b] \to \mathbb{R}$ so that $\mu = \lambda_g$, $g(a) = 0$ and g is right-continuous on $(a,b]$.
Hint. Use Exercise 20-24e.

(c) Let μ be a Borel measure on $[a,b]$ that is infinite on some subset of $[a,b]$. Prove that $\varphi(f) := \int_a^b f \, d\mu$ is undefined for some $f \in C^0[a,b]$.

Part III

Applied Analysis

Chapter 21

Physics Background

Although for the finer mathematical points we need the concepts developed in Parts I and II, the descriptions of physical phenomena in this chapter are accessible with a good background in calculus. Thus, this chapter could be read at any time to provide further motivation for the concepts encountered in the other chapters.

One reason analysis is so well developed is because it is useful to model physical phenomena. For example, the reader probably knows that in physics derivatives are typically interpreted as velocities (also see Proposition 17.26). The last part of this text focuses on some applications of analysis. The connection to applications was delayed until now, because in the final three chapters we will need almost every result in the text, either directly or as a lemma.[1] Consequently, it is safe to say that the closer we get to describing certain real life phenomena, the more mathematical detail is needed. We have seen this once already in our tour de force through manifolds. (Recall that manifolds can be used to describe surfaces, such as the body of an airplane.) But these details are what allows us to model many complicated phenomena. For example, the behavior of an airplane can be modeled so well that building prototypes of new commercial airplanes is no longer necessary. So, although the underlying abstract mathematics is challenging, although plenty of details need to be considered, and although the computational demands are steep indeed, this is where the fun really starts!

This chapter showcases some fundamental equations from physics. These equations gain their relevance through the encoding of physical quantities as mathematical objects and they are still at the heart of increasing numbers of sophisticated models. Some theoretical analysis and solution techniques are presented in the following chapters. Because this chapter describes physical phenomena, we will use notation similar

[1] It might be an interesting scavenger hunt to find the results in this text that are *not* used in Part III. The only rule the author would suggest is to count a one-dimensional result in Part I as used if a more abstract version from Part II is used.

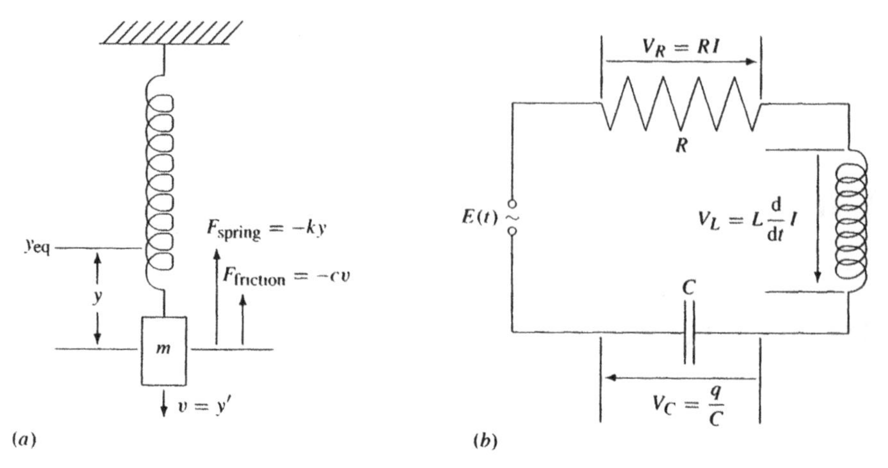

Figure 57: Forces in a spring mass system (a) and voltages in an RLC-circuit (b).

to that in physics and applications. Mainly, this means that vectors will be denoted as letters with arrows on top, inner products with a dot, determinants with absolute values and integrals over closed curves or surfaces with a circle in the integral sign.

21.1 Harmonic Oscillators

Newton's second law states that the net force acting on a body is equal to the product of the body's mass and its acceleration. Consider a mass attached to a spring (see Figure 57(a)). To make things simple, we will assume the motion of the mass is one-dimensional. That is, the mass moves in the direction in which the spring most naturally expands and compresses. The following forces act on the mass.

The spring force is proportional to the displacement $y(t)$ of the mass from the equilibrium position y_{eq}. As is customary, the variable t denotes time. Because the spring force tries to restore the body to the equilibrium position, it is $F_{spring} = -ky$ for some proportionality constant $k > 0$. The negative sign expresses that the spring force points toward the equilibrium point at all times. The constant k depends on the spring. It is also called the **spring constant**. It is common in the sciences to drop the arguments of functions and we will do so, too. The equations are often complicated enough without the arguments.

Friction is usually modeled as being proportional to the velocity y' of the mass. Because friction slows an object down, the friction force must be $F_{friction} = -cy'$, for some proportionality constant $c \geq 0$. The negative sign expresses that the friction force points opposite the direction of motion.

Finally, some forces may not depend on the position or the velocity of the mass, say, for example, gravity. These forces are usually combined into one term F. Overall, because the sum of all the forces acting on a body will equal its mass times its acceleration (Newton's second law), we obtain $my'' = -cy' - ky + F$. The above can be summarized as follows.

Theorem 21.1 *Equation of the* **forced harmonic oscillator with damping**. *The equation $my'' + cy' + ky = F$ describes the motion of a mass m attached to a spring with spring constant k with friction proportional to the velocity and further forces F acting on the mass.* ■

If friction is not too strong and if there are no further forces, the solutions to this equation are sine waves with exponentially decaying amplitude (see Exercise 21-1), which explains the name of the equation. This kind of equation is frequently encountered in science. Essentially, it arises whenever oscillations are involved. For example, in electric circuits (see Figure 57(b)), the voltage across a resistor with resistance R is $V_R = RI$, where I is the current through the resistor. The voltage across an inductor with inductance L is LI'. The voltage across a capacitor with capacitance C is $\frac{Q}{C}$, where Q is the charge of the capacitor. Now consider what happens when we place these three elements in a serial circuit and impose an external voltage $E(t)$. Because in a serial circuit the sum of the voltages across the individual elements equals the external voltage we obtain $E(t) = LI' + RI + \frac{Q}{C}$. Because the current is the derivative of the charge, with $Q' = I$ we get the equation $LQ'' + RQ' + \frac{1}{C}Q = E(t)$. That is, the circuit obeys the same equation of a forced harmonic oscillator with damping as the mechanical system described above. In fact, in physics one often uses the analogy between inductance and inertia or mass (which keep the system from stopping), between resistance and friction (which account for the energy losses), and between the reciprocal of the capacitance and the spring constant (because of which the system is attracted to its original equilibrium) to describe electrical systems in mechanical terms.

Of course, real life electrical and mechanical systems are not as simple as the lab systems described above. To describe real life systems, one often uses combinations of the above systems, which leads to systems of differential equations. For an introduction to mechanics, consider [11].

Exercises

21-1 Let $A, \varphi \in \mathbb{R}$ and let $m, k > 0$, $c \geq 0$ be so that $4km - c^2 > 0$.

(a) Prove that $y(t) := Ae^{-\frac{c}{2m}t} \sin\left(\sqrt{\frac{k}{m} - \frac{c^2}{4m^2}}\, t + \varphi\right)$ satisfies $my'' + cy' + ky = 0$.

(b) Explain why we need the inequality $4km - c^2 > 0$.

(c) Explain what happens (physically) for $c = 0$.

21-2. Prove that the gravitational force acting on a spring-mass-system can be ignored in the mathematical model by shifting the equilibrium point.

(a) Find the coordinates of the equilibrium point for the mass in the gravitational field of earth by solving $ky = -mg$ for y, where g denotes the constant gravitational acceleration.

(b) Let $\tilde{y}(t) := y(t) - \frac{mg}{k}$ and find a differential equation for $\tilde{y}$

Hint Write the force F as $F = -mg + \tilde{F}(t)$.

(c) Explain why we use a negative term for the gravitational force.

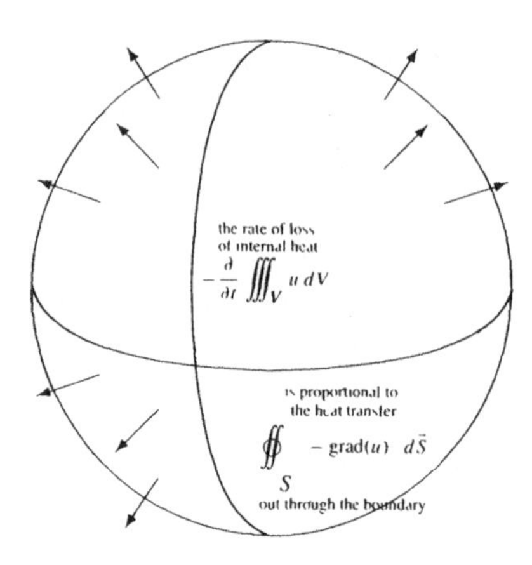

Figure 58: A test volume to analyze heat transfer. The rate of change of thermal energy inside the volume is proportional to the integral of the thermal flux across the surface.

21.2 Heat and Diffusion

To describe phenomena that depend on more than one space variable or on time and space, we need to use functions of several variables. Let $\Omega \subseteq \mathbb{R}^3$ be open. A function $u : \Omega \times [0, \infty) \to \mathbb{R}$ describes the value of a quantity $u(x, y, z, t)$ at the point $(x, y, z) \in \mathbb{R}^3$ at time t.

For this section, let $u = u(x, y, z, t)$ be the temperature at a point (x, y, z) at time t. This temperature is proportional to the thermal energy density at the point (x, y, z) at time t. Physical observations show that if a region has hot and cold parts and is not heated or cooled from the outside, then heat will diffuse so that all parts ultimately reach the same temperature. We can analyze this phenomenon mathematically. To make things simple, we assume that the medium conducts heat at equal rates in all places and that there are no heat sources or sinks. More sophisticated models will take these things into account, but we are first concerned with the fundamental principle.

Thermal energy follows the path that promises the fastest possible equalization of temperature between hot and cold regions. For example, if an ice cube is put in liquid, heat will diffuse through the surface of the ice cube. It will not in any way transfer around the ice cube without entering the ice cube and all thermal energy that enters the ice cube must do so through the surface. The negative gradient of the temperature u is the direction of steepest descent of u and the temperature is proportional to the thermal energy density. Therefore, $-\operatorname{grad}(u)$ gives the direction in which thermal energy flows as temperatures equalize. Consider a volume V of space, bounded by the closed surface S (see Figure 58). Formally, the volume V is a compact oriented three dimensional embedded manifold with boundary or corners (mathematical details were discussed in Chapters 16 and 19). To obtain an equation we analyze the net energy transfer between the volume V and the surrounding area. Positive transfer shall mean that energy is leaving the volume V, negative transfer shall mean that energy is entering the volume V. The transfer of energy out of the volume V can be measured in two ways.

- Thermal energy can only be interchanged through the surface $S := \partial V$. Thus the surface integral $\oiint_S -\text{grad}(u) \cdot d\vec{S}$ of the negative gradient of u over S is proportional to the energy transfer out of the volume V per time unit. (Formally, the integral is an integral of the appropriate form over $S = \partial V$.)
- Thermal energy transfer into and out of the volume V implies a change in the net thermal energy contained in V. Because $\iiint_V u \, dV$ is proportional to the net thermal energy content of V, the derivative $-\frac{\partial}{\partial t} \iiint_V u \, dV$ is proportional to the energy transfer out of the volume V per time unit. The negative sign is needed because transfer *out* of the volume V, which we wanted to be indicated by a positive number, means that the time derivative is negative. (Formally, the integral is a Lebesgue integral in $\mathbb{R}^3$.)

Because both integrals are proportional to the same quantity, there must be a number $k > 0$ such that $-k \oiint_S \text{grad}(u) \cdot d\vec{S} = -\frac{\partial}{\partial t} \iiint_V u \, dV$. The value of k depends on the heat conductivity properties of the region. The vector $\vec{q} = -k \, \text{grad}(u)$ is also called the **thermal flux vector**. It measures energy transfer per time and area. As is customary in physics, we assume that u is sufficiently often differentiable (mathematical details were discussed in Chapter 17, twice continuously differentiable suffices here). Then, because the integral equation must be valid for all volumes V, we obtain

$$-k \oiint_S \text{grad}(u) \cdot d\vec{S} = -\frac{\partial}{\partial t} \iiint_V u \, dV$$

Under mild hypotheses we can interchange integral and derivative (see Exercise 21-3).

$$\oiint_S k \, \text{grad}(u) \cdot d\vec{S} = \iiint_V \frac{\partial u}{\partial t} \, dV$$

Apply the Divergence Theorem at time t.

$$\iiint_V k \, \text{div}\big(\text{grad}(u)\big) \, dV = \iiint_V \frac{\partial u}{\partial t} \, dV.$$

Because this equation applies to all volumes V, we can pick a fixed point with position vector $\vec{r}$ and apply the equation to a small ball $V = B_a(\vec{r})$ of radius a around the point (see Figure 59).

$$\iiint_{B_a(\vec{r})} k \, \text{div}\big(\text{grad}(u)\big) \, dV = \iiint_{B_a(\vec{r})} \frac{\partial u}{\partial t} \, dV.$$

$$\lim_{a \to 0} \frac{1}{\frac{4}{3}\pi a^3} \iiint_{B_a(\vec{r})} k \, \text{div}\big(\text{grad}(u)\big) \, dV = \lim_{a \to 0} \frac{1}{\frac{4}{3}\pi a^3} \iiint_{B_a(\vec{r})} \frac{\partial u}{\partial t} \, dV.$$

As long as all functions involved are continuous, the limits are the values at the respective points.

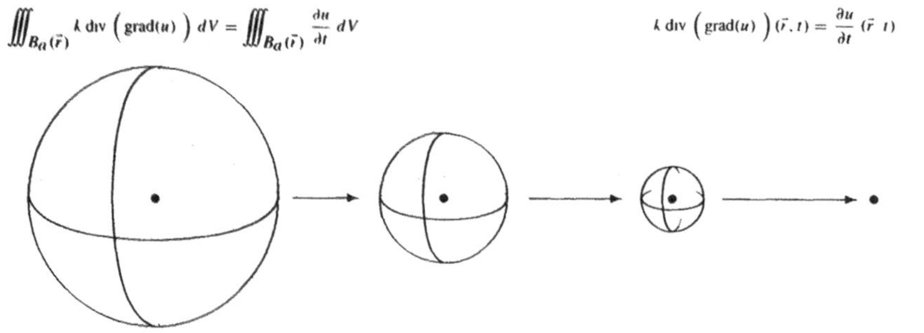

Figure 59: To derive a partial differential equation for heat transfer, we consider a small ball centered at $\vec{r}$. Shrinking this control volume to zero turns the integral equation into a differential equation.

$$k \operatorname{div}\big(\operatorname{grad}(u)\big)\,(\vec{r}, t) \quad = \quad \frac{\partial u}{\partial t}\,(\vec{r}, t)$$

The above argument shows that the divergence of the gradient of u determines the rate of change of the temperature u over time.

Definition 21.2 *The operator*

$$\Delta u := \operatorname{div}\big(\operatorname{grad}(u)\big) = \frac{\partial^2}{\partial x^2}u + \frac{\partial^2}{\partial y^2}u + \frac{\partial^2}{\partial z^2}u = \nabla \cdot \nabla u =: \nabla^2 u,$$

is called the **Laplace operator**.

Definition 21.3 *The partial differential equation* $k\Delta u = \frac{\partial}{\partial t}u$*, with* $k > 0$ *being constant, which is also often written as* $\frac{\partial}{\partial t}u - k\Delta u = 0$ *is called the* **heat equation**. *It is also known as the* **Fourier equation** *or the* **diffusion equation**. *The situation in which* $\frac{\partial}{\partial t}u = 0$ *is called the* **steady state** *and in this case the equation* $\Delta u = 0$ *is also referred to as the* **Laplace equation**.

Although we have derived the heat equation under the assumptions that there are no heat sources or sinks and that the material conducts heat equally in all places, the equation also works when sources and sinks are present and for inhomogeneous media. Because the derivation happened at individual points, the equation is valid at any point that is not on the boundary of a heat source or sink. Moreover, in inhomogeneous media, where k is not a constant and where heat conduction may be faster in certain directions, the heat equation is often written as $\frac{\partial}{\partial t}u - \nabla\big(A\nabla u\big) = 0$, where the function $A : \mathbb{R}^3 \to \mathbb{R}^3$ is a linear function describing the heat conducting properties of the medium, which may vary over space and time.

The initial state of a real system can be incorporated into the model by demanding that the solution u of the heat equation satisfies $u(x, y, z, 0) = i(x, y, z)$ for all (x, y, z)

in the domain, where the function $i(x, y, z)$ is the initial heat distribution at time $t = 0$. This type of condition is also called an **initial condition**. Heat sources and sinks can be modeled with **boundary conditions**, that is, by demanding that on certain surfaces in space the function u must equal certain values at all times. Initial and boundary conditions can be imposed not just on the function u, but also on its partial derivatives. The remarks at the end of Section 23.3 show that initial and boundary conditions can be incorporated into the equation using a nonzero right side.

Note that the heat equation applies to any transfer/diffusion phenomenon in which energy (like thermal energy) or a substance (say, salt dissolved in water) travels in the direction of the negative gradient of the density (or concentration) function u. The argument is exactly the argument we used to derive the heat equation. Just substitute the new interpretations for "heat" and "temperature".

In terms of rewriting the equation, it should be said that for ordinary as well as for partial differential equations it is common to write the equation so that all derivatives are on one side and all remaining terms are on the other. This is because an equation $Du = f$, where D is a differential operator acting on u, is accessible to methods like the finite element method (see Chapter 23).

Exercises

21-3. Let $\Omega \subseteq \mathbb{R}^d$ be a bounded open set and let $q : \Omega \times (a, b) \to \mathbb{R}$ be differentiable with bounded derivative. Prove that $\frac{d}{dt}\int_{\Omega} q(x, t)\, d\lambda(x) = \int_{\Omega} \frac{\partial}{\partial t} q(x, t)\, d\lambda(x)$

Hint. Use the Mean Value Theorem and the Dominated Convergence Theorem or use a similar procedure as for Exercise 17-58.

21-4. Equality of functions through equality of integrals. Let $\Omega \subseteq \mathbb{R}^d$ be open. Prove that if $f, g : \Omega \to \mathbb{R}$ are two continuous functions with equal domains and $\int_V f\, d\lambda = \int_V g\, d\lambda$ over all open sets V that are contained in the domain, then $f = g$

Hint. Mimic the argument in the derivation of the heat equation.

21-5. Equality of vector fields through equality of surface integrals. Prove that if $\vec{F}$ and $\vec{G}$ are two continuous vector fields in $\mathbb{R}^3$ with equal open domains and $\iint_S \vec{F} \cdot d\vec{S} = \iint_S \vec{G}\ \, d\vec{S}$ over all surfaces S (compact oriented two dimensional embedded manifolds with boundary or corners) that are contained in the domain, then $\vec{F} = \vec{G}$.

Hint Mimic the argument in the derivation of the heat equation Use disks instead of balls.

21-6 Prove the following differentiation formulas. $\vec{F}, \vec{G}$, etc denote three dimensional vector fields, a, b, etc. denote scalar functions. Always assume that the requisite derivatives exist.

(a) $\text{grad}\,(a + b) = \text{grad}\,(a) + \text{grad}\,(b)$

(b) $\text{div}\left(\vec{F} + \vec{G}\right) = \text{div}\left(\vec{F}\right) + \text{div}\left(\vec{G}\right)$

(c) $\text{curl}\left(\vec{F} + \vec{G}\right) = \text{curl}\left(\vec{F}\right) + \text{curl}\left(\vec{G}\right)$

(d) $\text{grad}\,(ab) = a\,\text{grad}\,(b) + b\,\text{grad}(a)$

(e) $\text{div}\left(a\vec{F}\right) = a\,\text{div}\left(\vec{F}\right) + \text{grad}(a) \cdot \vec{F}$

(f) $\text{curl}\left(a\vec{F}\right) = a\,\text{curl}\left(\vec{F}\right) + \text{grad}(a) \times \vec{F}$

(g) $\text{div}\left(\vec{F} \times \vec{G}\right) = \vec{G} \cdot \text{curl}\left(\vec{F}\right) - \vec{F} \cdot \text{curl}\left(\vec{G}\right)$

(h) $\text{div}\left(\text{grad}(a) \times \text{grad}(b)\right) = 0$

(i) $\text{curl}\left(\text{curl}\left(\vec{F}\right)\right) = \text{grad}\left(\text{div}\left(\vec{F}\right)\right) - \Delta\vec{F}$, where the Laplace operator Δ acts componentwise.

(j) $\operatorname{curl}\left(\vec{F}\times\vec{G}\right)=\vec{F}\operatorname{div}\left(\vec{G}\right)-\vec{G}\operatorname{div}\left(\vec{F}\right)+\left(\vec{G}\cdot\nabla\right)\vec{F}-\left(\vec{F}\cdot\nabla\right)\vec{G}$

(k) $\Delta(ab)=a\Delta b+b\Delta a+2\operatorname{grad}(a)\cdot\operatorname{grad}(b)$

21-7. Let $\Omega\subseteq\mathbb{R}^3$ be open and let $u:\Omega\to\mathbb{R}$ be twice differentiable. Prove the following representations of the **Laplace operator** in cylindrical and spherical coordinates

(a) $\Delta u=\frac{\partial^2 u}{\partial r^2}+\frac{1}{r}\frac{\partial u}{\partial r}+\frac{1}{r^2}\frac{\partial^2 u}{\partial\theta^2}+\frac{\partial^2 u}{\partial z^2}$, where $x=r\cos(\theta)$, $y=r\sin(\theta)$ and $z=z$

Hint. Use the Chain Rule as in Exercise 17-56.

(b) $\Delta u=\frac{\partial^2 u}{\partial\rho^2}+\frac{2}{\rho}\frac{\partial u}{\partial\rho}+\frac{1}{\rho^2}\frac{\partial^2 u}{\partial\phi^2}+\frac{\cos(\phi)}{\rho^2\sin(\phi)}\frac{\partial u}{\partial\phi}+\frac{1}{\rho^2\sin^2(\phi)}\frac{\partial^2 u}{\partial\theta^2}$,

where $x=\rho\cos(\theta)\sin(\phi)$, $y=\rho\sin(\theta)\sin(\phi)$ and $z=\rho\cos(\phi)$.

21-8 **Green's Identities** Let $V\subseteq\mathbb{R}^3$ be a compact oriented three dimensional embedded manifold with boundary or corners with $S=\partial V$ being the boundary surface of V with positive (outward) orientation. Let u and v be functions with continuous second partial derivatives on an open region that contains V. Prove the following using Exercise 21-6e and the Divergence Theorem

(a) $\iiint_V \nabla u\cdot\nabla v+u\Delta v\,dV=\oint\!\!\oint_S u\nabla v\cdot d\vec{S}$ (Green's First Identity)

(b) $\iiint_V \left(u\Delta v-v\Delta u\right)\,dV=\oint\!\!\oint_S \left(u\nabla v-v\nabla u\right)\cdot d\vec{S}$ (Green's Second Identity)

21.3 Separation of Variables, Fourier Series, and Ordinary Differential Equations

To predict real life phenomena, we need to solve the equations that model them. Because these equations are rather complicated, it is common to impose some simplifying assumptions. Consider, for example, a straight metal rod that is held at constant temperature at the ends. Because heat conducts much faster through metal than through the metal–air interface, for an initial investigation we can assume that heat is only interchanged at the ends of the rod. Moreover, because it is unlikely that the temperature distribution of the rod varies significantly over its cross section, we can assume the rod is one dimensional with its temperature u only depending on the position x on the rod and the time t.

In this section, we will solve the one dimensional heat equation $k\frac{\partial^2 u}{\partial x^2}=\frac{\partial u}{\partial t}$ on the interval $[0,\pi]$ with the boundary condition $u(0,t)=u(\pi,t)=0$ and initial condition $u(\cdot,0)=f(\cdot)$. This equation describes the rod from the previous paragraph. The ends are kept at constant temperature (we chose this constant to be zero because any solution of the heat equation can be rescaled by adding a constant), the initial temperature distribution is f and heat transfer occurs only within the rod and at the ends. Of course, the rod could have any length l. We chose the interval $[0,\pi]$ so that our solutions will be easier to read (this will become clear below). Exercise 21-9 shows that if we can solve the problem for rods of length π, then we can solve it for rods of arbitrary length.

The idea behind **separation of variables** is to write the function of two variables $u(x,t)$ as a product of single variable functions. With $u(x,t)=X(x)T(t)$ the equation $k\frac{\partial^2 u}{\partial x^2}=\frac{\partial u}{\partial t}$ reduces to $kX''T=XT'$, and then to $\frac{X''}{X}=\frac{1}{k}\frac{T'}{T}$, assuming that all functions are sufficiently often differentiable. Because the left side depends only on x and the right side depends only on t, we conclude that both sides are constant (see Exercise 21-10). That is, there must be a real number c so that $\frac{X''}{X} = \frac{1}{k}\frac{T'}{T} = c$.

Physical reality dictates that $\frac{T'}{T} \leq 0$, because if $T > 0$, the rod will cool ($T' < 0$), while if $T < 0$ the rod will warm up ($T' > 0$). Because $k > 0$ we can assume that $c = -\lambda^2$. Negative numbers are typically represented as negative squares. That way we can assume throughout that the symbolic constant λ is nonnegative. From $\frac{X''}{X} = \frac{1}{k}\frac{T'}{T} = -\lambda^2$, we obtain that $T' = -k\lambda^2 T$ and $X'' + \lambda^2 X = 0$. This means that $X(x) = a\cos(\lambda x) + b\sin(\lambda x)$ (if this does not ring a bell, Example 22.13 will establish this claim once more) and $T(t) = e^{-k\lambda^2 t}$ (Exercise 22-12 will provide the general solution; this particular solution is chosen so that $T(0) = 1$).

At this point, we appear to be stuck, because any solution of the form $X(x)T(t)$ satisfies an initial condition that looks like a sine or a cosine wave. Obviously, heat distributions can also assume other shapes. To resolve this dilemma, note that any linear combination of solutions of the heat equation also solves the heat equation (Exercise 21-11). With what we know about Fourier series (see Section 20.2) we should be able to sum solutions of the form $X(x)T(t)$ so that the initial condition $u(\cdot, 0) = f(\cdot)$ is satisfied. Because $T(0) = 1$, independent of the value of λ, it is reasonable to first approximate $f(\cdot)$ with functions of the form $X(x) = a\cos(\lambda x) + b\sin(\lambda x)$.

To assure that each $X(x)$ satisfies the boundary condition, it is sensible to demand that there is no cosine function in $X(x)$ (after all, $\cos(0) = 1 \neq 0$) and to demand that the part $\sin(\lambda x)$ of $X(x)$ satisfies the boundary condition. Because $\sin(\lambda\pi) = 0$ is only possible for $\lambda \in \mathbb{N}$, it is sensible to demand that λ is a natural number n. Now that $\lambda = n \in \mathbb{N}$, we represent f as a Fourier series. Let a_n, b_n be the Fourier coefficients of the odd function $g(x) := \begin{cases} f(x); & \text{for } x \in (0, \pi], \\ 0; & \text{for } x = 0, \\ -f(-x); & \text{for } x \in [-\pi, 0). \end{cases}$ Because g is odd, all the a_n are zero. For each n, we choose $X_n(x) := b_n \sin(nx)$. We automatically get $T_n(t) = e^{-kn^2 t}$ and our solution should be $u(x,t) := \sum_{n=1}^{\infty} b_n \sin(nx) e^{-kn^2 t}$.

It remains to be shown that the infinite sum makes sense and that the function actually is differentiable. Let $f \in L^2[0, \pi]$. By the Cauchy Schwarz inequality, the Fourier coefficients of g are uniformly bounded by $\|g\|_2$. For fixed $k > 0$ and $t > 0$, the sum $\sum_{n=1}^{\infty} e^{-kn^2 t}$ converges, so the series for u converges absolutely for all places $x \in [0, \pi]$ and all times $t > 0$. Similarly, the Fourier series of all termwise partial derivatives (of any order) converge absolutely for all $x \in [0, \pi]$ and all $t > 0$. Moreover, for each point $(x, t) \in (0, \pi) \times (0, \infty)$ the series converge uniformly in a neighborhood of (x, t). This means by Theorem 11.11 that u has partial derivatives of any order on $(0, \pi) \times (0, \infty)$ and the partial derivatives are taken termwise. By Theorem 17.41, this means that u as defined above is in $C^\infty\big((0, \pi) \times (0, \infty)\big)$ and because the derivatives are taken termwise and the terms satisfy the heat equation, $u|_{(0,\pi)\times(0,\infty)}$ is a solution of the heat equation.

To establish that u has the right initial and boundary values, first note that u trivially satisfies $u(0,t) = u(\pi,t) = 0$ for all $t > 0$. For the initial conditions by Theorem 20.12, the equality $u(\cdot,0) = f(\cdot)$ "holds in the L^2 sense." But *how* exactly are the initial values assumed? Is the function u continuous on $[0,\pi] \times [0,\infty)$? If it is not continuous, in what sense are the initial values assumed (if at all)? These questions are one of the inspirations of harmonic analysis. For our purposes, we can say that under mild hypotheses (such as f being continuously differentiable with bounded derivative and zero on the boundary, see Exercise 21-13) that are often satisfied in realistic situations, the function u is continuous and satisfies the initial and boundary conditions.

In summary, separation of variables together with series representation of initial values is a useful tool to reduce partial differential equations to ordinary differential equations whose solutions can be combined to solve the partial differential equation for given initial and boundary values. The differential equations that arise in this approach (also see Exercise 21-14), their solutions, and the properties of their solutions are investigated in the branch of mathematics called "special functions of mathematical physics" (see [1] for a collection of standard functions).

As a historical note it should be said that Fourier's investigation of the heat equation had far reaching consequences in mathematics. Fourier did not have our modern formalisms at his disposal. Yet the occurrence of trigonometric functions and the utility of the summation representations prompted people to rethink their definition of a function. Before Fourier, functions were only those entities that could be represented with a "closed expression" like $f(x) = x^2$ or finite sums. Fourier's investigation started turning the tide toward the more modern definition of a function, first leading to the inclusion of representations that were not "closed" and ultimately to the set theoretical definition that is the standard today.

Exercises

21-9 Let $l > 0$ and let $f : [0,l] \to \mathbb{R}$ be a function. Prove that if $u(\cdot) : [0,\pi] \times [0,\infty) \to \mathbb{R}$ solves the equation $\frac{l^2}{\pi^2} k \frac{\partial^2 u}{\partial x^2} = \frac{\partial u}{\partial t}$ with $u(0,t) = u(\pi,t) = 0$ and initial condition $u(x,0) = f\left(\frac{l}{\pi}x\right)$, then $v(x,t) := u\left(\frac{\pi}{l}x, t\right) : [0,l] \times [0,\infty) \to \mathbb{R}$ solves the equation $k\frac{\partial^2 v}{\partial x^2} = \frac{\partial v}{\partial t}$ with boundary condition $v(0,t) = v(l,t) = 0$ and initial condition $v(x,0) = f(x)$.

21-10. Let $D_f, D_g \subseteq \mathbb{R}$ and let $f : D_f \to \mathbb{R}$ and $g : D_g \to \mathbb{R}$ be functions. Prove that if for all $x \in D_f$ and all $y \in D_g$ we have $f(x) = g(y)$, then there must be a constant λ such that $f(x) = g(y) = \lambda$ for all $x \in D_f$ and all $y \in D_g$.

21-11. Prove that if u, v solve the heat equation $k\frac{\partial^2 u}{\partial x^2} = \frac{\partial u}{\partial t}$ and $a, b \in \mathbb{R}$, then $au + bv$ also solves the heat equation.

21-12. Prove that if the function f is continuous on $[0,\pi]$, continuously differentiable on $(0,\pi)$ with bounded derivative and $f(0) = f(\pi) = 0$, then the series $\sum_{n=1}^{\infty} |b_n|$ converges, where the b_n are the Fourier sine coefficients of the function g as defined in this section.

Hint Use integration by parts over $[\delta, \pi - \delta]$ (with $\delta \to 0$) to prove that $b_n^g = \frac{1}{n} a_n^{g'}$ and then use Exercise 20-12b.

Note This result also works for f being continuous and piecewise smooth with $f(0) = f(\pi) = 0$.

21-13. Prove that if the function f is continuous on $[0, \pi]$, continuously differentiable on $(0, \pi)$ with bounded derivative and $f(0) = f(\pi) = 0$, then the solution u of the heat equation as constructed in this section is continuous on $[0, \pi] \times [0, \infty)$.
Hint. Use Exercise 21-12.

21-14 Use the separation of variables $u = R(r)D(\theta)T(t)$ for the heat equation $k\Delta u = \frac{\partial u}{\partial t}$ on a domain that is a closed disk of radius a around the origin in $\mathbb{R}^2$ to prove that R must solve the **parametric Bessel equation** $r^2 R'' + rR' + \left(\lambda^2 r^2 - \nu^2\right) R = 0$ with $\lambda \in \mathbb{R}$ and ν being an integer
Hint. The heat equation in polar coordinates is $\frac{\partial^2 u}{\partial r^2} + \frac{1}{r}\frac{\partial u}{\partial r} + \frac{1}{r^2}\frac{\partial^2 u}{\partial \theta^2} = \frac{1}{k}\frac{\partial u}{\partial t}$, where $x = r\cos(\theta)$, $y = r\sin(\theta)$. λ and ν are squared to indicate nonnegativity, which is a consequence of physical considerations.

21-15. Use a separation of variables similar to the one in this section to compute a function u that should solve the partial differential equation $k\frac{\partial^2 u}{\partial x^2} = \frac{\partial^2 u}{\partial t^2}$ with initial values $u(\cdot, 0) = f(\cdot) \in L^2[0, \pi]$ and boundary values $u(0, t) = u(\pi, t) = 0$. Prove that if f is continuous on $[0, \pi]$, continuously differentiable on $(0, \pi)$ with bounded derivative and $f(0) = f(\pi) = 0$, then the function u is in $C^\infty\left((0, \pi) \times (0, \infty)\right) \cap C^0\left([0, \pi) \times (0, \infty)\right)$, it solves the equation on $(0, \pi) \times (0, \infty)$ and it satisfies the boundary conditions.

21-16. Another approach to the one dimensional heat equation.

(a) Let $k \in \mathbb{R}$. Prove that $H_k(x, t) := \frac{1}{\sqrt{4\pi k}\sqrt{t}} e^{-\frac{x^2}{4kt}}$ solves $k\frac{\partial^2}{\partial x^2}u = \frac{\partial}{\partial t}u$

(b) Prove that $\int_{-\infty}^{\infty} H_k(x, t)\, d\lambda(x) = 1$. *Hint.* Exercise 18-44c.

(c) Now let $k \in \mathbb{R}$, $g \in L^2(\mathbb{R})$ and let $u(x, t) := \int_{-\infty}^{\infty} g(z) H_k(x - z, t)\, d\lambda(z)$.

i. Prove that u solves the heat equation $k\frac{\partial^2}{\partial x^2}u = \frac{\partial}{\partial t}u$.
Hint. Use the Mean Value Theorem and the Dominated Convergence Theorem.

ii. Prove that $\int_{-\infty}^{\infty} u(x, t)\, d\lambda(x) = \int_{-\infty}^{\infty} g\, d\lambda$.
Hint. Apply Proposition 14.65 to the positive and negative parts of the integrand

iii. Prove that if g is continuous at x, then $\lim_{(z,t)\to(x,0)} u(z, t) = g(x)$

Note. If g is continuous, then the function u constructed in part 21-16c solves the heat equation on the real line so that the initial heat distribution is given by g and so that the total amount of thermal energy is constant over time.

21.4 Maxwell's Equations

Maxwell's equations describe the behavior of electric and magnetic fields. As such, they should be considered to be physical observations. We state them as theorems, because Theorems 21.4 and 21.5 are indeed mathematically equivalent. As a way of justifying that Maxwell's equations truly describe real phenomena, the reader should consider that every law in Theorem 21.4 can be verified experimentally. The verification is possible, because the integral of a physical density corresponds to a macroscopic measurement of the underlying quantity and because the surface integral of a quantity can be macroscopically obtained by measuring the quantity at enough points and computing a Riemann sum to obtain an approximation. Thus, the equivalent Theorems 21.4

and 21.5 describe real electric and magnetic fields because Theorem 21.4 does. The solid regions mentioned are, mathematically speaking, compact oriented three dimensional embedded manifolds with boundary or corners in $\mathbb{R}^3$. All surfaces mentioned are compact oriented two dimensional embedded manifolds with boundary or corners in $\mathbb{R}^3$ and any loops or closed curves mentioned are boundary curves of compact oriented two dimensional embedded manifolds with boundary or corners.

Theorem 21.4 **Maxwell's equations** *in integral form. Let $\vec{E}$ denote the electric field, $\vec{B}$ the magnetic field, ρ the charge density and $\vec{j}$ the current density. Let c be the speed of light, let ε_0 be the* **permittivity constant** $\varepsilon_0 = 8.8542 \cdot 10^{-12} \frac{As}{Vm}$ *and let $\mu_0 := \frac{1}{\varepsilon_0 c^2} = 1.2566 \cdot 10^{-6} \frac{Vs}{Am}$ be the* **permeability constant**. *Then the following can be observed experimentally (when no dielectric or magnetic materials are present).*

1. **Gauss' Law**. *The net electrical flux across a closed surface is proportional to the charge inside the surface.* $\oiint_S \vec{E} \cdot d\vec{S} = \frac{1}{\varepsilon_0} \iiint_V \rho \, dV.$

2. *The net flux of the magnetic field $\vec{B}$ across a closed surface is 0.* $\oiint_S \vec{B} \cdot d\vec{S} = 0.$

3. **Ampère-Maxwell Law**. *The line integral of a magnetic field $\vec{B}$ along a closed curve C is equal to $\mu_0\varepsilon_0$ times the rate of change of the electric flux through the surface S bounded by C plus μ_0 times the current that goes through the loop defined by C.* $\oint_C \vec{B} \cdot d\vec{s} = \mu_0\varepsilon_0 \frac{d}{dt} \iint_S \vec{E} \cdot d\vec{S} + \mu_0 \iint_S \vec{j} \cdot d\vec{S}.$

4. **Faraday's Law**. *The integral of the electric field $\vec{E}$ along a closed loop is equal to the rate at which the magnetic flux through the loop changes with time.* $\oint_C \vec{E} \cdot d\vec{s} = -\frac{d}{dt} \iint_S \vec{B} \cdot d\vec{S}.$

The integral forms of Maxwell's equations can be translated into partial differential equations with arguments similar to what was done for the heat equation in Section 21.2. Moreover, the integral and the differential forms of Maxwell's equations actually are equivalent. The reader can produce the details in Exercise 21-18.

Theorem 21.5 **Maxwell's equations** *in differential form.*

1. **Poisson Equation**. *(Differential form of Gauss' Law.)* $\operatorname{div}\vec{E} = \frac{\rho}{\varepsilon_0}.$

2. $\operatorname{div}\vec{B} = 0.$

3. **Ampère-Maxwell Law**. $\operatorname{curl}\vec{B} = \frac{1}{c^2}\frac{\partial}{\partial t}\vec{E} + \mu_0\vec{j}.$

4. **Induction Law**. *(Differential form of Faraday's Law.)* $\operatorname{curl}\vec{E} = -\frac{\partial}{\partial t}\vec{B}.$

We conclude this section with two consequences of Maxwell's equations. The first is a simple rewriting of the Poisson equation. Because in a static field all time derivatives are zero and no currents flow, the static electric field satisfies $\text{curl}\vec{E} = 0$. Hence (see Exercise 19-48), the static electric field $\vec{E}$ can be expressed as the negative gradient of a potential function u. That is, $\vec{E} = -\text{grad}(u)$. Thus the Poisson equation can be rewritten as follows.

Proposition 21.6 *The* **Poisson Equation** *for the electric potential u is* $-\Delta u = \frac{\rho}{\varepsilon_0}$. ■

Proposition 21.6 shows that the steady state heat equation (with sources and sinks encoded as a nonzero right side) and the equation for the static electric potential are very similar.

The next consequence of Maxwell's equation demonstrates the predictive power of mathematical models. In a vacuum, the current and charge densities are both zero, that is, $\vec{j} = 0$ and $\rho = 0$. Hence, $\text{div}\vec{E} = 0$ and $\text{curl}\vec{B} = \frac{1}{c^2}\frac{\partial}{\partial t}\vec{E}$ in a vacuum. With Exercise 21-6i, the Laplace operator applied componentwise, and Clairaut's Theorem we obtain the following:

$$\begin{aligned} -\Delta\vec{E} &= \text{grad}\,\text{div}\vec{E} - \Delta\vec{E} = \text{curl}\,\text{curl}\,\vec{E} = \text{curl}\left(-\frac{\partial}{\partial t}\vec{B}\right) \\ &= -\frac{\partial}{\partial t}\text{curl}\vec{B} = -\frac{\partial}{\partial t}\frac{1}{c^2}\frac{\partial}{\partial t}\vec{E} = -\frac{1}{c^2}\frac{\partial^2}{\partial^2 t}\vec{E}, \end{aligned}$$

or $\Delta\vec{E} = \frac{1}{c^2}\frac{\partial^2}{\partial^2 t}\vec{E}$. This equation is also called a wave equation.

Definition 21.7 *The partial differential equation* $a^2\Delta u = \frac{\partial^2}{\partial t^2}u$, *with* $a > 0$ *a constant, which is also often written as* $\frac{\partial^2}{\partial t^2}u - a^2\Delta u = 0$, *is called the* **wave equation.**

Maxwell observed that the wave equation has nonzero solutions that look like traveling waves (see Exercise 21-19) and he concluded that there must be such a thing as an electric wave. These electric waves were first observed by H. Hertz, years after Maxwell's death. That is, the prediction may have been purely theoretical, but it was undeniably true and beneficial. Electromagnetic waves are of course the foundation for the multitude of wireless communication techniques that surround us today.

The wave equation also governs wave phenomena such as vibrating strings, drum membranes, or vibration of parts in motors. Like the heat equation, Maxwell's equations and their consequences are used in modeling real life phenomena by imposing initial and boundary conditions and then finding solutions that satisfy these conditions. For more on electrodynamics, [19] is a standard reference.

Exercises

21-17. Derive the wave equation for $\vec{B}$ in a vacuum.

21-18 Prove that the integral and differential forms of Maxwell's equations are equivalent That is,

(a) Prove that part 1 of Theorem 21 4 is equivalent to part 1 of Theorem 21 5

(b) Prove that part 2 of Theorem 21.4 is equivalent to part 2 of Theorem 21.5.

(c) Prove that part 3 of Theorem 21 4 is equivalent to part 3 of Theorem 21 5

(d) Prove that part 4 of Theorem 21.4 is equivalent to part 4 of Theorem 21.5.

Hint. Each derivation from integral form to differential form is similar to the derivation in Section 21 2. Use Exercises 21-4 and 21-5 as appropriate. Each converse is a more direct application of the Divergence Theorem or Stokes' Theorem, as appropriate

21-19. Let $f : \mathbb{R}^3 \to \mathbb{R}$ be a twice differentiable function of three variables such that all mixed second partial derivatives of f are zero Let $\vec{v} = \begin{pmatrix} v_1 \\ v_2 \\ v_3 \end{pmatrix}$ and $g(x, y, z, t) = f(x - v_1 t, y - v_2 t, z - v_3 t)$.

(a) Prove that g solves the three dimensional wave equation $\|\vec{v}\|^2 \Delta g = \frac{\partial^2 g}{\partial^2 t}$.

(b) Explain why we need to demand that the mixed partial derivatives of f are zero.

(c) Explain why the function g describes a wave form shaped like the function f that is traveling at velocity $\vec{v}$.

21-20 Prove that the following two assertions are equivalent

(a) The net flux of the gravitational field $\vec{F}$ across a closed surface is $-4\pi G$ times the mass enclosed in the surface, where G is the gravitational constant

(b) If $\vec{F}$ is the gravitational field, then $\operatorname{div}\vec{F} = -4\pi G\rho$, where G is the gravitational constant and ρ denotes regular (mass) density

21-21 Use the following steps to derive Gauss' Law from **Coulomb's Law**, which states that at a point $\vec{r}$ the electrostatic field of a point charge q located at $\vec{a}$ is $\vec{E}_q = \frac{q}{4\pi\varepsilon_0}\frac{1}{\|\vec{r}-\vec{a}\|^2}\frac{\vec{r}-\vec{a}}{\|\vec{r}-\vec{a}\|}$.

(a) Prove that the integral of $\vec{E}_q$ over the surface $\delta B_\varepsilon(\vec{a})$ of any sphere of radius $\varepsilon > 0$ centered at a is $\oint\!\!\oint_{\delta B_\varepsilon(\vec{a})} \frac{q}{4\pi\varepsilon_0}\frac{1}{\|\vec{r}-\vec{a}\|^2}\frac{\vec{r}-\vec{a}}{\|\vec{r}-\vec{a}\|} \cdot d\vec{S} = \frac{q}{\varepsilon_0}$

(b) Prove that $\operatorname{div}\left(\frac{q}{4\pi\varepsilon_0}\frac{1}{\|\vec{r}-\vec{a}\|^2}\frac{\vec{r}-\vec{a}}{\|\vec{r}-\vec{a}\|}\right) = 0$ for $\vec{r} \neq \vec{a}$

Hint. Componentwise computation

(c) Now let $q_1, \ldots, q_n$ be point charges at $\vec{a}_1, \ldots, \vec{a}_n$, let V be an embedded oriented connected compact three dimensional manifold with boundary or corners so that no point charge is on the boundary and let $\vec{E}$ be the net electrical field generated by the point charges. Prove that the integral of E over the surface of V is the sum of the charges contained in V divided by ε_0.

Hint. Around each point charge in V consider a sphere of radius small enough so that each sphere is contained in V°, contains exactly one point charge and no two spheres intersect Apply the Divergence Theorem to V with the solid spheres removed.

21.5 The Navier Stokes Equation for the Conservation of Mass

The equations governing **fluid flow** are called the **Navier Stokes equations**. Fundamentally, the Navier Stokes equations are nothing but the principles of conservation of mass, momentum, and energy. Because the equation for the mass requires the least amount of specialized knowledge of physics and because the relevant analysis can be explained with it, we will focus exclusively on the conservation of mass in this section.

There are two approaches to setting up the conservation principles. In the **control volume approach** (also called the **Eulerian approach**), we fix a given volume in space. We must then make sure that the difference between what enters the volume and what exits the volume equals the change within the volume. This approach is similar to our derivation of the heat equation in Section 21.2. In the **systems approach** (also called the **Lagrangian approach**), we track a fixed set of particles. As these particles travel, no matter enters or exits the volume they occupy, but the volume gets deformed.

Because the Navier Stokes equations are challenging, many possible (and reasonable) simplifying assumptions can be made. Some of these assumptions are introduced in Exercises 21-22–21-26. As should by now be expected, throughout we assume that all volumes, surfaces, etc., are so that the integrals, etc., are defined.

21.5.1 The Continuum Hypothesis

We will assume that our fluid is a continuous medium, not a collection of individual particles. As long as we do not use the resulting equations at too small a scale, this assumption is entirely appropriate and it gives experimentally verifiable results. Thus our underlying assumptions will be the following.

- Our macroscopic time scale is considerably larger than the largest molecular time scale (time between collision of molecules).
- Our macroscopic length scale is considerably larger than the largest molecular length scale (distance between collisions of molecules).

Under these assumptions the macroscopic continuum effects are not overridden by microscopic kinetic effects. These assumptions are also called the (mechanical) **Continuum Hypothesis**, which is not to be confused with the Continuum Hypothesis from set theory. Because of its wide applicability to continua of all kinds (liquids, gases, even solids), the analysis we provide in this section is part of an area called **continuum mechanics** or **mechanics of continua**.

21.5.2 Control Volume or Eulerian Approach

In the control volume approach to fluid dynamics, we consider a fixed volume of space V within a fluid flow and determine how the mass behaves in relation to the fixed volume V. The total mass contained in our volume V is $m = \iiint_V \rho \, dV$, where ρ denotes the density of the fluid at a point (x, y, z) at time t. This mass is time dependent because the density depends on time. The only way the mass contained in V can change is by matter entering or exiting the control volume through the surface. The rate at which volume (of new matter) enters or exits through each surface element dS is $\vec{v} \cdot d\vec{S}$, where $\vec{v}$ is the velocity field of the fluid flow. This means that mass enters or exits through the surface element dS at a rate of $\rho\vec{v} \cdot d\vec{S}$. Therefore the overall rate of

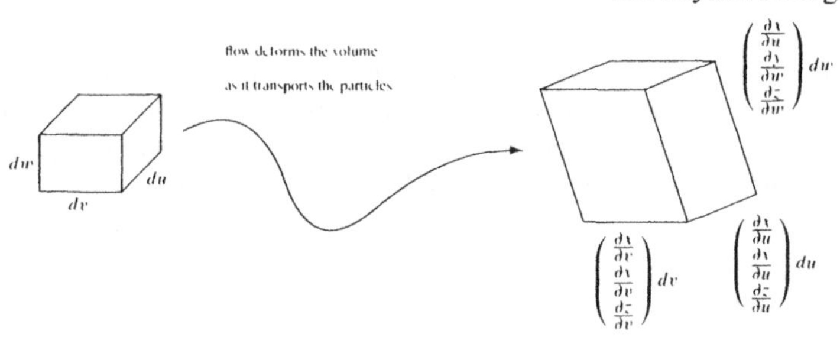

Figure 60: When tracking a set of particles over time, in first approximation, the flow deforms a box into a parallelepiped whose volume is given by a Jacobian (compare with the proof of the Multidimensional Substitution Formula). In actuality, the box can deform into arbitrarily complicated shapes.

change of the mass inside the volume V is $-\oint\!\!\oint_S \rho\vec{v}\cdot d\vec{S}$. The negative sign is needed because otherwise, with an outward pointing normal vector, a net entry of mass into the volume would be counted negatively instead of positively.

Theorem 21.8 *The integral form of the principle of* **conservation of mass** *for a fixed control volume is* $\frac{d}{dt}\iiint_V \rho\, dV = -\oint\!\!\oint_S \rho\vec{v}\cdot d\vec{S}$. *This is equivalent to the partial differential equation* $\frac{\partial}{\partial t}\rho + \vec{v}\cdot \mathrm{grad}(\rho) + \rho\mathrm{div}\,(\vec{v}) = 0$.

Proof. The integral equation was explained before the theorem was stated. The argument that the integral equation is equivalent to the partial differential equation is similar to the argument given for the heat equation in Section 21.2. The reader will provide this argument in Exercise 21-27. ■

21.5.3 Systems or Lagrangian Approach

In the systems approach to fluid dynamics, we consider a set of particles occupying a volume V_0 at time $t = 0$ and we analyze how these particles travel with the flow (see Figure 60). As the particles travel with the flow, the volume they occupy changes. Throughout our discussion, $V(t)$ will denote the volume occupied by the particles we track. The resulting integrals may look intimidating, but by choosing the right surfaces, etc. they can provide the simplest approach to the problem.

The mass $m = \iiint_{V(t)} \rho\, dV$ of the traveling volume $V(t)$ remains unchanged, because no particles enter or exit. This makes the integral equation very simple. Theorem 21.9 shows that both formulations of the principle of conservation of mass are indeed equivalent, because they are both equivalent to the same differential equation.

Theorem 21.9 *The integral form of the principle of* **conservation of mass** *when tracking a fixed set of particles occupying the volume $V(t)$ is $\frac{d}{dt}\iiint_{V(t)} \rho\, dV = 0$. This is equivalent to the partial differential equation $\frac{\partial}{\partial t}\rho + \vec{v}\cdot \text{grad}(\rho) + \rho \text{div}(\vec{v}) = 0$.*

Proof. The introduction has already verified the integral form of the principle of conservation of mass. We need to prove that the integral form translates into the given differential equation. To do this, we must take the derivative of a triple integral whose integrand ρ, as well as its domain $V(t)$ depend on time. This task is best achieved by first removing the time dependence of the domain with an appropriate, time-dependent change of variables. Time independent changes of variable were considered in Theorem 18.37 and the time dependence simply requires that Theorem 18.37 is applied at fixed times. The variables x, y, z will denote the "real" coordinates at time t of a particle whose initial coordinates at time $t = 0$ were u_1, u_2, u_3. That is, the particle that initially is at position (u_1, u_2, u_3) is at time t at the position given by $g(u_1, u_2, u_3, t) := \big(x(u_1, u_2, u_3, t), y(u_1, u_2, u_3, t), z(u_1, u_2, u_3, t)\big)$. The time derivatives of x, y, and z are the components of the flow velocity $\vec{v}$. The volume integral gives the mass for an arbitrary fixed time t. Hence, Theorem 18.37 can be applied at every time t. But the Jacobian $\left|\frac{\partial(x, y, z)}{\partial(u_1, u_2, u_3)}\right| := \det\big(Dg(\cdot, \cdot, \cdot, t)\big)$ (we will assume the determinant is nonnegative) now formally also depends on time.

$$0 = \frac{d}{dt}\iiint_{V(t)} \rho(x, y, z, t)\, dV = \frac{d}{dt}\iiint_{V_0} \rho(x, y, z, t)\left|\frac{\partial(x, y, z)}{\partial(u_1, u_2, u_3)}\right| dV$$

Now that the volume does not depend on time, we can move the time derivative inside the integral. We assume that the mild differentiability conditions needed for this interchange are satisfied. See Exercise 21-3 for details.

$$= \iiint_{V_0} \frac{\partial}{\partial t}\left[\rho(x, y, z, t)\left|\frac{\partial(x, y, z)}{\partial(u_1, u_2, u_3)}\right|\right] dV$$

$$= \iiint_{V_0} \frac{\partial}{\partial t}\left[\rho(x, y, z, t)\right]\left|\frac{\partial(x, y, z)}{\partial(u_1, u_2, u_3)}\right| + \rho(x, y, z, t)\frac{\partial}{\partial t}\left|\frac{\partial(x, y, z)}{\partial(u_1, u_2, u_3)}\right| dV$$

The derivative of the density function is

$$\frac{d}{dt}\rho\big(x(t), y(t), z(t), t\big) = \frac{\partial \rho}{\partial x}\frac{dx}{dt} + \frac{\partial \rho}{\partial y}\frac{dy}{dt} + \frac{\partial \rho}{\partial z}\frac{dz}{dt} + \frac{\partial \rho}{\partial t} = \text{grad}(\rho)\cdot\vec{v} + \frac{\partial \rho}{\partial t},$$

which is also called the **Lagrangian derivative**. For the rather technical derivative of the Jacobian $\left|\frac{\partial(x, y, z)}{\partial(u_1, u_2, u_3)}\right|$ we prove a separate lemma.

Lemma 21.10 $\frac{\partial}{\partial t}\left|\frac{\partial(x, y, z)}{\partial(u_1, u_2, u_3)}\right| = \text{div}(\vec{v})\left|\frac{\partial(x, y, z)}{\partial(u_1, u_2, u_3)}\right|.$

Proof of Lemma 21.10. By Exercise 18-34, the determinant of a matrix A is equal to the determinant of its transpose A^T. Thus we shall freely switch between the Jacobian and its transpose during this proof. The Jacobian $\left|\frac{\partial(x, y, z)}{\partial(u_1, u_2, u_3)}\right|$ is a determinant of a 3×3 matrix. Because (use transposes) determinants are linear in each row, the Chain Rule shows that the derivative of the Jacobian is obtained by differentiating each row and adding the results.

$$\frac{\partial}{\partial t}\left|\frac{\partial(x, y, z)}{\partial(u_1, u_2, u_3)}\right|$$

$$= \frac{\partial}{\partial t}\begin{vmatrix} \frac{\partial x}{\partial u_1} & \frac{\partial x}{\partial u_2} & \frac{\partial x}{\partial u_3} \\ \frac{\partial y}{\partial u_1} & \frac{\partial y}{\partial u_2} & \frac{\partial y}{\partial u_3} \\ \frac{\partial z}{\partial u_1} & \frac{\partial z}{\partial u_2} & \frac{\partial z}{\partial u_3} \end{vmatrix}$$

$$= \begin{vmatrix} \frac{\partial}{\partial t}\frac{\partial x}{\partial u_1} & \frac{\partial}{\partial t}\frac{\partial x}{\partial u_2} & \frac{\partial}{\partial t}\frac{\partial x}{\partial u_3} \\ \frac{\partial y}{\partial u_1} & \frac{\partial y}{\partial u_2} & \frac{\partial y}{\partial u_3} \\ \frac{\partial z}{\partial u_1} & \frac{\partial z}{\partial u_2} & \frac{\partial z}{\partial u_3} \end{vmatrix} + \begin{vmatrix} \frac{\partial x}{\partial u_1} & \frac{\partial x}{\partial u_2} & \frac{\partial x}{\partial u_3} \\ \frac{\partial}{\partial t}\frac{\partial y}{\partial u_1} & \frac{\partial}{\partial t}\frac{\partial y}{\partial u_2} & \frac{\partial}{\partial t}\frac{\partial y}{\partial u_3} \\ \frac{\partial z}{\partial u_1} & \frac{\partial z}{\partial u_2} & \frac{\partial z}{\partial u_3} \end{vmatrix} + \begin{vmatrix} \frac{\partial x}{\partial u_1} & \frac{\partial x}{\partial u_2} & \frac{\partial x}{\partial u_3} \\ \frac{\partial y}{\partial u_1} & \frac{\partial y}{\partial u_2} & \frac{\partial y}{\partial u_3} \\ \frac{\partial}{\partial t}\frac{\partial z}{\partial u_1} & \frac{\partial}{\partial t}\frac{\partial z}{\partial u_2} & \frac{\partial}{\partial t}\frac{\partial z}{\partial u_3} \end{vmatrix}$$

Now we note that the derivatives of x, y, and z with respect to time are the components of the velocity vector $\vec{v}$.

$$= \begin{vmatrix} \frac{\partial v_1}{\partial u_1} & \frac{\partial v_1}{\partial u_2} & \frac{\partial v_1}{\partial u_3} \\ \frac{\partial y}{\partial u_1} & \frac{\partial y}{\partial u_2} & \frac{\partial y}{\partial u_3} \\ \frac{\partial z}{\partial u_1} & \frac{\partial z}{\partial u_2} & \frac{\partial z}{\partial u_3} \end{vmatrix} + \begin{vmatrix} \frac{\partial x}{\partial u_1} & \frac{\partial x}{\partial u_2} & \frac{\partial x}{\partial u_3} \\ \frac{\partial v_2}{\partial u_1} & \frac{\partial v_2}{\partial u_2} & \frac{\partial v_2}{\partial u_3} \\ \frac{\partial z}{\partial u_1} & \frac{\partial z}{\partial u_2} & \frac{\partial z}{\partial u_3} \end{vmatrix} + \begin{vmatrix} \frac{\partial x}{\partial u_1} & \frac{\partial x}{\partial u_2} & \frac{\partial x}{\partial u_3} \\ \frac{\partial y}{\partial u_1} & \frac{\partial y}{\partial u_2} & \frac{\partial y}{\partial u_3} \\ \frac{\partial v_3}{\partial u_1} & \frac{\partial v_3}{\partial u_2} & \frac{\partial v_3}{\partial u_3} \end{vmatrix}$$

For these three matrices, it is possible to prove that the first matrix is equal to $\frac{\partial v_1}{\partial x}\left|\frac{\partial(x, y, z)}{\partial(u_1, u_2, u_3)}\right|$, the second is equal to $\frac{\partial v_2}{\partial y}\left|\frac{\partial(x, y, z)}{\partial(u_1, u_2, u_3)}\right|$, and the third is equal to $\frac{\partial v_3}{\partial z}\left|\frac{\partial(x, y, z)}{\partial(u_1, u_2, u_3)}\right|$. We prove the first equation, leaving the remaining equations to Exercise 21-28.

$$\begin{vmatrix} \frac{\partial v_1}{\partial u_1} & \frac{\partial v_1}{\partial u_2} & \frac{\partial v_1}{\partial u_3} \\ \frac{\partial y}{\partial u_1} & \frac{\partial y}{\partial u_2} & \frac{\partial y}{\partial u_3} \\ \frac{\partial z}{\partial u_1} & \frac{\partial z}{\partial u_2} & \frac{\partial z}{\partial u_3} \end{vmatrix}$$

We will work with the transpose so that the formulas fit the breadth of the page.

$$= \begin{vmatrix} \frac{\partial v_1}{\partial u_1} & \frac{\partial y}{\partial u_1} & \frac{\partial z}{\partial u_1} \\ \frac{\partial v_1}{\partial u_2} & \frac{\partial y}{\partial u_2} & \frac{\partial z}{\partial u_2} \\ \frac{\partial v_1}{\partial u_3} & \frac{\partial y}{\partial u_3} & \frac{\partial z}{\partial u_3} \end{vmatrix}$$

The velocity $\vec{v}(u_1, u_2, u_3, t)$ gives the velocity at time t of the particle that started at time $t = 0$ in position (u_1, u_2, u_3). Its derivative with respect to the u-coordinates can be computed with the Chain Rule, using x, y, z and t as intermediaries.

$$= \left| \begin{matrix} \left[\frac{\partial v_1}{\partial x}\frac{\partial x}{\partial u_1} + \frac{\partial v_1}{\partial y}\frac{\partial y}{\partial u_1} + \frac{\partial v_1}{\partial z}\frac{\partial z}{\partial u_1} + \frac{\partial v_1}{\partial t}\frac{\partial t}{\partial u_1}\right] & \frac{\partial y}{\partial u_1} & \frac{\partial z}{\partial u_1} \\ \left[\frac{\partial v_1}{\partial x}\frac{\partial x}{\partial u_2} + \frac{\partial v_1}{\partial y}\frac{\partial y}{\partial u_2} + \frac{\partial v_1}{\partial z}\frac{\partial z}{\partial u_2} + \frac{\partial v_1}{\partial t}\frac{\partial t}{\partial u_2}\right] & \frac{\partial y}{\partial u_2} & \frac{\partial z}{\partial u_2} \\ \left[\frac{\partial v_1}{\partial x}\frac{\partial x}{\partial u_3} + \frac{\partial v_1}{\partial y}\frac{\partial y}{\partial u_3} + \frac{\partial v_1}{\partial z}\frac{\partial z}{\partial u_3} + \frac{\partial v_1}{\partial t}\frac{\partial t}{\partial u_3}\right] & \frac{\partial y}{\partial u_3} & \frac{\partial z}{\partial u_3} \end{matrix} \right|$$

Because time t does not depend on any of the u_i, the last summand in each term in the first column is zero. Moreover, by linearity in the columns we can break the determinant up into the sum of three separate determinants and we can pull out a factor that is common to all entries of a column.

$$= \left| \begin{matrix} \frac{\partial v_1}{\partial x}\frac{\partial x}{\partial u_1} & \frac{\partial y}{\partial u_1} & \frac{\partial z}{\partial u_1} \\ \frac{\partial v_1}{\partial x}\frac{\partial x}{\partial u_2} & \frac{\partial y}{\partial u_2} & \frac{\partial z}{\partial u_2} \\ \frac{\partial v_1}{\partial x}\frac{\partial x}{\partial u_3} & \frac{\partial y}{\partial u_3} & \frac{\partial z}{\partial u_3} \end{matrix} \right| + \left| \begin{matrix} \frac{\partial v_1}{\partial y}\frac{\partial y}{\partial u_1} & \frac{\partial y}{\partial u_1} & \frac{\partial z}{\partial u_1} \\ \frac{\partial v_1}{\partial y}\frac{\partial y}{\partial u_2} & \frac{\partial y}{\partial u_2} & \frac{\partial z}{\partial u_2} \\ \frac{\partial v_1}{\partial y}\frac{\partial y}{\partial u_3} & \frac{\partial y}{\partial u_3} & \frac{\partial z}{\partial u_3} \end{matrix} \right| + \left| \begin{matrix} \frac{\partial v_1}{\partial z}\frac{\partial z}{\partial u_1} & \frac{\partial y}{\partial u_1} & \frac{\partial z}{\partial u_1} \\ \frac{\partial v_1}{\partial z}\frac{\partial z}{\partial u_2} & \frac{\partial y}{\partial u_2} & \frac{\partial z}{\partial u_2} \\ \frac{\partial v_1}{\partial z}\frac{\partial z}{\partial u_3} & \frac{\partial y}{\partial u_3} & \frac{\partial z}{\partial u_3} \end{matrix} \right|$$

$$= \frac{\partial v_1}{\partial x}\left| \begin{matrix} \frac{\partial x}{\partial u_1} & \frac{\partial y}{\partial u_1} & \frac{\partial z}{\partial u_1} \\ \frac{\partial x}{\partial u_2} & \frac{\partial y}{\partial u_2} & \frac{\partial z}{\partial u_2} \\ \frac{\partial x}{\partial u_3} & \frac{\partial y}{\partial u_3} & \frac{\partial z}{\partial u_3} \end{matrix} \right| + \frac{\partial v_1}{\partial y}\left| \begin{matrix} \frac{\partial y}{\partial u_1} & \frac{\partial y}{\partial u_1} & \frac{\partial z}{\partial u_1} \\ \frac{\partial y}{\partial u_2} & \frac{\partial y}{\partial u_2} & \frac{\partial z}{\partial u_2} \\ \frac{\partial y}{\partial u_3} & \frac{\partial y}{\partial u_3} & \frac{\partial z}{\partial u_3} \end{matrix} \right| + \frac{\partial v_1}{\partial z}\left| \begin{matrix} \frac{\partial z}{\partial u_1} & \frac{\partial y}{\partial u_1} & \frac{\partial z}{\partial u_1} \\ \frac{\partial z}{\partial u_2} & \frac{\partial y}{\partial u_2} & \frac{\partial z}{\partial u_2} \\ \frac{\partial z}{\partial u_3} & \frac{\partial y}{\partial u_3} & \frac{\partial z}{\partial u_3} \end{matrix} \right|$$

The last two matrices have two equal columns each, which means that their determinants are zero.

$$= \frac{\partial v_1}{\partial x}\left| \begin{matrix} \frac{\partial x}{\partial u_1} & \frac{\partial y}{\partial u_1} & \frac{\partial z}{\partial u_1} \\ \frac{\partial x}{\partial u_2} & \frac{\partial y}{\partial u_2} & \frac{\partial z}{\partial u_2} \\ \frac{\partial x}{\partial u_3} & \frac{\partial y}{\partial u_3} & \frac{\partial z}{\partial u_3} \end{matrix} \right| = \frac{\partial v_1}{\partial x}\left|\frac{\partial(x,y,z)}{\partial(u_1,u_2,u_3)}\right|.$$

This proves the first of the three promised equalities, with the other two left to Exercise 21-28. Overall we obtain

$$\frac{\partial}{\partial t}\left|\frac{\partial(x,y,z)}{\partial(u_1,u_2,u_3)}\right|$$

$$= \left| \begin{matrix} \frac{\partial v_1}{\partial u_1} & \frac{\partial v_1}{\partial u_2} & \frac{\partial v_1}{\partial u_3} \\ \frac{\partial y}{\partial u_1} & \frac{\partial y}{\partial u_2} & \frac{\partial y}{\partial u_3} \\ \frac{\partial z}{\partial u_1} & \frac{\partial z}{\partial u_2} & \frac{\partial z}{\partial u_3} \end{matrix} \right| + \left| \begin{matrix} \frac{\partial x}{\partial u_1} & \frac{\partial x}{\partial u_2} & \frac{\partial x}{\partial u_3} \\ \frac{\partial v_2}{\partial u_1} & \frac{\partial v_2}{\partial u_2} & \frac{\partial v_2}{\partial u_3} \\ \frac{\partial z}{\partial u_1} & \frac{\partial z}{\partial u_2} & \frac{\partial z}{\partial u_3} \end{matrix} \right| + \left| \begin{matrix} \frac{\partial x}{\partial u_1} & \frac{\partial x}{\partial u_2} & \frac{\partial x}{\partial u_3} \\ \frac{\partial y}{\partial u_1} & \frac{\partial y}{\partial u_2} & \frac{\partial y}{\partial u_3} \\ \frac{\partial v_3}{\partial u_1} & \frac{\partial v_3}{\partial u_2} & \frac{\partial v_3}{\partial u_3} \end{matrix} \right|$$

$$= \frac{\partial v_1}{\partial x}\left|\frac{\partial(x,y,z)}{\partial(u_1,u_2,u_3)}\right| + \frac{\partial v_2}{\partial y}\left|\frac{\partial(x,y,z)}{\partial(u_1,u_2,u_3)}\right| + \frac{\partial v_3}{\partial z}\left|\frac{\partial(x,y,z)}{\partial(u_1,u_2,u_3)}\right|$$

$$= \operatorname{div}(\vec{v})\left|\frac{\partial(x,y,z)}{\partial(u_1,u_2,u_3)}\right|.$$

■

Completing the proof of Theorem 21.9. We can now finish the proof of Theorem 21.9. The goal is to factor out the Jacobian, so we can change the integral back to an integral over our time dependent volume $V(t)$.

$$\frac{d}{dt}\iiint_{V(t)} \rho(x,y,z,t)\,dV$$

$$= \iiint_{V_0} \frac{\partial}{\partial t}[\rho(x,y,z,t)] \left| \frac{\partial(x,y,z)}{\partial(u_1,u_2,u_3)} \right| + \rho(x,y,z,t) \frac{\partial}{\partial t} \left| \frac{\partial(x,y,z)}{\partial(u_1,u_2,u_3)} \right| dV$$

$$= \iiint_{V_0} \left(\frac{\partial \rho}{\partial t} + \operatorname{grad}(\rho) \cdot \vec{v} \right) \left| \frac{\partial(x,y,z)}{\partial(u_1,u_2,u_3)} \right| + \rho \operatorname{div}(\vec{v}) \left| \frac{\partial(x,y,z)}{\partial(u_1,u_2,u_3)} \right| dV$$

$$= \iiint_{V_0} \left[\frac{\partial \rho}{\partial t} + \operatorname{grad}(\rho) \cdot \vec{v} + \rho \operatorname{div}(\vec{v}) \right] \left| \frac{\partial(x,y,z)}{\partial(u_1,u_2,u_3)} \right| dV$$

$$= \iiint_{V(t)} \frac{\partial \rho}{\partial t} + \operatorname{grad}(\rho) \cdot \vec{v} + \rho \operatorname{div}(\vec{v})\, dV.$$

The back-and-forth conversion to the partial differential equation is now exactly as for the heat equation. For one direction, we simply note that the integral must be zero when the partial differential equation is satisfied. For the other direction, we fix a time t and a position (x,y,z). We then set $V(t)$ equal to a ball of radius a about (x,y,z) and take the limit of the quotient of the integral and the volume of the ball. The reader will fill in the limit argument in Exercise 21-29. ■

21.5.4 Reynolds' Transport Theorem and Leibniz' Rule

The main part of the proof of Theorem 21.9 was rewriting the time derivative of the traveling volume integral. During the proof we did not use the fact that we were working with a mass density. Thus the formula for the derivative of the volume integral must hold in general. This is Reynolds' Transport Theorem. We state it here with the mathematical hypotheses that are needed to formally execute the proof. The function f is the flow, $\vec{v}$ is the velocity and the quantity q takes the role of the density.

Theorem 21.11 Reynolds' Transport Theorem. *Let $\Omega \subseteq \mathbb{R}^3$ be an open set, let $\delta, b > 0$ and let $f : \Omega \times (-\delta, b) \to \mathbb{R}^3$ be continuously differentiable, bounded, and so that for each $t \in (-\delta, b)$ the function $g_t := f(\cdot,\cdot,\cdot,t)$ is bijective with $\det(Dg_t(u_1,u_2,u_3)) > 0$ for all $(u_1,u_2,u_3) \in \Omega$. Let $\Omega' \subseteq \mathbb{R}^3 \times (-\delta, b)$ be an open set that contains $\bigcup_{-\delta<t<b} g_t[\Omega] \times \{t\}$ and let $q : \Omega' \to \mathbb{R}$ be continuously differentiable. Let $\vec{v} = \frac{\partial f}{\partial t}$, let $V_0 \subseteq \Omega$ be a compact oriented three dimensional embedded manifold with boundary or corners and let $V(t) := g_t[V_0]$. Then*

$$\frac{d}{dt} \iiint_{V(t)} q\, dV = \iiint_{V(t)} \frac{\partial}{\partial t} q + \vec{v} \cdot \operatorname{grad}(q) + q \operatorname{div}(\vec{v})\, dV.$$

Proof. Each function g_t satisfies the hypotheses of the Multidimensional Substitution Formula and the boundedness hypotheses allow us to exchange integration over Ω and differentiation with respect to t (see Exercise 21-3). Therefore, the transformation of the integral in the proof of Theorem 21.9 proves Reynolds' Transport Theorem. ■

An alternative formulation of Reynolds' Transport Theorem is Leibniz' Rule.

Theorem 21.12 Leibniz' Rule. *Let $\Omega \subseteq \mathbb{R}^3$ be an open set, let $\delta, b > 0$ and let the function $f : \Omega \times (-\delta, b) \to \mathbb{R}^3$ be continuously differentiable, bounded, and* *so that for each* $t \in (-\delta, b)$ *the function* $g_t := f(\cdot,\cdot,\cdot,t)$ *is bijective and satisfies* $\det\left(Dg_t(u_1,u_2,u_3)\right) > 0$ *for all* $(u_1,u_2,u_3) \in \Omega$. *Let* $\Omega' \subseteq \mathbb{R}^3 \times (-\delta, b)$ *be an open set that contains* $\bigcup_{-\delta<t<b} g_t[\Omega] \times \{t\}$ *and let* $q : \Omega' \to \mathbb{R}$ *be continuously differentiable. Let* $\vec{v} = \frac{\partial f}{\partial t}$, *let* $V_0 \subseteq \Omega$ *be a compact oriented three dimensional embedded manifold with boundary or corners, let* $V(t) := g_t[V_0]$ *and let* $S(t) := g_t[\partial V_0]$. *Then*

$$\frac{d}{dt}\iiint_{V(t)} q\, dV = \iiint_{V(t)} \frac{\partial}{\partial t} q\, dV + \oint\!\!\!\oint_{S(t)} q\vec{v}\cdot d\vec{S}.$$

Proof. Because g_t is an orientation preserving C^1-diffeomorphism, $V(t)$ is a compact oriented three dimensional embedded C^1-manifold with boundary or corners and $g_t[\partial V_0] = \partial g_t[V_0]$. Recall that $\vec{v}\cdot\text{grad}(q) + q\,\text{div}(\vec{v}) = \text{div}(q\vec{v})$ (see Exercise 21-6e). The remainder of the proof is an application of Reynolds' Transport Theorem and the Divergence Theorem.

$$\begin{aligned}
\frac{d}{dt}\iiint_{V(t)} q\, dV &= \iiint_{V(t)} \frac{\partial}{\partial t} q + \vec{v}\cdot\text{grad}(q) + q\,\text{div}(\vec{v})\, dV \\
&= \iiint_{V(t)} \frac{\partial}{\partial t} q + \text{div}(q\vec{v})\, dV \\
&= \iiint_{V(t)} \frac{\partial}{\partial t} q\, dV + \iiint_{V(t)} \text{div}(q\vec{v})\, dV \\
&= \iiint_{V(t)} \frac{\partial}{\partial t} q\, dV + \oint\!\!\!\oint_{S(t)} q\vec{v}\cdot d\vec{S}.
\end{aligned}$$

■

Leibniz' Rule and Reynolds' Transport Theorem are actually equivalent. The reader will show in Exercise 21-30 that Leibniz' Rule implies Reynolds' Transport Theorem. Moreover, Exercise 21-31 shows that both results can be generalized to space domains $\Omega \subseteq \mathbb{R}^d$ and flows $f : \Omega \times (-\delta, b) \to \mathbb{R}^d$.

For an explanation of the Navier Stokes equations for the conservation of linear momentum and for the conservation of energy, the reader should consult a text in continuum mechanics, such as [3] or [31].

Exercises

21-22. A flow is called **solenoidal** if and only if $\text{div}(\vec{v}) = 0$.

(a) For a solenoidal flow, state the partial differential equation for the conservation of mass (see Theorems 21.8 and 21.9).

(b) In what way does the equation simplify for a solenoidal flow?

(c) Explain why a particle traveling in a solenoidal flow does not experience any density changes.

Hint Use the Chain Rule to compute the derivative $\frac{d}{dt}\rho\left(x(t), y(t), z(t), t\right)$ Then reconsider part 21-22a

21-23. A flow is called **irrotational** or **lamellar** if and only if $\text{curl}(\vec{v}) = \vec{0}$ and the flow is defined on a convex or simply connected subset of $\mathbb{R}^3$ (The hypothesis on the domain is needed to apply Exercise 19-48.)

(a) Explain why an irrotational flow is called "irrotational".

(b) Explain why an irrotational flow has a potential function P such that $\vec{v} = \text{grad}(P)$.

(c) For an irrotational flow, state the partial differential equation for the conservation of mass using the potential function of the flow.

(d) In what way does the equation simplify for the potential function of an irrotational flow?

21-24. A flow is called **Laplacian** if and only if it is both solenoidal and irrotational

(a) For a Laplacian flow, state the partial differential equation for the conservation of mass.

(b) In what way does the equation simplify for a Laplacian flow?

21-25. A flow is called **complex lamellar** if and only if $\vec{v} \cdot \text{curl}(\vec{v}) = 0$ Prove that every irrotational flow is complex lamellar.

21-26 A flow is called **Beltrami** if and only if $\vec{v} \times \text{curl}(\vec{v}) = \vec{0}$

(a) Prove that every irrotational flow is Beltrami.

(b) Prove that the curl of a Beltrami flow is again Beltrami.

21-27 Finish the proof of Theorem 21.8

21-28. Finish the proof of Lemma 21 10 by proving the remaining two equations for the determinants on page 500.

21-29 Finish the proof of Theorem 21.9 by proving that the integral and differential equations are equivalent

21-30. Prove that Leibniz' Rule implies Reynolds' Transport Theorem.

21-31. Formulate Reynolds' Transport Theorem and Leibniz' Rule in $\mathbb{R}^d$ and explain why the proof given here would only need to be changed slightly to prove these higher dimensional versions.

21-32. Let $\Omega \subseteq \mathbb{R}^3$ be a bounded open set and let $q : \Omega \times (a, b) \to \mathbb{R}$ be continuously differentiable and bounded Use Leibniz' Rule to prove that $\frac{d}{dt} \iiint_{\Omega} q \, dV = \iiint_{\Omega} \frac{\partial}{\partial t} q \, dV$.

This shows that the result of Exercise 21-3 is a special case of Leibniz' rule Nonetheless, the result of Exercise 21-3 is formally not a corollary of Leibniz' Rule because it is used in the proof of Reynolds' Transport Theorem.

21-33 Let $\Omega \subseteq \mathbb{R}^3$ be a bounded open set, let $\delta, b > 0$ and let $f : \Omega \times (-\delta, b) \to \mathbb{R}^3$ be continuously differentiable, bounded and so that for each $t \in (-\delta, b)$ the function $g_t := f(\cdot, \cdot, \cdot, t)$ is bijective with $\det\left(Dg_t(u_1, u_2, u_3)\right) > 0$ for all $(u_1, u_2, u_3) \in \Omega$ Let $\Omega' \subseteq \mathbb{R}^3 \times (-\delta, b)$ be an open set that contains $\bigcup_{-\delta<t<b} g_t[\Omega] \times \{t\}$ and let $\rho, h : \Omega' \to \mathbb{R}$ be continuously differentiable and so that ρ satisfies the Navier Stokes equation for the conservation of the mass. Let $\vec{v} = \frac{df}{dt}$, let $V_0 \subseteq \Omega$ be a compact oriented three dimensional embedded manifold with boundary or corners and let $V(t) := g_t[V_0]$. Prove that $\frac{d}{dt} \iiint_{V(t)} \rho h \, dV = \iiint_{V(t)} \rho \frac{\partial h}{\partial t} + \rho \vec{v} \cdot \text{grad}(h) \, dV$.

Chapter 22

Ordinary Differential Equations

Equations involving the derivatives of a function of one variable arise as physical models (see Section 21.1), as well as in the solution of more complicated equations (see Section 21.3). In this chapter, we briefly touch the theory of these equations. Section 22.1 shows some standard transformations and Section 22.2 presents a well-known existence and uniqueness result. The important special case of linear equations is treated in Section 22.3.

22.1 Banach Space Valued Differential Equations

We start by exactly defining the equations we want to investigate.

Definition 22.1 *Let X be a Banach space, let $a < b$ and let $g : [a, b) \to X$. Then g is called* **differentiable** *on $[a, b)$ iff g is differentiable on (a, b) and the* **right-sided derivative** $g'(a) := \lim_{z \to a^+} \frac{g(z) - g(a)}{z - a}$ *exists. We define n times differentiable functions on $[a, b)$ similarly.*

Definition 22.2 *Let X be a Banach space, let $D_0, D_1, \ldots, D_n \subseteq X$, let $a < b$ and let $F : [a, b) \times \prod_{j=0}^{n} D_j \to X$ be a function so that there are $y_0, \ldots, y_{n-1}, y, z \in X$ and $t \in [a, b)$ with $F(t, y_0, \ldots, y_{n-1}, y) \neq F(t, y_0, \ldots, y_{n-1}, z)$. Then the equation $F\left(t, y, y', \ldots, y^{(n-1)}, y^{(n)}\right) = 0$ is called an* **(ordinary) n^{th} order differential equation**. *A function $f : [a, b) \to X$ is called a* **solution** *of the equation iff it is n times differentiable on $[a, b)$ and $F\left(t, f(t), f'(t), \ldots, f^{(n-1)}(t), f^{(n)}(t)\right) = 0$ for all $t \in [a, b)$.*

Finally, an **initial value problem** *consists of a differential equation as above and a specification of (initial) values for $y(a), y'(a), \ldots, y^{(n-1)}(a)$. A* **solution** *of the initial value problem solves the differential equation and it has the specified values at a.*

The condition $F(t, y_0, \ldots, y_{n-1}, y) \neq F(t, y_0, \ldots, y_{n-1}, z)$ assures that F genuinely depends on its last variable, which guarantees that the nth derivative occurs in a nontrivial way in the differential equation. The term "*ordinary* differential equations" indicates explicitly that the equations only involve ordinary, not partial, derivatives. We often simply say "differential equations," because equations involving partial derivatives are usually called "partial differential equations." The independent variable in a differential equation is typically interpreted as time. This interpretation also explains the consideration of differential equations for $t \in [a, b)$. The situation at time $t = a$ is specified by the initial conditions and we are interested in what happens in the future.

In physical terms, a second order differential equation often describes the motion of a particle (see Section 21.1). It is well known in physics that to predict the motion of a classical particle it is enough to know the (second order differential) equation of motion plus the particle's initial position and velocity. Of course, we could also define everything for intervals $(a - \delta, b)$ and thus extrapolate into the past.

When an initial value problem and a function are given, it is easy to verify if the function solves the initial value problem (see Exercise 22-1). Many symbolic and numerical methods have been developed to find solutions of differential equations. Yet certain initial value problems are so bad, that they do not even have a solution (see Exercise 22-2) or they have several solutions (see Exercise 22-8). Theorem 22.6 will provide a criterion which guarantees unique solutions for certain initial value problems.

Systems of differential equations arise when investigating coupled systems whose behavior is governed by ordinary differential equations (think of two spring mass systems that are connected to each other). The formalism of Definition 22.2 includes systems of differential equations, because if $f_1, \ldots, f_m$ solve a system of differential equations, then $(f_1, \ldots, f_m)$ solves an appropriate single differential equation with values in a product of Banach spaces. Therefore by considering Banach space valued differential equations we also build the theory of systems of differential equations. Moreover, any Banach space valued differential equation is equivalent to a first order Banach space valued differential equation.

Proposition 22.3 *Let X be a Banach space, let $D_0, D_1, \ldots, D_n \subseteq X$, let $a < b$ and let $F : [a, b) \times \prod_{j=0}^{n} D_j \to X$ be a function so that $F\left(t, y, y', \ldots, y^{(n-1)}, y^{(n)}\right) = 0$ is an nth order differential equation. If $f : [a, b) \to D_0$ is a solution of the differential equation then $g : [a, b) \to \prod_{j=0}^{n-1} D_j$ defined by $g(t) := \left(f(t), f'(t), \ldots, f^{(n-1)}(t)\right)$ is a solution of the first order X^n-valued differential equation*

$$\begin{pmatrix} y_1 - y_0' \\ \vdots \\ y_{n-1} - y_{n-2}' \\ F\left(t, y_0, y_1, \ldots, y_{n-1}, y_{n-1}'\right) \end{pmatrix} = \begin{pmatrix} 0 \\ \vdots \\ 0 \\ 0 \end{pmatrix}.$$

Conversely, if g solves the above X^n-valued differential equation, then its first component solves the original n^{th} order differential equation.

Proof. For both claims, the proof is a direct substitution (Exercise 22-3). ∎

The equivalence in Proposition 22.3 greatly simplifies the theory of (systems of) differential equations, because we only need to consider first order Banach space valued differential equations. Among the first order equations, those which can be explicitly solved for the derivative occupy a special place.

Definition 22.4 *Let X be a Banach space, let $D \subseteq X$ be a subset of X, let $a < b$ and let $f : [a, b) \times D \to X$ be a function. Then the equation $y' = f(t, y)$ is called an* **explicit first-order differential equation**.

For explicit differential equations, solutions of initial value problems are equivalent to solutions of certain integral equations. This is useful because integrals, unlike derivatives, usually define continuous operators. The proof of Theorem 22.6 will show how useful this change to integral equations can be. The integral below is the Riemann integral of Banach space valued functions discussed in Exercise 17-41.

Proposition 22.5 *Let X be a Banach space, let $a < b$, let $D \subseteq X$ be a subset of X and let $f : [a, b) \times D \to X$. Let $s : [a, b) \to D$ be a function such that $f\big(t, s(t)\big)$ is continuous on $[a, b)$. Then s is a solution of the initial value problem $y' = f(x, y)$, $y(t_0) = y_0$ iff for all $t \in [a, b)$ we have $s(t) = y_0 + \int_{t_0}^{t} f\big(\tau, s(\tau)\big)\, d\tau$.*

Proof. This application of the Fundamental Theorem of Calculus as given in Exercises 17-41c and 17-41e is left to the reader in Exercise 22-4. ∎

In conclusion, this section has shown how some simple translations turn differential equations into equations involving operators on Banach spaces. Despite their relative simplicity, the steps provided here are crucial to bring the machinery of analysis to bear on differential equations. Chapter 23 will show how similar ideas make partial differential equations accessible to analysis.

Exercises

22-1. For each differential equation and initial value problem, determine if the given function is a solution.

(a) $y' = y$, $y(0) = 1$, $s(x) = e^x$

(b) $y' = y$, $y(0) = 2$, $s(x) = e^x$

(c) $y' = y$, $y(0) = 1$, $s(x) = e^{2x}$

(d) $y' = \dfrac{1}{\sqrt{1 - x^2}}$, $y(0) = 0$, $s(x) = \arcsin(x)$

22-2. Prove that the differential equation $y' = \begin{cases} 1; & \text{for } x \in \mathbb{Q}, \\ 0; & \text{for } x \in \mathbb{R} \setminus \mathbb{Q}, \end{cases}$ has no solution on any interval $[0, \delta)$ with $\delta > 0$

Hint. Exercise 4-25.

22-3. Prove Proposition 22.3.

22-4 Prove Proposition 22.5.

22-5 Consider the second order differential equation $my'' + cy' + ky = 0$ with initial value problem $y(0) = y_0$ and $y'(0) = v_1$.

(a) Find a first order Banach space valued differential equation with initial value problem that is equivalent (in the sense of Proposition 22.3) to this problem

(b) Find an integral equation that is equivalent (in the sense of Proposition 22.5) to this problem.

22.2 An Existence and Uniqueness Theorem

It is natural to ask when differential equations have solutions and under what conditions these solutions are unique. Picard and Lindelöf's Existence and Uniqueness Theorem shows that unique solutions exist under relatively mild conditions. By Proposition 22.5, the solution of the differential equation $y' = f(t, y)$ is actually a fixed point of the operator T that maps functions $g : [a, b) \to X$ to functions $Tg : [a, b) \to X$ so that $Tg(y) = y_0 + \int_a^t f\big(\tau, g(\tau)\big)\, d\tau$. The proof of Theorem 22.6 exploits the properties of this operator.

Theorem 22.6 Picard and Lindelöf's Existence and Uniqueness Theorem. *Let $[a, b) \subseteq \mathbb{R}$, let X be a Banach space, let $r > 0$, let $z \in X$, and let the function $f : [a, b) \times B_r(z) \to X$ be continuous and such that there is a $C \in \mathbb{R}$ so that for all $t \in [a, b)$ and all $y_1, y_2 \in B_r(z)$ we have $\big\|f(t, y_1) - f(t, y_2)\big\| \leq C\|y_1 - y_2\|$. Then for each $y_0 \in B_r(z)$ there is a $\delta > 0$ so that the initial value problem $y' = f(t, y)$, $y(a) = y_0$ has exactly one solution on $[a, a + \delta)$.*

Proof. Let $y_0 \in B_r(z)$ and let $\varepsilon := \frac{1}{2}\big(r - \|y_0 - z\|\big)$. The supremum

$$M := \sup\left\{\big\|f(t, y)\big\| : (t, y) \in \left[a, \frac{a + b}{2}\right] \times \overline{B_\varepsilon(y_0)}\right\}$$

exists because f satisfies a Lipschitz condition in the second coordinate. Let $\delta := \min\left\{\frac{b - a}{2}, \frac{\varepsilon}{M + 1}, \frac{1}{2C + 1}\right\}$. Let $C^0\left([a, a + \delta], \overline{B_\varepsilon(y_0)}\right)$ be the space of continuous functions $g : [a, a+\delta] \to \overline{B_\varepsilon(y_0)}$. It is easy to see that $C^0\left([a, a + \delta], \overline{B_\varepsilon(y_0)}\right)$ with the metric induced by the uniform norm $\|g\|_\infty := \max\big\{\|g(t)\| : t \in [a, a + \delta]\big\}$ of $C^0\big([a, a + \delta], X\big)$ is a complete metric space (see Exercise 22-7).

For all functions $g \in C^0\left([a, a + \delta], \overline{B_\varepsilon(y_0)}\right)$, we define the function Tg pointwise by $Tg(t) := y_0 + \int_a^t f\big(\tau, g(\tau)\big)\, d\tau$. Then Tg is continuous on $[a, a + \delta]$ and for all $t \in [a, a + \delta]$ we infer

$$\big\|Tg(t) - y_0\big\| = \left\|\int_a^t f\big(\tau, g(\tau)\big)\, d\tau\right\| \leq \int_a^t \big\|f\big(\tau, g(\tau)\big)\big\|\, d\tau \leq M(t - a) \leq M\delta < \varepsilon.$$

Hence, T maps the space $C^0\left([a, a+\delta], \overline{B_\varepsilon(y_0)}\right)$ to itself. Moreover, for all functions $g_1, g_2 \in C^0\left([a, a+\delta], \overline{B_\varepsilon(y_0)}\right)$ and all $t \in [a, a+\delta]$ we obtain the inequality

$$\begin{aligned}
\left|Tg_1(t) - Tg_2(t)\right| &= \left|\int_a^t f\big(\tau, g_1(\tau)\big)\, d\tau - \int_a^t f\big(\tau, g_2(\tau)\big)\, d\tau\right| \\
&\leq \int_a^t \left|f\big(\tau, g_1(\tau)\big) - f\big(\tau, g_2(\tau)\big)\right| d\tau \\
&\leq \int_a^t C\left|g_1(\tau) - g_2(\tau)\right| d\tau \leq \delta C \|g_1 - g_2\|_\infty \\
&\leq \frac{1}{2}\|g_1 - g_2\|_\infty .
\end{aligned}$$

This means that for all $g_1, g_2 \in C^0\left([a, a+\delta], \overline{B_\varepsilon(y_0)}\right)$ the operator T satisfies the Lipschitz condition $\|Tg_1 - Tg_2\|_\infty \leq \frac{1}{2}\|g_1 - g_2\|_\infty$. Thus by Banach's Fixed Point Theorem (Theorem 17.64) T has a fixed point $s \in C^0\left([a, a+\delta], \overline{B_\varepsilon(y_0)}\right)$. But then for all $t \in [a, a+\delta]$ the equality $s(t) = Ts(t) = y_0 + \int_a^t f\big(\tau, s(\tau)\big)\, d\tau$ holds and by Proposition 22.5 the restriction of s to $[a, a+\delta)$ solves the initial value problem. ■

Note how the Picard-Lindelöf Theorem shows why it is appropriate to specify initial values for $y, y', \ldots, y^{(n-1)}$. These specifications guarantee that the initial value problem has a unique solution when the hypotheses of the Picard-Lindelöf Theorem are satisfied.

The transformation between differential and integral equations and the interpretation of differential equations as operator equations as exhibited here are both fundamental for work in differential equations.

Exercises

22-6 The supremum M in the proof of Theorem 22 6.

(a) Prove that the supremum M in the second line of the proof of Theorem 22.6 is finite.

(b) Find a function $f : [0, 1) \times \mathbb{R}$ that satisfies $\left|f(t, y_1) - f(t, y_2)\right| \leq |y_1 - y_2|$ for all $t \in [0, 1)$ and all $y_1, y_2 \in \mathbb{R}$ and for which $\sup\left\{\left|f(t, y)\right| : (t, y) \in [0, 1) \times \overline{B_\varepsilon(y_0)}\right\}$ is infinite for all $\varepsilon > 0$ and all $y_0 \in \mathbb{R}$.

(c) Explain why part 22-6b does not contradict part 22-6a.

22-7. Let $a \in \mathbb{R}$, let $\delta > 0$, let $[a, a+\delta] \subseteq \mathbb{R}$, let X be a Banach space, and let $\overline{B_\varepsilon(y_0)} \subseteq X$. Prove that the space $C^0\left([a, a+\delta], \overline{B_\varepsilon(y_0)}\right)$ equipped with the metric that is induced by the uniform norm $\|f\|_\infty := \max\left\{\|f(t)\| : t \in [a, a+\delta]\right\}$ of $C^0\left([a, a+\delta], X\right)$ is complete

22-8. Initial value problems need not have unique solutions

(a) Prove that for any $c \geq 0$ the function $f_c(x) = \begin{cases} 0; & \text{for } 0 \leq x < c, \\ (x-c)^2; & \text{for } x \geq c \end{cases}$ solves the initial value problem $y' = 2\sqrt{y};\ y(0) = 0$.

(b) Explain why this example does not contradict Theorem 22.6.

22.3 Linear Differential Equations

Existence and uniqueness theorems are useful in determining all solutions of a given type of differential equation. If existence is proved, then we know that there are solutions and if uniqueness is proved, we know for how many solutions we should look. In this section, we illustrate this use of existence and uniqueness by analyzing linear differential equations like the ones we encountered in Sections 21.1 and 21.3. Specifically, Example 22.13 and Exercise 22-12 will establish the claims about the general solutions of the differential equations that arose in the separation of variables in Section 21.3.

Definition 22.7 *Let $a < b$ be real numbers and let $g, a_0, \ldots, a_n : [a, b) \to \mathbb{R}$ be functions with $a_n \neq 0$. Then the equation $a_n(t)y^{(n)} + \cdots + a_1(t)y' + a_0(t)y = g(t)$ is called a* **linear** *n^{th} order differential equation. In case $g = 0$, the equation is called* **homogeneous** *and otherwise it is called* **inhomogeneous**.

Linear differential equations are called linear, because the solutions of homogeneous linear equations form a linear space (a vector space) and because the sum of a solution of the inhomogeneous equation and a solution of the homogeneous equation solves the inhomogeneous equation.

Proposition 22.8 *If $y_1, y_2 : [a, b) \to \mathbb{R}$ are solutions of the linear homogeneous n^{th} order differential equation $a_n(t)y^{(n)} + \cdots + a_1(t)y' + a_0(t)y = 0$ and $c_1, c_2 \in \mathbb{R}$, then $h(t) := c_1y_1(t) + c_2y_2(t)$ solves the equation, too.*

Proof. Exercise 22-9. ■

Proposition 22.9 *If the function $y_h : [a, b) \to \mathbb{R}$ is a solution of the linear homogeneous n^{th} order differential equation $a_n(t)y^{(n)} + \cdots + a_1(t)y' + a_0(t)y = 0$ and the function $y_p : [a, b) \to \mathbb{R}$ is a solution of the linear inhomogeneous n^{th} order differential equation $a_n(t)y^{(n)} + \cdots + a_1(t)y' + a_0(t)y = g(t)$, then $h(t) := y_p(t) + y_h(t)$ solves the inhomogeneous equation, too.*

Proof. Exercise 22-10. ■

The Picard-Lindelöf Theorem guarantees that for certain linear equations every initial value problem has a unique solution, where the uniqueness is understood to mean that if f_1 and f_2 both solve the same initial value problem for the linear differential equation, then one function is an extension of the other.

Theorem 22.10 *Let $a < b$, let $g, a_0, \ldots, a_n : [a, b) \to \mathbb{R}$ be continuous functions with $a_n(t) \neq 0$ for all $t \in [a, b)$ and let $y_0, \ldots, y_{n-1} \in \mathbb{R}$. Then the initial value problem $a_n(t)y^{(n)} + \cdots + a_1(t)y' + a_0(t)y = g(t)$, $y(a) = y_0$, $y'(a) = y_1, \ldots,$ $y^{(n-1)}(a) = y_{n-1}$ has a unique solution.*

Proof. Because we could consider the problem on an interval $[a, c)$ for some c with $a < c < b$ and because all a_j are continuous on $[a, c]$, we can assume that all functions a_j are bounded and $a_n > \varepsilon$ for some $\varepsilon > 0$. Then the equivalent $\mathbb{R}^n$ valued linear inhomogeneous differential equation

$$\begin{pmatrix} y_0' \\ \vdots \\ y_{n-2}' \\ y_{n-1}' \end{pmatrix} = \begin{pmatrix} y_1 \\ \vdots \\ y_{n-1} \\ \frac{g(t)}{a_n(t)} - \left(\frac{a_{n-1}(t)}{a_n(t)} y_{n-1} + \cdots + \frac{a_0(t)}{a_n(t)} y_0 \right) \end{pmatrix}$$

satisfies the hypotheses of the Picard-Lindelöf Theorem with the Lipschitz condition on y being valid on all of $\mathbb{R}^n$. (This is most easily seen by using $\|\cdot\|_1$ on $\mathbb{R}^n$.) Therefore every initial value problem for the $\mathbb{R}^n$ valued differential equation has a unique solution on some interval $[a, a+\delta)$.

Now consider an initial value $y_{\text{init}} \in \mathbb{R}^n$ and two functions $f_1 : [a, c_1) \to \mathbb{R}^n$ and $f_2 : [a, c_2) \to \mathbb{R}^n$ that satisfy the $\mathbb{R}^n$ valued differential equation with initial value $f_1(a) = y_{\text{init}} = f_2(a)$. Without loss of generality assume that $c_1 \leq c_2$. Let $d := \sup\left\{x \in [a, b) : f_1|_{[a,x)} = f_2|_{[a,x)}\right\}$ and suppose for a contradiction that $d < c_1$. Then $f_1|_{[d,c_1)}$ and $f_2|_{[d,c_2)}$ both solve the $\mathbb{R}^n$ valued differential equation and continuity implies $f_1(d) = f_2(d)$, but the two functions are not equal on any interval $[d, d+\delta)$, contradicting the Picard-Lindelöf Theorem. Hence, $d = c_1$, which means that f_2 is an extension of f_1. Consequently, each initial value problem for the $\mathbb{R}^n$ valued differential equation has a (globally) unique solution, and hence the same holds for the original n^{th} order differential equation. ■

With existence and uniqueness of the solutions of initial value problems established, it is possible to state the form of all solutions of linear homogeneous and linear inhomogeneous differential equations.

Theorem 22.11 *Let $a < b$ and let $a_0, \ldots, a_n : [a, b) \to \mathbb{R}$ be continuous functions with $a_n(t) \neq 0$ for all $t \in [a, b)$. Then the linear homogeneous differential equation $a_n(t)y^{(n)} + \cdots + a_1(t)y' + a_0(t)y = 0$ has n linearly independent solutions $y_1, \ldots, y_n$ and every solution of the differential equation is of the form $y = \sum_{j=1}^{n} c_j y_j$, where $c_1, \ldots, c_n \in \mathbb{R}$.*

Proof. By Theorem 22.10, the function F that maps each solution y of the differential equation to the vector $\left(y(a), y'(a), \ldots, y^{(n-2)}(a), y^{(n-1)}(a)\right)$ is bijective and it is easy to prove that it is linear, too. Hence, there is a linear isomorphism between the vector space of solutions of the differential equation and $\mathbb{R}^n$. The result is now proved by choosing a base in $\mathbb{R}^n$ and using the inverse images of the base vectors as the solutions $y_1, \ldots, y_n$. ■

Theorem 22.12 *Let $a < b$ and let $g, a_0, \ldots, a_n : [a, b) \to \mathbb{R}$ be continuous functions with $a_n(t) \neq 0$ for all $t \in [a, b)$. Let y_p be a particular solution of the linear inhomogeneous differential equation $a_n(t)y^{(n)} + \cdots + a_1(t)y' + a_0(t)y = g(t)$ and let $y_1, \ldots, y_n$ be linearly independent solutions of the linear homogeneous differential equation $a_n(t)y^{(n)} + \cdots + a_1(t)y' + a_0(t)y = 0$. Then every solution of the linear* *inhomogeneous differential equation* $a_n(t)y^{(n)} + \cdots + a_1(t)y' + a_0(t)y = g(t)$ *is of the form* $y = y_p + \sum_{j=1}^{n} c_j y_j$, *where* $c_1, \ldots, c_n \in \mathbb{R}$.

Proof. Exercise 22-11. ■

Knowing what a general solution of a differential equation looks like allows us to write an expression that encodes all solutions of the differential equation. This expression is also called the **general solution**.

Example 22.13 *The general solution of the differential equation* $X'' + \lambda^2 X = 0$ *is* $X(x) = a\cos(\lambda x) + b\sin(\lambda x)$.

By Theorem 22.11, every solution of the differential equation is a linear combination of two linearly independent solutions. Thus we are done if we can find two linearly independent solutions. It is easily checked that $\cos(\lambda x)$ and $\sin(\lambda x)$ are linearly independent solutions of $X'' + \lambda^2 X = 0$, which proves the claim. ■

Finally, note that because all results proved in this section are valid for linear differential equations with nonconstant coefficients, the results in this section apply to a wide variety of differential equations that arise in mathematical physics. For an example, consider Exercise 22-13.

Exercises

22-9. Prove Proposition 22.8.

22-10. Prove Proposition 22.9.

22-11. Prove Theorem 22.12.

Hints. First prove that each y as stated is a solution. Then use that if y is any solution, then $y - y_p$ solves the homogeneous equation.

22-12. Prove that the general solution of $T' = -k\lambda^2 T$ is $T(t) = ce^{-k\lambda^2 t}$.

22-13. Prove that for fixed λ and ν the **Bessel equation** (see Exercise 21-14) has two linearly independent solutions R_1 and R_2 defined on $(0, \infty)$ and that every solution R of the Bessel equation is a linear combination of the form $R = c_1 R_1 + c_2 R_2$.

Hint. Argue around an initial point $a \neq 0$ and use a substitution $u := -r$ to go right to left in the independent variable r.

22-14. Consider the differential equation $my'' + cy' + ky = 0$.

(a) Prove that if $4km - c^2 > 0$, then the general solution of the differential equation is

$$y(t) := Ae^{-\frac{c}{2m}t}\cos\left(\sqrt{\frac{k}{m} - \frac{c^2}{4m^2}}\,t\right) + Be^{-\frac{c}{2m}t}\sin\left(\sqrt{\frac{k}{m} - \frac{c^2}{4m^2}}\,t\right).$$

(b) Prove that if $c^2 - 4km > 0$, then the general solution of the differential equation is

$$y(t) := Ae^{\left(-\frac{c}{2m} + \sqrt{\frac{c^2}{4m^2} - \frac{k}{m}}\right)t} + Be^{\left(-\frac{c}{2m} - \sqrt{\frac{c^2}{4m^2} - \frac{k}{m}}\right)t}.$$

(c) Prove that if $c^2 - 4km = 0$, then the general solution of the differential equation is $y(t) := Ae^{-\frac{c}{2m}t} + Bte^{-\frac{c}{2m}t}$.

(d) Physically interpret the results of parts 22-14a ("underdamped oscillator"), 22-14b ("overdamped oscillator"), and 22-14c ("critically damped oscillator").

Chapter 23

The Finite Element Method

The finite element method uses a deep and beautiful combination of analysis, geometry, linear algebra and computation to provide approximations to solutions of partial differential equations. As with each chapter in Part III, we will be able to highlight the main ideas, leaving deeper study to the reader as desired. In particular, this chapter will address the theoretical background by introducing the requisite spaces and the method. At the end, a brief outline indicates how the method is used in practice.

Simplistically speaking, the finite element method provides something like a best approximation within a finite dimensional subspace of a Hilbert space of functions. This approximation is pieced together from functions whose domains are small subsets of the overall domain under consideration. Though it is technically not quite correct, these functions can be considered to be the "finite elements."

The key to obtaining the mentioned best approximation is to interpret the left side of the partial differential equation as an operator. For operators with sufficiently nice properties, we can then derive theoretical existence and convergence theorems (see Section 23.1). Unfortunately, differential operators need not be well behaved. Spaces on which the results of Section 23.1 can be applied to differential operators are defined in Sections 23.2 and 23.3 and some operators within the scope of the theory are introduced in Section 23.4. The main practical challenges lie in the choice of the subspace (see Section 23.5), in the computation of the best approximation and in the estimation of the error. These practical challenges are addressed in the extensive literature on the finite element method.

23.1 Ritz-Galerkin Approximation

A partial differential equation can be written in the form $Du = f$, where Du is some combination of partial derivatives of u and f is a function. Any function u that satisfies $Du = f$ is a solution of the partial differential equation. The examples in Chapter 21 show that in applications the left side of a partial differential equation is often linear, that is, $D(\alpha u + \beta v) = \alpha Du + \beta Dv$ for all numbers α, β and all sufficiently often differentiable functions u, v. Thus it is sensible to consider linear differential operators and we will do so throughout this chapter. If our functions are elements of a Hilbert space H *(all Hilbert spaces in this chapter are assumed to be real)*, then $Du = f$ iff u satisfies the system of equations $\langle Du, h\rangle = \langle f, h\rangle$ for all $h \in H$ (see Exercise 20-3). Ultimately, the systems of equations for the finite element method are defined slightly differently, but the above will serve as a good motivation until we are ready to discuss details. The right side $F := \langle f, \cdot\rangle$ of the equation is a functional in the dual space of H and the left side $B(\cdot,\cdot) := \big\langle D(\cdot), \cdot\big\rangle$ is a bilinear function. Therefore in this section we will focus on systems of equations of the form

$$B(\cdot, h) = F(h) \quad \text{for all} \quad h \in H \qquad (H\text{–PDE}),$$

where $B : H \times H \to \mathbb{R}$ is bilinear and $F \in H^*$. These are the appropriate systems of equations to consider, because in the eventual formulation, we will also have a bilinear form and a linear functional. An element $u \in H$ will be called a **solution** of (H-PDE) iff $B(u, h) = F(h)$ for all $h \in H$. Partial differential equations typically come with boundary conditions attached. Section 23.3 will show how boundary conditions can be absorbed into (H-PDE) and into the definition of the underlying space. Thus we will not be concerned with boundary conditions for now. Aside from B being continuous, we need the following property to build the theory.

Definition 23.1 *Let H be a real Hilbert space and let $B : H \times H \to \mathbb{R}$ be a bilinear function. Then B is called* **elliptic** *or* **coercive** *iff there is a $\lambda > 0$ so that for all $u \in V$ we have $B(u, u) \geq \lambda\|u\|^2$. The parameter λ is also called the* **coercivity coefficient**.

In general, a bilinear function $B(\cdot,\cdot) = \big\langle D(\cdot), \cdot\big\rangle$ need not be continuous or elliptic. However, we will see that both properties can be obtained if the spaces and bilinear functions are chosen appropriately. Formally, we note that continuous, elliptic, and symmetric bilinear functions introduce an inner product that could well replace the original inner product on the space.

Proposition 23.2 *Let H be a real Hilbert space and let $B : H \times H \to \mathbb{R}$ be a symmetric, continuous, elliptic bilinear function. Then $B(\cdot,\cdot)$ defines an inner product on H so that the norm $\|\cdot\|_B$ induced by B is equivalent to the norm of H.*

Proof. Exercise 23-1. ■

Proposition 23.2 shows that if the bilinear function B is symmetric, continuous, and elliptic, then replacing the original inner product $\langle\cdot,\cdot\rangle$ with the inner product $B(\cdot,\cdot)$ would not affect the notion of convergence. The only change would be that angles are measured differently. Most importantly, if $B(\cdot,\cdot)$ is an inner product, then by Riesz' Representation Theorem there must be a unique $u \in H$ so that $B(u, h) = F(h)$ for all $h \in H$. Therefore, in this case the existence and uniqueness of a solution of the system (H-PDE) would already be established. Unfortunately, the bilinear functions $B(\cdot,\cdot) = \big\langle D(\cdot), \cdot\big\rangle$ need not be symmetric, even for very simple operators.

Example 23.3 Consider the space $C_0^\infty(0, 1)$ as an inner product subspace of $L^2(0, 1)$. The derivative $\frac{d}{dx} : C_0^\infty(0, 1) \to C_0^\infty(0, 1)$ is a linear function on $C_0^\infty(0, 1)$. Hence,

$B(f,g) := \left\langle \frac{d}{dx} f, g \right\rangle$ is a bilinear function on $C_0^\infty(0,1) \times C_0^\infty(0,1)$. But for all functions $f, g \in C_0^\infty(0,1)$ integration by parts leads to the following.

$$\begin{aligned} B(f,g) &= \left\langle \frac{d}{dx} f, g \right\rangle = \int_0^1 f' g \, d\lambda \\ &= \lim_{u\to 1^-} f(u)g(u) - \lim_{b\to 0^+} f(b)g(b) - \int_0^1 f g' \, d\lambda \\ &= -\int_0^1 f g' \, d\lambda = -\left\langle f, \frac{d}{dx} g \right\rangle = -\left\langle \frac{d}{dx} g, f \right\rangle = -B(g,f). \end{aligned}$$

Therefore B is not symmetric. □

Typically the bilinear functions associated with partial differential equations cannot be made symmetric. However, on the positive side, under the right circumstances our bilinear functions will have all other properties of an inner product. Hence, we should be able to obtain results similar to those for inner products. First we establish that for continuous, elliptic bilinear functions B the system (H-PDE) has a unique solution.

Lemma 23.4 Lax-Milgram Lemma. *Let H be a real Hilbert space, let $F \in H^*$ and let $B : H \times H \to \mathbb{R}$ be a continuous elliptic bilinear function. Then there is exactly one $u \in H$ so that $B(u,h) = F(h)$ for all $h \in H$. Moreover, for this u the inequality $\|u\| \leq \frac{\|F\|}{\lambda}$ holds, where λ is the coercivity coefficient.*

Proof. Because B is bilinear and continuous, for every $w \in H$ the function $B(w, \cdot)$ is linear and continuous. Thus by Riesz' Representation Theorem there is a $T(w) \in H$ so that for all $h \in H$ the equality $B(w,h) = \langle T(w), h \rangle$ holds.

Note that if we can prove that $T : H \to H$ is surjective, then we have proved that a function u as claimed exists. Indeed, by Riesz' Representation Theorem for every $F \in H^*$ there is an $f \in H$ so that $F(\cdot) = \langle f, \cdot \rangle$. If T is surjective, then $B\left(T^{-1}(f), \cdot\right) = \langle f, \cdot \rangle = F(\cdot)$ for all $F \in H^*$, as claimed. In the remainder of this proof, we show that T is surjective.

First we show that T is linear and continuous. For all $x, y \in H$, $\alpha, \beta \in \mathbb{R}$ and $h \in H$ we obtain

$$\begin{aligned} \langle T(\alpha x + \beta y), h \rangle &= B(\alpha x + \beta y, h) = \alpha B(x,h) + \beta B(y,h) \\ &= \alpha \langle T(x), h \rangle + \beta \langle T(y), h \rangle = \langle \alpha T(x) + \beta T(y), h \rangle, \end{aligned}$$

and hence $T(\alpha x + \beta y) = \alpha T(x) + \beta T(y)$, because $h \in H$ was arbitrary (see Exercise 20-3). Therefore T is linear. Moreover, T is continuous, because for all $w \in H$ the inequality $\|T(w)\|^2 = \langle T(w), T(w) \rangle = B(w, T(w)) \leq \|B\|\|w\|\|T(w)\|$ holds, which means that $\|T(w)\| \leq \|B\|\|w\|$ for all $w \in H$.

Second, we prove that T is injective and the inverse of T is continuous. Suppose for a contradiction that T was not injective. Then there is an element $w \neq 0$ with $T(w) = 0$. But then $0 = \langle T(w), w \rangle = B(w,w) \geq \lambda \|w\| > 0$, a contradiction. Now let

$T^{-1} : T[H] \to H$ be the inverse of T. Then for all $y \in T[H]$ we infer the inequalities $\|y\| \left\|T^{-1}(y)\right\| \geq \left\langle y, T^{-1}(y)\right\rangle = B\left(T^{-1}(y), T^{-1}(y)\right) \geq \lambda \left\|T^{-1}(y)\right\|^2$, which means that $\left\|T^{-1}(y)\right\| \leq \frac{\|y\|}{\lambda}$, and hence T^{-1} is continuous. Note that once we have proved that T is surjective, this inequality also establishes the inequality claimed at the end of the Lax-Milgram Lemma.

Third, we prove that $T[H]$ is a closed subspace of H. Let $\{y_n\}_{n=1}^{\infty}$ be a sequence in $T[H]$ that converges to $y \in H$. Then, because T^{-1} is continuous, the sequence $\left\{T^{-1}(y_n)\right\}_{n=1}^{\infty}$ is a Cauchy sequence. Because H is a Hilbert space, there is an $x \in H$ so that $\lim_{n\to\infty} T^{-1}(y_n) = x$. But then because T is continuous, we infer that $T(x) = T\left(\lim_{n\to\infty} T^{-1}(y_n)\right) = \lim_{n\to\infty} T\left(T^{-1}(y_n)\right) = \lim_{n\to\infty} y_n = y$. Hence, $y \in T[H]$ and because $\{y_n\}_{n=1}^{\infty}$ was arbitrary we conclude that $T[H]$ is closed.

Now we can finally prove that $T[H] = H$. Suppose for a contradiction that $T[H] \neq H$. Then there is a $b \in H \setminus T[H]$. By Theorem 20.22, we can assume without loss of generality that b is orthogonal to all $y \in T[H]$. But then we obtain the inequalities $0 = \left\langle T(b), b\right\rangle = B(b, b) \geq \lambda\|b\|^2 > 0$, a contradiction. ■

With unique solvability of the system (H-PDE) established, we can turn to the task of approximating the solutions. By Exercise 20-22, if V is a complete subspace of H then $u \in V$ is the best approximation of f in V iff $\langle u, v\rangle = \langle f, v\rangle$ for all $v \in V$. Because our bilinear functions are not inner products, we cannot formally talk about best approximations, but the system of equations

$$B(\cdot, v) = F(v) \quad \text{for all} \quad v \in V \subseteq H \qquad (V\text{−PDE})$$

is well-defined. Ultimately, the right choice of a finite dimensional subspace V of H will produce a solution of a system of equations (V-PDE) that is close to the solution of (H-PDE). Before we can address such details, we need to assure that the system (V-PDE) actually has a unique solution with the right properties.

Definition 23.5 *Let H be a real Hilbert space, let V be a subspace, let $F \in H^*$ and let $B : H \times H \to \mathbb{R}$ be a continuous elliptic bilinear function. A* **Ritz-Galerkin approximation** *of the solution u of $B(u, h) = F(h)$ for all $h \in H$ (that is, of the solution of H-PDE) is defined to be a $u_V \in V$ so that $B(u_V, v) = F(v)$ for all $v \in V$ (that is, u_V is a solution of V-PDE) if such an element exists.*

Because continuous elliptic bilinear maps B share so many properties with the inner product of H it is not surprising that unique Ritz-Galerkin approximations exist and that if a sequence of spaces "fills" H in the right way, then the Ritz-Galerkin approximations will converge to the solution of (H-PDE).

Lemma 23.6 *Let H be a real Hilbert space, let V be a closed subspace, let $F \in H^*$ and let $B : H \times H \to \mathbb{R}$ be a continuous elliptic bilinear function. Then V contains a unique Ritz-Galerkin approximation u_V for the solution of (H-PDE).*

Proof. Let $C := B|_{V\times V}$ and let $G := F|_V$. Then C is bilinear, continuous, and elliptic and $G \in V^*$. By the Lax-Milgram Lemma there is a unique $u_V \in V$ so that for all $v \in V$ we have $C(u_V, v) = G(v)$, that is, $B(u_V, v) = F(v)$, which was to be proved. ■

Lemma 23.7 Céa's Lemma. *Let H be a real Hilbert space, let V be a closed subspace, let $F \in H^*$, let $B : H \times H \to \mathbb{R}$ be a continuous elliptic bilinear function, let u be the solution of the equations $B(u,h) = F(h)$ for all $h \in H$ and let u_V be the Ritz-Galerkin approximation of u in V. Then $\|u - u_V\| \leq \frac{\|B\|}{\lambda} \inf_{v\in V} \|u - v\|$, where $\|B\|$ is the tensor norm of B and λ is its coercivity coefficient.*

In particular, if $\{V_n\}_{n=1}^{\infty}$ is a sequence of closed spaces so that $\lim_{n\to\infty} \operatorname{dist}(u, V_n) = 0$, then $\lim_{n\to\infty} \left\|u - u_{V_n}\right\| = 0$.

Proof. Note that $B(u - u_V, w) = B(u, w) - B(u_V, w) = F(w) - F(w) = 0$ for all $w \in V$. Hence, for arbitrary $v \in V$ we infer

$$\begin{aligned}
\lambda\|u - u_V\|^2 &= \lambda\langle u - u_V, u - u_V\rangle \leq B(u - u_V, u - u_V) \\
&= B(u - u_V, u) - B(u - u_V, u_V) = B(u - u_V, u) \\
&= B(u - u_V, u) - B(u - u_V, v) = B(u - u_V, u - v) \\
&\leq \|B\|\|u - u_V\|\|u - v\|,
\end{aligned}$$

so $\|u - u_V\| \leq \frac{\|B\|}{\lambda}\|u - v\|$ for all $v \in V$, which implies the desired inequality. ■

The above shows that if we can define Hilbert spaces of functions and continuous and elliptic bilinear functions so that the solution of a partial differential equation $Du = f$ is also the solution of a system of equations (H-PDE), then the solution can be approximated with functions u_V taken from appropriately defined subspaces V of H. To make the problem accessible to computation, it is sensible to find Ritz-Galerkin approximations for the solution of (H-PDE) in appropriately chosen finite dimensional subspaces V. Using finite dimensional subspaces reduces the Ritz-Galerkin approximation to solving a system of linear equations.

Theorem 23.8 *Let H be a real Hilbert space, let S be a finite dimensional subspace with base $\{s_1, \ldots, s_d\}$, let $B : H \times H \to \mathbb{R}$ be a continuous elliptic bilinear function and let $F \in H^*$. Then $u_1, \ldots, u_d$ are the coefficients of the unique Ritz-Galerkin approximation $u_S = \sum_{j=1}^{d} u_j s_j$ in S for the solution of (H-PDE) iff $u_1, \ldots, u_d$ solve the system of equations given by*

$$\sum_{j=1}^{d} B(s_j, s_i)u_j = F(s_i) \quad \text{for all} \quad i = 1, \ldots, d \qquad (S\text{–PDE}).$$

Proof. Exercise 23-2. ■

In summary, because finite dimensional subspaces are closed our abstract considerations have, under the right circumstances, reduced the task of approximating the solution of a partial differential equation to the task of solving a system (S-PDE) of linear equations. We will now show how to translate many partial differential equations into the framework presented in this section.

Exercises

23-1. Prove Proposition 23.2.

23-2. Prove Theorem 23.8.

23.2 Weakly Differentiable Functions

The first step toward using the results of Section 23.1 (specifically Theorem 23.8) is to encode derivatives so that we can define operators that represent the left side of a partial differential equation on a Hilbert space of functions. We start by introducing notation to simplify the details.

Definition 23.9 *A* **multiindex** *is a d-tuple $\alpha := (\alpha_1, \ldots, \alpha_d)$ of nonnegative integers. We define $|\alpha| := \sum_{j=1}^{d} |\alpha_j|$. The set of* **nonnegative integers** *will be denoted $\mathbb{N}_0$.*

Definition 23.10 *Let $\Omega \subseteq \mathbb{R}^d$ be open, let $f : \Omega \to \mathbb{R}$ be k times differentiable and let $\alpha \in \mathbb{N}_0^d$ be a multiindex with $|\alpha| \leq k$. We define $D^\alpha f := \frac{\partial^{|\alpha|}}{\partial^{\alpha_1} \ldots \partial^{\alpha_d}} f$. $D^\alpha f$ will also be called a* **partial derivative of $|\alpha|^{\text{th}}$ order**.

Consider an open set $\Omega \subseteq \mathbb{R}^d$. The elements of $L^2(\Omega)$ can have discontinuities, so they are not all differentiable. But Theorem 18.12 shows that the space $C_0^\infty(\Omega)$ is dense in $L^2(\Omega)$. By Exercise 20-3, we know that $u = f$ iff $\langle u, x \rangle = \langle f, x \rangle$ for all x in a dense subspace of H. Definition 23.11 below defines a weak derivative via this property. The inspiration comes from integration by parts. Consider a continuously differentiable function $f : \Omega \to \mathbb{R}$ and let $g \in C_0^\infty(\Omega)$. Then for all $j \in \{1, \ldots, d\}$ Fubini's Theorem implies the following.

$$\int_\Omega \frac{\partial f}{\partial x_j} g \, d\lambda_d = \int_{\mathbb{R}^{d-1}} \int_{\mathbb{R}} \mathbf{1}_{\Omega_j^{(x_1, \ldots, x_d)}} \frac{\partial f}{\partial x_j} g \, dx_j \, d\lambda_{d-1}(x_1, \ldots, \widehat{x_j}, \ldots, x_d)$$

With $x := (x_1, \ldots, x_d)$, by Proposition 17.38, the section Ω_j^x is open in $\mathbb{R}$, so by Exercise 16-103d it is the union of at most countably many pairwise disjoint open intervals (a_k^x, b_k^x). If Ω_j^x is the union of finitely many open intervals, choose $a_k^x = b_k^x \in \Omega_j^x$ for the remaining k.

$$= \int_{\mathbb{R}^{d-1}} \sum_{k=1}^{\infty} \int_{a_k^x}^{b_k^x} \frac{\partial f}{\partial x_j} g \, dx_j \, d\lambda_{d-1}$$

Now use that the sections of g are zero in intervals $\left[a_k^x, a_k^x + \delta_k\right]$ and $\left[b_k^x - \delta_k, b_k^x\right]$ and integrate by parts (formally over $\left[a_k^x + \delta_k, b_k^x - \delta\right]$). Then undo the decomposition.

$$= -\int_{\mathbb{R}^{d-1}} \sum_{k=1}^{\infty} \int_{a_k^x}^{b_k^x} f \frac{\partial g}{\partial x_j}\, dx_j\, d\lambda_{d-1} = -\int_{\Omega} f \frac{\partial g}{\partial x_j}\, d\lambda_d.$$

Therefore, an easy induction shows that if $D^\alpha f$ is continuous and $g \in C_0^\infty(\Omega)$, then $\int_\Omega (D^\alpha f)\, g\, d\lambda = (-1)^{|\alpha|} \int_\Omega f\, (D^\alpha g)\, d\lambda$. This is the motivation for defining weak derivatives in Definition 23.11 below. Proposition 23.12 then shows that the weak derivative is unique if it exists and Proposition 23.13 shows that regular derivatives are also weak derivatives.

Definition 23.11 *Let $\Omega \subseteq \mathbb{R}^d$ be open, let $1 \leq p \leq \infty$, let $f \in L^p(\Omega)$, and let $\alpha \in \mathbb{N}_0^d$ be a multiindex. The function $w \in L^p(\Omega)$ is called a* **weak derivative** *of f of order $|\alpha|$ iff for all test functions $g \in C_0^\infty(\Omega)$ we have $\langle w, g\rangle_{L^2(\Omega)} = (-1)^{|\alpha|} \left\langle f, D^\alpha g\right\rangle_{L^2(\Omega)}$. (Hölder's inequality guarantees that $\langle w, g\rangle_{L^2(\Omega)}$ exists for all $w \in L^p(\Omega)$ and all $g \in C_0^\infty(\Omega)$.) The function f is called $|\alpha|$ times* **weakly differentiable** *iff all weak derivatives up to order $|\alpha|$ exist. For $|\alpha| = 1$, we say f is weakly differentiable.*

Proposition 23.12 *Let $\Omega \subseteq \mathbb{R}^d$ be an open subset of $\mathbb{R}^d$, let $1 \leq p \leq \infty$, let $f \in L^p(\Omega)$, and let $\alpha \in \mathbb{N}_0^d$ be a multiindex. If the functions $v, w \in L^p(\Omega)$ are both weak derivatives of f, that is, if $\langle v, g\rangle_{L^2(\Omega)} = (-1)^{|\alpha|} \left\langle f, D^\alpha g\right\rangle_{L^2(\Omega)} = \langle w, g\rangle_{L^2(\Omega)}$ for all test functions $g \in C_0^\infty(\Omega)$, then $v = w$.*

Proof. Use the fact that $C_0^\infty(\Omega)$ is dense in $L^2(\Omega)$. (Exercise 23-3.) ■

Proposition 23.13 *Let $\Omega \subseteq \mathbb{R}^d$ be an open subset of $\mathbb{R}^d$, let $\alpha \in \mathbb{N}_0^d$ be a multiindex, let $1 \leq p \leq \infty$, and let $f \in L^p(\Omega) \cap C^{|\alpha|}(\Omega)$ be so that the regular α^{th} partial derivative $D^\alpha f$ is in $L^p(\Omega)$. Then for all test functions $g \in C_0^\infty(\Omega)$ the equality $\left\langle D^\alpha f, g\right\rangle_{L^2(\Omega)} = (-1)^{|\alpha|} \left\langle f, D^\alpha g\right\rangle_{L^2(\Omega)}$ holds, which means that $D^\alpha f$ is a weak α^{th} derivative of f.*

Proof. Use integration by parts in the coordinates affected by D^α (similar to what was done to motivate the weak derivative), and then apply Proposition 23.12. (Exercise 23-4.) ■

With weak derivatives coinciding with regular derivatives when regular derivatives exist, it is sensible to extend the notation to weak derivatives.

Definition 23.14 *Let $\Omega \subseteq \mathbb{R}^d$ be an open subset of $\mathbb{R}^d$, let $1 \leq p \leq \infty$, let $f \in L^p(\Omega)$, and let $\alpha \in \mathbb{N}_0^d$ be a multiindex. Then a weak derivative $w \in L^p(\Omega)$ of f so that for all test functions $g \in C_0^\infty(\Omega)$ we have $\langle w, g\rangle_{L^2(\Omega)} = (-1)^{|\alpha|} \left\langle f, D^\alpha g\right\rangle_{L^2(\Omega)}$ is denoted $D^\alpha f := w$. For a weak derivative of order 1 in the ith coordinate direction, we also write $D^{(i)} f$ instead of using a less easy to read d-tuple.*

We should note that there are functions that are not differentiable, but which have a weak derivative. Recall that the absolute value function is not differentiable at zero.

Example 23.15 *Let $\Omega := (-1, 1) \subseteq \mathbb{R}$ and consider $f(x) = |x|$. The weak derivative of f is $D^{(1)}f(t) = \begin{cases} 1; & \text{for } t > 0, \\ -1; & \text{for } t \leq 0. \end{cases}$*

Let $g \in C_0^\infty(-1, 1)$ and let $a \in (0, 1)$ be so that $\mathrm{supp}(g) \subseteq (-a, a)$. Then

$$\begin{aligned} (-1)^1 \langle f, g' \rangle &= -\int_{-1}^{1} fg'\,d\lambda = -\int_{-a}^{a} fg'\,d\lambda = -\int_{0}^{a} xg'\,d\lambda + \int_{-a}^{0} xg'\,d\lambda \\ &= \int_{0}^{a} x'g\,d\lambda - \int_{-a}^{0} x'g\,d\lambda = \int_{0}^{a} 1g\,d\lambda + \int_{-a}^{0} (-1)g\,d\lambda \\ &= \int_{-a}^{a} \left(D^{(1)}f\right) g\,d\lambda = \int_{-1}^{1} \left(D^{(1)}f\right) g\,d\lambda = \left\langle D^{(1)}f, g \right\rangle. \end{aligned}$$

Because $g \in C_0^\infty(-1, 1)$ was arbitrary this proves the claim. □

Unfortunately, not all functions in $L^2(\Omega)$ have weak derivatives (see Exercise 23-5 for a specific example). The following results give an exact characterization of the weakly differentiable functions on open subintervals of $\mathbb{R}$ and Theorem 23.17 also is a key element of the proof of the beautiful Antiderivative Form of the Fundamental Theorem of Calculus for the Lebesgue integral (see Exercise 23-8). We first need to prove that if two functions have equal weak first partial derivatives, then the functions must differ (a.e.) by a constant.

Theorem 23.16 *Let $\Omega \subseteq \mathbb{R}^d$ be open and connected and let $f, g \in L^p(\Omega)$ be so that all weak derivatives of order 1 of f and g exist. If $D^{(i)}f = D^{(i)}g$ a.e. for all $i \in \{1, \ldots, d\}$, then there is a $c \in \mathbb{R}$ so that $f = g + c$ a.e.*

Proof. Let $\varphi \in C_0^\infty(\Omega)$. Then for all $i \in \{1, \ldots, d\}$ we infer

$$\int_\Omega (f - g)\frac{\partial \varphi}{\partial x_i}\,d\lambda = -\int_\Omega \left(D^{(i)}f - D^{(i)}g\right)\varphi\,d\lambda = 0.$$

This condition implies that $f - g$ is equal to a constant almost everywhere. The details are a bit technical and the reader can produce them in Exercise 23-6. ■

Theorem 23.17 *A function $f \in L^p(a, b)$ is weakly differentiable iff f is equal a.e. to a function $g : (a, b) \to \mathbb{R}$ that is absolutely continuous on every closed subinterval of (a, b) and that satisfies $g' \in L^p(a, b)$. In this case, the weak derivative is a.e. equal to the pointwise derivative.*

Proof. For the "$\Leftarrow$" part, we can let $f : (a, b) \to \mathbb{R}$ be absolutely continuous on every closed subinterval of (a, b). By Exercise 23-7b on every closed subinterval of (a, b), f is the difference of two absolutely continuous nondecreasing functions. Therefore, it is enough to prove the result for absolutely continuous nondecreasing f. By Exercise 10-7, f is differentiable a.e. on every closed subinterval of (a, b), and hence f is differentiable a.e. on (a, b) itself. We first prove the result for an absolutely continuous nondecreasing f so that $\left\{\frac{f\left(x+\frac{1}{n}\right)-f(x)}{\frac{1}{n}}\right\}_{n=1}^{\infty}$ is bounded on every closed subinterval $[c, d] \subseteq (a, b)$. (We assume that we discarded any early terms from the sequence for which $d+\frac{1}{n} \notin (a, b)$ or $c-\frac{1}{n} \notin (a, b)$.) Let $\varphi \in C_0^{\infty}(a, b)$ and let $[c, d] \subseteq (a, b)$ be a closed interval so that $\mathrm{supp}(\varphi) \subseteq (c, d)$. Because $\varphi \in C_0^{\infty}(a, b)$ all difference quotients for φ are bounded on $[c, d]$ by the maximum of φ' on $[a, b]$. Thus, because $[c, d]$ is bounded, we can apply the Dominated Convergence Theorem to the integrals below.

$$
\begin{aligned}
\int_a^b f'\varphi \, d\lambda &= \int_c^d \lim_{n\to\infty} \frac{f\left(x+\frac{1}{n}\right)-f(x)}{\frac{1}{n}} \varphi(x)\, d\lambda(x) \\
&= \lim_{n\to\infty} \int_c^d \frac{f\left(x+\frac{1}{n}\right)-f(x)}{\frac{1}{n}} \varphi(x)\, d\lambda(x) \\
&= \lim_{n\to\infty} \int_a^b f(x) \frac{\varphi\left(x-\frac{1}{n}\right)-\varphi(x)}{\frac{1}{n}} \, d\lambda(x) \\
&= -\int_a^b f(x) \lim_{n\to\infty} \frac{\varphi\left(x-\frac{1}{n}\right)-\varphi(x)}{-\frac{1}{n}} \, d\lambda(x) \\
&= -\int_a^b f(x)\varphi'(x)\, d\lambda(x),
\end{aligned}
$$

which proves $D^{(1)}f = f'$ a.e. In case $\left\{\frac{f\left(x+\frac{1}{n}\right)-f(x)}{\frac{1}{n}}\right\}_{n=1}^{\infty}$ is unbounded on $[c, d] \subseteq (a, b)$, for each $m \in \mathbb{N}$ let $B_m := \left\{x : \limsup_{n\to\infty} \frac{f\left(x+\frac{1}{n}\right)-f(x)}{\frac{1}{n}} > m\right\}$.

Because f is differentiable a.e., the set $\bigcap_{m=1}^{\infty} B_m$ is a null set. For each $m \in \mathbb{N}$, let $\left\{I_j^m\right\}_{j=1}^{\infty}$ be a countable family of pairwise disjoint open intervals that contains B_m and satisfies $\sum_{j=1}^{\infty}\left|I_j^m\right| \leq \lambda(B_m)+\frac{1}{m}$ and $\bigcup_{j=1}^{\infty} I_j^m \subseteq \bigcup_{j=1}^{\infty} I_j^{m-1}$ for $m > 1$. For each $m \in \mathbb{N}$ the function $f_m := f - \sum_{j=1}^{\infty}\left[f|_{I_j^m} + \mathbf{1}_{\left[\sup I_j^m, b\right)}\left(f\left(\sup I_j^m\right) - f\left(\inf I_j^m\right)\right)\right]$

is nondecreasing, absolutely continuous and $\left\{ \dfrac{f_m\left(x+\frac{1}{n}\right) - f_m(x)}{\frac{1}{n}} \right\}_{n=1}^{\infty}$ is bounded.
Moreover, because f is absolutely continuous, $\{f_m\}_{m=1}^{\infty}$ converges uniformly from below to f and $\{f'_m\}_{m=1}^{\infty}$ converges a.e. from below to f'. Let $\varphi \in C_0^{\infty}(a,b)$. By the Dominated Convergence Theorem we conclude

$$\begin{aligned}\int_a^b f'\varphi \, d\lambda &= \int_a^b \lim_{m\to\infty} f'_m\varphi \, d\lambda = \lim_{m\to\infty} \int_a^b f'_m\varphi \, d\lambda = \lim_{m\to\infty} -\int_a^b f_m\varphi' \, d\lambda \\ &= -\int_a^b \lim_{m\to\infty} f_m\varphi' \, d\lambda = -\int_a^b f\varphi' \, d\lambda,\end{aligned}$$

which establishes the "$\Leftarrow$" part as well as the claim about the derivatives.

For the "$\Rightarrow$" part, let the function $f:(a,b)\to\mathbb{R}$ be weakly differentiable and let $[c,d]\subset(a,b)$. Consider the function $\tilde{f}(x) := \int_c^x D^{(1)}f(t)\,d\lambda(t)$. Then by Exercise 14-36c, $\tilde{f}$ is absolutely continuous and by the Derivative Form of the Fundamental Theorem of Calculus (see Exercise 18-6) the equality $\frac{d}{dx}\tilde{f}(x) = D^{(1)}f(x)$ holds a.e. on (c,d). But with what we already proved, this means that $\tilde{f}$ is weakly differentiable on (c,d) and $D^{(1)}f = D^{(1)}\tilde{f}$ a.e. on (c,d). By Theorem 23.16, we conclude that f and $\tilde{f}$ differ at most by a constant on (c,d), and hence f is a.e. equal to an absolutely continuous function on $[c,d]$. ■

Exercises

23-3 Prove Proposition 23.12.

23-4. Prove Proposition 23.13

23-5 Let $\Omega := (-1,1)\subseteq\mathbb{R}$ and consider the function $g\in L^2(\Omega)$ defined by $g(t) = \begin{cases} 1; & \text{for } t>0, \\ -1; & \text{for } t\le 0. \end{cases}$
Prove directly (that is, without using Theorem 23 17) that g does not have a weak derivative.
Hint. Prove that if $w\in L^2(-1,1)$ was a weak derivative of g it must satisfy $w(x)=0$ a.e. Then show that for some even functions $e\in C_0^{\infty}(-1,1)$ the inner product $\langle e,w\rangle$ is not zero

23-6. Let $\Omega\subseteq\mathbb{R}^d$ be a connected open set. Use the steps below to prove that if $f\in L^1(\Omega)$ is so that $\int_\Omega f\frac{\partial\varphi}{\partial x_i}\,d\lambda = 0$ for all $\varphi\in C_0^{\infty}(\Omega)$ and all $i\in\{1,\ldots,d\}$, then there is a $c\in\mathbb{R}$ so that $f=c$ a.e
We will argue by contradiction, so in the following we assume that f is not constant a.e. Also, we will first establish the result in case $\Omega = \prod_{j=1}^{d}(a_j,b_j)$ is a bounded open box.

(a) Prove that if $f:\Omega\to\mathbb{R}$ is so that there is no $c\in\mathbb{R}$ with $f=c$ a.e., then there is an $a\in\mathbb{R}$ so that $\lambda\left(\left\{x\in\mathbb{R}^d : f(x)>a\right\}\right)>0$ and $\lambda\left(\left\{x\in\mathbb{R}^d : f(x)<a\right\}\right)>0$.

(b) Let a be as in part 23-6a, let $v>0$ be so that $A = \left\{x\in\Omega : f(x)<a-v\right\}$ is not a null set. Prove that there must be a $\delta\in\mathbb{R}$ and an $i\in\{1,\ldots,d\}$ so that $\lambda\left((\Omega\cap(A+\delta e_i))\setminus A\right)>0$.
Hint. Suppose for a contradiction that all these measures are zero. First prove that for all translations $x+S\subseteq\Omega$ of subsets S of A we have $\lambda\left((x+S)\cap A\right)=\lambda(x+S)$ Then let $\varepsilon>0$ and let B be an open box in Ω so that $\lambda(A\cap B)>(1-\varepsilon)\lambda(B)$. Prove that then for all translations $x+B$ of B that are contained in Ω we have $\lambda\left(A\cap(x+B)\right)>(1-\varepsilon)\lambda(x+B)$. Use that B,ε were arbitrary to conclude that $\lambda(A)>(1-\varepsilon)\lambda(\Omega)$ and then $\lambda(A)=\lambda(\Omega)$, a contradiction.

(c) Prove that there must be a $\delta \in \mathbb{R}$, an $i \in \{1, \ldots, d\}$ and a Lebesgue measurable bounded subset C of A so that $\delta e_i + C \subseteq \Omega$ and $\lambda\left((\delta e_i + C) \setminus A\right) = \lambda(\delta e_i + C) > 0$.

(d) Let C, i be as in part 23-6c, let $\{\psi_n\}_{n=1}^{\infty}$ be a sequence of bounded functions in $C_0^{\infty}(\Omega)$ so that $\lim_{n\to\infty} \|\psi_n - \mathbf{1}_C\|_1 = 0$ and $\lim_{n\to\infty} \psi_n = \mathbf{1}_C$ a.e. Prove that

$$\varphi_n(x) := \int_{a_i}^{x_i} \psi_n(x_1, \ldots, x_{i-1}, t, x_{i+1}, \ldots, x_d) - \psi_n(x_1, \ldots, x_{i-1}, t - \delta, x_{i+1}, \ldots, x_d)\, dt$$

defines a function in $C_0^{\infty}(\Omega)$ with $\frac{\partial \varphi_n}{\partial x_i}(x) = \psi_n(x) - \psi_n(x - \delta e_i)$

(e) Prove that $\lim_{n\to\infty} \int_{\Omega} f \frac{\partial \varphi_n}{\partial x_i}\, d\lambda = \int_C f\, d\lambda - \int_{\delta e_i + C} f\, d\lambda \neq 0$. Conclude that there is a function $\varphi \in C_0^{\infty}(\Omega)$ with $\int_{\Omega} f \frac{\partial \varphi}{\partial x_i}\, d\lambda \neq 0$, which establishes the contradiction for bounded open boxes.

(f) Prove that the result holds for all connected open sets Ω.

23-7 Let $f : [a, b] \to \mathbb{R}$ be an **absolutely continuous** function.

(a) Prove that f is of bounded variation.

(b) Prove that the two nondecreasing functions in Exercise 8-12a are both absolutely continuous.

(c) Use the above and Exercises 10-7 and 14-38b to conclude that f is differentiable almost everywhere and that the derivative is integrable over $[a, b]$.

23-8 **Fundamental Theorem of Calculus, Antiderivative Form** Prove that $f : [a, b] \to \mathbb{R}$ is **absolutely continuous** iff f is differentiable a.e., f' is Lebesgue integrable and for all $x \in [a, b]$ the equality $f(x) = f(a) + \int_a^x f'(t)\, d\lambda(t)$ holds.

Hint. Use Exercise 14-36c, Exercise 23-7c, Exercise 18-6, Theorem 23.16, and Theorem 23.17.

Note. Unlike for the Riemann integral, the Antiderivative Form of the Fundamental Theorem of Calculus for the Lebesgue integral is a biconditional and it has no artificial-looking hypotheses. In particular, the integrability of the derivative does not need to be demanded explicitly, because it follows from absolute continuity.

23-9 Use the Antiderivative Form of the Fundamental Theorem of Calculus to prove that Lebesgue's singular function is uniformly continuous, but not absolutely continuous.

23-10 **Integration by Parts** Let $f, g : [a, b] \to \mathbb{R}$ be **absolutely continuous**. Prove that the product fg is absolutely continuous and that if fg' and $g'f$ are integrable, then $\int_a^b f'g\, d\lambda = fg\,\big|_a^b - \int_a^b g'f\, d\lambda$.

Hint. Use Exercise 23-8.

23-11 Let $\Omega \subseteq \mathbb{R}^d$ be open. A function $f : \Omega \to \mathbb{R}$ is called **absolutely continuous** iff for every $\varepsilon > 0$ there is a $\delta > 0$ so that for all $x_1, \ldots, x_n, z_1, \ldots, z_n \in \Omega$ with $\sum_{j=1}^{n} \|z_j - x_j\| < \delta$ we have $\sum_{j=1}^{n} \left| f(z_j) - f(x_j) \right| < \varepsilon$. Prove that if $f \in L^p(\Omega)$ is absolutely continuous and all first partial derivatives (which exist a.e.) are in $L^p(\Omega)$, then f is weakly differentiable and each weak first partial derivative equals the regular first partial derivative a.e.

Hint. Use Exercise 23-10.

23-12 Let $\Omega \subseteq \mathbb{R}^d$ be open. Prove that if a function $f \in L^p(\Omega)$ is weakly differentiable, then on every compact box B that is contained in Ω, and on which the weak partial derivatives are in $L^{\infty}(B)$, f is equal a.e. to a function $g : B \to \mathbb{R}$ that is **absolutely continuous** on B and so that $\frac{\partial g}{\partial x_j} = D^{(j)} f$ a.e. on B° for all $j \in \{1, \ldots, d\}$.

Hint Without loss of generality assume that $B = [0, 1]^d$ Use Fubini's Theorem to prove that for each $j \in \{1, \ldots, d\}$ there is a null set $N \subseteq \mathbb{R}^{d-1}$ so that for all $(x_1, \ldots, \widehat{x_j}, \ldots, x_d) \notin N$ we have

$$\int_0^1 f(x_1, \ldots, x_{j-1}, t, x_{j+1}, \ldots, x_d)\varphi'(t)\, dt = \int_0^1 D^{(j)} f(x_1, \ldots, x_{j-1}, t, x_{j+1}, \ldots, x_d)\varphi(t)\, dt$$

for φ in a countable dense subset of $C_0^\infty(0, 1)$ (so that for every $\psi \in C_0^\infty(0, 1)$ there is a sequence with $\varphi_n \to \psi$ and $\varphi_n' \to \psi'$). Then the equation holds for all $\varphi \in C_0^\infty(0, 1)$. Conclude that $f(x_1, \ldots, x_d) = c_j + \int_a^{x_j} D^{(j)} f(x_1, \ldots, x_{j-1}, t, x_{j+1}, \ldots, x_d)\, d\lambda(t)$ for almost all $x_j \in [0, 1]$ and for all $(x_1, \ldots, \widehat{x_j}, \ldots, x_d) \notin N$. Conclude that because j was arbitrary and the weak first partial derivatives are bounded, f is equal a.e. on B to an absolutely continuous function.

Note $f(x) := \ln \|x\|$ on the unit ball of $\mathbb{R}^3$ shows that boundedness of the $D^{(j)} f$ is needed

23-13 Let $\Omega \subseteq \mathbb{R}^d$ be open, let $f \in L^p(\Omega)$ and let $g \in C^\infty(\Omega)$ be bounded

(a) Let $i \in \{1, \ldots, d\}$ Prove that if the weak derivative $D^{(i)} f$ in the ith coordinate direction exists and $D^{(i)} g$ is bounded, then fg is weakly differentiable in the ith coordinate direction and the product rule $D^{(i)}(fg) = \left(D^{(i)} f\right) g + f\frac{\partial g}{\partial x_i}$ holds

(b) Let $\alpha \in \mathbb{N}_0^d$ be a multiindex Prove that if all weak derivatives $D^\beta f$ of f of order $|\beta| \leq |\alpha|$ exist and $D^\alpha g$ is bounded, then the weak derivative $D^\alpha(fg)$ exists.

Hint Induction on $|\alpha|$.

23.3 Sobolev Spaces

As we define the spaces underlying the finite element method, we should recall that part 2 of Example 17.3 shows that differential operators need not be continuous. On the other hand, Exercise 17-13 shows that differential operators *can* be continuous if the space is chosen appropriately. Definition 23.18 assures that the functions in our spaces have sufficiently many weak derivatives. Theorem 23.19 equips the spaces with norms (similar to that of Exercise 17-13) that turn them into Banach spaces on which (weak) differential operators can be used to define continuous bilinear functions (see Theorem 23.22).

Definition 23.18 *Let $\Omega \subseteq \mathbb{R}^d$ be open, let $m \in \mathbb{N}_0$ and let $1 \leq p \leq \infty$. Then we define $W^{m,p}(\Omega) := \left\{u \in L^p(\Omega) : D^\alpha u \in L^p(\Omega) \text{ for all } |\alpha| \leq m\right\}$ and call it a* **Sobolev space**.

It is easy to see that Sobolev spaces are vector spaces (see Exercise 23-14) and Proposition 23.13 shows that Sobolev spaces contain the m times continuously differentiable functions whose partial derivatives up to order m are p-integrable. Moreover, for $m = 0$, the space $W^{0,p}(\Omega)$ is $L^p(\Omega)$, because the zeroth derivative of a function is the function itself.

Theorem 23.19 *Let $\Omega \subseteq \mathbb{R}^d$ be open, let $m \in \mathbb{N}_0$ and let $1 \leq p < \infty$ Then the function $\|u\|_{W^{m,p}(\Omega)} := \left(\sum_{|\alpha| \leq m} \left\|D^\alpha u\right\|_p^p\right)^{\frac{1}{p}}$ defines a norm on the Sobolev space $W^{m,p}(\Omega)$ so that $\left(W^{m,p}(\Omega), \|\cdot\|_{W^{m,p}(\Omega)}\right)$ is a Banach space. Moreover, the function*

$\|u\|_{W^{m,\infty}(\Omega)} := \sum_{|\alpha| \leq m} \left\| D^\alpha u \right\|_\infty$ *defines a norm on the Sobolev space* $W^{m,\infty}(\Omega)$ *so that* $\left(W^{m,\infty}(\Omega), \|\cdot\|_{W^{m,\infty}(\Omega)}\right)$ *is a Banach space.*

Proof. The properties of a norm are easily checked for $\|\cdot\|_{W^{m,p}(\Omega)}$. For completeness, note that if $\{u_n\}_{n=1}^\infty$ is a Cauchy sequence in $W^{m,p}(\Omega)$, then u and all its weak partial derivatives are Cauchy sequences in $L^p(\Omega)$. Then use Hölder's inequality to prove that the limits of the weak partial derivatives are the weak partial derivatives of the limit of $\{u_n\}_{n=1}^\infty$ (Exercise 23-15). ■

For $p = 2$, we can equip the Sobolev space $W^{m,2}(\Omega)$ with a Hilbert space structure.

Theorem 23.20 *Let* $\Omega \subseteq \mathbb{R}^d$ *be an open subset of* $\mathbb{R}^d$ *and let* $m \in \mathbb{N}_0$. *The function* $\langle u, v\rangle_{H^m(\Omega)} := \sum_{|\alpha| \leq m} \left\langle D^\alpha u, D^\alpha v\right\rangle_{L^2(\Omega)}$ *defines an inner product on the Sobolev space* $H^m(\Omega) := W^{m,2}(\Omega)$ *so that* $\left(H^m(\Omega), \langle\cdot,\cdot\rangle_{H^m(\Omega)}\right)$ *is a Hilbert space.*

Proof. The verification that $\langle\cdot,\cdot\rangle_{H^m(\Omega)}$ is an inner product on $H^m(\Omega)$ is a straightforward computation (Exercise 23-16).

The claim that $H^m(\Omega)$ is a Hilbert space follows directly from Theorem 23.19. ■

Similar to $L^p(\Omega)$ the infinitely differentiable functions are dense in Sobolev spaces as long as we do not insist on compact supports.

Theorem 23.21 *Let* $\Omega \subseteq \mathbb{R}^d$ *be an open subset of* $\mathbb{R}^d$, *let* $m \in \mathbb{N}_0$ *and let* $1 \leq p < \infty$. *Then* $C^\infty(\Omega) \cap W^{m,p}(\Omega)$ *is dense in* $W^{m,p}(\Omega)$.

Proof. Let $g : \mathbb{R}^d \to \mathbb{R}$ be the function $g(x) := \begin{cases} e^{\frac{1}{\sum_{j=1}^d x_j^2 - 1}}; & \text{for } \|x\|_2 < 1, \\ 0; & \text{for } \|x\|_2 \geq 1, \end{cases}$ from Exercise 18-15 and for $a > 0$ define $g_a(x) := \frac{1}{\int_{\mathbb{R}^d} g\left(\frac{y}{a}\right) \, d\lambda(y)} g\left(\frac{x}{a}\right)$. For every function $f \in L^p(\Omega)$ define $f_a(x) := \int_\Omega f(z) g_a(x-z) \, d\lambda(z)$. Functions such as f_a are also called **regularizations** of f and the construction of f_a by integrating the product of $f(z)$ and $g_a(x-z)$ with respect to z is called the **convolution** of f and g_a. By Exercise 18-17d each f_a is infinitely differentiable and by Exercise 18-17b the partial derivatives of f_a can be obtained as the convolution of f with the appropriate partial derivative of g_a.

To prove that $C^\infty(\Omega) \cap W^{m,p}(\Omega)$ is dense in $W^{m,p}(\Omega)$, first let $f \in W^{m,p}(\Omega)$ be so that $\operatorname{supp}(f)$ is compact and let $a < \operatorname{dist}\left(\operatorname{supp}(f), \mathbb{R}^d \setminus \Omega\right)$ be positive. Let α be a multiindex with $|\alpha| \leq m$. Then with $D_z^\alpha\left(g_a(x-z)\right)$ denoting the αth partial derivative of $g_a(x-\cdot)$, where x is held constant, we obtain the following for all $x \in \Omega$.

$$\begin{aligned} D^\alpha f_a(x) &= \int_\Omega f(z) \left(D^\alpha g_a\right)(x-z) \, d\lambda(z) = (-1)^{|\alpha|} \int_\Omega f(z) D_z^\alpha\left(g_a(x-z)\right) d\lambda(z) \\ &= \int_\Omega D^\alpha f(z) g_a(x-z) \, d\lambda(z). \end{aligned}$$

Now let $\varepsilon > 0$ and let $p \in (1, \infty)$. (The case $p = 1$ is left to Exercise 23-17.) We will prove that there is an $A > 0$ so that for all $a \in (0, A)$ the inequality $\|f - f_a\|_{W^{m,p}(\Omega)} < \varepsilon$ holds. To do so, we will estimate all terms of the Sobolev norm separately. So let α be a multiindex with $|\alpha| \leq m$ and let $\nu > 0$.

By Exercise 18-16, there is an $A_\alpha < \operatorname{dist}\left(\operatorname{supp}(f), \mathbb{R}^d \setminus \Omega\right)$, so that for all $y \in \mathbb{R}^d$ with $\|y\| < A_\alpha$ the inequality $\int_\Omega \left|D^\alpha f(x) - D^\alpha f(x+y)\right|^p \, d\lambda(x) < \nu$ holds. Then for all $a \in (0, A_\alpha)$ we obtain the following:

$$\begin{aligned}
&\int_\Omega \left|D^\alpha f(x) - D^\alpha f_a(x)\right|^p \, d\lambda(x) \\
&= \int_\Omega \left|\int_\Omega D^\alpha f(x) g_a(x-z) \, d\lambda(z) - \int_\Omega D^\alpha f(z) g_a(x-z) \, d\lambda(z)\right|^p \, d\lambda(x) \\
&\leq \int_\Omega \left(\int_\Omega \left|D^\alpha f(x) - D^\alpha f(z)\right| g_a(x-z) \, d\lambda(z)\right)^p \, d\lambda(x)
\end{aligned}$$

Use Hölder's inequality.

$$\begin{aligned}
&\leq \int_\Omega \left(\left(\int_\Omega \left(\left|D^\alpha f(x) - D^\alpha f(z)\right| \left(g_a(x-z)\right)^{\frac{1}{p}}\right)^p \, d\lambda(z)\right)^{\frac{1}{p}} \times \right. \\
&\qquad \left. \times \left(\int_\Omega \left(\left(g_a(x-z)\right)^{1-\frac{1}{p}}\right)^q \, d\lambda(z)\right)^{\frac{1}{q}}\right)^p \, d\lambda(x) \\
&= \int_\Omega \int_\Omega \left|D^\alpha f(x) - D^\alpha f(z)\right|^p g_a(x-z) \, d\lambda(z) \left(\int_\Omega g_a(x-z) \, d\lambda(z)\right)^{\frac{p}{q}} \, d\lambda(x) \\
&\leq \int_\Omega \int_{B_a(0)} \left|D^\alpha f(x) - D^\alpha f(x+y)\right|^p g_a(-y) \, d\lambda(y) \, d\lambda(x)
\end{aligned}$$

Use Proposition 14.65.

$$\begin{aligned}
&= \int_{B_a(0)} \int_\Omega \left|D^\alpha f(x) - D^\alpha f(x+y)\right|^p \, d\lambda(x) g_a(y) \, d\lambda(y) \\
&< \int_{B_a(0)} \nu g_a(y) \, d\lambda(y) = \nu.
\end{aligned}$$

Let $\nu := \dfrac{\varepsilon^p}{\left|\left\{\alpha \in \mathbb{N}_0^d : |\alpha| \leq m\right\}\right|}$ and let $A := \min\left\{A_\alpha : |\alpha| \leq m\right\}$, where the A_α are chosen so that $\left\|D^\alpha f - D^\alpha f_a\right\|_{L^p(\Omega)}^p < \nu$ for all $a \in (0, A_\alpha)$. Then for all $a \leq A$, we obtain the inequality $\|f - f_a\|_{W^{m,p}(\Omega)} = \left(\sum_{|\alpha| \leq m} \left\|D^\alpha f - D^\alpha f_a\right\|_{L^p(\Omega)}^p\right)^{\frac{1}{p}} < \varepsilon.$

Now consider an arbitrary $f \in W^{m,p}(\Omega)$. By Theorem 16.112 (applied to an open cover of Ω that consists of open balls contained in Ω), there is a locally finite cover $\{U_n\}_{n=1}^\infty$ of Ω consisting of open sets with compact closure $\overline{U_n} \subseteq \Omega$ and by Theorem

16.115 (using functions as in Lemma 18.11) there is a C_0^∞-partition of unity $\{\varphi_n\}_{n=1}^\infty$ subordinate to $\{U_n\}_{n=1}^\infty$. Therefore the equality $f = \sum_{n=1}^\infty f\varphi_n$ holds pointwise, but it need not imply convergence of the sum in $W^{m,p}(\Omega)$. Hence, we continue the argument as follows. By Exercise 23-13b, because all partial derivatives of φ_n are bounded, for each $n \in \mathbb{N}$ the function $f\varphi_n$ is in $W^{m,p}(\Omega)$ and it is compactly supported inside Ω. Let $\varepsilon > 0$. By what we have already proved, there is a $u_n \in C_0^\infty(\Omega)$ so that $\mathrm{supp}(u_n) \subseteq U_n$ and $\|u_n - f\varphi_n\|_{W^{m,p}(\Omega)} < \frac{\varepsilon}{2^n}$. Then $\sum_{n=1}^\infty \|u_n - f\varphi_n\|_{W^{m,p}(\Omega)} < \sum_{n=1}^\infty \frac{\varepsilon}{2^n} = \varepsilon$, which means $h := \sum_{n=1}^\infty (u_n - f\varphi_n) \in W^{m,p}(\Omega)$, because the series converges absolutely in the Sobolev space $W^{m,p}(\Omega)$. Therefore $h + f \in W^{m,p}(\Omega)$ and the sum satisfies $\left\|(h+f) - f\right\|_{W^{m,p}(\Omega)} = \|h\|_{W^{m,p}(\Omega)} \leq \sum_{n=1}^\infty \|u_n - f\varphi_n\|_{W^{m,p}(\Omega)} < \varepsilon$. But now (pointwise) $h + f = \sum_{n=1}^\infty (u_n - f\varphi_n) + f = \left(\sum_{n=1}^\infty u_n\right) - f + f = \sum_{n=1}^\infty u_n$ and because we started with a locally finite cover, the sum $\sum_{n=1}^\infty u_n$ is infinitely differentiable. Therefore $C^\infty(\Omega) \cap W^{m,p}(\Omega)$ is dense in $W^{m,p}(\Omega)$. ■

We next turn our attention to the bilinear functions we want to use in our investigation of the systems of equations (H-PDE), (V-PDE) and (S-PDE). We first note that these bilinear functions will be continuous on Sobolev spaces.

Theorem 23.22 *Let $\Omega \subseteq \mathbb{R}^d$ be open and let $m \in \mathbb{N}_0$. Then for all multi-indices α, β with $|\alpha| \leq m$ and $|\beta| \leq m$ the function $B(f, g) := \langle D^\alpha f, D^\beta g\rangle_{L^2(\Omega)}$ is continuous and bilinear on $H^m(\Omega)$.*

Proof. Use the Cauchy-Schwarz inequality. (Exercise 23-18.) ■

Ellipticity of the bilinear functions is not proved as easily. In fact, to prove ellipticity for the bilinear form that we will associate with the Poisson equation we will first need to consider the boundary values of functions in Sobolev spaces.

Physically there is a significant difference between boundary values, which dictate values at fixed places in space as time progresses, and initial values, which dictate values at a fixed time for all places in space. For an abstract investigation, time is just another variable, so initial values can be treated like boundary values (on the "time boundary," if you will). Hence, in the remainder of this section we will concentrate on boundary values. As soon as we talk about the boundary values of functions, the geometric shape of the boundary becomes an issue.

For sets whose boundary has the right properties, we will define the boundary values as an operator from the functions on $\overline{\Omega}$ to the functions on $\delta\Omega$. To do this, we first need to prove that $W^{m,p}(\Omega)$ has a dense subset of functions that can be evaluated on

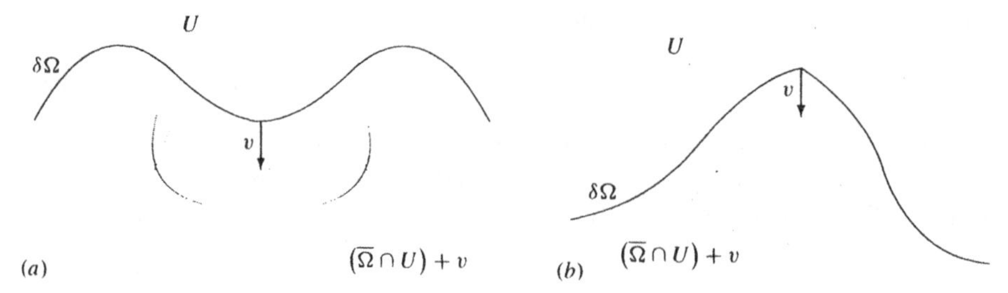

Figure 61: The segment property guarantees that every point on the boundary has a neighborhood U that (in a certain direction v) can be translated an arbitrarily small distance into the interior of the set. Image (a) shows the process for a smooth boundary, image (b) shows the process for a corner.

the boundary. Recall that the set $C^\infty\left(\overline{\Omega}\right)$ is defined in Definition 19.11. Although formally the domain of the functions in $C^k\left(\overline{\Omega}\right)$ is greater than that of the functions in $W^{m,p}(\Omega)$, for $k \geq m$ we will consider $C^k\left(\overline{\Omega}\right)$ to be a subspace of $W^{m,p}(\Omega)$, implicitly assuming that the functions in $C^k\left(\overline{\Omega}\right)$ have been restricted to Ω.

Definition 23.23 *The open set $\Omega \subseteq \mathbb{R}^d$ has the* **segment property** *iff there is a locally finite open cover $\{U_i\}_{i=1}^\infty$ of $\delta\Omega$ and a sequence $\{v_i\}_{i=1}^\infty$ of vectors in $\mathbb{R}^d$ so that if $x \in \overline{\Omega}\cap U_i$ for some $i \in \mathbb{N}$, then for all $t \in (0,1)$ we have $x + tv_i \in \Omega$ (see Figure 61).*

The segment property will allow us to approximate pieces of $W^{m,p}(\Omega)$-functions with shifted versions that are defined on a set that contains the boundary.

Theorem 23.24 *Let $\Omega \subseteq \mathbb{R}^d$ be a bounded open set that has the segment property, let $m \in \mathbb{N}_0$ and let $1 \leq p < \infty$. Then $C^\infty\left(\overline{\Omega}\right) \cap W^{m,p}(\Omega)$ is dense in $W^{m,p}(\Omega)$.*

Proof. For each $x \in \delta\Omega$, let U be a neighborhood of x and $v \in \mathbb{R}^d$ be as in the definition of the segment property. Without loss of generality we can assume that v is so that there is another neighborhood $W \subseteq U$ of x so that $\overline{(W\cap\delta\Omega)} - tv \subseteq U$ for all $t \in (0,1)$. Then for all $t \in (0,1)$ we infer $\left(\overline{(W\cap\delta\Omega)} - tv\right)\cap\overline{\Omega} = \emptyset$, because otherwise there is a $z = y - tv \in \overline{\Omega}$ with $y \in \delta\Omega$, $t \in (0,1)$ and then (via the segment property) we would obtain $y = z + tv \in \Omega$, a contradiction.

By the above, for each $x \in \overline{\Omega}$ there is an open neighborhood W_x of x and a vector $k_x \in \mathbb{R}^d$ so that if $W_x \cap \delta\Omega \neq \emptyset$ then $\left[\overline{(W_x\cap\delta\Omega)} + tk_x\right]\cap\overline{\Omega} = \emptyset$ for all $t \in (0,1)$. Because $\overline{\Omega}$ is compact we can find a finite open cover with these properties. Call this cover $\{W_i\}_{i=1}^n$ and the corresponding vectors $\{k_i\}_{i=1}^n$.

Let $\{\psi_i\}_{i=1}^n$ be a C_0^∞-partition of unity on $\overline{\Omega}$ subordinate to $\{W_i\}_{i=1}^n$. That is, $\sum_{i=1}^n \psi_i(x) = 1$ for all $x \in \overline{\Omega}$, $\psi_i \in C_0^\infty(W_i)$ and $0 \leq \psi_i \leq 1$. This partition of unity is constructed as in Theorem 16.115 using functions that are in $C_0^\infty\left(\overline{\Omega}\right)$ (these functions are compactly supported *in a neighborhood* of $\overline{\Omega}$).

Let $f \in W^{m,p}(\Omega)$. By Exercise 23-13b, each $f_i := f\psi_i$ is in $W^{m,p}(\Omega)$. Now fix an index $i \in \{1, \ldots, n\}$ and a $t \in (0,1)$ and let H_i^t be a neighborhood of $\overline{\Omega}$ so that $\left[\overline{(\delta\Omega \cap W_i)} + tk_i\right] \cap H_i^t = \emptyset$. Define $g_i^t(x) := f_i(x - tk_i)$. Then $g_i^t \in W^{m,p}\left(H_i^t\right)$ with $D^\alpha g_i^t(x) = D^\alpha f_i(x - tk_i)$ for all $x \in H_i^t$ (see Exercise 23-19). In particular, $g_i^t \in W^{m,p}(\Omega)$ because $\Omega \subseteq H_i^t$ and the equality for the derivatives also holds in Ω.

By Exercise 18-16, for all $\alpha \in \mathbb{N}_0^d$ with $|\alpha| \leq m$ the limit of $\left\| D^\alpha g_i^t - D^\alpha f_i \right\|_{L^p(\Omega)}$ is zero as $t \to 0$. Hence, $\lim_{t\to 0} \left\| g_i^t - f_i \right\|_{W^{m,p}(\Omega)} = 0$. By considering $f = \sum_{i=1}^n f_i$ and $g := \sum_{i=1}^n g_i^t$ for small enough t we conclude that for every $\varepsilon > 0$ there is a neighborhood $H := \bigcap_{i=1}^n H_i^t$ of $\overline{\Omega}$ and a $g \in W^{m,p}(H)$ so that $\|g - f\|_{W^{m,p}(\Omega)} < \frac{\varepsilon}{2}$. By Theorem 23.21, there is a function $h \in C^\infty(H) \cap W^{m,p}(H)$ with $\|h - g\|_{W^{m,p}(H)} < \frac{\varepsilon}{2}$, which means $\|h - f\|_{W^{m,p}(\Omega)} < \varepsilon$. Hence, the space $C^\infty\left(\overline{\Omega}\right) \cap W^{m,p}(\Omega)$ is dense in $W^{m,p}(\Omega)$. ■

Theorem 23.25 Trace Theorem. *Let $\Omega \subseteq \mathbb{R}^d$ be open so that $\overline{\Omega}$ is a d-dimensional compact connected embedded oriented manifold with boundary or corners and let $1 \leq p < \infty$. Then there is a continuous linear operator $\gamma : W^{1,p}(\Omega) \to L^p(\delta\Omega)$ so that for all functions $u \in C^1\left(\overline{\Omega}\right) \cap W^{1,p}(\Omega)$ we have $\gamma\left(u|_\Omega\right) = u|_{\delta\Omega}$.*

Proof. First note that by Theorem 17.13 any continuous linear function on a dense subspace can be extended to a continuous linear function on the whole space. Thus we are done if we can show that the function $\gamma : C^1\left(\overline{\Omega}\right) \cap W^{1,p}(\Omega) \to L^p(\delta\Omega)$ defined by $\gamma\left(u|_\Omega\right) := u|_{\delta\Omega}$ is continuous. (It obviously is linear.)

By Exercise 23-20, Ω has the segment property. Let $\{\varphi_i\}_{i=1}^n$ be a partition of unity for $\overline{\Omega}$ so that for each φ_i with $\text{supp}(\varphi_i) \cap \delta\Omega \neq \emptyset$ there is a $v_i \in \mathbb{R}^d$ and a neighborhood $N_i \subseteq \delta\Omega$ of $\text{supp}(\varphi_i) \cap \delta\Omega$ so that for some $\varepsilon_i > 0$ and all unit normal vectors $n(x)$ of $\delta\Omega$ at $x \in N_i$ we have $\left|\langle v_i, n(x)\rangle\right| > \varepsilon_i$, so that for all $t \in (0,1]$ the containment $N_i + tv_i \subseteq \Omega$ holds and so that $(N_i + v_i) \cap \text{supp}(\varphi_i) = \emptyset$. For $u \in C^1\left(\overline{\Omega}\right) \cap W^{1,p}(\Omega)$ we infer $\left\| u|_{\delta\Omega} \right\|_{L^p(\delta\Omega)} = \left\| u|_{\delta\Omega} \sum_{i=1}^n \varphi_i|_{\delta\Omega} \right\|_{L^p(\delta\Omega)} \leq \sum_{i=1}^n \left\| (u\varphi_i)|_{\delta\Omega} \right\|_{L^p(\delta\Omega)}$. Therefore, the result is proved if we can prove that there is a $c \geq 0$ so that for all $i = 1, \ldots, n$ the inequality $\left\| (u\varphi_i)|_{\delta\Omega} \right\|_{L^p(\delta\Omega)} \leq c\|u\|_{W^{1,p}(\Omega)}$ holds.

So let $\varphi \in \{\varphi_1, \ldots, \varphi_n\}$ be so that $\text{supp}(\varphi) \cap \delta\Omega \neq \emptyset$, let N be the corresponding neighborhood N_i, let v be the corresponding vector v_i and let ε be the ε_i from above. Moreover, let $S \subseteq \mathbb{R}^{d-1}$ be open and let $b : S \to \delta\Omega$ be a parametrization of N. Then for each $z \in S$ we obtain

$$(u\varphi)(b(z)) = \int_{-1}^{0} \frac{d}{dt}\left[(u\varphi)(b(z) - tv)\right] dt = \int_{-1}^{0} \left\langle \nabla(u\varphi)(b(z) - tv), -v \right\rangle dt$$

$$= \int_{-1}^{0} \left\langle \varphi \nabla u\big(b(z) - tv\big), -v \right\rangle dt + \int_{-1}^{0} \left\langle u \nabla \varphi\big(b(z) - tv\big), -v \right\rangle dt.$$

The integral of the pth power of the second term over N can be bounded above as follows. Recall that if N contains points in corners, then the integral over N is a sum of integrals over smooth pieces, so without loss of generality we can assume that N does not intersect any corners. Also recall that the surface volume element for $\delta\Omega$ at $b(z)$ is $\det\left(n(b(z)), \frac{\partial b}{\partial z_1}(z), \ldots, \frac{\partial b}{\partial z_{d-1}}(z)\right)$ and note that the vector v can be represented as $v = \langle v, n(b(z))\rangle n(b(z)) + \big(v - \langle v, n(b(z))\rangle n(b(z))\big)$, where the term in parentheses is in the span of the vectors $\frac{\partial b}{\partial z_1}(z), \ldots, \frac{\partial b}{\partial z_{d-1}}(z)$. This means that $\det\left(n(b(z)), \frac{\partial b}{\partial z_1}(z), \ldots, \frac{\partial b}{\partial z_{d-1}}(z)\right) = \frac{1}{\langle v, n(b(z))\rangle} \det\left(v, \frac{\partial b}{\partial z_1}(z), \ldots, \frac{\partial b}{\partial z_{d-1}}(z)\right)$.

Hence,

$$\int_N \left| \int_{-1}^{0} \left\langle u \nabla \varphi\big(b(z) - tv\big), -v \right\rangle dt \right|^p d\omega_V^{\delta\Omega}$$

First use Cauchy-Schwarz in $\mathbb{R}^d$, then use Hölder.

$$\leq \int_N \left| \int_{-1}^{0} \left| u\big(b(z) - tv\big)\right| \left\| \nabla \varphi\big(b(z) - tv\big)\right\| \|v\| \, dt \right|^p d\omega_V^{\delta\Omega}$$

$$\leq \int_N \int_{-1}^{0} \left| u\big(b(z) - tv\big)\right|^p dt \left\| \nabla \varphi\big(b(z) - tv\big)\right\|_{L^q(-1,0)}^p d\omega_V^{\delta\Omega} \|v\|^p$$

$$\leq \underbrace{\left(\|v\|^p \max\left\{ \left\| \nabla \varphi\big(b(z) - tv\big)\right\|_{L^q(-1,0)}^p : z \in S \right\} \right)}_{=: \varepsilon c_{\varphi,v}} \times$$

$$\times \int_S \int_{-1}^{0} \left| u\big(b(z) - tv\big)\right|^p dt \det\left(n(b(z)), \frac{\partial b}{\partial z_1}(z), \ldots, \frac{\partial b}{\partial z_{d-1}}(z)\right) dz$$

$$\leq \varepsilon c_{\varphi,v} \int_S \int_{-1}^{0} \left| u\big(b(z) - tv\big)\right|^p \frac{1}{\varepsilon} \left| \det\left(v, \frac{\partial b}{\partial z_1}(z), \ldots, \frac{\partial b}{\partial z_{d-1}}(z)\right)\right| dt \, dz$$

$$= c_{\varphi,v} \int_{\{x = b + tv \in \Omega : b \in N, t \in (0,1)\}} |u(x)|^p \, d\lambda \leq c_{\varphi,v} \|u\|_{W^{1,p}(\Omega)}^p.$$

The term in front of the norm of u does not depend on u. We estimate the first term similarly and adding the estimates for the all the functions in our finite partition of unity gives the desired estimate. ■

Theorem 23.25 shows that the sensible way to assign boundary values to $W^{1,p}(\Omega)$-functions that have boundary values in the conventional sense (limits as we approach the boundary) can be extended to all of $W^{1,p}(\Omega)$. Hence, we can assign the following notation.

Definition 23.26 *An open set $\Omega \subseteq \mathbb{R}^d$ that has the segment property, for which there is an operator $\gamma : W^{1,p}(\Omega) \to L^p(\delta\Omega)$ as in the Trace Theorem and for which the Divergence Theorem holds will be called a* **Trace Theorem Domain**. *If Ω is a Trace Theorem Domain, then for $u \in W^{1,p}(\Omega)$ the function $\gamma(u)$ is called the* **generalized boundary value** *of u.*

The name "Trace Theorem Domain," which is not standard nomenclature, is introduced because the Trace Theorem can be proved with a variety of different hypotheses. The results for Trace Theorem Domains that we present in the following are valid for any domain with the segment property on which the Trace Theorem and the Divergence Theorem hold. In this fashion, the reader can easily adapt the results to domains that may be defined differently, as long as the two theorems hold for the domain.

Note that if we try to solve partial differential equations with systems of equations like (H-PDE), the Hilbert space computations from Section 23.1 can affect the boundary values. Hence, we will consider the problem (H-PDE) in spaces of functions that are zero on the boundary as defined below.

Definition 23.27 *Let the set $\Omega \subseteq \mathbb{R}^d$ be a Trace Theorem Domain, let $m \in \mathbb{N}_0$ and let $1 \leq p < \infty$. We define the* **Sobolev spaces** *of functions that are zero on the boundary as $W_0^{m,p}(\Omega) := \{f \in W^{m,p}(\Omega) : \gamma(u)=0\}$ and $H_0^m(\Omega) := \{f \in H^m(\Omega) : \gamma(u)=0\}$.*

To solve boundary value problems $Du = f$ with $u|_{\delta\Omega} = g|_{\delta\Omega}$ for some function $g \in H^m(\Omega)$, we simply solve the problem $Dw = f - Dg$ in $H_0^m(\Omega)$. (The boundary values of $f - Dg$ must the zero or the problem is not solvable.) If w solves this transformed problem, then, because $w|_{\delta\Omega} = 0$, the function $u := w + g$ solves the original problem. *Therefore, from here on we will assume that the functions we seek vanish on the boundary of the domain.* Boundary value problems involving derivatives are handled with similar transformations, using spaces in which the weak derivatives vanish on the boundary.

Exercises

23-14 Prove that every Sobolev space $W^{m,p}(\Omega)$ is a vector space

23-15. Prove Theorem 23.19. (Norm properties and completeness.)

23-16 Prove Theorem 23.20.

23-17. Prove Theorem 23.21 for $p = 1$

23-18 Prove Theorem 23.22.

23-19. Let $\Omega \subseteq \mathbb{R}^d$ be a bounded open set, let $U \subseteq \mathbb{R}^d$ be open and let the vector $v \in \mathbb{R}^d$ be so that $\left[\overline{(\delta\Omega \cap U)} + v\right] \cap \overline{\Omega} = \emptyset$. Prove that if $f \in W^{m,p}(\Omega)$ is compactly supported in U and H is a neighborhood of $\overline{\Omega}$ with $\left[\overline{(\delta\Omega \cap U)} + v\right] \cap H = \emptyset$, then $g(x) := f(x - v)$ defines a function in $W^{m,p}(H)$ so that $D^\alpha g(x) = D^\alpha f(x - v)$ for all $x \in H$.

23-20 Sets with the segment property

(a) Let $\Omega \subseteq \mathbb{R}^d$ be a connected open set so that $\overline{\Omega}$ is a d-dimensional embedded connected compact oriented manifold with boundary. Prove that Ω has the segment property.
Hint. By Exercise 19-5d, we know that $\delta\Omega = \partial\Omega$.

(b) Let $\Omega \subseteq \mathbb{R}^d$ be a connected open set so that $\overline{\Omega}$ is a d-dimensional embedded connected compact oriented manifold with corners Prove that Ω has the segment property

23-21. Let $\Omega \subseteq \mathbb{R}^d$ be a connected open set with the segment property so that $\overline{\Omega}$ is a finite union of d-dimensional embedded connected compact oriented manifolds with corners whose interiors are pairwise disjoint. Prove that Ω is a Trace Theorem Domain

23-22. Let $\Omega \subseteq \mathbb{R}^d$ be open and let $f \in L^p(\Omega)$ and $\alpha, \beta \in \mathbb{R}^d$ be so that (componentwise) $\beta \leq \alpha$ and the weak derivatives $D^\alpha f$ and D^β exist Prove that $D^{\alpha-\beta}\left(D^\beta f\right) = D^\alpha f$.

23-23. Would it be easy or hard to prove a Trace Theorem for $W^{m,p}(\Omega)$?

23-24 **Containment relations of Sobolev spaces.** Let $\Omega \subseteq \mathbb{R}^d$ be open

(a) Prove that if $m \leq n$ and $1 \leq p \leq \infty$, then $W^{n,p}(\Omega) \subseteq W^{m,p}(\Omega)$.

(b) Prove that if $\lambda(\Omega) < \infty$, $m \in \mathbb{N}$ and $1 \leq p \leq q \leq \infty$, then $W^{m,q}(\Omega) \subseteq W^{m,p}(\Omega)$

23-25. Let $\Omega \subseteq \mathbb{R}^d$ be a Trace Theorem Domain. Prove that $C_0^\infty(\Omega)$ is dense in $H_0^1(\Omega)$.
Hint. Use the proof of Theorem 23 24 as guidance, but push the functions "inward" instead of "outward"

23.4 Elliptic Differential Operators

Now it is time to connect the theory developed so far to concrete differential equations. The most frequently investigated differential operators are given below. The reason for the negative sign before the first sum will become clear after Proposition 23.31.

Definition 23.28 *Let $\Omega \subseteq \mathbb{R}^d$ be open and let $a_{ij}, b_k, c \in C^\infty(\Omega)$. Consider the* **differential operator**

$$Du(x) := -\sum_{i,j=1}^{d} a_{ij}(x)\frac{\partial^2}{\partial x_i \partial x_j}u(x) + \sum_{k=1}^{d} b_k(x)\frac{\partial}{\partial x_k}u(x) + c(x)u(x)$$

and let $A(x) := \left(a_{ij}(x)\right)_{\substack{i=1,\ldots,d \\ j=1,\ldots,d}}$.

1. *D is called* **elliptic** *iff A is positive definite at every $x \in \Omega$, that is, for all $x \in \Omega$ and all $z \in \mathbb{R}^d \setminus \{0\}$ the inequality $\langle A(x)z, z\rangle > 0$ holds.*

2. *D is called* **parabolic** *iff A is positive semidefinite at every $x \in \Omega$, that is, for all $x \in \Omega$ and all $z \in \mathbb{R}^d \setminus \{0\}$ the inequality $\langle A(x)z, z\rangle \geq 0$ holds.*

3. *D is called* **hyperbolic** *iff A is indefinite at every $x \in \Omega$, that is, for all $x \in \Omega$ there are $z, z' \in \mathbb{R}^d \setminus \{0\}$ so that $\langle A(x)z, z\rangle > 0$ and $\langle A(x)z', z'\rangle < 0$.*

The naming convention is inspired by geometry. Positive definite 2×2-matrices A define ellipses via $x^T Ax = 1$. In the same fashion, indefinite matrices define hyperbolas and in similar fashion positive semidefinite matrices can be used to define parabolas.

Example 23.29 Some examples of elliptic and nonelliptic differential operators.

1. The negative Laplace operator $-\Delta = -\sum_{j=1}^{d}\frac{\partial^2}{\partial^2 x_j}$ is elliptic.

2. The operator $-k\Delta + \frac{\partial}{\partial t} = -k\left(\sum_{j=1}^{d} \frac{\partial^2}{\partial^2 x_j}\right) + \frac{\partial}{\partial t}$ is parabolic. (The matrix of the a_{ij} has zeroes in the row and the column corresponding to t.)

3. The operator $-k\Delta + \frac{\partial^2}{\partial^2 t} = -k\left(\sum_{j=1}^{d} \frac{\partial^2}{\partial^2 x_j}\right) + \frac{\partial^2}{\partial^2 t}$ is hyperbolic. □

It would be nice if elliptic differential operators D would induce elliptic bilinear forms $\langle D(\cdot), \cdot\rangle$. Unfortunately this is not the case (also see Exercise 23-26). To make some elliptic differential equations accessible to the ideas of Section 23.1, we need to formulate our bilinear form a bit differently, we need to use a stronger property than ellipticity for the operator, and we need to explicitly use that the boundary values are zero. First we rewrite the differential operator D.

Proposition 23.30 *Let* $Du(x) = -\sum_{i,j=1}^{d} a_{ij}(x)\frac{\partial^2 u(x)}{\partial x_i \partial x_j} + \sum_{k=1}^{d} b_k(x)\frac{\partial u(x)}{\partial x_k} + c(x)u(x)$ *be a differential operator as in Definition 23.28. Then, with the scalar product of* $a, b \in \mathbb{R}^d$ *denoted by* $a \cdot b$*,* D *can be rewritten as*

$$\begin{aligned} Du(x) &= -\sum_{i,j=1}^{d} \frac{\partial}{\partial x_i}\left(a_{ij}(x)\frac{\partial}{\partial x_j}u(x)\right) + \sum_{k=1}^{d} \tilde{b}_k(x)\frac{\partial}{\partial x_k}u(x) + c(x)u(x) \\ &= -\nabla\big(A\nabla u\big) + \tilde{b}\cdot\nabla u + cu. \end{aligned}$$

Proof. Exercise 23-27. ■

For operators as in Proposition 23.30, which satisfy $\tilde{b} = 0$, we can now rewrite the bilinear form $\langle D(\cdot), \cdot\rangle$ as a more symmetric entity. This will allow us to associate with the partial differential equation $Du = f$ a system of equations (H-PDE) for which the bilinear form is elliptic. The rewriting in Proposition 23.31 explains why we wanted a negative sign in front of the second derivatives. The results we obtain will be practically useful, because if $D = \Delta$, then clearly $\tilde{b} = 0$.

Proposition 23.31 *Let* $\Omega \subseteq \mathbb{R}^d$ *be a Trace Theorem Domain. Then for all functions* $u, v \in C_0^\infty(\Omega)$ *we have*

$$\int_\Omega \big(-\nabla\big(A\nabla u\big) + cu\big)\, v\, dV = \int_\Omega \big(A\nabla u\big)\cdot\nabla v\, dV + \int_\Omega cuv\, dV.$$

Proof. Because the second term is unaffected, we can concentrate on the first term. By the Divergence Theorem and Exercise 21-6e, we obtain

$$\int_\Omega \nabla\big(A\nabla u\big)v\, dV + \int_\Omega \big(A\nabla u\big)\cdot\nabla v\, dV = \int_\Omega \nabla\big(A(\nabla u)v\big)\, dV = \int_{\delta\Omega} \big(A\nabla u\big)v\cdot d\vec{S},$$

and the latter term is zero because u and v are zero on the boundary of Ω. Hence, $\int_\Omega -\nabla\big(A\nabla u\big)v\, dV = \int_\Omega \big(A\nabla u\big)\cdot\nabla v\, dV.$ ■

We can now summarize the results of the first two sections of this chapter in more concrete terms for partial differential equations.

Definition 23.32 *Let $\Omega \subseteq \mathbb{R}^d$ be a Trace Theorem Domain, let $Du := -\nabla(A\nabla u)+cu$ and let $f \in H_0^1(\Omega)$. Then $u \in H_0^1(\Omega)$ is called a* **weak solution** *of the equation $Du = f$ iff u solves the system of equations (H-PDE) with the bilinear form B defined by $B(u,v) := \int_\Omega (A\nabla u)\cdot\nabla v + cuv\,dV$ and with $F(v) = \int_\Omega fv\,dV$. The equation $\int_\Omega (A\nabla u)\cdot\nabla v + cuv\,dV = \int_\Omega fv\,dV$ for all $v \in H_0^1(\Omega)$ is also called the weak* **variational formulation** *of the equation $-\nabla(A\nabla u) + cu = f$.*

Because $C_0^\infty(\Omega)$ is dense in $H_0^1(\Omega)$ (Exercise 23-25), Proposition 23.31 shows that if $u \in H_0^2(\Omega)$ solves the equation in the regular sense, then it will also be a weak solution.

By Theorem 23.22, the bilinear function in Definition 23.32 is continuous on the space $H_0^1(\Omega)$. To prove that the bilinear function B is elliptic, we need the differential operator to satisfy the property below.

Definition 23.33 *An elliptic differential operator is called* **uniformly elliptic** *iff the there is a constant $C_A > 0$ so that for all $x \in \Omega$ and all $z \in \mathbb{R}^d \setminus \{0\}$ the inequality $\langle A(x)z, z\rangle \geq C_A\|z\|^2$ holds.*

To prove that uniformly elliptic operators induce elliptic bilinear functions, we proceed as follows.

Theorem 23.34 Poincaré-Friedrichs inequality. *Let $\Omega \subseteq \mathbb{R}^d$ be a Trace Theorem Domain that is contained in a cube of side length $C > 0$. Then for all $u \in H_0^1(\Omega)$ the inequalities $\|u\|_{L^2(\Omega)} \leq C\sum_{j=1}^d \left\|\frac{\partial u}{\partial x_j}\right\|_{L^2(\Omega)}$ and $\|u\|_{L^2(\Omega)}^2 \leq C^2\sum_{j=1}^d \left\|\frac{\partial u}{\partial x_j}\right\|_{L^2(\Omega)}^2$ hold.*

Proof. Because $C_0^\infty(\Omega)$ is dense in $H_0^1(\Omega)$ (see Exercise 23-25) it is enough to prove the inequality for functions in $C_0^\infty(\Omega)$. Moreover, without loss of generality we can assume that $\Omega \subseteq [0, C]^d$. Let $u \in C_0^\infty(\Omega)$ and set $u(x) := 0$ for all $x \in [0, C]^d \setminus \Omega$. Then for all $(x_1, \ldots, x_d) \in \Omega$ we infer

$$\int_0^C \cdots \int_0^C \left|u(x_1, x_2, \ldots, x_d)\right|^2 dx_1 \cdots dx_d$$

$$= \int_0^C \cdots \int_0^C \left|u(0, x_2, \ldots, x_d) + \int_0^{x_1} \frac{\partial u}{\partial x_1}(t, x_2, \ldots, x_d)\,dt\right|^2 dx_1 \cdots dx_d$$

The boundary term is zero and we can apply the Cauchy-Schwarz inequality to the integral of the partial derivative.

$$\leq \int_0^C \cdots \int_0^C \int_0^{x_1} 1\,dt \int_0^{x_1} \left|\frac{\partial u}{\partial x_1}(t, x_2, \ldots, x_d)\right|^2 dt\,dx_1 \cdots dx_d$$

$$
\begin{aligned}
&\le \int_0^C \cdots \int_0^C c \int_0^C \left| \frac{\partial u}{\partial x_1}(t, x_2, \ldots, x_d) \right|^2 dt\, dx_1 \cdots dx_d \\
&= C \int_0^C \cdots \int_0^C \left(\int_0^C \int_0^C \left| \frac{\partial u}{\partial x_1}(t, x_2, \ldots, x_d) \right|^2 dx_1\, dt \right) dx_2 \cdots dx_d
\end{aligned}
$$

Note that the integrand does not depend on x_1.

$$
\begin{aligned}
&= C^2 \int_0^C \cdots \int_0^C \int_0^C \left| \frac{\partial u}{\partial x_1}(t, x_2, \ldots, x_d) \right|^2 dt\, dx_2 \cdots dx_d \\
&= C^2 \left\| \frac{\partial u}{\partial x_1} \right\|_{L^2(\Omega)}^2 \le C^2 \sum_{j=1}^d \left\| \frac{\partial u}{\partial x_j} \right\|_{L^2(\Omega)}^2 .
\end{aligned}
$$

and moreover $\|u\|_{L^2(\Omega)} \le C \left\| \frac{\partial u}{\partial x_1} \right\|_{L^2(\Omega)} \le C \sum_{j=1}^d \left\| \frac{\partial u}{\partial x_j} \right\|_{L^2(\Omega)}$. ■

Theorem 23.35 *Let $\Omega \subseteq \mathbb{R}^d$ be a bounded Trace Theorem Domain, let the differential operator $Du := -\nabla(A\nabla u) + cu$ be uniformly elliptic and let $f \in H_0^1(\Omega)$. Then the equation $Du = f$ has a unique weak solution u. Moreover, if $\{V_n\}_{n=1}^\infty$ is a sequence of subspaces so that $\lim_{n\to\infty} \operatorname{dist}(w, V_n) = 0$ for all $w \in H_0^1(\Omega)$, then with u_{V_n} being the solution of the system of equations (V_n-PDE) we have $\lim_{n\to\infty} \|u - u_{V_n}\| = 0$.*

Proof. By Theorem 23.22, the bilinear form $B(u, v) := \int_\Omega (A\nabla u) \cdot \nabla v + cuv\, dV$ from Definition 23.32 is continuous. Moreover, with C being the side length of a cube that contains Ω, by the Poincaré-Friedrichs inequality we obtain the following for all $v \in H_0^1(\Omega)$.

$$
\begin{aligned}
B(v, v) &= \int_\Omega (A\nabla v) \cdot \nabla v + cv^2\, dV \ge \int_\Omega C_A \nabla v \cdot \nabla v\, dV \\
&= \frac{C_A}{1 + C^2} \left(\sum_{j=1}^d \left\| \frac{\partial v}{\partial x_j} \right\|_{L^2(\Omega)}^2 + C^2 \sum_{j=1}^d \left\| \frac{\partial v}{\partial x_j} \right\|_{L^2(\Omega)}^2 \right) \\
&\ge \frac{C_A}{1 + C^2} \left(\sum_{j=1}^d \left\| \frac{\partial v}{\partial x_j} \right\|_{L^2(\Omega)}^2 + \|v\|_{L^2(\Omega)}^2 \right) = \frac{C_A}{1 + C^2} \|v\|_{H_0^1(\Omega)}^2 .
\end{aligned}
$$

Therefore B is elliptic and the result follows from the Lax-Milgram Lemma and Céa's Lemma. ■

Theorem 23.35 and part 1 of Example 23.29 tell us that the Poisson equation can be solved with the finite element method. That is, the potentials of static electrical fields and the temperature/density distributions of the steady state of heat/diffusion phenomena can be approximated by solving large systems of linear equations. Unfortunately, parts 2 and 3 of Example 23.29 show that we cannot directly apply the results developed so far to the heat and wave equations. There are ways to apply the finite element method to parabolic and hyperbolic equations. For our introduction, we shall be satisfied having proved that the method can be applied to certain elliptic equations, including the Laplace and Poisson equations.

Exercises

23-26 Consider the differential operator $D := -\frac{d^2}{dt^2} + \frac{d}{dt}$ on $H^2(-1, 1)$ Prove that D is uniformly elliptic, but that the bilinear form $\langle D(\cdot), \cdot \rangle$ is not elliptic. Then explain why this is not a contradiction to what was done in the proof of Theorem 23.35.

23-27. Prove Proposition 23.30.

23-28. Prove that there is no reversal of the Poincaré-Friedrichs inequality, that is, prove that there is *no* $c > 0$ so that for all $u \in H_0^1(\Omega)$ we have $\|u\|_{L^2(\Omega)} \geq c \sum_{j=1}^{d} \left\| \frac{\partial u}{\partial x_j} \right\|_{L^2(\Omega)}$.

Hint. Consider nonnegative functions in $C_0^\infty(-1, 1)$ whose maximum value is 1 and whose L^2 norms go to zero

23.5 Finite Elements

So far, we have established that an elliptic partial differential equation $Du = f$ with uniformly elliptic left side of the form $Du := -\nabla(A\nabla u) + cu$ has a unique weak solution. Moreover, with the right sequence $\{V_n\}_{n=1}^{\infty}$ of finite dimensional subspaces, the weak solution can be approximated with the solutions of the corresponding systems of equations (V_n-PDE). This simplification is significant, because the infinite dimensional problem of solving the partial differential equation is now reduced to solving the finite dimensional systems of linear equations given by (V_n-PDE). From a theoretical point of view, all we need are the right spaces V_n and we will get an approximation of any given quality. From a practical point of view, we therefore need to address how to construct such spaces. We will build the approximation of the solution on small subsets of Ω. These subsets and the functions on them are called finite elements.

Definition 23.36 *A* **finite element** *in* $\mathbb{R}^d$ *is a triple* $(\tau, P_\tau, \Sigma_\tau)$ *such that the following hold.*

1. $\tau \subseteq \mathbb{R}^d$ *is compact and* $\tau^\circ \neq \emptyset$ *is a connected Trace Theorem Domain.*
2. P_τ *is a finite dimensional subspace of* $C^m(\tau)$.
3. *With* $n := \dim(P_\tau)$ *the set* Σ_τ *consists of linearly independent continuous linear functionals* $B_1, \ldots, B_n : C^m(\tau) \to \mathbb{R}$ *so that for all* $\alpha_1, \ldots, \alpha_n \in \mathbb{R}$ *there is a* $p \in P_\tau$ *so that* $B_i(p) = \alpha_i$.

The functionals B_i *are called the* **degrees of freedom** *of the finite element. Functions* $p_1, \ldots, p_n$ *so that* $B_i(p_j) = \begin{cases} 1; & \text{if } i = j, \\ 0; & \text{if } i \neq j, \end{cases}$ *are called the* **base functions** *of the finite element. Finite elements are also often denoted by* τ *only.*

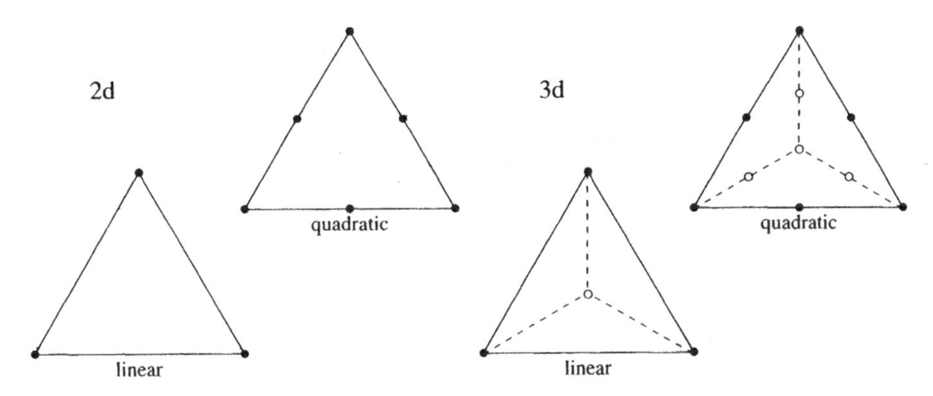

Figure 62: Left to right. A 2-simplex with the evaluation points and a base function for linear Lagrangian finite elements (dotted), a 2-simplex with the evaluation points and a base function for quadratic Lagrangian finite elements (dotted), a 3-simplex with the evaluation points for linear Lagrangian finite elements, and a 3-simplex with the evaluation points for quadratic Lagrangian finite elements.

Basically, the definition of a finite element provides a set τ on which we can build an approximation to the solution, and a space of functions P_τ with which to build the approximation. The demand that P_τ is finite dimensional assures that our space is not too large. The degrees of freedom in Σ_τ assure that the space is large enough to reach a certain set of functions. We will see below that the degrees of freedom also are used merge the pieces into a function on all of Ω.

Example 23.37 Some simple finite elements. Let $P^u(\Omega)$ be the space of **polynomials** $p : \Omega \to \mathbb{R}$ of degree at most u. For points $a_1, \ldots, a_{d+1} \in \mathbb{R}^d$, so that the set $\{a_1 - a_{d+1}, \ldots, a_d - a_{d+1}\}$ is linearly independent, define the d-**simplex** spanned by $a_1, \ldots, a_{d+1}$ to be $S := \left\{ \sum_{j=1}^{d+1} \lambda_j a_j : \lambda_1, \ldots, \lambda_{d+1} \in [0,1], \sum_{j=1}^{d+1} \lambda_j = 1 \right\}$. Geometrically, the linear independence of $\{a_1 - a_{d+1}, \ldots, a_d - a_{d+1}\}$ assures that the points $a_1, \ldots, a_{d+1}$, also called the **vertices** of S, are not all in the same hyperplane of $\mathbb{R}^d$. Figure 62 shows some simplices. Some properties of simplices are highlighted in Exercise 23-29.

1. **Linear Lagrangian finite elements** in $\mathbb{R}^d$. For a simplex τ in $\mathbb{R}^d$ with vertices $a_1, \ldots, a_{d+1}$, let $P_\tau := P^1(\tau)$. Then $\dim(P_\tau) = d + 1$. For the degrees of freedom, we choose $B_j(p) := p(a_j)$ for $j = 1, \ldots, d+1$. For the base functions, recall that $\{a_1 - a_{d+1}, \ldots, a_d - a_{d+1}\}$ was a base. Therefore, for each $y \in \mathbb{R}^d$ there are unique $y_1, \ldots, y_d$ so that $y = \sum_{j=1}^{d} y_j(a_j - a_{d+1})$. For $j = 1, \ldots, d$ we define $p_j(x) := (x - a_{d+1})_j$ (the jth coordinate of x with respect to the aforementioned base) and for $j = d+1$ we define $p_{d+1}(x) := 1 - \sum_{j=1}^{d} p_j(x)$. Note

that a simplex has exactly the right number of vertices to define polynomials of degree 1 by specifying the values of the polynomial at the vertices.

2. **Quadratic Lagrangian finite elements** in $\mathbb{R}^d$. For a simplex τ in $\mathbb{R}^d$ with vertices $a_1, \ldots, a_{d+1}$, let $P_\tau := P^2(\tau)$. Then, counting second order terms first and taking symmetry into account, $\dim(P_\tau) = \frac{d}{2}(d+1) + d + 1 = \frac{(d+1)(d+2)}{2}$. The simplex τ has $d+1$ vertices. For $k = 1, \ldots, d$ and each vertex a_{k+1}, there are k line segments connecting a_{k+1} to $a_1, \ldots, a_k$. The vertices of τ and the centers of these segments give us $(d+1) + \sum_{k=1}^{d} k = \frac{(d+1)(d+2)}{2}$ points $a_1, \ldots, a_{\frac{(d+1)(d+2)}{2}}$. We choose $B_j(p) := p(a_j)$ as the degrees of freedom. The coefficients of the base functions $p_1, \ldots, p_{\frac{(d+1)(d+2)}{2}}$ are obtained by solving the system of equations that is implicitly given in the definition of finite elements for the coefficients of each p_j. Exercise 23-30 gives an impression of the computations.

3. **Cubic Hermitian finite elements** in $\mathbb{R}^d$. For a simplex τ in $\mathbb{R}^d$ with vertices $a_1, \ldots, a_{d+1}$, let $P_\tau := P^3(\tau)$. Then, counting first the third powers of the coordinates, then cubic terms with all factors distinct and then the remaining terms, taking symmetry into account for the cubic summands,

$$\dim(P_\tau) = d + \binom{d}{3} + \frac{1}{3}\left(d^3 - d - d(d-1)(d-2)\right) + \frac{(d+1)(d+2)}{2}.$$

Regarding finite elements, we only consider two dimensions. For $d = 2$, we obtain $\dim(P_\tau) = 10$. Regarding the degrees of freedom note that evaluation of the polynomial at the vertices and the center plus evaluation of the partial derivatives $\frac{\partial p}{\partial x}$ and $\frac{\partial p}{\partial y}$ at the vertices yields 10 equations for the coefficients of the polynomial. These equations can be used to determine the base functions, so the above mentioned evaluations can be chosen as the degrees of freedom. □

Definition 23.38 *A finite element is called* **Lagrangian** *iff the degrees of freedom consist of evaluation operators that evaluate the function at points. A finite element is called* **Hermitian** *iff the degrees of freedom consist of evaluation operators that evaluate the function and its directional derivatives.*

With finite elements available to locally approximate the solution, we need to determine how to approximate the solution overall. This is done by partitioning Ω into subsets on which we have finite elements.

Definition 23.39 *Let $\Omega \subseteq \mathbb{R}^d$ be a bounded set. A set T of subsets of $\overline{\Omega}$ is called a* **triangulation** *(also see Figure 63) of Ω iff the following hold.*

1. All sets $\tau \in T$ are closed and each τ° is a nonempty Trace Theorem Domain.

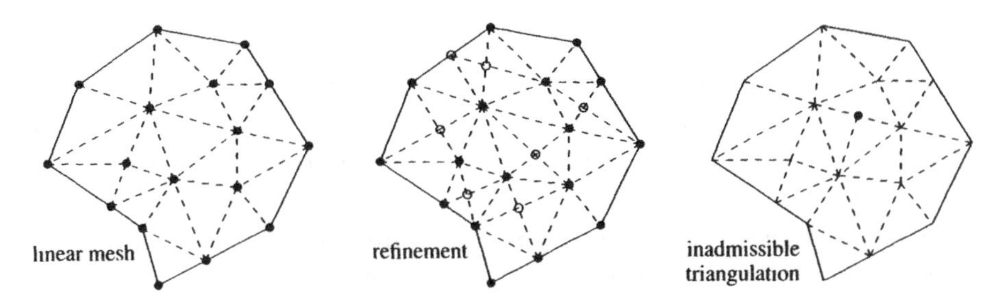

Figure 63: An admissible triangulation with nodes for linear Lagrangian finite elements marked, a refinement of the admissible triangulation with new nodes marked with unfilled circles and an inadmissible triangulation with the "hanging node" marked.

2. $\overline{\Omega} = \bigcup_{\tau \in T} \tau.$

3. *For all distinct* $\tau_1, \tau_2 \in T$ *we have* $\tau_1^\circ \cap \tau_2^\circ = \emptyset$.

Moreover, similar to partitions of intervals, we define $\|T\| := \sup\{\text{diam}(\tau) : \tau \in T\}$ *and we say that the triangulation R is a* **refinement** *of the triangulation T iff each element of R is contained in an element of T and each element of T is the union of finitely many elements of R.*

To construct functions on Ω from finite elements, we must merge the functions on different finite elements so that the resulting function is at least in $H^1(\Omega)$. For the remainder of this chapter, we will focus on Lagrangian finite elements.

Definition 23.40 *Let* $\Omega \subseteq \mathbb{R}^d$ *be an open, connected polyhedron and let T be a triangulation into Lagrangian finite elements using simplices. (Formally, T is triangulated by sets that are themselves parts of Lagrangian finite elements, but this is quite cumbersome to state. So we assume that "triangulate into finite elements" says just that.) Then T is called an* **admissible triangulation** *of* Ω *iff every face of a* $\tau_1 \in T$ *is also a face of exactly one* $\tau_2 \in T$ *or it is a part of* $\delta\Omega$. *Two elements of T that share a face are also called* **neighbors** *of each other. The set of points where the degrees of freedom of each* τ *are evaluated is called the set of* **nodes** *of the elements of T.*

The finite element space is now the space $\prod_{\tau \in T} P_\tau$ of $|T|$-tuples of P_τ functions on the finite elements so that any two functions that share a node agree at their common nodes. These $|T|$-tuples need not turn into functions, because two functions in P_{τ_1} and P_{τ_2} may be equal at their common nodes and still different somewhere else on the shared part $\delta\tau_1 \cap \delta\tau_2$ of their boundary.

Definition 23.41 *Let* $\Omega \subseteq \mathbb{R}^d$ *be an open, connected polyhedron and let T be a triangulation into Lagrangian finite elements. Let N be the set of nodes of the elements of T. For each* $b \in N$, *let* $T(b)$ *be the set of all finite elements* $\tau \in T$ *so that b is a* *node of τ. For any node b of τ, let $B_{b,\tau}$ be the degree of freedom that evaluates each P_τ-function at b. The* **finite element space** *X is defined to be*

$$X = \left\{ (v_\tau)_{\tau\in T} \in \prod_{\tau\in T} P_\tau : \left(\forall b \in N : \forall \tau_1, \tau_2 \in T(b) : B_{b,\tau_1}(v_{\tau_1}) = B_{b,\tau_2}(v_{\tau_2})\right)\right\}.$$

If for all $v \in X$ and for all neighboring τ_1, τ_2 we have $v_{\tau_1}|_{\delta\tau_1\cap\delta\tau_2} = v_{\tau_2}|_{\delta\tau_1\cap\delta\tau_2}$, then v can be considered to be a function on $\overline{\Omega}$ and we also write

$$\begin{aligned} X = \Big\{ v : \overline{\Omega} \to \mathbb{R} : \Big(\forall \tau \in T : v|_\tau \in P_\tau \text{ and} \\ \forall b \in N : \forall \tau_1, \tau_2 \in T(b) : B_{b,\tau_1}(v|_{\tau_1}) = B_{b,\tau_2}(v|_{\tau_2})\Big)\Big\}. \end{aligned}$$

Clearly, every finite element space is finite dimensional.

Example 23.42 For the linear and quadratic Lagrangian finite elements introduced in Example 23.37, equality of the elements at the nodes implies equality of the elements on the boundaries of the simplices. Therefore the finite element spaces associated with linear and quadratic Lagrangian finite elements are subspaces of $C^0\left(\overline{\Omega}\right)$. □

Finally, Proposition 23.44 below shows that if the triangulation and the finite elements are chosen appropriately, the associated finite element space is a finite dimensional subspace of a Sobolev space. Moreover, Theorem 23.45 shows that if we choose an appropriate sequence of such spaces, then the Ritz-Galerkin approximations of the solution of $Du = f$ converge to the actual (weak) solution.

Lemma 23.43 Green's Theorem *for H^1 functions. Let $\Omega \subseteq \mathbb{R}^d$ be a Trace Theorem Domain. Then $\int_\Omega \frac{\partial u}{\partial x_i} v \, d\lambda = -\int_\Omega u \frac{\partial v}{\partial x_i} \, d\lambda + \int_{\delta\Omega} u v e_i \cdot d\vec{S}$ for all $u, v \in H^1(\Omega)$, where formally the values of u and v on the boundary are given by $\gamma(u)$ and $\gamma(v)$ with γ as in Theorem 23.25 and the partial derivatives are weak partial derivatives.*

Proof. For all $u, v \in C^1\left(\overline{\Omega}\right)$ we obtain via the Divergence Theorem

$$\begin{aligned} \int_\Omega \frac{\partial u}{\partial x_i} v \, d\lambda + \int_\Omega u \frac{\partial v}{\partial x_i} \, d\lambda &= \int_\Omega \frac{\partial u}{\partial x_i} v + u \frac{\partial v}{\partial x_i} \, d\lambda = \int_\Omega \frac{\partial}{\partial x_i}(uv) \, d\lambda \\ &= \int_\Omega \operatorname{div}(u v e_i) \, d\lambda = \int_{\delta\Omega} u v e_i \cdot d\vec{S}. \end{aligned}$$

Now let $u, v \in H^1(\Omega)$. Because $C^1\left(\overline{\Omega}\right)$ is dense in $H^1(\Omega)$, we can choose sequences $\{u_n\}_{n=1}^\infty$ and $\{v_n\}_{n=1}^\infty$ in $C^1\left(\overline{\Omega}\right)$ with $\|u - u_n\|_{H^1} \to 0$ and $\|v - v_n\|_{H^1} \to 0$ as $n \to \infty$. Because

$$\begin{aligned} &\left| \int_\Omega \frac{\partial u}{\partial x_i} v \, d\lambda - \int_\Omega \frac{\partial u_n}{\partial x_i} v_n \, d\lambda \right| \\ \leq\ &\left| \int_\Omega \frac{\partial u}{\partial x_i} v \, d\lambda - \int_\Omega \frac{\partial u_n}{\partial x_i} v \, d\lambda \right| + \left| \int_\Omega \frac{\partial u_n}{\partial x_i} v \, d\lambda - \int_\Omega \frac{\partial u_n}{\partial x_i} v_n \, d\lambda \right| \\ \leq\ &\int_\Omega \left| \left(\frac{\partial u}{\partial x_i} - \frac{\partial u_n}{\partial x_i} \right) v \right| \, d\lambda + \int_\Omega \left| \frac{\partial u_n}{\partial x_i} (v - v_n) \right| \, d\lambda \\ \leq\ &\|u - u_n\|_{H^1} \|v\|_{H^1} + \|u_n\|_{H^1} \|v - v_n\|_{H^1}, \end{aligned}$$

we conclude $\lim_{n\to\infty}\int_{\Omega}\frac{\partial u_n}{\partial x_i}v_n\,d\lambda=\int_{\Omega}\frac{\partial u}{\partial x_i}v\,d\lambda$. Similar limiting statements hold for the other terms (use continuity of γ for the surface integral). Therefore the claimed equality holds for all $u, v\in H^1(\Omega)$. ■

For the remainder of this section, we will work with compact polyhedra. Note that by Exercises 23-21 and 23-29d (the interiors of) compact polyhedra are Trace Theorem Domains.

Proposition 23.44 *Let $\Omega\subseteq\mathbb{R}^d$ be an open, connected polyhedron, let T be an admissible triangulation into Lagrangian finite elements and let X be the associated finite element space If $P_\tau\subseteq H^1(\tau^\circ)$ for all $\tau\in T$ and $X\subseteq C^0\left(\overline{\Omega}\right)$, then $X\subseteq H^1(\Omega)$.*

Proof. Let $f\in X$ and for $\tau\in T$, let $i\in\{1,\ldots,d\}$ and let $D_\tau^{(i)}f$ be the weak partial derivative of $f|_{\tau^\circ}$ in the direction of e_i. We claim that $D^{(i)}f:=\sum_{\tau\in T}D_\tau^{(i)}f$ (with the $D_\tau^{(i)}$ being zero outside τ°) is the weak partial derivative of f in the direction of e_i. To prove this claim, let $g\in C_0^\infty(\Omega)$. By Lemma 23.43, we obtain the following.

$$\begin{aligned}\int_{\Omega}fD^{(i)}g\,d\lambda &= \sum_{\tau\in T}\int_{\tau}f|_{\tau}D^{(i)}g\,d\lambda=\sum_{\tau\in T}-\int_{\tau}D_\tau^{(i)}f\,g\,d\lambda+\int_{\delta\tau}f|_{\tau}ge_i\cdot d\vec{S}\\ &= -\int_{\Omega}D^{(i)}f\,g\,d\lambda+\sum_{\tau\in T}\int_{\delta\tau}f|_{\tau}ge_i\cdot d\vec{S}=-\int_{\Omega}D^{(i)}f\,g\,d\lambda,\end{aligned}$$

where the sum of the boundary terms vanishes because g is zero on $\delta\Omega$ and all interior boundary terms occur exactly twice and with opposite signs. Because the function $g\in C_0^\infty(\Omega)$ was arbitrary, $D^{(i)}f$ is the weak i^{th} partial derivative of f. Because $f|_{\tau^\circ}\in H^1(\tau^\circ)$, for each $\tau\in T$ we have that $D_\tau^{(i)}f\in L^2(\Omega)$. Thus $D^{(i)}f\in L^2(\Omega)$. Because $i\in\{1,\ldots,d\}$ was arbitrary, all weak first partial derivatives of f exist and are in $L^2(\Omega)$. Because $f\in L^2(\Omega)$, too, we infer that $f\in H^1(\Omega)$. Because $f\in X$ was arbitrary this establishes the claim. ■

Theorem 23.45 *Let $\Omega\subseteq\mathbb{R}$ be the interior of a compact connected polyhedron, let $Du:=-\nabla(A\nabla u)+cu$ be a uniformly elliptic differential operator, let $f\in H^1(\Omega)$ and let $\{S_n\}_{n=1}^\infty$ be a sequence of finite element subspaces of $H^1(\Omega)$ associated with admissible triangulations T_n of Ω into linear or quadratic Lagrangian finite elements on simplices so that all elements of S_n are in $H_0^1(\Omega)$, so that $\lim_{n\to\infty}\|T_n\|=0$ and so that T_{n+1} refines T_n Then the unique weak solution of the equation $Du=f$ in $H_0^1(\Omega)$ is the H_0^1-limit of the solutions of the problems (S_n-PDE)*

Proof. Because each T_{n+1} refines T_n, the containment $S_n\subseteq S_{n+1}$ holds. Hence, it is enough to prove that $\bigcup_{n=1}^{\infty}S_n$ is dense in $H_0^1(\Omega)$. To prove this claim, it is enough to prove that for every function $f\in C_0^\infty(\Omega)$ there is a sequence $\{f_n\}_{n=1}^\infty$ with $f_n\in S_n$ so that $\lim_{n\to\infty}\|f-f_n\|_{H_0^1}=0$. So let $f\in C_0^\infty(\Omega)$. For each $n\in\mathbb{N}$, let $f_n\in S_n$ be the unique function so that for all nodes b of T_n we have $f_n(b) = f(b)$. Then, because f is infinitely differentiable, the sequence $\{f_n\}_{n=1}^{\infty}$ converges uniformly to f and all partial derivatives converge uniformly where they are defined (which is everywhere outside a null set). Because the domain of f is bounded, this means that $\{f_n\}_{n=1}^{\infty}$ converges to f in $H_0^1(\Omega)$. ∎

Theorem 23.45 establishes that the (weak) solutions of certain elliptic partial differential equations, including the Laplace and Poisson equations, can be approximated with the finite element method. While this is theoretically satisfying, it is still not enough for practitioners. Just as it was stated at the beginning of Chapter 13, in numerical analysis it is important to know *how fast* convergence happens. In this regard, Theorem 23.45 falls short, because it does not say anything about how close we can get to the solution of a given problem in the subspaces we mention.

Moreover, the finite element method is computationally intensive, because large systems of linear equations need to be solved. The size of the systems is proportional to the number of elements and the constant of proportionality involves the degree of the elements. Therefore, the practical application of the finite element method involves many steps, some of which are outlined below.

- To start a finite element approach to a problem, we must obtain a variational formulation (compare with Definition 23.32).

- If the domain is not a polyhedron, then a polyhedron (or another domain that is accessible with the finite element method) must be found so that the solution on the approximate domain is close to the solution on the actual domain.

- The finite elements need to be chosen so that the resulting (large) systems of equations are well-behaved numerically and so that good error estimates are available for the approximation of the solution of the variational problem.

- If the solution is to be approximated successively, the degree of the elements as well as the size and shape (use squares, hexagons, etc., instead of triangles) of the mesh elements can be adjusted.

- To reduce the computational effort, one can refine the mesh more where the solution is expected to fluctuate greatly and less where it is expected to be nearly constant.

- The triangulations themselves can be modified. A finite element method based on admissible triangulations is also called **conforming**, while a method involving nonadmissible triangulations with "hanging nodes" is called **nonconforming**.

- It is also possible to combine methods and approach a problem using a mixed finite element–finite difference scheme.

- Error bounds need to be established. Generally speaking, convergence in the L^p norm with larger p is better and L^∞ convergence would be ideal. On the other hand, if L^2 estimates are hard, one can try to establish L^p estimates with $p < 2$.

- The approximation is not solely judged by how close it is to the solution with respect to an L^p norm, but also by how its properties relate to the modeled phenomenon. If an approximation has nonphysical properties (like oscillations when we solve the heat equation), then the approximation must be discarded as physically meaningless, no matter how "close" it is in the L^2 sense.

The considerable amount of detail needed here is beyond the aim of this text, which was to provide the theoretical foundation for such investigations. The text [25] could be picked up at this point to expose the reader to more details. Also, for those who read German, the freely available notes [21] are recommended.

Exercises

23-29. Let $S \subseteq \mathbb{R}^d$ be a simplex with vertices $a_1, \ldots, a_{d+1}$.

(a) Prove that for any $x \in S$ the numbers $\lambda_1, \ldots, \lambda_{d+1}$ with $\sum_{j=1}^{d+1} \lambda_j = 1$ so that $x = \sum_{j=1}^{d+1} \lambda_j a_j$ are unique.
Hint. Write $x - a_{d+1}$ as a linear combination of $a_1 - a_{d+1}, \ldots, a_d - a_{d+1}$.

(b) Prove that S is closed.

(c) Prove that S is convex.

(d) Prove that S is a d-dimensional manifold with corners (and hence S° is a Trace Theorem Domain).
Hint. There is linear function that maps the standard base to $\{a_1 - a_{d+1}, \ldots, a_d - a_{d+1}\}$.

23-30. Let $\hat{\tau}$ be the triangle in $\mathbb{R}^2$ with vertices $(0,0)$, $(0,1)$, and $(1,0)$

(a) For quadratic Lagrangian finite elements, use a computer to compute the base functions $p_1, \ldots, p_6$.
Hint For each $p_i(x,y) = ax^2 + bxy + cy^2 + dx + ey + f$, set up a 6×6 system of linear equations

(b) Let τ be an arbitrary triangle in $\mathbb{R}^2$ with vertices (a_x, a_y), (b_x, b_y) and (c_x, c_y). Find a bijective, affine linear function $f : \mathbb{R}^2 \to \mathbb{R}^2$ (that is, a sum of a constant and a linear function) that maps $\hat{\tau}$ to τ

(c) Explain why the base functions in part 23-30a are sufficient to construct base functions for quadratic Lagrangian finite elements on arbitrary triangles in $\mathbb{R}^2$.

23-31. Prove $\int_\Omega u \Delta v \, d\lambda = -\int_\Omega \nabla u \cdot \nabla v \, d\lambda + \int_{\delta\Omega} u \nabla v \cdot d\vec{S}$ for all $u \in H^1(\Omega)$ and $v \in H^2(\Omega)$, where formally the values of u and ∇v on the boundary are given by $\gamma(u)$ and $\gamma(\nabla v)$ with γ as in Theorem 23.25.

23-32. Prove $\int_\Omega \operatorname{div}(u) v \, d\lambda = -\int_\Omega u \cdot \nabla v \, d\lambda + \int_{\delta\Omega} vu \cdot d\vec{S}$ for all $u \in \left(H^1(\Omega)\right)^d$ and $v \in H^1(\Omega)$, where formally the values of u and v on the boundary are given by $\gamma(u_1), \ldots, \gamma(u_d)$ and $\gamma(v)$ with γ as in Theorem 23.25

Conclusion and Outlook

It was mentioned in the preface that the text is meant to lay a foundation for a number of topics in mathematics. We can now take a quick look at these topics.

Complex analysis investigates the analytical properties of functions from $\mathbb{C}$ to $\mathbb{C}$. It turns out that if such a function is differentiable, it is locally equal to a power series.

(Ordinary) differential equations. Theoretical approaches focus on results similar to the Picard-Lindelöf Theorem. Applied approaches focus, for example, on special functions of mathematical physics or stability theory (continuous dependence of solutions on input parameters). Numerical approaches focus on numerical schemes to approximate solutions.

Differential geometry investigates the geometric properties of manifolds. An important application here is the general theory of relativity.

Functional analysis investigates the properties of Banach and Hilbert spaces as well as the properties of linear and nonlinear operators on these spaces. These ideas can then be applied, for example, to solve ordinary and partial differential equations, to approximate solutions, and also to model quantum mechanical phenomena.

Harmonic analysis investigates the properties of harmonic functions (solutions of the steady state heat equation or real parts of differentiable complex functions), Fourier series and integral operators.

Mathematical physics draws on all branches of mathematics to model phenomena in all branches of physics.

Measure theory investigates properties of measures and integrable functions.

Numerical analysis provides numerical approximation schemes for solutions of equations and systems of equations. Often the focus is on the application of the method, say, for the finite element method the focus would be error estimates and the choice of mesh, step sizes, and degrees of the elements.

Partial differential equations. Topics in this area can reach from theoretical investigations about existence and stability of (weak) solutions to solution schemes with possible overlaps into numerical analysis.

Probability theory investigates phenomena governed by chance. It ultimately draws on measure theory, because probability spaces are special measure spaces.

Topology investigates properties defined in terms of open sets (point-set topology). Low-dimensional topology focuses on the properties of three dimensional space.

You are ready for the topics above. Choose wisely and enjoy.

Appendix A

Logic

Sets and Logic are the foundation of mathematics. All mathematical results are ultimately derived from the axioms of set theory using the rules of logic. A start into mathematics from set theory, constructing the real numbers, is almost a course in itself. This being a text on *analysis*, Appendices A, B, and C are used to outline the necessary background in and connections to the foundations. Appendix A establishes the notation for logic and some fundamental techniques. Appendix B does the same for set theory. Appendix C presents a construction of the rational numbers from the axioms of set theory. In particular, together with the remarks after Theorem 16.89, Appendix C shows that the real numbers can indeed be constructed from the axioms of set theory.

Specifics of set theory and logic are only rarely used in analysis. Yet when they are needed, they are essential. In the preface of [13], Paul Halmos stated the fundamental importance of set theory by saying one should "read it, absorb it and forget it." The author wholeheartedly agrees. Fundamental ideas that are frequently used will become second nature. The remaining details often fade from conscious memory without any loss of mathematical ability.

Logic provides the language of mathematics and set theory provides the objects. Of course, the two are intertwined. Without language it is not possible to communicate anything about the objects. On the other hand, without objects, what would there be to talk about? We choose to start the fundamentals with logic.

A.1 Statements

In mathematics, there are absolute notions of "true" and "false." These notions are used to full effect by mostly working with statements.

Definition A.1 *A* **statement** *is a sentence that is either true or false.*

Once statements are given, more statements can be formed. Definition A.2 applies to arbitrary statements, Definition A.3 applies to statements with variables.

Definition A.2 *Let p and q be statements.*

1. *The statement $p \wedge q$ ("p **and** q") is true iff p is true and q is true.*
2. *The statement $p \vee q$ ("p **or** q") is true iff p is true or q is true, where the "or" also allows for both statements to be true.*
3. *The statement $p \Rightarrow q$ ("p **implies** q") is false iff p is true and q is false.*
4. *The statement $p \Leftrightarrow q$ ("p **if and only if** q" or "p **iff** q") is true iff p and q are both true or both false.*
5. *The statement $\neg p$ ("**not** p") is true iff p is false.*

Definition A.3 *Let $P(x)$ be a statement that depends on the variable x and let S be a set.*

1. *The statement $\forall x \in S : P(x)$ ("for all x in S we have $P(x)$") is true iff $P(x)$ holds for all elements x in the set S.*
2. *The statement $\exists x \in S : P(x)$ ("there is an x in S so that $P(x)$") is true iff $P(x)$ holds for at least one element x in the set S.*

The symbols $\forall$ and $\exists$ are called **quantifiers**. *$\forall$ is the* **universal quantifier** *and $\exists$ is the* **existential quantifier**.

Proposition A.4 *Let p and q be statements. The* **contrapositive** *of the statement $p \Rightarrow q$ is $(\neg q) \Rightarrow (\neg p)$. An implication and its contrapositive are either both true or both false. That is, the contrapositive says the same as the original implication.*

A.2 Negations

To learn more about what it means that a statement is true, it is often helpful to investigate what it means that the statement is false. That is, it is helpful to investigate the **negation** of the statement. Negations are also used in the contrapositive.

Theorem A.5 *Let p, q be statements.*

1. *The negation of the statement $p \wedge q$ is $\neg(p \wedge q) = (\neg p) \vee (\neg q)$.*
2. *The negation of the statement $p \vee q$ is $\neg(p \vee q) = (\neg p) \wedge (\neg q)$.*
3. *The negation of the statement $p \Rightarrow q$ is $\neg(p \Rightarrow q) = p \wedge (\neg q)$.*

Theorem A.6 *Let $P(x)$ be a statement that depends on the variable x and let S be a set.*

1. *The negation of $\forall x \in S : P(x)$ is $\neg\big(\forall x \in S : P(x)\big) = \exists x \in S : \big(\neg P(x)\big)$.*
2. *The negation of $\exists x \in S : P(x)$ is $\neg\big(\exists x \in S : P(x)\big) = \forall x \in S : \big(\neg P(x)\big)$.*

Appendix B

Set Theory

This appendix presents the Zermelo-Fraenkel axioms of set theory and it defines relations and functions. Note that products are defined in Definition 7.8.

B.1 The Zermelo-Fraenkel Axioms

Axiom B.1 *The Zermelo-Fraenkel Axioms for Set Theory*

1. *For every object x and every set S, we can determine if $x \in S$ or $x \notin S$.*

2. *Axiom of Specification. If S is a set and $P(\cdot)$ is a meaningful statement for each element of S, then the set of all elements $x \in S$ that satisfy $P(x)$ is also a set. It is denoted as $\{x \in S : P(x)\}$ or also as $\{x \in S | P(x)\}$.*

3. *There is a set, or equivalently, there is a set $\emptyset$ that has no elements. (For every set, the set $\{x \in S : x \neq x\}$ is empty.)*

4. *Axiom of Extension. Two sets are equal if and only if they have the same elements.*

5. *Axiom of Pairing. For any two sets, there exists a set to which they both belong. That is, if A, B are sets, then $\{A, B\}$ also is a set.*

6. *Axiom of Unions. For every collection $\mathcal{C}$ of sets, there exists a set whose elements are all the elements that belong to at least one element of the collection. This set is denoted $\bigcup \mathcal{C}$ and it is called the* **union** *of $\mathcal{C}$.*

7. *Axiom of Powers. For each set S, there exists a set $\mathcal{P}(S)$, called the* **power set** *of S, whose elements are all the subsets of S.*

8. *Axiom of Infinity. There is a set I that contains $\emptyset$ and for each $a \in I$ the set $\{a, \{a\}\}$ is also in I.*

9. *Axiom of Substitution. If $S(a, b)$ is a sentence such that for each $a \in A$ the set $\{b : S(a, b)\}$ can be formed, then there exists a function F with domain A such that $F(a) = \{b : S(a, b)\}$ for all $a \in A$.*

Two more important axioms are independent of the Zermelo-Fraenkel axioms.

Axiom B.2 *The* **Axiom of Choice**. *Let $\{A_i\}_{i\in I}$ be an indexed family of sets. Then there is a function $f : I \to \bigcup_{i \in I} A_i$ so that $f(i) \in A_i$ for all $i \in I$.*

Axiom B.3 *The* **Continuum Hypothesis**. *With $\aleph_0$ and $\aleph_1$ being the first two infinite cardinal numbers, $\aleph_1$ is equivalent to the power set of $\aleph_0$.*

B.2 Relations and Functions

Relations and functions are fundamental to analysis. In set theory, they are defined as special subsets of the product of two sets.

Definition B.4 *Let A and B be sets. Then a* **relation** *ρ from A to B is a set $\rho \subseteq A \times B$. For $a \in A$ and $b \in B$ it is customary to write $a\rho b$ instead of $(a, b) \in \rho$.*

Definition B.5 *Let A and B be sets.*

1. *A relation $\rho \subseteq A \times B$ is called* **totally defined** *iff for all $a \in A$ there is a $b \in B$ with $a\rho b$.*

2. *A relation $\rho \subseteq A \times B$ is called* **well-defined** *iff for all $a \in A$ there is at most one $b \in B$ with $a\rho b$.*

Definition B.6 *Let A and B be sets. A* **function** *$f : A \to B$ is a relation $f \subseteq A \times B$ that is totally defined and well-defined. For $a \in A$ and $b \in B$, it is customary to write $b = f(a)$ instead of $(a, b) \in f$. Functions are also called* **maps** *or* **mappings**.

Definition B.7 *Let A and B be sets and let $f : A \to B$ be a function.*

1. *The function f is called* **injective** *or* **one-to-one** *iff $x \neq y$ implies $f(x) \neq f(y)$, for all $x, y \in A$.*

2. *The function f is called* **surjective** *or* **onto** *iff for all $b \in B$ there is an $a \in A$ such that $f(a) = b$.*

3. *The function f is called* **bijective** *iff it is injective and surjective.*

Appendix C

Natural Numbers, Integers, and Rational Numbers

A lot of mathematics seems as if it is not founded on sets, but actually on the number systems that we are familiar with. This appendix briefly indicates how the familiar number systems are all part of set theory.

C.1 The Natural Numbers

Axiom C.1 *The Peano Axioms for* $\mathbb{N}$.

1. *There is a natural number* $1 \in \mathbb{N}$,
2. *Each* $x \in \mathbb{N}$ *has a (unique) successor* x',
3. *For all* $x, y \in \mathbb{N}$ *if* $x' = y'$, *then* $x = y$,
4. *The element* 1 *is not the successor of any natural number,*
5. *The only natural numbers are those given by 1 and 2.*

Proposition C.2 *We can construct a model of* $\mathbb{N}$ *in set theory by setting* $1 := \{\emptyset\}$, *and by setting* $x' := \{x, \{x\}\}$ *for every* x *that is already defined.*

Arithmetic on the natural numbers can also be defined.

Definition C.3 *A (binary)* **operation** *on a set* S *is a function* $\circ : S \times S \to S$. *For elements* $a, b \in S$ *we set* $a \circ b := \circ(a, b)$.

Definition C.4 *We define the operation* $+ : \mathbb{N} \times \mathbb{N} \to \mathbb{N}$ *by* $n + 1 := n'$ *for all* $n \in \mathbb{N}$ *and* $n + m' := (n + m)'$ *for all* $m, n \in \mathbb{N}$. *The operation* $\cdot : \mathbb{N} \times \mathbb{N} \to \mathbb{N}$ *is defined by* $n \cdot 1 := n$ *and* $n \cdot m' := n \cdot m + n$.

C.2 The Integers

Definition C.5 *Let X be a set. A relation $\sim\subseteq X \times X$ is called an* **equivalence relation** *iff*

1. *$\sim$ is* **reflexive**. *That is, for all $x \in X$ we have $x \sim x$.*
2. *$\sim$ is* **symmetric**. *That is, for all $x, y \in X$ we have $x \sim y$ iff $y \sim x$.*
3. *$\sim$ is* **transitive**. *That is, for all $x, y, z \in X$ we have that $x \sim y$ and $y \sim z$ implies $x \sim z$.*

For each $x \in X$, the set $[x] := \{y \in X : y \sim x\}$ is called the **equivalence class** *of x.*

Proposition C.6 *The relation $(a, b) \sim (c, d)$ defined by $a + d = b + c$ is an equivalence relation on the set $\mathbb{N} \times \mathbb{N}$.*

Definition C.7 *The* **integers** *$\mathbb{Z}$ are defined to be the set of equivalence classes $\big[(a, b)\big]$ of elements of $\mathbb{N} \times \mathbb{N}$ under the equivalence relation $\sim$ of Proposition C.6. Addition of integers is defined by $\big[(a, b)\big] + \big[(c, d)\big] := \big[(a + b, c + d)\big]$ and multiplication is defined by $\big[(a, b)\big] \cdot \big[(c, d)\big] := \big[(ac + bd, bc + ad)\big]$. Both operations are well-defined and $\mathbb{N}$ is isomorphic to the subset $\big\{\big[(n, 1)\big] : n \in \mathbb{N} \setminus \{1\}\big\}$. This set will also be called $\mathbb{N}$.*

C.3 The Rational Numbers

Proposition C.8 *The relation $(a, b) \sim (c, d)$ defined by $a \cdot d = b \cdot c$ is an equivalence relation on the set $\mathbb{Z} \times \big(\mathbb{Z} \setminus \{0\}\big)$.*

Definition C.9 *The* **rational numbers** *$\mathbb{Q}$ are defined to be the set of equivalence classes $\big[(a, b)\big]$ of elements of $\mathbb{Z} \times \big(\mathbb{Z} \setminus \{0\}\big)$ under the equivalence relation $\sim$ of Proposition C.8. Addition is defined by $\big[(a, b)\big] + \big[(c, d)\big] := \big[(ad + bc, bd)\big]$ and multiplication is defined by $\big[(a, b)\big] \cdot \big[(c, d)\big] := \big[(ac, bd)\big]$. Both operations are well-defined.*

Theorem C.10 *With operations as defined above, the rational numbers are an ordered field. That is, $\mathbb{Q}$ satisfies all the properties outlined in Axioms 1.1 and 1.6 for the real numbers at the beginning of the text. The set $\mathbb{Q}^+$ is $\big\{\big[(a, b)\big] : a, b \in \mathbb{N} \subseteq \mathbb{Z}\big\}$.*

Bibliography

[1] M. Abramowitz and I. Stegun (1965), *Handbook of mathematical functions: with formulas, graphs, and mathematical tables*, Dover, New York.

[2] R. Adams (1978), *Sobolev Spaces*, Academic Press, Boston.

[3] R. Aris (1962), *Vectors, Tensors, and the Basic Equations of Fluid Mechanics*, Prentice-Hall, Englewood Cliffs, NJ.

[4] R. Bjork (1994), Memory and Metamemory Considerations in the Training of Human Beings, in J. Metcalfe and A. Shimamura (eds.), Metacognition: Knowing about knowing, MIT Press, Cambridge, MA, 185–205.

[5] J.Bransford, R. Sherwood, N. Vye, and J. Rieser (1986), Teaching Thinking and Problem Solving, American Psychologist, October issue.

[6] A. C. Chapman (1987), *Fundamentals of Heat Transfer*, MacMillan, New York.

[7] D. Cohn (1980), *Measure Theory*, Birkhäuser, Boston, MA.

[8] J. Dieudonné (1960), *Foundations of Modern Analysis*, Academic Press, New York, London.

[9] C. Dodge (1969), *Sets, Logic and Numbers*, Prindle, Weber & Smith, Incorporated, Boston, London, Sydney.

[10] D. Ferguson (1973), Sufficient conditions for Peano's kernel to be of one sign, *SIAM J. Numer. Anal.* 10, 1047–1054.

[11] H. Goldstein (1950), *Classical Mechanics*, Addison-Wesley, Cambridge, MA.

[12] D. Halliday, R. Resnick and J. Walker (2001), *Fundamentals of Physics*, J. Wiley & Sons, Hoboken, NJ.

[13] P. R. Halmos (1974), *Naive set theory*, Undergraduate Texts in Mathematics, Springer Verlag, New York.

[14] E. Hewitt and K. Stromberg (1965), *Real and Abstract Analysis*, Graduate Texts in Mathematics, Springer Verlag, New York, Heidelberg, Berlin.

[15] H. Heuser (1986), *Lehrbuch der Analysis, Teil 1 (4. Auflage)*, B. G. Teubner, Stuttgart.

[16] H. Heuser (1983), *Lehrbuch der Analysis, Teil 2 (2. Auflage)*, B. G. Teubner, Stuttgart.

[17] H. Heuser (1986), *Funktionalanalysis (2. Auflage)*, B. G. Teubner, Stuttgart.

[18] A. Hurd and P. Loeb (1985), *An Introduction to Nonstandard Real Analysis*, Academic Press, Orlando, FL.

[19] J.D. Jackson (1999), *Classical Electrodynamics (Third Edition)*, John Wiley & Sons, Inc., New York.

[20] R. Johnsonbaugh and W. Pfaffenberger (2002), *Foundations of Mathematical Analysis*, Dover, Mineola, NY.

[21] A. Jüngel (2004), *Das kleine Finite-Elemente-Skript*, Vorlesungsskript, Johannes Gutenberg Universität Mainz.

[22] M. Lehn (2003), *Analysis III*, Vorlesungsskript, Johannes Gutenberg Universität Mainz.

[23] M. Renardy and R. Rogers (1993), *An Introduction to Partial Differential Equations*, Springer, New York.

[24] J. T. Sandefur (1990), *Discrete Dynamical Systems*, Clarendon Press, Oxford.

[25] P. Šolín (2006), *Partial Differential Equations and the Finite Element Method*, J. Wiley and Sons, Inc., Hoboken, NJ.

[26] M. Spivak (1965), *Calculus on Manifolds*, W. A. Benjamin Inc., New York.

[27] M. Spivak (1979), *A Comprehensive Introduction to Differential Geometry, vol. I, second ed.*, Publish or Perish, Houston, TX.

[28] J. Stoer and R. Bulirsch (1980), *Introduction to Numerical Analysis*, Springer Verlag, New York, Heidelberg, Berlin.

[29] K. Stromberg (1981), *An Introduction to Classical Real Analysis*, Wadsworth International, Belmont, CA.

[30] A. Torchinsky (1986), *Real-Variable Methods in Harmonic Analysis*, Academic Press, San Diego, CA.

[31] J.Welty, C. Wicks, and R.Wilson (1969), *Fundamentals of Momentum, Heat and Mass Transfer*, John Wiley & Sons, Inc., New York, London, Sydney, Tokyo.

[32] S. Willard (1970), *General Topology*, Addison-Wesley, Reading, MA.

[33] K. Yosida (1968), *Functional Analysis (Second Edition)*, Springer Verlag, New York, Heidelberg, Berlin.

[34] E. Zeidler (1990), *Nonlinear Functional Analysis and its Applications II/A*, Springer Verlag, New York, Berlin, Heidelberg.

Index

Printed and bound by CPI Group (UK) Ltd, Croydon, CR0 4YY
16/04/2025
14658521-0005